BIOCHEMISTRY

Edited by **Deniz Ekinci**

Biochemistry
Edited by Deniz Ekinci

Published by InTech

Janeza Trdine 9, 51000 Rijeka, Croatia

Publishing Process Manager Masa Vidovic

Technical Editor Teodora Smiljanic

Cover Designer InTech Design Team

Additional hard copies can be obtained from orders@intechopen.com

Biochemistry, Edited by Deniz Ekinci
p. cm.
ISBN 978-953-51-0076-8

For EU product safety concerns:
IN TECH d.o.o., Prolaz Marije Krucifikse Kozulić 3, 51000 Rijeka, Croatia,
info@intechopen.com or visit our website at intechopen.com.

Meet the editor

Dr. Deniz Ekinci obtained a BSc in Chemistry in 2004, an MSC in Biochemistry in 2006 and a PhD in Biochemistry in 2009 from Atatürk University, Turkey. He studied at Stetson University, USA, in 2007-2008 and at the Max Planck Institute of Molecular Cell Biology and Genetics, Germany, in 2009-2010. Dr. Ekinci currently works as an Assistant Professor of Biochemistry in the Faculty of Agriculture, and is the Head of the Enzyme and Microbial Biotechnology Division, Ondokuz Mayıs University, Turkey. He is a member of Turkish Biochemical Society, American Chemical Society and German Genetics society. Dr. Ekinci published over fifty scientific papers, reviews and book chapters and presented several conferences to scientists. He has received numerous publication awards from several scientific councils. Dr. Ekinci serves as the Editor in Chief of three international books and is involved in the Editorial Board of several international journals.

Contents

Preface

Biochemistry includes the chemical processes in living systems which govern all living organisms and living processes. It deals with the structures and functions of biomolecules. Over the recent years, biochemistry has become responsible for explaining living processes such that many scientists in the life sciences from agronomy to medicine are engaged in biochemical research. The main focus of biochemistry is in understanding how biomolecules give rise to the chemical processes that occur within living cells. Although extensive research has been performed on biochemistry for many years, there is still deep need of understanding the biochemical reactions as well as the structures of biomolecules.

This book titled "*Biochemistry*" contains a selection of chapters focused on the research area of proteins, enzymes, cellular mechanisms and chemical compounds used in relevant approaches. The book provides an overview on basic issues and some of the recent developments in biochemical science and technology. Particular emphasis is devoted to both theoretical and experimental aspect of modern biochemistry. The primary target audience for the book includes students, researchers, biologists, chemists, chemical engineers and professionals who are interested in biochemistry, molecular biology and associated areas.

The textbook is written by international scientists with expertise in protein biochemistry, enzymology, molecular biology and genetics many of which are active in biochemical and biomedical research. I would like to acknowledge the authors for their contribution to the book. We hope that the textbook will enhance the knowledge of scientists in the complexities of some biochemical approaches; it will stimulate both professionals and students to dedicate part of their future research in understanding relevant mechanisms and applications.

Dr. Deniz Ekinci
Assistant Professor of Biochemistry
Ondokuz Mayıs University
Turkey

Part 1

Proteins and Hormones

1

Protein Flexibility and Coiled-Coil Propensity: New Insights Into Type III and Other Bacterial Secretion Systems

Spyridoula N. Charova[1,2], Anastasia D. Gazi[1,2], Marianna Kotzabasaki[2], Panagiotis F. Sarris[1,2], Vassiliki E. Fadouloglou[2,3], Nickolas J. Panopoulos[1,2] and Michael Kokkinidis[1,2]

[1]*Institute of Molecular Biology & Biotechnology, Foundation of Research & Technology*
[2]*Department of Biology, University of Crete, Vasilika Vouton, Heraklion, Crete*
[3]*Department of Molecular Biology and Biotechnology*
Democritus University of Thrace, Alexandroupolis
Greece

1. Introduction

Secretion in unicellular species is the transport or translocation of molecules, for example proteins, from the interior of the cell to its exterior. In bacteria secretion is a very important mechanism, either modulating their interactions with their environment for adaptation and survival or establishing interactions with their eukaryotic hosts for pathogenesis or symbiosis. To overcome the physical barriers of membranes, Gram-negative bacteria use a variety of molecular machines which have been elaborated to secrete a wide range of proteins and other molecules; their functions include biogenesis of organelles (e.g. pili and flagella), virulence, efflux of toxins etc. As in some cases the secreted proteins are destined to enter host cells (effectors, toxins), some of the secretion systems include extracellular appendices to translocate proteins across the plasma membrane of the host.

With the rapid accumulation of bacterial genome sequences, our knowledge of the complexity of bacterial protein secretion systems has expanded and several secretion systems have been identified. Gene Ontology has been very useful for describing the components and functions of these systems, and for capturing the similarities among the diverse systems (Tseng et al., 2009). These analyses along with numerous biochemical studies have revealed the existence of at least six major mechanisms of protein secretion. These pathways are highly conserved throughout the Gram-negative bacterial species and are functionally independent with respect to outer membrane translocation; commonalities exist in the inner membrane transport steps of some systems, with most of them being terminal branches of the general secretion pathway (Sec). The pathways have been numbered Type I, II, III, IV, V and VI.

In Gram-negative bacteria, some secreted proteins are exported across the inner and outer membranes in a single step via the Type I, III, IV or VI pathways. Other proteins are first exported into the periplasmic space using the universal Sec or two-arginine (Tat) pathways and then translocated across the outer membrane via the Type II, V or less commonly, the Type I or IV machinery. In Gram-positive bacteria, secreted proteins are commonly translocated across the single membrane by the Sec pathway, the two-arginine (Tat) pathway, or the recently identified type VII secretion system (Abdallah et al., 2007). In the following we will briefly survey the six Gram-negative bacterial secretion systems known to modulate interactions with host organisms:

Type I secretion system: This system (T1SS) forms a contiguous channel traversing the inner and outer membranes of Gram-negative bacteria. It is a simple system, which consists of only three major components: ATP-binding cassette transporters, Outer Membrane Factors, and Membrane Fusion Proteins (Holland et al., 2005). T1SS transports ions and various molecules including proteins of various sizes (20-900 kDa) and non-proteinaceous substrates like cyclic β-glucans and polysaccharides.

Type II secretion system: This system (T2SS) is encoded by at least 12 genes and supports the transport of a group of seemingly unrelated proteins across the outer membrane. In order for these proteins to enter the type II secretion pathway, they have to first translocate across the cytoplasmic membrane via the Sec-system and then fold into a translocation competent conformation in the periplasm. Proteins secreted by T2SS include proteases, cellulases, pectinases, phospholipases, lipases, and toxins which contribute to cell damage and disease. Although Sec-dependent translocation is universal (Cao & Saier, 2003), the T2SS is found only in Gram-negative proteobacteria phylum (Cianciotto, 2005; Filloux, 2004). A bacterial species may have more than one T2SS (Cianciotto, 2005; Filloux, 2004).

Type III secretion system: These systems (T3SS) are essential mediators of the interaction of many Gram-negative pathogenic proteobacteria (α, β, γ and δ subdivisions) with their human, animal, or plant hosts and are evolutionarily related to bacterial flagella. (Dale & Moran, 2006; Tampakaki et al., 2004; Troisfontaines & Cornelis, 2005). The machinery of the T3SS, termed the injectisome, appears to have a common evolutionary origin with the flagellum and translocates a diverse repertoire of effector proteins either to extracellular locations or directly into eukaryotic cells, in a Sec-independent manner (interkingdom protein transfer device). The T3SS effectors (T3EPs) modulate the function of crucial host regulatory molecules and trigger a range of highly dynamic cellular responses which determine pathogen-host recognition, pathogen/symbiont accommodation and elicitation or suppression of defense responses by the eukaryotic hosts. In some cases however, effector proteins are simply secreted out of the cell. T3SS have evolved into seven families (Troisfontaines & Cornelis, 2005). Some bacteria may harbor more than one T3SS, usually from different families. T3SS genes are encoded in pathogenicity islands and/or are located on plasmids, and are commonly subject to horizontal gene transfer.

Type IV secretion system: In comparison to other secretion systems, T4SS is unique in its ability to transport nucleic acids in addition to proteins into plant and animal cells, as well as into yeast and other bacteria. Usually T4SS comprises 12 proteins that can be identified as homologs of the VirB1–11 and VirD4 proteins of the *Agrobacterium tumefaciens* Ti plasmid transfer system (Christie & Vogel, 2000). T4SS spans both membranes of Gram-negative bacteria, using a specific transglycosylase, VirB1, to digest the intervening murein (Koraimann, 2003; Baron et al., 1997). While many organisms have homologous type IV secretion systems, not all systems contain the same sets of genes. The only common protein is VirB10 (TrbI) among all T4SS systems (Cao & Saier, 2003).

Type V secretion system: T5SS is the simplest protein secretion mechanism. Proteins are secreted via the autotransporter system (type Va or AT-1), the two-partner secretion pathway (type Vb), and the oligomeric autotransporters (type Vc or AT-2 system) (Yu et al., 2008; Desvaux et al., 2004). Proteins secreted via these pathways have similarities in their primary structures as well as striking similarities in their modes of biogenesis.There are three sub-classes of T5SS. The archetypal bacterial proteins secreted via the T5SS (T5aSS subclass) consist of a N-terminal passenger domain of 40-400 kD in size and a conserved C-terminal domain (Henderson et al., 2004). The proteins are synthesized with a N-terminal signal peptide that directs their export into the periplasm via the Sec machinery.

Type VI secretion system: In T6SS 13 genes are thought to constitute the minimal number needed to produce a functional apparatus (Boyer et al., 2009). TheT6SS gene clusters (T6SS loci) often occur in multiple, non-orthologous copies per genome and have probably been acquired via horizontal gene transfer (Sarris & Skoulica, 2011; Sarris at al., 2011). Each T6SS probably assumes a different role in the interactions of the harbouring organism with others. Although the T6SS has been studied primarily in the context of pathogenic bacteria-host interactions, it has been suggested that it may also function to promote commensal or mutualistic relationships between bacteria and eukaryotes, as well as to mediate cooperative or competitive interactions between bacterial species. The T6SS machinery constitutes a phage-tail-spike-like injectisome that has the potential to introduce effector proteins directly into the cytoplasm of host cells, analogous to the T3SS and T4SS machineries.

Genetic, structural and biochemical studies of the above bacterial secretion systems along with massive *in silico* analyses of microbial genomes have been used to distinguish pathogens from their non-pathogenic relatives. These studies have established the presence of characteristic conserved features within individual types of secretion systems (e.g. Tampakaki et al., 2004), along with considerable sequence and structural diversities within each system at the level of specific components and effector proteins.

Despite the complexity of these systems however, the problem of identifying conserved features and properties within each secretion system type, or across several types of systems is of particular importance, going beyond a fundamental understanding of how bacterial secretion works. Even for well studied pathogens, not all virulence factors have been identified, making it possible that e.g. effector proteins that are associated with different diseases are still unknown. In less well characterized bacterial species there is certainly a wide spectrum of unknown effectors. This situation may be now changing through new approaches that use advanced machine learning algorithms to identify within individual types of secretion systems common themes for effectors and other system components that go beyond simple amino acid motifs (Arnold et al., 2009; Samudrala et al., 2009), or through the identification of important structural and physicochemical properties as universal signatures of virulence factors (Gazi et al., 2008; 2009).

This review will focus on the well-characterized T3SS proteins where the prevalence of coiled-coil domains along with pronounced structural flexibility/disorder have been proposed to be characteristic properties associated with a protein-protein interaction mode within T3SS and as essential requirements for secretion (Delahay and Frankel, 2002; Pallen et al., 1997; Gazi et al., 2008; 2009). Common themes with other secretion systems (T4SS, T6SS) will be also discussed.

2. Overview of the T3SS system: Architecture, conserved features and protein structures

Pathogenic bacterial strains are distinguished from non-pathogenic ones by the presence of specific set of genes that code for toxins, secretion systems, effectors that are meant to act extracellularly or effectors that should be delivered inside the host cell cytoplasm. These genes are usually tightly organized in operones that are located in chromosomal areas with a high distribution of mobile elements or can be found in virulence plasmids. Usually these chromosomal areas are called pathogenicity islands as they possess a different GC content from the rest of the genome, which implies recent acquisition through horizontal gene transfer events. One of the most profound cases was a set of approximately 20-25 genes which together encode one of the best characterized pathogenic mechanisms termed "type III secretion". By this mechanism extracellularly located bacteria that are in a close contact with a eukaryotic cell deliver proteins into the host cell cytosol. While the T3S apparatus is conserved in pathogens across the plant/animal phyllogenetic divide, the secreted proteins differ considerably. The genes coding for what are now recognized as structural T3SS components were first described as a contiguous cluster, designated "*hrp*" (hypersensitive response and pathogenicity) in plant pathogens. Important insights into fundamental questions of bacterial pathobiology came with the recognition, in subsequent years, of the T3SS as a complex multiprotein channel dedicated to translocate the effectors from the pathogen to the host. Although originally linked to pathogenesis, T3SS are also found in members of the phylum proteobacteria that are symbiotic, commensal or otherwise associated with insects, nematodes, fishes, plants, as well as in obligatory bacterial parasites of the phylum *Chlamydiae* (Dale and Moran, 2006; Marie et al., 2001).

T3SS is a multicomponent apparatus with the following characteristics: i) when fully developed it spans both bacterial membranes and the periplasmic space; ii) it possesses a large extracellular appendage (termed 'pilus' in plant pathogenic bacteria or 'needle' in animal pathogenic ones) that reaches the eukaryotic host cell contributing to bacterial adherence; iii) it forms the translocation pore in the host cell membrane to efficiently deliver proteins of bacterial origin inside the host cell; iv) a large number of T3SS cytosolic components form the export gate into the bacterial cytoplasm which sorts and prepares the substrates for secretion (Fig. 1).

The integral bacterial membrane part of the T3S apparatus consists of a series of rings. The protein that oligomerizes and forms the outer membrane and periplasmic rings (yellow parts in Fig. 1) belongs to the secretin family of proteins (which is also common to T2SS) and has a crucial role in T3S biogenesis (Diepold et al., 2010; Korotkov et al., 2011). Secretins consist of various domains with the C-terminal one integrated in the outer membrane. The N-terminal domains are less conserved among secretion systems and are responsible for the formation of the periplasmic rings. An N-terminal signal targets secretins to the periplasmic space through the Sec pathway. From there they are delivered to the outer membrane through a specific small lipidated protein, pilotin (Okon et al., 2008). Pilotins from various secretion systems possess different structures despite their common function, probably due to their interaction with the non-conserved C-terminal tail of various secretins. Thus, for example, the T3SS pilotin of *Shigella flexneri* possess an overall fold which differs from the fold of the T3SS pilotin of *Pseudomonas aeruginosa* or the T2SS pilotins of *Neisseria meningitis* and *P. aeruginosa* (Izore et al., 2011).

The T3SS inner membrane (IM) rings are formed by the proteins SctD and SctJ [orange parts in Fig. 1; the unified nomenclature is followed here as proposed by Hueck (1998)]. SctD is a single-pass inner membrane protein that oligomerizes to form the most external inner membrane ring of the T3SS. Its N-terminal domain is facing the bacterial cytoplasm and its structure is homologous to forkhead-associated (FHA) domains (McDowell et al., 2011). The inner membrane part of the *Salmonella typhimurium* injectisome has been studied by EM (Marlovits et al., 2006; Marlovits et al., 2004). The inner membrane topology of six conserved components ($HrcD^{SctD}$, $HrcR^{SctR}$, $HrcS^{SctS}$, $HrcT^{SctT}$, $HrcU^{SctU}$ and $HrcV^{SctV}$) of the T3SS from *Xanthomonas campestris* was recently studied (Berger et al., 2010) by translational fusions to a dual alkaline phosphatase–β-galactosidase reporter protein. Full IM rings have been modeled for $PrgH^{SctD}$ and $PrgK^{SctJ}$ [the species-specific name is followed by the standard T3SS nomenclature as proposed by Hueck (1998) in superscript] based on docking of atomic structures of individual domains to cryo electron microscopy maps (Schraidt & Marlovits, 2011). The central density observed in the inner membrane rings (socket region) of a T3SS needle complex cryo electron microscopy reconstruction map from *Salmonella enterica* sv. typhimurium (Fig. 1, red parts) is attributed to the $SpaP^{SctR}$, $SpaQ^{SctT}$, $SpaR^{SctS}$, $SpaS^{SctU}$ and $InvA^{SctV}$ proteins (Schraidt & Marlovits, 2011; Wagner et al., 2010).

In the socket region numerous cytosolic components are recruited to orchestrate the secretion of various T3SS substrates, like the ATPase SctN and its various subunits SctO, SctL. As biogenesis of the T3SS must take place before the secretion of the effectors, the first T3SS substrates to be secreted are the proteins that build the needle or pilus (SctF) and the inner rod (SctI), (green part in Fig. 1). The proteins that form the translocator pore in the eukaryotic membrane along with the proteins found in the needle tip are the next substrates to be secreted prior to effector proteins secretion.

An additional cytoplasmic ring is believed to be formed around the T3SS export gate as in the case of the flagellum (Thomas et al., 2006). Although never really observed by electron microscopy, recently Lara-Tejero and colleagues have reported the presence of a large platform in the T3SS of *S. enterica* sv. typhimurium that can sort substrates prior to secretion (Lara-Tejero et al., 2011). This platform consists of $SpaO^{SctQ}$, $OrgA^{SctK}$ and $OrgB^{SctL}$.

Numerous crystal structure determinations of T3SS components have been reported: The structures of the C-terminal domain of $HrcQ_B{}^{SctQ}$ (Fadouloglou et al., 2004; Fadouloglou et al., 2009), the C-terminal domain of FliN (Brown et al., 2005) and the central part of FliM (Park et al., 2006), all members of the SctQ/FliN,Y family and components of the cytoplasmic ring of the T3SS apparatus (C-ring) have been determined. Extended mutational and cross linking studies support a donut-shaped tetramer organization for the FliN protein which is localized at the bottom of the C-ring (Paul and Blair, 2006). A model where the FliN tetramers alterates with the C-terminal domain of FliM ($FliM_C$) seems to be in agreement with the major features observed in electron microscopic reconstructions. The side-wall of C-ring above the $FliN_4FliM_C$ array is formed by the middle domain of FliM while the N-terminal domain interacts with the FliG which is localised in proximity with the inner membrane and is the connection unit between the C-ring and the inner membrane, MS-ring (Sarkar et al., 2010; Paul et al., 2011). FliG has no homolog in non-flagellar T3SS and the homolog SctQ proteins are interacting to the T3SS injectisome through the SctD proteins.

The structures of $EscU^{SctU}$ and $YscU^{SctU}$, *EPEC* and *Yersinia* homologs of $HrcU^{SctU}$ respectively (Zarivach et al., 2008; Lountos et al., 2009; Thomassin et al., 2011) provide insights into the properties of conserved core components. The periplasmic domain of PrgH[SctC] from *Salmonella* (Spreter et al., 2009) and the cytoplasmic domain of MxiD[SctC] from *Shigella* (McDowell et al., 2011) have been recently determined. Structures of the periplasmic domains of the membrane components EscJ[SctJ] from the enteropathogenic *Escherichia coli* (EPEC) are also available (Yip et al., 2005b; Spreter et al., 2009).

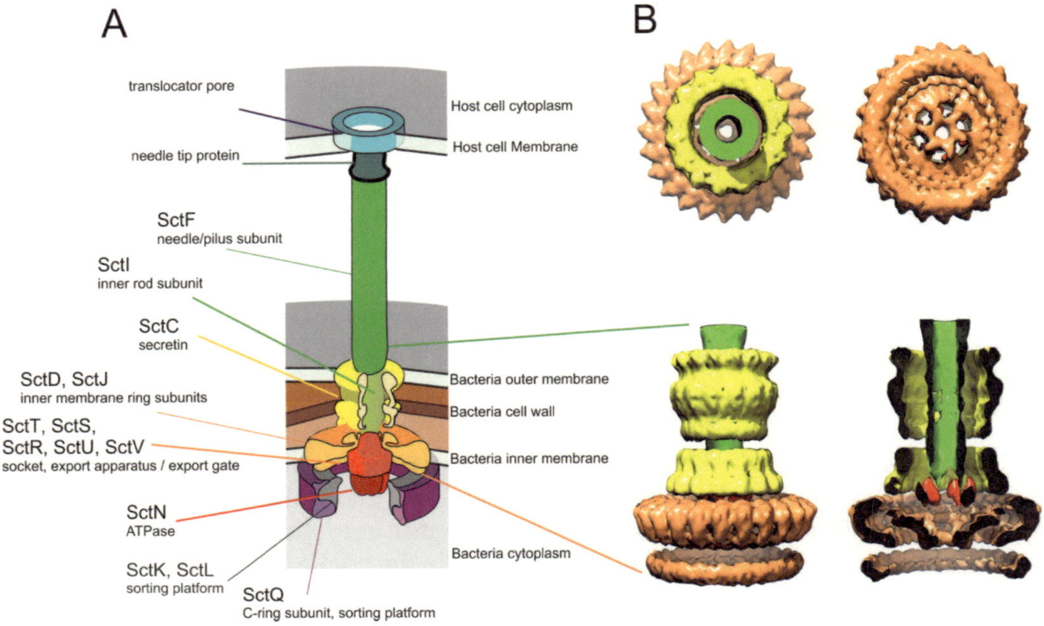

Fig. 1. (A) Overview of the T3S injectisome. (B) Different views of the *S. enterica* T3SS needle complex (Schraidt & Marlovits, 2011). Various parts of the needle complex are depicted in different colors using UCSF Chimera (Goddard et al., 2007). The colouring scheme used in (A) is followed. Top left: The T3SS needle complex viewed from top, Top right: View from the bottom, Lower left: side view, Lower right: a cross section of the side view. The inner membrane rings (orange) possess a 24-fold symmetry axis while the secretin rings (yellow) possess a 15-fold symmetry axis resulting in an overall 3-fold symmetry for the needle complex. The socket region (red parts and part of the orange area under the red parts) has a 6-fold symmetry (top right, icenter of the bottom view), which is also the symmetry of the T3SS ATPase that presumably docks in this area.

The T3SS utilizes an ATPase dedicated to drive secretion substrates through the central channel of the apparatus. Members of the SctN family (HrcN/FliI/YscN homologs) have a demonstrated ATP-hydrolysis activity, and exhibit extensive sequence and structural similarities with the F_0F_1-ATPase α/β subunits. Biochemical and electron microscopy data suggest that as it is the case with F_0F_1-ATPases, also the T3SS ATPases are hexamers anchored at the bacterial inner membrane. The crystal structure of FliI has been determined in the ADP-bound state and extensive structural similarities with to the α and β subunits of the F_0F_1-ATPsynthase have been found (Imada et al., 2007). The catalytic domain of EscN[SctN] also shows structural similarity with F_0F_1-ATPases (Zarivach et al., 2007). Recently the structure of FliJ, member of the SctO family (HrpO/FliJ/YscO homologs) has been reported as an α-helical coiled coil (Ibuki et al., 2011). Its structural similarity to a subunit of the F_0F_1-ATPsynthase and its interactions with FliI will be presented in section 6.1.1. Proteins from the SctL family (HrpE/FliH/YscL homologs) interact both with the T3SS ATPase and with structural proteins from the YscQ/FliN family located at the cytoplasm/inner membrane i.e. $YscQ^{SctQ}$, $EscQ^{SctQ}$ and FliN (Blaylock et al., 2006; McMurry et al., 2006; Biemans-Oldhinkel et al., 2011).

The outer supramolecular structure of the needle has been studied (Cordes et al., 2003), while structures of needle subunits from various bacteria have been recently determined (Wang et al., 2007; Deane et al., 2006; Zhang et al., 2006). At the tip of the T3SS needle resides an adaptor structure which mediates the interaction between the needle and the translocation pore at the eukaryotic membrane. The adaptor is formed through polymerization of a single protein. Information is available for the following needle tip proteins from three T3SS families: IpaD (*Shigella flexneri*), SipD (*Salmonella* spp.) and BipD (*Burkholderia pseudomallei*) from the *Inv-Mxi-Spa* T3SS family; LcrV (*Yersinia* spp.), PcrV (*Pseudomonas aeruginosa*, Sato et al., 2011; Sato & Frank, 2011) and AcrV (*Aeromonas salmonicida*) from the *Ysc* T3SS family; EspA (EPEC) from the *Ssa-Esc* family. The structures of IpaD, BipD, LcrV and part of the EspA structure have been elucidated (Espina et al., 2007; Johnson et al., 2007; Yip et al., 2005a; Derewenda et al., 2004), while a 3D-reconstruction of the MxiH filament is available (Deane et al., 2006).

Effectors are a large and structurally diverge group of virulence proteins which usually comprise a domain or a motif with a significant and proven role whithin the host cell during infection (for a review see Dean, 2011). Structures of several T3SS effectors from plant and animal pathogens are known (Desveaux et al., 2006; Stebbins, 2005). In addition, several structures for chaperones and chaperone-substrate complexes have been determined, including class I, class II and class III chaperones (Lilic et al., 2006; Buttner et al., 2008; Quinaud et al., 2007; Sun et al., 2008). Chaperones will be also presented in section 2.1.

2.1 The T3SS secretion signal

Type III effector proteins (T3EPs) possess non-cleavable secretion signals in the N-terminal protein regions, but no discernible amino acid or peptide similarities (Buttner and He, 2009) can be found. Three different types of potential secretion signals have been discussed: i) the N-terminus of the effector protein, ii) the ability of a chaperone to bind the effector before secretion, and iii) the 5′-end region of the mRNA; this hypothesis is very controversial (Gauthier et al., 2003; Anderson & Schneewind, 1997; Ramamurthi et al., 2002)

The prevailing view, supported by extensive biocomputing analyses, is that the amino acid composition of the N-terminal region of the effectors serves as secretion signal (Lloyd et al., 2001; Buttner and He, 2009; Arnold et al., 2009; Samudrala et al., 2009). The required N-terminal peptide length for secretion is usually 10–15 residues, whereas the minimum length needed for translocation is 50–60 residues. Additional targeting information is contained within the first 200 residues which provide binding sites for secretion chaperones (Lilic et al., 2006). T3SS chaperones of mammal pathogens interact with their cognate effectors through a chaperone-binding domain (CBD) located within the first 100 amino acids of the effector, after the N-terminal export signal (Cornelis, 2006).

Analyses of effectors from pathogenic bacteria revealed that the 25 N-terminal residues are enriched in Ser and lack Leu (Buttner and He, 2009; Arnold et al., 2009; Samudrala et al.,

2009). The N-terminal regions of T3EPs are probably unfolded, which is an important prerequisite for their transport through the narrow inner T3SS channel of presumably only 2.8 nm in diameter as was previously shown for the T3SS of several animal pathogenic bacteria (Marlovits et al., 2004, 2006; Galan and Wolf-Watz, 2006; Gazi et al., 2009).

For some effectors however, the N-terminal secretion signal is not sufficient for maximal secretion (Buttner & He, 2009) and specific chaperone proteins are needed; these are usually located adjacent to the cognate effector genes, suggesting strong selection for their coexistence in the genome. T3S chaperones are proposed to play a role in targeting secretory cargo to the injectisome, either by providing targeting information (Birtalan et al., 2002), or facilitating the exposure of the N-terminal export signal (Cornelis et al., 2006). Some chaperones are involved in the translocation of many substrate proteins, e.g. the global HpaB chaperone from *Xanthomonas campestris* pv. *vesicatoria* or Spa15 of *S. flexneri* (Hachani et al., 2008; Parsot et al., 2003; 2005; Buttner et al., 2004 ; 2006). Class I chaperones (the chaperones of effectors) are soluble small, usually homodimeric proteins that bind effector proteins. Although diverse in their sequences, they belong to the structural class of α/β proteins with a two-layer-sandwich architecture. For the chaperone-effector interaction a β-strand of the effector is added to extend the β-sheet layer of the chaperone (Lilic et al., 2006). Class I chaperones have been further subclassified depending on whether they associate with one (class Ia) or several (class Ib) effectors (Page & Parsot, 2002). Class II chaperones are T3SS chaperones of the translocators (Neyt & Cornelis, 1999). Experimental determinations of their structures (Buttner et al., 2008; Lunelli et al., 2009; Job et al., 2010; Priyadarshi & Tang, 2011) have confirmed earlier sequence analyses (Pallen et al., 2003) predicting an all-α-helical domain structure, with the bulk of the protein consisting of three tandem tetratricopeptide repeats (TPRs) which are involved in protein-protein interactions. Their substrate is recognised and bound into a concave site of the chaperone. Class III chaperones prevent the premature polymerization of needle components in the bacterial cytoplasm. They are predicted to adopt extended α-helical structures; this was confirmed by the crystal structure of the CesA which binds the EspA filament protein (Yip et al., 2005a).

Many functions have been attributed to T3SS chaperones, but the exact role(s) of the entire family of chaperones remain to be determined. However, it has been proposed that one of the main roles of the T3SS chaperones is the stabilization of at least some effector proteins inside bacterial cell, as well as their maintainance in a secretion-competent state, i.e. a partially folded or unfolded conformation.

3. The coiled-coil motif in proteins and α-helical bundles

The coiled-coil motif in protein structures consists of amphipathic α-helices that twist around each other to form a supercoiled bundle (Burkhard et al., 2001). It represents one of the efficient geometric solutions to packing helices in a stable way. The motif was one of the earliest protein structures discovered, first described for the hair protein alpha keratin (Crick, 1952). Coiled-coils are associated with all types of protein structure (globular, fibrous, membrane) and frequently provide a structural scaffold linked with molecular recognition interactions and oligomerization. Coiled-coil interactions play a major role in the formation of protein complexes in transcription, cell divisions, host-pathogen interactions etc. (Rackham et al., 2010). Coiled-coil helices may run parallel or antiparallel, and may form homo- or heterocomplexes (Grigoryan and Keating, 2008). The structures range from simple dimers through pentamers to more complex assemblies of many helices or bundles of bundles. Sequences of regular, left-hand twisted coiled-coils are characterized by a seven-residue periodicity (heptad repeat). If the heptad positions are labeled *a-g*, then positions *a* and *d* are hydrophobic and form the core of the bundle (Fig. 2). Positions *b, c, e, f* and *g* are more solvent-exposed and their amino acid preferences reflect constraints which are specific to each type of helical bundle (Paliakasis & Kokkinidis, 1992; Lupas et al., 1991). The hydrophobic residues *a* and *d* from one helix form 'knobs' that pack into 'holes' formed by residues *g, a, d, e* on neighbouring helices. Coiled-coil helices are distinguished from other amphipathic helices by the periodicity of hydrophobic residues (3.5 vs. 3.65 residues per turn), the length (long vs. short) and the packing interactions. Some proteins are induced to form coiled-coils upon association with a binding partner (Lupas, 1996).

Coiled-coil predictions at the genome level have explored the 'coilomes' of individual organisms (Barbara et al., 2007; Newman et al., 2000, Rose et al., 2004). In addition, genomewide analyses of coiled-coils evolution have been performed (Rackham et al., 2010) and shown that coiled-coils do not change their oligomeric state over evolution, and do not evolve from rearrangements of α-helices in protein structures. An analysis of proteomes (Liu and Rost, 2001) showed that twice as many coiled-coils are found in eukaryotes (10%) as in prokaryotes and archaea (4%-5%). The size of coiled-coil proteins ranges from short domains of 6-7 heptad repeats, e.g. Leucine zippers, serving as homo-/heterodimerization motifs in transcription factors (Jakoby et al., 2002; Vinson et al., 2002), to long domains of several hundreds amino acids found in functionally distinct proteins, often involved in attaching protein complexes to larger cellular structures (e.g. the Golgi, centrosomes, centromers, or the nuclear envelope).

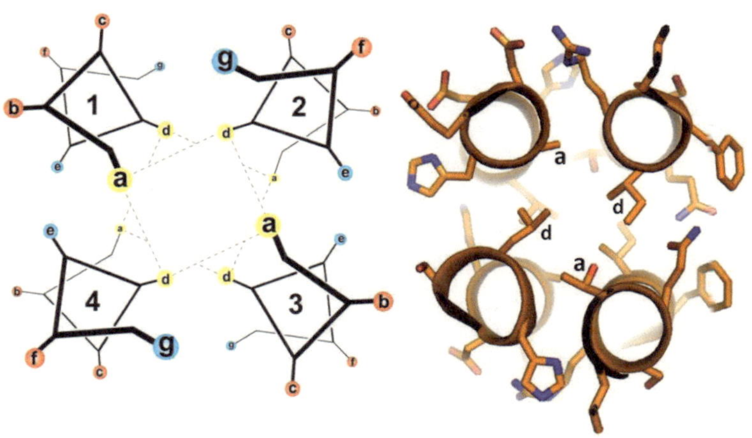

Fig. 2. Left: Antiparallel coiled-coil structure (4-α-helical bundle) and assignment of heptad positions a-g. Hydrophobic and hydrophilic positions are coloured yellow and red respectively. Right: Positions a, d and their packing in the core of the 4-α-helical ColE1 Rop (Banner et al., 1987). Positions a, d form slices perpendicularly to the bundle axis.

4. Flexibility and disorder in proteins

Until recently the classical structure-function paradigm which states that protein function is dependent on a defined, if flexible, three-dimensional polypeptide structure was widely accepted in protein science (Anfinsen, 1973). However, even in the early days of structural biology, with only approx. 20 protein crystal structures determined, some protein segments were known which yield weak or non-detectable electron density and yet they may be essential for function (Bloomer et al., 1978; Bode et al., 1978). A common reason (apart from crystal defects) for missing electron density is that the unobserved region fails to scatter X-rays coherently due to variation in position from one atom to the next, i.e. the unobserved atoms are disordered. In addition, during the last decade, many proteins have been described that fail to adopt a stable tertiary structure under physiological conditions and yet display biological activity (Dunker et al., 2008a; Uversky & Dunker, 2010). This state of the proteins, defined as intrinsic disorder, has been found to be rather widespread; disordered regions lacking stable secondary and tertiary structure are often a prerequisite for biological activity, suggesting that structure-function relationships can be frequently only understood in a dynamic context in which function arises from conformational freedom. Fully or partly nonstructured proteins are described as intrinsically disordered (IDPs) or intrinsically unstructured proteins. The term natively unfolded proteins indicates that protein function is associated with a dynamic ensemble of different conformations (Gazi et al., 2008).

Structural plasticity and flexibility is believed to represent a key functional feature of IDPs (Dunker et al., 2008a, 2008b; Dunker & Uversky, 2008; Xie et al., 2007; Cortese et al., 2008), enabling them to interact with numerous binding partners, e.g. proteins, membranes, nucleic acids and small molecules (Durand et al., 2008; Uversky et al., 2009). Because of their functional importance, intrinsically disordered domains are very common in proteomes and play crucial roles in signaling, recognition, regulation and self-assembly (Namba, 2001). The extreme flexibility of IDPs has been suggested to represent a strategy for optimizing the search and interaction with their targets (Sugase et al., 2007). Intrinsically disordered proteins are substantially depleted in W,C,F,Y,V,L,N (order-promoting) and enriched in A,R,G,Q,S,P,E,K (disorder-promoting residues) (Dunker et al., 2002; Uversky, 2010). These biases in the amino acid compositions of IDPs (which result in low overall hydrophobicity and low net charge) are used in various methods for the prediction of the ID propensities (Prilusky et al., 2005). Such analyses suggest that approx. 45% of proteins within a eukaryotic proteome contain a disordered region (Pentony & Jones, 2010). As a result of their frequent node positions in interactoms, many disordered proteins are tightly regulated at the levels of their synthesis, degradation and posttranslational modifications (Gsponer, 2008). It is noteworthy that extreme structural plasticity and ensembles of different conformations has been occasionally observed for coiled-coils and α-helical bundles (Glykos et al., 1999, 2004); as is the case with other proteins, the plasticity of coiled coils may have functional implications, e.g. in the establishment of macromolecular assemblies based on coiled-coil interactions (Gazi et al., 2008).

5. Tools for the analysis of coiled-coils and intrinsic disorder

5.1 *In silico* prediction and analysis of coiled-coil domains

Prediction of coiled coils from sequence: The 'COILS' webserver assesses the probability that a residue in a sequence is part of a coiled-coil structure by comparison of its flaking sequences with sequences of known coiled-coil proteins (Lupas et al., 1991) (http://www.ch.embnet.org/software/COILS_form.html). In the 'Paircoil2' algorithm

(McDonnell et al., 2006), pairwise residue probabilities are used to detect coiled-coil motifs in protein sequences (http://groups.csail.mit.edu/cb/paircoil2/paircoil2.html). 'Matcher' (http://cis.poly.edu/~jps/) determines whether a given sequence contains heptads and assigns heptad positions to residues (Fischetti et al., 1993). To predict the oligomerization states of coiled coils 'Multicoil2' (Trigg et al., 2011) uses pairwise correlations and Hidden Markov Models (HMMs). For distinguishing dimers, trimers and non-coiled-coil oligomerization states the algorithm integrates sequence features through a multinomial logistic regression and devises an optimized scoring function that incorporates pairwise correlations localized in the sequence. A database comprising 2015 sequences with reliable structural annotation from experimental data is used (http://multicoil2.csail.mit.edu). 'SCORER' (Armstrong et al., 2011) also provides predictions of coiled-coil oligomerization (http://coiledcoils.chm.bris.ac.uk/Scorer) .

Assignment of the coiled coil packing: COILCHECK (Alva et al., 2008) can be used for analysis and validation of coiled-coil structures through calculation of the strength of interhelical interactions in coiled coils; it can be used to rationalize the behaviour of single residue mutations and to design mutations (http://caps.ncbs.res.in/coilcheck/). SOCKET (Walshaw & Woolfson, 2001) can be used to identify coiled coils through an analysis of the knobs-into-holes side chain packing (http://coiledcoils.chm.bris.ac.uk/socket/).

Databases: For genomewide predictions the 'SpiriCoil' algorithm (Rackham et al., 2010) is employed which uses hundreds of HMMs representing coiled-coil-containing domain families. Their results are available through the SpiriCoil Database (http //supfam.org/SUPERFAMILY/spiricoil). It includes results from all completely sequenced genomes. The CC+ database is a detailed, searchable repository accessible via the SOCKET program (Testa et al., 2009) (http://coiledcoils.chm.bris.ac.uk/ccplus/).

Several of the above tools have been used in sections 6 and 7 of this chapter. In addition, protein sequences were retrieved from the NCBI/GenBank and specialized databases e.g. PPI: *P.syringae* Genome Resources (www.pseudomonas-syringae.org) and the Kyoto Encyclopedia for Genes and Genomes (KEGG) (Kanehisa & Goto, 2000). Secondary structure predictions were performed with 'PSIPRED' (Jones, 1999). Protein structures were retrieved from the Protein Data Bank (PDB).

5.2 *In silico* analysis of T3SS effectors and secretion signals

A selection of bioinformatics tools is available for T3SS effector and secretion signal prediction: 'Effective' is an on-line tool for sequence-based prediction of secreted proteins available from the TUM Genome Oriented Bioinformatics, University of Vienna (Arnold et al., 2009; Jehl et al., 2011), which can be used for the effector prediction in bacterial protein-sequences (http://www.effectors.org/). 'Effective' provides pre-calculated predictions on bacterial effectors in all publicly available pathogenic and symbiotic genomes or using sequence data provided by the user. T3SS secretion signal predictions from amino acid sequences, is available from 'moblab' (http://gecco.org.chemie.uni-frankfurt.de/T3SS_prediction/T3SS_prediction.html). The basic concepts of this tool are described by Lower & Schneider (2009). The 'SIEVE' Server (http://www.sysbep.org/sieve/) for the prediction of type III secreted effectors was originally described by Samudrala et al., (2009) and recently reviewed by McDermott et al., (2011). Potential T3SS effectors are scored using a computational model developed via Machine-Learning Methodologies.

5.3 Experimental and *in silico* analysis of disordered domains

Disordered regions may be detected in protein structures determined by X-ray crystallography through missing electron density. Heteronuclear multidimensional NMR is a powerful tool for the characterization of protein disorder and provides direct measurement of the mobility of unstructured regions (Eliezer, 2007). Loss of secondary structure may be detected (among other methods) by far-UV CD (Kelly & Price, 1997) and Fourier transform infra-red spectroscopy (FTIR) (Uversky et al., 2000). Hydrodynamic parameters obtained from techniques such as gel filtration, SAXS (Gazi et al., 2008), dynamic and static light scattering provide information on whether a protein is unfolded since the unfolding results in an increase in protein hydrodynamic volume. The degree of globularity, which reflects the presence of a well-packed hydrophobic core may be estimated by a special analysis of small angle X-ray scattering (SAXS) data in form of a Kratky plot. Kratky plots are obtained by plotting $I(s) x s^2$ against s (scattering intensity: I; momentum transfer: $s=4\pi \sin(\theta)/\lambda$; 2θ: scattering angle; wavelength of X-rays: λ). They are used to judge the folding of the protein, as the shape of the curve is sensitive to the conformational state of the scattering molecules (Gazi et al., 2008).

Several algorithms have been developed to predict protein disorder on the basis of specific biochemical properties and biased amino acid compositions. These tools include PONDR (Romero et al., 2001; Peng et al., 2005), DisEMBL (Linding et al., 2003), IUPred (Dosztanyi et al., 2005), FoldUnfold (Galzitskaya et al., 2006) and PrDOS (Ishida & Kinoshita, 2007).

The main tool used in sections 6 and 7 for the *in silico* prediction of protein disorder from sequences is FoldIndex© (Prilusky et al., 2005). The propensity of N-termini of proteins for disorder was analyzed on the basis of their biased content of order-/disorder- promoting residues (Dunker et al., 2002).

6. The occurrence of coiled-coils and intrinsic disorder in T3SS proteins

Analyses of T3SS protein sequences (Table 1) reveal an unusually frequent occurrence of predicted heptad repeats, which is indicative of a high propensity for coiled-coil formation (Delahay and Frankel, 2002; Pallen et al., 1997; Gazi et al., 2009; Knodler et al., 2011). Structural studies have confirmed the unsual prevalence of coiled-coils among T3SS proteins (Gazi et al., 2009; Ibuki et al., 2009; Lorenzini et al., 2010). In addition, coiled-coil interactions occur frequently in crystal structures of T3SS protein complexes, e.g. in the a macromolecular assembly TyeA-YopN that regulates type III secretion in *Yersinia pestis* (Schubot et al., 2005) or in the complex of the filament protein EspA from the enteropathogenic *E. coli* T3SS with its chaperone CesA (Yip et al., 2005a). In a recent report, the interactions of the *Salmonella typhimurium* needle protein PrgI, an α-helical hairpin, with the tip protein SipD which comprises a long, central coiled coil (Rathinavelan et al., 2011) were studied using NMR paramagnetic relaxation enhancement. A specific region on the SipD coiled-coil was identified as the binding site for the α-helix of PrgI. Crystallographic studies of the PrgI-SipD complex have revealed coiled-coil interactions via the formation of an intermolecular 4-α-helical bundle structure (Lunelli et al., 2011). These studies also showed the importance of the structural flexibility of SipD (introduced by a π-bulge structure) in complex formation. Coiled-coil interactions of HrpO and FliJ with their cognate protein targets have been also reported (Gazi et al., 2008).

Predicted coiled-coil domains have been shown by mutagenesis to enhance membrane association of *Salmonella* T3SS effectors (Knodler et al., 2011). T3SS proteins and coiled-coil domains are frequently predicted to be structurally disordered (Table 1, 2). For many T3SS effectors disorder in their N-terminal region, as well as an increased overall flexibility have been also noted (Table 1, Gazi et al., 2009). In the following, these aspects of T3SS proteins will be elaborated with specific examples from various protein families. Structures of T3SS proteins with increased coiled-coil content are shown in Fig. 3

Protein	% heptad repeats	% order-promoting aa among the 50 N-terminal residues	% disorder-promoting aa among the 50 N- terminal residues	% overall disorder
EFFECTORS				
AvrPto1	27 (~70)	32	52	58
HopE1	23	24	44	50
HopH1	24	26	49	61
HopY1	26	30	62	47
PILUS				
HrpA1	43	40	46	26
OTHER SECRETED/PUTATIVE SECRETED				
HrpJ	41	32	54	30
HrpF	60	28	48	57
HrpB	23	26	58	18
CYTOPLASMIC				
HrpO	79	22	58	85
HrpE	51	30	54	17
HrcN	21	36	54	21
$HrcQ_B$	53	14	54	46

Table 1. Heptad repeats prediction and disorder analysis for selected proteins from the T3SS of *P. syringae* pv. tomato DC3000. Only proteins with coiled-coil content above 20% are given. For the AvrPto1 protein the crystalographically determined coiled-coil content is given in parentheses. The overall disorder was calculated using FOLDINDEX. N-terminal protein disorder calculations used Dunker's et al. (2002) definition of order-/ disorder-promoting residues. $HrcQ_B$ does not include the disordered N-terminal domain.

6.1 Cytoplasmic proteins

Several cytoplasmic T3SS proteins exhibit a significant coiled coil propensity and intrinsic disorder (Table 1, Fig. 3, 4). Evidence from some cytoplasmic proteins (see section 6.1.1) suggests that these properties might be essential elements in the establishment of key protein-protein interaction networks required for T3SS function (Gazi et al., 2008).

6.1.1 The SctO family (HrpO/FliJ/YscO homologs)

The most extensive heptad repeat pattern occurs in the HrpO/FliJ/YscO family of T3SS proteins (Gazi et al., 2008). Despite the absence of significant homologies, the family members share specific characteristics, e.g. increased propensity for coiled coil formation and intrinsic disorder (Gazi et al., 2008). The extreme flexible nature of $HrpO^{SctO}$ from

Pseudomonas syringae pv. phaseolicola (Gazi et al., 2008), a property shared with FliJ from *S. typhimurium,* has prevented its crystallization and determination of its 3D-structure by X-ray crystallography. A variant form of the FliJ protein from *S. enterica* sv. typhimurium was crystallized however (Ibuki et al., 2009), and its structure was found to be remarkably similar (Fig. 5) to that of the two-stranded α-helical coiled-coil part of the γ subunit of F_0F_1-ATP synthase (Ibuki et al., 2011). A similar coiled coil structure (Fig. 5) consisting of two long α-helices was also reported for the crystal structure of the CT670^{SctO} protein (a YscO homolog) from *Chlamydia trachomatis* (Lorenzini et al., 2010).

Protein	% total disorder	Protein	% total disorder	Protein	% total disorder
NEEDLE		**CHAPERONES**		**TIP**	
MxiH	52	SycD	12	LcrV	45
PrgI	21	PscE	0	IpaD	41
BsaL	30	PscG	7	BipD	26
		YscE	27	EspA	6
		YscG	39		
		CesA	66		

Table 2. Disorder analysis for T3SS proteins of known 3D-structures with coiled-coil content exceeding 30%. The overall protein disorder was calculated from sequence data using the FoldIndex program with a window of 21 residues.

Small angle X-ray scattering (SAXS) and circular dichroism (CD) characterization of HrpO^{SctO} from *P. syringae* pv. phaseolicola revealed a high α-helical content with coiled-coil characteristics and molten globule-like properties (Gazi et al., 2008). HrpO^{SctO} like its flagellar counterpart FliJ is essential for export, but its function remains obscure. HrpO^{SctO} interacts, probably via intermolecular coiled-coil formation, with HrpE, a highly α-helical T3SS protein which belongs to the HrpE/FliH/YscL family. FliH, the flagellar counterpart of HrpE is a regulator of the FliI ATPase (Lane et al., 2006). Evidence from HrpO^{SctO} and its analogs in various flagellar or non-flagellar T3S systems suggests that the extreme flexibility (Fig. 4) and propensity for coiled-coil interactions observed in members of the HrpO/FliJ/YscO family might be important factors for increased interactivity and the establishment of functional protein-protein interaction networks in T3SS. This is consistent with the observation that several members of the HrpO/FliJ/YscO family were found to interact with other cytosolic T3SS components or self-associate via coiled-coil interactions: The flagellar FliJ protein, a key player in a chaperone escort mechanism that recruits unloaded chaperones for the minor filament-class subunits of the filament cap and hook-filament junction substructures (Evans et al., 2006) binds to the same chaperone site as the cognate export substrate of the chaperone, albeit with a much lower affinity. Similarly, YscO^{SctO} from *Yersinia enterocolitica* and InvI^{SctO} from *Salmonella typhimurium* do not bind to export substrates but recognize a subset of export chaperones that are specialized to deliver the T3SS translocators to the export apparatus (Evans & Hughes, 2009). In all of these cases the interaction partners of the HrpO/FliJ/YscO family members exhibit a very high α-helical/coiled-coil content (Fig. 5). FliJ was also found to interact with structural cytoplasmic components of the T3SS like the FliM^{SctQ} protein, even in the absence of FliH, suggesting a docking mechanism for export substrates, chaperones and the ATPase to the T3SS machinery (Gonzalez-Pedrajo et al., 2006). The CT670^{SctO} protein exists in monomeric and dimeric forms, with the monomeric form dominating at low protein concentrations. For self-association and dimer formation the involvement of coiled-coil interactions is predicted (Lorenzini et al., 2010). CT670^{SctO} interacts with CT671^{SctP}, a T3SS protein with a predicted coiled-coil domain in its C-terminal region. CT671^{SctP} is a homolog of the YscP protein which has been characterized as a molecular ruler and as a switch for T3SS substrate specificity in *Yersinia* species (Agrain et al., 2005). The two coiled-coil containing proteins CT670^{SctO} and CT671^{SctP} have been suggested to form a chaperone-effector-like pair with CT670^{SctO} acting as chaperone (Lorenzini et al., 2010).

The HrpO/FliJ/YscO family members are encoded by genes located always downstream of the gene coding for T3SS ATPases (the SctN family of T3SS proteins which includes HrcN/FliI/YscN homologs); this implies a close connection between these proteins and the ATPase. In flagellar T3SS the FliI protein is an ATPase that has extensive structural similarity to the α- and β- subunits of the F_oF_1-ATP synthase (Imada et al., 2007), while also the structure of FliJ from *S. enterica* sv. typhimurium (Fig. 4, 5) is remarkably similar to that of the two-stranded α-helical coiled-coil part of the γ-subunit of F_oF_1-ATP synthase (Ibuki et al., 2001). FliJ promotes the formation of FliI hexamer rings by binding to the center of the ring in a similar way to the γ-subunit penetrating into the central channel of the $\alpha_3\beta_3$ ring in F_oF_1-ATPase. Moreover, the HrpE/FliH/YscL family of proteins (interaction partners of the HrpO/FliJ/YscO family) are distant homologs to both β- and δ- subunits of the F_oF_1-ATP synthase (Pallen et al., 2006). In flagellar systems the docking of the ATPase to the T3S machinery is mediated by the FliJ/FliH pair (Minamino et al., 2009). These results strongly suggest that T3SS and F- and V-type ATPases share a similar mechanism and an evolutionary relationship. It is thereby striking that extensive coiled-coil domains (e.g. FliJ, FliH) have been conserved between the two systems.

Overall, the above remarkable findings support our earlier suggestions (Gazi et al., 2008, 2009) that T3SS proteins, and in particular members of the SctO family, with long disordered/flexible coiled coil structures occupy node positions in the T3SS interactome, being capable of interacting with different partners and possess various roles in the secretion mechanism. These roles are to a large extent poorly understood and remain to be elucidated experimentally.

6.1.2 The SctL family (HrpE/FliH/YscL homologs)

In terms of predicted heptad repeats content (Gazi et al., 2008) the HrpE/FliH/YscL family of proteins comes second after the HrpO/FliJ/YscO family. These proteins are distant homologs of the second-stalk components of the F_0-F_1 ATPases (Pallen et al., 2006). FliH is a regulator of the FliI ATPase (Evans et al., 2006) and was found to interact with the 18 N-terminal residues of FliI that are predicted to form an amphipathic α-helix upon interaction with FliH (Lane et al., 2006).

The HrpE/FliH/YscL family members possess glycine-rich repeats of the form $AxxxG(xxxG)_mxxxA$ with m representing a non constant value between FliH proteins from different bacteria and x standing for any residue. The amino acid sequence distribution of each of the three x positions was found to differ significantly from the overall amino acid composition of the HrpE/FliH/YscL proteins. The high frequency of Glu, Gln, Lys and Ala residues in the repeat positions suggests the presence of α-helical structure for this motif (Trost et al., 2009). When the Protein Data Bank was searched for GxxxG repeats similar in length to those found in FliH, no helices containing more than three contiguous glycine repeat segments were found implying that long GxxxG repeats are presumably quite rare in nature.

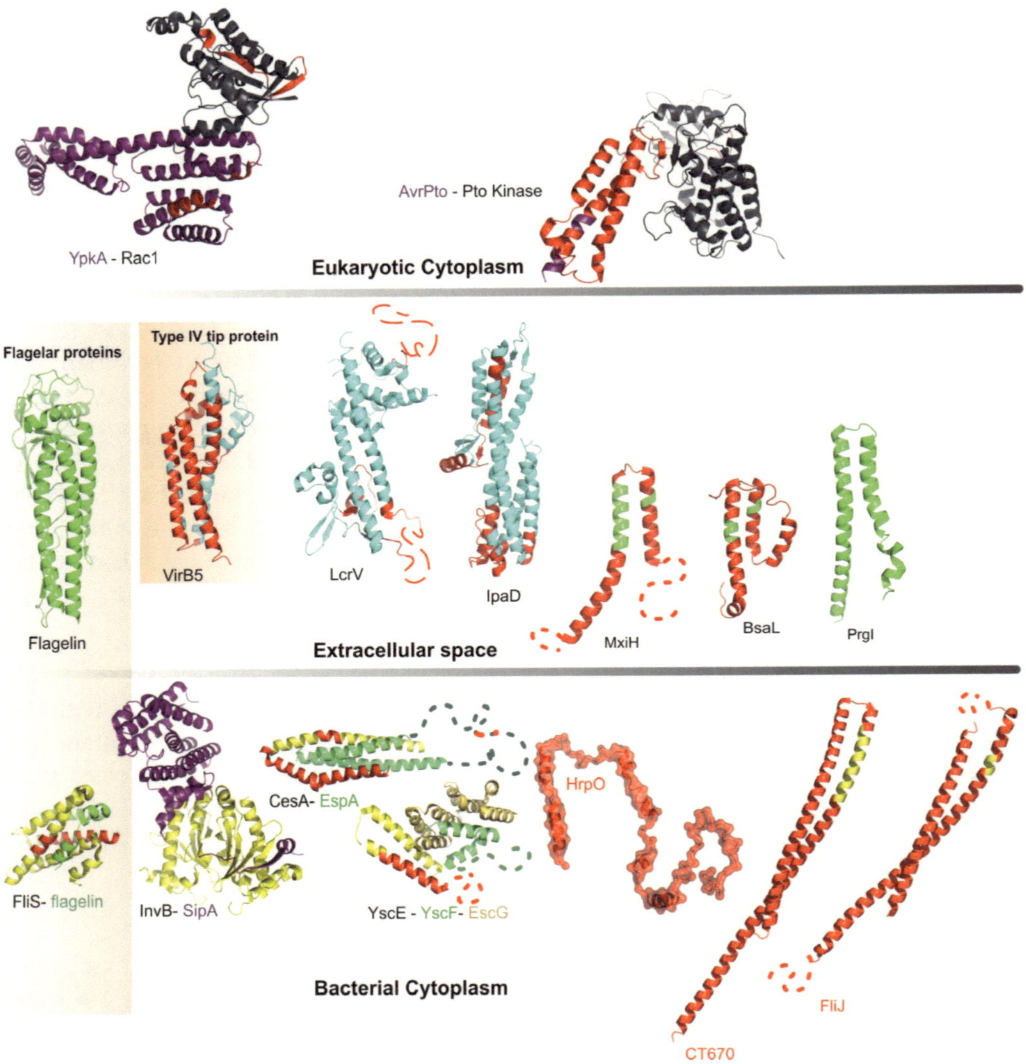

Fig. 3. 3D-structures of T3SS proteins with significant coiled-coil content and their locations. The structure of the intrinsically disordered HrpO is based on SAXS data (Gazi et al., 2008).

6.2 T3SS needle and pilus proteins

The major extracellular T3SS component is the needle with a length of 60 nm and an external diameter of 7 nm for animal pathogens; a much longer structure (up to 2μm) named the Hrp pilus is the needle counterpart in phytopathogenic bacteria (Barrett et al., 2008; Cordes et al., 2003; He and Jin, 2003; Alfano & Collmer, 1997; Roine, 1997). The needle appears to play a major role in host sensing and signal transmission from the distal to the basal end of T3SS (Deane et al., 2006). The needle structures are formed through the helical assembly of multiple copies of a small α-helical protein. Along the needle axis runs a narrow (2.5 nm) conduit which is used for the passage of needle components, tip proteins, translocators and effectors, whereby a partially unfolded form of the substrate is required. Structures are available (Wang et al., 2007; Deane et al., 2006; Zhang et al., 2006) for three needle components from animal pathogens: MxiH (*S. flexneri*), BsaL (*B. pseudomallei*) and PrgI (*S. typhimurium*). These structures are highly α-helical, with a central coiled-coil which is essential for needle assembly. Outside this coiled coil, all three proteins have highly mobile N-termini and C-termini, although these regions may retain some degree of helical structure in solution (Blocker et al., 2008). The sequences of the coiled-coil parts of the needle proteins show strong similarities, suggesting that they all share a common fold and pattern of interactions (Wang et al., 2007; Zhang et al., 2006). Analysis of structures and sequences of needle components from various pathogens suggests that the majority has a propensity for structural disorder (Table 2). An analysis of needle/pilus components predicts a mean overall disorder of approx. 30%. The predicted disorder has been confirmed by CD and thermal unfolding studies of MxiH, BsaL and PrgI which reveal that under conditions resembling the physiological ones, all three C-terminally truncated proteins adopt a molten globule-like state; at temperatures above 37° C their tertiary structure collapses while the secondary structure is largely retained (Barrett et al., 2008); this behaviour is strongly reminescent to the one observed for the cytoplasmic $HrpO^{SctO}$ protein (Gazi et al., 2008). A partially unfolded state of the needle components could be functionally important e.g. for transversing the needle channel and for the extracellular assembly. Signal transmission for host cell sensing is suspected to utilize the flexibility of needle subunits (Deane et al., 2006).

The major subunits of the Hrp pilus (HrpA) are generally predicted to be almost entirely α-helical, with the exception of the *Pseudomonas syringae* species, for which the 50 N-terminal amino acids are predicted to contain β-strands (He and Jin, 2003; Koebnik, 2001). The major subunits of the Hrp-dependent pili, like the needle structural proteins, are all small proteins (of 6 to 11 kDa), but their sequences are surprisingly hypervariable, even within *P. syringae* pathovars. This hypervariability may reflect the evolutionary adaptations to evade plant defense systems. The predicted secondary structures of the major pilus subunits however, are remarkably similar, almost entirely α-helical. Insights into the structure of Hrp pilus components have been obtained from the recent investigation of the HrpA protein from the *P. syringae* pv. phaseolicola T3SS pilus (Kotzabasaki & Kokkinidis, unpublished). The C-terminal part of the 11 kDa protein is responsible for the assembly of multiple HrpA copies in the pilus (Roine, 1997). No chaperons for HrpA have been identified. The secondary structure of HrpA is predicted to be highly α-helical, with a propensity for coiled-coil formation in its functionally important C-terminal region. Surprisingly, experimental characterization of the HrpA protein using CD, Raman, FTIR and SAXS provides strong evidence that HrpA does not adopt a helical structure, but rather a highly disordered state with β-strand features. High resolution transmission electron microscopy (TEM) of purified HrpA samples reveals a pronounced propensity for polymerization and formation of two types of fibrils with nano-to micro scale features, one of which has comparable geometrical parameters with the Hrp pilus.

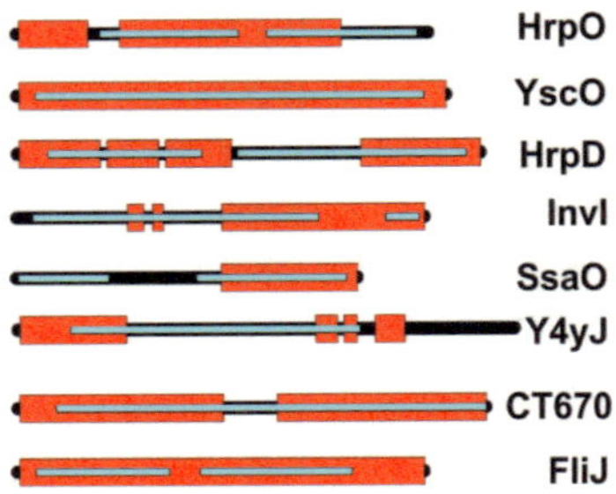

Fig. 4. Predicted coiled-coil regions and structural disorder for the HrpO/FliJ/YscO family. Disordered segments (predicted by FoldIndex) are colored red, coiled-coil domains cyan. All sequences were predicted to be almost entirely α-helical, which was confirmed for HrpO (Gazi et al., 2008), FliJ (Ibuki et al., 2011) and CT670 (Lorenzini et al., 2010). Sequences are drawn to scale, and vary from 166 (HrpD) to 125 residues (SsaO).

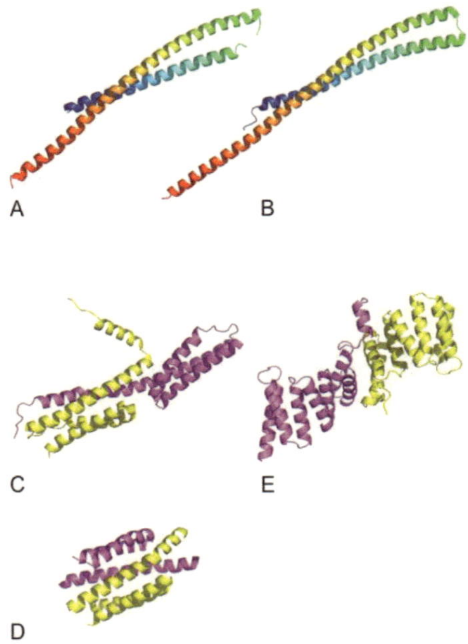

Fig. 5. T3SS proteins involved in coiled-coil interactions: (A) FliJ (Ibuki et al., 2011) from *S. typhimurium*, (B) CT670 ^{SctO} (Lorenzini et al., 2010) from *Ch. trachomatis* colored from N- (blue) to C-terminus (red). The dimeric FliT of *S. enterica* sv. typhimurium (C) and *Bordetella bronchiseptica* (D), each monomer differently coloured. In (C) the C-terminal helix adopts a different conformation in each monomer (Imada et al., 2010). (E) The *Y. enterocolitica* SycD (Buttner et al., 2008) which is recognized by YscO ^{SctO} (Evans & Hughes, 2009).

6.3 T3SS effector flexibility and coiled-coil propensity

Approximately a third of the T3SS effector structures known forms regular coiled-coils with knob-into-holes packing, while several others exhibit short heptad repeat patterns in their sequences which give rise to coiled-coil interactions and short, distorted α-helical bundles: A YpkA subdomain folds in two 3-helical bundles (Prehna et al., 2006). Coiled-coils are also observed in the N-terminal domains of YopH and SptP (Khandelwal et al., 2002; Stebbins and Galan, 2000) and in MxiC and YopN-TyeA (Deane et al., 2008; Schubot et al., 2005). The *S. flexneri* effector IpaH is a E3 ubiquitin ligase with a C-terminal which consists entirely of α-helical bundles and carries the catalytic activity for ubiquitin transfer (Singer et al., 2008). Helical bundle structures are also adopted by AvrPto (Wulf et al., 2004) and AvrPtoB (Dong et al., 2009), although both interact with their target protein kinase Pto via β-strand addition (Xing et al., 2007). AvrPto (PDB ids: 1R5E, 2QKW) displays considerable structural plasticity which is consistent with its predicted increased flexibility (Table 1). In a recent analysis, 49% of the *S. typhimurium* effectors have been predicted to posses at least one coiled-coil domain which enhances membrane association in mammalian cells (Knodler et al., 2011).

Apart from the frequent occurrence of coiled-coils, a further common feature among T3SS effectors are disorder effects established both as localized disorder of their extreme N-terminal peptide where the secretion signal resides, or frequently as an overall structural disorder. Usually, the 15-20 N-terminal residues of effectors are highly disordered. Truncation of N- and C-terminal residues was necessary for the NMR study of the AvrPto effector (Wulf et al., 2004), while in the crystal structure of the AvrPto-Pto kinase complex 28 N-terminal residues are missing (Xing et al., 2007). In the case of ExsE the N- terminal fifteen residues had to be omitted for crystallization and structure determination (Vogelaar et al., 2010). Although, the full length AvrB and AvrPphF ORF2 were crystallized, electron density was not observed for the 27 N-terminal residues due to disorder (Lee et al., 2004; Singer et al., 2004). The N-terminal region of the SipA effector in the complex with the chaperone InvB is highly disordered (Lilic et al., 2006). In the CBD of effectors bound to class IA chaperones, there is a prevalent localized disorder for the part of the effector that crosses the interface of the dimeric chaperone; thus, this part of the effector cannot be modelled, as can be seen not only for SipA but also for ExsE and YopN (Vogelaar et al., 2010). Moreover, a large majority (~75%) of *P.syringae* pv. tomato effectors show a significant propensity for structural disorder in the region of their 50 N-terminal residues based on the ratio of order- vs. disorder- promoting residues (selected effectors are shown in Table 1) which has an average value of 0.45 (M. Kokkinidis, unpublished). Other secreted proteins have an average value of 0.50. The ratio for cytoplasmic T3SS proteins is 0.55; the average ratio in proteomes is 0.58 based on the amino acid frequencies of order- and disorder-promoting residues (Brooks et al., 2002); significantly lower values, as in the case of the N-termini of T3SS effectors, indicate a propensity for disorder. The only structured N-terminal region of a T3SS effector is that of YopH. The 129 residue N-terminal domain has two functions: the first 70 residues contain the CBD domain for chaperone SycH, while the full 129 N-terminal domain binds to phosphotyrosine-containing proteins and adopts an overall globular fold (Khandelwal et al., 2002). The N-terminal region of VirA is partially disordered (Davis et al., 2008). Apart from the N-terminal disorder, a propensity for overall disorder is predicted for T3SS effectors, which may reflect an increased structural flexibility. An average value of 35% of disordered residues is predicted by FOLDINDEX for the T3SS effectors of *P. syringae* pv. tomato DC3000, which is to be contrasted with an average value of 28% for cytoplasmic T3SS proteins, if the extensively disordered members of the HrpO/FliJ/YscO family (some of which could be classified as IDPs) are excluded. For other secreted/putatively secreted T3SS proteins the values are 30-37%.

7. Comparison with other secretory systems: T2SS, T4SS and T6SS

To compare with T3SS, earlier analyses for T2SS and T4SS were updated and an analysis of T6SS protein sequences was performed. Coiled-coil predictions and disorder analysis were carried out for *Helicobacter pylori* (T4SS), *Legionella pneumophila* (T2SS and T4SS), the Type-4-pili (T4P) of *L. pneumophila* and *Pseudomonas aeruginosa* and *P. aeruginosa* strain PA14 (T6SS). For 2878 proteins encoded in the *L. pneumophila* genome (GenBank accession number NC_006369), the predicted coiled-coil content is 4%.

Using the Virulence Factor Database (Yang et al., 2008) for the classification of *L. pneumophila* proteins, a coiled-coil content of 14% is predicted for T2SS (11 proteins), 19% for T4SS (50 proteins), and 13% for T4P (3 proteins). For 1573 proteins of *H. pylori* (NC_000915) a coiled-coil content of 3% is predicted, with 26% for T4SS (24 proteins). For 5571 proteins of *P. aeruginosa* (NC_002516) the predicted content is 4%, 8% for T2SS (11 proteins) and 10% for T4P (32 proteins). A high coiled-coil content is predicted for the T4SS of *H. pylori* and the T4SS effectors of *L. pneumophila*. The latter (Table 3) exhibit a particularly high propensity for structural disorder (on the average 46% disordered regions) and coiled-coil content (30%), thus strongly resembling T3SS effectors. The analysis of T6SS protein sequences retrieved from the KEGG database is in Table 4. Hcp and Vgr proteins are proposed effectors (Mougous, et al., 2006), although they may also act as T6SS machine components (Zheng &Leung, 2007). Non-secreted components include ClpV an AAA+ Clp-like ATPase(Cascales, 2008) and various other core components. The analysis suggests a low coiled-coil content for most secreted proteins (on the average 6%) and a higher one (12% on the average) for core proteins. Interestingly, the mean overall disorder of secreted T6SS components is very high (41%) and comparable to T3SS effectors or to T4SS effectors of *L. pneumophila*. Core components of T6SS display a significantly lower degree disorder (28%).

T4SS effectors	% heptad repeats	% overall disorder	T4SS effectors	% heptad repeats	% overall disorder
Ceg9	0	56	SidJ	5	35
VipA	29	36	LaiA/SdeA	28	42
AnkX/AnkN/LegA8	15	35	YlfA/LegC7	60	40
LidA	47	68	DrrA/SidM	42	47
Ceg19	33	30	LepB	49	50
LegC3	62	60	SidC	36	62
LegC2	67	61	SidF	28	56
RalF	0	43	LepA	7	16
SetA	26	42	LubX/LegU2	10	46
AnkB/Cag27/LegAU13	20	51	VipD	42	35

Table 3. Predictions (using MATCHER) of the heptad repeats content and disorder analysis (using FOLDINDEX) for effectors from the T4SS of *L. pneumophila* strain Philadelphia 1.

Locus No (Protein)	% heptad repeats	% overall disorder	Locus No (Protein)	% heptad repeats	% overall disorder
SECRETED COMPONENTS			CORE COMPONENTS		
PA14_01030 (Hcp)	14	37	PA14_00875 (PpkA)	7	25
PA14_01110 (VgrG)	6	42	PA14_00890 (PppA)	0	14
PA14_01160 (VgrG)	6	59	PA14_00910 (ImpL)	18	22
PA14_33960 (VgrG)	0	34	PA14_00925 (ImpK)	10	29
PA14_34030 (Hcp)	0	20	PA14_00940 (ImpJ)	0	20
PA14_03220 (VgrG)	4	38	PA14_00960 (VasD)	0	35
PA14_03240 (Hcp)	0	52	PA14_00970 *unknown*	45	29
PA14_18985 (VgrG)	0	45	PA14_00980 (HFA)	0	30
PA14_21450 (VgrG)	0	41	PA14_00990 (ImpA)	12	20
PA14_29390 (VgrG)	8	45	PA14_01060 (ImpF)	14	31
PA14_44900 (VgrG)	5	42	PA14_01070 (ImpG)	5	28
PA14_67230 (VgrG)	0	46	PA14_01100 (ClpV)	18	21
PA14_69550 (VgrG)	0	40	PA14_34050 (ImpC)	8	19
PA14_29190 (tse2)	44	27	PA14_34070 (ImpB)	22	49

Table 4. T6SS proteins (T6S system HSI-I, HSI-III) of the *P. aeruginosa* strain PA14.

8. Conclusions

In conclusion, structural studies and *in silico* analyses of bacterial genomes have confirmed the occurrence of coiled-coil domains and protein flexibility in the T3SS and provide a more consolidated insight into the occurrence of such features in other secretory systems, e.g. T4SS and T6SS. In the case of T3SS the occurrence of coiled-coils is considerably higher than the average predicted occurrence in prokaryotic proteomes (Schubot et al., 2005). Coiled-coils occur in all types of T3SS proteins, including in proteins from the T3SSs of plant pathogens, for which in earlier studies no coiled-coils could be predicted (Delahay and Frankel, 2002). Apart from coiled-coils, a further widespread feature in T3SS proteins is a considerable structural flexibility which may range from localized to extensive disorder effects. At the level of experimental observations, disorder manifests itself as missing stretches of electron density in crystallographically determined structures (e.g. in the case of the N-termini of effectors), or occasionally as establishment of a molten-globule-like state at conditions resembling the physiological ones. Examples for the latter include the IDPs HrpO (Gazi et al., 2008) and HrpA from *P. syringae* pv. phaseolicola or the needle subunits MxiH, BsaL and PrgI (Barrett et al., 2008). The flexibility of T3SS proteins is frequently associated with a plasticity of coiled-coil domains; this becomes evident in the case of multiple structural studies of the same protein, e.g. AvrPto (PDB ids: 2QKW, 1R5E) or in differences between subunits of oligomeric proteins, e.g. in the FliT dimer (PDB id: 3A7M).

The combination of coiled-coiled interactions and structural plasticity are frequently essential prerequisites for the establishment of interaction networks within T3SS, as exemplified by the interactions of proteins of the HrpO/FliJ/YscO family with members of the HrpE/FliH/YscL family (Gazi et al., 2008), the SipD/PrgI (Lunelli et al., 2011; Rathinavelan et al., 2011) or the CT670/CT671 interaction (Lorenzini et al., 2010). In addition, the assembly of the T3SS supramolecular structures frequently requires a combination of coiled-coils and conformational flexibility: T3SS needle assembly occurs through the stepwise polymerization of a major subunit (e.g. MxiH, BsaL and PrgI) via a flexible or partially disordered C-terminal helix which exhibits a propensity for coiled-coil interactions. For the IDP HrpA polymerization into pilus-like fibrils has been observed, although no experimental evidence for the involvement of coiled-coil interactions could be obtained, despite the high α-helical content predicted by sequence analysis.

The propensity for disorder is frequently reflected the amino acid composition of T3SS protein sequences. The vast majority N-terminal sequences of T3SS effectors and other secreted proteins exhibits specific biases (Table 1) in their composition with respect to order- and disorder-promoting residues (Dunker et al., 2002; Uversky, 2010), from which a disorder propensity can be predicted, usually in agreement with experimental observations. Interestingly, these disorder-associated biases (as reflected in the ratio of order- vs. disorder- promoting residues), result in sequence preferences for the N-termini which are similar to those determined for T3SS effectors from various bacterial species (Greenberg and Vinatzer, 2003). The structural disorder of the N-termini may thus play a role as a secretion signal, a suggestion made earlier by Akeda & Galan (2005) and confirmed by subsequent analyses (Gazi et al., 2009). However, as N-terminal structural disorder does not ensure specificity of substrate recognition (e.g. the cytoplasmic $HrcQ_B$ protein is predicted to possesses a highly flexible N-terminus), it may be assumed that N-terminal flexibility could be one of multiple secretion signals (Marlovits et al., 2006), with other signals, e.g. chaperones, ensuring specificity. Analysis of effectors and other secreted/non-secreted T3SS components strongly suggests that the overall disorder of T3SS proteins is a further parameter strongly correlated with secretion (Table 1, 2). Flexible or disordered T3SS domains could facilitate rapid unfolding which is necessary for secretion. Both N-terminal and overall flexibility might be thus considered in prediction algorithms for the identification of universal T3SS effectors signatures; this would complement recent efforts based on machine learning approaches (Arnold et al., 2009; Samudrala et al., 2009). Interestingly, sequence stretches with coiled-coils propensities are suitable tertiary motifs to provide the necessary flexibility which is proposed to be associated with secretion. In fact, coiled-coil proteins are frequently viewed as a specific set of intrinsically disordered proteins (Gaspari & Nyitray, 2011) and occasionally they have been observed to display molten globule characteristics (Glykos & Kokkinidis, 2004). A further advantage of coiled-coils might be associated with specific features of their disordered state: As shown in the case of the HrpO protein (Gazi et al., 2008), proteins exhibiting coiled-coil propensity are capable of adopting highly non-globular conformations, while maintaining a considerable α-helical content. The geometrical dimensions of such non-globular helical conformations permit passage through the narrow needle/pilus channel if the appropriate secretion signal is present. It is intuitive to assume that after passing this conduit, such preformed and folding-competent helices encompassing a few turns may form a nucleation site which promotes fast assembly of a globular coiled-coil domain. Flexible coiled-coil domains are thus particularly suitable as secretion substrates as they can easily unfold into secretion-competent α-helices, which in turn may refold in the host cell into a native structure following a relatively fast pathway, and thus avoid degradation of the unfolded polypeptide by host defences. In addition, coiled-coils of effectors may also be a particularly suitable structural motif for interactions in the host cell, as many key processes in the eukaryotic cell involve coiled-coil domains, a fact already noted by Pallen (1997), and confirmed by recent experiments (Knodler et al., 2011). It might be thus hypothetized that the selective evolutionary pressure for optimization of bacterial effectors favours coiled-coil domains and increased flexibility, and this in turn creates a basis for the overall prevalence of coiled-coil domains in T3SS, as this helps establish interaction networks within the T3SS, which may be exploited by even partially unfolded effectors or other secretion substrates.

The predicted high occurrence of coiled-coil domains and structural disorder in T4SS effectors of *L. pneumophila* (Table 3) indicate that the concepts outlined above for T3SS effectors might also some validity in other Gram-negative secretory mechanisms. In addition, the analysis of T6SS secreted components (Table 4) strongly supports the concept of structural flexibility of proteins being an important prerequisite for bacterial secretion. We still have a long way to go to decipher the full complexity of bacterial secretion, even for extensively studied systems such as T3SS. However, the elegant genetic, biochemical, genetic and computational studies which were reviewed in this contribution may open ways to resolve this issue.

9. Acknowledgment

Work was partially funded by the GSRT, Joint Research and Technology Programmes for Greece-France (2010-11). Support for access and use of the SOLEIL synchrotron for SAXS studies is acknowledged. V.E.F is supported by a Marie Curie Reintegration grant. Figure 1 was produced with UCSF Chimera, University of California, San Francisco

10. References

Abdallah, A.M., Gey van Pittius, N.C., Champion, P.A., Cox, J., Luirink, J., Vandenbroucke-Grauls, C.M., Appelmelk, B.J., Bitter, W. (2007). Type VII secretion--mycobacteria show the way. *Nat Rev Microbiol.*, Vol. 5, No 11, pp: 883-91

Agrain, C., I. Sorg, I., C. Paroz, C. and Cornelis, G. R. (2005). Secretion of YscP from *Yersinia enterocolitica* is essential to control the length of the injectisome needle but not to change the type III secretion substrate specificity. *Mol. Microbiol., Vol.* 57, No. 5, pp.1415–27

Akeda, Y. and Galan, J.E. (2005) Chaperone release and unfolding of substrates in type III secretion. *Nature*, Vol. 437, No. 7060, pp. 911-15

Alfano, J.R., and Collmer, A. (1997). The type III (Hrp) secretion pathway of plant pathogenic bacteria: trafficking hairpins, Avr proteins and death. *J. Bacteriol., Vol.* 179, No. 18, pp. 5655-62

Alva, V., Syamala Devi, D. P., & Sowdhamini, R. (2008). COILCHECK: an interactive server for the analysis of interface regions in coiled coils. *Protein Pept Lett*, Vol. 15, No. 1, pp. 33-38

Anderson, D. M., Schneewind, O. (1997). A mRNA signal for the type III secretion of Yop proteins by *Yersinia enterocolitica. Science.* Vo. 278, No. 5340, pp. 1140-1143

Anfinsen, C.B. (1973). Principles that govern the folding of protein chains. *Science*, Vol. 181, No. 96, pp. 223–30

Armstrong, C. T., Vincent, T. L., Green, P. J., & Woolfson, D. N. (2011). SCORER 2.0: an algorithm for distinguishing parallel dimeric and trimeric coiled-coil sequences. *Bioinformatics*, Vol. 27, No. 14, pp. 1908-14

Arnold, R., Brandmaier, S., Kleine, F., Tischler, P., Heinz, E., Behrens, S., Niinikoski, A., Mewes, H-W., Horn, M. & Rattei, T. (2009). Sequence-based prediction of Type III secreted proteins. *PLoS Pathog., Vol.* 5, No. 4:e1000376

Banner, D.W., Kokkinidis, M. & Tsernoglou, D. (1987). Structure of the ColE1 Rop protein at 1.7 Å resolution. *J. Mol. Biol., Vol.* 196, No. 3, pp. 657–75

Barbara, K. E., Willis, K. A., Haley, T. M., Deminoff, S. J. & Santangelo, G. M. (2007). Coiled coil structures and transcription: an analysis of the *S.cerevisiae* coilome. *Mol. Genet.Genomics*, Vol. 278, No. 2, pp. 135–47

Baron, C., Llosa, M., Zhou, S. & Zambryski, P. C. (1997). VirB1, a component of the T-complex transfer machinery of *Agrobacterium tumefaciens*, is processed to a C-terminal secreted product, VirB1. *J. Bacteriol.*, Vol. 179, No. 4, pp. 1203–1210

Barrett, B.S., Picking, W.L., Picking, W.D., Middaugh, C.R. (2008). The response of type three secretion system needle proteins MxiH(Delta5), BsaL(Delta5), and PrgI(Delta5) to temperature and pH. *Proteins*, Vol. 73, No. 3, pp. 632-43

Berger, B., Wilson, D.B., Wolf, E., Tonchev, Th., Milla, M. & Kim, P.S. (1995) Predicting Coiled Coils by Use of Pairwise Residue Correlations. *Proc. National Academy of Science USA*, Vol 92, No. 18, pp. 8259-63

Berger, C., Robin,G. P., Bonas, U. and Koebnik, R. (2010). Membrane topology of conserved components of the type III secretion system from the plant pathogen Xanthomonas campestris pv. vesicatoria. Microbiology, Vol. 156, No. 7, pp. 1963-74

Biemans-Oldehinkel, E., Sal-Man, N., Deng, W., Foster, L.J. and Finlay, B.B. (2011). Quantitative proteomic analysis reveals formation of an EscL-EscQ-EscN type III complex in enteropathogenic *Escherichia coli. J. Bacteriol.*Vol. 193, No. 19, pp. 5514-19

Birtalan, S.C., Phillips, R.M. & Ghosh, P. (2002). Three-dimensional secretion signals in chaperone-effector complexes of bacterial pathogens. *Mol. Cell*, Vol. 9, No. 5, pp. 971–80

Blaylock, B., Riordan, K.E., Missiakas, D.M. & Schneewind, O. (2006). Characterization of the *Yersinia enterocolitica* type III secretion ATPase YscN and its regulator, YscL. *J. Bacteriol.*, Vol. 188, pp 3525-34

Blocker, A. J., Deane, J. E., Veenendaal, A. K. J., Roversi, P., Hodgkinson, J. L., Johnson, S., and Lea, S. M. (2008). What's the point of the type III secretion system needle? *Proc Natl Acad Sci USA*, Vol. 105, No. 18, pp. 6507-13

Bloomer, A.C., Champness, J.N., Bricogne, G., Staden, R. & Klug, A. (1978). Protein disk of tobacco mosiac virus at 2.8Å resolution showing the interactions within and between subunits. *Nature,* Vol. 276, No. 7363, pp. 362-68

Bode, W., Schwager, P. & Huber, R. (1978). The transition of bovine trypsinogen to a trypsin-like state upon strong ligand binding. The refined crystal structures of the bovine trypsinogen-pancreatic trypsin inhibitor complex and of its ternary complex with Ile-Val at 1.9Å resolution. *J. Mol. Biol.,* Vol. 118, No. 1, pp. 99-112

Boyer, F., Fichant, G., Berthod, J., Vandenbrouck, Y. & Attree, I. (2009). Dissecting the bacterial type VI secretion system by a genome wide *in silico* analysis: what can be learned from available microbial genomic resources? *BMC Genomics,* Vol. 10, No, pp 104

Brooks, D.J., Fresco, J.R., Lesk, A.M. & Singh. M. (2002). Evolution of amino acid frequencies in proteins over deep time: inferred order of introduction of amino acids into the genetic code. *Mol. Biol. Evol.,* Vol. 19, No. 10, pp. 1645-55

Brown, P.N., Mathews, M.A., Joss, L.A., Hill, C.P. & Blair, D.F. (2005). Crystal structure of the flagellar rotor protein FliN from *Thermotoga maritime. J. Bacteriol.,* Vol. 187, No 8, pp. 2890-2902

Burkhard, P., Stetefeld, J. & Strelkov, S.V. (2001). Coiled coils: a highly versatile protein folding motif. *Trends Cell Biol.,* Vol. 11, No. 2, pp. 82–88

Buttner, D., Gurlebeck, D., Noel, L. D., Bonas, U. (2004). HpaB from *Xanthomonas campestris* pv. vesicatoria acts as an exit control protein in type III dependent protein secretion. *Mol Microbiol.,* Vol. 54, No. 3, pp. 755–68

Buttner, D., Lorenz, C., Weber, E., Bonas, U. (2006). Targeting of two effector protein classes to the type III secretion system by a HpaC- and HpaBdependent protein complex from Xanthomonas campestris pv. vesicatoria. *Mol Microbiol.,* Vol. 59, No. 2, pp. 513–27

Buttner, C.R., Sorg, I., Cornelis, G.R., Heinz, D.W. & Niemann, H.H. (2008). Structure of the *Yersinia enterocolitica* type III secretion translocator chaperone SycD. *J. Mol. Biol., Vol.* 375, No. 4, pp. 997-1012

Buttner, D. & He, S-Y. (2009). Type III Protein Secretion in Plant Pathogenic Bacteria. *Plant Physiology,* Vol. 150, No. 4, pp. 1656–64

Cao, T.B. & Saier, M.H. Jr. (2003). The general protein secretory pathway phylogenetic analyses leading to evolutionary conclusions. *Biochim Biophys Acta,* Vol. 1609, No. 1, pp. 115-25

Cascales, E. (2008). The type VI secretion toolkit. *EMBO Reports,* Vol. 9, No 8, pp 735-41

Christie, P. J. & Vogel, J. P. (2000). Bacterial type IV secretion: conjugation systems adapted to deliver effector molecules to host cells. *Trends Microbiol.,* Vol. 8, No. 8, pp. 354-60

Cianciotto, N.P. (2005). Type II secretion: a protein secretion system for all seasons. *Trends Microbiol.,* Vol. 13, No. 12, pp. 581-88

Cordes, F.S., Komoriya, K., Larquet, E., Yang, S., Egelman, E.H., Blocker, A., Lea, S.M. (2003). Helical structure of the needle of the type III secretion system of *Shigella flexneri. J Biol Chem.,* Vol. 278, No. 19, pp. 17103-107

Cornelis, G. R. (2006). The type III secretion injectisome. *Nature Reviews. Microbiology,* Vol. 4, No.11, pp. 811-25

Cortese, M. S., Uversky, V. N. & Dunker, A.K. (2008). Intrinsic disorder in scaffold proteins: getting more from less. *Progress Biophys. Mol. Biol.*, vol. 98, no. 1, pp. 85–106

Crick, F.H. (1952). Is alpha-keratin a coiled coil? *Nature* Vol. 170, No. 4334, pp. 882–83

Dale, C. & Moran, N.A. (2006). Molecular interactions between bacterial symbionts and their hosts. *Cell*, Vol. 126, No. 3, pp. 453-65

Davis, J., Wang, J., Tropea, J. E., Zhang, D., Dauter, Z., Waugh, D. S., Wlodawer, A. (2008). Novel fold of VirA, a type III secretion system effector protein from *Shigella flexneri*. *Protein Sci.*, Vol. 17, No. 12, pp. 2167-73

Dean, P. (2011). Functional domains andmotifs of bacterial type III effector proteins and their roles in infection. *FEMS Microbiol Rev* pp. 1–26

Deane, J.E., Roversi, P., Cordes, F.S., Johnson, S., Kenjale, R., Daniell, S., Booy, F., Picking, W.D., Picking, W.L., Blocker, A.J. & Lea, S.M. (2006). Molecular model of a type III secretion system needle: Implications for host-cell sensing. *Proc Natl Acad Sci USA, Vol.* 103, No. 33, pp. 12529–33

Deane, J.E., Roversi, P., King, C., Johnson, S., Lea, S.M. (2008). Structures of the *Shigella flexneri* type 3 secretion system protein Mxic reveal conformational variability amongst homologues. *J.Mol.Biol., Vol.* 377, No. 4, pp. 985-92

Delahay, R.M. & Frankel, G. (2002) Coiled-coil proteins associated with type III secretion systems: a versatile domain revisited. *Mol Microbiol.*, Vol. 45, No. 4, pp. 905-16

Derewenda, U., Mateja, A., Devedjiev, Y., Routzahn, K.M., Evdokimov, A.G., Derewenda, Z.S. & Waugh, D.S. (2004) The structure of *Yersinia pestis* V-antigen, an essential virulence factor and mediator of immunity against plague. *Structure*, Vol. 12, No. 2, pp. 301-306

Desvaux, M., Parham, N.J. & Henderson, I.R. (2004). Type V protein secretion: simplicity gone awry? *Curr. Issues Mol. Biol.*, Vol. 6, No. 2, pp. 111–24

Desveaux, D., Singer, A.U. & Dangl, J.L. (2006). Type III effector proteins: doppelgangers of bacterial virulence. *Current Opinion in Plant Biology*, Vol. 9, No. 4, 376–82

Diepold, A., Amstutz, M., Abel, S., Sorg, I., Jenal, U., & Cornelis, G. R. (2010). Deciphering the assembly of the *Yersinia* type III secretion injectisome. *Embo J.*, Vol. 29, No. 11, pp. 1928-40

Dong, J., Xiao, F., Fan, F., Gu, L., Cang, H., Martin, G. B., Chai. (2009). Crystal structure of the complex between *Pseudomonas* effector AvrPtoB and the tomato Pto kinase reveals both a shared and a unique interface compared with AvrPto-Pto. *Plant Cell,* Vol. 21, No. 6, pp. 1846-59

Dosztanyi, Z. , Csizmok, V., Tompa, P. & Simon, I. (2005). IUPred: web server for the prediction of intrinsically unstructured regions of proteins based on estimated energy content. *Bioinformatics*, Vol. 21, No. 16, pp. 3433–34

Dunker, A.K., Brown, C.J., Lawson, J.D., Iakoucheva, L.M. & Obradovic, Z. (2002). Intrinsic disorder and protein function. *Biochemistry, Vol.* 41, No. 21: 6573-82

Dunker, A.K., Oldfield, C.J., Meng, J., Romero, P., Yang, J.Y., Chen, J.W., Vacic, V., Obradovic, Z. & Uversky, V.N. (2008a). The unfoldomics decade: an update on intrinsically disordered proteins. *BMC Genomics*, Vol. 9, suppl. 2, article S1

Dunker, A.K., Silman, I., Uversky, V.N. & Sussman, J. L. (2008b). Function and structure of inherently disordered proteins. *Current Opinion in Structural Biology*, Vol. 18, No. 6, pp. 756–64

Dunker, A.K. & Uversky, V.N. (2008). Signal transduction via unstructured protein conduits. *Nature Chemical Biology*, Vol. 4, No. 4, pp. 229–230

Durand, F., Dagkessamanskaia, A., Martin-Yken, H., Graille, M., Van Tilbeurgh, H., Uversky, V.N. & Francois, J.M. (2008). Structure-function analysis of Knr4/Smi1, a newly member of intrinsically disordered proteins family, indispensable in the absence of a functional PKC1-SLT2 pathway in Saccharomyces cerevisiae. *Yeast*, Vol. 25, No. 8, pp. 563–76

Eliezer, D. (2007). Characterizing residual structure in disordered protein States using nuclear magnetic resonance. *Methods Mol. Biol.*, Vol. 350 pp. 49–67

Espina, M., Ausar, F., Middaugh, C.R., Baxter, M.A., Picking, W.D. & Picking, W.L. (2007) Conformational stability and differential structural analysis of LcrV, PcrV, BipD and SipD from type III secretion systems. *Protein Sci., Vol.* 16, No. 4, pp. 704-714

Evans, L.D., Stafford, G.P., Ahmed, S., Fraser, G.M. & Hughes, C. (2006). An escort mechanism for cycling of export chaperones during flagellum assembly. *Proc. Natl. Acad. Sci. USA*, Vol. 103, No. 46, pp. 17474-79

Evans, L.D. & Hughes, C. (2009). Selective binding of virulence type III export chaperones by FliJ escort orthologues InvI and YscO. *FEMS Microbiol. Lett.*, Vol. 293, No. 2, pp. 292–97

Fadouloglou, V.E., Tampakaki, A.P., Glykos, N.M., Bastaki, M.N., Hadden, J.M, Phillips, S.E., Panopoulos, N.J. & Kokkinidis, M. (2004). Structure of HrcQ$_B$-C, a conserved component of the bacterial type III secretion systems. *Proc. Natl. Acad. Sci. USA, Vol.* 101, No. 1, pp. 70-75

Fadouloglou, V.E., Bastaki, M.N., Ashcroft, A.E., Phillips, S.E., Panopoulos, N.J., Glykos, N.M. & Kokkinidis, M. (2009). On the quaternary association of the type III secretion system HrcQ$_B$-C protein: experimental evidence differentiates among the various oligomerization models. *J. Struct. Biol., Vol.* 166, No. 2, pp. 214-25

Filloux, A. (2004). The underlying mechanisms of type II protein secretion. *Biochimica et Biophysica Acta*, Vol. 1694, No. 1–3, pp. 163-79

Fischetti, V. A., Landau, G. M., Schmidt, J. P., & Sellers, P. (1993). Identifying periodic occurences of a template with applications to protein structure. *Inform Process Let*, Vol. 45, No. 1993, pp. 11-18

Galan, J. E., Wolf-Watz, H. (2006). Protein delivery into eukaryotic cells by type III secretion machines. *Nature*, Vol. 444, No. 7119, pp. 567–73

Galzitskaya, O. V., Garbuzynskiy, S. O. & Lobanov, M. Y. (2006). FoldUnfold: web server for the prediction of disordered regions in protein chain. *Bioinformatics*, Vol. 22, No. 23, pp. 2948–49

Gaspari, Z. & Nyitray, L. (2011). Coiled-coils as possible models of protein structure evolution. *Biomol. Conc.*, Vol. 2, No. 3, pp. 199-210

Gauthier, A., Thomas, N.A. & Finlay, B.B. (2003). Bacterial injection machines. *J. Biol. Chem.*, Vol. 278, No. 28, pp. 25273-76

Gazi, A.D., Bastaki, M., Charova, S.N., Gkougkoulia, E.A., Kapellios, E.A., Panopoulos, N.J. & Kokkinidis, M. (2008). Evidence for a widespread interaction mode of disordered proteins in bacterial type III secretion systems *J. Biol. Chem., Vol.* 283, No. 49, pp. 34062-68

Gazi, A., Charova, S.N., Panopoulos, N.J. & Kokkinidis, M. (2009). Coiled-coils in type III secretion systems: structural flexibility, disorder and biological implications. *Cell. Microbiol., Vol.* 11, No. 5, pp. 719–29

Glykos, N.M., Cesareni, G. & Kokkinidis, M. (1999). Protein plasticity to the extreme: changing the topology of a 4-α-helical bundle with a single amino acid substitution. *Structure*, Vol. 7, No. 6, pp. 597-603

Glykos, N.M. & Kokkinidis, M. (2004). Structural polymorphism of a marginally stable 4-alpha-helical bundle. Images of a trapped molten globule? *Proteins*, Vol. 56, No. 3, pp. 420-25

Glykos, N.M., Papanikolau, Y., Vlassi, M., Kotsifaki, D., Cesareni, G. & Kokkinidis, M. (2006). Loopless Rop: Structure and dynamics of an engineered homotetrameric variant of the repressor of primer protein. *Biochemistry*, Vol. 45, No. 36, pp. 10905-19

Goddard, T. D., Huang, C. C., & Ferrin, T. E. (2007). Visualizing density maps with UCSF Chimera. *J Struct Biol*, Vol. 157, No. 1, pp. 281-287

Gonzalez-Pedrajo, B., Minamino, T., Kihara, M., Namba, K. (2006). Interactions between C ring proteins and export apparatus components: a possible mechanism for facilitating type III protein export. *Mol Microbiol.*, Vol. 60, No. 4, pp. 984-98

Greenberg, J.T. & Vinatzer, B.A. (2003). Identifyng type III efectors of plant pathogens and analysing their interaction with plant cell. *Curr. Opin. Microb.*, Vol. 6, No. 1, pp. 20-28

Grigoryan, G. & Keating, A. E. (2008). Structural Specificity in Coiled-coil Interactions. *Curr. Opin. Struct. Biol.*, Vol. 18, No. 4, pp. 477-483

Gsponer, J., Futschik, M.E., Teichmann, S.A. & Babu, M.M. (2008). Tight regulation of unstructured proteins: from transcript synthesis to protein degradation. *Science*, Vol. 322, No. 5906, pp. 1365–68

Hachani, A., Biskri, L., Rossi, G., Marty, A., Menard, R., Sansonetti, P., Parsot, C., Van Nhieu, G.T., Bernardini, M.L. & Allaoui, A. (2008). IpgB1 and IpgB2, two homologous effectors secreted via the Mxi-Spa type III secretion apparatus, cooperate to mediate polarized cell invasion and inflammatory potential of *Shigella flexenri*. *Microbes Infect.*, Vol. 10, No. 3, pp. 260-68

He, S.Y. & Jin, Q. (2003). The Hrp pilus: learning from flagella. *Current Opinion in* Microbiology, Vol. 6, No. 1, pp. 15-19

Henderson, I.R., Navarro-Garcia, F., Desvaux, M., Fernandez, R.C. & Ala'Aldeen, D. (2004) Type V protein secretion pathway: the autotransporter story. *Microbiol. Mol. Biol. Rev., Vol.* 68, No. 4, pp. 692-744

Hienonen, E., Romantschuk, M. & Taira, S. (2006). Stabilization of heterologous transcripts with hrpA, mRNA of a type III secretion system component. *Microbial Cell Factories*, Vol. 5 (Suppl 1), P72

Holland, I.B., Schmitt, L. & Young J. (2005). Type 1 protein secretion in bacteria, the ABC-transporter dependent pathway. *Mol. Membr. Biol.*, Vol. 22, No. 1–2, pp. 29-39

Hubber, A., Roy, C.R. (2010). Modulation of host cell function by *Legionella pneumophila* type IV effectors. *Ann. Rev. Cell. Dev. Biol.* Vol. 26, pp. 261-83

Hueck, C.J. (1998). Type III protein secretion systems in bacterial pathogens of animals and plants. *Microbiol. Mol. Biol. Rev.*, Vol. 62, No. 2, pp. 379-433

Ibuki, T., Shimada, M., Minamino, T., Namba, K. & Imada, K. (2009). Crystallization and preliminary X-ray analysis of FliJ, a cytoplasmic component of the flagellar type III protein-export apparatus from Salmonella sp. *Acta. Crystallogr. Sect. F Struct. Biol. Cryst. Commun.*, Vol. 65, No. 1, pp. 47-50

Ibuki, T., Imada, K., Minamino, T., Kato, T., Miyata, T. & Namba, K. (2011). Common architecture of the flagellar type III protein export apparatus and F- and V-type ATPases. *Nat. Struct. Mol. Biol., Vol.* 18, No. 3, pp. 277-82

Imada, K., Minamino, T., Tahara, A. & Namba, K. (2007). Structural similarity between the flagellar type III ATPase FliI and F1-ATPase subunits. *Proc. Natl. Acad. Sci. USA* Vol 104, No2, pp 485-90

Imada, K., Minamino, T., Kinoshita, M., Furukawa, Y. & Namba, K. (2010). Structural insight into the regulatory mechanisms of interactions of the flagellar type III chaperone FliT with its binding partners. *Proc.Natl.Acad. Sci. USA*, Vol. 107, No. 19, pp. 8812-17

Ishida, T. & Kinoshita, K. (2007). PrDOS: prediction of disordered protein regions from amino acid sequence. *Nucl. Acids Res., Vol.* 35, Web server issue, pp. W460–W464

Izore, T., Perdu, C., Job, V., Attree, I., Faudry, E., & Dessen, A. (2011). Structural characterization and membrane localization of ExsB from the type III secretion system (T3SS) of *Pseudomonas aeruginosa. J. Mol. Biol.*, Vol. 413, No. 1, pp. 236-46

Jakoby, M., Weisshaar, B., Droge-Laser, W., Vincente-Carbajosa, J,. Tiedemann, J., Kroj, T., Parcy, F. (2002). bZIP transcription factors in *Arabidopsis. Trends Plant Sci., Vol.* 7, No. 3, pp. 106–11

Jehl, M.A., Arnold, R. & Rattei, T. (2011). Effective--a database of predicted secreted bacterial proteins. *Nucl. Acids Res., Vol.* 39, Database issue, pp. D591-95

Jin, Q., Thilmony, R., Zwiesler-Vollick, J. & He, S.-Y. (2003). Type III protein secretion in *Pseydomonas syringae. Microbes Infect., Vol.* 5, No. 4, pp. 301-10

Job, V., Matteï, P-J., Lemaire, D., Attree, I. & Dessen, A. (2010). Structural Basis of Chaperone Recognition of Type III Secretion System Minor Translocator Proteins *J. Biol. Chem., Vol.* 285, No. 30, pp. 23224-23232

Jones, D.T. (1999) Protein secondary structure prediction based on position-specific scoring matrices. *J. Mol. Biol.*, Vol. 292, No. 2, pp. 195-202

Kanehisa, M. & Goto, S. (2000). KEGG: Kyoto encyclopedia of genes and genomes. *Nucl. Acids Res.*, Vol. 28, No. 1, pp. 27-30

Kelly, S.M. & Price, N.C. (1997). The application of circular dichroism to studies of protein folding and unfolding. *Biochim. Biophys. Acta*, Vol. 1338, No. 2, pp. 161–85

Khandelwal, P., Kellikuli, K., Smith, C.L., Saper, M.A., Zuiderweg, E. R.P. (2002). Solution structure and phopsphopeptide binding to the N-terminal domain of *Yersinia* YopH: comparison with a crystal structure. *Biochemistry*, Vol. 41, No. 38, pp. 11425-437

Knodler, L.A., Ibarra, J.A., Perez-Rueda, E., Yip, C.K., & Steele-Mortimer, O. (2011). Coiled-coil domains enhance the membrane association of *Salmonella* type III effectors. *Cell. Microbiol.*, Vol. 13, No. 10, pp. 1497-1517

Koebnik, R. (2001). The role of bacterial pili in protein and DNA translocation. *Trends Microbiol.*, Vol. 9, No. 12, pp. 586-90

Koraimann, G. (2003). Lytic transglycosylases in macromolecular transport systems of Gram-negative bacteria. *Cell Mol. Life Sci.*, Vol. 60, No. 11, pp. 2371–88

Korotkov, K. V., Gonen, T., & Hol, W. G. (2011). Secretins: dynamic channels for protein transport across membranes. *Trends Biochem Sci*, Vol. 36, No. 8, pp. 433-43

Lane, M,C, O'Toole, P.W. & Moore, S.A. (2006). Molecular basis of the interaction between the flagellar export proteins FliI and FliH from Helicobacter pylori. *J. Biol. Chem.*, Vol. 281, No. 1, pp. 508-17

Lara-Tejero, M., Kato, J., Wagner, S., Liu, X., & Galan, J. E. (2011). A sorting platform determines the order of protein secretion in bacterial type III systems. *Science*, Vol. 331, No. 6021, pp. 1188-91

Lee, C.C., Wood, M.D., Ng, K., Luginbuhl, P., Spraggon, G., Katagiri, F. (2004). Crystal Structure of the Type III Effector AvrB from *Pseudomonas syringae*. *Structure*, Vol. 12, No. 3, pp. 487-94

Lilic, M., Vujanac, M., Stebbins, C.E. (2006). A common structural motif in the binding of virulence factors to bacterial secretion chaperones. *Mol.Cell*, Vol. 21, No. 5, pp. 653-64

Linding, R., Jensen, L. J., Diella, F., Bork, P., Gibson, T. J. & Russell, R. B. (2003). Protein disorder prediction: implications for structural proteomics. *Structure*, Vol. 11, No. 11, pp. 1453– 1459

Liu, J. & Rost, B. (2001). Comparing function and structure between entire proteomes. *Protein Sci., Vol.* 10, No. 10, pp. 1970-79

Lloyd, S. A., Norman, M., Rosqvist, R., and Wolf-Watz, H. (2001). *Yersinia* YopE is targeted for type III secretion by N-terminal, not mRNA, signals. *Mol. Microbiol.*, Vol. 39, No. 2, pp. 520–31

Lorenzini, E., Singer, A., Singh, B., Lam, R., Skarina, T., Chirgadze, N.Y., Savchenko, A. & Gupta, R.S. (2010). Structure and protein-protein interaction studies on *Chlamydia trachomatis* protein CT670 (YscO Homolog). *J. Bacteriol., Vol.* 192, No. 11, pp. 2746-56

Lountos, G.T., Austin, B.P., Nallamsetty, S., Waugh, D.S. (2009). Atomic resolution structure of the cytoplasmic domain of *Yersinia pestis* YscU, a regulatory switch involved in type III secretion. *Protein Sci., Vol.* 18, No. 2, pp. 467-74

Löwer, M. & Schneider, G. (2009). Prediction of Type III Secretion Signals in Genomes of Gram-Negative Bacteria. *PLoS ONE, Vol.* 4, No. 6: e5917.doi:10.1371/journal.pone.0005917

Lunelli, M., Lokareddy, R.M., Zychlinsky, A. & Kolbe M. (2009). IpaB-IpgC interaction defines binding motif for type III secretion translocator. *Proc. Natl. Acad. Sci. USA, Vol.* 106, No. 24 pp. 9661-66

Lunelli M, Hurwitz R, Lambers J, Kolbe M. (2011). Crystal Structure of PrgI-SipD: Insight into a Secretion Competent State of the Type Three Secretion System Needle Tip and its Interaction with Host Ligands. *PLoS Pathog.*, Vol. 7, No. 8: e1002163

Lupas, A. (1996). Coiled coils: new structures and new functions. *TRENDS in Biochemical Sciences*, Vol. 21, No. 10, pp. 375-82

Lupas, A., Van Dyke, M. & Stock, J. (1991). Predicting coiled coils from protein sequences. *Science*, Vol. 252, No. 5010, pp. 1162-64

Marie, C., Broughton, W.J. and Deakin, W.J. (2001) *Rhizobium* type III secretion systems: legume charmers or alarmers? *Curr Opin Plant Biol.*, Vol. 4, No. 4, pp. 336-42

Marlovits, T. C., Kubori, T., Sukhan, A., Thomas, D. R., Galan, J. E., Unger, V. M. (2004). Structural insights into the assembly of the type III secretion needle complex. *Science*, Vol. 306, No. 5698, pp. 1040–42

Marlovits, T. C., Kubori, T., Lara-Tejero, M., Thomas, D., Unger, V. M., Galan, J. E. (2006). Assembly of the inner rod determines needle length in the type III secretion injectisome. *Nature*, Vol. 441, No. 7093, pp. 637–40

McDermott, J. E., Corrigan, A., Peterson, E., Oehmen, C., Niemann, G., Cambronne, E. D., Sharp, D., Adkins, J. N., Samudrala, R., Heffron, F. (2011). Computational prediction of type III and IV secreted effectors in gram-negative bacteria. *Infect Immun.*, Vol. 79, No. 1, pp. 23-32

McDonnell, A. V., Jiang, T., Keating, A. E., & Berger, B. (2006). Paircoil2: improved prediction of coiled coils from sequence. *Bioinformatics*, Vol. 22, No. 3, pp. 356-58

McDowell, M.A., Johnson, S., Deane, J.E., Cheung, M. Roehrich, D. A., Blocker, A.J., McDonnell, J. M. & Lea, S. M. (2011). Structural and Functional Studies on the N-terminal Domain of the *Shigella* Type III Secretion Protein MxiG. *J. Biol. Chem., Vol.* 286, No. 35, pp. 30606-14

McMurry, J.L., Murphy, J.W. & Gonzalez-Pedrajo, B. (2006). The FliN-FliH interaction mediates localization of flagellar export ATPase FliI to the C ring complex. *Biochemistry* Vol 45, pp. 11790-98

Minamino, T., Yoshimura, S. D., Morimoto, Y. V., Gonzalez-Pedrajo, B., Kami-Ike, N., Namba, K. (2009). Roles of the extreme N-terminal region of FliH for efficient localization of the FliH-FliI complex to the bacterial flagellar type III export apparatus. *Mol Microbiol., Vol.* 74, No. 6, pp. 1471-83

Mougous, J. D., Cuff, M. E., Raunser, S., Shen, A., Zhou, M., Gifford, C. A., Goodman, A. L., Joachimiak, G., Ordonez, C. L., Lory, S., Walz, T., Joachimiak, A. & Mekalanos, J. J. (2006). A virulence locus of *Pseudomonas aeruginosa* encodes a protein secretion apparatus. *Science*, Vol. 312, No 5779, pp 1526-30

Newman, J. R. S., Wolf, E. & Kim, P. S. (2000). A computationally directed screen identifying interacting coiled coils from *Saccharomyces cerevisiae*. *Proc. Natl. Acad. Sci. USA*, Vol. 97, No. 24, pp. 13203–08

Namba, K. (2001). Roles of partly unfolded conformations in macromolecular self-assembly. *Genes to Cells*, Vol. 6, No. 1, pp. 1-12

Neyt, C. & Cornelis, G.R. (1999). Role of SycD, the chaperone of the *Yersinia* Yop translocators YopB and YopD. *Mol. Microbiol.*, Vol. 31, No. 1, pp. 143–156

Okon, M., Moraes, T. F., Lario, P. I., Creagh, A. L., Haynes, C. A., Strynadka, N. C. & MacIntosh, L.P. (2008). Structural characterization of the type-III pilot-secretin complex from *Shigella flexneri*. *Structure*, Vol. 16, No. 10, pp. 1544-54

Page, A. & Parsot, C. (2002). Chaperones of the type III secretion pathway: jacks of all trades. *Mol. Microbiol.*, Vol. 46, No. 1, pp. 1–11

Paliakasis, C.D. & Kokkinidis, M. (1992). Relationships between sequence and structure for the four-alpha-helix bundle tertiary motif in proteins. *Prot Engineering*, Vol. 5, No. 8, pp. 739-49

Pallen, M.J., Dougan, G. & Frankel, G. (1997). Coiled-coil domains in proteins secreted by type III secretion systems. *Mol. Microbiol.*, Vol. 25, No. 2, pp. 423-25

Pallen, M.J., Francis, M.S. & Futterer, K. (2003). Tetratricopeptide-like repeats in type-III-secretion chaperones and regulators. *FEMS Microb.. Lett., Vol.* 223, No. 1, pp. 53–60

Pallen, M.J., Bailey, C.M. & Beatson, S.A. (2006). Evolutionary links between FliH/YscL-like proteins from bacterial type III secretion systems and second-stalk components of the FoF1 and vacuolar ATPases. *Protein Sci.*, Vol. 15, No. 4, pp. 935-41

Park, S. Y., Lowder, B., Bilwes, A. M., Blair, D. F. & Crane, B. R. (2006). Structure of FliM provides insight into assembly of the switch complex in the bacterial flagella motor. *Proc. Natl. Acad. Sci. U S A, Vol.* 103, No. 32, pp. 11886–91

Parsot, C., Hamiaux, C., Page, A. L. (2003). The various and varying roles of specific chaperones in type III secretion systems. *Curr. Opin. Microb.*, Vol. 6, No. 1, pp. 7-14

Parsot, C., Ageron, E., Penno, C., Mavris, M., Jamoussi, K., d'Hauteville, H., Sansonetti, P. & Demers, B. (2005). A secreted anti-activator, OspD1, and its chaperone, Spa15, are involved in the control of transcription by the type III secretion apparatus activity in *Shigella flexneri*. *Mol Microbiol*, Vol. 56, No. 6, pp. 1627-35

Paul, K. & Blair, D.F. (2006). Organization of FliN subunits in the flagellar motor of Escherichia coli. *J. Bacteriol.*, Vol. 188, No. 7, pp. 2502-11

Paul, K., Gonzalez-Bonet, G., Bilwes, A.M., Crane, B.R. & Blair, D. (2011). Architecture of the flagellar rotor. *EMBO J.*, Vol. 30, No. 14, pp. 2962-71

Peng, K., Vucetic, S., Radivojac, P., Brown, C. J., Dunker, A. K. & Obradovi´c, Z. (2005). Optimizing long intrinsic disorder predictors with protein evolutionary information. *J. Bioinf. Comp. Biol.*Vol. 3, No. 1, pp. 35–60

Pentony, M. M. & Jones, D. T. (2010). Modularity of intrinsic disorder in the human proteome. *Proteins, Vol.* 782010, No. 178, pp. 212–221

Prehna, G., Ivanov, M.I., Bliska, J.B., Stebbins, C.E. (2006). *Yersinia* virulence depends on mimicry of host rho-family nucleotide dissociation inhibitors. *Cell*, Vol. 126, No. 5, pp. 869-80

Prilusky, J., Felder, C.E., Zeev-Ben Mordehai, T., Rydberg, E.H., Man, O., Beckmann, J.S., Silman, I. & Sussman J.L. (2005) FoldIndex©: a simple tool to predict whether a given protein sequence is intrinsically unfolded. *Bioinf.*, Vol. 21, No. 16, pp. 3435-38

Priyadarshi, A. & Tang, L. (2010). Crystallization and preliminary crystallographic analysis of the type III secretion translocator chaperone SicA from *Salmonella enterica*. *Acta Cryst. F, Vol.* 66, No. 11, pp. 1533-35

Quinaud, M., Ple, S., Job, V., Contreras-Martel, C., Simorre, J.P., Attree, I. & Dessen, A. (2007). Structure of the heterotrimeric complex that regulates type III secretion needle formation. *Proc. Natl. Acad. Sci. U S A*, Vol. 104, No. 19, pp. 7803–08

Rackham, O. J., Madera, M., Armstrong, C. T., Vincent, T. L., Woolfson, D. N., & Gough, J. (2010). The evolution and structure prediction of coiled coils across all genomes. *J Mol Biol*, Vol. 403, No. 3, pp. 480-93

Ramamurthi, K. S., and Schneewind, O. (2002). *Yersinia enterocolitica* type III secretion: mutational analysis of the yopQ secretion signal. *J. Bacteriol., Vol.* 184, No. 12,pp. 3321–28

Rathinavelan, T., Tang, Ch. & De Guzman, R.N. (2011). Characterization of the Interaction between the *Salmonella* Type III Secretion System Tip Protein SipD and the Needle Protein PrgI by Paramagnetic Relaxation Enhancement. *J. Biol. Chem.*, Vol. 286, No. 6, pp. 4922-30

Roine, E., Wei, W., Yuan, J., Nurmiaho-Lassila, E. L., Kalkkinen, N., Romantschuk, M., He, S-Y. (1997). Hrp pilus: an *hrp*-dependent bacterial surface appendage produced by Pseudomonas syringae pv. Tomato DC3000. *Proc Natl Acad Sci USA, Vol.* 94, No. 7, pp. 3459-64

Romero, P., Obradović, Z., Li, X., Garner, E. C., Brown, C. J. & Dunker, A. K. (2001). Sequence complexity of disordered protein. *Proteins*, Vol. 42, No. 1, pp. 38–48

Rose, A., Manikantan, S., Schraegle, S. J., Maloy, M. A., Stahlberg, E. A. & Meier, I. (2004). Genome-wide identification of *Arabidopsis* coiled-coil proteins and establishment of the ARABI-COIL database. *Plant Physiol.*, Vol. 134, No. 3, pp. 927–39.

Samudrala, R., Heffron, F., McDermott, J. E. (2009). Accurate prediction of secreted substrates and identification of a conserved putative secretion signal for type III secretion systems. *PLoS Pathog., Vol.* 5, No. 4:e1000375

Sarris, P. F. & Scoulica, E. V. (2011). *Pseudomonas entomophila* and *Pseudomonas mendocina*: Potential models for studying the bacterial type VI secretion system. *Infection, Genetics and Evolution*, Vol. 11, No 6, pp 1352-60

Sarris, P. F., Zoumadakis, C., Panopoulos, N. J. & Scoulica, E. (2011). Distribution of the putative type VI secretion system core genes in *Klebsiella* spp. *Infection Genetics and Evolution*, Vol. 11, No 1, pp 157–166

Sato, H., Hunt, M.L., Weiner, J.J., Hansen, A.T. & Frank, D.W. (2011). Modified needle-tip PcrV proteins reveal distinct phenotypes relevant to the control of type III secretion and intoxication by *Pseudomonas aeruginosa*. *PLoS One*, Vol 6, No. 3:e18356

Sato, H. & Frank, D.W. (2011). Multi-functional characteristics of the *Pseudomonas aeruginosa* type III needle-tip protein, PcrV; comparison to orthologs in other Gram-negative bacteria. *Front Microbiol.*, Vol. 2:142

Sarkar, M.K., Paul, K. & Blair, D.F. (2010). Subunit organization and reversal-associated movements in the flagellar switch of E. coli. *J Biol Chem.*, Vol. 285, No. 1, pp. 675-84

Schraidt, O. & Marlovits, T. C. (2011). Three-dimensional model of Salmonella's needle complex at subnanometer resolution. *Science*, Vol. 331, No. 6021, pp. 1192-95

Schubot, F.D., Jackson, M.W., Penrose, K.J., Cherry, S., Tropea, J.E., Plano, G.V., Waugh, D.S. (2005). Three-dimensional structure of a macromolecular assembly that regulates type III secretion in *Yersinia pestis*. *J. Mol.Biol.*, Vol 346, No. 4, pp. 1147-61

Singer, A.U., Desveaux, D., Betts, L., Chang, J.H., Nimchuk, Z., Grant, S.R., Dangl, J.L. & Sondek, J. (2004). Crystal structures of the type III effector protein AvrPphF and its chaperone reveal residues required for plant pathogenesis. *Structure*, Vol. 12, No. 9, pp. 1669-81

Singer, A. U., Rohde, J. R., Lam, R., Skarina, T., Kagan, O., Dileo, R., Chirgadze, N. Y., Cuff, M. E., Joachimiak, A., Tyers, M., Sansonetti, P. J., Parsot, C., Savchenko, A. (2008). Structure of the *Shigella* T3SS effector IpaH defines a new class of E3 ubiquitin ligases. *Nat Struct Mol Biol.*, Vol. 15, No. 12, pp. 1293-1301

Spreter, T., Yip, C. K., Sanowar, S., Andre, I., Kimbrough, T. G., Vuckovic, M., Pfuetzner, R. A., Deng, W., Yu, A. C., Finlay, B. B., Baker, D., Miller, S. I., and Strynadka, N. C. (2009). A conserved structural motif mediates formation of the periplasmic rings in the type III secretion system. *Nat. Struct. Mol. Biol., Vol.* 16, pp. 468-76

Stebbins, C.E., Galan, J.E. (2000). Modulation of host signaling by a bacterial mimic: structure of the Salmonella effector SptP bound to Rac1. *Mol.Cell*, Vol. 6, No. 6, pp. 1449-60

Stebbins, C.E. (2005) Structural microbiology at the pathogen–host interface. *Cell. Microbiol.*, Vol. 7, No. 9, pp. 1227–36

Sugase, K., Dyson, H.J. & Wright, P.E. (2007) Mechanism of coupled folding and binding of an intrinsically disordered protein. *Nature*, Vol. 447 , No. 7356, pp. 1021-25

Sun, P., Austin, B.P., Tropea, J.E. & Waugh, D.S. (2008). Structural characterization of the *Yersinia pestis* type III secretion system needle protein YscF in complex with its heterodimeric chaperone YscE/YscG. *J.Mol.Biol., Vol.* 377, No. 3, pp. 819-30

Tampakaki, A. P., Fadouloglou, V. E., Gazi, A. D., Panopoulos, N. J. & Kokkinidis, M. (2004). Conserved features of type III secretion. *Cell. Microbiol.*, Vol. 6, No. 9, pp. 805-16

Testa, O. D., Moutevelis, E., & Woolfson, D. N. (2009). CC+: a relational database of coiled-coil structures. *Nucleic Acids Res*, Vol. 37, Database issue, pp. D315-322

Thomas, D. R., Francis, N. R., Xu, C., & DeRosier, D. J. (2006). The three-dimensional structure of the flagellar rotor from a clockwise-locked mutant of Salmonella enterica serovar Typhimurium. *J. Bacteriol.*, Vol. 188, No. 20, pp. 7039-48

Thomassin, J.L., He, X. & Thomas, N.A. (2011). Role of EscU auto-cleavage in promoting type III effector translocation into host cells by Enteropathogenic Escherichia coli. *BMC Microbiol.*, Vol. 11, No. 205.Troisfontaines, P. & Cornelis, G. R. (2005). Type III secretion: more systems than you think. *Physiol.*, Vol. 20, No. 5, pp. 326-39

Trigg, J., Gutwin, K., Keating, A. E., & Berger, B. (2011). Multicoil2: predicting coiled coils and their oligomerization States from sequence in the twilight zone. *PLoS One*, Vol. 6, No. 8, pp. e23519

Troisfontaines, P. & Cornelis, G. R. (2005). Type III secretion: more systems than you think. *Physiol.*, Vol. 20, No. 5, pp. 326-339Trost, B. & Moore, S.A. (2009). Statistical characterization of the GxxxG glycine repeats in the flagellar biosynthesis protein FliH and its Type III secretion homologue YscL. *BMC Microbiol.*, Vol. 9, No. 1, p. 72

Tseng, T-T.; Tyler, B.M. & Setubal, J.C. (2009). Protein secretion systems in bacterial-host associations, and their description in the Gene Ontology. *BMC Microbiology*, Vol. 9, Suppl 1, article S2

Uversky, V.N., Gillespie, J.R. & Fink, A.L. (2000). Why are "natively unfolded" proteins unstructured under physiologic conditions? *Proteins*. Vol. 41, No. 3, pp. 415–27

Uversky, V.N., Oldfield, C.J., Midic, U., Xie, H., Xue, B., Vucetic, S., Iakoucheva, L.M., Obradovic, Z. & Dunke, A.K. (2009). Unfoldomics of human diseases: linking protein intrinsic disorder with diseases. *BMC Genomics*, Vol. 10, Suppl. 1, article S7

Uversky, V.N. & Dunker, A.K. (2010) Understanding protein non-folding. *Biochim Biophys Acta*.Vol. 1804, No. 6, pp. 1231–64

Uversky, V.N. (2010). The mysterious unfoldome: structureless, underappreciated, yet vital part of any given proteome. *J. Biomed. Biotechnolol.*, Vol. 2010: 568068

Vinson, C., Myakishev, M., Acharya, A., Mir, A.A., Moll, J.R., Bonovich, M. (2002). Classification of human B-ZIP proteins based on dimerization properties. *Mol Cell Biol.*, Vol. 22, No. 18, pp. 6321–35

Vogelaar, N. J., Jing, X., Robinson, H. H., Schubot, F. D. (2010). Analysis of the crystal structure of the ExsC.ExsE complex reveals distinctive binding interactions of the *Pseudomonas aeruginosa* type III secretion chaperone ExsC with ExsE and ExsD. *Biochemistry* Vol. 49, No. 28, pp. 5870-79

Wagner, V.E., Bushnell, D., Passador, L., Brooks, A.I. & Iglewski, B.H. (2003) Microarray analysis of Pseudomonas aeruginosa quorum-sensing regulons: effects of growth phase and environment. *J Bacteriol.*, Vol. 185, No. 7, pp. 2080-95

Wagner, S., Konigsmaier, L., Lara-Tejero, M., Lefebre, M., Marlovits, T. C., & Galan, J. E. (2010). Organization and coordinated assembly of the type III secretion export apparatus. *Proc. Natl. Acad. Sci. U S A*, Vol. 107, No. 41, pp. 17745-50

Walshaw J. & Woolfson D.N. (2001) SOCKET: a program for identifying and analysing coiled-coil motifs within protein structures *J. Mol. Biol., Vol.* 307, No 5, pp. 1427-50

Wang, Y., Ouellette, A.N., Egan, C.W., Rathinavelan, T., Im, W., De Guzman, R.N. (2007). Differences in the Electrostatic Surfaces of the Type III Secretion Needle Proteins PrgI, BsaL, and MxiH. *J.Mol.Biol.*, Vol. 371, No. 5, pp. 1304-14

Wolf, E., Kim, P.S. & Berger, B. (1997). MultiCoil: A program for predicting two- and three-stranded coiled coils. *Protein Science*, Vol. 6, No. 6, pp. 1179-89

Wulf, J., Pascuzzi, P.E., Martin, G.B., Nicholson, L.K. (2004). The solution structure of type III effector protein AvrPto reveals conformational and dynamic features important for plant pathogenesis. *Structure*, Vol. 12, No. 7, pp. 1257-68

Xie, H., Vucetic, S., Iakoucheva, L.M., Oldfield, C.J., Dunker, A.K., Uversky, V.N., Obradovic, Z. (2007). Functional anthology of intrinsic disorder. 1. Biological processes and functions of proteins with long disordered regions. *Journal of Proteome Research*, Vol. 6, No. 5, pp. 1882–98

Xing, W.M., Zou, Y., Liu, Q., Hao, Q., Zhou, J.M., Chai, J.J. (2007). The structural basis for activation of plant immunity by bacterial effector protein AvrPto. *Nature*, Vol. 449, No. 7159, pp. 243-47

Yang, J., Chen, L.H., Sun, L.L., Yu, J., Jin, Q. (2008). VFDB 2008 release: an enchanced web-based resource for comparative pathogenomics. *Nucleic Acids Res.*, Vol. 36, Database issue, pp. D539-D542

Yip, C.K., Finlay, B.B. & Strynadka, N.C. (2005a). Structural characterization of a type III secretion system filament protein in complex with its chaperone. Nature Struct. Mol. Biol., Vol. 12, No. 1, pp. 75-81

Yip, C.K., Kimbrough, T.G., Felise, H.B., Vuckovic, M., Thomas, N.A., Pfuetzner, R.A., Frey, E.A., Finlay, B.B., Miller, S.I., Strynadka, N.C. (2005b). Structural characterization of the molecular platform for type III secretion system assembly. *Nature* Vol 435, No. 7042, pp. 702-07

Yu, C., Ruiz, T., Lenox, C., & Mintz, K. P. (2008). Functional mapping of an oligomeric autotransporter adhesin of *Aggregatibacter actinomycetemcomitans*. *J Bacteriol*, Vol. 190, No. 9, pp. 3098-3109

Zarivach, R., Vuckovic, M., Deng, W., Finlay, B.B. & Strynadka, N.C.J. (2007). Structural analysis of a prototypical ATPase from the type III secretion system. *Nat. Struct. Biol.*, Vol 14, No. 2, pp. 131-137

Zarivach, R., Deng, W., Vuckovic, M., Felise, H.B., Nguyen, H.V., Miller, S.I., Finlay, B.B. & Strynadka, N.C. (2008). Structural analysis of the essential self-cleaving type III secretion proteins EscU and SpaS. *Nature*, Vol. 453, No 1, pp. 124-127

Zhang, L., Wang, Y., Picking, W. L., Picking, W. D. & De Guzman R. N. (2006). Solution structure of monomeric BsaL, the type III secretion needle protein of Burkholderia pseudomallei. *J Mol Biol.*, Vol. 359, No. 2, pp. 322–330

Zheng, J. & Leung, K. Y. (2007). Dissection of a type VI secretion system in *Edwardsiella tarda*. *Molecular Microbiology*, Vol. 66, No 5, pp 1192-1206

2

Peptides and Peptidomimetics as Tools to Probe Protein-Protein Interactions – Disruption of HIV-1 gp41 Fusion Core and Fusion Inhibitor Design

Lifeng Cai, Weiguo Shi and Keliang Liu
Beijing Institute of Pharmacology & Toxicology, Beijing China

1. Introduction

Protein-protein interactions play important roles in many critical processes in the life sciences, such as signal transduction, lipid membrane fusion, receptor recognition, *etc.*, and many of them are important targets for drug development and design (Wilson 2009; Tavassoli 2011). Unlike an enzyme-substrate interaction which usually has a deep binding pocket in the protein for substrate binding, protein-protein interactions usually involve a large interacting interface; as a result, it is a big challenge for small molecule drugs to efficiently competitively occupy the interface and disrupt protein-protein interactions that modulate these life processes. Proteins are natural ligands that can modulate protein-protein interactions; however, they are not ideal therapeutic agents because of their expensive production costs and the fact that they are not able to be administered orally. In protein-protein interactions, energy is not always equally distributed throughout the binding interface; a couple of focused areas may account for the main protein-protein interaction energy, called a hot spot, which can be the target for a small molecule protein-protein interaction inhibitor (PPII).

Peptides, with suitable molecular size, provide a bridge between protein and small molecule drugs. Similar to proteins, many peptides are natural ligands that modulate protein-protein interactions in important life processes; they are used as drug leads and/or modified to increase potency and selectivity. Compared with small molecules, peptides are more efficient PPIIs due to their relatively large size, and can be useful tools to probe protein-protein interactions for PPII design.

HIV-1 gp41 mediated virus-cell membrane fusion is critical for HIV-1 infection and *in vivo* propagation (Eckert & Kim 2001; Caffrey 2011), and the mechanism is shared by many other viruses using a class 1 fusion protein as membrane fusion machinery, including some life threatening pathogens such as influenza virus, respiratory syncytial virus (RSV), Ebola virus, and severe acute respiratory syndrome (SARS) virus (Harrison 2008). A critical step in HIV-1 infection is a protein-protein interaction between the gp41 N- and C-terminal heptad repeats (NHR and CHR), that form a coiled-coil six-helical bundle (6-HB), providing energy for virus-cell membrane fusion (Fig. 1). Peptides derived from CHR or NHR can interact with their counterparts in gp41 to prevent fusogenic 6-HB formation and inhibit HIV-1-cell membrane fusion, thus preventing HIV-1 infection and replication. T20 (Fuzeon, enfuvirtide), a 36-mer peptide from HIV-1 gp41 CHR, was approved by the USA FDA in 2003 as the first fusion inhibitor for salvage therapy in HIV/AIDS patients unresponsive to common antiretroviral therapy. Its application has been limited by i) the high cost of peptide synthesis, ii) rapid *in vivo* proteolysis, and iii) poor efficacy against emerging T20-resistant strains. These drawbacks have called for a new generation of fusion inhibitors with improved antiviral and pharmacokinetic profiles.

In this chapter, we will focus on the development of HIV-1 fusion inhibitors, concentrating on C-peptide fusion inhibitors and their peptidomimetics, which have been used as probes and tools to elucidate gp41 NHR-CHR interactions for future fusion inhibitor design and improve, and in the long run, the development of small molecule inhibitors that can disrupt this important protein-protein interaction.

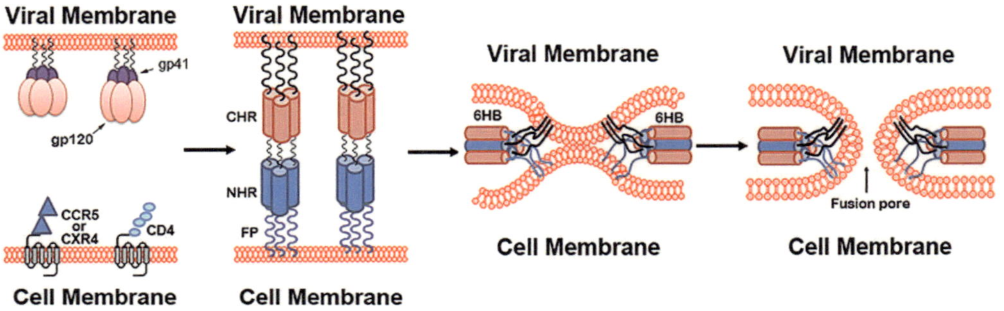

Fig. 1. HIV-1 gp41 mediated virus-cell membrane fusion.

2. Peptides as a model to build the HIV-1 gp41 fusion core

HIV-1 uses an envelope protein (ENV) mediated virus-cell membrane fusion to enter host cells for infection (Eckert & Kim 2001). HIV-1 ENV is composed of noncovalently associated gp120/gp41 trimers that form spikes and decorate the viral surface, in which the metastable transmembrane subunit gp41 is sequestered by the cell surface subunit gp120. During HIV-1 infection, gp120 first interacts with the T-cell receptor CD4, ensuring the viruses approach the target cells; then, the coreceptor binding sites in gp120 are sequentially exposed and gp120-coreceptor (CCR5 or CXR4) binding follows (Fig. 1). The resulting dissociation of the gp120-gp41 complex and the release of the unstable gp41 subunit trigger virus-cell membrane fusion. First, gp41 inserts into the target cell membrane using its fusion peptide, resulting in a pre-hairpin intermediate (PHI) in which its C-terminus anchors to the viral membrane and its N-terminus inserts into the host cell membrane, bridging the viral and cellular membranes (Fig. 1). The gp41 PHI automatically undergoes structure rearrangement with its NHR and CHR folding towards each other to form the fusogenic 6-HB. The energetic 6-HB formation drives the juxtaposition of the viral and cellular membrane, and finally results in virus-cell membrane fusion (Fig. 1). Agents that target the presumed gp41 PHI to prevent fusogenic 6-HB formation can terminate the virus-cell membrane fusion processes and be used as fusion inhibitors for antiretroviral therapy (Cai & Jiang 2010).

The discovery of potent anti-HIV peptides from HIV-1 gp41 NHR and CHR sequences suggests that gp41 is a target for fusion inhibitors (Wild et al. 1992; Jiang et al. 1993; Wild et al. 1994); these exogenous HIV-1 gp41 peptides interact with their counterparts in the gp41 6-HB, forming an unproductive complex that prevents gp41 fusion core formation. During the membrane fusion process, HIV-1 gp41 progressively undergoes a conformational change, and the gp41 PHI target exists for only a couple of minutes and then rapidly folds into a 6-HB; therefore, gp41 and its ectodomain are not suitable targets for a fusion inhibitor. Efforts to obtain a whole structure of the gp41 ectodomain also have been unsuccessful. So, the identification of a stable target in the PHI or gp41 fusion core is necessary for understanding the mechanism of gp41 mediated virus-cell membrane fusion for fusion inhibitor design and development.

The HIV-1 gp41 fusogenic 6-HB core has been constructed using synthesized peptides from the related gp41 wild-type sequences. Typical resolved crystal structures of the 6-HB fusogenic core include the N36/C34 complex (Chan et al. 1997), the IQNgp41/C43 complex (Weissenhorn et al. 1997), and the N34(L6)C28 trimer (Tan et al. 1997). These crystal structures provide atomic resolution of the interactions between NHR and CHR, verifying that NHR and CHR can be both a target and ligand from which a pharmacophore model can be deduced for fusion inhibitor design and optimization.

The crystal structures show that a parallel coiled-coil trimerized NHR forms the interior core, which is antiparallel packed with three CHR helices, to form a 6-HB (Fig. 2a,2b) (Chan et al. 1997). In the NHR interior core, the N-peptide uses its amino acid residues at the *a* and *d* positions of the heptads for self trimerization to stabilize the core; while the *e* and *g* residues of two adjacent helices form three hydrophobic grooves along the whole NHR trimer, which serve as targets that interact with the *a* and *d* residues of the C-peptides. Each groove contains a particularly deep cavity: Val-570, Lys-574, and Gln-577 from the left N36 ($gp41_{546-581}$) helix form the left side; Leu-568, Trp- 571, and Gly-572 from the right N36 helix form the right side; and Thr-569, Ile-573, and Leu-576 form the floor, resulting in a pocket of ~16 Å long, 7 Å wide, and 5–6 Å deep (Fig. 2d). With the exception of Ile-573, all of the residues forming the cavity are identical between HIV-1 and SIV. The NHR deep pocket accommodates three hydrophobic residues from the abutting C34 ($gp41_{628-661}$) helix: Ile-635, Trp-631, and Trp-628 constitute a WWI motif (Fig. 2c). The interaction between the NHR pocket and the WWI motif is predominately hydrophobic. A salt bridge between Lys-574 of NHR and Asp-632 of CHR immediately to the left of the cavity is also important for the NHR-CHR interaction (Chan et al. 1997). In addition to be the main binding sites for the C-peptide, the deep NHR pocket is also an attractive target for small molecule fusion inhibitors. Besides the deep pocket, the rest of the groove along the NHR helices also makes extensive contact with CHR, providing additional energy to stabilize the 6-HB. The N36/C34 complex shows striking structural similarity to the low-pH-induced conformation of the influenza HA2 subunit (TBHA2) and the TM subunit of Mo-MLV, both of which have been proposed to be in a fusogenic conformation, suggesting a common mechanism of virus-cell membrane fusion among enveloped viruses (Chan et al. 1997).

During 6-HB formation, NHR and CHR are mutual target and ligand, so either can be the target for fusion inhibitor design. In a 6-HB, NHRs form a trimerized interior core that contains three grooves, and each with a deep pocket, which is more like a target, especially for small molecule fusion inhibitors. An electrostatic potential map of the N36 coiled-coil trimer shows that its surface is largely uncharged; and the grooves that are the sites for C34 interaction are aligned with predominantly hydrophobic residues that would be expected to lead to aggregation upon exposure to solvent. In contrast, the N36/C34 complex shows a much more highly charged surface due to acidic residues on the outside of the C34 helices, resulting in greater solubility of the heterodimeric complex (Chan et al. 1997). As a result, N-peptides are prone to aggregate in the absence of C-peptides under physiological conditions. This also accounts for a much weaker inhibitory potency for N-peptides compared to C-peptides, since they must form a stable discrete trimerized inner core to efficiently interact with the CHR. Thus, construction of a stable and soluble discrete trimerized gp41 NHR core as a target is important for fusion inhibitor design and development.

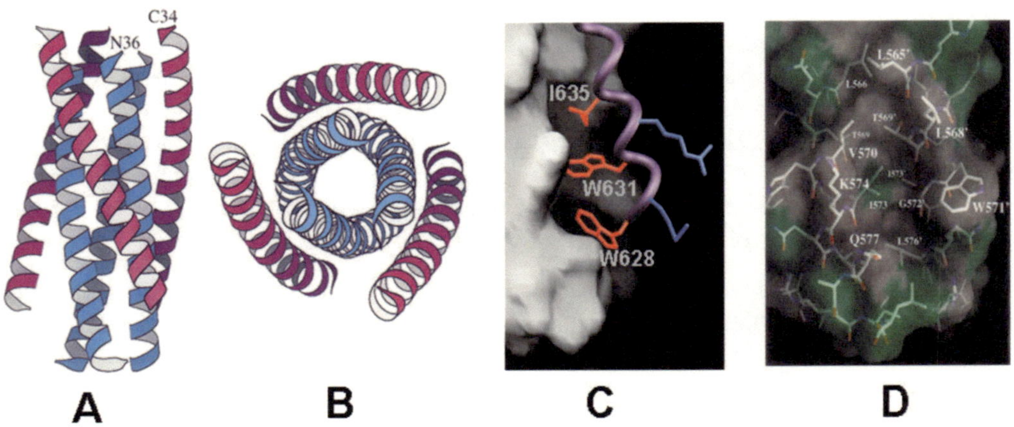

Fig. 2. Crystal structures of the HIV-1 gp41 fusion core. (A) 6-HB structure of the gp41 N36/C34 fusion core; (B) the top to bottom view of the N36/C34 6-HB structure (Chan et al. 1997); (C) the deep pocket in the NHR groove interacts with the WWI motif of CHR (Chan et al. 1998); (D) the NHR deep pocket.

The key for constructing an efficient NHR target is to promote trimerization of N-peptides without changing their native binding sites and conformation. Addition of physicochemical restraints in N-peptides has been shown to be an efficient way to construct a stable and discrete NHR trimer. Typical NHR constructs include: IQN17 (**3**) and IZN17 (**4**) (Eckert & Kim 2001), 5-helix (Root et al. 2001; Frey et al. 2006), and Env2.0 (**5**) and Env5.0 (**6**) (Cai & Gochin 2007; Cai et al. 2009). These stable NHR-trimers can be efficient targets for fusion inhibitor discovery and development. Through forming discrete and stable trimers, they are also highly potent HIV-1 fusion inhibitors by themselves. The sequences of the N-peptide targets are shown in Fig. 3.

IQN17/IZN17 (Fig. 4): A trimeric coiled-coil GCN4 isoleucine zipper was used to construct the first HIV-1 gp41 fusion core for an x-ray crystallographic study (Weissenhorn et al. 1997). IQN17 was constructed by fusing a modified GCN4-pI$_Q$I peptide sequence to the 17-mer N-peptide gp41$_{565-581}$ (N17) that comprises the gp41 hydrophobic pocket (Eckert et al. 1999). The resulting peptide, IQN17, is a fully helical discrete trimer in solution, as determined by circular dichroism (CD) and sedimentation equilibrium experiments. The crystal structure of the IQN17/D10-p1 complex, a cyclic D-peptide fusion inhibitor, showed that the overall architecture of the HIV-1 gp41 hydrophobic pocket in the complex is almost identical to that in the wild-type HIV-1 gp41 N36/C34 structure, with a $C_α$ root mean square deviation (rmsd) of 0.65 Å. In follow-up studies, a new version, IZN17, was designed using the same strategy. IZN17 is more thermally stable than IQN17, with a T_m > 100 °C, compared with ~100 °C for IQN17; the enhancement of thermal stability was further confirmed by measuring the T_m in 2 M guanidine chloride, with a T_m of 66 °C and 74 °C for IQN17 and QZN17, respectively. IZN17 is also more soluble than IQN17 under physiological conditions (Eckert & Kim 2001). Both IQN17 and IZN17 were used as targets in a mirror-image phage display experiment to identify D-peptide fusion inhibitors (Eckert et al. 1999; Welch et al. 2007; Welch et al. 2010).

DP107	(1)	NNLLRAIEAQQHLLQLTVWGIKQLQARILAVERYLKDQ
N36	(2)	SGIVQQQNNLLRAIEAQQHLLQLTVWGIKQLQARIL
IQN17	(3)	RMKQIEDKIEEIESKQKKIENEIARIKKLLQLTVWGIKQLQARIL
IZN17	(4)	IKKEIEAIKKEQEAIKKKIEAIEKLLQLTVWGIKQLQARIL
Env2.0	(5)	Bpy-GQAVEAQQHLLQLTVWGIKQLQARILAVEKK
Env5.0	(6)	Bpy-GQAVSGIVQQQNNLLRAIEAQQHLLQLTVWGIKQLQARILAVEKK

Fig. 3. NHR target sequences. The sequences and groups responsible for physicochemical constraint are shown in grey.

5-Helix (Fig. 4): 5-Helix was designed using the 6-HB as a motif (Root et al. 2001). In 5-helix, five of the six helices that make up the 6-HB core structure are connected by short peptide linkers. The 5-helix protein lacks a third C-peptide helix, and this vacancy is expected to create a high-affinity binding site for the gp41 CHR. Under physiological conditions, 5-helix is soluble and a well folded protein that adopts >95% helical content, as expected from the design, and is extremely stable. In addition, denaturation was not observed, even at 96 °C or in 8 M guanidine chloride. 5-Helix interacts strongly and specifically with C-peptides, inducing a helical conformation in the bound C-peptide as judged by CD. 5-Helix was successfully used as the target in a fluorescence polarization assay to identify small molecule fusion inhibitors (Frey et al. 2006).

Env2.0/Env5.0 (Fig. 4): A trivalent coordination metal complex was used to fortify the gp41 NHR trimer (Gochin et al. 2003). 5-Carboxy-2,2′-bipyridine (BPY) was attached to an N-peptide that contains a deep pocket. Addition of a metal ion such as Fe^{2+} or Ni^{2+} resulted in the formation of a tris-BPY metal complex, which stabilizes the coiled-coil structure. The resulting magenta $Fe^{2+}(BPY)_3$ complex solution was due to a Fe^{2+}–BPY charge transfer band at 545 nm and confirmed Fe^{2+}-BPY binding. The apo-Env2.0 displayed 40% α-helical structure that increased to 89% upon the addition of Fe^{2+} ions, as measured by CD. The integrity of the binding grooves in $Fe^{2+}(Env2.0)_3$ was confirmed by its efficient binding with a matched C-peptide, as shown by CD and NMR (Gochin et al. 2006). The 545 nm absorbance agrees well with the emission maxima of the fluorophores fluorescein and Lucifer yellow. Fluorescence quenching by fluorescence resonance energy transfer (FRET) should occur if the fluorophore is brought close to the Fe^{2+}–BPY center. This enables direct determination of binding by using a fluorophore labeled C-peptide as the probe. Compounds which are able to bind to the NHR target and displace the probe can be measured with a competitive inhibition assay by following the recovery of probe fluorescence intensity (Cai & Gochin 2007). The BPY-metal complex FRET strategy is generally applicable to different interacting peptide pairs, as long as the two peptide sequences are matched. This has been confirmed by the development of a longer gp41 N-peptide/C-peptide pair, Env5.0 and CP5; the peptide pair showed nanomolar binding affinity and can be used to screen more potent fusion inhibitors. Env5.0 contains the whole groove, and it has been used to identify ligands that interact with the range of the groove outside of the deep pocket by designing suitable probes (Cai et al. 2009). In addition, Env2.0 has been successfully used as a target for a screening assay to identify small molecule fusion inhibitors (Cai & Gochin 2007; Zhou et al. 2010).

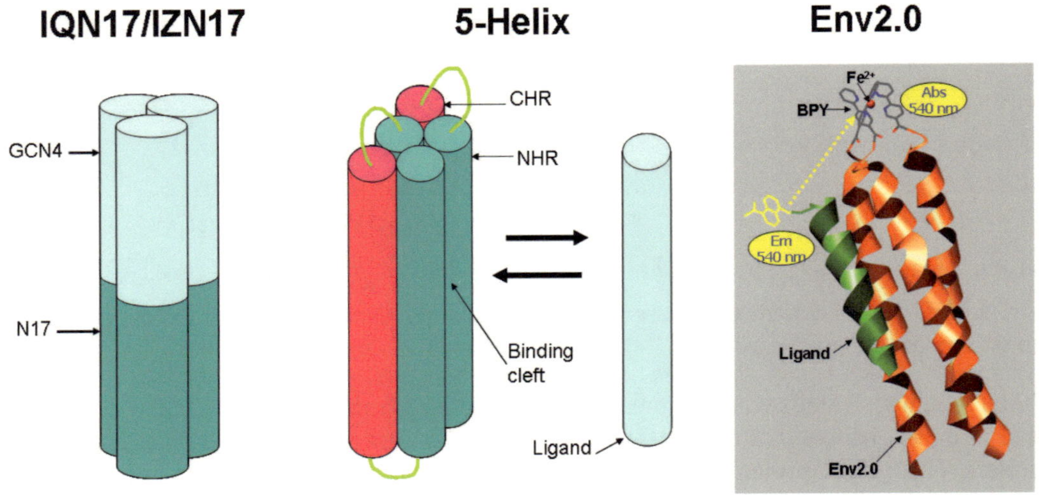

Fig. 4. Designed soluble and discrete NHR target.

In summary, peptides have been used to construct the HIV-1 gp41 fusion core, which is a 6-HB. Crystal structure analysis showed that the gp41 NHR trimer forms an interior core, which contains three hydrophobic grooves as the binding site for C-peptide. A deep pocket in the groove is a hot spot for the NHR-CHR interaction, and can be a target for small molecule fusion inhibitors. The NHR target can be constructed by adding physicochemical constraints in the N-peptides to promote the formation of a soluble and discrete NHR trimer, which can be used for screening HIV-1 fusion inhibitors targeting gp41 NHR or the deep pocket.

3. Peptide fusion inhibitors target the gp41 NHR core

Peptides, especially C-peptides (sequence see Fig. 5), can efficiently block the gp41 NHR-CHR interaction to inhibit HIV-cell membrane fusion and infection. They act in a dominant-negative manner by binding to the transiently exposed coiled-coil N-peptide region in the PHI (Eckert & Kim 2001). The wild-type C-peptide sequences have been shown to have low nanomolar IC_{50} values for HIV-1 ENV mediated membrane fusion and viral infection. Peptide engineering has been employed on wild-type C-peptide sequences to obtain structure activity relationship (SAR) data for the peptide fusion inhibitors, resulting in peptides with an improved anti-HIV profile and a better understanding of the mechanism of gp41 mediated virus-cell membrane fusion (Otaka et al. 2002; Dwyer et al. 2007). The insight gained from these works was finally tested by the artificial design of peptide fusion inhibitors with few sequence homologies to natural peptides or protein sequences (Qi et al. 2008; Shi et al. 2008).

3.1 Peptides from the wild-type gp41 sequence

The first highly potent HIV-1 fusion inhibitors were independently discovered by two groups in the early 1990s, including SJ-2176 ($gp41_{630\text{-}659}$) (Jiang et al. 1993; Jiang et al. 1993) and DP178 ($gp41_{638\text{-}673}$, later named T20) (Wild et al. 1994), which were both derived from the gp41 CHR wild-type sequence. Due to their stronger anti-HIV activity compared with N-peptides, most of the exploited fusion inhibitors were C-peptides, among them, T20 (**7**) and C34 ($gp41_{628\text{-}661}$, **8**) were extensively studied. C-peptide fusion inhibitors are usually unstructured in solution by themselves, and form α-helical structures in a 6-HB after interaction with NHR.

T20, originally named DP178, was developed into the first HIV-1 fusion inhibitor with the brand name Enfuvirtide (Lazzarin et al. 2003; Walmsley et al. 2003; Su et al. 2004). It has low nanomolar antiretroviral activity. Under physiological conditions, it is unstructured and cannot form a stable 6-HB with N-peptide; however, it is highly soluble, making it a good drug candidate. Its mechanism of action has been controversial until now, since it cannot form a 6-HB with N-peptide, which is an established interaction model of HIV-1 peptide fusion inhibitors that has been supported by x-ray crystallography (Chan et al. 1997). T20 does not contain the WWI motif necessary to bind with the primary NHR deep pocket. This may account for its relatively weak binding with NHR and the resulting loss of activity against emerging drug resistant HIV-1 isolates. The 8-residue C-terminus of T20 contains three Trp residues and is highly hydrophobic, which enables T20 to bind with the lipid membrane; thus, this 8-residue motif is called the lipid binding domain (LBD). The hydrophobic residues in the LBD are critical for T20 to maintain high anti-HIV activity, although the LBD elicits no anti-HIV activity by itself. It seems that T20 may interact with both the gp41 NHR groove and the lipid membrane to interfere with 6-HB formation, thus inhibiting HIV-1 infection (Liu et al. 2005; Liu et al. 2007).

C34 forms a stable 6-HB with NHR, thus preventing productive 6-HB formation, a mechanism well supported by x-ray crystallography (Chan et al. 1997). It also displays stronger antiretroviral activity than T20, while its poor solubility under physiological conditions hinders it as a promising drug candidate (Otaka et al. 2002). Like all other wild-type gp41 C-peptides, C34 is unstructured under physiological conditions, while it adopts a nearly full α-helical structure when interacting with N-peptide to form a 6-HB. C34 contains the WWI motif, so it can interact with the primary binding pocket in NHR to form a stable complex with N-peptides. N-PAGE has shown that C34 can form a stable 6-HB in the presence of N36 or N46; and thermal denaturation has shown that the N36/C34 complex displays typical two-state denaturation behavior with a T_m value of ~61 °C (Pan et al. 2009). In a 6-HB, C34 uses residues *a* and *d* to interact with NHR. The *a* and *d* residues in the N-terminal half of C34 are uniformly hydrophobic and elicit a predominantly hydrophobic interaction with residues *e* and *g* in NHR and bury these residues in the 6-HB; and the *a* and *d* residues in the C-terminal half of C34 form a hydrophilic layer spanning four α-helical turns, which is assumed to match the similar hydrophilic layer in the related NHR sequence. Thus, C34 is widely used as a tool to study the mechanism of HIV fusion inhibitors, as well as the lead or template for next generation fusion inhibitor design, and will be discussed in Section 3.2.

CP32 ($gp41_{621\text{-}652}$, **14**) is another identified highly potent wild-type gp41 C-peptide fusion inhibitor that targets NHR sequences other than T20 and C34 (He et al. 2008). It contains a 7-residue motif upstream of the C34 sequence. Interestingly, the CP32 sequence matches the

T21 sequence, the first identified peptide HIV-1 fusion inhibitor from gp41 under the name DP107 (Wild et al. 1992); the match is expected based on the N36/C34 complex and the anti-parallel interactions between gp41 NHR and CHR. CP32 contains the WWI motif, so it can interact with the NHR deep pocket to form a stable 6-HB. The CP32/T21 complex is ~100% α-helical with a T_m of 82 °C, which is more stable than the N36/C34 complex. The discrete 6-HB conformation of the CP32/T21 complex was supported by N-PAGE, size exclusion chromatography, as well as analytical ultracentrifugation. CP32 showed an IC_{50} value of 4.2 nM against HIV-1 ENV mediated cell-cell fusion and an IC_{50} value of 4.6 nm against HIV-$1_{NL4\text{-}3wt}$ infection of MT-2 cells. Although it has a similar potency as C34 against wild-type HIV-1 isolates, CP32 is ~20-fold and >500-fold more potent than C34 and T20, respectively, against the drug-resistant HIV-$1_{NL4\text{-}3\text{-}V38SE/N42S}$ isolate, possibly due to the fact that it targets a different sequence in gp41 NHR.

There are also longer C-peptides fusion inhibitors, such as C43 and C52, which include both partial or complete sequences of C34 and T20; however, none of these longer peptides have improved anti-HIV potency compared with T20 and C34 (Deng et al. 2007). In the PHI, a long groove may expand throughout the gp41 NHR and beyond, and it may be targeted by its CHR counterpart. The C-terminal half of the gp41 ectodomain may make contact with the N-terminal half at a certain time during fusion processes, so the C-peptide sequence may expand to the whole C-terminal half of the gp41 ectodomain and interact with the PHI to inhibit gp41 mediated virus-cell membrane fusion. The energy along the gp41 NHR-CHR interface is not evenly distributed; the WWI motif and the LBD serve as hot spots in the gp41 NHR-CHR interaction. A highly potent C-peptide fusion inhibitor must contain at least the WWI motif or the LBD; in addition, a suitable length of total peptide sequence is required to provide additional interactions in the NHR groove to stabilize the C-peptide-NHR interaction. Though they form stable α-helical structures in the 6-HB, C-peptides and N-peptides from the wild-type gp41 sequence are usually unstructured in solution; thus, they are prone to proteolysis. The viral strains resistant to T20 also required the development of a highly potent fusion inhibitor to overcome drug resistance. The use of protein/peptide engineering to improve the physicochemical properties of the wild-type C-peptide sequence and to increase the stability of the C-peptide-NHR complex is discussed below.

3.2 Engineered peptides

New generations of peptide fusion inhibitors have been developed by engineering C34-related sequences in order to increase the *in vivo* stability and NHR binding affinity, and to overcome T20 resistance. It is well accepted that increasing the helicity of the peptide fusion inhibitor will increase its antiretroviral potency by increasing its binding affinity with NHR and the *in vivo* stability (Otaka et al. 2002). In a 6-HB, C-peptides interact with NHR with their *a* and *d* residues, which are considered to be critical for molecular recognition between CHR and NHR; while amino acid residues at the *b*, *c*, *f*, and *g* positions are exposed to solution and are not considered to be critical for the gp41 NHR and CHR interaction (Chan et al. 1997). However, the solvent exposed residues have a global effect on the solubility, stability, and other physicochemical properties of the C-peptides, so they affect the *in vivo* activity and the druggability of peptide fusion inhibitors. Salt bridges and helical enhancers have been engineered by replacing the solvent exposed residues with the desired residues in order to get more potent HIV-1 fusion inhibitors.

T1249 (**13**) was developed by Trimeris as a second generation peptide HIV-1 fusion inhibitor after T20 (Miralles et al. 2003; Eggink et al. 2008; Pan et al. 2009) It was designed to include both hot spots, the WWI motif of C34 and the LBD of T20. To keep the peptides a suitable length, the seven residues following the WWI motif were considered to be not critical for the NHR interaction and were deleted; thus, the WQEWEQKI motif remained. It also contained the conserved amino acid residues from SIV and HIV-2 that are essential for fighting contains the T20 resistant virus. In addition, alanine substitutions and salt bridges were added to increase the α-helicity, resulting in a 39-mer highly mutated peptide based on the wild-type HIV-1 gp41 sequence. T1249 showed enhanced antiretroviral activity against the T20 resistant virus, and ~50% α-helicity compared to the unstructured character of the wild-type gp41 peptide. It entered into phase II clinical trials, but it was terminated due to side effects (Miralles et al. 2003).

T1144 (**9**) and T2635 (**31**) are third generation peptide fusion inhibitors developed by Trimeris (Pan et al. 2011). They fully exploited the strategy used in the development of T1249, however, they are based on the $gp41_{626-663}$ sequence (Dwyer et al. 2007). T1144 and T2635 showed strong activity against both native and highly T20 resistant HIV-1 strains. They form stable α-helices in solution with a helical content of 97% and 75% for T1144 and T2635, respectively. In addition, they both form a very stable 6-HB with NHR under physiological conditions. Analytical ultracentrifugation showed that these highly helical peptides form trimers in solution, which may make them more resistant to proteolysis and increase their *in vivo* stability. Ultra stable C-peptides from the same CHR sequence were also obtained, which showed ~100% helical content in solution and formed ultra stable 6-HBs with N-peptide with T_m values >100 °C, even in 8 M urea solutions. However, these peptides showed very weak antiretroviral activity. This indicated that it required a suitable degree of stability and α-helical content for the C-peptide to efficiently inhibit 6-HB formation to stop the HIV-1-cell fusion process.

Sifuvirtide (SFT, **12**) was developed by FusoGen and was based on the C34-related sequence $gp41_{627-662}$ (He et al. 2008; Liu et al. 2011). It was derived from the HIV-1 subtype E sequence and was engineered to mutate the exposed residues to salt bridges to increase the helical content and solubility. Like C34, Sifuvirtide is featureless under physiological conditions, while it forms a nearly full α-helical 6-HB with N36, with a T_m of 72 °C, 10 °C higher than that of N36/C34. As expected from its sequence origin, Sifuvirtide does not interact with the lipid membrane. Sifuvirtide showed low nanomolar inhibitory activity against HIV-1 ENV mediated cell-cell fusion and HIV-1 infection, including T20-resistant HIV-1 isolates. It showed an *in vivo* half-life of 20 h in a single dose administration in 12 healthy volunteers, which is much more stable than T20 and suitable for a once daily administration. It has finished phase IIb clinical trials in China and has shown promising antiretroviral profiles against both T20 resistant and T20 sensitive HIV-1 strains (Wang et al. 2009). The same strategy was applied to CP32 and resulted in CP32M with an improved anti-HIV profile (He et al. 2008).

SC35EK (**10**), also based on C34, was developed by Fujii's group. Most of the *b*, *c*, *f*, and *g* residues were substituted with glutamic acid and lysine residues in order to form EE-KK double salt bridges to fortify the α-helical structure (Otaka et al. 2002). SC35EK showed a little bit more potency than C34 in a multinuclear activation of galactosidase indicator (MAGI) assay (IC_{50} from 0.68 to 0.39 nM), while the salt bridge greatly enhanced its solubility and made it a suitable drug candidate. Its structure is still largely random in solution, while the T_m of its 6-HB formed with N36 increased from 57 °C to 77 °C, which is 20 °C higher than C34. SC35EK was further shortened to SC29EK (**11**), with similar potency (Naito et al. 2009). The same strategy was applied to T20, the resulting T20EK (**16**) showed eight times more potency than T20 and can efficiently inhibit T20 resistant HIV-1 strains (Oishi et al. 2008).

In summary, the C-peptide fusion inhibitor could be engineered to improve the anti-HIV profile. Exposed residues in the 6-HB were substituted to build salt-bridges to significantly stabilize the C-peptide-NHR complex. This type of substitution can improve the solubility of the peptide fusion inhibitor to improve its druggability. The substitution also improved the pharmacokinetic profile, resulting in a longer *in vivo* half life. The helicity of isolated C-peptides were greatly increased by replacing the *a* and *d* residues in the hydrophilic layer, resulting in thermally stable C-peptide fusion inhibitors with high α-helical content; they formed an extremely stable complex with NHR. Some of these structured C-peptides showed high anti-HIV potency, especially against highly drug-resistant HIV-1 isolates; while too thermally stable C-peptides of this type caused abolishment of their inhibitory activities.

T20	(**7**)	YTSLIHSLIEESQNQQEKNEQELLELDKWASLWNWF
C34	(**8**)	WMEWDREINNYTSLIHSLIEESQNQQEKNEQELL
T1144	(**9**)	TTWEAWDRAIAEYAARIEALLRALQEQQEKNEAALREL
SC35EK	(**10**)	WEEWDKKIEEYTKKIEELIKKSEEQQKKNEEELKK
SC29KE	(**11**)	WEEWDKKIEEYTKKIEELIKKSEEQQKKN
Sifuvirtide	(**12**)	SWETWEREIENYTRQIYRILEESQEQQDRNERDLLE
T1249	(**13**)	WQEWEQKITALLEQAQIQQEKNEYELQKLDKWASLWEWF
CP32	(**14**)	QIWNNMTWMEWDREINNYTSLIHSLLEESQNQ
CP32M	(**15**)	VEWNEMTWMEWEREIENYTKLIYKILEESQEQ
T20EK	(**16**)	YTSLIEELIKKSEEQQKKNEEELKKLEEWAKKWNWF

Fig. 5. C-peptide fusion inhibitors

3.3 Artificially designed peptides

Artificial design was employed to design unknown peptide sequences with few homologies to natural peptide sequences (Qi et al. 2008; Shi et al. 2008). Based on the crystal structures of the HIV-1 gp41 fusion core, C-peptide uses it hydrophobic *a* and *d* residues to interact with the NHR. An EEYTKKI heptad unit (HR) was designed, with the heptad repeat 'bcdefga', as the building block. The *d* and *a* positions in the HR were hydrophobic Tyr and Ile residues, respectively, which were expected to form a hydrophobic face to interact with the hydrophobic NHR grooves. The residues at the *b* and *c* positions in the HR were negatively charged Glu, which were expected to form an intrahelical salt bridge with positively charged Lys at the *f* and *g* positions to stabilize the helical structure; these highly polar residues also form a highly hydrophilic face that increases the solubility of the peptides.

A 35-mer 5HR (**17**) (Fig. 6), which contains five copies of the HR, based on the length of most highly potent HIV-1 fusion inhibitors, was used as a template to build peptides to disrupt the HIV-1 gp41 NHR-CHR interaction. The interaction between 5HR and N46 ($gp41_{536-581}$) was modeled by using a PyMOL program based on the crystal structure of the N36/C34 6-HB, and compared with that of C34. The binding between the residues of 5HR and N46 was less complementary than that of the residues between C34 and N46. 5HR showed weak anti-HIV-1 activity (IC_{50} = 156 ± 8 μg/mL), as measured by a dye transfer HIV-1-mediated cell-cell fusion assay. The WWI motif and LBD were used to replace the HR unit at the N- or C-terminus of 5HR, respectively, or both, based on the SAR of the C-peptide fusion inhibitors, resulting in PBD-4HR (**18**), 4HR-LBD (**19**), and PBD-3HR-LBD (**20**). Inserting a LBD or WWI motif in the 5HR sequences resulted in 2-fold and 6-fold increased potency, with an IC_{50} value of 74 ± 4 and 26 ± 0.4 μg/mL for 4HR-LBD and PBD-4HR, respectively. The increasing potency was synergistic and PBD-3HR-LBD had an IC_{50} value of 4.8 ± 0.3 μg/mL, a striking 33-fold increase over 5HR. As expected from the design, peptides containing PBD, e.g. PBD-4HR and PBD-3HR-LBD, could form a stable 6-HB with the N-peptide N46 and effectively blocked gp41 core formation, as measured by CD spectroscopy and N-PAGE; peptides containing the LBD, including 4HRLBD and PBD-3HR-LBD, were bound tightly to lipid vehicles, with an association constant of 6.80 × 10^4 and 1.27 × 10^5 M^{-1}, respectively, as determined by isothermal titration calorimetry (ITC). These results suggest that the HR sequence can be efficiently docked into the NHR groove and act as a structural domain; and the interaction can be greatly increased by including the WWI motif and LBD in the sequence. Thus, 4HR-LBD, PBD-4HR, and PBD-3HR-LBD are artificial fusion inhibitors that mimic T20, C34, and T1249 – the three typical highly potent HIV-1 fusion inhibitors target different sites of gp41 NHR, respectively.

The anti-HIV-1 activities of 4HR-LBD and PBD-4HR are lower than those of T20 and C34, which may be due to less sequence complementarity between the artificially designed HR and HIV-1 gp41 NHR. The resulting less tight binding suggests that a specific interaction should be uncovered and be addressed for the design of highly potent fusion inhibitors targeting specific viruses.

5HR (17) EEYTKKIEEYTKKIEEYTKKIEEYTKKIEEYTKKI
4HR-LBD (18) EEYTKKIEEYTKKIEEYTKKIEEYTKKIWASLWNWF
PBD-4HR (19) WMEWDREIEEYTKKIEEYTKKIEEYTKKIEEYTKKI
PBD-4HR-LBD (20) WMEWDREIEEYTKKIEEYTKKIEEYTKKIWASLWNWF

Fig. 6. Artificially designed peptide fusion inhibitors

In summary, C-peptide fusion inhibitors interact with gp41 NHR to prevent fusogenic 6-HB formation, and thus terminal HIV-1 ENV mediated virus-cell membrane fusion. A WWI motif in C-peptide that interacts with the NHR deep pocket is critical to the C-peptide-NHR interaction, and an extended interaction between C-peptide and the rest of the groove in the NHR trimer provides additional energy to stabilize the 6-HB. Artificial peptide design, based on the knowledge learned from SAR studies of the C-peptides, provides an alternative for peptide fusion inhibitor design; it also provides a stringent test for the knowledge gained and sets a new starting point for fully understanding the fundamentals of virus-cell membrane fusion in order to guide future fusion inhibitor design against HIV and other viruses with class I fusion proteins.

4. Peptidomimetics as probes and inhibitors to study the gp41 NHR-CHR interaction

Several SAR studies of highly potent peptide fusion inhibitors have provided an efficient way to disrupt the HIV-1 gp41 NHR-CHR interaction for anti-HIV therapy; they have also deepened our understanding of the gp41 NHR-CHR interaction. Peptide drugs have their intrinsic weaknesses, however, such as high-cost, and unsuitability for oral administration due to *in vivo* proteolysis. Peptidomimetics that use unnatural building blocks may overcome the *in vivo* instability of peptide drugs, leading to orally bioavailable drugs. Peptidomimetics are more like small molecules than peptide drugs, so highly potent peptidomimetic fusion inhibitor studies can be useful for guiding small molecule fusion inhibitor design. Peptidomimetic fusion inhibitors that target gp41 NHR, including D-peptides, foldamers, and covalently linked restrained α-helical peptides (sequences or structures see Fig. 7), are discussed in this section.

4.1 D-peptides

As enantiomers of natural L-peptides, D-peptides are not degraded by proteases and have the potential for oral bioavailability. D-Peptides that target a specific protein or peptide target can be discovered by mirror-image phage display (Eckert et al. 1999). The target is synthesized chemically with D-amino acids, resulting in a product that is the mirror image of the natural L-amino acid form, which is used to screen phage that expresses a peptide library of phage coat proteins, to select phage clones with L-peptide sequences that specifically bind to the D-target. The mirror images of the phage-expressed L-peptide sequences are chemically synthesized with D-amino acids. By symmetry, these D-peptides should bind to the natural L-amino acid target.

Cyclic D-peptide HIV-1 fusion inhibitors targeting IQN17 have been identified by mirror-image phage display (Eckert et al. 1999). The phage-expressed peptide library contained ten random amino acid residues flanked by either a cysteine or a serine on both sides. Of the 12 identified IQN17-specific phage clones, nine were pocket specific binders, and eight contained the consensus sequence CXXXXXEWXWLC. The corresponding D-peptides were synthesized and were oxidized to form disulfide bonds. Lysines were added to improve the solubility. An intramolecular disulfide bond was critical for pocket binding and viral inhibition by these D-peptides, since cysteines were selected from an initial phage library containing either Cys or Ser at these positions. Replacing the Cys with Ala in the most potent derivative D10-p5-2K (**22**, IC_{50} of 3.6 μM) caused complete loss of inhibitory activity in a gp41 mediated cell/cell fusion assay.

A IQN17/D10-p1 (**21**) co-crystal was obtained and resolved to 1.5 Å resolution by x-ray crystallography. Structural superposition showed that the overall architecture of the gp41 NHR deep pocket in the IQN17/D10-p1 complex is almost identical to that in the wild-type N36/C34 structure (Chan et al. 1997), with a C_{α} rmsd of 0.65 Å. D10-p1 forms a circular structure and binds only to the gp41 region of IQN17. Ala-2 to Ala-5 and Ala-11 to Ala-16 form short left-handed α-helices, and the middle region is unstructured. The overall positions of the D10-p1 and C34 helices closely overlap, but most of the side chains are significantly different, corresponding to the opposite handedness of the inhibitors. Of the 16 residues in D10-p1, only six interact directly with the gp41 pocket of IQN17, including Trp-10, Trp-12, and Leu-13 in the conserved EWXWL sequence, and Gly-1, Ala-2, and Ala-16 in the invariant original flanking phage sequence. The side chains of Trp-10, Trp-12, Leu-13, and Ala-16 are deeply buried in the hydrophobic pocket of IQN17. A hydrogen bond is formed between a pocket residue Gln-577 and Trp-12 in D10-p1. The packing difference between the Trp-12 and Leu-13 side chains in D10-p1 and Trp-631 and Ile-635 in C34 results in slight changes in the shape of the pocket. Overall, however, the hydrophobic pocket maintains its integrity between the N36/C34 and IQN17/D10-p1 structures. NHR chemical shift differences showed that, for all of the identified D-peptides, Trp-10, Trp-12, and Leu-13 are buried in the IQN17 pocket, validating the pocket as a target for drug development.

In follow-up work, the consensus residues in the sequence ($CX_5EWXWLC$) reported above were fixed so that a constrained library was constructed in which the other six positions were randomized (Welch et al. 2007). The mirror-image phase display using IQN17 as the target identified, incidentally, the potent 8-mer D-peptide 2K-PIE1 (**23**) in the 10-mer template phage library. The x-ray crystal structure showed that 2K-PIE1 interacts in a similar manner as D10-p1 to IQN17, and 2K-PIE1 forms a more compact structure with IQN17. So, a comprehensive 1.5×10^8 member 8-mer phage library of the form CX4WXWLC (3.4×10^7 possible sequences) was generated, and was screened using IZN17 as the target (Eckert & Kim 2001). The resulting PIE7 (**24**) was the most potent inhibitor (IC_{50} = 620 nM) and is 15-fold more potent than the best first-generation D-peptide (D10-p5). Comparison of the crystal structures of 2K-PIE1 and PIE7 complexed with IQN17 reveals several interesting differences. First, an intramolecular polar contact between the hydroxyl of D-Ser7 and the carbonyl of D-Gly3 in 2K-PIE1 is lost in PIE7 but is replaced with a new interaction between the side chain carboxylate of D-Asp6 and the amide of D-Gly3. Second, new hydrophobic interactions are created in PIE7 between the ring carbons of D-Tyr7 and the pocket residue Trp-571. Third, the carbonyl of D-Lys2 of PIE7, although somewhat flexible in orientation, forms a direct hydrogen bond with the ε nitrogen of Trp-571 in some of the structures. Fourth, in some of the structures the hydroxyl of D-Tyr7 in PIE7 forms a new water-mediated hydrogen bond with the pocket residue Gln-575, and this interaction cannot be formed in the 2K-PIE1 structure. Dimerized or trimerized PIE7 was constructed via PEG cross-linkers. The resulting $(PIE7)_2$ and $(PIE7)_3$ have IC_{50} values of 1.9 nM and 250 pM against HXB2, respectively. In contrast, PIE7 inhibits both JRFL, a primary R5-tropic strain, (IC_{50} = 24 μM) and BaL (IC_{50} = 2.2 μM) entry, although ~40- and 4-fold less potently than HXB2 entry, respectively; the PIE7 trimer is a moderately potent inhibitor of this strain (IC_{50} = 220 nM) and an extremely potent inhibitor against BaL (IC_{50} = 650 pM).

Structure-guided phage display was used to optimize the flanking residues for further improvement of PIE7 (Welch et al. 2010). The crystal structure shows significant contacts between the presumed inert flanking residues (Gly-Ala on the N-terminus and Ala-Ala on the C-terminus) and the NHR deep pocket. A new phage library was designed using XXCDYPEWQWLCXX as the template. PIE12 (**25**) was identified as the most potent (40-fold more potent than PIE7 against the JRFL strain). The x-ray crystal structure showed similarity between PIE12/IZN17 and PIE7/IZN17 structures with a RMSD of 0.6 to 1.2 Å on all C_α atoms. In PIE12/IZN17, new N-terminal flank residues (His1 and Pro2) form favorable ring stacking interactions with the pocket (IQN17-Trp571), the substitution of Leu for Ala in the C-terminal flank sequence causes it to be buried an additional 50 Å into the hydrophobic surface area of the pocket, and the new interactions with the flanking sequence do not perturb the pocket-binding structure of the core PIE7 residues. These differences may account for the improved activity of PIE12 over PIE7. CD thermal denaturation showed that the PIE12-trimer forms the more stable complex with IZN17 with a T_m of 81 °C in 2 M Guanidine chloride (Gua.HCl), 8 °C higher than that of the PIE7-trimer complex. The anti-HIV-1 breadths of the PIE7-trimer, PIE12-trimer, and PIE12 were tested by a pseudovirion assay against a panel of 23 pseudotyped viruses representing clades A to D, several CRFs, and enfuvirtide-resistant strains. Both PIE7 and PIE12-trimers potently inhibited all strains tested, though PIE12-trimer was generally a superior inhibitor.

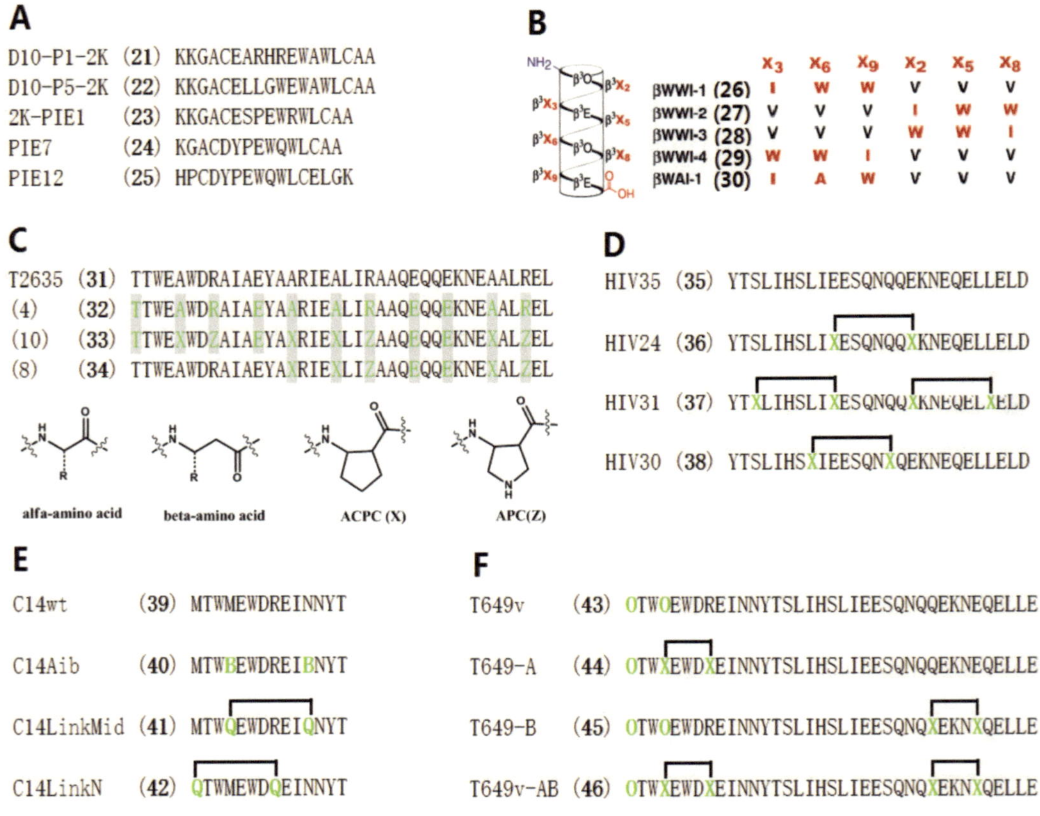

Fig. 7. Peptidomimetics used to disrupt the HIV-1 gp41 NHR-CHR interaction. (A) D-peptides; (B) β-foldamers; (C) α/β-foldamers; (D)/(E) linked peptides; (F) stapled peptides.

Viral passage studies were conducted to select for resistant strains. A strain bearing E560K/V570I mutations, which conferred a 400-fold resistance to PIE7-dimer, was selected with 20 weeks of propagation. These mutations dramatically weaken the binding of D-peptides to the gp41 pocket but not the C-peptide inhibitor C37. Despite this loss of affinity, the escape mutations had a minimal effect on the potencies of PIE12-dimer and PIE12-trimer. PIE12-dimer and PIE12-trimer resistant virus were identified after 40 and 65 weeks of propagation, respectively, using a much slower escalation strategy; only a Q577R single substitution was identified. Interestingly, this substitution is present in nearly all group O isolates but is rare among group M isolates. Examination of the PIE12 crystal structure shows that Q577 makes hydrogen bonds with Glu7 and Trp10 in PIE12, which may explain the disruptive effects of this mutation.

4.2 Foldamers

A foldamer is a discrete chain molecule or oligomer that adopts a secondary structure stabilized by noncovalent interactions. Foldamers use unnatural building blocks instead of natural amino acids or nucleotides; as a result, they are more resistant to enzymatic degradation and show enhanced *in vivo* stability. They can mimic the ability of proteins, nucleic acids, and polysaccharides to fold into well-defined conformations, such as helices and β-sheets.

Short β^3-foldamers have been designed that mimic the WWI motif of C-peptide fusion inhibitors (Stephens et al. 2005). A β-amino acid contains an additional methylene unit between the amine and carboxylic acid (Fig. 7c), and the amide bonds in β-peptides can resist *in vivo* proteolysis. A set of β^3-decapeptides, *β*-WWI-1-4 (**26-29**), in which the WWI motif is presented on one face of a short 1,4-helix (Fig. 7b), were designed. Each *β*-peptide was fluorescently labeled at the N-terminus and was used in direct fluorescence polarization experiments to determine its binding affinity to IZN17. *β*-WWI-1-4-Flu bound IZN17 well, with equilibrium affinities of 0.75 ± 0.1, 1.0 ± 0.3, 2.4 ± 0.7, and 1.5 ± 0.4 μM, respectively. A WWI-1 analog *β*-WAI-1-Flu (**30**), containing Ala in place of the central Trp of the WWI motif, bound IZN17 with lower affinity (K_d > 20 μM), suggesting the WWI motif is critical for pocket binding. The binding affinities are consistent with the cell-cell fusion assay results; *β*-WWI-1-4 inhibited cell-cell fusion with EC_{50} values of 27 ± 2.5, 15 ± 1.6, 13 ± 1.9, and 5.3 ± 0.5 μM, respectively, whereas *β*-WAI-1 was inactive under the same conditions.

In a follow-up study (Bautista et al. 2009), the second Trp in *β*-WWI-4 (**29**) was replaced with unnatural residues to probe steric and electronic effects on the NHR deep pocket binding. Most of the new *β*-peptides (EC_{50} 8.2–19 μM) are more potent than *β*WWI-1 (**26**) (EC_{50} = 56 μM) at promoting the survival of HIV-infected cells. However, high cytotoxicities, with a selective index (CC_{50}/EC_{50}) <10, render these short β-peptides unsuitable as drug leads.

An α/β foldamer with partial β-amino acid replacement was used to modify a highly potent C-peptide fusion inhibitor to increase its *in vivo* stability (Horne et al. 2009). A two-stage design strategy was employed to modify T2635 (**31**), a highly potent third generation peptide fusion inhibitor (Dwyer et al. 2007). A fluorescent polarization binding assay using 5-helix as the target, a cell-cell fusion assay, and a protease K assay were used to assess the peptide and designed foldamer. In the first stage, one amino acid residue in each α-helix turn at the same position was replaced by a β-amino acid. The optimized α/β foldamer (**32**), containing systematic β-amino acid substitutions at positions *c* and *f*, showed weak binding affinity (K_i = 3800 nM) and cell-cell fusion inhibitory activity (IC_{50} = 390 nM), compared with those of T2635 (K_i < 0.2 nM and IC_{50} = 9 nM). However, **32** showed a 20-fold increase in half-life (14 min) in the protease K assay, compared with that for T2635. The β-amino acid, with one additional methylene unit, may make the backbone of the α/β foldamer too flexible to adapt a suitable conformation and results in the loss of activity. Accordingly, in the second stage of design, β-amino acids at certain positions were replaced with cyclic β-amino acids to restore the rigidity of the backbone. The resulting α/β foldamer (**33**) had an IC_{50} value of 5 nM in the cell-cell fusion assay, similar to that of T2635; its binding affinity to 5-helix was 9 nM, similar to its cell-cell fusion inhibitory potency, despite its significantly weaker binding affinity to 5-helix than T2635. The α/β foldamer (**33**) showed a half-life of 200 min in the protease K assay, a 280-fold increase from T2635. The anti-HIV-1 infection activity of the α/β foldamer was also similar to T2635, as measured by HIV-1 infection of TZM-bl (JC53BL) cells using both R5 and X4 HIV-1 strains.

X-ray crystallography was used to characterize the structures of N36/T2635 and N36/**33** complexes. The N36/T2635 6-HB structure is almost identical to that of the wild-type N36/C34 6-HB, with a rmsd of 0.73 Å for the $C_α$ atoms. However, the N36/**33** complex showed large structure distortion in the N-terminus (4.2 Å $C_α$ rmsd for residues 2–15); the side chains of Trp^3 and Trp^5 were not resolved in electron density, suggesting a high degree of disorder, indicating that the N-terminal segment of **33** does not engage the NHR binding pocket in the complex. However, removal of the first ten residues of **33**, where the WWI motif is located, causes the loss of binding to 5-helix (K_i >10 μM), indicating that the N-terminal segment of **33** is essential for high-affinity 5-helix binding. The N36/**34** complex, maintaining the intact WWI motif in the foldamer sequence (**8**), was crystallized and resolved to 2.8 Å resolution. Relative to **33**, **34** tracks much more closely with T2635, with a 1.4 Å $C_α$ rmsd for residues 2–33 between the two structures. The side chains of the WWI motif in the N-terminal segment of **34** show the expected packing into the binding pocket on the NHR core trimer. The above results suggest that the lack of direct contact between the N-terminal portion of **33** and the NHR trimer in the N36/**33** complex may be an artifact of crystal packing.

4.3 Covalent-linked constrained peptides

Helical structure is critical for C-peptide fusion inhibitors to make proper contacts with the NHR binding sites to elicit potent inhibition. Constraining methods that add structural constraints into the peptide sequence by covalently cross-linking amino acid residues at suitable positions can promote the formation of the α-helical conformation, even in short peptides. The covalent linker can be a longer linker between the *i* and *i* + 7 residues, or a short linker called a stapler between the *i* and *i* + 4 residues.

The first selected gp41 C-peptide was truncated T20 that lacks the LBD sequence, called HIV35 ($gp41_{638-665}$, **35**) (Judice et al. 1997). A covalent cross-linker between the *i* and *i* + 7 residues of the polypeptide chain locks the intervening residues into a α-helical conformation. Residues at adjacent *f* positions on the opposite face of the helix were selected for cross-linking to enforce the residues at positions *a* and *d* to adopt a suitable conformation for target binding. Analogs of HIV35 were prepared containing either one, HIV24 (**36**), or two, HIV31 (**37**), tethers to impart increasing helicity. A control peptide, HIV30 (**38**), was prepared in which a tether was introduced between successive *d* residues to stabilize the helicity while blocking potential binding interactions across the *a-d* face. HIV24 and HIV30 were partially α-helical as measured by CD. By contrast, the doubly constrained analog HIV31 was mostly α-helical. HIV35 showed very weak inhibitory activity against HIV-1 in primary infectivity assays by using peripheral blood mononuclear cells with the virus JRCSF, a nonsyncytium-inducing strain, and BZ167, a syncytium-inducing HIV-1 strain. Single restrained HIV24 is more potent than HIV35, partially restoring the inhibitory activity of T20. Doubly constrained HIV31 shows dramatically higher potency, and its activity was comparable with T20 in both HIV-1 infection assays.

In another report, a 14-residue C-peptide C14 ($gp41_{626-639}$, **39**) was selected (Sia et al. 2002). A cell–cell fusion assay was used to evaluate the biological activity of the peptides. Two strategies were employed, substitution with 2-aminoisobutyric acid (Aib) or a diaminoalkane crosslinker, to stabilize the helical conformation of C14. Six peptides were designed and produced, C14linkmid (**41**) was the most potent inhibitor against syncytia formation (IC_{50} = 35 μM), followed by C14Aib (**40**) (IC_{50} = 144 μM). C14linkmid and C14Aib bind to IQN17 with a K_d of 1.2 μM, respectively, as measured by ITC. The efficacy of the cross-linker on the inhibitory activities depends on its position in the peptide sequence, N-terminal cross-linked C14linkN does not inhibit cell–cell fusion, whereas the middle cross-linked C14linkmid inhibits cell–cell fusion at micromolar concentrations. The cell–cell fusion inhibitory activities of the peptides generally correlated with their NHR binding affinities, although the cell–cell fusion activities were consistently ~10-fold less potent than the K_d of NHR binding. Additional factors, other than binding affinity to the target, may be necessary for blocking viral entry. The crystal structure of the C14linkmid/IQN17 complex showed that C14linkmid binds to the gp41 hydrophobic pocket in essentially the same conformation as the pocket-binding region of C34, demonstrating that the crosslink imparts no detectable distortion on the backbone of the C14 peptide in the bound conformation.

Chemical staples have been used to fortify peptides to overcome the proteolytic shortcomings of highly potent peptide HIV fusion inhibitors as therapeutics. As an example, chemical staples were inserted at the N- or C-termini of T649v (**43**) by substituting (*S*)-2-(((9H-fluoren-9-yl)methoxy)carbonylamino)-2-methyl-hept-6-enoic acid at select (i and $i + 4$) positions, followed by ruthenium-catalyzed olefin metathesis (Bird et al. 2010). Sites for unnatural amino acid insertion were carefully selected to avoid disruption of the critical hydrophobic interface between NHR and CHR helices as delineated by the crystal structure of N36/C34. Three stapled peptides were designed by inserting single or double staples at selected positions. The activities of related peptides were measured using a luciferase-based HIV-1 infectivity assay, using viruses derived from HXBc2 and the neutralization-resistant primary R5 isolate, YU2. All of the peptides showed low nanomolar IC_{50} values against HXBc2 strains, suggesting that chemical modification in the stapled peptides does not disrupt its NHR interaction. Moreover, all of the stapled peptides showed higher inhibitory activities against drug resistant HIV-1 isolates, such as YU2 and the HIV-1 HXBc2 virus bearing the T20-resistant V38A/N42T or V38E/N42S double mutations in gp41 NHR, with a rank order of SAH-$gp41_{626-662}$ (**46**) > **44** > **45** > T649v > enfuvirtide. SAH-$gp41_{626-662}$ (**46**) displayed medium to low nanomolar IC_{50} values for all of the viruses tested, including T20 and the T649v-resistant YU2 isolate.

The pharmacokinetic properties of **44** were evaluated in a mouse model (Bird et al. 2010). The total body clearance of **44** (1.0 mL/min/kg) was 10-fold more slow than that of the unmodified T649v peptide (9.5 mL/min/kg). A proteolysis assay using both chymotrypsin and pepsin suggested that the striking protease resistance of stapled peptides is conferred by a combination of (1) decreased rate of proteolysis due to induction of α-helical structure and (2) complete blockage of peptidase cleavage at sites localized within or immediately adjacent to the (i, $i + 4$)-crosslinked segment. In addition, a pilot study was undertaken to compare the oral absorption of T649v and **44** using a mouse model. Measurable concentrations of the full-length peptide were found in plasma samples from all **44** treated animals after oral dosing, and the concentration was dose dependent; no T649v was detected in plasma under the same conditions. The hydrocarbon double-stapling confers striking protease resistance of the peptide fusion inhibitor, which translates into markedly improved pharmacokinetic properties, including oral absorption, thus unlocking the therapeutic potential of natural bioactive polypeptides.

In summary, highly potent peptidomimetic HIV-1 fusion inhibitors have been discovered based on peptide fusion inhibitors, including: D-peptide fusion inhibitors discovered by mirror-image phage display using a D-amino acid form of the HIV-1 gp41 target; foldamers constructed from highly potent C-peptide fusion inhibitors by proper substitution of selected residues with β-amino acid residues; and structurally constrained peptides by covalently linking two residues at the same positions in a helical turn to promote α-helical structure formation. More like small molecule drugs, these peptidomimetics are potentially orally bioavailable and also provide clues for small molecule fusion inhibitor design.

5. Small molecule helix mimetics

The ultimate goal for drug development is small molecule drugs; it is also the main challenge in PPII development. The NHR deep pocket is a hot spot for the NHR-CHR interaction; it has an internal volume of roughly 400 $Å^3$, and could be filled by a molecule with a molecular weight of approximately 500 Da, raising the possibility that it could be targeted by small molecule drugs (Chan et al. 1997). Several groups have identified small molecules that show low micromolar inhibitory potency against HIV-1 ENV mediated cell-cell fusion and virus infection (Debnath et al. 1999; Frey et al. 2006; Cai & Gochin 2007; Zhou et al. 2010), however no direct evidence supports that these small molecule fusion inhibitors bind to the deep pocket (Gochin & Cai 2009; Cai & Jiang 2010). Therefore, providing direct structural evidence that a small molecule can bind to the NHR deep pocket, so that a small molecule pharmacophore model can be deduced, is highly desired for small molecule HIV fusion inhibitor design and development.

5.1 Small molecule-peptide conjugates

To identify small molecule ligands that specifically bind to the gp41 NHR deep pocket, Harrison's group has synthesized a biased peptide conjugate library (Ferrer et al. 1999). It contained ~60,000 compounds and used three small molecule building blocks to replace the WWI motif in C-peptide and links to the same peptide sequence. The library was synthesized and screened against 5-helix (Weissenhorn et al. 1997) using an on-bead affinity-based assay. A small molecule moiety was identified, which sequentially contained cyclopentyl propionic acid–ε-glutamic acid–p-(N-carboxyethyl) aminomethyl benzoic acid (Fig. 8) (**47**). The moiety alone had no activity based on an HIV-1 ENV mediated cell-cell fusion assay. However, when conjugated to a 30-mer C-peptide C30 ($gp41_{636-665}$) without the PBD sequence, the resulting conjugate peptide showed an IC_{50} value of 0.3 μM, which was 20-fold increase compared with the IC_{50} value of 7 μM for C30. The conjugated peptide still had a much lower potency than a 38-mer ($gp41_{628-665}$) C-peptide containing the PBD that showed an IC_{50} value of 3 nM. The conjugated peptide could form a stable complex with N-peptide, as shown by size exclusion chromatography and native N-PAGE. This indicated that the small molecule moiety could partially mimic the WWI motif of C-peptide to occupy the deep binding pocket of the NHR, while structure modification is needed to optimize the binding.

Crystallography was used to characterize the interaction between the conjugated peptide and gp41 NHR (Zhou et al. 2000). The full length of the non-peptide moiety is visible in electron density maps, but unexpectedly in two orientations, each with about 50% occupancy. The two binding modes share the same aminobenzoic acid position (F1) but diverge at the two more distal building blocks. Also, the electron density for the amino acid at the connection to the non-peptide moiety is poor, suggesting disorder in the peptide linkage to the non-peptide moiety.

5.2 Small molecules

A small molecule α-helical mimetic based on a substituted terphenyl scaffold was designed to inhibit the assembly of the 6-HB core (Ernst et al. 2002). Tris-functionalized 3,2′,2″-terphenyl derivatives can serve as effective mimics of the surface functionality projected along one face of an α-helix. Compound 1a (**48**) was designed to mimic the side chains of an *i*, *i*+4, *i*+7 hydrophobic surface in an α-helix, using the branched alkyl substituents isobutyl and isopropyl (to avoid complications from chirality in a sec-butyl group) to mimic the side chains of the most prevalent Leu and Ile in the *a* and *d* positions of a 3-4 heptad repeat. Terminal carboxylate groups were also added to mimic the anionic character of the C-peptide and to improve the aqueous solubility. The ability of 1a (**48**) to disrupt the gp41 core was studied by CD spectroscopy, using a N36/C34 6-HB model (T_m = 66 °C). Titration of 1a into a 10 μM solution of N36/C34 resulted in a decrease of the CD signal at 222 and 208 nm, which corresponds to a reduction in the helicity of the 6-HB. A plot of θ_{222} versus inhibitor concentration shows saturation at approximately three equivalents of 1a. The CD spectrum with excess 1a was similar to the theoretical addition of the individual N36 and C34 spectra at the same concentration; and the thermal denaturation curve of the gp41 core in the presence of 50 μM 1a shows a significant 18 °C drop in the T_m value and closely resembles the melting transition of N36 alone at the same concentration. These data suggest that the 6-HB structure is completely disrupted by helix mimetic 1a. Both the hydrophobic and electrostatic features of 1a are important for its ability to disrupt the bundle. Analogs lacking the key alkyl side chains or carboxylic groups have little effect on the CD spectrum of the protein, even at high concentrations. Mimetic 1a effectively disrupts N36/C34 complexation with an IC_{50} value of 13.18 ± 2.54 μg.mL^{-1}, as measured by an ELISA assay using NC-1 (Jiang et al. 1998). Compound 1a inhibits HIV-1 mediated cell-to-cell fusion with an IC_{50} value of 15.70 ± 1.30 μg.mL^{-1}, using a dye-transfer cell fusion assay. In comparison, analogs lacking hydrophobic side chains or carboxylic groups had no inhibitory activity and proved to be cytotoxic at similar concentrations. Compound 1b (**49**), with larger hydrophobic groups than 1a, showed marginally enhanced activity than 1a.

Cai and Gochin identified a set of small molecule fusion inhibitors from a peptidomimetic library using a fluorescent biochemistry assay using Env2.0 as the target (Cai & Gochin 2007). Compounds **54** [3,5] and **55** [6,11] showed K_i values of 1.51 ± 0.16 and 1.34 ± 0.19 μM, respectively, in a competitive binding assay using Env2.0 as the target; and an IC_{50} value of ~8 μM in an HIV-1 gp41 mediated cell-cell fusion assay using a CCR5/CXR4 dual dependent target cell line (JI et al. 2006). These compounds contain two units that are covalently linked by an amide bond, and each unit contains two aromatic rings that may bind into the gp41 NHR hydrophobic pocket. A carboxyl group provides electrostatic interaction with K574 in the binding pocket and is critical for the activity of these small molecules; methylation of the carboxyl group resulted in loss of activities of the compounds in both the biochemical assay and cell-cell fusion assay. Three-unit compounds are prone to form aggregates under the assay conditions used and showed no activity, while single-unit compounds, such as M1 (**56**), display submillimolar inhibitory activity (Cai et al. 2009). Compound 1 (**57**), based on M1, was developed, which displayed an IC_{50} value of 4.5 ± 0.5 and 3.2 ± 0.5 µM in a fluorescence biochemical assay and a cell-cell fusion assay, respectively (Zhou et al. 2010). Compound 1 (**57**) showed very low cytotoxcity ($IC_{50} > 500$ µM); with a relatively small size, it is a promising lead for fusion inhibitor design.

Others have reported well-characterized small molecule fusion inhibitors targeting gp41, including SDS-J1 (**50**) (Debnath et al. 1999), NB64 (**51**), NB2 (**52**) (Jiang et al. 2004), and 4M041 (**53**) (Frey et al. 2006). These fusion inhibitors were selected from an active compound library by visual screening, then identified by high-throughput screening, and finally verified by a cell-cell fusion assay or HIV-1 infection assay. They usually showed low micromolar IC_{50} values for fusion inhibition; however, the following work to optimize the structures to obtain more potent fusion inhibitors were less fruitful, resulting in the identification of more small molecules with similar activity (Jiang et al. 2011). Also, their exact binding model with the gp41 NHR deep pocket still needs to be verified.

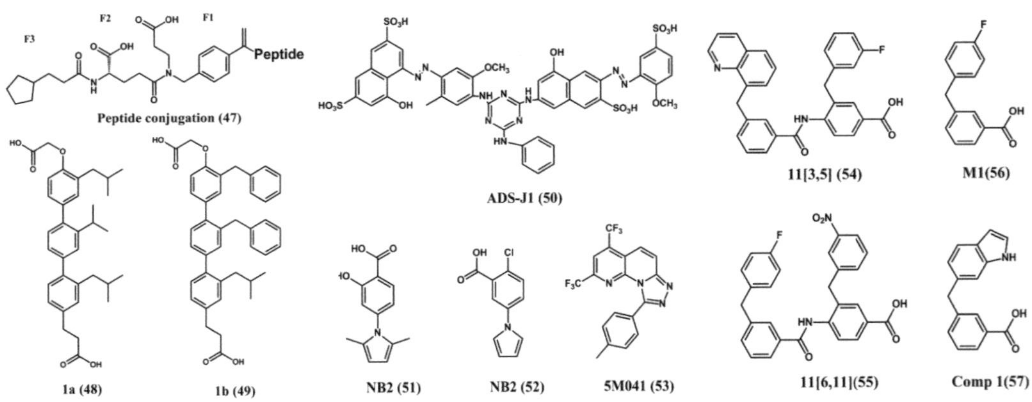

Fig. 8. Small molecule fusion inhibitors.

6. Conclusion

Peptides and peptidomimetics are efficient tools to study the HIV-1 gp41 NHR-CHR interaction, a key protein-protein interaction for HIV-1 gp41 mediated virus-cell membrane fusion, which enables HIV-1 enters and ultimately infects host cells. Peptides derived from wild-type HIV-1 gp41 sequences have been used to model the HIV-1 gp41 fusogenic core, a 6-HB formed by the NHR trimer as the inner core, and anti-parallel bind with three CHRs. Crystallographic structure analysis of the 6-HB has uncovered structure details for the gp41 NHR-CHR interaction. A deep pocket in the surface of NHR is a hot spot for the NHR-CHR interaction and a potential target for small molecule fusion inhibitors. N-peptides can be efficient targets for screening fusion inhibitors targeting the gp41 deep pocket by adding structural modulators to promote the trimeration of N-peptide.

Natural C-peptides can efficiently inhibit the gp41 NHR-CHR interaction by interacting with their counterpart in the gp41 6-HB; therefore, they can be used as fusion inhibitors against HIV-1 ENV mediated virus-cell fusion. They use residues at the *a* and *d* positions in heptad registration to bind the NHR hydrophobic grooves. The WWI motif of C-peptide provides a critical interaction with the NHR deep pocket, and an additional interaction between the C-peptide and NHR groove is required for a highly potent peptide fusion inhibitor of 30–40 residues. The *b*, *c*, *f*, and *g* residues in the C-peptide that form the predominantly hydrophilic surface in the 6-HB can be modified for increasing the secondary structure and solubility of the C-peptide, in order to increase its anti-HIV potency. The knowledge gained has been tested by artificial design of highly potent peptide fusion inhibitors with few similarities from known peptide sequences.

Peptidomimetics using unnatural building blocks have been successfully employed to mimic the molecular structures involved in the gp41 NHR-CHR interaction, resulting in highly potent HIV-1 fusion inhibitors with extraordinary *in vivo* stability to overcome the weakness of peptide drugs with potential oral administration possibilities. The achievements of the high potency peptidomimetic fusion inhibitors can also be used to guide small molecule fusion inhibitor design to disrupt this important protein-protein interaction.

In summary, HIV-1 fusion inhibitor development provides a model for using peptides as tools to probe protein-protein interactions for small molecule PPII design and development. The methods and results described in this chapter not only provide clues for future HIV-1 fusion inhibitor design, but also can be used for other viruses using a familiar virus-cell membrane fusion mechanism, as well as to guide other PPII design and development.

7. Acknowledgment

This work was supported by National Key Technologies R&D Program for New Drugs Grant 2009ZX09301-002 and National Natural Science Foundation of China Grant 81072581.

8. References

Bautista, A. D., et al. (2009). Identification of a beta(3)-peptide HIV fusion inhibitor with improved potency in live cells. *Bioorganic & Medicinal Chemistry Letters,* Vol. 19, No. 14, pp. 3736-3738, ISSN 0960-894X.

Bird, G. H., et al. (2010). Hydrocarbon double-stapling remedies the proteolytic instability of a lengthy peptide therapeutic. *Proceedings of the National Academy of Sciences of the United States of America,* Vol. 107, No. 32, pp. 14093-14098, ISSN 0027-8424.

Caffrey, M. (2011). HIV envelope: challenges and opportunities for development of entry inhibitors. *Trends in Microbiology,* Vol. 19, No. 4, pp. 191-197, ISSN 0966-842X.

Cai, L. & S. Jiang (2010). Development of peptide and small-molecule HIV-1 fusion inhibitors that target gp41. *Chemmedchem,* Vol. 5, No. 11, pp. 1813-1824, ISSN 1860-7187.

Cai, L. F., E. Balogh & M. Gochin (2009). Stable Extended Human Immunodeficiency Virus Type 1 gp41 Coiled Coil as an Effective Target in an Assay for High-Affinity Fusion

Inhibitors. *Antimicrobial Agents and Chemotherapy,* Vol. 53, No. 6, pp. 2444-2449, ISSN 0066-4804.

Cai, L. F. & M. Gochin (2007). A novel fluorescence intensity screening assay identifies new low-molecular-weight inhibitors of the gp41 coiled-coil domain of human immunodeficiency virus type 1. *Antimicrobial Agents and Chemotherapy,* Vol. 51, No. 7, pp. 2388-2395, ISSN 0066-4804.

Chan, D. C., C. T. Chutkowski & P. S. Kim (1998). Evidence that a prominent cavity in the coiled coil of HIV type 1 gp41 is an attractive drug target. *Proceedings of the National Academy of Sciences of the United States of America,* Vol. 95, No. 26, pp. 15613-15617, ISSN 0027-8424.

Chan, D. C., D. Fass, J. M. Berger & P. S. Kim (1997). Core structure of gp41 from the HIV envelope glycoprotein. *Cell,* Vol. 89, No. 2, pp. 263-273, ISSN 0092-8674.

Debnath, A. K., L. Radigan & S. B. Jiang (1999). Structure-based identification of small molecule antiviral compounds targeted to the gp41 core structure of the human immunodeficiency virus type 1. *Journal of Medicinal Chemistry,* Vol. 42, No. 17, pp. 3203-3209, ISSN 0022-2623.

Deng, Y. Q., Q. Zheng, T. J. Ketas, J. P. Moore & M. Lu (2007). Protein design of a bacterially expressed HIV-1 gp41 fusion inhibitor. *Biochemistry,* Vol. 46, No. 14, pp. 4360-4369, ISSN 0006-2960.

Dwyer, J. J., et al. (2007). Design of helical, oligomeric HIV-1 fusion inhibitor peptides with potent activity against enfuvirtide-resistant virus. *Proceedings of the National Academy of Sciences of the United States of America,* Vol. 104, No. 31, pp. 12772-12777, ISSN 0027-8424.

Eckert, D. M. & P. S. Kim (2001). Design of potent inhibitors of HIV-1 entry from the gp41 N-peptide region. *Proceedings of the National Academy of Sciences of the United States of America,* Vol. 98, No. 20, pp. 11187-11192, ISSN 0027-8424.

Eckert, D. M. & P. S. Kim (2001). Mechanisms of viral membrane fusion and its inhibition. *Annual Review of Biochemistry,* Vol. 70, pp. 777-810, ISSN 0066-4154.

Eckert, D. M., V. N. Malashkevich, L. H. Hong, P. A. Carr & P. S. Kim (1999). Inhibiting HIV-1 entry: Discovery of D-peptide inhibitors that target the gp41 coiled-coil pocket. *Cell,* Vol. 99, No. 1, pp. 103-115, ISSN 0092-8674.

Eggink, D., et al. (2008). Selection of T1249-resistant human immunodeficiency virus type 1 variants. *Journal of Virology,* Vol. 82, No. 13, pp. 6678-6688, ISSN 0022-538X.

Ernst, J. T., et al. (2002). Design of a protein surface antagonist based on alpha-helix mimicry: Inhibition of gp41 assembly and viral fusion. *Angewandte Chemie-International Edition,* Vol. 41, No. 2, pp. 278-+, ISSN 1433-7851.

Ferrer, M., et al. (1999). Selection of gp41-mediated HIV-1 cell entry inhibitors from biased combinatorial libraries of non-natural binding elements. *Nature Structural Biology,* Vol. 6, No. 10, pp. 953-960, ISSN 1072-8368.

Frey, G., et al. (2006). Small molecules that bind the inner core of gp41 and inhibit HIV envelope-mediated fusion. *Proceedings of the National Academy of Sciences of the United States of America,* Vol. 103, No. 38, pp. 13938-13943, ISSN 0027-8424.

Gochin, M. & L. F. Cai (2009). The Role of Amphiphilicity and Negative Charge in Glycoprotein 41 Interactions in the Hydrophobic Pocket. *Journal of Medicinal Chemistry,* Vol. 52, No. 14, pp. 4338-4344, ISSN 0022-2623.

Gochin, M., R. K. Guy & M. A. Case (2003). A metallopeptide assembly of the HIV-1 gp41 coiled coil is an ideal receptor in fluorescence detection of ligand binding. *Angewandte Chemie-International Edition,* Vol. 42, No. 43, pp. 5325-5328, ISSN 1433-7851.

Gochin, M., R. Savage, S. Hinckley & L. F. Cai (2006). A fluorescence assay for rapid detection of ligand binding affinity to HIV-1 gp41. *Biological Chemistry,* Vol. 387, No. 4, pp. 477-483, ISSN 1431-6730.

Harrison, S. C. (2008). Viral membrane fusion. *Nature Structural & Molecular Biology,* Vol. 15, No. 7, pp. 690-698, ISSN 1545-9985.

He, Y. X., et al. (2008). Identification of a critical motif for the human immunodeficiency virus type 1 (HIV-1) gp41 core structure: Implications for designing novel anti-HIV fusion inhibitors. *Journal of Virology,* Vol. 82, No. 13, pp. 6349-6358, ISSN 0022-538X.

He, Y. X., et al. (2008). Potent HIV fusion inhibitors against Enfuvirtide-resistant HIV-1 strains. *Proceedings of the National Academy of Sciences of the United States of America,* Vol. 105, No. 42, pp. 16332-16337, ISSN 0027-8424.

He, Y. X., et al. (2008). Design and evaluation of sifuvirtide, a novel HIV-1 fusion inhibitor. *Journal of Biological Chemistry,* Vol. 283, No. 17, pp. 11126-11134, ISSN 0021-9258.

Horne, W. S., et al. (2009). Structural and biological mimicry of protein surface recognition by alpha/beta-peptide foldamers. *Proceedings of the National Academy of Sciences of the United States of America,* Vol. 106, No. 35, pp. 14751-14756, ISSN 0027-8424.

JI, C., J. ZHANG, N. CAMMACK & S. SANKURATRI (2006). Development of a Novel Dual CCR5-Dependent and CXCR4-Dependent Cell-Cell Fusion Assay System with Inducible gp160 Expression. *Journal of Biomolecular Screening,* Vol. 11, No. 1, pp. 65-74, ISSN 1087-0571.

Jiang, S., K. Lin & M. Lu (1998). A conformation-specific monoclonal antibody reacting with fusion-active gp41 from the human immunodeficiency virus type 1 envelope glycoprotein. *Journal of Virology,* Vol. 72, No. 12, pp. 10213-10217, ISSN 0022-538X.

Jiang, S., et al. (2011). Design, Synthesis, and Biological Activity of Novel 5-((Arylfuran/1H-pyrrol-2-yl)methylene)-2-thioxo-3-(3-(trifluoromethyl)p henyl)thiazolidin-4-ones as HIV-1 Fusion Inhibitors Targeting gp41. *Journal of Medicinal Chemistry,* Vol. 54, No. 2, pp. 572-579, ISSN 0022-2623.

Jiang, S. B., K. Lin, N. Strick & A. R. Neurath (1993). HIV-1 inhibition by a peptide. *Nature,* Vol. 365, No. 6442, pp. 113-113, ISSN 0028-0836.

Jiang, S. B., K. Lin, N. Strick & A. R. Neurath (1993). Inhibition of hiv-1 infection by a fusion domain binding peptide from the HIV-1 envelope glycoprotein-gp41. *Biochemical and Biophysical Research Communications,* Vol. 195, No. 2, pp. 533-538, ISSN 0006-291X.

Jiang, S. B., et al. (2004). N-substituted pyrrole derivatives as novel human immunodeficiency virus type 1 entry inhibitors that interfere with the gp41 six-helix bundle formation and block virus fusion. *Antimicrobial Agents and Chemotherapy,* Vol. 48, No. 11, pp. 4349-4359, ISSN 0066-4804.

Judice, J. K., et al. (1997). Inhibition of HIV type 1 infectivity by constrained alpha-helical peptides: Implications for the viral fusion mechanism. *Proceedings of the National Academy of Sciences of the United States of America,* Vol. 94, No. 25, pp. 13426-13430, ISSN 0027-8424.

Lazzarin, A., et al. (2003). Efficacy of enfuvirtide in patients infected with drug-resistant HIV-1 in Europe and Australia. *New England Journal of Medicine,* Vol. 348, No. 22, pp. 2186-2195, ISSN 0028-4793.

Liu, S. W., et al. (2007). HIV gp41 C-terminal heptad repeat contains multifunctional domains - Relation to mechanisms of action of anti-HIV peptides. *Journal of Biological Chemistry,* Vol. 282, No. 13, pp. 9612-9620, ISSN 0021-9258.

Liu, S. W., et al. (2005). Different from the HIV fusion inhibitor C34, the anti-HIV drug fuzeon (T-20) inhibits HIV-1 entry by targeting multiple sites in gp41 and gp120. *Journal of Biological Chemistry,* Vol. 280, No. 12, pp. 11259-11273, ISSN 0021-9258.

Liu, Z., et al. (2011). In Vitro Selection and Characterization of HIV-1 Variants with Increased Resistance to Sifuvirtide, a Novel HIV-1 Fusion Inhibitor. *Journal of Biological Chemistry,* Vol. 286, No. 5, pp. 3277-3287, ISSN 0021-9258.

Miralles, G. D., et al. (2003). Baseline and on-treatment gp41 genotype and susceptibility to enfuvirtide (ENF) and T-1249 in a 10-day study of T-1249 in patients failing an ENF-containing regimen (T1249-102). *Antiviral Therapy,* Vol. 8, No. 3, pp. 21, ISSN 1359-6535.

Naito, T., et al. (2009). SC29EK, a Peptide Fusion Inhibitor with Enhanced alpha-Helicity, Inhibits Replication of Human Immunodeficiency Virus Type 1 Mutants Resistant to Enfuvirtide. *Antimicrobial Agents and Chemotherapy,* Vol. 53, No. 3, pp. 1013-1018, ISSN 0066-4804.

Oishi, S., et al. (2008). Design of a novel HIV-1 fusion inhibitor that displays a minimal interface for binding affinity. *Journal of Medicinal Chemistry,* Vol. 51, No. 3, pp. 388-391, ISSN 0022-2623.

Otaka, A., et al. (2002). Remodeling of gp41-C34 peptide leads to highly effective inhibitors of the fusion of HIV-1 with target cells. *Angewandte Chemie-International Edition,* Vol. 41, No. 16, pp. 2938-2940, ISSN 1433-7851.

Pan, C., L. Cai, H. Lu, L. Lu & S. Jiang (2011). A novel chimeric protein-based HIV-1 fusion inhibitor targeting gp41 with high potency and stability. *Journal of Biological Chemistry,* Vol. 286, No. 32, pp. 28425-28434, ISSN 0021-9258.

Pan, C. G., L. F. Cai, H. Lu, Z. Qi & S. B. Jiang (2009). Combinations of the First and Next Generations of Human Immunodeficiency Virus (HIV) Fusion Inhibitors Exhibit a Highly Potent Synergistic Effect against Enfuvirtide-Sensitive and -Resistant HIV Type 1 Strains. *Journal of Virology,* Vol. 83, No. 16, pp. 7862-7872, ISSN 0022-538X.

Qi, Z., et al. (2008). Rationally Designed Anti-HIV Peptides Containing Multifunctional Domains as Molecule Probes for Studying the Mechanisms of Action of the First and Second Generation HIV Fusion Inhibitors. *Journal of Biological Chemistry,* Vol. 283, No. 44, pp. 30376-30384, ISSN 0021-9258.

Root, M. J., M. S. Kay & P. S. Kim (2001). Protein design of an HIV-1 entry inhibitor. *Science,* Vol. 291, No. 5505, pp. 884-888, ISSN 0036-8075.

Shi, W. G., et al. (2008). Novel anti-HIV peptides containing multiple copies of artificially designed heptad repeat motifs. *Biochemical and Biophysical Research Communications,* Vol. 374, No. 4, pp. 767-772, ISSN 0006-291X.

Sia, S. K., P. A. Carr, A. G. Cochran, V. N. Malashkevich & P. S. Kim (2002). Short constrained peptides that inhibit HIV-1 entry. *Proceedings of the National Academy of Sciences of the United States of America,* Vol. 99, No. 23, pp. 14664-14669, ISSN 0027-8424.

Stephens, O. M., et al. (2005). Inhibiting HIV Fusion with a β -Peptide Foldamer. *Journal of the American Chemical Society,* Vol. 127 pp. 13126-13127, ISSN 0002-7863.

Su, C., et al. (2004). Substitutions within HIV gp41 amino acids 36-45 are identified as the primary determinants for loss of in vitro susceptibility to enfuvirtide: results of data mining analyses of genotypic changes in gp41 in TORO 1 and TORO 2 that associate with changes in phenotypic susceptibility to enfuvirtide. *Antiviral Therapy,* Vol. 9, No. 4, pp. U120-U121, ISSN 1359-6535.

Tan, K. M., J. H. Liu, J. H. Wang, S. Shen & M. Lu (1997). Atomic structure of a thermostable subdomain of HIV-1 gp41. *Proceedings of the National Academy of Sciences of the United States of America,* Vol. 94, No. 23, pp. 12303-12308, ISSN 0027-8424.

Tavassoli, A. (2011). Targeting the protein-protein interactions of the HIV lifecycle. *Chemical Society Reviews,* Vol. 40, No. 3, pp. 1337-1346, ISSN 0306-0012.

Walmsley, S., et al. (2003). Enfuvirtide (T-20) cross-reactive glycoprotein 41 antibody does not impair the efficacy or safety of enfuvirtide. *Journal of Infectious Diseases,* Vol. 188, No. 12, pp. 1827-1833, ISSN 0022-1899.

Wang, R. R., et al. (2009). Sifuvirtide, a potent HIV fusion inhibitor peptide. *Biochemical and Biophysical Research Communications,* Vol. 382, No. 3, pp. 540-544, ISSN 0006-291X.

Weissenhorn, W., A. Dessen, S. C. Harrison, J. J. Skehel & D. C. Wiley (1997). Atomic structure of the ectodomain from HIV-1 gp41. *Nature,* Vol. 387, No 6631, pp. 426-430, ISSN 0028-0836.

Welch, B. D., et al. (2010). Design of a Potent D-Peptide HIV-1 Entry Inhibitor with a Strong Barrier to Resistance. *Journal of Virology,* Vol. 84, No. 21, pp. 11235-11244, ISSN 0022-538X.

Welch, B. D., A. P. VanDemark, A. Heroux, C. P. Hill & M. S. Kay (2007). Potent D-peptide inhibitors of HIV-1 entry. *Proceedings of the National Academy of Sciences of the United States of America,* Vol. 104, No. 43, pp. 16828-16833, ISSN 0027-8424.

Wild, C., T. Oas, C. McDanal, D. Bolognesi & T. Matthews (1992). A synthetic peptide inhibitor of human-immunodeficiency-virus replication - correlation between solution structure and viral inhibition. *Proceedings of the National Academy of Sciences of the United States of America,* Vol. 89, No. 21, pp. 10537-10541, ISSN 0027-8424.

Wild, C. T., D. C. Shugars, T. K. Greenwell, C. B. McDanal & T. J. Matthews (1994). Peptides corresponding to a predictive alpha-helical domain of human-immunodeficiency-virus type-1 gp41 are potent inhibitors of virus-infection. *Proceedings of the National*

Academy of Sciences of the United States of America, Vol. 91, No. 21, pp. 9770-9774, ISSN 0027-8424.

Wilson, A. J. (2009). Inhibition of protein–protein interactions using designed molecules. *Chemical Society Reviews,* Vol. 38, pp. 3289-3300, ISSN 0306-0012.

Zhou, G. F., et al. (2000). The structure of an HIV-1 specific cell entry inhibitor in complex with the HIV-1 gp41 trimeric core. *Bioorganic & Medicinal Chemistry,* Vol. 8, No. 9, pp. 2219-2227, ISSN 0968-0896.

Zhou, G. Y., D. Wu, E. Hermel, E. Balogh & M. Gochin (2010). Design, synthesis, and evaluation of indole compounds as novel inhibitors targeting Gp41. *Bioorganic & Medicinal Chemistry Letters,* Vol. 20, No. 5, pp. 1500-1503, ISSN 0960-894X.

3

Profilin, and Vascular Diseases

Mohammad T. Elnakish and Hamdy H. Hassanain
Department of Anesthesiology, Dorothy M. Davis Heart & Lung Research Institute
Molecular, Cellular, and Developmental Biology Program
The Ohio State University, Columbus, OH
USA

1. Introduction

Actin is a highly dynamic network. It is essential for several important activities, such as muscle contraction and transmembrane signaling (Luna & Hitt, 1992; Salmon, 1989). Actin consists of actin filaments and a variety of associated proteins (Schmidt & Hall, 1998). Many proteins associated with the actin cytoskeleton control actin assembly and disassembly. These proteins regulate actin assembly at multiple levels, including the organization of actin monomers into actin polymers (Schmidt & Hall, 1998).

One key actin-regulatory protein is profilin, which associates with polymerization of actin. Profilin is a ubiquitous small (12–15 kDa) actin-binding protein expressed in eukaryotes (Kwiatkowski & Bruns, 1988; Magdolen et al., 1988; Sonobe et al., 1986; Tseng et al., 1984; Valenta et al., 1991a, b; Widada et al., 1989) and some viruses (Machesky et al., 1994). Profilin plays an important role in the regulation of actin polymerization in a number of motility functions (Haarer & Brown, 1990). The ability of profilin to bind to many ligands suggests that profilin is involved in signal transduction and may link transmembrane signaling to the control of the microfilament system (Korenbaum et al., 1998; Pantaloni & Carlier, 1993).

Early biochemical studies indicated that profilin interacts with actin in a 1:1 ratio and participates in the addition of monomers at the free barbed end of the filament then disassociates at the barbed end (Pantaloni & Carlier, 1993). Latest work has suggested several more functions of profilin aside from its monomer-sequestering ability. Profilin promotes the exchange of adenine nucleotide bound to actin monomer and also effectively lowers the critical concentration of monomer actin for polymerization of actin (Borisy & Svitkina, 2000; Theriot & Mitchison, 1993). It also promotes nucleotide exchange on an actin monomer by lowering the affinity of the actin monomer for its bound nucleotide by 1000-fold (Goldschmidt-Clermont et al., 1991).

It became progressively clear that profilins are vital constituents of the cytoskeleton. Additionally, the role of profilins in several cytoskeleton-based processes of clinical relevance has been proven. Several studies showed abnormal profilin levels in some pathological conditions. For example, high levels of profilin expression have been reported in human gastric cancer (Tanaka et al., 1992). On the contrary, profilin-I has been described as a tumor suppressor in some other types of cancer such as breast cancer (Das et al., 2009;

Zuo et al., 2007). Another clinical problem in which profilins may be involved is the lateral spreading of some infectious diseases (Pistor et al., 1995; Smith et al., 1996; Zeile et al., 1996). Moreover, profilins got a clinical consideration in other unexpected milieu. In this regard, profilins have been reported as major allergens implicated in pollen and food allergies in approximately 20% of type I allergy patients (Ebner et al., 1995; Valenta et al., 1991c, 1992). Furthermore, we (Hassona et al., 2010, 2011; Moustafa-Bayoumi et al., 2007) and others (Caglayan et al., 2010; Romeo & Kazlauskas, 2008; Romeo et al., 2004, 2007) have shown that profilin-I is an unexpectedly novel molecule that plays a highly significant role in vascular problems that predict a higher risk for developing arteriosclerosis, hypertension, stroke, heart failure, and finally death. Therefore, the aim of this chapter is to shed light on the significance of profilin-I via understanding the molecular and cellular aspects of this molecule, and its role in the vascular diseases.

2. Profilin family

2.1 Gene expression, products & intracellular localization of profilins

So far, there are four profilin genes that have been identified in the mouse and humans. Normally, the isoforms are expressed by diverse genes; nevertheless, differentially spliced isoforms are known to be present as well. It has been reported that in human, bovine, mouse, and rat, profilin-II is alternatively spliced into profilin-IIA and -IIB (Di Nardo et al., 2000; Lambrechts et al., 2000). In humans, profilin-I is expressed in every cell, while other isoforms are expressed in specific tissues. For example, profilin-IIA and -IIB are found to be brain specific and they are essential for neuronal development (Witke et al., 2001). Profilin-II complexes with other proteins such as synapsin and dynamin-I, well- known proteins that implicated in membrane trafficking. In addition, in humans and mouse profilin-III has been shown to be expressed in the testis and kidney and entirely in developing spermatids (Braun et al., 2002). At the amino acid level profilin-III and -IV exhibited only 30% identity among themselves and with other mammalian profilins (Obermann et al., 2005). Profilin-IV plays a key role in acrosome production and sperm morphogenesis. The same study by Obermann et al. proposes that profilin-III and -IV are transcribed in the germ cells. Yet, the expression timing was different during the rat testis post-natal development and in the rat spermatogenetic cycle. In the human testis, there is a correlation between profilin-IV mRNA expression and the presence of germ cells. Profilin-III and -IV may control testicular actin cytoskeleton dynamics and be a factor in acrosome production and spermatid nuclear shaping (Obermann et al. 2005).

Additionally, in *Caenorhabditis elegans* three profilin isoforms, profilin-I, profilin-II, and profilin-III, have been reported, among them profilin-I is crucial; however, profilin-II and profilin-III are not (Polet et al., 2006). As evident by immunostaining expression patterns for the profilin isoforms was different. At the early stages of embryogenesis, profilin-I confines to the cytoplasm and to the cellular contacts, while at the later stages of embryogenesis it confines to the nerve ring. At the late stages of embryogenesis, it has been shown that profilin-III expresses exclusively in the muscle cell walls. On the other hand, during adulthood, profilin-I is expressed in the neurons, the vulva, and the somatic gonad, profilin-II in the intestinal wall, the spermatheca, and the pharynx, and profilin-III, as dots, in the muscle cells of the body wall (Polet et al., 2006). Furthermore, two profilin isoforms (I and II) have been identified in *Dictyostelium amoebae;* profilin-I is fundamental for growth and development, where profilin-II is not. Moreover, it has been reported that *Saccharomyces cerevisiae* and *S. pombe* have only a single profilin isoform (Ezezika et al., 2009; Magdolen et al., 1988).

Based on the small sizes of profilin (15 kDa) and the profilactin complex (57 kDa) one might expect that they can easily diffuse to the nucleus. Nonetheless, profilin ordinarily is excluded from the nucleus and can be found only in the cytoplasm. Either the most part of profilin is bound in the cytoplasm and only a small portion can diffuse freely or there is a particular export mechanism that can actively take the profilin out of the nucleus (Witke, 2004). Recently, Stuven et al., (2003) reported a profilin-specific exportin present in the mammalian cells. Exportin 6 identifies the actin-bound profilin only, as a cargo and moves it out of the nucleus. The reasons for the existence of this profilactin-specific exportin still unclear, but this finding proposes that the nuclear levels of profilin and actin should be strictly regulated (Witke, 2004). Conversely, there are numerous reports about a nuclear fraction of profilin. For example, it has been reported that profilin-I is linked with subnuclear structures such as ribonuclear particles and Cajal bodies, and anti-profilin antibodies interfere with splicing *in vitro*. This implies a role for profilin-I in pre-mRNA processing (Skare et al., 2003). Also, it has been proposed that in the nucleus profilin-I and profilin-II interact with the survival of motor neuron (SMN) protein, a nuclear factor that is mutated in spinal muscular atrophy (Giesemann et al., 1999). SMN is important for splicing regulation yet it is not known whether this requires profilin binding or not. Still, in cell culture, co-localization of profilin-I and profilin-II with SMN in nuclear gems has been established (Giesemann et al., 1999).

To date the nuclear localization of profilins is a mystifying finding. Only a role for profilin and actin in splicing, chromatin remodeling or transcriptional regulation can be speculated. A more detailed understanding of the dynamics and properties of nuclear profilin and actin is required. It is possible that in the nucleus these proteins are considered necessary momentarily during the cell cycle or, particularly, in cells experiencing transcriptional activity changes (Witke, 2004).

2.2 Structural aspects of profilin

All recognized profilins share common structural and biochemical properties, though the amino acid sequences of the analogous isoforms in distantly related species may demonstrate less than 25% homology (Schlüter et al., 1997). Numerous studies on profilins from different origins demonstrate that they have highly similar tertiary structures (Fedorov et al., 1994, 1997; Metzler et al., 1993; Schutt et al., 1993; Thorn et al., 1997; Vinson et al., 1993) (Figure. 1). The profilin polypeptide consists of 100-131amino acids (Krishnan & Moens, 2009) and it is folded into a central β-pleated sheet formed of 5–7 antiparallel β-strands (Schlüter et al., 1997). On one side, this core is flanked by N- and C-terminal α-helices, with both termini next to each other, and on the opposed side by an extra α -helix attached to either additional α -helix or a small β-strand (Schlüter et al., 1997) (Figure. 1).

It has been reported that there are three groups of ligands characterize profilins: (1) G-actin and actin-related proteins (Machesky et al., 1994; Schutt et al., 1989; Tobacman et al., 1983) (2) polyphosphoinositides (Lassing & Lindberg, 1985, 1988) (3) poly-L-proline (PLP) with the exception of *Vaccinia* profilin (Kaiser et al., 1989; Lindberg et al., 1988; Tanaka & Shibata 1985), existing either as a peptide or as a sequence motif in particular proteins.

In this context, Gieselmann et al., (1995) showed that human profilin-I exhibits about five folds higher affinity for actin than profilin-II. Radiography analyses of the structures of human profilin isoforms imply that the substitution of profilin-I S29 by Y29 in profilin-II participates in the higher affinity of profilin-II for proline-rich sequences (Nodelman et al., 1999). In spite of the similarity in the 3D structures of human profilin-I and -II, the surface characteristics, such as exposure of hydrophobic patches (Figure 2), and biochemical properties of each isoform are different (Krishnan & Moens, 2009).

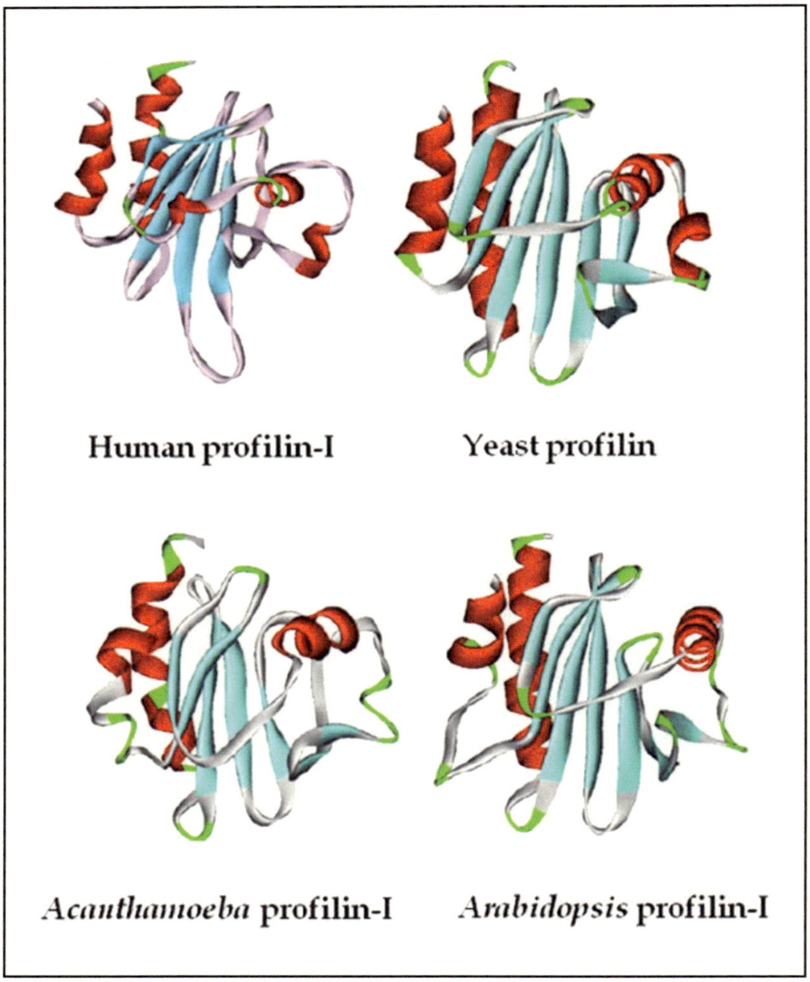

Fig. 1. Profilin-I isoforms from different organisms showing a similar helix (red) and strand (cyan) structure (PDB database: 1PFL, 1KOK, 2PRF, and 3NUL) with the loops highlighted in green, adapted from Krishnan & Moens, (2009) with permission.

2.3 Profilin ligands

Despite its relatively small size, many profilin ligands have by now been recognized, such as actin and actin-related proteins, polyphosphoinositides, PLP, annexin-I, and the list still increasing (Schlüter et al., 1997). Recently, there are more than 50 described profilin-binding ligands from diverse origins. However, this represents only a part of the real number of profilin-binding partners. Figure 3 shows the identified profilin ligands in mammalian cells. These do not include only molecules of focal contacts that could link profilin directly to actin polymerization such as VASP (vasodilator-stimulated Phosphoprotein) or Mena (mouse homolog of Drosophila enabled) (Gertler et al., 1996; Parast & Otey, 2000; Reinhard et al., 1995) but also include other molecules such as nuclear-export receptors (Boettner et al., 2000; Camera et al., 2003), regulators of endocytosis and membrane trafficking (Witke et al., 1998), Rac and Rho effectors molecules (Alvarez-Martinez et al., 1996; Miki et al., 1998; Ramesh et al., 1997; Suetsugu et al., 1998; Watanabe et al., 1997; Witke et al., 1998; Yayoshi-Yamamoto et al., 2000) and synaptic scaffold proteins (Mammoto et al., 1998; Miyagi et al., 2002; Wang et al., 1999). While a small number of these interactions demonstrated a physiological relevance, the recognition of profilin-interacting proteins could explain the unpredicted roles of profilin in mammalian cells. The profilin-ligands binding might help in linking different pathways to cytoskeletal dynamics via a mechanism that still unknown. Instead, the profilin–ligand interaction might work independently of actin to control the ligands directly (Witke, 2004).

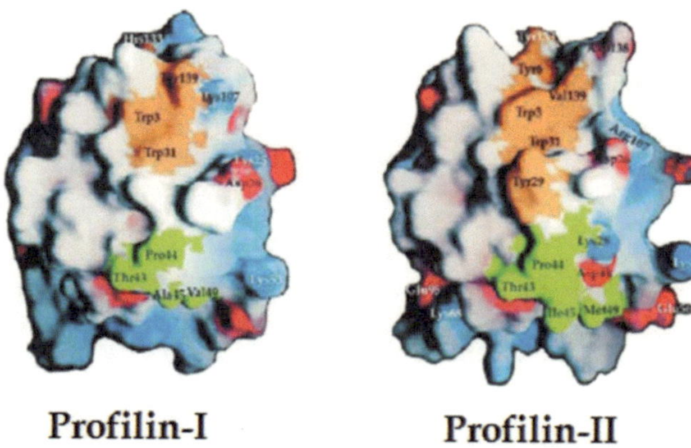

Fig. 2. Structure of human profilin-I and –II: differences in the surface-charge distribution might account for the ligand-binding specificity of profilin-I and -II. Colored regions highlight amino acid residues that are different in profilin-I and -II. Non-conserved residues are shown in blue; conserved residues are shown in brown, adapted from Witke, (2004) with permission.

Among this large number of profilin ligands we will focus on the binding of profilin to some of those ligands believed to be of relevant role in vascular problems such as actin and ligands in Rho/Rac pathway.

2.3.1 Profilin, actin & cytoskeleton

In vitro, Profilins can interact with and sequester actin monomers, in that way diminishing the concentration of free actin monomers that are accessible for filament elongation (Carlsson et al., 1977). They refill the pool of ATP-actin monomers via rising the nucleotide exchange rate by 1000-fold in comparison with that rate obtained from simple diffusion (Goldschmidt-Clermont et al., 1992). The profilin–ATP–actin complex can bind to the fast growing, barbed, or plus end of the actin filament and liberate the ATP–actin monomer, which is after that added to the filament (Figure 4). As a result, the elongating filament is made of ATP-actin. Down the filament, the ATP is slowly hydrolyzed via the actin intrinsic

ATPase activity. This produces ADP–actin in the older part of the filament. ADP–actin can be liberated gradually from the pointed or minus end of the filament by depolymerization or at faster rate by actin-depolymerizing proteins (Witke, 2004).

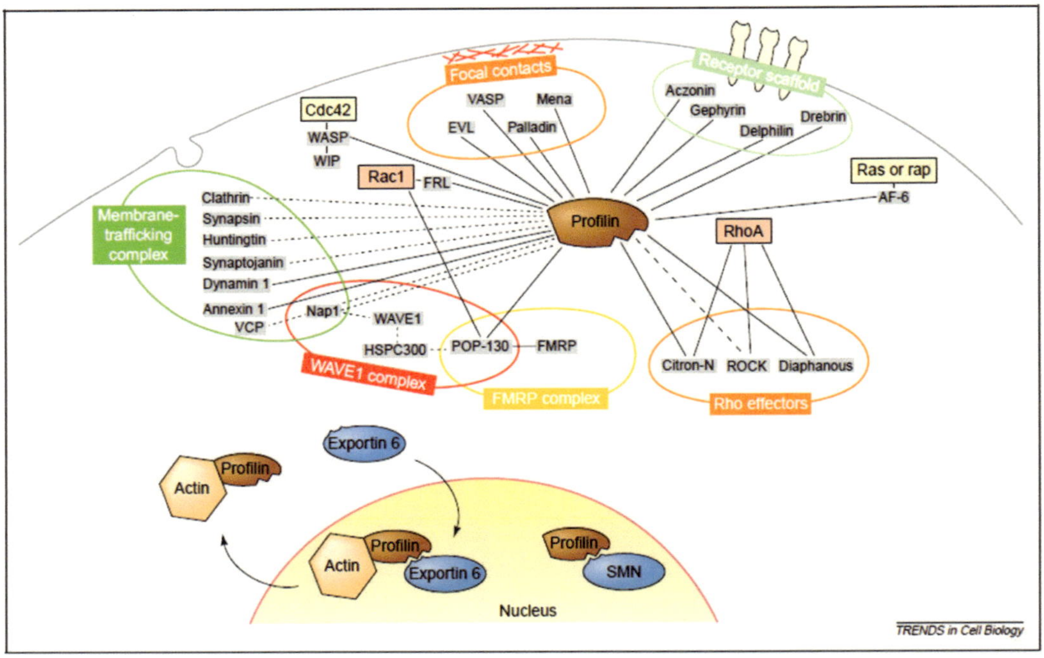

Fig. 3. Network of molecular interactions of profilin. Proteins that are known to interact with profilin are grouped according to their cellular location or the complexes in which they are found. Some of the profilin ligands are shared among different complexes (indicated by the intersecting fields), which suggests a crosstalk among signaling platforms, with profilin as the common denominator. Several links exist to small GTPases such as Rac1, RhoA, cdc42, Ras and Rap that are part of pathways that signal to the actin cytoskeleton. For simplicity, the term profilin is commonly used for profilin-I and profilin-II. Direct interactions between profilin and the ligands are indicated by unbroken lines, whereas potentially direct interactions are indicated by broken lines. Abbreviations: AF-6, All-1 fusion partner from chromosome 6; EVL, Ena VASP like; FMRP, fragile X mental retardation protein; FRL, formin-related gene in leukocytes; HSP, heat-shock protein; Mena, mouse homolog of Drosophila enabled; POP, partner of profilin; SMN, survival of motor neuron; VASP, vasodilator-stimulated phosphoprotein; VCP, valosine-containing protein; WASP, Wiskott-Aldrich syndrome protein; WAVE, WASP family verprolin-homologous protein; WIP, WASP-interacting protein, adapted from Witke, (2004) with permission.

It is worth noting that the presence of other G-actin binding proteins, such as thymosin β4 or any of the ADF family members can alter these processes (Pantaloni & Carlier, 1993). Additionally, capping the plus end of the filaments inhibits the addition of the profilin-actin complexes and consequently limits the activity of profilin to a simple sequestering effect (Pantaloni & Carlier, 1993; Perelroizen et al., 1996; Pring et al., 1992). Thus, the presence of other G-actin binding and/or capping proteins could regulate the profilin effect on cellular actin (Schlüter et al., 1997).

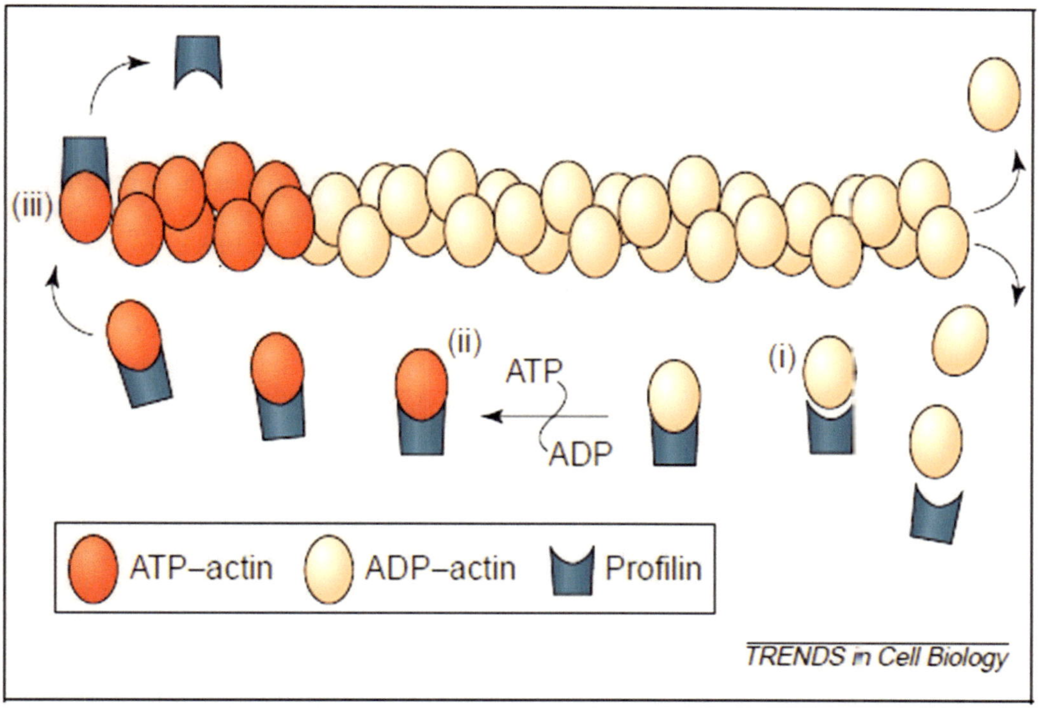

Fig. 4. Role of profilins in actin polymerization. An actin filament consists of two α-helical protofilaments. *In vitro*, three major functions have been identified for profilins in the regulation of actin polymerization. (i) Profilins can bind to and sequester actin monomers, thereby decreasing the concentration of free actin monomers that are available for filament elongation. (ii) Profilins replenish the pool of ATP–actin monomers (red) by increasing the rate of nucleotide exchange on the bound actin monomer 1000-fold compared with the rate of exchange based on simple diffusion. (iii) The profilin–ATP–actin complex can interact with the fast growing end of the actin filament and release the ATP–actin monomer, which is then added to the filament. Consequently, the elongating filament consists of ATP–actin. Along the filament, the ATP is slowly hydrolyzed by the intrinsic ATPase activity of actin, which generates ADP–actin (orange) in the older part of the filament. ADP–actin can be released slowly from the end of the filament by depolymerization or at an accelerated rate by 'actin-depolymerizing proteins' (not shown), adapted from Witke, (2004) with permission.

Previously, it has been considered that the profilins effect on nucleotide exchange on actin directly regulates their ability to promote filament assembly at the plus end. Polymerization of filament is coupled with the actin-bound ATP hydrolysis and thus far, profilins are unique microfilament associated proteins that can work as nucleotide exchange factors. Polymerization of ATP-actin occurs more rapidly and at a lower critical concentration than ADP-actin (Pollard, 1986). Nonetheless, profilin isoforms I and III in *Arabidopsis* are unable to speed up the rate of nucleotide exchange on G-actin yet still reduce the critical concentration at the plus ends of filaments, similar to vertebrate profilin (Perelroizen et al., 1996). These data demonstrate that the major effect of profilins on actin polymerization cannot be linked with their capacity to work as nucleotide exchange factors.

In vivo, the global view that the main biological function of profilin was observed in its actin sequestering effect became debatable, principally due to the finding that the concentration of profilin in cells and its actin-binding affinity are inadequate to stabilize the G-actin pool (Babcock & Rubenstein 1993; Goldschmidt-Clermont et al., 1991; Machesky & Pollard, 1993; Sohn & Goldschmidt-Clermont, 1994). Generally, the data obtained from cells with different profilin levels are in harmony with the notion that in lower eukaryotes the central role of profilin is to sequester G-actin, whereas in higher eukaryotes this is mainly done via other G-actin binding proteins such as thymosin β4 (Safer et al., 1991), and profilins are mostly implicated in the actin filament dynamics control (Sohn & Goldschmidt-Clermont, 1994).

Based on this notion, lower eukaryotes deficient in profilins should exhibit an increase in F-actin, however in higher eukaryotes this would not be the principal outcome. Compatible with this model, *S. pombe* cells with profilin overproduction showed undetectable amount of actin filaments, and are incapable of forming a contractile ring (Balasubramanian et al., 1994). In *S. cerevisiae* cells harmful effects due to actin overexpression, could be compensated by profilin overexpression (Magdolen et al., 1993). Conversely, several studies reported about filament-stabilizing or -regulating functions of profilin in higher eukaryotes. For example, the overall F-actin content and stability were elevated whereas; a considerable amount of F-actin was shifted from stress fibers to the cortical actin network in Chinese hamster ovary cells overexpressing profilin (Finkel et al., 1994). Likewise, actin filaments were stabilized against cytochalasin D and latrunculin in baby hamster kidney cells overexpressing birch profilin (Rothkegel et al., 1996). In addition, a shift in F-actin from stress fibers to thick peripheral actin filament bundles with a corresponding increase in cellular adhesion to fibronectin has been reported in cultured human endothelial cells overexpressing profilin (Moldovan et al., 1997).

Although these findings indicated a differential role of profilins between lower and higher eukaryotes, a few studies showed contradictory data to these reports (Cao et al., 1992; Edwards et al., 1994; Staiger et al., 1994). Consequently, a final conclusion on the validity of the assumption regarding differential functions of profilins in higher and lower eukaryotes needs to be confirmed with further experimentations (Schlüter et al., 1997).

2.3.2 Profilin & Rho/Rac pathway

Rho/Rac signaling pathway represents one of the well-known pathways in the regulation of actin cytoskeleton, as indicated by the Rac1-dependent membrane ruffling and RhoA-stimulated stress-fiber formation (Nobes & Hall, 1995). Although there is no any report about the direct interaction between profilins and Rho and/or Rac or any other small GTPases, many of the profilin ligands are well-recognized Rho/Rac effector molecules (Witke, 2004). In this regard, our recent data showed that profilin overexpression in vascular smooth muscle cells (VSMC) of transgenic mice results in vascular remodeling and hypertension. These were associated with increased Rho-GTPase activity and Rho-dependent coiled-coil kinase (ROCK) expression (Hassona et al., 2010; Moustafa-Bayoumi et al., 2007). As well, it has been reported that ROCK is a part of the profilin-II complex in the brain (Witke et al., 1998) and this binding is significant in the regulation of neurite outgrowth by ROCK (Da Silva et al., 2003). Furthermore, two other proteins that connect profilin to the Rac pathway were recognized in the profilin-II complex in the brain, Nck-associated protein (Nap 1) and partner of profilin (POP)-130 (Witke et al., 1998). GTP-Rac1 interacts with POP-130 and can detach the tetrameric WAVE 1 [Wiskott–Aldrich syndrome protein (WASP) family verprolin-homologous protein]1 complex, resulting in the activation of actin polymerization by WAVE1. Yet, the role of profilin binding to POP-130 is not apparent however it is possible that profilin might manage the complex formation between WAVE1 and POP-130 and between FMRP (fragile X mental retardation protein) and POP-130, in the same way as Rac1 (Witke, 2004).

Additional small-GTPase-binding molecules that can interact with profilin are the Rho-binding molecules, mouse homologs of the *Drosophila* gene diaphanous (mDia1, mDia2 and mDia3) which are known as potent nucleators of actin polymerization (Wallar & Alberts, 2003). Generally, the diaphanous protein exists in an inactive conformation due to folding back of its N terminal GTPase-binding domain onto its C-terminal Dia-autoregulatory domain resulting in association and autoinhibition. RhoA binding to the N terminus releases the autoinhibition and activates actin nucleation (Alberts, 2002). Profilin binding occurs through the proline-rich formin homology domain that present in the core of that diaphanous molecule (Watanabe et al., 1997). Yet, the significance of that binding is not clear. One interesting possibility is that diaphanous can move actin after it has been sequestered by profilin and activate actin polymerization (Li & Higgs, 2003). However, this is limited by the argument that the studies of profilin-diaphanous binding used truncated versions of diaphanous, rather than the full-length protein. *In vivo*, it has been suggested that large complex of diaphanous oligomers is present as well (Li & Higgs, 2003), which via diaphanous monomers can interact with profilin and/or profilactin molecules. Nevertheless, the structure and regulation of this enormous signaling platform for actin nucleation need to be understood (Witke, 2004).

2.3.3 Ligands binding sites

In this section we will discuss the binding sites of the main profilin ligands, actin, phatidylinositol 4,5-bisphosphate (PtdIns 4,5-P), and PLP. Initially, profilin binds with high affinity (micromolar range) to G-actin in a 1:1 stoichiometric complex (Schlüter et al., 1997). The amino acid motif LADYL in the C-terminal α-helix was first proposed to be implicated in actin binding depending on (1) the presence of this motif in most of profilins, (2) the presence of homologous sequences in a range of actin-binding proteins such as DNase I, fragmin, gelsolin, severin, villin and the vitamin D-binding protein (Binette et al., 1990; Tellam et al., 1989; Vandekerckhove, 1989; Vinson et al., 1993). Nevertheless, this hypothesis was neglected due to (1) the absence of this sequence in mammalian profilins (2) the ability of Saccharomyces profilin to interact with actin even after deletion of this motif (Haarer et al., 1993). Now, the LADYL-motif is believed to be a central element in the dense structure of these proteins (Ampe & Vandekerckhove, 1994; Fedorov et al., 1994; Haarer et al., 1993; McLaughlin & Weeds, 1995). Studies on bovine profilin-I and β-actin showed that the actin binding sites on profilin are localized in the α-helix 3, the proximal part of α-helix 4, and in the β-strands 4, 5 and 6 (Schutt et al., 1993) (Figure 5). These residues bind to subdomains 1 and 3 on the actin molecule; however, they do not exhibit a conserved sequence motif (Thorn et al., 1997). In the bovine complex, Phe375 appears to be a key residue that interacts with Ile73, His119, Gly121 and Asn124 on the profilin side (Schutt et al., 1993). Similarly, other studies on *Acanthamoeba* reported that actin-related proteins such as Arp2 interact with profilin using the same binding site (Kelleher et al., 1995; Machesky, 1997).

Fig. 5. Topographical relation of the main ligand binding domains as seen on the X-ray structure of bovine profilin (Schutt et al., 1993). The binding domains of actin and actin related proteins (blue; Schutt et al., 1993) and PtdIns 4,5-P_2 (red; Sohn et al., 1995) overlap, while that for proline-cluster sequences (green; Metzler et al., 1994) is located at the opposite side of the profilin molecule, adapted from Schlüter et al., (1997) with permission.

On the other hand, studies on *Acanthamoeba* described a positively charged area opposing both termini, placed in the G-actin binding site as the binding motif of the second key ligand of profilin, PtdIns 4,5-P_2 (Fedorov et al., 1994) (Figure 5). This was supported by mutation studies on *Saccharomyces* profilin and human profilin-I. Point mutations in this region diminished the binding affinity of profilin to PtdIns 4,5-P_2 (Haarer et al., 1993; Sohn et al., 1995). In line with the observation that the binding sites of G-actin and PtdIns 4,5-P_2 on profilin overlap (Figure 5), it has been reported that these ligands compete with each other for binding to profilin (Lassing & Lindberg, 1985; 1988; Machesky et al., 1990). In addition, other reports showed that binding of PtdIns 4,5-P2 results in a conformational change in profilin and disrupts the profilin-actin complex (Raghunathan et al., 1992). Also, it has been revealed that profilin can bind a variety of phosphatidylinositol and the binding affinity of human profilin-I to phosphatidylinositol 3,4-bisphosphate (PtdIns 3,4-P_2), and phosphatidylinositol 3,4,5-trisphosphate (PtdIns 3,4,5-P_3) is higher than its affinity to PtdIns 4,5-P_2 (Lu et al., 1996). Furthermore, phosphoinositide (PI) 3-kinase activity may be regulated by profilin through direct binding to the p85 subunit of this enzyme (Singh et al., 1996). PI 3-kinase has no effect on the binding of actin to profilin (Singh et al., 1996), signifying that the binding sites of actin and p85 on profilin are different.

Conversely, the profilin-PtdIns 4,5-P2 complex can be hydrolyzed only via phospholipase Cγ1 (PLCγ1). Phosphorylation and activation of this lipase as a result of transmembrane signaling (Goldschmidt-Clermont et al., 1990; 1991) leads to the conclusions that (1) profilins are implicated in the metabolism of phosphoinositide and (2) hydrolysis of PtdIns 4,5-P2 causes profilin to move out from the membrane to the cytosol where it can bind to actin or other ligands. These conclusions propose that profilin–phosphoinositide binding plays a vital role *in vivo* (Janmey et al., 1995; Ostrander et al., 1995).

Mutation (Björkegren et al., 1993; Haarer et al., 1993) and NMR (Archer et al., 1994; Metzler et al., 1994) analyses described the binding site of profilin to the third main ligand, PLP as a hydrophobic patch including the NH- and COOH-terminal α-helices and the upper face of the antiparallel β-sheet, opposing to the actin/PtdIns 4,5-P2 binding region (Figure 5). The binding of PLP to profilins has no effect on the interaction with G-actin or PtdIns 4,5-P_2 (Archer et al., 1994; Kaiser et al., 1989), indicating that PLP has a distinct binding site (Figure 4). Expediently, this specific PLP-profilin binding is used for profilins purification (Kaiser et al., 1989; Lindberg et al., 1988). For effective profilin binding, it has been proposed that 6 continuous prolines would be sufficient (Metzler et al., 1994). Nevertheless, other reports demonstrated that at least 8–10 prolines are required for efficient binding (Domke et al., 1997; Machesky & Pollard, 1993; Perelroizen et al., 1994; Petrella et al., 1996).These proline stretches may be interrupted by single glycine residues (Domke et al., 1997; Lambrechts et al., 1997) and may be capable of simultaneous binding of two profilins (Lambrechts et al., 1997), depending on the ability of profilin to oligomerize (Babich et al., 1996).

The first recognized ligand for PLP was VASP, a focal adhesion molecule that was reported to interact directly with F-actin (Jockusch et al., 1995; Reinhard et al., 1995), and it also described as a substrate of both cGMP- and cAMP-dependent protein kinases in platelets (Halbrügge et al., 1990). VASP has a central proline-rich domain with a single copy and a 3-fold tandem repeat of a remarkable (G)P_5 motif (Haffner et al., 1995). This motif is both required and sufficient for profilin binding (Domke et al., 1997; Lambrechts et al., 1997; Reinhard et al., 1995). Another PLP-binding ligand similar to VASP is a VASP-related mouse protein, Mena (Gertler et al., 1996). Additional PLP-biding ligands are the formin-related proteins, *S. cerevisiae* Bni1p and Bnr1p, *S. pombe* Cdc12p, *Drosophila* cappuccino and p140mDia, the mammalian homologue of the *Drosophila* protein diaphanous (Chang et al., 1997; Evangelista et al., 1997; Imamura et al., 1997; Manseau et al., 1996; Watanabe et al., 1997). These proteins have a proline-rich domain with numerous proline stretches consisting of 5–13 residues and a C-terminal consensus sequence of approximately 100 amino acids (Castrillon & Wasserman, 1994). Due to the high specific binding of Bni1p, Bnr1p and p140mDia to the GTP-bound form of Rho family members (Kohno et al., 1996; Imamura et al., 1997; Watanabe et al., 1997); they perhaps represent significant connectors between signal transduction, profilin and the cytoskeleton. Furthermore, adenylyl cyclase-associated protein (CAP) has been described as PLP-binding ligand. CAP has a $G(P)_6$ $G(P)_5$ motif and it can bind to profilin (Domke et al., 1997; Lambrechts et al., 1997). Nevertheless, other studies demonstrated that CAP exists in a folded configuration (Lambrechts et al., 1997) and hence its binding to profilin may be firmly regulated.

2.3.4 Regulation of profilin-ligands binding

The important factors that could help in understanding the process of profilin-ligand binding regulation include the structural requirements for the binding of profilin to the ligand, the binding specificity of ligands to different profilins and the mechanisms of ligand release. Initially, the structural requirements for the profilin-ligand binding are not completely understood. In spite of binding of profilin to an extremely diverse group of ligands either directly or as part of a larger complex, the binding sites on both profilin and ligands appear to be well conserved. The majority of ligands are believed to interact with the PLP domain of profilin that contains the N- and C-terminal helices. The only exception, so far, to this model is gephyrin, which appears to bind to a special profilin domain (Giesemann et al., 2003). All profilin ligands are characterized by the presence of stretched or nearly stretched proline-rich domains that are required for profilin binding. Still, a contiguous prolines stretch is insufficient. Depending on the data obtained from *in vitro* studies using synthetic PLP peptides of different length high-affinity binding requires a decamer as a minimum, (Perelroizen et al., 1994) however this cannot be extended to cover proteins or to be used for recognizing or evaluating the ability of profilin to bind to a ligand. A lot of profilin ligands contain in their proline-rich domains proline repeats of no more than three or four successive prolines. Further amino acids, mostly glycines, appear to be capable of replacing proline, and an efficient profilin-binding domain appears to include numerous repeats that have the consensus sequence ZPPX (where Z=P, G or A; and X= any hydrophobic amino acid) (Witke et al., 1998).

The second important factor in regulating the binding of profilins to their ligands is the binding specificity of ligands to the different profilins. Previous reports showed that the interaction of ligands with profilin-I and profilin-II occurs in a highly specific manner (Witke et al., 1998) and it looks likely that it is not only the PLP-binding domains but also other complex binding parameters have to be considered. Comparative studies on the structures of mammalian profilin-I and profilin-II indicated that they are approximately superimposable (Nodelman et al., 1999). Nevertheless, the distribution of surface charges in profilin-I and profilin-II is significantly different and this perhaps participates in the ligand-binding specificity (Figure 2). Eventually, identifying the structural features of different profilin complexes will be helpful to understand the basis of specificity.

Finally, the profilin- ligands binding should be a dynamic process and the mechanisms of ligand release under physiological conditions have to be determined. For example, actin can be released from profilactin complex via PtdIns(4,5)P_2, and an analogous mechanism might be used for ligand binding regulation. For instance, PtdIns(4,5)P_2 can regulate the interaction between dynamin 1 and profilin-II, but not the Mena–profilin or VASP–profilin complexes. Regulation of Mena, VASP, and other ligands binding might be achieved in different ways such as profilin or ligand phosphorylation (Witke, 2004).

2.4 Role of profilin in signal transduction

Profilins bind to several ligands, and a lot of these ligands are part of various complexes or interact with each other as well (Figure 3). This results in an intimate crosstalk among these complexes that can substitute and distribute components and, thus, could assimilate signals from other signaling pathways such as small-GTPase and phosphoinositide pathways. In these signaling platforms profilins appear to be a common denominator (Witke, 2004). Figure 6 is a schematic representation demonstrating various interactions between profilin, the microfilament system and different signaling pathways.

Profilins are linked to the phosphatidylinositol cycle and in turns to the receptor tyrosine kinase pathway through their binding to PtdIns 4,5-P_2. Profilin-bound PtdIns 4,5-P_2 is resistant to hydrolysis by phospholipase Cγ1 (Goldschmidt-Clermont et al., 1990). However, this resistance can be overcome after activating phospholipase via receptor tyrosine kinases-dependant phosphorylation (Goldschmidt-Clermont et al., 1991). This activation process results in PtdIns 4,5-P_2 hydrolysis with subsequent formation of other two second messengers, diacylglycerol and inositol 1,4,5-trisphosphate. Additionally, profilin releases from the membrane, which might initiate fast, local actin polymerization. Conversely, activated phospholipase Cγ1 cannot hydrolyze other PI 3-kinase activity products such as PtdIns 3,4-P_2 and PtdIns 3,4,5-P_3 ,that bind to profilin with a higher affinity than PtdIns 4,5-P_2. Consequently, it has been revealed that PtdIns 3,4-P_2 and PtdIns 3,4,5-P_3 may regulate phospholipase Cγ1-controlled turnover of PtdIns 4,5-P_2 (Lu et al., 1996).

In addition, the profilin ligands of the formin-related proteins such as p140mDia connect the GTPase-related signaling cascade, which is also coupled with the PtdIns 4,5-P_2 signaling pathway to the microfilament system. The small GTPases of the Rho family are active members that are involved in regulating the cytoskeleton-based processes such as cell morphology, adhesion and cytokinesis (Tapon & Hall, 1997). Most likely, these formin-related proteins are down-stream effectors of Rho in this cascade (Evangelista et al., 1997; Watanabe et al., 1997).

Also, the microfilament system is linked to the adenylyl cyclase-related pathway via substrates of the cAMP/cGMP-dependent protein kinases such as VASP/Mena family (Butt et al., 1994; Gertler et al., 1996) and the putative profilin ligand CAP, which is an adenylyl cyclase activator (Fedor-Chaiken et al., 1990; Field et al., 1990; Toda et al., 1985). This linking can be executed through either direct binding of CAP and VASP proteins to actin (Freeman et al., 1995; Gieselmann & Mann, 1992; Gottwald et al., 1996; Hubberstey et al., 1996; Reinhard et al., 1992) or recruiting profilin and profilin-actin complexes to areas of dynamic actin remodelling via the interaction of VASP proteins with cell contact proteins such as zyxin and vinculin (Brindle et al., 1996; Gertler et al., 1996; Reinhard et al., 1995, 1996).

Furthermore, annexin I could be involved in this crosstalk depending on previous reports that described the sensitivity of annexin I-profilin binding to PtdIns 4,5-P_2 and actin (Alvarez-Martinez et al., 1996). On top of that the annexins activity is controlled by the free Ca^{2+} level, which is adjusted via PtdIns 4,5-P_2 hydrolysis upon the action of the activated phospholipase Cγ 1 (Figure 6). In addition to annexin I, Ca^{2+} level will affect various Ca^{2+}-actin-binding and –severing proteins which slice the actin filaments and create new plus ends to which profilin-actin complexes can be added (Schlüter et al., 1997) (Figure 6).

Interestingly, in mesangial cells extracellular profilin was shown to bind specifically to a putative receptor and stimulates AP-1, a key element in signal transduction that is involved in the regulation of the transcription of several genes and cell growth (Tamura et al, 2000).

With the current large number of profilins ligands the future challenge is to determine their role in this complicated signaling crosstalk. One possibility is that profilins may act as regulators for the composition of the complexes and facilitate entrance or exit of certain ligands. Additionally, they might act as direct regulators for the ligands activities. Identification of all profilins molecular interactions, their ligands, and recognizing the structure of these complexes will be helpful to understand the mechanisms by which profilins can control this diverse signaling complexes (Witke, 2004).

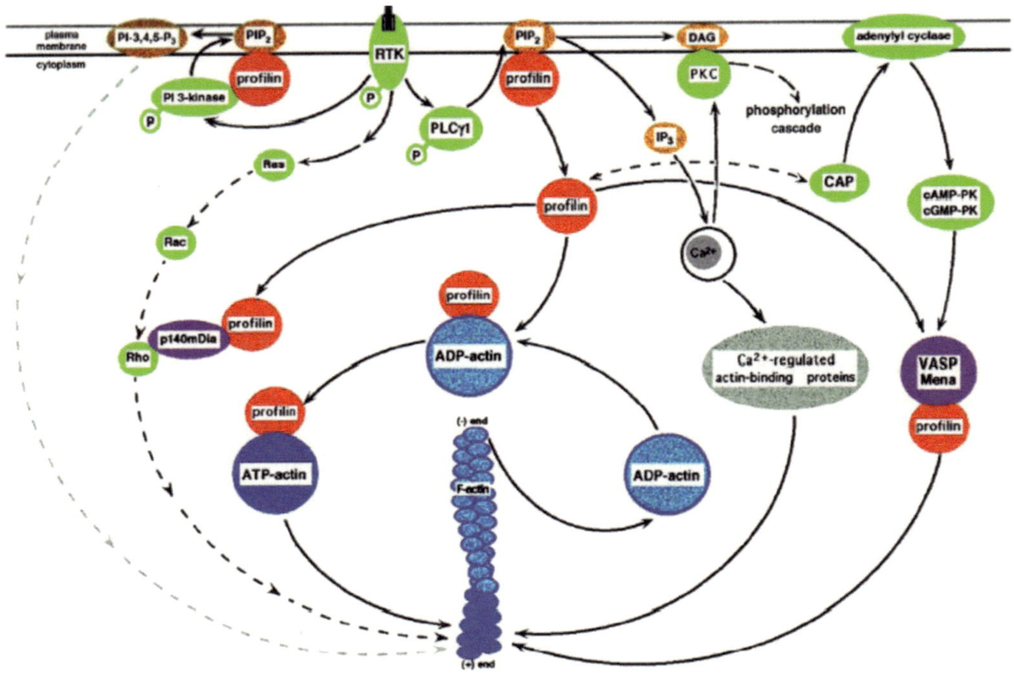

Fig. 6. The involvement of profilin (red) in different signaling routes. This schematic drawing depicts only the main connections established so far. Molecules and second messengers of the polyphosphoinositide signaling pathway are indicated in yellow, protein members of signaling routes are marked green, proline-cluster proteins identified as profilin ligands are marked purple, the actin cycle is seen in blue, Ca^{2+} in intracellular stores and Ca^{2+} regulated microfilament proteins are marked grey. For simplicity, the solid arrows indicate either direct interactions between components, as shown by biochemical assays, or point to pathways. Broken arrows indicate suspected or indirect interactions. Abbreviations: PI-3,4,5-P_3: phosphatidylinositol 3,4,5-trisphosphate; PIP_2: phosphatidylinositol 4,5-bisphosphate; RTK: receptor tyrosine kinase; DAG: diacylglycerol; PLCγ 1: phospholipase Cγ 1; cAMP/cGMP- PK: cAMP/cGMP dependent protein kinase; IP_3: inositol 1,4,5-trisphosphate, adapted from Schlüter et al., (1997) with permission.

3. Profilin & vascular diseases

3.1 Role of profilin in vascular smooth muscle & endothelial cells

3.1.1 Profilin & vascular smooth muscle cells migration & proliferation

Migration of smooth muscle cell takes place throughout vascular development, as a result of vascular injury, and throughout atherogenesis. Throughout vascular development, platelet-derived growth factor promotes migration of pericyte or other precursors of smooth muscle that is required for the formation of correct vessel wall (Hellstrom et al., 1999). Clinically, vascular injury takes place after angioplasty, vascular stent implantation, or organ transplantation. In vascular injury in animals, thickening of intima and media has been attributed to VSM proliferation and migration from media to intima (Clowes et al., 1989; Majesky & Schwartz, 1990; Reidy, 1992). Throughout atherogenesis, VSMCs migrate to occupy the intima, either from the media (Murry et al., 1997) or from the circulation via $CD34^+$ hematopoietic progenitor cells migration, resulting in smooth muscle progenitor cells (Yeh et al., 2003). Figure 7 shows the inner lining of a normal artery.

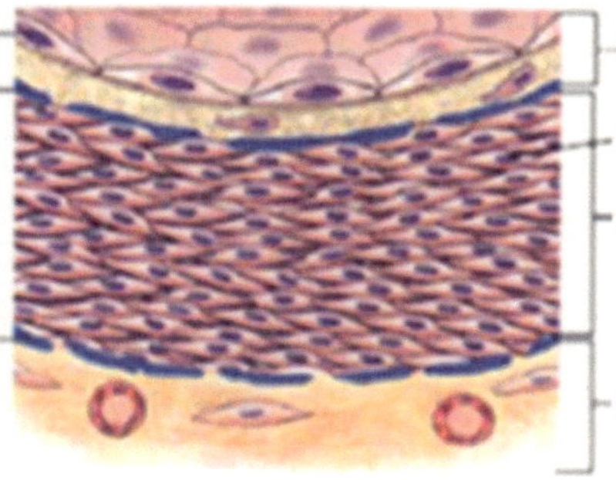

Fig. 7. Inner lining of normal artery

VSMC migration requires the extension of lamellipodia toward the stimulus via actin polymerization, trailing edge detachment via focal contacts degradation, and force generation via myosin II in the cellular body to drive the cell forward (Gerthoffer, 2007). Initiation of new filaments nucleation is achieved by actin-related proteins such as ARP2/3 complex interacting with the minus end and elimination at the plus end of capping proteins that are PIP_2-sensitive. Extension of new actin filaments is improved by formin-related proteins such as mDia1 and mDia2 that operate along with profilin on the plus end. Activation of The formins mDia1 and mDia2 is achieved by RhoA and Cdc42, respectively. Profilin released from the binding sites of membrane phospholipid enhances nucleotide exchange on G-actin monomers and promotes actin polymerization. Stimulation of filament branching is accomplished via activating WAVE complex and WASP by Rac and Cdc42, respectively. WAVE and WASP increase nucleation and branching through activating actin-related proteins such as ARP2/3 complex. Severing of Actin-filament by gelsolin is stimulated by Ca^{2+}, and nucleation is favored via liberating gelsolin from plus ends of F-actin by PtdIns 4,5-P2. Stimulation of actin depolymerization is executed by cofilin at the minus end. Cofilin acts to limit the filaments length and to induce the existing filaments turnover. These operations have been reported to be sufficient for force generation to expand the leading edge of the cell toward the stimulus (Mogilner & Oster, 2003; Prass et al., 2006). Consistent with these findings our recent data confirmed the significant role of profilin-I in VSMC migration. Migration assays performed on VSMC isolated from the aorta of transgenic mice that overexpress the cDNA of profilin-I or profilin-I-dominant negative mutant (88R/L) and nontransgenic controls showed that the rate of cell migration of profilin-I VSMCs is significantly higher than that of the control and 88R/L. Conversely, 88R/L mice exhibited a significantly lower rates compared to nontransgenic controls (Figure 8) (Hassanain HH, unpublished).

On the other hand, it has been shown that profilin plays a vital role in the proliferation and differentiation of normal cell. Disruption in the profilin results in embryonic lethality due to gross impairment in growth, motility, and cytokinesis in single cells (Haugwitz et al., 1994; Witke, 2004; Witke et al., 2001). Also, profilin-1 was demonstrated to exert cellular responses such as DNA synthesis and increasing the binding activity of AP-1 DNA in mesangial cells via activating putative cell surface receptors (Tamura et al., 2000).

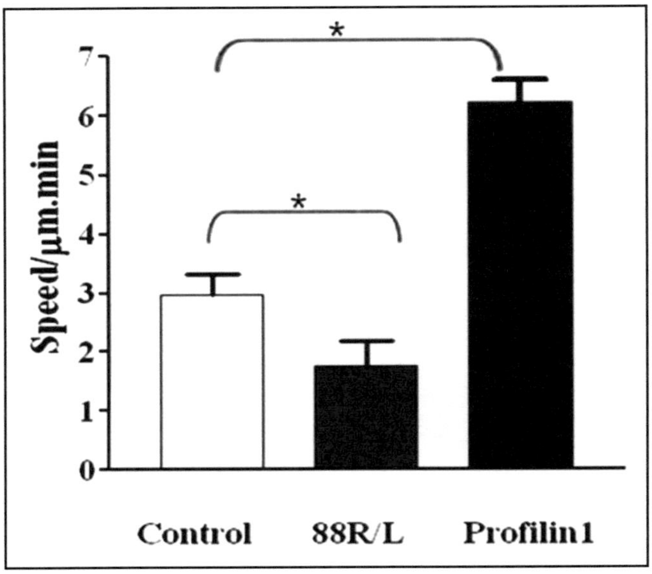

Fig. 8. The MetaMorph image analysis of the mean speed of the individual cells (µm/min) of profilin-I, 88 R/L and control VSMCs. The differences in mean were determined by ANOVA. *P < 0.05, compared with corresponding control, is considered to be significant (Hassanain HH, unpublished).

In line with the established role of profilin in cellular migration and proliferation, it has been shown that recombinant profilin-I stimulates DNA-synthesis and migration of both rat and human VSMCs in a concentration-dependent manner (Caglayan et al., 2010). The same study indicated that profilin-induced VSMCs migration is dependent on PI3K activity (Caglayan et al., 2010). Moreover, Cheng et al., (2011) found that profilin-I plays a key role in Angiotensin (Ang) II-induced VSMCs proliferation. They also suggested that Ang-II increases profilin-I expression and promotes VSMCs proliferation via activating AT1 receptor/JAK2/STAT3 pathway (Cheng et al., 2011). On the contrary, other studies described the involvement of phospho-extracellular signal-regulated kinase1/2 (P-ERK1/2) and phospho-c-Jun NH2-terminal kinase (P-JNK) in Ang-II-induced profilin-I expression (Zhong et al., 2011), and that PI3-kinase, Src, and, to a lesser extent, P-ERK1/2 are required for profilin-I-dependent VSMCs proliferation (Caglayan et al., 2010). Consequently, Cheng et al., (2011) proposed that the interaction of these signaling pathways mediating the role of profilin-I in VSMCs proliferation requires further investigation. Consistent with these data, we observed that the treatment of mouse aortic VSMCs with Ang-II (100 nM/10 min) resulted in increased profilin-I expression (Hassanain HH, unpublished)

3.1.2 Profilin & vascular smooth muscle contraction

Regulation of smooth muscle contraction has been thought to be only dependent on the 20-kDa regulatory light chain of myosin (MLC20) that in turn modulates cross-bridge cycling of actomyosin. Numerous studies showed that contractile stimulation promotes actin polymerization in vascular and airway smooth muscle tissues (Cipolla & Osol, 1998; Jones et al., 1999; D. Mehta & Gunst, 1999) and in cultured smooth muscle cells (An et al., 2002; Barany, et al., 2001; Hirshman & Emala, 1999). In addition, inhibition of actin polymerization by specific inhibitors such as latrunculin decreases the contractile stimuli- activated force development in smooth muscle (Cipolla & Osol, 1998; D. Mehta & Gunst, 1999; Youn et al., 1998). However, this does not affect contractile stimulation-induced MLC20 phosphorylation (34), suggesting that actin polymerization plays a central role during smooth muscle contraction. Tang & Tan, (2003) investigated the effect of profilin, the main actin-regulatory protein on the regulation of smooth muscle contraction. They demonstrated that profilin downregulation with antisense repressed force generation, without affecting MLC20 phosphorylation, signifying that profilin is crucial for smooth muscle contraction and that it does not regulate the activation of contractile protein. Yet, profilin downregulation repressed increases in the F-actin/G-actin ratio in return to agonist stimulation, showing that profilin is essential for actin dynamics during contractile stimulation of smooth muscle (Tang & Tan, 2003). In harmony with these finding our results showed higher expression of stress fibers and membrane ruffling in vascular smooth muscle cells from profilin-I transgenic mice compared with nontransgenic control and 88R/L. The 88R/L cells, however, showed lower expression of stress fiber formation and ruffling than the nontransgenic controls (Figure 9*A*) (Moustafa-Bayoumi et al., 2007). In addition, we confirmed these findings by assessing the ratio of F-actin/G-actin in the aortic smooth muscle cells from profilin-I. Our results showed a significant increase in F/G actin ratio in the aortic smooth muscle cells from profilin-I mice compared with the nontransgenic controls (Figure 9B) (Moustafa-Bayoumi et al., 2007). Furthermore, we showed that profilin-I plays a significant role in increased contractility and force development in the mesenteric arteries of profilin-I mice via activating Rho/ROCK pathway and MLC20 (Hassona et al., 2010). Activated Rho elevates MLC20 phosphorylation by *1*) directly phosphorylating MLC20 and *2*) phosphorylation and inhibition of the MBS of MLC20 phosphatase (Higgs & Pollard, 2001; Pollard & Borisy, 2003). This increases myosin contractility and tension contributing to stress fibers. In conclusion, our results indicate that overexpression of profilin-I in smooth muscle cells leads to increased contractility and force development via increasing actin polymerization (Moustafa-Bayoumi et al., 2007) and MLC20 activation(Hassona et al., 2010), which in turn induce mechanical stress that is considered as the main initiator for arterial stiffness and hypertension observed in these mice.

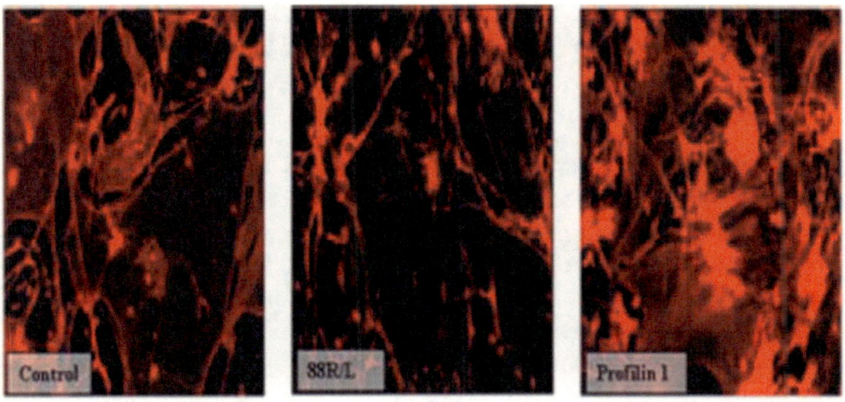

(a)

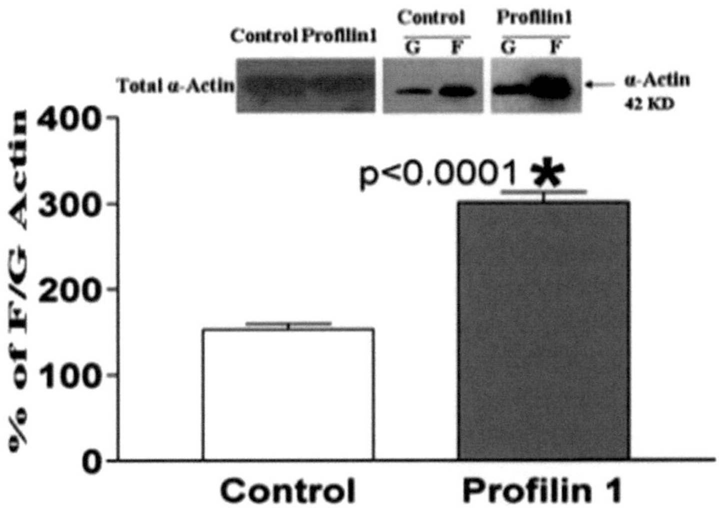

(b)

Fig. 9. Overexpression of profilin-I induced actin polymerization in vascular smooth muscle cells. Rhodamine-phalloidin staining of smooth muscle cell confluent monolayers shows increased stress fibers in vascular smooth muscle cells from profilin-I transgenic mouse as compared with nontransgenic control (*a*). The 88R/L cells, however, show lower expression of stress fiber formation than the control (*a*). Analysis of the F-actin/G-actin ratio shows significant increase in F-actin/G-actin (*F/G*) ratio in the aortic smooth muscle cells from profilin-I mice compared with the nontransgenic controls (*b*) (Moustafa-Bayoumi et al., 2007).

3.1.3 Profilin & vascular endothelial cells adhesion

Endothelial cells survival neatly depends on their ability to anchor to extracellular matrix proteins. Suppression of endothelial cell adhesion has been shown to induce apoptosis in these cells (Meredith et al., 1993; Re et al., 1994; Zang et al., 1995). It has been found that transient overexpression of profilin in cultured human aortic endothelial cells using replication-incompetent adenovirus enhances endothelial cells adhesion to the extracellular matrix via promoting the binding of extracellular fibronectin to its receptor on the surface of these cells. Additionally, it was revealed that profilin-mediated enhancement of endothelial cell adhesion has a protective role in situations of focal contacts disruption due to shear, stretch or other focal injuries (Moldovan et al., 1997).

Moreover, the authors, Moldovan et al., (1997) proposed that the profilin-mediated effect seems to be stimulated via recruiting integrins $\alpha_5\beta_1$ to the endothelial cell surface. Numerous mechanisms may explain this later effect. One possibility is that profilin might cause improvement in the access of receptor molecules to the cell surface. Instead, profilin might cause impairment in the internalization of membrane receptors. These effects may be achieved in 1) actin-dependent manner, where profilin might decrease receptor internalization via disrupting actin stress fibers or it might offer a stronger anchor for fibronectin receptor molecules in focal contacts via stabilizing actin filaments that are not stress fibers (Finkel et al., 1994), or 2) actin-independent manner, where profilin interacts with PtdIns 4,5-P2 and inhibits its hydrolysis by phospholipase C (Goldschmidt-Clermont et al., 1990, 1991; Lassing & Lindberg, 1985). Increased concentrations of PtdIns 4,5-P2 could stimulate the stabilization of newly formed focal contacts including the fibronectin receptor via an unknown mechanism or profilin overexpression could overcome other actin-binding proteins for interacting with PtdIns 4,5-P2 and thus enhance their binding to actin filaments (Hartwig et al., 1995).

3.1.4 Role of profilin in vascular endothelial cells migration, proliferation & capillary morphogenesis

Vascular endothelial cell (VEC) migration is vital for capillary outgrowth from pre-existing blood vessels during angiogenesis (Bauer, et al., 2005). During cell migration, actin cytoskeleton reorganization is a dynamic process that includes both actin polymerization and depolymerization in an accurate spatiotemporal manner. Regulation of this actin remodeling process is achieved by a large number of actin binding proteins such as those involved in monomer sequestering, nucleating, elongating, severing, depolymerizing, and capping of actin filaments (Pollard & Borisy, 2003). Expression profiles in VEC experiencing capillary morphogenesis identified some of the key actin-binding proteins that have been previously involved in angiogenesis such as thymosin β4, profilin, gelsolin and VASP. Among these proteins, as a minimum thymosin β4 has been established as a proangiogenic molecule *in vivo* (Philp et al., 2004; Salazar et al., 1999). In addition, it has been reported that silencing profilin-I expression in human umbilical vein endothelial cells significantly decreases their capability of forming planar cord-like structures on matrigel (a commonly adopted *in vitro* representation for angiogenesis). These findings proposed for the first time that profilin-I might play a key role in VEC capillary morphogenesis (Ding et al., 2006).

In a more recent report for the same group they adopted a knockdown–knockin experimental system to stably express either fully functional form or mutants of profilin-I that are deficient in binding to actin and proteins containing polyproline domains, in a human dermal microvascular cell line. They showed that silencing endogenous profilin-I expression in this cell line results in slow rate of random migration, decreased membrane protrusion velocity and a significant reduction in matrigel-induced cord formation. These defects were rescued only via re-expression of fully functional but not any of the two ligand-binding deficient mutants of profilin-I. They also showed that loss of profilin-I expression in VEC inhibits three dimensional capillary morphogenesis, MMP2 secretion and ECM invasion. Disruption of actin and polyproline interactions of profilin-I inhibited VEC invasion through ECM, as well. They concluded that profilin-I regulates VEC migration, invasion and capillary morphogenesis through its binding to both actin and proline-rich ligands (Ding et al., 2009). Furthermore, they indicated that these *in vitro* findings pave the way for future *in vivo* studies to investigate the role of profilin-I in angiogenesis.

Interestingly, cutaneous wound healing experiments in our profilin-I and 88R/L transgenic mice showed a significant increase in blood vessel density in profilin-I transgenic mice compared to 88R/L transgenic mice and nontransgenic control at post wound day 7 (Figure 10) (Hassanain HH, unpublished). These data could indicate the importance of profilin-I in angiogenic reponse in VEC.

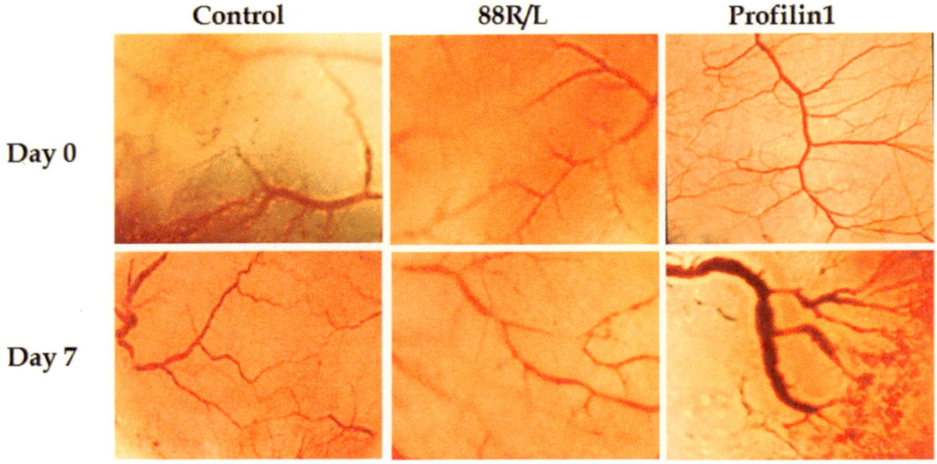

Fig. 10. Stimulation of angiogenesis in the wound area of profilin-I mice. Distribution of capillaries along the margin of the excision wound in Profilin-I, 88R/L and nontransgenic control mice at post wound days 0 and 7. High magnification of capillaries in the skin was obtained with a 2X objective lens light microscope (Hassanain HH, unpublished).

3.2 Role of profilin in vascular remodeling & hypertension

Hypertension represents a major risk factor for cardiovascular events such as stroke and myocardial infarction. It is well established that hypertension leads to remodeling of large and small arteries (Folkow, 1982; Simon, 2004). Remodeling of the vasculature is an active process of structural changes that involves alterations in cellular processes, including growth and changes in the extracellular matrix integrin-cytoskeleton axis, resulting in an increase in the media-to-lumen ratio (Gimbrone et al., 1997; Intengan & Schiffrin, 2001). Physiological remodeling is an adaptive process occurring in response to hemodynamic changes and aging. However, when this process becomes maladaptive, it plays a role in hypertension's complications (Ming et al, 2002; Touyz, 2007). Increased mechanical strain/hypertension in the vessel wall triggers the hypertrophic signaling pathway resulting in structural remodeling of vasculature. Increased actin polymerization and stress fiber formation generate mechanical force that represents an important modulator of cellular morphology and function in a variety of tissues and is an important contributor to hypertrophy in the cardiovascular system (Ruwhof &van der Laarse, 2000). Also, it has been shown that actin polymerization within VSMCs in response to increased intravascular pressure is a novel mechanism underlying arterial myogenic behavior. The cytosolic concentration of G-actin is significantly reduced by an elevation in intravascular pressure, demonstrating the dynamic nature of actin within VSMCs and implying a shift in the F:G equilibrium in favor of F-actin. Profilin-I which is a key actin-regulatory protein that plays an essential role in regulating de novo actin polymerization, particularly actin treadmilling (Carlier & Pantaloni, 2007; Suetsugu et al., 1999) could be vital in regulating all of these vascular events. Indeed, our report in the Journal of Biological Chemistry (Moustafa-Bayoumi et al., 2007) established the feasibility of our proposal. We showed that elevated expression of profilin-I gene in VSMCs of profilin-I mice favoring F-actin induces stress fiber formation (Figure 11) and plays an important role in vascular hypertrophy by inducing internal mechanical stress and triggering the hypertrophic signaling pathways, integrins-$\alpha_1\beta_1$/Rho-ROCK/MAPKs e.g. P-ERK and P-JNK, leading to vascular remodeling in both large (e.g. aorta) and small (e.g. mesenteric) arteries (Figure 12A, B) of profilin transgenic mice (Hassona et al., 2010; Moustafa-Bayoumi et al., 2007).

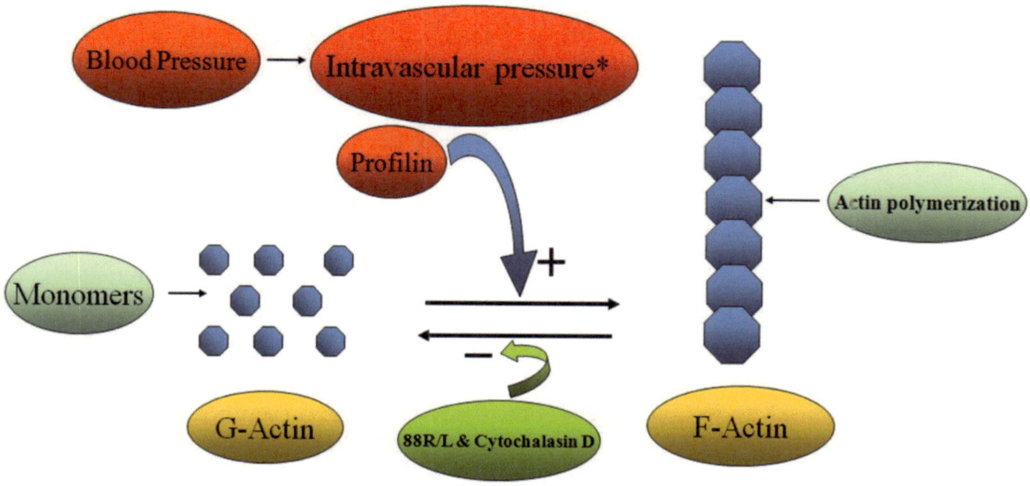

Fig. 11. Hypertension or increased profilin-I expression in VSMCs leads to a shift in the F:G equilibrium in favor of F-actin and an elevation in intravascular pressure. This pathway can be reversed by F-actin inhibitor, cytochalasin D or profilin-I mutant, 88R/L.

Consistent with our finding, very recent studies showed increased profilin-I expression in hypertrophic thoracic aorta and mesenteric arteries of spontaneously hypertensive rats with subsequent elevation in both P-ERK and P-JNK, suggesting that profilin-I may contribute to the vascular remodeling in these rats (Cheng et al., 2011; Zhong et al., 2011). In this context, previous studies suggested that mechanical stretch is closely related to JNK and ERK1/2 activation (Hu et al., 1997; Pyles et al., 1997). These cascades play an important role in remodeling of blood vessels, as well. In addition, this pathway is activated by Ang-II and has been implicated in the pathogenesis of cardiovascular diseases (P.K. Mehta & Griendling, 2007). Interestingly, it has been recently reported that profilin-I is a key component in the Ang-II-induced vascular remodeling (Cheng et al., 2011; Zhong et al., 2011).

As it was mentioned above that hypertension is a major cause of vascular remodeling. The primary aim of anti-hypertensive drugs, particularly Ang-converting enzyme inhibitors and Ang receptor subtype 1 antagonists, is to lower the blood pressure with the hope of reversing this remodeling (Schiffrin, 2001). Importantly, In our profilin-I model we demonstrate that the reverse can be true as well, i.e. alteration in cytoskeleton dynamics favoring increased actin polymerization can contribute to vascular adaptations with aging resulting in increased systolic blood pressure by the time the profilin-I mice were six months old (Figure 12C) (Moustafa-Bayoumi et al., 2007). The blood pressure in the profilin-I mice was elevated 25-30 mm Hg higher than nontransgenic controls. In agreement with our findings, it has been demonstrated that profilin speeds up the actin remodeling and accordingly improves the growth and invasion force of VSMCs resulting in increased vascular resistance and accelerated formation of pulmonary hypertension (Dai et al., 2006).

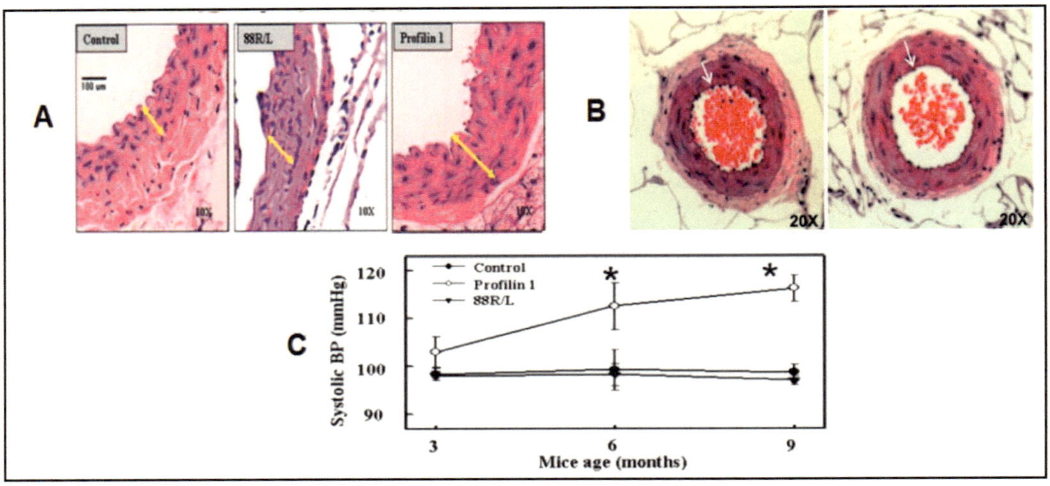

Fig. 12. Profilin overexpression induced vascular hypertrophy and hypertension. Hematoxylin and eosin staining shows clear signs of remodeling and vascular hypertrophy in the aorta of profilin-I transgenic mice (yellow arrows; **A**) and mesenteric arteries (white arrows; **B**). There are no differences, however, between 88R/L and nontransgenic control aortic sections (**A**). Tail cuff measurements of blood pressure show significant increase in the systolic blood pressure (BP) in profilin-I transgenic mice at 6 months and older compared with nontransgenic control mice (C) (Hassona et al., 2010; Moustafa-Bayoumi et al., 2007).

On the other hand, the blood pressure in 88R/L mice was below the control littermates; however, it did not reach statistical significance. The absence of a hypotensive phenotype in the 88R/L mice could be due to the lack of significant vascular remodeling as a result of decreased actin polymerization. Our results showed a decrease in stress fibers formation in 88R/L mice (Figure 9); however, these changes did not translate into significant alterations in the vasculature. This might be due to an activation of a compensatory mechanism to maintain the integrity of vessel structure and thus keep the blood pressure at a survival level. Additionally, our preliminary data showed that inhibition of profilin-I-induced stress fibers by cytochalasin D lowered blood pressure in profilin-I mice. As a pilot study the profilin-I mice were injected with a single dose of cytochalasin D (0.5 µg/gram body weight) which led to lowered blood pressure within 10 minutes in these mice from 140 mmHg to 70 mmHg and the effect was sustained for more than 1.5 hours. Then the mice were recovered without any sign of sickness. To make sure that cytochalasin D had no damaging effect on the endothelium, we assessed the functionality of the endothelium using Ach and wire-myography. Our results showed no damage in the endothelium after cytochalasin D treatment (Hassanain HH, unpublished). We should note that cytochalasin D was used before by other investigators in different studies with much higher doses and no toxicity was observed (Speirs & Kaufman, 1989).

Furthermore, stress fiber formation could affect the relaxation/contraction process of the smooth muscles, making it more constrictive and/or less responsive to vasodilators such as nitric oxide. That could be an important factor contributing to hypertension besides the vascular hypertrophy in the profilin-I transgenic mice. Our recent report in the American Journal of Physiology confirmed this proposal. We showed that vascular hypertrophy-associated hypertension of profilin-I transgenic mice led to functional remodeling of peripheral arteries. Our results showed a significant increase in the contraction response of profilin-I mesenteric arteries toward phenylephrine and significant decreases in the relaxation response toward ACh and sodium nitrite compared with nontransgenic controls (Hassona et al, 2010). Additionally, inhibiting stress fibers formation with cytochalasin D significantly relaxes the phenylephrine-contracted mesenteric arteries, suggesting that the increased constriction of mesenteric arteries to phenylephrine could be because of the increased F- to G actin ratio; however, cytochalasin D treatment reduced this ratio (Hassona et al., 2010).

Moreover, it has been reported that in addition to the role of hypertension in vascular remodeling, there are pressure-independent genes that play a key role in vascular remodeling. This concept is supported by the observation that despite blood pressure control in hypertensive patients, the rate of restenosis (attributable to remodeling) remains high (Gurlek et al., 1995). In harmony with this concept we recently showed that normalization of blood pressure by selected anti-hypertensive agents is not enough to correct the structural and functional remodeling of profilin-I transgenic mice (Hassona et al., 2011). Our results demonstrated that there is only correction in the functional remodeling and signaling cascades of the mesenteric arteries of losartan- and amlodipine-treated, but not those of atenolol-treated profilin-I transgenic mice, where losartan and amlodipine decrease the F-actin and stress fibers formation, proposing that the stress fibers seem to play a major role in the development and progression of the vascular remodeling-associated hypertension. We finally concluded that profilin-I gene, which is the key player controlling stress fiber formation may be a good target to treat not only hypertension but also the vascular remodeling in hypertensive patients (Hassona et al., 2011).

3.3 Role of profilin in atherosclerosis & vascular complication in diabetes

Vascular endothelium dysfunction goes before, and may participate in atheroma formation in return to various cardiovascular risk factors such as diabetes (Johnstone et al., 1993; Tesfamariam et al., 1990;), hyperlipidemia (Chikani et al., 2004; Steinberg et al., 1997), and both local and systemic inflammatory mediators (Libby, 2002). Interestingly, Romeo et al., (2004) revealed that profilin-I levels are improved in the endothelium of diabetic aorta of both human and experimental animals. They also demonstrated that profilin overexpression in primary aorta EC was capable of triggering indicators of endothelial dysfunction such as apoptosis, ICAM-1 up-regulation, and decreased VASP phosphorylation. In addition, profilin was found to be required for LDL-mediated ICAM-1 up-regulation and it can be regulated by LDL/cholesterol signaling, but not high glucose (Romeo et al., 2004). Although, Clarkson et al., (2002) reported that exposure to high glucose was able to increase profilin-I mRNA in mesangial cells and in the diabetic rat kidney. Romeo et al., (2004) suggested that the inability of high glucose to enhance profilin-I protein levels in EC is in line with a multifactorial etiology of endothelial dysfunction coupled with the metabolic syndrome and may reveal the inadequate effect of glucose-lowering monotherapy to prevent macrovascular complications in type 2 diabetic patients (U.K. Prospective Diabetes Study (UKPDS) Group, 1998). On the other hand, our preliminary data showed that mouse aortic VSMCs treated with glucose (25 mM/24 hours) increased profilin-I expression (Hassanain HH, unpublished).

Furthermore, Romeo et al., (2004) showed that profilin was clearly increased in EC and macrophages within atherosclerotic lesions of apoE null mice. In a more recent report, the same group specified the significance of profilin-I for atherogenesis *in vivo* as profilin-I heterozygosity resulted in protection from atherosclerosis in LDL receptor-null mice (Romeo et al., 2007). In this report, a variety of atheroprotective indicators were recognized in mice with heterozygous deficiency of profilin-I, as compared to profilin-I wild-type mice. Aortas from these heterozygous mice exhibited preserved activation of endothelial nitric oxide synthase (eNOS) and nitric oxide-dependent signaling, decreased expression of vascular cell adhesion molecule (VCAM)-1 and decreased accumulation of macrophage at the sites of injury. Correspondingly, profilin-I knockdown in cultured aortic ECs was able to protect against endothelial dysfunction induced by oxidized low-density lipoproteins (oxLDL). Additionally, macrophages from bone marrow of profilin-I-deficient heterozygous mice exhibited diminished internalization of oxLDL and oxLDL-induced inflammation. These studies concluded that profilin-I plays a vital role in early atheroma formation and that decreasing profilin-I levels is atheroprotective. Finally, profilin-I atheroprotective effect is mediated via combined mechanisms that depend on both endothelium and macrophages (Romeo et al., 2007).

Moreover, the same group addressed the pathways responsible for profilin-I gene expression in 7-ketocholesterol (oxysterol)-stimulated endothelial cells and in the diabetic aorta. They showed that oxysterol-binding protein-1 (OSBP1) is required for oxysterol-dependent nucleation and activation of the JAK2/STAT3 pathway, which in turn regulates profilin-I gene expression in endothelial cells. Similarly, diabetes increases the activation of STAT3 and its recruitment to the profilin-I promoter in large vessels *in vivo* (Romeo et al., 2008)

Very recently, it has been reported that profilin-I expression is markedly increased in human atherosclerotic plaques compared to the normal vessel wall (Caglayan et al., 2010). A correlation was found between profilin-I serum levels and the degree of atherosclerosis, as well. The atherogenic effects of profilin-I on VSMCs imply an auto-/paracrine role within the plaque. In addition, it was found that profilin-I acts as an extracellular ligand and triggers atherogenic effects in VSMCs including DNA synthesis and migration. Besides, profilin-1 stimulates typical signaling pathways such as the PI3K/AKT and RAS-RAf-MEK-ERK pathways. These findings revealed that profilin-I might play a critical role in atherogenesis and may represent a novel therapeutic target in human patients (Caglayan et al., 2010).

3.4 Role of profilin in age-associated vascular problems

Aging is a major risk factor for the development of vascular diseases, such as hypertension and arteriosclerosis, which lead to stroke and heart failure (Spagnoli et al., 1991). Aging is also linked with decreased stress tolerance. Susceptibility to a variety of physiological stresses such as infection, inflammation, and oxidative damage enhances with age and is causally coupled with clinical problems in the elderly (Starr et al., 2011). So far, the mechanism of age-related changes in vasculature has not been completely understood. On the top of that, the role of profilin in these age related changes remains largely unstudied.

Recently, it has been reported that protein nitration levels increased in aged mice compared to young mice. Also, particularly strong nitration was found in the pulmonary vascular endothelium during systemic inflammatory response syndrome (SIRS). Age- and SIRS-dependent increased protein nitration was evident in proteins related to the actin cytoskeleton that are responsible for maintaining pulmonary vascular permeability such as transgelin-2, LASP 1, tropomyosin, myosin and profilin-I. Recognizing the nitrated proteins indicated important modifications to the vascular endothelial cytoskeleton, which potentially participates in the barrier dysfunction, enhanced vascular permeability, and pulmonary edema (Starr et al., 2011).

It has been established that deficiency in plasma fibronectin increases lung vascular permeability (Wheatley et al., 1993); consequently, as adhesion of endothelial cell to fibronectin depends on profilin expression (Moldovan et al., 1997), lack of functional profilin may be to some extent responsible for vascular permeability as a result of inefficient barrier integrity. These data can fairly elucidate the age-associated enhancement in susceptibility to systemic inflammation, acute lung injury, and respiratory failure (Starr et al., 2011).

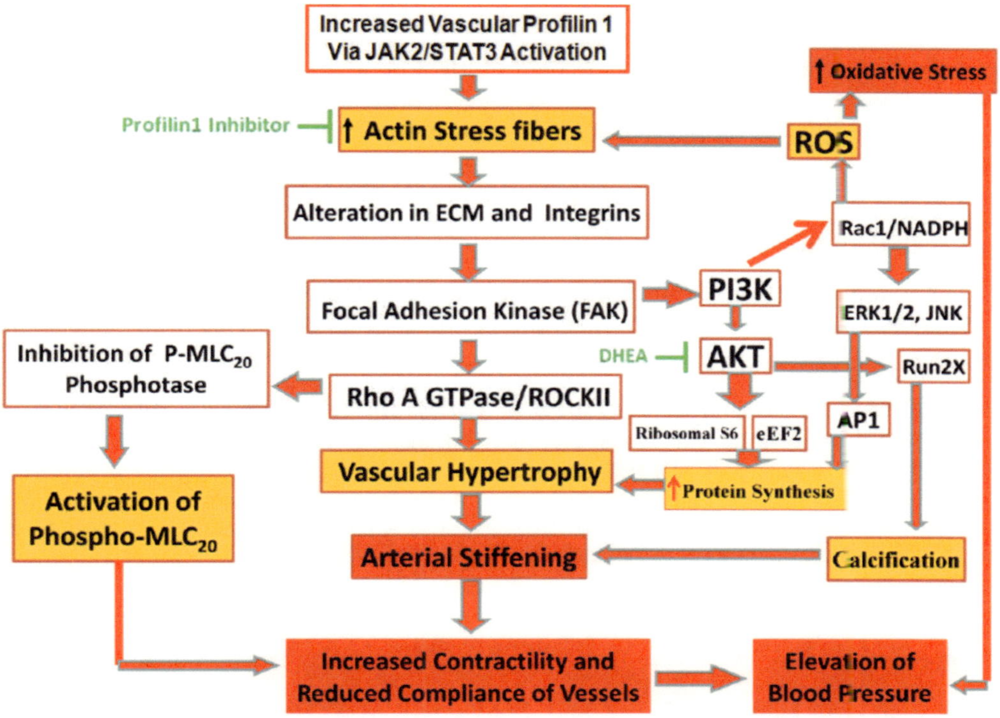

Fig. 13. JAK2/STAT3 pathway activation increases profilin-I (Romeo et al., 2008; Cheng et al., 2011) in the vessel media induced stress fiber formation and increased internal mechanical stress in the vessel walls (Moustafa-Bayoumi et al., 2007) which modulates changes in ECM and integrins (Abouelnaga et al., 2009; Hassona et al., 2010). These changes led to activation of FAK (Abouelnaga et al., 2009) that in turn activate Rho/ROCKII (Hassona et al, 2010; Moustafa-Bayoumi et al., 2007), PI3 kinase and AKT (Caglayan et al., 2010). Activation of Rac1/NADPH pathway (Abouelnaga et al., 2009) results in increased superoxide production and increases oxidative stress (Hassanain HH, unpublished) in vessel walls which could contribute to hypertension. The activation of Rho/ROCKII and AKT result in activation of MLC20 (Hassona et al., 2010), and increases in protein synthesis (Gingras et al., 1998; Kitamura et al., 1998; Ushio-Fukai et al., 1999) and calcification (Byon et al., 2008), respectively. These changes in the media of the vessel walls result in arterial stiffening and hypertension (Moustafa-Bayoumi et al., 2007). Profilin-I inhibitor can block the stress fiber formation in this pathway (Moustafa-Bayoumi et al., 2007) and dehydroepiandrosterone (DHEA) can inhibit AKT kinase pathway (Bonnet et al., 2009).

Conversely, other indirect evidence showed that profilin-I increased with age; a recent study using proteomic and genomic analyses of hippocampus from young and old rats showed a significant increase in profilin-I expression in aged rat hippocampus (Weinreb et al., 2007). Another study investigating differential protein expression profiles in chronically stimulated T cell clones found that profilin-I was widely and highly expressed in cytoplasm (Mazzatti et al., 2007). The study concluded that differential expression of profilin-I in aging may contribute directly to immunosenescence via disrupting the intracellular signaling and intercellular communication (Egerton et al., 1992; Witke et al., 1998). Consistent with these findings our preliminary data showed an increase in profilin-I expression in the aortic medial layers of older wild-type mice compared with young mice (Hassanain HH, unpublished).

Taken together, this review shed some light on the important role of profilin-I in vascular diseases. However, more studies need to be done in order to fully understand the profilin-I signaling pathway and its mechanism(s) of regulation. Figure 13 summarize some of the proposed signaling molecules involved in profilin-induced vascular complications.

4. Acknowlegements

Figure 1 is reprinted from Krishnan, K. & Moens, P.D.J. (2009). Structure and functions of profilins. *Biophysical reviews*, Vol.1, No.2, pp. 71–81, ISSN 1867-2450, with permission from Springer.

Figures 2, 3 & 4 are reprinted from Witke, W. (2004). The role of profilin complexes in cell motility and other cellular processes. *Trends in cell biology*, Vol.14, No.8, pp. 461-469, ISSN 0962-8924, with permission from Elsevier.

Figures 5 & 6 are reprinted from Schlüter, K.; Jockusch, B.M. & Rothkegel, M. (1997). Profilins as regulators of actin dynamics. *Biochimica et biophysica acta*, Vol.1359, No.2, pp. 97-109, ISSN 0006-3002, with permission from Elsevier.

5. References

Abouelnaga, Z.A.; Hassona, M.D.; Awad, M.M; Alhaj, M.A.; Badary, O.A.; Hamada, F.A.; Bergese, S.D. & Hassanain, H.H. (2009). Mechanical Strain in VSM Cells Triggers Vascular Remodeling and Hypertension and Activates Integrins in Profilin1 Transgenic Model. *The FASEB journal*, April 2009 23 (Meeting Abstract Supplement) 704.6, ISSN 0892-6638.

Alberts, A.S. (2002). Diaphanous-related Formin homology proteins. *Current biology*, Vol.12, No. 23, pp. R796, ISSN 0960-9822.

Alvarez-Martinez, M.T.; Mani, J.C.; Porte, F.; Faivre-Sarrailh, C.; Liautard, J.P.; Sri Widada, J. (1996). Characterization of the interaction between annexin I and profilin. *European journal of biochemistry*, Vol.238, No.3, pp. 777–784, ISSN 0014-2956.

Ampe, C. & Vandekerckhove, J. (1994). Actin-actin binding protein interfaces. *Seminars in cell biology*, Vol.5, No.3, pp. 175-182, ISSN 1043-4682.

An, S.S.; Laudadio, R.E.; Lai, J.; Rogers, R.A. & Fredberg, J.J. (2002). Stiffness changes in cultured airway smooth muscle cells. *American journal of physiology. Cell physiology*, Vol.283, No.3, pp. C792–C801, ISSN 0363-6143.

Archer, S.J.; Vinson, V.K.; Pollard, T.D. & Torchia, D.A. (1994). Elucidation of the poly-L-proline binding site in Acanthamoeba profilin I by NMR spectroscopy. *FEBS Letters*, Vol.337, No.2, pp. 145-151, ISSN 0014-5793.

Babcock, G. & Rubenstein, P.A. (1993). Control of profilin and actin expression in muscle and nonmuscle cells. *Cell motility and the cytoskeleton*, Vol.24, No.3, pp. 179-188, ISSN 0886-1544.

Babich, M.; Foti, L.R.; Sykaluk, L.L. & Clark, C.R. (1996). Profilin forms tetramers that bind to G-actin. *Biochemical and biophysical research communications*, Vol.218, No.1, pp. 125-131, ISSN 0006-291X.

Balasubramanian, M.K.; Hirani, B.R.; Burke, J.D. & Gould, K.L. (1994). The Schizosaccharomyces pombe cdc3+ gene encodes a profilin essential for cytokinesis. *The Journal of cell biology*, Vol.125, No.6, pp. 1289-1301, ISSN 0021-9525.

Barany, M.; Barron, J.T.; Gu, L. & Barany, K. (2001). Exchange of the actin-bound nucleotide in intact arterial smooth muscle. *The Journal of biological chemistry*, Vol.276, No.51, pp. 48398–48403, ISSN 0021-9258.

Bauer, S.M; Bauer, R.J. & Velazquez, O.C. (2005). Angiogenesis, vasculogenesis, and induction of healing in chronic wounds. *Vascular and endovascular surgery*, Vol.39, No.4, pp. 293–306, ISSN 1538-5744.

Binette, F.; Bénard, M.; Laroche, A.; Pierron, G.; Lemieux, G. & Pallotta, D. (1990). Cell-specific expression of a profilin gene family. *DNA and cell biology*, Vol.9, No.5, pp. 323-324, ISSN 1044-5498.

Björkegren, C.; Rozycki, M.; Schutt, C.E.; Lindberg, U. & Karlsson, R. Mutagenesis of human profilin locates its poly(L-proline)-binding site to a hydrophobic patch of aromatic amino acids. FEBS Letters, Vol.333, No.(1-2), PP.123-126, ISSN 0014-5793.

Boettner, B.; Govek, E.E.; Cross, J. & Van Aelst, L. (2000). The junctional multidomain protein AF-6 is a binding partner of the Rap1A GTPase and associates with the actin cytoskeletal regulator profilin. *Proceedings of the National Academy of Sciences of the United States of America*, Vol.97, No.16, pp. 9064–9069, ISSN 0027-8424.

Bonnet, S.; Paulin, R.; Sutendra, G.; Dromparis, P.; Roy, M.; Watson, K.O.; Nagendran, J.; Haromy, A.; Dyck, J.R. & Michelakis, E.D. (2009). Dehydroepiandrosterone reverses systemic vascular remodeling through the inhibition of the Akt/GSK3-β/NFAT axis. Circulation. Vol.120, No.13, pp. 1231-1240, ISSN 0009-7322.

Borisy, G.G. & Svitkina, T.M. (2000). Actin machinery: pushing the envelope. *Current opinion in cell biology*, Vol.12, No.1, pp. 104-112, ISSN 0955-0674.

Braun, A.; Aszódi, A.; Hellebrand, H.; Berna, A.; Fässler, R. & Brandau, O. (2002). Genomic organization of profilin-III and evidence for a transcript expressed exclusively in testis. *Gene*, Vol.283, No.1-2, pp. 219–225, ISSN 0378-1119.

Brindle, N.P.; Holt, M.R.; Davies, J.E.; Price, C.J. & Critchley, D.R. (1996). The focal-adhesion vasodilator-stimulated phosphoprotein (VASP) binds to the proline-rich domain in vinculin. *The Biochemical journal*, Vol.318, Pt.3, pp. 753-757, ISSN 0264-6021.

Butt, E.; Abel, K.; Krieger, M.; Palm, D.; Hoppe, V.; Hoppe, J. & Walter, U. (1994). cAMP- and cGMP-dependent protein kinase phosphorylation sites of the focal adhesion vasodilator-stimulated phosphoprotein (VASP) in vitro and in intact human platelets. *The Journal of biological chemistry*, Vol.269, No.20, pp. 14509-14517, ISSN 0021-9258.

Byon, C.H.; Javed, A.; Dai, Q.; Kappes, J.C.; Clemens, T.L.; Darley-Usmar, V.M.; McDonald, J.M. & Chen, Y. (2008). Oxidative stress induces vascular calcification through modulation of the osteogenic transcription factor Runx2 by AKT signaling. *The Journal of biological chemistry*, Vol.283, No.22, pp. 15319-15327, ISSN 0021-9258.

Caglayan, E.; Romeo, G.R.; Kappert, K.; Odenthal, M.; Südkamp, M.; Body, S.C.; Shernan, S.K.; Hackbusch, D.; Vantler, M.; Kazlauskas, A. & Rosenkranz, S. (2010). Profilin-1 is expressed in human atherosclerotic plaques and induces atherogenic effects on vascular smooth muscle cells. *PLoS one*, Vol.5, No.10, pp. e13608, ISSN 1932-6203.

Camera, P.; da Silva, J.S.; Griffiths, G.; Giuffrida, M.G.; Ferrara, L.; Schubert, V.; Imarisio, S. Silengo, L., Dotti, C.G. & Di Cunto, F. (2003). Citron-N is a neuronal Rho-associated protein involved in Golgi organization through actin cytoskeleton regulation. *Nature cell biology*, Vol.5, No.12, pp. 1071–1078, ISSN 1465-7392.

Cao, L.G.; Babcock, G.G.; Rubenstein, P.A. & Wang, Y.L. (1992). Effects of profilin and profilactin on actin structure and function in living cells. *The Journal of cell biology*, Vol.117, No.5, pp. 1023-1029, ISSN 0021-9525.

Carlier, M.F. & Pantaloni, D. (2007). Control of actin assembly dynamics in cell motility. *The Journal of biological chemistry*, Vol.282, No.32, pp. 23005–23009, ISSN 0021-9258.

Castrillon, D.H. & Wasserman, S.A. (1994). Diaphanous is required for cytokinesis in Drosophila and shares domains of similarity with the products of the limb deformity gene. *Development*. Vol.120, No.12, pp. 3367-3377, ISSN 0950-1991.

Chang, F.; Drubin, D. & Nurse, P. (1997). cdc12p, a protein required for cytokinesis in fission yeast, is a component of the cell division ring and interacts with profilin. *The Journal of cell biology*, Vol.137, No.1, pp. 169-182, ISSN 0021-9525.

Cheng, J.F.; Ni, G.H.; Chen, M.F.; Li, Y.J.; Wang, Y.J.; Wang, C.L.; Yuan, Q.; Shi, R.Z.; Hu, C.P. & Yang, T.L. (2011). Involvement of profilin-1 in angiotensin II-induced vascular smooth muscle cell proliferation. *Vascular Pharmacology*, Vol.55, No.1-3, pp. 34-41, ISSN 1537-1891.

Chikani, G.; Zhu, W. & Smart, E.J. (2004). Lipids: potential regulators of nitric oxide generation. *American journal of physiology. Endocrinology and metabolism*, Vol.287, No.3, pp. E386-E389, ISSN 0193-1849.

Cipolla, M.J. & Osol, G. (1998). Vascular smooth muscle actin cytoskeleton in cerebral artery forced dilatation. *Stroke*, Vol.29, No.6, pp. 1223– 1228, ISSN 0039-2499.

Cipolla, M.J; Gokina, N.I. & Osol, G. (2002). Pressure-induced actin polymerization in vascular smooth muscle as a mechanism underlying myogenic behavior. *The FASEB Journal*, Vol.16, No.1, pp. 72-76, ISSN 892-6638.

Clarkson, M. R., Murphy, M., Gupta, S., Lambe, T., Mackenzie, H. S., Godson, C., Martin, F., and Brady, H. R. (2002) High glucose-altered gene expression in mesangial cells. Actin-regulatory protein gene expression is triggered by oxidative stress and cytoskeletal disassembly. *The Journal of biological chemistry*, Vol.277, No.12, pp. 9707-9712, ISSN 0021-9258.

Clowes, A.W.; Clowes, M.M.; Fingerle, J. & Reidy, M.A. (1989). Regulation of smooth muscle cell growth in injured artery. *Journal of cardiovascular pharmacology*, Vol.14, Suppl.6, pp. S12–S15, ISSN 0160-2446.

Da Silva, J.S.; Medina, M.; Zuliani, C.; Di Nardo, A.; Witke, W. & Dotti, C.G. (2003). RhoA/ROCK regulation of neuritogenesis via profilin IIa-mediated control of actin stability *The Journal of cell biology*, Vol.162, No.7, pp. 1267-1279, ISSN 0021-9525.

Dai, Y.P.; Bongalon, S.; Tian, H.; Parks, S.D.; Mutafova-Yambolieva, V.N. & Yamboliev, I.A. (2006). Upregulation of profilin, cofilin-2 and LIMK2 in cultured pulmonary artery smooth muscle cells and in pulmonary arteries of monocrotaline-treated rats. *Vascular Pharmacology*, Vol.44, No.5, pp. 275–282, ISSN 1537-1891.

Das, T.; Bae, Y.H.; Wells, A. & Roy, P. (2009). Profilin-1 overexpression upregulates PTEN and suppresses AKT activation in breast cancer cells. *Journal of cellular physiology*, Vol.218, No.2, pp. 436-443, ISSN 0021-9541.

Di Nardo, A.; Gareus, R.; Kwiatkowski, D. & Witke, W. (2000). Alternative splicing of the mouse profilin II gene generates functionally different profilin isoforms. *Journal of cell science*, Vol.113, No.21, pp. 3795–3803, ISSN 0021-9533.

Ding, Z.; Gau, D.; Deasy, B.; Wells, A. & Roy, P. (2009). Both actin and polyproline interactions of profilin-1 are required for migration, invasion and capillary morphogenesis of vascular endothelial cells. *Experimental cell research*, Vol.315, No.17, pp. 2963-2973, ISSN 0014-4827.

Ding, Z.; Lambrechts, A.; Parepally, M. & Roy, P. (2006). Silencing profilin-1 inhibits endothelial cell proliferation, migration and cord morphogenesis. *Journal of cell science*, Vol.119, Pt.19, pp. 4127–4137, ISSN 0021-9533.

Domke, T.; Federau, T.; Schlüter, K.; Giehl, K.; Valenta, R.; Schomburg, D. & Jockusch, B.M. (1997). Birch pollen profilin: structural organization and interaction with poly-(L-proline) peptides as revealed by NMR. *FEBS Letters*, Vol.411, No.2-3, pp. 291-295, ISSN 0014-5793.

Ebner, C.; Hirschwehr, R.; Bauer, L.; Breiteneder, H.; Valenta, R.;, Ebner, H.;, Kraft, D. & Scheiner, O. (1995). Identification of allergens in fruits and vegetables: IgE cross-reactivities with the important birch pollen allergens Bet v 1 and Bet v 2 (birch profilin). *The Journal of allergy and clinical immunology*, Vol.95, No.5 Pt.1, pp. 962-969, ISSN 0091-6749.

Edwards, K.A.; Montague, R.A.; Shepard, S.; Edgar, B.A.; Erikson, R.L. & Kiehart, D.P. Identification of Drosophila cytoskeletal proteins by induction of abnormal cell shape in fission yeast. *Proceedings of the National Academy of Sciences of the United States of America*, Vol.91, No.10, 4589-4593, ISSN 0027-8424.

Egerton, M.; Ashe, O.R.; Chen, D.; Druker, B.J.; Burgess, W.H. & Samelson, L.E. (1992). VCP, the mammalian homolog of cdc48, is tyrosine phosphorylated in response to T cell. *The EMBO Journal*, Vol.11, No.10, pp. 3533-3540, ISSN 0261-4189.

Evangelista, M.; Blundell, K.; Longtine, M.S.; Chow, C.J.; Adames, N.; Pringle, J.R.; Peter, M. & Boone, C. (1997). Bni1p, a yeast formin linking cdc42p and the actin cytoskeleton during polarized morphogenesis. *Science*, Vol.276, No 5309, pp. 118-122, ISSN 0036-8075.

Ezezika, O.C.; Younger, N.S.; Lu, J.; Kaiser, D.A.; Corbin, Z.A.; Nolen, B.J.; Kovar, D.R. & Pollard, T.D. (2009). Incompatibility with Formin Cdc12p prevents human profilin from substituting for fission yeast profilin: insights from crystal structures of fission yeast profilin. *The Journal of biological chemistry*, Vol.284, No.4, pp. 2088–2097, ISSN 0021-9258.

Fedor-Chaiken, M.; Deschenes, R.J. & Broach, J.R. (1990). SRV2, a gene required for RAS activation of adenylate cyclase in yeast. *Cell*, Vol.61, No.2, pp. 329-340, ISSN 0092-8674.

Fedorov, A.A.; Ball, T.; Mahoney, N.M.; Valenta, R. & Almo, S.C. (1997). The molecular basis for allergen cross-reactivity: crystal structure and IgE-epitope mapping of birch pollen profilin. *Structure*, Vol.5, No.1, pp. 33-45, ISSN 0969-2126.

Fedorov, A.A.; Magnus, K.A.; Graupe, M.H.; Lattman, E.E.; Pollard, T.D. & Almo, S.C. (1994). X-ray structures of isoforms of the actin-binding protein profilin that differ in their affinity for phosphatidylinositol phosphates. *Proceedings of the National Academy of Sciences of the United States of America*, Vol.91, No.18, pp. 8636-8640, ISSN 0027-8424.

Field, J.; Vojtek, A.; Ballester, R.; Bolger, G.; Colicelli, J.; Ferguson, K.; Gerst, J.; Kataoka, T.; Michaeli, T.; Powers, S.; Riggs, M.; Rodgers, L.; Wieland, I.; Wheland, B.; & Wigler, M. (1990). Cloning and characterization of CAP, the S. cerevisiae gene encoding the 70 kd adenylyl cyclase-associated protein. *Cell*, Vol.61, No.2, pp. 319-327, ISSN 0092-8674.

Finkel, T.; Theriot, J.A.; Dise, K.R.; Tomaselli, G.F. & Goldschmidt-Clermont, P.J. (1994). Dynamic actin structures stabilized by profilin. *Proceedings of the National Academy of Sciences of the United States of America*, Vol.91, No.4, pp. 1510-1514, ISSN 0027-8424.

Folkow, B. (1982). Physiological aspects of primary hypertension. Physiological reviews, Vol.62, No.2, pp. 347-504, ISSN 0031-9333.

Freeman, N.L.; Chen, Z.; Horenstein, J.; Weber, A. & Field, J. An actin monomer binding activity localizes to the carboxyl-terminal half of the Saccharomyces cerevisiae cyclase-associated protein. *The Journal of biological chemistry*, Vol.270, No.10, pp. 5680-5685, ISSN 0021-9258.

Gerthoffer, W.T. (2007). Mechanisms of vascular smooth muscle cell migration. *Circulation Research*, Vol.100, No.5, 607-621, ISSN 0009-7330.

Gertler, F.B.; Niebuhr, K.; Reinhard, M.; Wehland, J. & Soriano, P. (1996). Mena, a relative of VASP and Drosophila Enabled, is implicated in the control of microfilament dynamics. *Cell*, Vol.87, No.2, pp. 227-239, ISSN 0092-8674.

Gieselmann, R. & Mann, K. (1992). ASP-56, a new actin sequestering protein from pig platelets with homology to CAP, an adenylate cyclase-associated protein from yeast. *FEBS Letters*, Vol.298, No.2-3, pp. 149-153, ISSN 0014-5793.

Gieselmann, R.; Kwiatkowski, D.J.; Janmey, P.A. & Witke, W. (1995). Distinct biochemical characteristics of the two human profilin isoforms. *European journal of biochemistry*, Vol.229, No.3, pp. 621–628, ISSN 0014-2956.

Giesemann, T.; Rathke-Hartlieb, S.; Rothkegel, M.; Bartsch, J.W.; Buchmeier, S.; Jockusch, B.M. & Jockusch, H. (1999). A role for polyproline motifs in the spinal muscular atrophy protein SMN. Profilins bind to and colocalize with SMN in nuclear gems. *The Journal of biological chemistry*, Vol.274, No.53, pp. 37908–37914, ISSN 0021-9258.

Giesemann, T.; Schwarz, G.; Nawrotzki, R.; Berhörster, K.; Rothkegel, M.; Schlüter, K.; Schrader, N.; Schindelin, H.; Mendel, R.R.; Kirsch, J. & Jockusch, B.M. (2003). Complex formation between the postsynaptic scaffolding protein gephyrin, profilin, and Mena: a possible link to the microfilament system. *The Journal of neuroscience*, Vol.23, No.23, pp. 8330-8339, ISSN 0270-6474.

Gimbrone, M.A.; Nagel, T. & Topper, J.N. (1997). Biomechanical activation: an emerging paradigm in endothelial adhesion biology. *The Journal of clinical investigation*, Vol.99, No.8, pp. 1809–1813, ISSN 0021-9738.

Gingras, A.C.; Kennedy, S.G.; O'Leary, M.A.; Sonenberg, N. & Hay, N. (1998). 4E-BP1, a repressor of mRNA translation, is phosphorylated and inactivated by the Akt(PKB) signaling pathway. *Genes & development*, Vol.12, No.4, pp. 502-513, ISSN 0890-9369.

Goldschmidt-Clermont, P.J.; Kim, J.W.; Machesky, L.M.; Rhee, S.G. & Pollard, T.D. (1991) Regulation of phospholipase C-gamma 1 by profilin and tyrosine phosphorylation. *Science*, Vol.251, No.4998, pp. 1231–1233, ISSN 0036-8075.

Goldschmidt-Clermont, P.J.; Machesky, L.M.; Baldassare, J.J. & Pollard, T.D. (1990). The actin-binding protein profilin binds to PIP2 and inhibits its hydrolysis by phospholipase C. *Science*, Vol.247, No.4950, pp. 1575-1578, ISSN 0036-8075.

Goldschmidt-Clermont, P.J.; Mendelsohn, M.E. & Gibbs, J.B. (1992). Rac and Rho in control. *Current Biology*, Vol.2, No.12, pp. 669-671, ISSN 0960-9822.

Gottwald, U.; Brokamp, R.; Karakesisoglou, I.; Schleicher, M. & Noegel, A.A. (1996). Identification of a cyclase-associated protein (CAP) homologue in Dictyostelium discoideum and characterization of its interaction with actin. *Molecular biology of the cell*, Vol.7, No.2, pp. 261-272, ISSN 1059-1524.

Gürlek, A.; Dağalp, Z.; Oral, D.; Omürlü, K.; Erol, C.; Akyol, T. & Tutar, E. (1995). Restenosis after transluminal coronary angioplasty: a risk factor analysis. *Journal of cardiovascular risk*, Vol.2, No.1, pp. 51-55, ISSN 1350-6277.

Haarer, B.K. & Brown, S.S. (1990). Structure and function of profilin. *Cell motility and the cytoskeleton* Vol.17, No.2, pp. 71-74, ISSN 0886-1544.

Haarer, B.K.; Petzold, A.S. & Brown, S.S. (1993). Mutational analysis of yeast profilin. *Molecular and cellular biology*, Vol.13, No.12, pp. 7864-7873, ISSN 0270-7306.

Haffner, C.; Jarchau, T.; Reinhard, M.; Hoppe, J.; Lohmann, S.M. & Walter, U. (1995). Molecular cloning, structural analysis and functional expression of the proline-rich focal adhesion and microfilament-associated protein VASP. *The EMBO journal*, Vol.14, No.1, pp. 19-27, ISSN 0261-4189.

Halbrügge, M.; Friedrich, C.; Eigenthaler, M.; Schanzenbächer, P. & Walter, U. (1990). Stoichiometric and reversible phosphorylation of a 46-kDa protein in human platelets in response to cGMP- and cAMP-elevating vasodilators. *The Journal of biological chemistry*, Vol.265, No.6, pp. 3088–3093, ISSN 0021-9258.

Hartwig, J.H.; Bokoch, G.M.; Carpenter, C.L.; Janmey, P.A.; Taylor, L.A.; Toker, A. & Stossel, T.P. (1995). Thrombin receptor ligation and activated Rac uncap actin filament barbed ends through phosphoinositide synthesis in permeabilized human platelets. *Cell*, Vol.82, No.4, pp. 643–653, ISSN 0092-8674.

Hassona, M.D.; Abouelnaga, Z.A.; Elnakish, M.T.; Awad, M.M.; Alhaj, M., Goldschmidt-Clermont, P.J. & Hassanain, H.H. (2010). Vascular hypertrophy-associated hypertension of profilin1 transgenic mouse model leads to functional remodeling of peripheral arteries. *American journal of physiology. Heart and circulatory physiology*, Vol.298, No.6, pp. 2112-2120, ISSN 0363-6135

Hassona, M.D.; Elnakish, M.T.; Abouelnaga, Z.A.; Alhaj, M.; Wani, A.A. & Hassanain, H.H. (2011). The Effect of Selective Antihypertensive Drugs on the Vascular Remodeling-associated Hypertension: Insights from a Profilin1 Transgenic Mouse Model. *Journal of Cardiovascular Pharmacology*, Vol.57, No.5, pp. 550-558, ISSN 0160-2446.

Haugwitz, M.; Noegel, A.A.; Karakesisoglou, J. & Schleicher, M. (1994). Dictyostelium amoebae that lack G-actin-sequestering profilins show defects in F-actin content, cytokinesis, and development. *Cell*, Vol.79, No.2, pp. 303–314, ISSN 0092-8674.

Hellstrom, M.; Kalen, M.; Lindahl, P.; Abramsson, A. & Betsholtz, C. (1999). Role of PDGF-B and PDGFR-beta in recruitment of vascular smooth muscle cells and pericytes during embryonic blood vessel formation in the mouse. *Development*, Vol.126, No.14, pp. 3047–3055, ISSN 0950-1991.

Higgs, H.N. & Pollard, T.D. (2001). Regulation of actin filament network formation throughARP2/3 complex: activation by a diverse array of proteins. *Annual review of biochemistry*, Vol.70, pp. 649–662, ISSN 0066-4154.

Hirshman, C.A. & Emala, C.W. (1999). Actin reorganization in airway smooth muscle cells involves Gq and Gi-2 activation of Rho. *American journal of physiology. Lung cellular and molecular physiology*, Vol.277, No.3 Pt.1, pp. L653–L661, ISSN 0002-9513.

Hu, Y.; Cheng, L.; Hochleitner, B.W. & Xu, Q. (1997). Activation of mitogen-activated protein kinases (ERK/JNK) and AP-1 transcription factor in rat carotid arteries after balloon injury. *Arteriosclerosis, thrombosis, and vascular biology*, Vol.17, No.11, pp. 2808-2816, ISSN 1079-5642.

Hubberstey, A.; Yu, G.; Loewith, R.; Lakusta, C. & Young, D. (1996). Mammalian CAP interacts with CAP, CAP2, and actin. *Journal of cellular biochemistry*, Vol.61, No.3, pp. 459-466, ISSN 0730-2312.

Imamura, H.; Tanaka, K.; Hihara, T.; Umikawa, M.; Kamei, T.; Takahashi, K.; Sasaki, T. & Takai, Y. (1997). Bni1p and Bnr1p: downstream targets of the Rho family small G-proteins which interact with profilin and regulate actin cytoskeleton in Saccharomyces cerevisiae. *The EMBO journal*, Vol.16, No.10, pp. 2745-2755, ISSN 0261-4189.

Intengan, H.D. & Schiffrin, E.L. (2001). Vascular remodeling in hypertension: roles of apoptosis, inflammation, and fibrosis. *Hypertension*, Vol.38, No.3 Pt.2, pp. 581–587, ISSN 0194-911X.

Janmey, P.A. (1995). Protein regulation by phosphatidylinositol lipids. *Chemistry & biology*, Vol.2, No. 2, pp. 61–65, ISSN 1074-5521.

Jockusch, B.M.; Bubeck, P.; Giehl, K.; Kroemker, M.; Moschner, J.; Rothkegel, M.; Rüdiger, M.; Schlüter, K.; Stanke, G. & Winkler, J. (1995). The molecular architecture of focal adhesions. *Annual review of cell and developmental biology*, Vol.11, pp. 379-416, ISSN 1081-0706.

Johnstone, M.T.; Creager, S.J.; Scales, K.M.; Cusco, J.A.; Lee, B.K. & Creager, M.A. (1993). Impaired endothelium-dependent vasodilation in patients with insulin-dependent diabetes mellitus. *Circulation*, Vol.88, No.6, pp. 2510-2516, ISSN 0009-7322.

Jones, K.A.; Perkins, W.J.; Lorenz, R.R.; Prakash, Y.S.; Sieck, G.C. & Warner, D.O. (1999). F-actin stabilization increases tension cost during contraction of permeabilized airway smooth muscle in dogs. *The Journal of physiology*, Vol.519, Pt.2, pp. 527–538, ISSN 0022-3751.

Kaiser, D.A.; Goldschmidt-Clermont, P.J.; Levine, B.A. & Pollard, T.D. (1989). Characterization of renatured profilin purified by urea elution from poly-L-proline agarose columns. *Cell motility and the cytoskeleton*, Vol.14, No.2, pp. 251-262, ISSN 0886-1544.

Kelleher, J.F.; Atkinson, S.J. & Pollard TD. (1995). Sequences, structural models, and cellular localization of the actin-related proteins Arp2 and Arp3 from Acanthamoeba. *The Journal of cell biology*, Vol.131, No.2, pp. 385-397, ISSN 0021-9525.

Kitamura, T.; Ogawa, W.; Sakaue, H.; Hino, Y.; Kuroda, S.; Takata, M.; Matsumoto, M.; Maeda, T.; Konishi, H.; Kikkawa, U. & Kasuga, M. (1998). Requirement for activation of the serine-threonine kinase Akt (protein kinase B) in insulin stimulation of protein synthesis but not of glucose transport. *Molecular and cellular biology*, Vol.18, No.7, pp. 3708-3717, ISSN 0270-7306.

Kohno, H.; Tanaka, K.; Mino, A.; Umikawa, M.; Imamura, H.; Fujiwara, T.; Fujita, Y.; Hotta, K.; Qadota, H.; Watanabe, T.; Ohya, Y. & Takai, Y. (1996).Bni1p implicated in cytoskeletal control is a putative target of Rho1p small GTP binding protein in Saccharomyces cerevisiae. *The EMBO journal*, Vol.15, No.22, pp. 6060-6068, ISSN 0261-4189.

Korenbaum, E.; Nordberg, P.; Björkegren-Sjögren, C.; Schutt, C.E.; Lindberg, U. & Karlsson, R. (1998). The role of profilin in actin polymerization and nucleotide exchange. *Biochemistry*, Vol.37, No.26, pp. 9274-9283, ISSN 0006-2960.

Krishnan, K. & Moens, P.D.J. (2009). Structure and functions of profilins. *Biophysical reviews*, Vol.1, No.2, pp. 71–81, ISSN 1867-2450.

Kwiatkowski, D.J. & Bruns, G.A. (1988). Human profilin: Molecular cloning, sequence comparison and chromosomal analysis. *The Journal of biological chemistry*, Vol.263, No.12, pp. 5910–5915, ISSN 0021-9258.

Lambrechts, A.; Braun, A.; Jonckheere, V.; Aszodi, A.; Lanier, L.M.; Robbens, J.; Van Colen, I.; Vandekerckhove, J.; Fässler, R. & Ampe, C. (2000). Profilin II is alternatively spliced, resulting in profilin isoforms that are differentially expressed and have distinct biochemical properties. *Molecular and cellular biology*, Vol.20, No.21, pp. 8209–8219, ISSN 0270-7306.

Lambrechts, A.; Verschelde, J.L.; Jonckheere, V.; Goethals, M.; Vandekerckhove, J. & Ampe, C. (1997). The mammalian profilin isoforms display complementary affinities for PIP2 and proline-rich sequences. *The EMBO journal*, Vol.16, No.3, pp. 484-94, ISSN 0261-4189.

Lassing, I. & Lindberg, U. (1985). Specific interaction between phosphatidylinositol 4,5-bisphosphate and profilactin. Nature, Vol.314, No.6010, pp. 472-474, ISSN 0028-0836.

Lassing, I. & Lindberg, U. (1988). Specificity of the interaction between phosphatidylinositol 4,5-bisphosphate and the profilin:actin complex. *Journal of cellular biochemistry*, Vol. 37, No.3, pp. 255-267, ISSN 0730-2312.

Li, F. & Higgs, H.N. (2003). The mouse Formin mDia1 is a potent actin nucleation factor regulated by autoinhibition. *Current biology*, Vol.13, No.15, 1335-1340, ISSN 0960-9822.

Libby, P. (2002). Inflammation in atherosclerosis. *Nature*, Vol.420, No.6917, pp. 868-874, ISSN 0028-0836.

Lindberg, U.; Schutt, C.E.; Hellsten, E.; Tjäder, A.C. & Hult, T. (1988). The use of poly(L-proline)-Sepharose in the isolation of profilin and profilactin complexes. *Biochimica et biophysica acta*, Vol.967, No.3, pp. 391-400, ISSN 0006-3002.

Lu, P.J.; Shieh, W.R.; Rhee, S.G.; Yin, H.L. & Chen, C.S. (1996). Lipid products of phosphoinositide 3-kinase bind human profilin with high affinity. Biochemistry, Vol.35, No.44, pp. 14027–14034, ISSN 0006-2960.

Luna, E.J. & Hitt, A.L. (1992) Cytoskeleton--plasma membrane interactions. *Science*, Vol.258 No.5084, pp. 955-964, ISSN 0036-8075.

Machesky, L.M. & Pollard, T.D. (1993). Profilin as a potential mediator of membrane-cytoskeleton communication. *Trends in cell biology*, Vol.3, No.11, pp. 381–385, ISSN 0962-8924

Machesky, L.M. (1997). Cell motility: complex dynamics at the leading edge. *Current biology*, Vol.7, No.3, pp. R164-R167, ISSN 0960-9822.

Machesky, L.M.; Atkinson, S.J.; Ampe, C.; Vandekerckhove, J. & Pollard, T.D. (1994). Purification of a cortical complex containing two unconventional actins from *Acanthamoeba* by affinity chromatography on profilin-agarose. *The Journal of cell biology*, Vol.127, No.1, pp. 107–115, ISSN 0021-9525.

Machesky, L.M.; Cole, N.B.; Moss, B. & Pollard, T.D. (1994). Vaccinia virus expresses a novel profilin with a higher affinity for polyphosphoinositides than actin. *Biochemistry*, Vol.33, No.35, pp. 10815–10824, ISSN 0006-2960.

Machesky, L.M.; Goldschmidt-Clermont, P.J. & Pollard, T.D. (1990). The affinities of human platelet and Acanthamoeba profilin isoforms for polyphosphoinositides account for their relative abilities to inhibit phospholipase C. *Cell regulation*, Vol.1, No.12, pp. 937-950, ISSN 1044-2030.

Magdolen, V.; Drubin, D.G.; Mages, G.; Bandlow, W. (1993). High levels of profilin suppress the lethality caused by overproduction of actin in yeast cells. *FEBS Letters*, Vol.316, No.1, pp.41-47, ISSN 0014-5793.

Magdolen, V.; Oechsner, U.; Müller, G. & Bandlow, W. (1988). The introncontaining gene for yeast profilin (PFY) encodes a vital function. *Molecular and cellular biology*, Vol.8, No.12, pp. 5108–5115, ISSN 0270-7306.

Majesky, M.W. & Schwartz, S.M. (1990). Smooth muscle diversity in arterial wound repair. *Toxicologic Pathology*, Vol.18, No.4 Pt. 1, pp. 554 –559, ISSN 0192-6233.

Mammoto, A.; Sasaki, T.; Asakura, T.; Hotta, I.; Imamura, H.; Takahashi, K.; Matsuura, Y.; Shirao, T. & Takai, Y. (1998). Interactions of drebrin and gephyrin with profilin. *Biochemical and biophysical research communications*, Vol.243, No.1, pp. 86–89, ISSN 0006-291X.

Manseau, L.; Calley, J. & Phan, H. (1996). Profilin is required for posterior patterning of the Drosophila oocyte. *Development*, Vol.122, No.7, pp. 2109-2116, ISSN 0950-1991.

Mazzatti, D.J.; Pawelec, G.; Jonath, R. & Forsey, R.J. (2007). SELDI-TOF-MS protein Chip array profiling of T-cell propagated in long-term culture identifies human P potential bio-marker of immunosenescence. *Proteome Science*, 5:7, ISSN 1477-5956.

McLaughlin, P.J. & Weeds, A.G. (1995). Actin-binding protein complexes at atomic resolution. *Annual review of biophysics and biomolecular structure*, Vol.24, pp. 643-675, ISSN 1056-8700.

Mehta, D. & Gunst, S.J. (1999). Actin polymerization stimulated by contractile activation regulates force development in canine tracheal smooth muscle. *The Journal of physiology*, Vol.519, Pt.3, pp. 829–840, ISSN 0022-3751.

Mehta, P.K. & Griendling, K.K. (2007). Angiotensin II cell signaling: physiological and pathological effects in the cardiovascular system. *American journal of physiology. Cell physiology*, Vol.292, No.1, pp. C82-97, ISSN 0363-6143.

Meredith, J.E. Jr.; Fazeli, B. & Schwartz, M.A. (1993). The extracellular matrix as a cell survival factor. *Molecular biology of the cell*, 1993, Vol.4, No.9, pp. 953–961, ISSN 1059-1524.

Metzler, W.J.; Bell, A.J.; Ernst, E.; Lavoie, T.B. & Mueller, L. (1994). Identification of the poly-L-proline-binding site on human profilin. *The Journal of biological chemistry*, Vol.269, No.6, pp. 4620-4625, ISSN 0021-9258.

Metzler, W.J.; Constantine, K.L.; Friedrichs, M.S.; Bell, A.J.; Ernst, E.G.; Lavoie, T.B. & Mueller, L. (1993). Characterization of the three-dimensional structure of human profilin: ^{1}H, ^{13}C and ^{15}N NMR assignments and global folding pattern. *Biochemistry*, Vol.32, No.50, pp. 13818-13829, ISSN 0006-2960.

Miki, H.; Suetsugu, S. & Takenawa, T. (1998). WAVE, a novel WASP-family protein involved in actin reorganization induced by Rac. *The EMBO journal*, Vol.17, No.23, pp. 6932–6941, ISSN 0261-4189.

Ming, X.F.; Viswambharan, H.; Barandier, C.; Ruffieux, J.; Kaibuchi, K.; Rusconi, S. & Yang, Z. (2002). Rho GTPase/Rho kinase negatively regulates endothelial nitric oxide synthase phosphorylation through the inhibition of protein kinase B/Akt in human endothelial cells. *Molecular and cellular biology*, Vol.22, No.24, pp. 8467–8477, ISSN 0270-7306.

Miyagi, Y.; Yamashita, T.; Fukaya, M.; Sonoda, T.; Okuno, T.; Yamada, K. Watanabe, M.; Nagashima, Y.; Aoki, I.; Okuda, K.; Mishina, M. & Kawamoto S. (2002). Delphilin: a novel PDZ and formin homology domain-containing protein that synaptically colocalizes and interacts with glutamate receptor d2 subunit. *The Journal of neuroscience*, Vol.22, No.3, pp. 803–814, ISSN 0270-6474.

Mogilner, A. & Oster, G. (2003). Polymer motors: pushing out the front and pulling up the back. *Current biology*, Vol.13, No.18, pp. R721–R733, ISSN 0960-9822.

Moldovan, N.I.; Milliken, E.E.; Irani, K.; Chen, J.; Sohn, R.H.; Finkel, T. & Goldschmidt-Clermont, P.J. (1997). Regulation of endothelial cell adhesion by profilin. *Current biology*, Vol.7, No.1, pp. 24-30, ISSN 0960-9822.

Moustafa-Bayoumi, M.; Alhaj, M.A.; El-Sayed, O.; Wisel, S.; Chotani, M.A.; Abouelnaga, Z.A.; Hassona, M.D.H.; Rigatto, K.; Morris, M.; Nuovo, G. Zweier, J.L.; Goldschmidt-Clermont, P. & Hassanain, H.H. (2007). Vascular hypertrophy and hypertension caused by transgenic overexpression of profilin1. *The Journal of biological chemistry*, Vol.282, No.52, pp. 37632–7639, ISSN 0021-9258.

Murry, C.E.; Gipaya, C.T.; Bartosek, T.; Benditt, E.P. & Schwartz, S.M. (1997). Monoclonality of smooth muscle cells in human atherosclerosis. *The American journal of pathology*, Vol.151, No.3, pp. 697–705, ISSN 0002-9440.

Nobes, C.D. & Hall, A. (1995). Rho, rac, and cdc42 GTPases regulate the assembly of multimolecular focal complexes associated with actin stress fibers, lamellipodia, and filopodia. *Cell*, Vol.81, No.1, pp. 53–62, ISSN 0092-8674.

Nodelman, I.M.; Bowman, G.D.; Lindberg, U. & Schutt, C.E. (1999). X-ray structure determination of human Profilin II: a comparative structural analysis of human profilins. *Journal of molecular biology*, Vol.294, No.5, pp. 1271– 1285, ISSN 0022-2836.

Obermann, H.; Raabe, I.; Balvers, M.; Brunswig, B.; Schulze, W. & Kirchhoff, C. (2005). Novel testis-expressed profilin IV associated with acrosome biogenesis and spermatid elongation. *Molecular human reproduction*, Vol.11, No.1, pp. 53–64, ISSN 1360-9947.

Ostrander, D.B.; Gorman, J.A. & Carman, G.M. (1995). Regulation of profilin localization in Saccharomyces cerevisiae by phosphoinositide metabolism. *The Journal of biological chemistry*, Vol.270, No.45, pp. 27045–27050, ISSN 0021-9258.

Pantaloni, D. & Carlier, M.F. (1993). How profilin promotes actin filament assembly in the presence of thymosin beta 4. *Cell*, Vol.75, No.5, pp. 1007-1014, ISSN 0092-8674.

Parast, M.M. & Otey, C.A. (2000). Characterization of palladin, a novel protein localized to stress fibers and cell adhesions. *The Journal of cell biology*, Vol.150, No.3, pp. 643–656, ISSN 0021-9525.

Perelroizen, I.; Didry, D.; Christensen, H.; Chua, N.H. & Carlier, M.F. (1996). Role of nucleotide exchange and hydrolysis in the function of profilin in action assembly. *The Journal of biological chemistry*, Vol.271, No.21, pp. 12302-12309, ISSN 0021-9258.

Perelroizen, I.; Marchand, J.B.; Blanchoin L.; Didry D. & Carlier, M.F. (1994). Interaction of profilin with G-actin and poly(L-proline). *Biochemistry*, Vol.33, No.28, pp. 8472–8478, ISSN 0006-2960.

Petrella, E.C.; Machesky, L.M.; Kaiser, D.A. & Pollard TD. (1996). Structural requirements and thermodynamics of the interaction of proline peptides with profilin. *Biochemistry*, Vol.35, No.51, pp. 16535-16543, ISSN 0006-2960.

Philp, D.; Goldstein, A.L. & Kleinman, H.K. (2004). Thymosin beta4 promotes angiogenesis, wound healing, and hair follicle development. *Mechanisms of ageing and development*, Vol.125, No.2, pp. 113–115, ISSN 0047-6374.

Pistor, S.; Chakraborty, T.; Walter, U. & Wehland, J. The bacterial actin nucleator protein ActA of Listeria monocytogenes contains multiple binding sites for host microfilament proteins. *Current biology*, Vol.5, No.5, pp. 517-525, ISSN 0960-9822.

Polet, D.; Lambrechts, A.; Ono, K.; Mah, A.; Peelman, F.; Vandekerckhove, J.; Baillie, D.L.; Ampe, C. & Ono, S. (2006). Caenorhabditis elegans expresses three functional profilins in a tissue-specific manner. *Cell motility and the cytoskeleton*, Vol.63, No.1, pp. 14–28, ISSN 0886-1544.

Pollard, T.D & Borisy, G.G. (2003). Cellular motility driven by assembly and disassembly of actin filaments. *Cell*, Vol.112, No.4, pp. 453–465, ISSN 0092-8674.

Pollard, T.D. (1986). Rate constants for the reactions of ATP- and ADP-actin with the ends of actin filaments. *The Journal of cell biology*, Vol.103, No.6 Pt.2, pp. 2747-54, ISSN 0021-9525.

Prass, M.; Jacobson, K.; Mogilner, A. & Radmacher, M. (2006). Direct measurement of the lamellipodial protrusive force in a migrating cell. *The Journal of cell biology*. Vol.174, No.6, pp. 767–772, ISSN 0021-9525.

Pring, M.; Weber, A. & Bubb, M.R. (1992). Profilin-actin complexes directly elongate actin filaments at the barbed end. *Biochemistry*, Vol.31, No.6, pp. 1827-1836, ISSN 0006-2960.

Pyles, J.M.; March, K.L.; Franklin, M.; Mehdi, K.; Wilensky, R.L. & Adam, L.P. (1997). Activation of MAP kinase in vivo follows balloon overstretch injury of porcine coronary and carotid arteries. *Circulation research*, Vol.81, No.6, pp. 904-910, ISSN 0009-7330.

Raghunathan, V.; Mowery, P.; Rozycki, M.; Lindberg, U. & Schutt, C. Structural changes in profilin accompany its binding to phosphatidylinositol, 4,5-bisphosphate. *FEBS Letters*, Vol.297, No.1-2, pp. 46-50, ISSN 0014-5793.

Ramesh, N.; Antón, I.M.; Hartwig, J.H. & Geha, R.S. (1997). WIP, a protein associated with Wiskott–Aldrich syndrome protein, induces actin polymerization and redistribution in lymphoid cells. *Proceedings of the National Academy of Sciences of the United States of America*, Vol.94, No.26, pp. 14671–14676, ISSN 0027-8424.

Re, F.; Zanetti, A.; Sironi, M.; Polentarutti, N.; Lanfrancone, L.; Dejana, E. & Colotta, F. (1994). Inhibition of anchorage-dependent cell spreading triggers apoptosis in cultured human endothelial cells. *The Journal of cell biology*, 1994, Vol.127, No.2, pp. 537–546, ISSN 0021-9525.

Reidy, M.A. (1992). Factors controlling smooth-muscle cell proliferation. *Archives of pathology & laboratory medicine*, Vol.116, No.12, pp. 1276 –1280, ISSN 0003-9985.

Reinhard, M.; Giehl, K.; Abel, K.; Haffner, C.; Jarchau, T.; Hoppe, V.; Jockusch, B.M. & Walter, U. (1995). The proline-rich focal adhesion and microfilament protein VASP is a ligand for profilins. *The EMBO journal*, Vol.14, No.8, pp. 1583–1589, ISSN 0261-4189.

Reinhard, M.; Halbrügge, M.; Scheer, U.; Wiegand, C.; Jockusch, B.M. & Walter, U. (1992). The 46/50 kDa phosphoprotein VASP purified from human platelets is a novel protein associated with actin filaments and focal contacts. *The EMBO Journal*, Vol.11, No.6, pp. 2063-2070, ISSN 0261-4189.

Reinhard, M.; Rüdiger, M.; Jockusch, B.M. & Walter U. (1996). VASP interaction with vinculin: a recurring theme of interactions with proline-rich motifs. *FEBS Letters*, Vol.399, No.1-2, pp. 103-107, ISSN 0014-5793.

Romeo, G.R. & Kazlauskas, A. (2008). Oxysterol and diabetes activate STAT3 and control endothelial expression of profilin-1 via OSBP1. *The Journal of biological chemistry*, Vol.283, No.15, pp.9595-9605, ISSN 0021-9258.

Romeo, G.R.; Frangioni, J.V. & Kazlauskas A. (2004). Profilin acts downstream of LDL to mediate diabetic endothelial cell dysfunction. *The FASEB journal*, Vol.18, No.6, pp. 725–727, ISSN 0892-6638.

Romeo, G.R.; Moulton, K.S. & Kazlauskas, A. (2007). Attenuated Expression of Profilin-1 Confers Protection from Atherosclerosis in the LDL Receptor–Null Mouse. *Circulation research*, Vol.101, No.4, pp. 357-367, ISSN 0009-7330.

Rothkegel, M.; Mayboroda, O.; Rohde, M.; Wucherpfennig, C.; Valenta, R. & Jockusch, B.M. (1996). Plant and animal profilins are functionally equivalent and stabilize microfilaments in living animal cells. *Journal of cell science*, Vol.109, Pt.1, pp. 83-90, ISSN 0021-9533.

Ruwhof, C. & van der Laarse, A. (2000). Mechanical stress-induced cardiac hypertrophy: mechanisms and signal transduction pathways. *Cardiovascular research*, Vol.47, No.1, pp. 23-37, ISSN 0008-6363.

Safer, D.; Elzinga, M. & Nachmias, V.T. (1991). Thymosin beta 4 and Fx, an actin-sequestering peptide, are indistinguishable. *The Journal of biological chemistry*, Vol.266, No.7, pp. 4029-4032, ISSN 0021-9258.

Salazar, R.; Bell, S.E. & Davis, G.E. (1999). Coordinate induction of the actin cytoskeletal regulatory proteins gelsolin, vasodilator-stimulated phosphoprotein, and profilin during capillary morphogenesis in vitro. *Experimental cell research*, Vol.249, No.1, pp. 22–32, ISSN 0014-4827.

Salmon, E.D. (1989). Cytokinesis in animal cells. *Current opinion in cell biology*, Vol.1, No.3, pp. 541-547, ISSN 0955-0674.

Schiffrin, E.L. (2001). Effects of antihypertensive drugs on vascular remodeling: do they predict outcome in response to antihypertensive therapy? *Current opinion in nephrology and hypertension*, Vol.10, No.5, pp. 617-624, ISSN 1062-4821.

Schlüter, K.; Jockusch, B.M. & Rothkegel, M. (1997). Profilins as regulators of actin dynamics. *Biochimica et biophysica acta*, Vol.1359, No.2, pp. 97-109, ISSN 0006-3002.

Schmidt, A. & Hall, M. N. (1998). Signaling to the actin cytoskeleton. *Annual review of cell and developmental biology*, Vol.14, pp. 305-338, ISSN 1081-0706.

Schutt, C.E.; Lindberg, U.; Myslik, J. & Strauss, N. (1989). Molecular packing in profilin: actin crystals and its implications. *Journal of molecular biology*, Vol.209, No.4, pp. 735-746, ISSN 0022-2836.

Schutt, C.E.; Myslik, J.C.; Rozycki, M.D.; Goonesekere, N.C. & Lindberg, U. (1993). The structure of crystalline profilin-beta-actin. *Nature*, Vol.365, No.6449, 810-816, ISSN 0028-0836.

Simon, G. (2004). Pathogenesis of structural vascular changes in hypertension. *Journal of hypertension*, Vol.22, No.1, pp. 3-10, ISSN 0263-6352.

Singh, S.S.; Chauhan, A.; Murakami, N. & Chauhan, V.P. (1996). Profilin and gelsolin stimulate phosphatidylinositol 3-kinase activity. *Biochemistry*, Vol.35, No.51, pp. 16544-16549, ISSN 0006-2960.

Skare, P.; Kreivi, J.P.; Bergström, A. & Karlsson R. (2003) Profilin I colocalizes with speckles and Cajal bodies: a possible role in pre-mRNA splicing. *Experimental cell research*, Vol.286, No.1, pp. 12–21, ISSN 0014-4827.

Smith, G.A.; Theriot, J.A. & Portnoy, D.A. (1996). The tandem repeat domain in the Listeria monocytogenes ActA protein controls the rate of actin-based motility, the percentage of moving bacteria, and the localization of vasodilator-stimulated phosphoprotein and profilin. *The Journal of cell biology*, Vol.135, No.3, pp. 647-660, ISSN 0021-9525.

Sohn, R.H. & Goldschmidt-Clermont, P.J. (1994). Profilin: at the crossroads of signal transduction and the actin cytoskeleton. *BioEssays*, Vol.16, No.7, pp. 465-472, ISSN 0265-9247.

Sohn, R.H.; Chen, J.; Koblan, K.S.; Bray, P.F. & Goldschmidt-Clermont, P.J. (1995). Localization of a binding site for phosphatidylinositol 4,5-bisphosphate on human profilin. *The Journal of biological chemistry*, Vol.270, No.36, pp. 21114-21120, ISSN 0021-9258.

Sonobe, S.; Takahashi, S.; Hatano, S. & Kuroda, K. (1986). Phosphorylation of amoeba G-actin and its effect on actin polymerization. *The Journal of biological chemistry*, Vol.261, No.31, pp. 14837–14843, ISSN 0021-9258.

Spagnoli, L.; Bonanno, L.; Sangiorgi, G. & Mauriello, A. (1991). Role of Inflammation in Atherosclerosis. *Journal of nuclear medicine*, Vol.48, No.11, pp. 1800-1815, ISSN 0161-5505.

Speirs, S. & Kaufman, M.H. (1989). Cytochalasin D-induced triploidy in the mouse. The *Journal of experimental zoology*, Vol.250, No.3, pp. 339-345, ISSN 0022-104X.

Staiger, C.J.; Yuan, M.; Valenta, R.; Shaw, P.J.; Warn, R.M. & Lloyd, C.W. (1994). Microinjected profilin affects cytoplasmic streaming in plant cells by rapidly depolymerizing actin microfilaments. *Current biology*, Vol.4, No.3, pp. 215-219, ISSN 0960-9822.

Starr, M.E.; Ueda, J.; Yamamoto, S.; Evers, B.M. & Saito, H. (2011). The effects of aging on pulmonary oxidative damage, protein nitration, and extracellular superoxide dismutase down-regulation during systemic inflammation. *Free radical biology & medicine*, Vol.50, No.2, 371-380, ISSN 0891-5849.

Steinberg, H.O.; Bayazeed, B.; Hook, G.; Johnson, A.; Cronin, J. & Baron, A.D. (1997). Endothelial dysfunction is associated with cholesterol levels in the high normal range in humans. *Circulation*, Vol.96, No.10, pp. 3287-3293, ISSN 0009-7322.

Stüven, T.; Hartmann, E. & Görlich, D. (2003). Export in 6: a novel nuclear export receptor that is specific for profilin–actin complexes. *The EMBO journal*, Vol.22, No.21, pp. 5928–5940, ISSN 0261-4189.

Suetsugu, S.; Miki, H. & Takenawa, T. (1999). Distinct roles of profilin1 in cell morphological changes: microspikes, membrane ruffles, stress fibers, and cytokinesis. *FEBS Letters*, Vol.457, No.3, pp. 470–474, 1999, ISSN 0014-5793.

Suetsugu, S.; Miki, H.; Takenawa, T. (1998). The essential role of profilin in the assembly of actin for microspike formation. *The EMBO journal*, 17, 6516–6526, ISSN 0261-4189.

Tamura, M.; Yanagihara, N.; Tanaka, H.; Osajima, A.; Hirano, T.; Higashi, K.; Yamada, K.M.; Nakashima, Y. & Hirano, H. (2000). Activation of DNA synthesis and AP-1 by profilin, an actin-binding protein, via binding to a cell surface receptor in cultured rat mesangial cells. *Journal of the American Society of Nephrology*, Vol.11, No.9, pp. 1620-1630, ISSN 1046-6673.

Tanaka, M. & Shibata, H. (1985). Poly(L-proline)-binding proteins from chick embryos are a profilin and a profilactin. *European journal of biochemistry*, Vol.151, No.2, pp. 291-297, ISSN 0014-2956.

Tanaka, M.; Sasaki, H.; Kino, I.; Sugimura, T. & Terada, M. (1992). Genes preferentially expressed in embryo stomach are predominantly expressed in gastric cancer. *Cancer research*, Vol.52, No.12, pp. 3372-3377, ISSN 0008-5472.

Tang, D.D. & Tan, J. (2003). Downregulation of profilin with antisense oligodeoxynucleotides inhibits force development during stimulation of smooth muscle. *American journal of physiology. Heart and circulatory physiology*, Vol.285, No.4, H1528-H1536, ISSN 0363-6135.

Tapon, N. & Hall, A. (1997). Rho, Rac and Cdc42 GTPases regulate the organization of the actin cytoskeleton. *Current opinion in cell biology*, Vol.9, No.1, pp. 86-92, ISSN 0955-0674.

Tellam, R.L.; Morton, D.J. & Clarke, F.M. (1989). A common theme in the amino acid sequences of actin and many actin-binding proteins? *Trends in biochemical sciences*, Vol.14, No.4, pp. 130-133, ISSN 0968-0004.

Tesfamariam, B.; Brown, M.L.; Deykin, D. & Cohen, R.A. (1990). Elevated glucose promotes generation of endothelium-derived vasoconstrictor prostanoids in rabbit aorta. *The Journal of clinical investigation*, Vol.85, No.3, pp. 929-932, ISSN 0021-9738.

Theriot, J.A. & Mitchison, T.J. (1993). The three faces of profilin. *Cell*, Vol.75, No.5, pp. 835-838, ISSN 0092-8674.

Thorn, K.S.; Christensen, H.E.; Shigeta, R., Huddler, D.; Shalaby, L.; Lindberg, U.; Chua, N.H. & Schutt, C.E. (1997). The crystal structure of a major allergen from plants. *Structure*, Vol.5, No.1, pp. 19-32, ISSN 0969-2126.

Tobacman, L.S.; Brenner, S.L. & Korn, E.D. (1983). Effect of Acanthamoeba profilin on the pre-steady state kinetics of actin polymerization and on the concentration of F-actin at steady state. *The Journal of biological chemistry*, Vol.258, No.14, pp. 8806-8812, ISSN 0021-9258.

Toda, T.; Uno, I.; Ishikawa, T.; Powers, S.; Kataoka, T.; Broek, D.; Cameron, S.; Broach, J.; Matsumoto, K. & Wigler, M. (1985). In yeast, RAS proteins are controlling elements of adenylate cyclase. *Cell*, Vol.40, No.1, pp. 27-36, ISSN 0092-8674.

Touyz, RM. (2007). Vascular remodeling, retinal arteries, hypertension. *Hypertension*, Vol.50, No.4, pp. 602–603, ISSN 0194-911X.

Tseng, P.C-H.; Runge, M.S.; Cooper, J.A.; Williams, R.C. Jr. & Pollard, T.D. (1984). Physical, immunochemical, and functional properties of Acanthamoeba proflin. *The Journal of cell biology*, Vol.98, No.1, pp. 214–221, ISSN 0021-9525.

U.K. Prospective Diabetes Study (UKPDS) Group. (1998). Effect of intensive blood-glucose control with metformin on complications in overweight patients with type 2 diabetes (UKPDS 34). *Lancet*, Vol.352, No.9131, pp. 854-865, ISSN 0140-6736.

Ushio-Fukai, M.; Alexander, R.W.; Akers, M.; Yin, Q.; Fujio, Y.; Walsh, K.& Griendling, K.K. (1999). Reactive oxygen species mediate the activation of Akt/protein kinase B by angiotensin II in vascular smooth muscle cells. *The Journal of biological chemistry*, Vol.274, No.32, pp. 22699-22704, ISSN 0021-9258.

Valenta, R.; Breiteneder, H.; Pettenburger, K.; Breitenbach, M.; Rumpold, H.; Kraft, D. & Scheiner, O. (1991a). Homology of the major birchpollen allergen, Bet v I, with the major pollen allergens of alder, hazel, and hornbeam at the nucleic acid level as determined by cross-hybridization. *The Journal of allergy and clinical immunology*, Vol.87, No.3, pp. 677–682, ISSN 0091-6749.

Valenta, R.; Duchene, M.; Breitenbach, M.; Pettenburger, K.; Koller, L.; Rumpold, H.; Scheiner, O. & Kraft, D. (1991b). A low molecular weight allergen of white birch (Betula verrucosa) is highly homologous to human profilin. *International archives of allergy and applied immunology*, Vol.94, No.(1-4), pp. 368–370, ISSN 0020-5915.

Valenta, R.; Duchene, M.; Ebner, C.; Valent, P.; Sillaber, C.; Deviller, P.; Ferreira, F.; Tejkl, M.; Edelmann, H.; Kraft, D. & Scheiner O. (1992). Profilins constitute a novel family of functional plant pan-allergens. *The Journal of experimental medicine*, Vol.175, No.2, pp. 377-385, ISSN 0022-1007.

Valenta, R.; Duchêne, M.; Pettenburger, K.; Sillaber, C.; Valent, P.; Bettelheim, P.; Breitenbach, M.; Rumpold, H.; Kraft, D. & Scheiner O. (1991c). Identification of profilin as a novel pollen allergen; IgE autoreactivity in sensitized individuals. *Science*, Vol.253, No.5019, pp. 557-560, ISSN 0036-8075.

Vandekerckhove, J. (1989). Structural principles of actin-binding proteins. *Current opinion in cell biology*, Vol.1, No.1, pp.15-22, ISSN 0955-0674.

Vinson, V.K.; Archer, S.J.; Lattman, E.E.; Pollard, T.D. & Torchia, D.A. (1993). Three-dimensional solution structure of Acanthamoeba profilin-I. *The Journal of cell biology*, Vol.122, No.6, pp. 1277-1283, ISSN 0021-9525.

Wallar, B.J. & Alberts, A.S. (2003). The formins: active scaffolds that remodel the cytoskeleton. Trends in cell biology, Vol.13, No.8, pp. 435–446, ISSN 0962-8924.

Wang, X.; Kibschull, M.; Laue, M.M.; Lichte, B.; Petrasch-Parwez, E. & Kilimann, M.W. (1999). Aczonin, a 550-kD putative scaffolding protein of presynaptic active zones, shares homology regions with Rim and Bassoon and binds profilin *The Journal of cell biology*, Vol.147, No.1, pp. 151–162, ISSN 0021-9525.

Watanabe, N.; Madaule, P.; Reid, T.; Ishizaki, T.; Watanabe, G.; Kakizuka, A.; Saito, Y.; Nakao, K.; Jockusch, B.M. & Narumiya, S. (1997). p140mDia, a mammalian homolog of Drosophila diaphanous, is a target protein for Rho small GTPase and is a ligand for profilin. *The EMBO journal*, Vol.16, No.11, pp. 3044–3056, ISSN 0261-4189.

Weinreb, O.; Drigues, N.; Yotam Sagi, Y.; Abraham Reznick, A.Z.Z.; Tamar Amit, T. & Youdim, M. (2007).The application of proteomics and genomics to the study of age-related neurodegeneration and neuroprotection. *Antioxidants & Redox Signaling*, Vol.9, No.2, pp. 169–179, ISSN 1523-0864.

Wheatley, E. M.; Vincent, P. A.; McKeown-Longo, P. J. & Saba, T. M. Effect of fibronectin on permeability of normal and TNF-treated lung endothelial cell monolayers. (1993). *The American journal of physiology*, Vol.264, No.1 Pt.2, pp. R90–R96; ISSN 0002-9513.

Widada, J.S.; Ferraz, C. & Liautard, J.P. (1989). Total coding sequence of profilin cDNA from Mus musculus macrophage. *Nucleic acids research*, Vol.17, No.7, pp. 2855, ISSN 0305-1048.

Witke, W. (2004). The role of profilin complexes in cell motility and other cellular processes. *Trends in cell biology*, Vol.14, No.8, pp. 461-469, ISSN 0962-8924.

Witke, W.; Podtelejnikov, A.V.; Di Nardo, A.; Sutherland, J.D.; Gurniak, C B.; Dotti, C. & Mann, M. (1998). In mouse brain profilin I and profilin II associate with regulators of the endocytic pathway and actin assembly. *The EMBO journal*, Vol.17, No.4, pp. 967–976, ISSN 0261-4189.

Witke, W.; Sutherland, J.D.; Sharpe, A.; Arai, M. & Kwiatkowski, D.J. (2001). Profilin I is essential for cell survival and cell division in early mouse development. *Proceedings of the National Academy of Sciences of the United States of America*, Vol.98, No.7, pp. 3832–3836, ISSN 0027-8424.

Yayoshi-Yamamoto, S.; Taniuchi, I. & Watanabe, T. (2000). FRL, a novel formin-related protein, binds to Rac and regulates cell motility and survival of macrophages. *Molecular and cellular biology*, Vol.20, No.18, pp. 6872–6881, ISSN 0270-7306.

Yeh, E.T.; Zhang, S.; Wu, H.D.; Korbling, M.; Willerson, J.T. & Estrov, Z. (2003). Transdifferentiation of human peripheral blood $CD34^{+}$-enriched cell population into cardiomyocytes, endothelial cells, and smooth muscle cells in vivo. *Circulation*, Vol.108, No.17, pp. 2070 –2073, ISSN 0009-7322.

Youn, T.; Kim, S.A. & Hai, C.M. (1998). Length-dependent modulation of smooth muscle activation: effects of agonist, cytochalasin, and temperature. *American journal of physiology. Cell physiology*. Vol.274, No.6 Pt.1, pp. C1601–C1607, ISSN 0002-9513.

Zang, Z.; Vuori, K.; Reed, J.C. & Ruoslahti, E. (1995). The alpha 5 beta 1 integrin supports survival of cells on fibronectin and up-regulates Bcl-2 expression. Proceedings of the National Academy of Sciences of the United States of America, Vol.92, No.13, pp. 6161–6165, ISSN 0027-8424.

Zeile, W.L; Purich, D.L. & Southwick, F.S. (1996). Recognition of two classes of oligoproline sequences in profilin-mediated acceleration of actin-based Shigella motility. *The Journal of cell biology*, Vol.133, No.1, pp. 49-59, ISSN 0021-9525.

Zhong, J.C.; Ye, J.Y.; Jin, H.Y.; Yu, X.; Yu, H.M.; Zhu, D.L.; Gao, P.J.; Huang, D.Y.; Shuster, M.; Loibner, H.; Guo, J.M.; Yu, X.Y.; Xiao, B.X.; Gong, Z.H.; Penninger, J.M. & Oudit, G.Y. (2011). Telmisartan attenuates aortic hypertrophy in hypertensive rats by the modulation of ACE2 and profilin-1 expression. *Regulatory Peptides*. Vol.166, No.1-3, pp. 90-97, ISSN 0167-0115

Zou, L.; Jaramillo, M.; Whaley, D.; Wells, A.; Panchapakesa, V.; Das, T. & Roy, P. (2007) Profilin-1 is a negative regulator of mammary carcinoma aggressiveness. *British journal of cancer*, Vol. 97, No.10, pp. 1361-1371, ISSN 0007-0920.

4

E. coli Alpha Hemolysin and Properties

Bakás Laura, Maté Sabina, Vazquez Romina and Herlax Vanesa
Instituto de Investigaciones Bioquímicas La Plata (INIBIOLP), CCT- La Plata, CONICET, Facultad de Ciencias Médicas. Universidad Nacional de La Plata. La Plata, Buenos Aires Argentina

1. Introduction

Protein toxins are prominent virulence factors in many pathogenic bacteria. While toxins of Gram-positive bacteria do not generally require activation, many toxins of the Gram-negatives are translated into an inactive form and require a processing step.

The most common such step involves a proteolytic cleavage to generate the active form, especially in those toxins with enzymatic activity. Toxins are activated by proteolysis in a variety of ways: As examples, the anthrax toxin is proteolyzed after its interaction with the receptor on the target cell to promote the formation of a prepore (van der Goot & Young, 2009); the toxic subunit of the *Vibrio cholerae* toxin (CT) is posttranslationally modified through the action of a *V. cholerae* protease that generates two fragments, one containing the toxic activity and the other serving to interact with the binding domain (Sanchez & Holmgren, 2011); finally, the toxins that are synthesized as a single polypeptides must be separated by proteolytic cleavage to generate a catalytic, a transmembrane, and a receptor-binding domain—a salient example here being the diphtheria toxin (Murphy, 1996).

Another processing step involves the acylation of proteins, which substitution is achieved by various mechanisms that differ according to the particular fatty acid transferred, the modified amino acid, and the fatty-acyl donor. Myristate and palmitate are the most common fatty acids cross-linked to proteins. Proteins sorted to the bacterial outer membrane or to the eukaryotic plasma membrane undergo processing in which an acyl group is attached to the N-terminal amino acid. In prokaryotes, acyltransferase, lipases, or esterases use catalytic mechanisms involving ester-linked acyl groups attached to serine and cysteine residues; while eukaryotic proteins utilize ester-linked palmitoylation and ether-linked prenylation of cysteine residues for membrane sorting and protein-protein interaction (Stanley *et al.*, 1998).

The pore-forming **α-hemolysin (HlyA)** of *Escherichia coli,* a member of the RTX toxins, represents a unique class of bacterial toxins that require for activation a posttranslational modification involving a covalent amide linkage of fatty acids to two internal lysine residues (Stanley *et al.*, 1998). In general, protein acylation is divided into labile modifications of internal regions and stable modifications at the N and C termini. By contrast, the mechanism of stable internal acylation of HlyA represents a unique example among prokaryotic proteins, thus generating interest in its study and discussion. After introducing HlyA, its synthesis, posttranslational modification, secretion, and activity; this chapter will focus on the role that covalently bound fatty acids play in the toxin's mechanism of action.

In recent decades, scientific advances have permitted the manipulation of toxins by using different strategies for directing toxic moieties to diseased cells and tissues. The end of the chapter will involve a discussion of this so-called *toxin-based therapy* and the potential use of HlyA in that modality.

2. The alpha-hemolysin (HlyA) of *E. coli*

Extraintestinal pathogenic *Escherichia coli* (ExPEC) is the causative agent of at least 80% of all uncomplicated urinary-tract infections (UTIs), which pathologies currently rank among the most common of infectious diseases worldwide (Marrs et al. 2005), (Foxman & Brown, 2003). ExPEC strains that cause a UTI are called uropathogenic *E. coli* (UPEC). This unique group of *E. coli* strains can reside in the lower gastrointestinal tract of healthy adults (Foxman *et al.*, 2002), (Yamamoto *et al.*, 1997), but upon entry into the urinary tract can ascend to and colonize the bladder, causing cystitis. The infection may be confined to the bladder, or bacteria may ascend into the ureters to infect the kidneys and cause pyelonephritis. In severe cases, bacteria can further disseminate across the proximal-tubular and capillary endothelia to the bloodstream, causing bacteremia (Mobley *et al.*, 2009.). A significant proportion of UTIs occur in patients with no known abnormalities of the urinary tract—the so-called *uncomplicated UTIs*. Certain host characteristics, however, such as a congenital defect in urinary-tract anatomy, are considered complicating factors for UTI and accordingly increase susceptibility to this infection as well as affect its diagnosis and management (Foxman, 2002.). Finally, colonization of the bladder in high numbers may occur without eliciting symptoms in the host, a condition known as asymptomatic bacteriuria (Hooton *et al.*, 2000.). In recent years, an enormous amount of information has accrued through sequencing the genomes of several ExPEC patients. These data, together with epidemiological analyses, have confirmed that different ExPEC pathotypes share many known as well as putative virulence factors. These latter include a number of secreted toxins, iron-acquisition systems, adhesins, and capsular antigens (Wiles *et al.*, 2008). Secreted toxins—which proteins include **HlyA**, the cytotoxic necrotizing factor-1 (CNF-1), and the secreted autotransporter—can alter host signaling cascades, disrupt inflammatory responses, and induce host-cell death; whereas bacterial siderophores like aerobactin, bacteriocin, and enterobactin allow the ExPEC to sequester iron away from the host (Guyer *et al.*, 2002), (Wiles *et al.*, 2008). Adhesive organelles can mediate ExPEC interaction with, and entry into, host cells and tissues; while the expression of encapsulation may enable ExPEC to more effectively avoid professional phagocytes (Wiles *et al.*, 2008), (Dhakal *et al.* 2008).

Experiments in murine and cell-culture model systems have demonstrated that high levels of HlyA can cause the osmotic lysis of host cells, while sublytic concentrations of this pore-forming toxin can modulate host-survival pathways by interfering with phagocyte chemotaxis (Wiles *et al*, 2008),(Jonas *et al.*, 1993), (Cavalieri & Snyder, 1982), (Chen *et al.*, 2006). Both HlyA and CNF-1 may in addition stimulate the breakdown of tissue barriers and the release of nutrients (Smith *et al.* 2008), (Bauer & Welch, 1996), but through the use of the zebrafish infection model phagocytes were found that appeared to be the primary targets of these toxins (Wiles *et al.*,2009).

HlyA represents the prototype of the first RTX family of proteins characterized by Rodney Welch (Welch 1991). Produced by a variety of Gram-negative bacteria, these proteins exhibit two common features: The first is the presence of arrays of glycine- and aspartate-rich nonapeptide repeats, which sequences are located at the C-terminal portion. The second is the unique mode of secretion via the type-I system (an ABC-binding–cassette transporter). This first group of RTX toxins consists of toxins—mostly exhibiting cytotoxic pore-forming activity—that often are first detected as a hemolytic halo surrounding bacterial colonies grown on blood-agar plates (Muller *et al.*, 1983), (Welch, 1991), *(Felmlee et al.*, 1985). Recently, a subgroup of very large RTX toxins (>3200 residues) were discovered with multiple activities, such as protease and lipase. These pathogens were named the multifunctional autoprocessing RTX toxins, with the *Vibrio cholerae* toxin being the prototype of this group. In summary, the RTX proteins form a large and diverse family with a broad spectrum of biological and biochemical activities (Linhartova, *et al.*, 2010).

2.1 Synthesis and structure of HlyA

The synthesis, maturation, and secretion of *E. coli* HlyA are determined by the *hlyCABD* operon ((Felmlee *et al*, 1985), (Issartel *et al.*, 1991), (Koronakis *et al.*, 1997), (Nieto *et al.* ,1996)). The membrane-associated export proteins are synthesized at a lower level than the cytosolic HlyC and pro-HlyA, in part because of transcription termination within the *hlyCABD* operon (Felmlee *et al*, 1985). This termination is suppressed by the elongation protein RfaH and a short 59-bp, *ops* (operon polarity suppressor) (Bailey *et al.*, 1992, 1996), (Cross *et al.*, 1990), (Nieto *et al.*, 1996) that act in concert to allow the transcription of long operons such as *hly*, *rfa*, and *tra* encoding the synthesis and export of extracellular components key in the virulence and fertility of Gram-negative bacteria (Bailey *et al.*, 1992, 1997).

The structural gene *hlyA* produces a single 110-kDa polypeptide. The estimated pI of the toxin is 4.5, with this characteristic being common among the RTX toxins. The N-terminal hydrophobic domain is predicted to contain nine amphipathic α-helices (Soloaga *et al.*, 1999). Using photoactivable liposomes, Hyland *et al* (2001) demonstrated that the region comprised between residues 177-411 is the one that becomes inserted into membranes. The C-terminal calcium-binding domain contains 11-17 of the glycine- and aspartate-rich nonapeptide β-strand repeats. Although the membrane interaction of HlyA is assumed to occur mainly through the amphipathic α-helical domain, that both major domains of HlyA are directly involved in the membrane interaction of HlyA has recently been proposed, with the calcium-binding domain in particular being responsible for the early stages of the HlyA's docking to the target membrane (Sanchez-Magraner *et al.*, 2007).

The topic of the existence of a receptor for the toxin in erythrocytes remains quite controversial. Nevertheless, Cortajarena *et al* (2003) observed that a short sequence from the C-terminal domain (between residues 914–936) was the main HlyA segment that bound to the glycophorin A on erythrocytes.

The last 60 C-terminal amino acids consist of 2 α-helices separated by 8-10 charged residues. This domain is implicated in the transport of the toxin to the extracellular medium (Hui *et al.*, 2000). Fig. 1 shows a scheme of the HlyA structure.

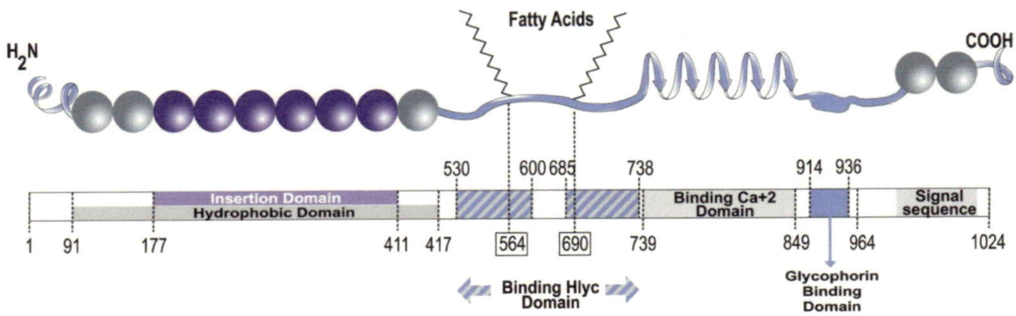

Fig. 1. A scheme of the HlyA structure.

The more relevant domains of HlyA are indicated.

2.2 The posttranslational activation of HlyA

The proHlyA protoxin is matured in the cytosol to the active form by HlyC-directed fatty acylation before export from the toxin-producing bacteria. This process consists in a posttranslational modification of the ε-amino groups of internal lysine residues by covalent attachment of amide-linked fatty-acyl residues. This reaction is catalyzed by the HlyC acyltransferases expressed together with the protoxins (Goebel & Hedgpeth, 1982). The mechanism of this novel type of protein acylation was extensively analyzed for HlyA (Issartel *et al.*, 1991), (Stanley *et al.*, 1994). HlyC uses the fatty-acyl residues carried by acyl-carrier protein (ACP) to form a covalent acyl-HlyC intermediate, which species then transfers the fatty-acyl residues to the ε-amino groups of the Lys 564 and Lys 690 residues of proHlyA (Worsham *et al.*, 2001, 2005). ACPs carrying various fatty-acyl residues—including palmitate (16:0) and palmitoleate (16:1), the most common in *E. coli*—could be efficiently used *in vitro* as acyl donors for the modification of HlyA ((Issartel *et al.*, 1991), (Trent *et al.*, 1998)). *In vivo*, however, HlyC exhibits a high selectivity for myristic acid (14:0), which species was found to constitute about 68% of the acyl chains covalently linked to Lys 564 and Lys 690 of the native HlyA (Lim *et al.*, 2000). Contrary to expectations, the extremely rare odd-carbon saturated fatty-acyl residues 15:0 and 17:0 were found to constitute the rest of the *in-vivo* acylation of HlyA in two different clinical *E. coli* isolates (Lim *et al.*, 2000). Both acylation sites in the HlyA genome function independently of one another with respect to the kinetics of their interaction with acyl-HlyC (Langston *et al.*, 2004). By using deleted protoxin variants and protoxin peptides as substrates in an *in-vitro* maturation reaction dependent on only HlyC and acyl-ACP, two independent HlyC-recognition domains were identified on the HlyA protoxin, each of which spanned one of the target lysine residues (Stanley *et al.*, 1996). Each domain required 15 to 30 amino acids for basal recognition and 50 to 80 for full wild-type acylation, but HlyC recognized a large topology rather than a linear sequence. The loss of the Lys 564 acylation site either by mutation or structural deletion affected the thermodynamics of the acylation reaction at Lys 690, implying an undefined connectivity between the two acylation sites (Worsham *et al.*, 2005). Nevertheless, the intact acylation at Lys 690 is essential for HlyA activity.

No other HlyA sequences are required for toxin maturation, including the immediately C-terminal Ca^{+2} binding repeats. Indeed, *in vitro,* Ca^{+2} ions prevent acylation at both sites (Stanley *et al.*, 1996). The extreme sensitivity of the proHlyA activation reaction to free Ca^{+2} supports the view that intracellular Ca^{+2} levels in *E. coli* are too low to affect toxin activity and that Ca^{+2} binding does not occur until the toxin is outside the cell.

This posttranslational modification is remarkable because the behavior of the protein is changed by lipid modification from a benign protein to a frank toxin—part of this transformation being an exclusive mechanism in prokaryotes since in only a few eukaryotic proteins is this type of acylation found (for example, in the nicotinic acetylcholine receptor; the insulin receptor; and cytokines such as TNF-α, IL-1α and IL-1β) (Stanley *et al.*, 1998). In the following section we discuss the role that these covalently bound fatty acids play in the toxin's mechanism of action.

2.3 The secretion of HlyA into the extracellular medium

Maturation increases the hydrophobicity of the protein, but that property is not required for export (Ludwig et al. 1987). *E. coli* HlyA-related toxins are all secreted across both membranes by the type-I export process employing an uncleaved C-terminal recognition signal (Nicaud *et al.*, 1986), (Stanley *et al.*, 1991), but no N-terminal leader peptide (Felmlee *et al.*, 1985) or periplasmic intermediate (Felmlee & Welch, 1988), (Koronakis *et al.*, 1989). The HlyA secretory apparatus comprises HlyB (an inner-membrane traffic ATPase, the ATP-binding cassette), HlyD (a membrane-fusion protein), and TolC (an outer-membrane protein) (Schulein *et al.*,1992), (Wandersman & Delepelaire, 1990), (Wang *et al.*, 1991). In *E. coli* and most other pathogens, TolC is encoded by a separate gene from *hlyCABD*. As mentioned before (*cf.* **Section 2.0**) the type-I-secretion–signal sequences have been located within the last 60 C-terminal amino acids, consisting of 2 α-helices separated by 8-10 charged residues (Hui *et al.*, 2000).

The mechanism of exportation of HlyA is as follows: The trimeric accessory protein HlyD has been proposed to form a substrate-specific complex with the inner-membrane protein HlyB, which latter species subsequently recognizes the C-terminal signal peptide of HlyA. Upon the binding of HlyA, the HlyD trimer interacts with the trimeric TolC protein of the outer membrane, inducing a conformational change and the consequent export of HlyA. This assembly between the complex HlyB-HlyD with TolC very likely occurs because, as has been demonstrated by X-ray crystallography, the trimeric complex of TolC is very similar in size to the trimeric structure of HlyD, thus facilitating the formation of a continuous transperiplasmic export channel through which HlyA can pass (Koronakis *et al.*, 2000). This complex appears to be transient, with it disengaging and reverting to a resting state once the substrate has been transported (Thanabalu *et al.*, 1998). The energy necessary for the secretion process depends not only on ATP hydrolysis mediated by HlyB but also on the proton motive force exerted on the inner membrane (Koronakis *et al.*, 1991, 1995). Type-I secretion is generally assumed to involve the translocation of unfolded proteins (Young & Holland, 1999), although Pimenta *et al* (2005) have suggested that contact with HlyD directly or indirectly affects the folding of HlyA either during the latter's transit through the translocator or afterwards.

In the last decade many researchers have been interested in this type of secretion machinery because of its potential use in the export of chimeric proteins and in vaccine production (Gentschev *et al.*,1996, 2002).

Although HlyA has its own machinery for export from the bacteria, the presence of a physiologically active HlyA in the outer-membrane vesicles (OMVs) of clinical-hemolytic (Balsalobre *et al.*, 2006) as well as laboratory-recombinant strains of *E. coli* (Herlax *et al.*, 2010) has recently been demonstrated.

OMVs are constantly being discharged from the surface of Gram-negative bacteria during bacterial growth. All Gram-negative bacteria studied to date, including *E. coli*, produce OMVs; and their release is increased when the bacteria are exposed to stressful conditions such as antibiotics or serum. Even though the release of OMVs could not be demonstrated *in vivo*, the presence of particles resembling those vesicles has, in fact, been detected in plasma from patients with different infectious processes (Beveridge, 1999). OMVs serve as secretory vehicles for the proteins and lipids of Gram-negative bacteria and in this manner play roles in establishing a colonization niche for carrying or transmitting virulence factors into host cells or otherwise modulating the host defense and response, thus acting as well as long-range virulence factors that can protect luminal cargo from extracellular host proteases and so penetrate into tissues more readily than the larger bacteria (Kuehn & Kesty, 2005). In addition to toxin-protein delivery, other roles have been characterized for OMVs—namely, interspecies interaction and communication during multispecies infections plus DNA uptake and transfer (Mashburn-Warren & Whiteley, 2006). In the particular example of HlyA, we have demonstrated that the toxin secreted in this way is transferred to the target cell in a concentrated manner and as such is more hemolytically efficient than the free HlyA (Herlax *et al.*, 2010). Moreover, Balsalobre *et al.* (2006) demonstrated that the HlyA associated with OMVs is protected from the attack of proteases, thus facilitating the survival of the toxin within the adverse medium of a patient's plasma.

2.4 The mechanism of action of HlyA

HlyA belongs to one class of a wide range of host-cell-specific toxins. HlyA acts on a variety of cell types from several species—*e. g.*, red blood cells, embryo and adult fibroblasts, granulocytes, lymphocytes, and macrophages (Cavalieri *et al.*, 1984)—and also binds to and disrupts protein-free liposomes (Ostolaza *et al.*, 1993).

The host environments encountered by the ExPEC are extremely nutrient-poor; and the function of HlyA has generally been thought to be primarily the destruction of host cells, thereby facilitating the release of nutrients and other factors, such as iron, that are critical for bacterial growth. The lytic mechanism of HlyA is a complex process. Three stages seem to be involved that ultimately lead to cell lysis: binding, insertion, and oligomerization of the toxin within the membrane.

Studies that have explored the binding of HlyA to membranes and the characterization of a putative toxin-specific receptor have produced contradictory results. First, the lymphocyte function-associated antigen (LFA-1) (CD11a/CD18; $\alpha_1\beta_2$ integrin), was reported to serve as the receptor for HlyA on polymorphonuclear neutrophils (Lally *et al.*, 1997) and HlyA was found to recognize and bind the N-linked oligosaccharides to their β2-integrin receptors (Morova *et al.*, 2008). This finding raises the possibility that the initial binding of the toxin to various cells might occur through the recognition of glycosylated membrane components, such as glycoproteins and gangliosides. Recently, Cortajarena *et al.* (2001) found that HlyA binds to the glycophorin of horse erythrocytes and that this binding was abolished by a trypsinization of the membranes. In addition, these authors found that the glycophorin purified from erythrocyte ghosts and reconstituted in liposomes significantly increased liposomal sensitivity to HlyA. Amino acids 914-936 of HlyA were subsequently hypothesized to be responsible for binding to the ghost receptor (Cortajarena *et al.*, 2003).

Other studies, however, indicated that the binding of HlyA to cells occurred in a nonsaturable manner and that the toxin did not interact with a specific protein receptor either on granulocytes or erythrocytes (Valeva *et al.*, 2005). Nevertheless, HlyA produces protein-free liposome disruption. Ostolaza *et al.* have reported that HlyA causes the release of fluorescent solutes following a so-called *all-or-none* mechanism. Using large unilamellar vesicles of different lipid compositions, the authors found that the vesicles composed of phosphatidylcholine, phosphatidylethanolamine, and cholesterol at a molar ratio of 2:1:1 were the most sensitive (Ostolaza *et al.*, 1993). These results demonstrated that the presence of a receptor was not necessary for hemolysis to occur. These contradictory findings regarding the presence or absence of a toxin-specific receptor might be related to the different amounts of toxin and/or the different types and animal species of target cells used in the various studies. At all events, the interaction of HlyA with a target-cell membrane devoid of any specific proteinaceous receptor appears to occur in two steps: an initial reversible adsorption of the toxin that is sensitive to electrostatic forces followed by an irreversible membrane insertion (Bakás *et al.*, 1996), (Ostolaza *et al.*, 1997). Studies with the isolated calcium-binding domain of HlyA revealed that that part of the protein may be adsorbed onto the membrane during the early stages of HlyA-membrane interaction (Sanchez-Magraner *et al.*, 2007).

The next step in the hemolytic process is the insertion of the toxin into the membrane. Hyland *et al.* (2001) demonstrated that the major region of HlyA that inserts into the membrane is located between residues 177 and 411. The insertion is furthermore independent of membrane lysis since HlyA-protein mutants that are completely nonlytic can insert into lipid monolayers (Sanchez-Magraner *et al.*, 2006). In addition, a binding of calcium to the toxin was shown to induce a protein conformational change that made the insertion process irreversible (Sanchez-Magraner *et al.*, 2006), (Bakás *et al.* 1998). Once the toxin is inserted, an oligomerization process occurs. We previously found that the fatty acids covalently bound to the toxin induce conformational changes that expose intrinsically disordered regions so as to promote protein-protein interactions. Thus, the oligomerization process of the toxin is facilitated by microdomains within the membrane (Herlax & Bakas, 2007), (Herlax *et al.*, 2009).

The HlyA pore that is formed is highly dynamic because the size depends on both the interaction time and the concentration of the toxin (Welch, 2001). We recently demonstrated that the pore is of a proteolipidic nature since the conductance and membrane lifetime are dependent on membrane composition (Bakas *et al.*, 2006).

Nevertheless, what is not clear is how often HlyA reaches levels that are high enough to lyse host target cells during the course of an infection. In fact, sublytic concentrations of HlyA may even be more physiologically relevant. Indeed, recent studies have demonstrated that sublytic concentrations of a number of pore-forming toxins can modulate a variety of host signaling pathways, including the transient stimulation of calcium oscillations, the activation of MAP-kinase signaling, and the alteration of histone-phosphorylation and -

acetylation patterns (Hamon *et al.*, 2007), (Ratner *et al.*, 2006). In addition, sublytic concentrations of HlyA have been recently found to potently stimulate the inactivation of the serine/threonine protein kinase B (PKB), which enzyme plays a central role in host cell-cycle progression, metabolism, vesicular trafficking, survival, and inflammatory-signaling pathways (Wiles *et al.*, 2008). These findings may help to explain previously published results implicating sublytic concentrations of HlyA in the inhibition of chemotaxis and in bacterial killing by phagocytes in addition to the HlyA-mediated stimulation of host apoptotic and inflammatory pathways (Cavalieri & Snyder, 1982), (Koschinski *et al.*, 2006), (Mansson *et al.*, 2007), (Tran Van Nhieu *et al.*,2004), (Uhlen *et al.*, 2000).

3. Role of the fatty acids covalently bound to HlyA

In general, lipid moieties play central roles in protein function—*e. g.*, the targeting into membranes, an increase in the affinity for biological membranes, and an enhancement of protein-protein interactions (Stanley *et al.*, 1998), (Chow *et al.*, 1992).

After a brief introduction to the general aspects of HlyA in the following section, we will describe the role that covalently bound fatty acids play in the mechanism of action of the toxin, from its initial activation to its final functioning in the target cell. This posttranslational modification must be critical since the presence of fatty acids transforms the innocuous proHlyA into the virulent toxin HlyA.

3.1 Exposure of intrinsically disordered regions

After the initial activation of HlyA by acylation, the toxin is exported into the extracellular medium by the type-I secretion system and by OMVs. None of the secretion routes are acylation-dependent, although the extracellular transport yield was found to be lower for proHlyA compared to that for HlyA. In addition, a high concentration of ProHlyA was found in inclusion bodies (Sanchez-Magraner *et al.*, 2006). For comparative studies where acylated and nonacylated proteins were used, proHlyA was obtained from *E. coli* DH1—it having been transformed by a recombinant plasmid, pSF4000Δ*BamHI*, in whose DNA a portion of the *hlyC* gene had been deleted. This strain secreted a full-length, but inactive hemolysin. Fatty acids were not necessary for the secretion of the toxin by OMVs, or by the bacteria's own export machinery; but they were essential for the toxin's hemolytic activity (Boehm *et al.*, 1990).

Several steps are involved in the lytic mechanism of the toxin: a binding of calcium previous to the toxin's interaction with membranes, the binding to and insertion into membranes, and the oligomerization of the toxin to form the final lytic pore. We will discuss below to what extent covalently bound fatty acids influence the different steps.

In the extracellular medium, HlyA must associate with calcium in order to bind to membranes in the lytically active form (Ostolaza & Goñi 1995), (Bakás *et al.*, 1998). This second activation step is acylation-dependent because the calcium-binding capacity is lower in the unacylated protein (Soloaga *et al.*, 1996). Once HlyA is calcium-activated, the toxin appears to have a two-stage interaction with membranes: first, a reversible adsorption that is sensitive to electrostatic forces; and second, an irreversible insertion (Bakás *et al.*, 1996). The inserted HlyA behaves as an integral protein because this form of the toxin cannot be extracted without the use of detergents (Soloaga *et al.*, 1999).

Nevertheless, proHlyA, though nonacylated, also interacts with membranes. This observation is not surprising because the amino-acid sequence of the polypeptide shows amphipathic helices in the 250–400 amino-acid region. Despite the amphipathic stretches known to be essential for lytic activity, however, proHlyA is unable to alter the bilayer permeability (Soloaga *et al.*, 1999). Experiments on protein adsorption at an air-water interface suggested that the fatty acids present in HlyA, unlike those in proHlyA, did not modify the surface-active properties of the protein and that the main difference between the precursor and the mature protein was that the proHlyA was virtually unable to insert itself into lipid monolayers (Sanchez-Magraner *et al.*, 2006). Furthermore, we found that the presence of two acyl chains in HlyA confers on this protein the property of irreversible binding to membranes, which feature is essential for the lytic process to take place (Herlax & Bakas, 2003). In summary, although fatty acids covalently bound to HlyA help the toxin to bind calcium in order to adopt a competitive conformation for interaction with membranes, the absence of these fatty acids does not modify that interaction of the toxin, so that these fatty acids must play some other relevant role. The answer is that the fatty acids expose intrinsically disordered regions of the toxin that are involved in a different step within the mechanism of action.

HlyA has a molten-globule conformation promoted by the presence of acyl chains, as demonstrated by a lower denaturing concentration of guanidinium-chloride. Other characteristics demonstrating this conformation were the binding of a higher number of 8-anilinonaphtalene-1-sulfonate (ANS) molecules to HlyA with a weaker affinity, a higher efficiency of energy transfer from tryptophan to the bound ANS, and a faster digestion of HlyA with trypsin compared to the same reactions with proHlyA (Herlax & Bakas, 2007).

The acylated protein was more stable in the absence of denaturant than the unacylated form, as demonstrated by the higher $\Delta G°H_2O$ value for HlyA compared to proHlyA. Acyl chains covalently bound to the protein, however, promote a steric hindrance that contributes to a more relaxed structure, which acylated form can thus be denatured at a lower guanidinium-chloride concentration.

ANS binding to ordered regions can be distinguished from the binding to molten-globule-like regions by differences in the apparent binding constant. The exceptionally high value of ANS bound to HlyA and proHlyA might result from amphipathic regions in both forms, but the presence of fatty acids has been observed to double this value because of the molten structure those lipids impart. The binding of a large number of ANS molecules in a weak manner is characteristic of the loose structure of the molten conformation. ANS binding to pockets in ordered or molten-globule proteins operationally gives apparent *Kd* values that differ by more than a factor of 5; thus, despite the uncertainties involved, these apparent *Kd* values serve as a diagnostic probe to distinguish ordered from molten proteins *(*Bailey *et al., 2001).* This structural difference was also observed between the HlyA and proHlyA *Kd* values, demonstrating by an independent means that the fatty acids on the former induce a molten structure. Moreover, the higher fluorescence-transfer efficiency for HlyA compared to that for proHlyA indicated that the quenching of tryptophan fluorescence was more effective when the binding of ANS to the molten-globule conformation took place, where the accessibility of both the surface and inner tryptophan residues was increased. Thus, the capability of ANS to quench tryptophan fluorescence was seen to be correlated with the ANS-binding behavior.

Proteins with molten-globule-like regions are included in the category of intrinsically disordered proteins, as recently reviewed elsewhere (Dunker *et al.*, 2001). Most of the disordered regions of proHlyA that were predicted through the use of the predictor of naturally disordered regions (PONDR) were located in the C-terminal half of the protein (Fig. 2). These domains could be related to the different steps in this toxin's mechanism of action from its export from the bacterium to pore formation in the target cell.

HlyA carries a carboxy-terminal–secretion signal located within the last 50–60 amino acids (Jarchau *et al.*, 1994). This region is predicted to be disordered; and although export of the toxin has been observed to be acylation-independent (Ludwig *et al.*, 1987), as mentioned above, the yield from extracellular transport for proHlyA was lower than that for HlyA. Consequently, covalently bound acyl chains can expose these signal regions and thus facilitate transport.

Intrinsically unstructured proteins can bind in several different patterns through a process termed *binding promiscuity*. The intrinsic lack of structure can confer functional advantages, including the ability to bind—perhaps in various conformations—to several different target cells. This binding promiscuity would furthermore explain the previously mentioned ambiguity in experimental determinations of the presence of a specific receptor for HlyA published to date (Lally *et al.*, 1997), (Cortajarena *et al.*, 2001), (Valeva *et al.*, 2005).

Many studies have searched for the presence of a receptor for HlyA in different target cells. For example, CD11a and CD18, the two subunits of β2-integrin, were identified as cell-surface receptors that mediate HlyA toxicity in the human target cells HL60 (Lally *et al.*, 1997). This receptor was found in most circulating leukocytes (lymphocytes, neutrophils, monocytes, and macrophages). Despite the absence of studies identifying the protein region responsible for the interaction with this receptor, studies on the adenylate-cyclase-containing hemolysin of *Bordetella pertusis* (CyaA)—another RTX toxin—revealed that the main integrin-interacting domain of CyaA is located in its glycine/aspartate-rich repeat region; which stretch is characteristic of all protein members of this family. These results allowed the identification of region 1166–1287 as a major CD11b-binding motif (Azami-El-Idrissi *et al.*, 2003). Because this domain is involved in calcium binding, the authors proposed that CyaA shifts from a disordered structure to an R-helical conformation upon calcium binding to the RTX motifs (Rose *et al.*, 1995); therefore, the speculation that the calcium-binding domain composed of glycine-rich tandem repeats corresponding to amino acids 550–850 of HlyA might be involved in the binding to β2-integrin is tempting. That these regions also match the disordered regions predicted and that acyl chains might be implicated in the exposure since the calcium-binding capacity of proHlyA is lower than that of HlyA, should also be borne in mind (Soloaga *et al.*, 1996).

As cited above in **Section 2.4**, another protein identified as a receptor of HlyA in horse erythrocytes is the glycoprotein glycophorin (Cortajarena *et al.*, 2001). A glycophorin-binding region between residues 914 and 936 accordingly has been identified (Cortajarena *et al.*, 2003). Previous sequence analyses of several RTX toxins had revealed that this stretch was a conserved region. If this region was deleted, the specific binding of HlyA to the cell-surface receptors on erythrocytes was lost without affecting its nonspecific binding (adsorption) to lipid bilayers. This region was also predicted to be intrinsically disordered.

The role of fatty acids in the exposure of disordered regions is supported by results published for the D12-monoclonal-antibody–epitope reactivity. The D12 epitope maps to amino acids 673–726. Since the D12 monoclonal antibody reacts with HlyA, but not with proHlyA; the acylation of the former is directly responsible for the exposure of the epitope within this region (Pellett *et al.*, 1990), (Rowe *et al.*, 1994).

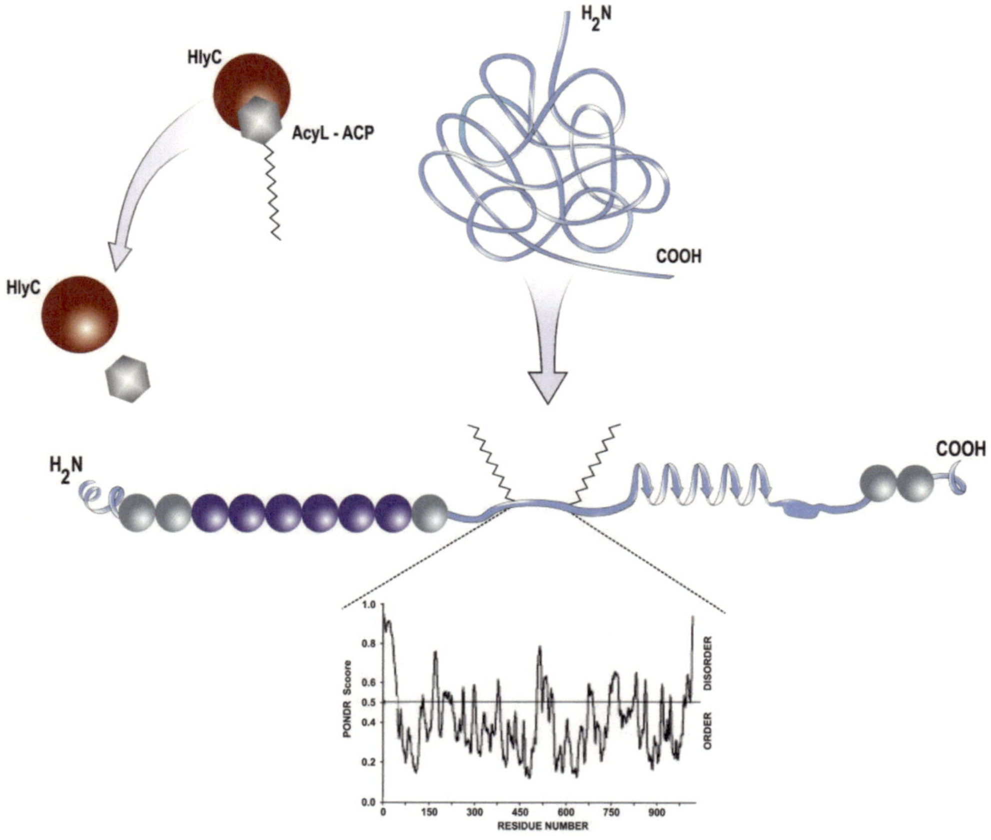

Fig. 2. ProHlyA is activated by acylation.

This process is catalyzed by HlyC, which species transfers a fatty acid from ACP to two internal lysines of ProHlyA (Lys 564 and Lys 690). Once covalently bound, these fatty acids induce a molten-globule conformation in HlyA that exposes intrinsically disordered regions, the existence of which zones was predicted by the predictor of naturally disordered regions PONDR. The amino-acid sequence is represented on the x-axis, and the prediction of disorder on the y-axis. Peaks >0.5 are strongly predicted to be disordered (Dunker *et al.*, 2005).

3.2 Promotion of protein oligomerization

Lipid binding to proteins can also be a determinant of specific protein-protein interactions such as the assembly of proteins into oligomeric complexes. This circumstance obtains for HlyA, where an oligomer was found at lytic concentrations in sheep-erythrocyte ghosts. In contrast, no oligomeric structure was found for proHlyA (Herlax *et al.*, 2009).

Fluorescence-Resonance-Energy Transfer (FRET) is a photochemical process whereby one fluorescent molecule or fluorophore, the "donor", upon excitation by an initial photon of light, spontaneously transfers its energy to another molecule, the "acceptor", by a nonradioactive dipole-dipole interaction (Forster, 1959). The distance over which energy can be transferred depends on the spectral characteristics of the fluorophores, but is generally within the 10–100-Å range. Hence, FRET can be used for measuring structure (Lakowicz *et al.*, 1990), conformational changes (Heyduk, 2002), and interactions between molecules (Parsons *et al.*, 2004). Since HlyA does not contain cysteine residues in its sequences, lysine 344 was replaced by a cysteine (HlyA K344C) and the same point mutation introduced into the unacylated protein (proHlyA K344C). The aim of this point mutagenesis was to permit the binding of only one fluorescent probe per protein, where that mutation—hopefully located in the insertion region of the toxin into membranes (Hyland *et al.*, 2001)—would not affect the hemolytic activity of the toxin. To carry out this study, two populations of HlyA K344C mutant proteins, one labelled with donor (Alexa-488) and the other with acceptor fluorophores (Alexa-546), were bound to sheep-erythrocyte ghosts. Our report showed that an oligomer was involved in the hemolytic mechanism of HlyA (Herlax *et al.*, 2009). FRET can be used to study the distribution of molecules in membranes because the average spacing between molecules of interest will depend primarily on their lateral arrangement. Molecules may be within FRET distance either because they are clustered or because they are randomly distributed at such high surface densities that a fraction of them is within FRET proximity. The latter possibility was avoided in our experiments by using a high lipid/protein molar ratio (10^9) to insure that the observed FRET corresponded to oligomerization of the toxin on the erythrocyte surface. In comparison, the absence of FRET in the mutant protein, proHlyA K344C confirmed the participation of the covalently bound fatty acids in the oligomerization process. Fig. 3 shows the fluorescence spectra obtained in the FRET experiments for both proteins. *Prima facie*, this absence of FRET could be attributed to a reduced binding of the mutant protein to the erythrocyte ghosts, but this possibility was discarded because the percentage of binding to the membranes of both proteins was similar. We need to underscore here that fatty acids are essential for hemolytic activity; and considering that they are needed for oligomerization, we can state that oligomerization is necessary for hemolysis. We thus feel tempted to propose that the presence of fatty acids covalently bound to the protein leads to the exposure of regions that are implicated in protein-protein interactions.

In addition, a critical role of acylation in the oligomerization process to form hemolytic pores has been proposed for the adenylate-cyclase toxin from *Bordetella pertussis* (*cf.* **Section 3.1**) (Hackett *et al.*, 1995).

Finally, if we consider that pores formed by HlyA are sensitive to proteases on the *cis* side of the planar lipid membranes (Menestrina *et al.*, 1987), we could propose the possibility that the part of the toxin remaining external to the membrane is involved in the protein-protein interaction responsible for oligomerization and thus participates in pore formation.

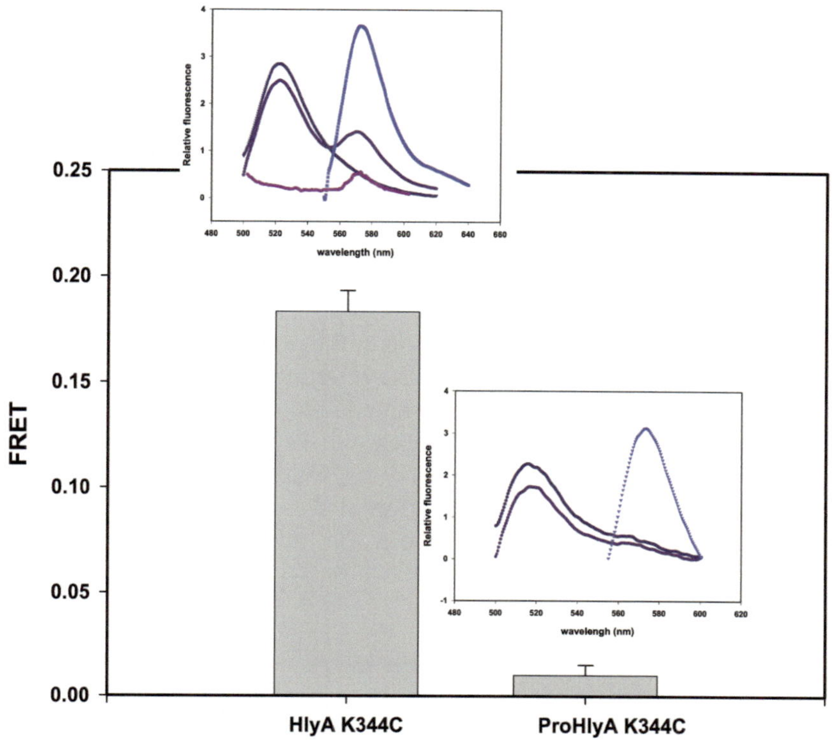

Fig. 3. Oligomerization of HlyA in erythrocyte ghosts. FRET calculated for HlyA K344C and proHlyA K344C bound to erythrocyte ghosts. The lipid/protein ratio was 10^3. *Left inset*: Example of spectra measured for HlyA K344C. Fluorescence-emission spectrum of erythrocyte ghosts containing donor/acceptor, $F_{D/A}^{(480,\lambda em)}$ (excited at 480 nm; *blue triangle*), emission spectrum of erythrocyte ghosts labelled only with donor, $F_D^{(480,\lambda em)}$ (*violet circle*), emission spectrum of erythrocyte ghosts containing D/A, $F_{D/A}^{(530,\lambda em)}$ (excited at 530 nm where only the acceptor absorbs; *light blue square*), and emission spectrum of erythrocyte ghosts labelled only with acceptor, $F_A^{(480,\lambda em)}$ (*purple square*). *Right inset*: The same emission spectrum as in the left inset but measured for ProHlyA K344C.

3.3 Contrary to expectations, fatty acids do not facilitate the interaction of HlyA with membrane microdomains

A variety of pathogens and toxins have been recognized as interacting with microdomains in the plasma membrane. These microdomains are enriched in cholesterol and sphingolipids and probably exist in a liquid-ordered phase, in which lipid acyl chains are extended and ordered (Brown & London, 1998). Many proteins are targeted to these membrane microdomains by their favorable association with ordered lipids. Interestingly, such proteins are linked to saturated acyl chains, which species partition well into those domains (Pike, 2003). Although covalently bound fatty acids had not been implicated in the targeting of HlyA to membranes, their involvement in the targeting to membrane microdomains was studied (Herlax *et al.*, 2009). For this purpose—and taking into account that these microdomains are enriched in cholesterol and sphingolipids—the hemolytic activity of the toxin on sheep erythrocytes was compared with the activity on cholesterol-depleted erythrocytes. The hemolysis rate of the cholesterol-poor erythrocytes was lower than that of the control erythrocytes at each HlyA concentration tested, thus pointing to the participation of cholesterol-enriched microdomains in the oligomerization process. For cholesterol-depleted erythrocytes, at low toxin concentrations, the kinetics of hemolysis seemed to be more complex, suggesting that toxin diffusion in membranes is the rate-limiting step. In order to determine if the decrease in the hemolytic rate observed in the cholesterol-depleted erythrocytes was caused by an impairment of toxin oligomerization, we repeated the FRET experiments comparing control and cholesterol-depleted sheep-erythrocyte ghosts. We demonstrated that cholesterol depletion led to a decrease in FRET of 75% compared to the control sheep ghosts. This result indicated that cholesterol-enriched microdomains played a significant role in the oligomerization process. To obtain more information about the effect of cholesterol-enriched microdomains within the oligomerization process, we performed FRET-kinetics experiments. The role of cholesterol was confirmed by the results of FRET kinetics, where the biphasic behavior of FRET suggested the initial formation of small oligomers, followed by their assembly to form multimeric structures (Fig. 4). The concentration of the small oligomers was favored by the cholesterol-enriched microdomains, where the diffusion time in the membrane became diminished. The number of HlyA molecules that became associated to form the

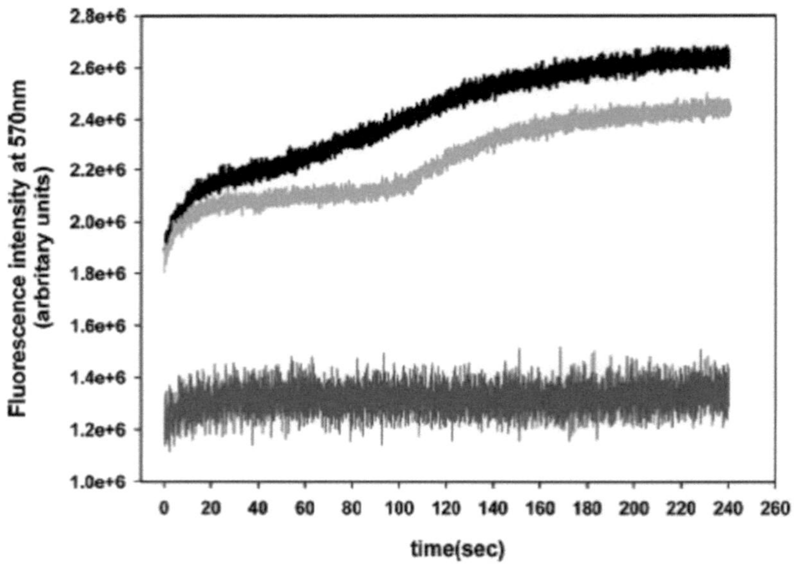

Fig. 4. *FRET kinetics.* Measurement of acceptor fluorescence at 570 nm as a function of time in a mixture composed of HlyA K344C labelled with fluorescent donor and acceptor plus either control erythrocytes (black line) or cholesterol-depleted erythrocytes (light gray line). Measurement of a mixture of unlabelled and labelled with acceptor HlyA K344C with control erythrocytes (*dark gray line*) was done as FRET-negative control. Assays were performed at a ratio of 5 µg of total toxin per 100 µg of phospholipids (as erythrocyte membranes). The excitation and emission monochromators were set at 480 nm and 570 nm, respectively. Alexa-546 emission was measured at a rate of 25 samples/s for 240 s, at 37°C. The *curves* represent the average value of three independent experiments containing five replicates each.

pore was uncertain; nevertheless, the assumption that several molecules could oligomerize to form a pore was not unreasonable. An extension of this reasoning suggested that at high doses a progressive oligomerization of HlyA leads to the fusion of the pore and rapid destruction of the cell membrane with little time for activation of the central apoptotic pathway. By contrast, at lower concentrations, the pores would be smaller and fewer in number so that the cells, though injured, would survive long enough for apoptosis to be observed (Lally *et al.*, 1997). These results can explain why toxin association with erythrocytes at 0–2°C is characterized as a prelytic state, whereas following a shift to 23°C—and after a lag period—lysis begins (Moayeri & Welch, 1997). In conclusion, the fusion of oligomers may be the rate-limiting step in pore formation, and the integrity of the cholesterol-enriched microdomains is necessary for the concentration of HlyA to induce hemolysis. This notion agrees with the findings of Moayeri and Welch (Moayeri & Welch, 1994), who observed that the degree of osmotic protection of erythrocytes afforded by protectants of varying sizes depended on the amount of toxin applied and the duration of the assay. These authors suggested that HlyA creates a lesion with a very small initial size that then increases in apparent diameter over time. Consequently, the larger the oligomer is, the bigger the pore size becomes.

That the terms "membrane microdomains" and "detergent-resistant microdomains" (DRMs) are not synonymous is essential to remember because the two have different origins and conceptual meanings (Lichtenberg *et al.*, 2005). The DRM technique, though, is widely used in the current literature to investigate the interaction between a protein and membrane microdomains. This technique takes advantage of the selective solubilization of different lipids that occurs when a biomembrane is submitted to the action of a nonionic detergent such as Triton X-100. When erythrocyte ghosts were incubated with HlyA and the DRMs were separated by sucrose-gradient ultracentrifugation, the immunoblot analysis revealed that most of the ghost-associated HlyA was localized in the DRMs, indicating that the binding of HlyA to the erythrocyte membranes was mediated by membrane microdomains that served as concentration platforms for the toxin's oligomerization. That proHlyA colocalizes with HlyA and flotillin (a microdomain protein marker) in DRMs emphasizes our hypothesis that the main role of the saturated acyl chain covalently bound to HlyA is a participation in the oligomerization process, and not the targeting to cholesterol-enriched membranes (Fig 5).

A key feature of cholesterol-enriched microdomains is the tight packing of lipid acyl chains in the liquid-ordered phase, where the lipid acyl chains are extended and ordered (Brown & London, 1998). Because of the difficulty in packing membrane-spanning helices into the ordered lipid environment, some proteins are linked to saturated acyl chains and partition well into those microdomains (de Planque & Killian, 2003). Shogomori *et al.* (2005) found, however, that acylation did not measurably enhance microdomain association, and they concluded that the acylated linker for the activation of T-cell transmembrane domains had a low inherent affinity for cholesterol-enriched microdomains. The possible inferrence is that acylation is not sufficient for the targeting of any transmembrane protein and that therefore a second mechanism—such as protein-protein interactions for microdomain associations—is required (Fragoso *et al.*, 2003), (Cherukuri *et al.*, 2004).

To conclude, we propose that fatty acids covalently bound to HlyA and membrane microdomains are implicated in the hemolysis process. Fatty acids are essential because they induce the exposure of intrinsic disordered regions in the toxin so as to enhance protein-protein interactions in order to form the oligomer, while the membrane microdomains act as platforms for the concentration of the toxin during the oligomerization process.

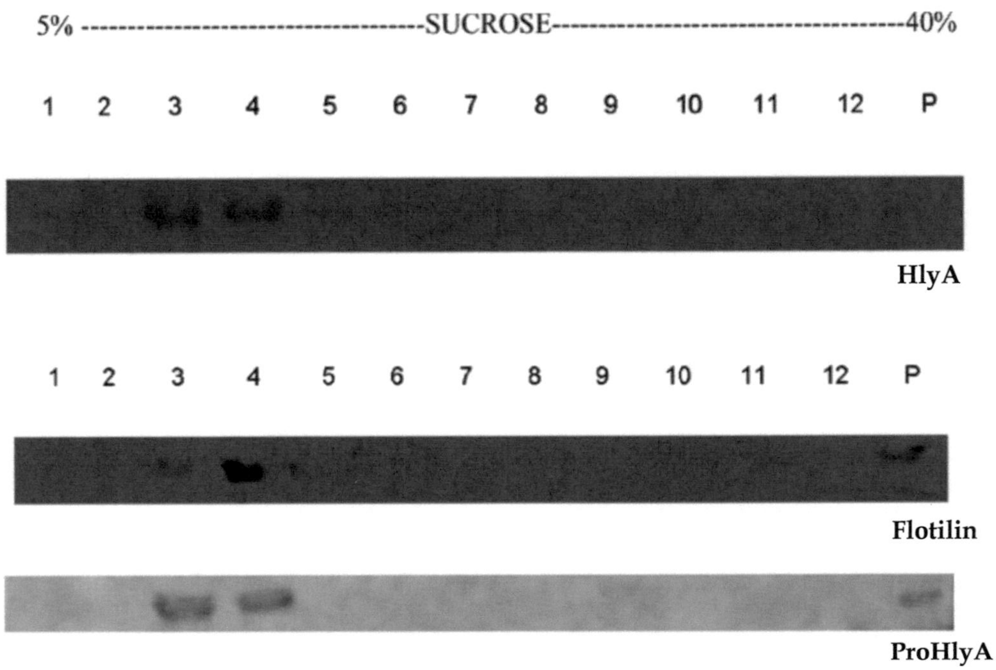

Fig. 5. *Interaction of HlyA with DRMs.*Thirty μg of HlyA were incubated with 100 μl of erythrocyte ghosts for 30 min at 37 °C. Cells were lysed with 1% (v/v) Triton X-100 and insoluble cell components separated by sucrose-density-gradient centrifugation. The gradient fractions were analyzed by immunoblotting with anti-HlyA antibodies. HlyA was present in fractions 3 and 4. Gradient fractions were also analyzed by immunoblotting with anti-Flotillin-1 antibodies. Flotillin-1 appears mainly in fractions 3 and 4. ProHlyA was incubated under the same conditions as HlyA. The gradient fractions were analyzed by immunoblotting with anti-HlyA antibodies. ProHlyA colocalizes with HlyA.

In summary, fatty acids covalently bound to HlyA induce a molten-globule structure in the toxin, exposing intrinsically disordered regions involved in the different steps in the toxin's mechanism of action. Fatty acids expose specific regions that induce protein-protein interaction in the oligomerization process that takes place within the membrane microdomains of erythrocytes. The irreversibility of the toxin's membrane binding promoted by fatty acids might result from the formation of the HlyA oligomeric structure (Herlax & Bakas, 2003,2007), (Herlax *et al.*, 2009).

4. Toxin-based theraphy

Bacterial toxins have been defined as "soluble substances that alter the normal metabolism of host cells with deleterious effects on the host" (Schlessinger & Schaechter, 1993). Nonetheless, during the last decade, taking advantage of advances in toxin research, investigators have sought ways of obtaining benefits from toxins. In the present section we will discuss these toxin-based therapies and the possible relevant use of HlyA.

4.1 Immunotoxins

Conventional cancer treatments such as surgery, chemotherapy, and radiotherapy often fail to achieve complete cancer remission. Moreover, radiotherapy and/or chemotherapy are almost always cause significant—and sometimes long-lasting—side effects. These considerations have prompted the development of many new approaches for the treatment of cancer. One such example involves the use of immunotoxins (Bernardes *et al.*, 2010).

The term "immunotoxin" classically refers to chimeric proteins with a cell-selective ligand chemically linked or genetically fused to a toxin moiety that can target cancer cells overexpressing tumor-associated antigens, membrane receptors, or carbohydrate antigens. In the 1970s the first therapeutic agents composed of toxins conjugated to antibodies against cell-surface antigens started to emerge as tumor-cell killers (Moolten & Cooperband, 1970), (Moolten *et al.*, 1976). Since then, many hybrid molecules consisting of a toxin coupled to a specific targeting antibody or ligand were developed, with most of these hybrids being directed against tumor cells (Pastan *et al.*, 2007).

First-generation immunotoxins were prepared by chemically conjugating antibodies to natural intact toxin units or to toxins with attenuated cell-binding capability. These constructs, however, were heterogeneous and nonspecific both because multiplicities of potential sites were available for chemical conjugation and since the presence of the cell-binding domain on the toxin led to an intoxication of nontumor cells as well. Immunotoxins of the second generation were also based on chemical conjugation between the targeting moiety and the toxin. Nevertheless, accumulated knowledge on the structure and function of the toxins enabled the removal of their native nonspecific cell-binding domain, thus generating immunotoxins that were much more target-specific when conjugated to monoclonal antibodies. Although more specific and thus better tolerated by animals, immunotoxins from this second generation were still chemically heterogeneous, and their large size hindered them from penetrating solid tumors. In order to avoid this heterogeneity, improve tumor penetration, and reduce production complexity and expense, recombinant-DNA techniques were applied in the production of the third-generation immunotoxins. In these constructs—mostly produced in the bacterium *E coli*—the cell-binding domain of the toxin is genetically replaced with a ligand or with the Fv portion of an antibody in which the immunoglobilin light- and heavy-chain variable regions are either genetically linked or held together by a disulfide bond (Shapira & Benhar, 2010).

Among the bacterial toxins that were used for the construction of immunotoxins, the most common were the diphtheria toxin and the pseudomonas exotoxin A, which toxins are naturally produced by the Gram-positive, aerobic *Corynebacterium diphtheria* and by the Gram-negative, aerobic *Pseudomonas aeruginosa*, respectively. Clinical trials with different exotoxin A-immunotoxins have already been performed with positive results in leukemia and bladder cancer (Kreitman *et al.*, 2001), (Kreitman *et al.*, 2005), (Biggers & Scheinfeld, 2008).

In spite of the promise shown by bacterial toxin-based chimeric proteins, these hybrids still present several obstacles that limit their clinical application. The toxin part of the fusion proteins elicits a high degree of humoral response in humans. In addition, in developed countries, where people have become immunized against diphtheria, the patient's serum will have circulating antibodies against the diphtheria toxin that will result in a neutralization of diphtheria toxin-based immunotoxins (Hall *et al.*, 2001). Both the *Pseudomonas* exotoxin and the diphtheria toxin are large molecules and are difficult to humanize. At sufficiently high concentrations these fusion proteins lead to symptoms like the vascular-leak syndrome and thus exhibit a certain degree of nonspecific toxicity.

HlyA as a possible candidate toxin for the synthesis of immunotoxins

Considering all the details of the structure and mode of action of HlyA discussed above, we can state that HlyA can be a good candidate for an effective immunotoxin. Although, certain additional details about the domains implicated in the binding of the toxin to target-cell receptors need to be clarified, we can consider that the domain that comprises amino acids 914–936 should be exchanged for the specific antibody (or ligand) chosen for interaction with the tumor cell of interest. Perhaps some amino acids within the repeat domain that might be involved in the interaction with the ß2-integrins should also be removed in the fusion protein.

The reason for using HlyA in an immunotoxin

An observation deserving emphasis is that the more relevant effects that HlyA produces during an infection are sublytic rather than cytolytic. Among these effects we must bear in mind the one related to the modulation of the host-signaling cascades, where HlyA was found to produce an inactivation of the PKB (*cf.* **Section 2.4**)—a key protein involved in several pathways related to host-cell survival, inflammatory responses, proliferation, and metabolism (Manning & Cantley, 2007), (Fayard *et al.*, 2005). By inactivating PKB, HlyA is able to fine-tune host responses related to the inflammatory- and apoptosis-signaling cascades that are initiated during the course of an infection. PKB inactivation is produced by an extracellular calcium-dependent, potassium-independent process requiring HlyA insertion into the host plasma membrane and subsequent pore formation. Calcium influx induces the activation of host-protein phosphatases that dephosphorylate PKB, inactivating it and finally inducing host-cell apoptosis. Thus, if a ligand directed at a specific tumor-cell receptor is fused with HlyA, that immunotoxin might induce the apoptosis of the desired cell.

The advantage of using HlyA is that the translocation of the immunotoxin into a tumor cell is not necessary, only its binding to the membrane where HlyA can insert itself and form the pore needed to execute its apoptotic action. Of course, these hypotheses are only possibilities that would warrant further investigation.

4.2 Vaccines

In recent years, an increase in the development of vaccination technology has taken place, but the ideal vaccine has not yet been found. In general terms, there are certain criteria that a vaccine must satisfy: it must be capable of eliciting the appropriate immune response; and it should be safe, stable, and reproducible (Perrie *et al.*, 2008).

UTIs caused by UPEC still represent an enormous challenge for the development of vaccines targeted to induce an immunity that can either prevent the infectious agent from attaching to and colonizing the mucosal epithelium and/or can block the binding and action of microbial toxins, such as HlyA (Holmgren *et al.*, 2003). Since the infections by these bacteria occur at, or take their departure from, a mucosal surface; a mucosal route of vaccination should be selected rather than a parenteral one.

A significant aspect of immune responses at mucosal surfaces is the production of a secretory IgA (S-IgA) and its transport across the epithelium. This S-IgA response represents the first line of defence against the invasion by bacterial pathogens. The mucosal immune system is an integrated network of tissues, cells, and effector molecules that functions to protect the host from those pathogens. Furthermore, mucosal lymphocytes exhibit unique homing receptors, the integrins, that recognize ligands expressed on mucosal endothelial cells so as to allow their retention within mucosal tissues for the delivery of cellular and humoral immune responses (Butcher & Picker, 1996). Because of the presence of specific interconnected mucosal induction and effector sites for eliciting the S-IgA antibody response, the mucosal immune system has been shown to be separated from the peripheral immune system. Thus, the induction of peripheral immune responses by parenteral immunization does not necessarily result in significant mucosal immunity; by contrast, mucosal immunization is capable of inducing immune protection in both the external secretions and the peripheral immune compartments (Kiyono *et al.*, 1992), (McGhee *et al.*, 1992).

The induction of immune responses following mucosal immunization is usually dependent upon the coadministration of the appropriate adjuvants that can initiate and support the transition from innate to adaptive immunity. While a number of substances of bacterial origin have been tested as mucosal adjuvants, the most widely used mucosal adjuvants in experimental animals are the cholera toxin (CT) and the closely related *E. coli* heat-labile enterotoxin (LT). Both CT and LT consist in homopentamers of cell-binding B subunits associated with a single toxically active A subunit. The A subunit enzymatically ADP-ribosylates the Gs protein of adenylate cyclase leading to an increased cAMP production in the affected cells (de Haan & Hirst, 2000). CT and LT can alter several steps in the induction of a mucosal immune response. These effects, which alone or in combination might explain their strong adjuvant action after oral immunization, include: (1) an increased permeability of the intestinal epithelium leading to an enhanced uptake of a coadministered antigen, (2) an enhanced antigen presentation by various antigen-presenting cells, (3) a promotion of isotype differentiation in B cells leading to increased IgA formation, and (4) complex stimulatory as well as inhibitory effects on T-cell proliferation and cytokine production. Finally, CT and LT have been shown not only to avoid an induction of oral tolerance but also to abrogate additional otherwise efficient regimens producing tolerance induction after oral antigen administration (Holmgren *et al.*, 2005).

A number of studies have examined the activity of LT and CT as mucosal adjuvants in vaccines against a variety of bacterial, fungal, and viral pathogens. Representative examples include the tetanus toxoid (Xu-Amano *et al.*, 1993), (Yamamoto *et al.*, 1996), (Cheng *et al.*, 1999), (Xu-Amano *et al.*, 1994), the inactivated influenza virus (Hashigucci *et al.*, 1996), (Tumpey *et al.*, 2001), a recombinant urease from *Helicobacter* spp. (Lee *et al.*, 1995), (Weltzin *et al.*, 1997), (Lee, 2001), and the pneumococcal surface protein A from *S. pneumoniae* (Wu *et al.*,1997). Many other examples have been reported, and all these studies clearly indicate that both LT and CT have significant potential for use as adjuvants for mucosally administered antigens. Nevertheless, both LT and CT are potent enterotoxins, and this property has seriously limited the practical use of these molecules (Freytag & Clements 2005). To avoid such toxicity, a number of nontoxic mutant derivatives of CT or LT have been proposed (Douce G, 1997) (Douce *et al.* 1998). In particular, single-amino-acid-substitution mutants of LT (R7K, S63K and R192G) that lack ADP-ribosyltransferase activity have been shown to retain their adjuvant properties (Douce et al. 1995).

In contrast, because of their size, plasticity, and safety profile in humans, OMVs are attractive vehicles for vaccine delivery. OMV vaccines for serogroup-B meningococcal disease, consisting in vesicles from *Salmonella thyphimurium* and *Pseudomonas aeruginosa* that contain surface antigens native to the pathogens have been shown to exhibit immunogenic properties (Alaniz *et al.*, 2007), (Bauman & Kuehn, 2006). Two vaccines for serogroup-B meningococcal disease currently exist that are formulations comprising bacterial surface antigens that have been naturally incorporated into OMVs (Oster *et al.*, 2005), (Feiring *et al.* , 2006). These OMV-based vaccines represent a novel system where both the antigen and delivery vehicle are derived from the *Neisseria meningitidis* pathogen itself (Claassen *et al.*, 1996), (Arigita *et al.*, 2004). Moreover, genetically engineered OMVs offer an attractive possibility for use as easily purified vaccine-delivery systems capable of greatly enhancing the immunogenicity of low-immunogenicity protein antigens without the need for an added adjuvant.

With the development of controlled-release technologies, the engineering of OMVs emerged as a promising strategy for antigen delivery because these vesicles are similar in geometry to naturally occurring pathogens and are readily internalized by antigen-presenting cells, thus avoiding the complex manufacturing steps required to purify and encapsulate antigens into particulate delivery systems such as polymer particles (Singh *et al.*, 2007), immune-stimulating complexes (Morein *et al.*, 1984), liposomes, proteosomes, and related vesicles (Lowell *et al.*, 1988), (Lowell *et al.*, 1988), (Felnerova *et al.*, 2004), (Copland *et al.*, 2005)—all of which processes render these approaches economically unfeasible (Ulmer *et al.*, 2006).

The genetic fusion of the green-fluorescent protein (GFP) as a model subunit antigen with the bacterial hemolysin ClyA resulted in a chimeric protein that elicited strong anti-GFP antibody titers in immunized mice, whereas immunization with GFP alone elicited no such titers. Similar to native unfused ClyA, the chimeric ClyA-fusion proteins were found localized in bacterial OMVs, where they retained the activity of the fusion partners, thus demonstrating for the first time that ClyA can be used to colocalize fully functional heterologous proteins directly in bacterial OMVs. The anti-GFP humoral response in mice immunized with the engineered OMV formulations was indistinguishable from the response to the purified ClyA-GFP fusion protein alone and was equal to the response to purified proteins adsorbed to aluminum hydroxide, a standard adjuvant. Engineered OMVs containing ClyA-GFP were easily isolated by ultracentrifugation, thus effectively eliminating the need for a laborious antigen purification from cell-culture expression systems (Chena *et al.*, 2010). The retention of hemolytic-protein activity indicated that ClyA-antigen fusions maintained their conformations. Although no pathologic effects were observed in mice immunized with ClyA, a detoxification of the toxin through mutation, truncation, or chemical methods may attenuate any possible toxicity while still retaining the hybrid's immunomodulatory capabilities.

On the basis of all these data, HlyA presents many properties that can be considered when designing a vaccine.

Anti- UPEC vaccine: The urinary tract is one of the most common sites of bacterial infection. As mentioned above, over half (53%) of all women along with 14% of men experience at least one UTI in their lifetime (Griebling, 2005), (Griebling, 2005). *E. coli* is the infectious agent in more than 80% of the uncomplicated UTIs (Marrs *et al.*, 2005), (Foxman & Brown, 2003). In addition, the upper UTIs of young children can cause permanent kidney damage. An estimated 57% of children with acute pyelonephritis develop renal scarring (Lin *et al.*, 2003).

In recent years, an increase in the antibiotic resistance of UPEC isolates has been observed (Bours *et al.*, 2010) that imposed an urgent need for alternative treatment and prevention strategies to combat this serious and widespread human pathogen. With this aim, much research has been focussed on the development of vaccines to stimulate protective immunity against UPEC. In those studies, surface-exposed molecules such as P fimbriae, the lipopolysaccharide core, α-hemolysin, and the salmochelin receptor IroN have been utilized as antigens for subunit vaccines (Goluszko *et al.*, 2005), (Russo *et al.*,2003), (O'Hanley *et al.*, 1991); but the limited success of these strategies prevented any vaccine from being currently available. One consideration is that this vaccine has to generate immune responses at the level of mucosal surfaces.

Large-scale reverse-vaccinology approaches offer an alternative to the traditional vaccine design through applying genomic and bioinformatic methods to identify novel vaccine targets (Pizza *et al.*, 2000). Using this technique, Alteri *et al* identified a class of molecules involved in iron acquisition as vaccine candidates and reported that intranasal immunization with this UPEC outer-membrane iron receptor generated an antigen-specific humoral response to provide protection from UTI (Alteri *et al.*, 2009). The authors proposed that the targeting of an entire class of molecules instead a single protein would permit the identification of components of a more generally protective UTI vaccine and that this strategy could be used in the development of vaccines to prevent infections caused by other pathogenic bacteria. During this present year, these same authors, using the *in-vivo*-induced-antigen technology, identified a novel UPEC virulence factor (*tosA*, a gene encoding a predicted repeat-in-toxin family member) that could be useful as a potential vaccine target (Vigil *et al.*, 2011). Although this methodology did not identify HlyA as a potential candidate for this vaccine, the introduction of that toxin would be beneficial. First of all, HlyA has been recognized as one of the main virulence factors associated with the pathogenicity caused by UPEC (Wiles, Kulesus and Mulvey 2008); second, the toxin induced an immunity response in host organisms (O'Hanley *et al.*, 1991) and thus is immunogenic in its native state; third, it can also produce focal leaks in intestinal epithelia (Troeger *et al.*, 2007). Focal leaks are small openings within the epithelium where bacterial penetration occurs. HlyA induces such focal leaks in a proinflammatory environment—those being also induced by the secretion of the cytokines TNFα and IL-13. Of relevance to highlight is that HlyA can increase the permeability of the intestinal epithelium so as to lead to an enhanced uptake of a coadministered antigen, thus acting as both a coadjuvant and an antigen in its own right. The dose that induces this effect would naturally have to be extensively investigated.

HlyA can also be used as adjuvants in any other vaccine design against another pathogen. For example, the toxin can be included in any liposomal vaccine in order to facilitate uptake though epithelia for the induction of immunity.

OMV vaccines: Balsalobre *et al* (2006) demonstrated that physiologically active HlyA is associated with the OMVs produced from *E. coli* laboratory strains and also from natural and clinical isolates. In our laboratory, we found that the unacylated toxin (proHlyA) can also be associated with OMVs (Herlax *et al.*, 2010). On the basis of this finding, OMV vaccines can be designed by effecting a fusion of the desired antigen with ProHlyA. In this way, ProHlyA would direct the exposure of the antigen on the surface of the OMVs without inducing any cytotoxic response. An advantage of OMV vaccines is that, because of their size and lipopolysaccharide content, they are able to induce an adequate immune response.

Finally, mention must be made that these hypotheses are just speculative on the basis of what is known about the structure and function of HlyA, whose application in toxin-based therapy still has to be exhaustively investigated and especially the immune response the toxin might evoke.

5. Conclusion

E. coli is one of the predominant species of facultative anaerobes in the human gut and in the majority of the cases is harmless to the host. These strains are mostly commensals but also contain a group called the extraintestinal pathogenic *E. coli* (ExPEC). Usually the ExPEC are also harmless colonizers but under certain circumstances can translocate and cause infection. The main virulence factor responsible for this translocation is the HlyA toxin, which pathogen is mainly associated with severe UTI but in addition with bacteremia and extraintestinal infections. In this chapter an exhaustive description of the toxin has been delineated; including its synthesis, maturation, and export from the bacteria. Effects produced by HlyA in different target organs have also been discussed. The significance of the maturation process for the toxin cannot be understated. The acylation of the protein at two internal lysines gives the toxin its virulence, by exposing intrinsic disordered regions that are essential to different steps of the toxin's mechanism of action. The further exposure of regions involved in the protein-protein interaction within the oligomerization process is responsible for the permeability induced in all the target cells, despite the intracellular signal pathway the toxin induces in each specific organ. This activation is unique to prokaryotic proteins.

Based on the already known structural and functional characteristics of HlyA, we might speculate about its use in toxin-based therapy. Such therapy is a versatile and dynamic research area with a great potential application. Further investigation, however, is required in order to improve the efficiency and safety of toxin-based agents. Investments in the development of delivery and targeting techniques are definitely needed in order to achieve this goal, though the basic research on the structure and mechanism of natural toxins should nevertheless not be abandoned. Topics related to HlyA have still to be clarified concerning the existence of a toxin-specific receptor in target cells and the domains of the toxin involved in its interaction with those putative binding sites. The deeper our knowledge becomes about this unique family of secreted polypeptides, the more easily will we be able to harness their great potential for our own benefit.

6. Acknowledgements

We thank Prof. Norma Tedesco for revising the English grammar; Mario Ramos for the graphic designs; and Dr. Donald F. Haggerty, a retired career biochemist and native English speaker, for editing the final version of the manuscript.

This work was supported by grants from the Comisión de Investigaciones Científicas de la Provincia de Buenos Aires, and Agencia Nacional de Promoción Científica Grant PICT N° 647, Argentina.

7. References

Alaniz, R. C., B. L. Deatherage, J. C. Lara and B. T. Cookson (2007). "Membrane vesicles are immunogenic facsimiles of Salmonella typhimurium that potently activate dendritic cells, prime B and T cell responses, and stimulate protective immunity *in vivo*." *J Immunol* 179: 7692–7701.

Alteri, C. J., E. C. Hagan, K. E. Sivick, S. N. Smith and H. L. Mobley (2009). "Mucosal immunization with iron receptor antigens protects against urinary tract infection." *PLoS Pathog.* 5(9): e1000586

Arigita, C., W. Jiskoot, J. Westdijk, C. van Ingen, W. E. Hennink, D. J. Crommelin and G. F. Kersten (2004). "Stability of mono- and trivalent meningococcal outer membrane vesicle vaccines." *Vaccine* 22: 629–642.

Azami-El-Idrissi, M. E., C. Bauche, J. Loucka, R. Osicka, P. Sebo, D. Ladant and C. Leclerc (2003). " Interaction of *Bordetella pertussis* Adenylate Cyclase with CD11b/CD18 : Role of Toxin acylation and identification of the main Integrin interaction domain." *J. Biol. Chem* 278: 38514-38521

Bailey, M., V. Koronakis, T. Schmoll and C. Hughes (1992). "*Escherichia coli* HlyT protein, a transcriptional activator of haemolysin synthesis and secretion, is encoded by *rfa*H (*srf*B) locus requierd for expression of sex factor and lipopolisaccharide genes." *Mol Microbiol.* 6: 1003-1012

Bailey, M. J., C. Hughes and V. Koronakis (1996). "Increased distal gene transcription by the elongation factor RfaH, a specialized homologue of NusG." *Mol Microbiol* 22(4): 729-37

Bailey, M. J., C. Hughes and V. Koronakis (1997). "RfaH and the ops element, components of a novel system controlling bacterial transcription elongation." *Mol Microbiol* 26(5): 845-51

Bailey, R., A. Dunker, C. Brown, E. Garner and G. M (2001). "Clusterin, a Binding Protein with a Molten Globule-like Region." *Biochemistry* 40: 11818-11840

Bakas, L., A. Chanturiya, V. Herlax and J. Zimmerberg (2006). "Paradoxical lipid dependence of pores formed by the *Escherichia coli* alpha-hemolysin in planar phospholipid bilayer membranes." *Biophys J* 91(10): 3748-55

Bakás, L., H. Ostolaza, W. L. Vaz and F. M. Goñi (1996). "Reversible adsorption and nonreversible insertion of *Escherichia coli* alpha-hemolysin into lipid bilayers." *Biophys J.* 71(4): 1869-1876

Bakás, L., M. Veiga, A. Soloaga, H. Ostolaza and F. Goñi (1998). "Calcium-dependent conformation of *E. coli* alpha-haemolysin. Implications for the mechanism of membrane insertion and lysis." *Biochim Biophys Acta* 1368(2): 225-234

Balsalobre, C., J. M. Silvan, S. Berglund, Y. Mizunoe, B. E. Uhlin and S. N. Wai (2006). "Release of the type I secreted alpha-haemolysin via outer membrane vesicles from *Escherichia coli.*" *Mol Microbiol* 59(1): 99-112

Bauer, M. E. and R. A. Welch (1996). "Association of RTX toxins with erythrocytes." *Infec Immun* 64(11): 4665-4672

Bauman, S. J. and M. J. Kuehn (2006). "Purification of outer membrane vesicles from *Pseudomonas aeruginosa* and their activation of an IL-8 response." *Microbes Infect* 8: 2400–2408.

Bernardes, N., R. Seruca, A. Chakrabarty and A. M. Fialho (2010). "Microbial-based therapy of cancer: Current progress and future prospects" *Bioengineered Bugs* 1(3): 178-190

Beveridge, T. J. (1999). "Structures of Gram-negative cell walls and their derived membrane vesicles." *J .Bacteriol* 181: 4725-4733

Biggers, K. and N. Scheinfeld (2008). "VB4-845, a conjugated recombinant antibody and immunotoxin for head and neck cancer and bladder cancer." *Curr Opin Mol Ther* 10: 176-186

Boehm, D., R. Welch and I. Snyder (1990). "Domains of *Escherichia coli* hemolysin (HlyA) involved in binding of calcium and erythrocyte membranes." *Infect Immunity* 58(6): 1959-1964

Bours, P. H., R. Polak, A. I. Hoepelman, E. Delgado, A. Jarquin and A. J. Matute (2010). " Increasing resistance in community-acquired urinary tract infections in Latin America, five years after the implementation of national therapeutic guidelines." *Int. J. Infect. Dis.* 4: e770–e774

Brown, D. A. and E. London (1998). "Structure and origin of ordered lipid domains in biological membranes." *J Membr Biol* 164: 103-14

Butcher, E. C. and L. J. Picker (1996). "Lymphocyte homing and homeostasis." *Science* 272: 60-66.

Cavalieri, S., G. A. Bohach and I. Synder (1984). "*Escherichia coli* _-hemolysin characteristics and probable role in pathogenicity." *Microbiol. Rev* 48: 326-343

Cavalieri, S. J. and I. S. Snyder (1982). "Effect of *Escherichia coli* alpha-hemolysin on human peripheral leukocyte viability in vitro." *Infect Immun* 36(2): 455-61

Chen, M., R. Tofighi, W. Bao, O. Aspevall, T. Jahnukainen, L. E. Gustafsson, S. Ceccatelli and G. Celsi (2006). "Carbon monoxide prevents apoptosis induced by uropathogenic *Escherichia coli* toxins." *Pediatr Nephrol* 21(3): 382-9

Chena, D. J., N. Osterriederb, S. M. Metzgerb, E. Bucklesd, A. M. Doodye, M. P. DeLisaa and D. Putnam (2010). "Delivery of foreign antigens by engineered outer membrane vesicle vaccines." *PNAS* 107(7): 3099–3104

Cheng, E., L. Cardenas-Freytag and J. D. Clements (1999). "The role of cAMP in mucosal adjuvanticity of *Escherichia coli* heat-labile enterotoxin (LT)." *Vaccine* 18: 38–49.

Cherukuri, A., T. Shoham, H. W. Sohn, S. Levy, S. Brooks, R. Carter and S. K. Pierce (2004). "The tetraspanin CD81 is necessary for partitioning of coligated CD19/CD21-B cell antigen receptor complexes into signaling-active lipid rafts." *J. Immunol.* 172: 370–380

Chow, M., C. J. Der and J. E. Buss (1992). "Structure and biological effects of lipid modifications on proteins." *Curr Opin Cell Biol* 4(4): 629-36

Claassen, I., J. Meylis, P. van der Ley, C. Peeters, H. Brons, J. Robert, D. Borsboom, A. van der Ark, I. van Straaten, P. Roholl, B. Kuipers and J. Poolman (1996). "Production, characterization and control of a *Neisseria meningitidis* hexavalent class 1 outer membrane protein containing vesicle vaccine." *Vaccine* 14: 1001–1008.

Copland, M. J., T. Rades, N. M. Davies and M. A. Baird (2005). "Lipid based particulate formulations for the delivery of antigen." *Immunol Cell Biol* 83: 97–105

Cortajarena, A., F. Goñi and H. Ostolaza (2001). "Glycophorin as a receptor for *Escherichia coli* α-hemolysin in erythrocytes." *J. Biol. Chem* 276(16): 12513-12519

Cortajarena, A. L., F. M. Goñi and H. Ostolaza (2003). "A receptor- binding region in *Escherichia coli* _-haemolysin." *J.Biol.Chem.* 278(21): 19159-19163

Cross, M. A., V. Koronakis, P. L. Stanley and C. Hughes (1990). "HlyB-dependent secretion of hemolysin by uropathogenic *Escherichia coli* requires conserved sequences flanking the chromosomal hly determinant." *J Bacteriol* 172(3): 1217-24

de Haan, L. and T. R. Hirst (2000). "Cholera toxin and related enterotoxins: a cell biological and immunological perspective." *J Nat. Toxins* 9: 281–97.

de Planque, M. R. and J. A. Killian (2003). "Protein-lipid interactions studied with designed transmembrane peptides: role of hydrophobic matching and interfacial anchoring." *Mol. Membr. Biol.* 20: 271–284

Dhakal, B. K., R. R. Kulesus and M. A. Mulvey (2008). "Mechanisms and consequences of bladder cell invasion by uropathogenic *Escherichia coli*." *Eur J Clin Invest* 38 Suppl 2: 2-11

Douce G, F. M., Pizza M, Rappuoli R, Dougan G. (1997). " Intranasal immunogenicity and adjuvanticity of site-directed mutant derivatives of cholera toxin." *Infect Immun* 65: 2821-2828.

Douce, G., M. M. Giuliani, V. Giannelli, M. Pizza, R. Rappuoli and G. Dougan (1998). "Mucosal immunogenicity of genetically detoxified derivatives of heat labile toxin from *Escherichia coli*." *Vaccine* 16: 1065-1073

Douce, G., C. Turcotte, I. Cropley and e. al. (1995). "Mutants of *Escherichia coli* heat-labile toxin lacking ADP-ribosyltransferase activity act as nontoxic, mucosal adjuvants." *Proc Natl Acad Sci USA* 92: 1644-1648.

Dunker, A. K., M. S. Cortese, P. Romero, L. M. Iakoucheva and V. Uversky (2005). "Flexible nets. The roles of intrinsic disorder in protein interaction networks." *FEBS J.* 272: 5129-5148.

Dunker, K., J. D. Lawson, R. Brown, P. Williams and J. Romero (2001). "Intrinsically disordered protein." *J. Mol. Graph. Model* 19: 26-59

Fayard, E., L. A. Tintignac, A. Baudy and B. A. Hemmings (2005). "Potein kinase B/Akt at a glance." *J.Cell Sci.* 118: 5675-5678

Feiring, B., J. Fuglesang, P. Oster, L. M. Naess, O. S. Helland, S. Tilman, E. Rosenqvist, M. A. Bergsaker, H. Nøkleby and I. S. Aaberge (2006). "Persisting immune responses indicating long-term protection after booster dose with meningococcal group B outer membrane vesicle vaccine." *Clin Vaccine Immunol* 13: 790–796

Felmlee, T., S. Pellet and R. Welch (1985). "Nucleotide sequence of en *Escherichia coli* chromosomal hemolysin." *J.Bacteriol.* 163(1): 94-105

Felmlee, T., S. Pellett, E. Y. Lee and R. A. Welch (1985). "*Escherichia coli* hemolysin is released extracellularly without cleavage of a signal peptide." *J Bacteriol* 163(1): 88-93

Felmlee, T. and R. A. Welch (1988). "Alterations of amino acid repeats in the *Escherichia coli* hemolysin affect cytolytic activity and secretion." *Proc Natl Acad Sci U S A* 85(14): 5269-73

Felnerova, D., J. F. Viret, R. Glück and C. Moser (2004). "Liposomes and virosomes as delivery systems for antigens, nucleic acids and drugs." *Curr Opin Biotechnol* 15: 518–529.

Forster, T. (1959). "Transfer mechanisms of electronic excitation." *Discuss. Faraday Soc.* 27: 7-17

Foxman, B. (2002.). "Epidemiology of urinary tract infections: incidence, morbidity, and economic costs." *Am. J. Med.* 113: 5S–13S

Foxman, B. and P. Brown (2003). "Epidemiology of urinary tract infections: transmission and risk factors, incidence, and costs." *Infect Dis Clin North Am* 17(2): 227-41

Foxman, B., S. D. Manning, P. Tallman, R. Bauer, L. Zhang, J. S. Koopman, B. Gillespie, J. D. Sobel and C. F. Marrs (2002). "Uropathogenic *Escherichia coli* are more likely than commensal *E. coli* to be shared between heterosexual sex partners." *Am. J. Epidemiol.* 156: 1133–1140.

Fragoso, R., D. Ren, X. Zhang, M. W. Su, S. J. Burakoff and Y. J. Jin (2003). "Lipid raft distribution of CD4 depends on its palmitoylation and association with Lck, and evidence for CD4-induced lipid raft aggregation as an additional mechanism to enhance CD3 signaling." *J. Immunol.* 170: 913–921

Freytag, L. C. and J. D. Clements (2005). "Mucosal adjuvants." *Vaccine.* 23: 1804-1813.

Gentschev, I., G. Dietrich and W. Goebel (2002). "The *E. coli* _-hemolysin secretion system and its use in vaccine development." *Trends Microbiol.* 10(1): 39-45

Gentschev, I., H. Mollenkopf, Z. Sokolovic, J. Hess, S. H. E. Kaufmann and W. Goebel (1996). "Development of antigen-delivery systems, based on the *Escherichia coli* haemolysin secretion pathway." *Genes* 179: 133-140

Goebel, W. and J. Hedgpeth (1982). "Cloning and functional characterization of the plasmid-encoded hemolysin determinant of Escherichia coli." *J Bacteriol* 151(3): 1290-8

Goluszko, P., E. Goluszko, B. Nowicki, S. Nowicki, V. Popov and e. al. (2005). "Vaccination with purified Dr Fimbriae reduces mortality associated with chronic urinary tract infection due to *Escherichia coli* bearing Dr adhesin." *Infect Immun* 73: 627–631.

Griebling, T. L. (2005). "Urologic diseases in america project: trends in resource use for urinary tract infections in men." *J Urol* 173: 1288–1294.

Griebling, T. L. (2005). "Urologic diseases in America project: trends in resource use for urinary tract infections in women." *J Urol* 173: 1281–1287.

Guyer, D. M., S. Radulovic, F. E. Jones and H. L. Mobley (2002). "Sat, the secreted autotransporter toxin of uropathogenic *Escherichia coli*, is a vacuolating cytotoxin for bladder and kidney epithelial cells." *Infect Immun* 70(8): 4539-46

Hackett, M., C. Walker, L. Guo, M. C. Gray, S. V. Cuyk, U. Ullmann, J. Shabanowitz, D. F. Hunt, E. L. Hewlett and P. Sebo (1995). "Hemolytic, but not cell-invasive activity of adenylate cyclase toxin is selectively affected by differential fatty acylation in *Escherichia coli*." *J Biol Chem* 270: 20250-20253

Hall, P. D., G. Virella, T. Willoughby, D. H. Atchley, R. J. Kreitman and A. E. Frankel (2001). "Antibody response to DT-GM, a novel fusion toxin consisting of truncated diphtheria toxin (DT) linked to human granulocyte-macrophage colonystimulating factor (GM), during a phase I trial of patients with relapsed or refractory acute myeloid leukemia." *Clin Immunol* 100: 191–197.

Hamon, M. A., E. Batsche, B. Regnault, T. N. Tham, S. Seveau, C. Muchardt and P. Cossart (2007). "Histone modifications induced by a family of bacterial toxins." *Proc Natl Acad Sci U S A* 104(33): 13467-72

Hashigucci, K., H. Ogawa, T. Ishidate and et. al. (1996). "Antibody responses in volunteers induced by nasal influenza vaccine combined with Escherichia coli heat-labile enterotoxin B subunit containing a trace amount of the holotoxin.' *Vaccine* 14: 113–119

Herlax, V. and L. Bakas (2003). "Acyl chains are responsible for the irreversibility in the *Escherichia coli* alpha-hemolysin binding to membranes." *Chem Phys Lipids.* 122: 185-190

Herlax, V. and L. Bakas (2007). "Fatty acids covalently bound to alpha-hemolysin of Escherichia coli are involved in the molten globule conformation: implication of disordered regions in binding promiscuity." *Biochemistry* 46(17): 5177-84

Herlax, V., M. F. Henning, A. M. Bernasconi, F. M. Goni and L. Bakas (2010). "The lytic mechanism of *Escherichia coli* _-hemolysin associated to outer membrane vesicles." *Health.* 2: 484-492

Herlax, V., S. Mate, O. Rimoldi and L. Bakas (2009). "Relevance of fatty acid covalently bound to *Escherichia coli* alpha-hemolysin and membrane microdomains in the oligomerization process." *J Biol Chem* 284(37): 25199-210

Heyduk, T. (2002). "Measuring protein conformational changes by FRET/LRET." *Curr. Opin. Biotechnol.* 13: 292-296

Holmgren, J., J. Adamsson, F. Anjuère, J. Clemens, C. Czerkinsky, K. Eriksson, C. F. Flach, A. George-Chandy, A. M. Harandi, M. Lebens, T. Lehner, M. Lindblad, E. Nygren, S. Raghavan, J. Sanchez, M. Stanford, J. B. Sun, A. M. Svennerholm and S. Tengvall (2005). "Mucosal adjuvants and anti-infection and anti-immunopathology vaccines based on cholera toxin, cholera toxin B subunit and CpG DNA." *Immunol Lett.* 97(2): 181-188.

Holmgren, J., C. Czerkinsky, K. Eriksson and A. Mharandi (2003). "Mucosal immunisation and adjuvants: a brief overview of recent advances and challenges " *Vaccine.* 21(2): S89-95.

Hooton, T. M., D. Scholes, A. E. Stapleton, P. L. Roberts, C. Winter, K. Gupta, M. Samadpour and W. E. Stamm (2000.). "A prospective study of asymptomatic bacteriuria in sexually active young women." *N. Engl. J. Med* 343: 992–997

Hui, D., C. Morden, F. Zhang and V. Ling (2000). "Combinatorial analysis of the structural requirements of the *Escherichia coli* hemolysin signal sequence." *J.Biol.Chem.* 275: 2713-2720

Hyland, C., L. Vuillard, C. Hughes and V. Koronakis (2001). "Membrane interaction of *Escherichia coli* Hemolysin: Flotation and Insertion-Dependent Labeling by Phospholipid Vesicles." *J. Bacteriol* 183: 5364-5370

Issartel, J., V. Koronakis and C. Hughes (1991). "Activation of *Escherichia coli* prohaemolysin to the mature toxin by acyl carrier protein-dependent fatty acylation." *Nature* 351: 759-761

Jarchau, T., T. Chakraborty, F. Garcia and W. Goebel (1994). "Selection for transport competence of C-terminal polypeptides derived from *Escherichia coli* hemolysin: the shortest peptide capable of autonomous HlyB/HlyD-dependent secretion comprises the C-terminal 62 amino acids of HlyA." *Mol Gen Genet* 245(1): 53-60

Jonas, D., B. Schultheis, C. Klas, P. Krammer and S. Bhadki (1993). "Cytocidal effects of *Escherichia coli* hemolysin on human T lymphocytes." *Infect Immun* 61 1715-1721

Kiyono, H., J. Bienenstock, J. R. McGhee and P. B. Ernst (1992). "The mucosal immune system: features of inductive and effector sites to consider in mucosal immunization and vaccine development." *Reg Immunol* 4: 54-62.

Koronakis, E., C. Hughes, I. Milisav and V. Koronakis (1995). "Protein exporter function and *in vivo* ATPase activity are correleted in ABC-domain mutants of HlyB." *Mol.Microbiol* 16: 87-96

Koronakis, V., C. Hughes and E. Koronakis (1991). "Energetically distinct early and late stages of HlyB/HlyD-dependent secretion across both *Escherichia coli* membranes." *EMBO J.* 10: 3263-3272

Koronakis, V., E. Koronakis and C. Hughes (1989). "Isolation and analysis of the C-terminal signal directing export of *Escherichia coli* hemolysin protein across both bacterial membranes." *EMBO J* 8(2): 595-605

Koronakis, V., J. Li, E. Koronakis and K. Stauffer (1997). "Structure of TolC, the outer membrane component of the bacterial type I efflux system, derived from two-dimensional crystals." *Mol Microbiol.* 23: 617-626

Koronakis, V., A. Sharff, E. Koronakis, B. Luisi and C. Hughes (2000). "Crystal structure of the bacterial membrane protein TolC central to multidrug efflux and protein export." *Nature* 405: 914-919

Koschinski, A., H. Repp, H. Unver, F. Dreyer, D. Brockmeier, A. Valeva, S. Bhakdi and I. Walev (2006). " Why Escherichia coli _-hemolysin induces calcium oscillations in mammalian cells–the pore is on its own." *FASEB J* E80-E87

Kreitman, R. J., D. R. Squires, M. Stetler-Stevenson, P. Noel, D. J. FitzGerald, W. H. Wilson and et. al (2005). " Phase I trial of recombinant immunotoxin RFB4 (dsFv)-PE38 (BL22) in patients with B-cell malignancies." *J Clin Oncol* 23: 6719-6729.

Kreitman, R. J., W. H. Wilson, K. Bergeron, M. Raggio, M. Stetler-Stevenson and D. J. FitzGerald (2001). "Efficacy of the anti-CD22 recombinant immunotoxin BL22 in chemotherapy-resistant hairy-cell leukemia." *N Engl J Med* 345: 241-247

Kuehn, M. J. and N. C. Kesty (2005). "Bacterial outer membrane vesicles and host-pathogen interaction." *Genes and Development* 19: 2645-2655

Lakowicz, J. R., I. Gryczynski, W. Wiczk, G. Laczko, F. C. Prendergast and M. L. Johnson (1990). "Conformational distributions of melittin in water/methanol mixtures from frequency-domain measurements of nonradiative energy transfer." *Biophys. Chem.* 36: 99-115

Lally, E. T., I. R. Kieba, A. Sato, C. L. Green, J. Rosenbloom, J. Korostoff, J. F. Wang, B. J. Shenker, S. Ortlepp, M. K. Robinson and P. C. Billings (1997). "RTX toxins recognize a beta2 integrin on the surface of human target cells." *J Biol Chem.* 272(48): 30463-9

Langston, K. G., L. M. Worsham, L. Earls and M. L. Ernst-Fonberg (2004). "Activation of hemolysin toxin: relationship between two internal protein sites of acylation." *Biochemistry* 43(14): 4338-46

Lee, C. K. (2001). "Vaccination against Helicobacter pylori in non-human primate models and humans." *Scand J Immunol* 53: 437–442.

Lee, C. K., R. Weltzin, W. D. Thomas and et. al. (1995). "Oral immunization with recombinant *Helicobacter pylori* urease induces secretory IgA antibodies and protects mice from challenge with Helicobacter felis." *J Infect Dis* 172: 161–171.

Lichtenberg, D., F. M. Goñi and H. Heerklotz (2005). "Detergent-resistant membranes should not be identified with membrane rafts." *Trends Biochem Sci* 30(8): 430-436

Lim, K. B., C. R. Bazemore Walker, L. Guo, S. Pellett, J. Shabanowitz, D. Hunt, E. L. Hewlett, A. Ludwig, W. Goebeli, R. A. Welch and Hackett.M. (2000). "*Escherichia coli* _-Hemolysin (HlyA) Is Heterogeneously Acylated in Vivo with 14-, 15-, and 17-Carbon Fatty Acids." *J Biol Chem* 275: 36698-36702

Lin, K. Y., N. T. Chiu, M. J. Chen, C. H. Lai, J. J. Huang and e. al. (2003). "Acute pyelonephritis and sequelae of renal scar in pediatric first febrile urinary tract infection." *Pediatr Nephrol* 18: 362–365.

Linhartova, I., L. Bumba, J. Masin, M. Basler, R. Osicka, J. Kamanova, K. Prochazkova, I. Adkins, J. Hejnova-Holubova, L. Sadilkova, J. Morova and P. Sebo (2010) "RTX proteins: a highly diverse family secreted by a common mechanism." *FEMS Microbiol Rev* 34(6): 1076-112

Lowell, G. H., W. R. Ballou, L. F. Smith, R. A. Wirtz, W. D. Zollinger and W. T. Hockmeyer (1988). "Proteosome-lipopeptide vaccines: Enhancement of immunogenicity for malaria CS peptides." *Science* 240: 800–802.

Lowell, G. H., L. F. Smith, R. C. Seid and W. D. Zollinger (1988). "Peptides bound to proteosomes via hydrophobic feet become highly immunogenic without adjuvants." *J Exp Med* 167: 658–663

Ludwig, A., M. Vogel and W. Goebel (1987). "Mutations affecting activity and transport of haemolysin in *Escherichia coli*." *Mol Gen Genet.* 206: 238-254

Manning, B. D. and L. C. Cantley (2007). "Akt/PKB signaling: navigating downstream." *Cell* 129: 1261-1274

Mansson, L. E., P. Kjall, S. Pellett, G. Nagy, R. A. Welch, F. Backhed, T. Frisan and A. Richter-Dahlfors (2007). "Role of the lipopolisacharide -CD14 complex for the activity of hemolysin from uropathogenic Escherichia coli." *Infec Immun* 75: 997-1004

Marrs, C. F., L. Zhang and B. Foxman (2005). "*Escherichia coli* mediated urinary tract infections: are there distinct uropathogenic *E. coli* (UPEC) pathotypes?" *FEMS Microbiol Lett* 252: 183–190

Mashburn-Warren, L. and M. Whiteley (2006). "Special delivery: Vesicle trafficking in prokaryotes." *Molecular Microbiology* 61(4): 839-846.

McGhee, J. R., J. Mestecky, M. T. Dertzbaugh, J. H. Eldridge, M. Hirasawa and H. Kiyono (1992). "The mucosal immune system: from fundamental concepts to vaccine development." *Vaccine* 10: 75-88

Menestrina, G., N. Mackman, I. Holland, B. and S. Bhadki (1987). "*Escherichia coli* haemolysin forms voltage-dependent ion channels in lipid membranes." *Biochem Biophys Acta* 905: 109-117

Moayeri, M. and R. Welch (1994). "Effects of temperature, time, and toxin concentration on lesion formation by the *Escherichia coli* hemolysin." *Infection and immunity* 62(10): 4124-4134

Moayeri, M. and R. Welch (1997). "Prelytic and lytic Conformation of erythrocyte- associated *Escherichia coli* hemolysin." *Infection and immunity* 65(6): 2233-2239

Mobley, H. L. T., M. S. Donnenberg and E. C. Hagan (2009.) *Uropathogenic Escherichia coli.* Washington, DC, ASM Press.

Moolten, F., S. Zajdel and S. Cooperband (1976). "Immunotherapy of experimental animal tumors with antitumor antibodies conjugated to diphtheria toxin or ricin." *Ann. NY Acad. Sci.* 277: 690–699.

Moolten, F. L. and S. R. Cooperband (1970). "Selective destruction of target cells by diphtheria toxin conjugated to antibody directed against antigens on the cells." *Science* 169: 68–70

Morein, B., B. Sundquist, S. Höglund, K. Dalsgaard and A. Osterhaus (1984). " Iscom, a novel structure for antigenic presentation of membrane proteins from enveloped viruses." *Nature* 308: 457–460

Morova, J., R. Osicka, J. Masin and P. Sebo (2008). "RTX cytotoxins recognize beta2 integrin receptors through N-linked oligosaccharides." *P Natl Acad Sci USA* 105: 5355–5360.

Muller, D., C. Hughes and W. Goebel (1983). "Relationship between plasmid and chromosomal hemolysin determinants of *Escherichia coli*." *J .Bacteriol.* 153: 846-851

Murphy, J. R. (1996). *Corynebacterium Diphtheriae.* University of Texas Medical Branch at Galveston.

Nicaud, J. M., N. Mackman, L. Gray and I. B. Holland (1986). "The C-terminal, 23 kDa peptide of *E. coli* haemolysin 2001 contains all the information necessary for its secretion by the haemolysin (Hly) export machinery." *FEBS Lett* 204(2): 331-5

Nieto, J., C. Hughes, M. Bailey and V. Koronakis (1996). "Suppression of transcription polarity in the *E.coli* hemolysin operon by a short upstream element shared by polysaccharide and DNA transfer determinants." *Mol.Microbiol* 19: 705-714

O'Hanley, P., G. Lalonde and G. Ji (1991). "Alpha-hemolysin contributes to the pathogenicity of piliated digalactoside-binding *Escherichia coli* in the kidney: efficacy of an alpha-hemolysin vaccine in preventing renal injury in the BALB/c mouse model of pyelonephritis." *Infect Immun* 59 1153–1161.

Oster, P., D. Lennon, J. O'Hallahan, K. Mulholland, S. Reid and D. Martin (2005). "MeNZB: A safe and highly immunogenic tailor-made vaccine against the New Zealand *Neisseria meningitidis* serogroup B disease epidemic strain." *Vaccine* 23: 2191–2196.

Ostolaza, H., L. Bakas and F. Goñi (1997). "Balance of electrostatic and hydrophobic interactions in the lysis of model membranes by *E. coli* alpha-haemolysin." *J Membr Biol.* 158(2): 137-145

Ostolaza, H., B. Bartolome, I. Ortiz de Zarate, F. de la Cruz and F. M. Goñi (1993). "Release of lipid vesicle contents by the bacterial protein toxin alpha-haemolysin." *Biochim Biophys Acta* 1147(1): 81-88

Ostolaza, H. and F. M. Goñi (1995). "Interaction of the bacterial protein toxin alpha-haemolysin with model membranes: protein binding does not always lead to lytic activity." *FEBS Lett.* 371(3): 303-306

Parsons, M., B. Vojnovic and S. Ameer-Beg (2004). "Imaging proteinprotein interactions in cell motility using fluorescence resonance energy transfer (FRET)." *Biochem. Soc. Trans.* 32: 431-433

Pastan, I., R. Hassan, D. J. FitzGerald and R. J. Kreitman (2007). "Immunotoxin treatment of cancer." *Annu.Rev. Med.* 58: 221–237.

Pellett, S., D. Boehm, I. Snyder, G. Rowe and R. Welch (1990). "Characterization of monoclonal anibodies against the *Escherichia coli* hemolysin." *Infection and immunity* 58(3): 822-827

Perrie, Y., A. R. Mohammed, D. J. Kirby, S. E. McNeil and V. W. Bramwell (2008). "Vaccine adjuvant systems: enhancing the efficacy of sub-unit protein antigens." *International Journal of Pharmaceutics* 364: 272–280

Pike, L. J. (2003). "Lipid rafts: bringing order to chaos." *J Lipid Res* 44(4): 655-67

Pimenta, A. L., K. Racher, L. Jamieson, M. A. Blight and I. B. Holland (2005). "Mutations in HlyD, part of the type 1 translocator for hemolysin secretion, affect the folding of the secreted toxin." *J Bacteriol* 187: 7471–7480.

Pizza, M., V. Scarlato, V. Masignani, M. M. Giuliani, B. Arico and et. al. (2000). "Identification of vaccine candidates against serogroup B meningococcus bywhole-genome sequencing." *Science* 287: 1816–1820

Ratner, A. J., K. R. Hippe, J. L. Aguilar, M. H. Bender, A. L. Nelson and J. N. Weiser (2006). "Epithelial cells are sensitive detectors of bacterial pore-forming toxins." *J Biol Chem* 281(18): 12994-8

Rose, T., P. Sebo, J. Bellalou and D. Ladant (1995). "Interaction of calcium with Bordetella pertussis adenylate cyclase toxin. Characterization of multiple calcium-binding sites and calcium-induced conformational changes." *J Biol Chem* 270(44): 26370-6

Rowe, G., S. Pellet and R. Welch (1994). "Analysis of toxinogenic functions associated with the RTX repeat region and monoclonal antibody D12 epitope of *Escherichia coli* hemolysin (HlyA)." *Infect. Immun.* 62: 579-588

Russo, T. A., C. D. McFadden, U. B. arlino-MacDonald, J. M. Beanan, R. Olson and e. al. (2003). "The Siderophore receptor IroN of extraintestinal pathogenic Escherichia coli is a potential vaccine candidate." *Infect Immun* 71: 7164–7169.

Sanchez-Magraner, L., A. Cortajarena, F. Goni and H. Ostolaza (2006). "Membrane insertion of *Escherichia coli* alpha-hemolysin is independent from membrane lysis." *J. Biol. Chem* 281: 5461-5467

Sanchez-Magraner, L., A. R. Viguera, M. Garcia-Pacios, M. P. Garcillan, J. L Arrondo, F. de la Cruz, F. M. Goni and H. Ostolaza (2007). "The calcium-binding C-terminal domain of *Escherichia coli* alpha-hemolysin is a major determinant in the surface-active properties of the protein." *J Biol Chem* 282(16): 11827-35

Sanchez, J. and J. Holmgren (2011). "Cholera toxin - a foe & a friend." *Indian J Med Res.* 133(2): 153-163

Schlessinger, D. and M. Schaechter (1993). *Bacterial toxins.* Baltimore, Williams and Wilkins.

Schulein, R., I. Gentschev, H. J. Mollenkopf and W. Goebel (1992). "A topolcgical model for the haemolysin translocator protein HlyD." *Mol Gen Genet* 234(1): 155-63

Shapira, A. and I. Benhar (2010). "Toxin-Based Therapeutic Approaches." *Toxins* 2: 2519-2583

Shogomori, H., A. T. Hammond, A. G. Ostermeyer-Fay, D. J. Barr, G. W. Feigenson, E. London and D. A. Brown (2005). "Palmitoylation and intracellular domain interactions both contribute to raft targeting of linker for activation of T cells." *J Biol Chem.* 280: 18931-42

Singh, M., A. Chakrapani and D. O'Hagan (2007). "Nanoparticles and microparticles as vaccine-delivery systems." *Expert Rev Vaccines* 6: 797–808

Smith, Y. C., S. B. Rasmussen, K. K. Grande, R. M. Conran and A. D. O'Brien (2008). "Hemolysin of uropathogenic *Escherichia coli* evokes extensive shedding of the uroepithelium and hemorrhage in bladder tissue within the first 24 hours after intraurethral inoculation of mice." *Infect Immun* 76(7): 2978-90

Soloaga, A., H. Ostolaza, F. Goñi and F. De la Cruz (1996). "Purification of *Escherichia coli* pro-haemolysin, and a comparison with the properties of mature _-haemolysin." *Eur. J. Biochem.* 238: 418-422

Soloaga, A., P. Veiga, L. García Segura, H. Ostolaza, R. Brasseur and G. F (1999). "Insertion of *Escherichia coli* _-haemolysin in lipid bilayer as a non-transmembrane integral protein:prediction and experiment." *Molecular Microbiology* 31: 1013-1024

Stanley, P., V. Koronakis, K. Hardie and C. Hughes (1996). "Independent interaction of the acyltransferase HlyC with two maturation domains of the *Escherichia coli* toxin HlyA." *Mol Microbiol* 20(4): 813-22

Stanley, P., V. Koronakis and C. Hughes (1991). "Mutational analysis supports a role for multiple structural features in the C-terminal secretion signal of *Escherichia coli* haemolysin." *Mol Microbiol* 5(10): 2391-403

Stanley, P., V. Koronakis and C. Hughes (1998). "Acylation of *Escherichia coli* hemolysin: A unique protein lipidation mechanism underlying toxin fuction." *Microbiology and Molecular Biology Reviews* 62: 309-333

Stanley, P., L. Packman, V. Koronakis and C. Hughes (1994). "Fatty acylation of two internal lysine residues required for the toxic activity of *Escherichia coli* hemolysin." *Science* 266: 1992-1996

Thanabalu, T., E. Koronakis, C. Hughes and V. Koronakis (1998). "Substrate-induced assembly of a contiguous channel for protein export from *E.coli* : reversible bridging of an inner-membrane translocase to an outer membrane exit pore." *The EMBO Journal* 17(22): 6487-6496

Tran Van Nhieu, G., C. Clair, G. Grompone and P. Sansonetti (2004). "Calcium signalling during cell interaction with bacterial pathogens." *Biology of the cell* 96: 93-101

Trent, M. S., L. M. Worsham and M. L. Ernst-Fonberg (1998). "The biochemistry of hemolysin toxin activation: characterization of HlyC, an internal protein acyltransferase." *Biochemistry* 37(13): 4644-52

Troeger, H., J. F. Richter, L. Beutin, D. Gunzel, U. Dobrindt, H. J. Epple, A. H. Gitter, M. Zeitz, M. Fromm and J. D. Schulzke (2007). "*Escherichia coli* alpha-haemolysin induces focal leaks in colonic epithelium: a novel mechanism of bacterial translocation." *Cell Microbiol* 9(10): 2530-40

Tumpey, T. M., M. Renshaw, J. D. Clements and J. M. Katz (2001). "Mucosal delivery of inactivated influenza vaccine induces B-cell-dependent heterosubtypic cross-protection against lethal influenza A H5N1 virus infection." *J Virol* 75: 5141–5150

Uhlen, P., A. Laestadius, T. Jahnukainen, T. Soderblom, F. Backhed, G. Celsi, H. Brisman, S. Normark, A. Aperia and A. Richter- Dahlfors (2000). "Alpha-haemolysin of uropathogenic *E. coli* induces Ca^{+2} oscillations in renal epithelial cells." *Nature* 405: 694–697

Ulmer, J. B., U. Valley and R. Rappuoli (2006). "Vaccine manufacturing: Challenges and solutions." *Nat Biotechnol* 24: 1377–1383

Valeva, A., I. Walev, H. Kemmer, S. Weis, I. Siegel, F. Boukhallouk, T. Wassenaar, T. Chavakis and S. Bhakdi (2005). "Binding of *Escherichia coli* Hemolysin and Activation of the Target Cells is Not Receptor-dependent." *J Biol Chem* 280: 36657-36663 van der Goot, G. and J. A. Young (2009). "Receptors of anthrax toxin and cell entry." *Mol Aspects Med.* 30(6): 406-412

Vigil, P. D., C. Alteri and H. L. Mobley (2011). "Identification of In Vivo-Induced Antigens Including an RTX Family Exoprotein Required for Uropathogenic *Escherichia coli* Virulence." *Infection and Immunity* 79: 2335–2344

Wandersman, C. and P. Delepelaire (1990). "TolC, an *Escherichia coli* outer membrane protein required for hemolysin secretion." *Proc Natl Acad Sci U S A.* 87: 4775-4780

Wang, R., S. Seror, M. Blight, J. Pratt, J. Broome-Smith and I. Holland (1991). "Analysis of the membrane organization of an *Escherichia coli* protein traslocator HlyB, a member of a large family of prokaryote and eukaryote surface transport proteins." *J.Mol.Biol* 217: 441-454

Welch, R. (2001). "RTX Toxin Structure and Function: A Story of Numerous Anomalies and Few Analogies in Toxin Biology." *Current Top Microbiol Immunol* 257: 85-111

Welch, R. A. (1991). "Pore-forming cytolisins of Gram-negative bacteria." *Mol. Microbiol* 5: 521-528

Weltzin, R., H. Kleanthous, F. Guirakhoo, T. P. Monath and C. K. Lee (1997). "Novel intranasal immunization techniques for antibody induction and protection of mice against gastric Helicobacter felis infection." *Vaccine* 4: 370–376.

Wiles, T. J., J. M. Bower, M. J. Redd and M. A. Mulvey (2009). "Use of zebrafish to probe the divergent virulence potentials and toxin requirements of extraintestinal pathogenic *Escherichia coli.*" *PLoS Pathog* 5(12): e1000697

Wiles, T. J., B. K.-. Dhakal, D. S. Eto and M. A. Mulvey (2008). "Inactivation of host Akt/protein kinase B signalling by bacterial pore- foming toxins." *Moleculas Biology of the cell* 19: 1427-1438

Wiles, T. J., R. R. Kulesus and M. A. Mulvey (2008). "Origins and virulence mechanisms of uropathogenic *Escherichia coli.*" *Exp Mol Pathol* 85(1): 11-9

Worsham, L. M., K. G. Langston and M. L. Ernst-Fonberg (2005). "Thermodynamics of a protein acylation: activation of *Escherichia coli* hemolysin toxin." *Biochemistry* 44: 1329-1337

Worsham, L. M., M. S. Trent, L. Earls, C. Jolly and M. L. Ernst-Fonberg (2001). "Insights into the catalytic mechanism of HlyC, the internal protein acyltransferase that activates *Escherichia coli* hemolysin toxin." *Biochemistry* 40(45): 13607-16

Wu, Y.-Y., M. H. Nahm, Y. Guo, M. W. Russell and D. E. Briles (1997). "Intranasal immunization of mice with PspA (pneumococcal surface protein A) can prevent intranasal carriage, pulmonary infection, and sepsis with *Streptococcus pneumoniae.*" *J Infect Dis* 175: 839–846.

Xu-Amano, J., R. J. Jackson, K. Fujihashi, H. Kiyono, H. F. Staats and J. R. McGhee (1994). "Helper Th1 and Th2 cell responses following mucosal or systemic immunization with cholera toxin." *Vaccine* 12: 903–911

Xu-Amano, J., H. Kiyono and R. J. Jackson (1993). "Helper T cell subsets for immunoglobulin A responses: oral immunization with tetanus toxoid and cholera toxin as adjuvant selectively induces Th2 cells in mucosa associated tissues." *J Exp Med* 178(4): 1309–1320.

Yamamoto, M., J. L. Vancott, N. Okahashi and e. al. (1996). "The role of Th1 and Th2 cells for mucosal IgA responses." *Ann NY Acad Sci* 778: 64–71.

Yamamoto, S., T. Tsukamoto, A. Terai, H. Kurazono, Y. Takeda and O. Yoshida (1997). "Genetic evidence supporting the fecal-perinealurethral hypothesis in cystitis caused by *Escherichia coli*." *J. Urol.* 157: 1127–1129.

Young, J. and I. B. Holland (1999). "ABC transporters: bacterial exporters-revisited five years on." *Biochim Biophys Acta* 1461: 177–200.

5

Human ERα and ERβ Splice Variants: Understanding Their Domain Structure in Relation to Their Biological Roles in Breast Cancer Cell Proliferation

Ana M. Sotoca[1,2], Jacques Vervoort[2],
Ivonne M.C.M. Rietjens[1] and Jan-Åke Gustafsson[3,4]
[1]*Toxicology section, Wageningen University, Wageningen*
[2]*Laboratory of Biochemistry, Wageningen University, Wageningen*
[3]*Department of Biology and Biochemistry, Center for Nuclear Receptors and Cell Signaling, University of Houston, Science & Engineering Research Center, Houston*
[4]*Department of Biosciences and Nutrition, Karolinska Institutet, Novum, Huddinge*
[1,2]*The Netherlands*
[3]*USA*
[4]*Sweden*

1. Introduction

ERs are members of the nuclear receptor superfamily and have a broad range of biological roles, such as growth, differentiation and physiology of the reproductive system (Pearce & Jordan, 2004). These enzymes also have roles in non-reproductive tissues such as bone, cardiovascular system, brain and liver (Heldring *et al.*, 2007). Until 1996, only one human estrogen receptor (ER) was known. That year Kuiper et al. discovered a novel nuclear estrogen receptor cloned from rat prostate. The known ER was renamed and called ERα to differentiate it from the novel ER, ERβ (Kuiper *et al.*, 1996). The complete human ERβ cDNA sequence was published in 1998 by Ogawa et al (Ogawa *et al.*, 1998a).

1.1 Estrogen receptors and signalling function

Estrogen receptors are products of distinct genes localized on different chromosomes; human ERα is encoded on chromosome 6q24-q27 (Gosden *et al.*, 1986), while the gene encoding human ERβ is localized on chromosome 14q22-q24 (Enmark *et al.*, 1997). Despite their distinct localization, the gene organization of the two receptors is well conserved. ESR1 (ERα) and ESR2 (ERβ) genes contain eight exons, separated by seven long intronic sequences. As members of the nuclear receptor superfamily, ERs contain 6 regions in their protein structure common for all nuclear receptors, namely: A, B, C, D, E and F which form functionally different but interacting domains (figure 1). Exon 1 encodes the A/B region in ERα and ERβ, exons 2 and 3 encode part of the C region. Exon 4 encodes the remaining part of region C, the whole of region D and part of region E. Exons 5 to 8 contain the rest of region E and region F is encoded by part of exon 8 [reviewed in (Ascenzi *et al.*, 2006)].

Although ERα and ERβ are encoded separately they share a high degree of homology. The most conserved domain among ERs is the DNA binding domain (DBD) corresponding to the C region, with 96% homology between α and β ER subtypes. The DBD is responsible for binding to specific DNA sequences (Estrogen Responsive Elements or EREs) in target gene promoter regions. High structure similarity in this region suggests similar target promoter sites for both receptors. The A/B region located in the N-terminus of the protein encompasses the AF-1 domain responsible for ligand independent transactivation. The AF-1 domain is the least conserved part among the two ERs with only 30% homology and it is functional only in the ERα subtype (Hall & McDonnell, 1999). The C-terminus of the protein contains the ligand dependent transactivation domain AF-2, the ligand binding domain (LBD) and the homo-/heterodimerization site. Homology between the E/F regions of both proteins is only 53%, explaining differences in ligand binding affinities between the two receptors. The hinge region localized in the D domain contains the nuclear localization signal of the ERs as well as post translational modification sites (Sentis *et al.*, 2005). Information on structure/function relationship of this region is very limited and it appears to be a variable and not well conserved part of the ERs (only 30% homology).

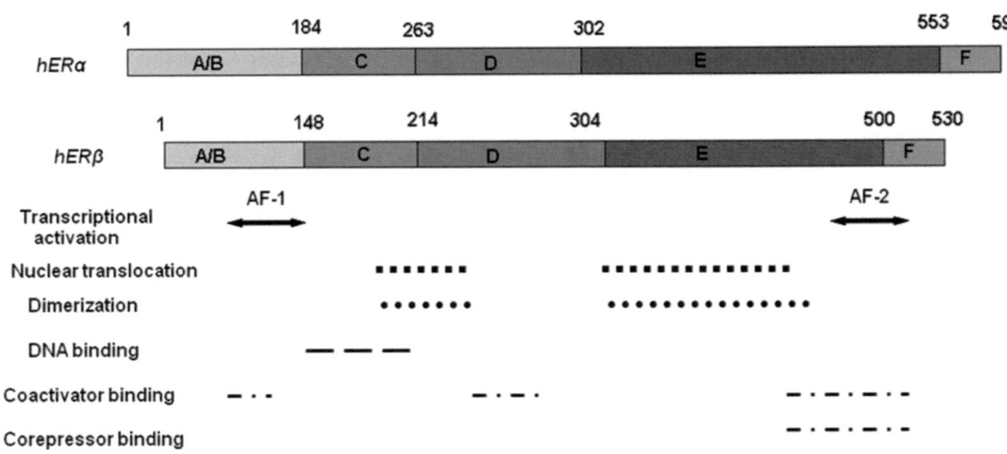

Fig. 1. Proteomic format, domain structure of human ERα (A) and ERβ (B). Based on Matthews and Gustafsson (Matthews & Gustafsson, 2003).

Estrogen (E2) binding to the receptor induces the LBD to undergo a conformational change, upon which the receptor dimerizes, binds to DNA, and stimulates gene expression (Cowley *et al.*, 1997; Katzenellenbogen & Katzenellenbogen, 2000).

1.2 Estrogen receptor distribution

The distribution of ERs varies both between and within human tissues (see Table 1). The cardiovascular system, brain, and bones express both receptors. ERβ is predominant in the male reproductive system. Expression of both ERα and ERβ has been found in all major human uterine cell types at every menstrual stage. However, expression varies from cell-type to cell-type with expression of ERα mRNA generally being higher than that of ERβ (Matsuzaki *et al.*, 1999). Changes in expression of estrogen receptors has been found in certain tumour types. Normal mammary tissue in man predominantly expresses ERβ mRNA, whereas most ER-positive breast tumours appear to exhibit increased ratios of ERα/ERβ (Leygue *et al.*, 1998).

Likewise, an increased ratio of ERα/ERβ mRNA has been demonstrated in ovarian carcinoma compared with normal tissue or cysts (Bardin *et al.*, 2004). High concentrations of ERβ have also been found within the human gut (Enmark *et al.*, 1997).

Therefore, the ultimate estrogenic effect of a certain compound on cells or tissues will be dependent on the receptor phenotype of these cells or tissues.

Organ/Tissue	Human ER subtype		Organ/Tissue	Human ER subtype	
	ERα	ERβ		ERα	ERβ
Heart	✓	✓	**Adrenal**	✓	-
Lung	-	✓	**Kidney**	✓	✓
Vascular	✓	✓	**Prostate**	-	✓
Bladder	-	✓	**Testes**	-	✓
Epididymus	-	✓	**Brain**	✓	✓
Pituitary	-	✓	**Thymus**	-	✓
Liver	✓	-	**Breast**	✓	✓
Muscle	-	-	**Uterus**	✓	✓
Fat	-	-	**Endometrium**	✓	✓
Gastrointestinal tract	-	✓	**Vagina**	✓	-
Colon	-	✓	**Fallopian tube**	-	✓
Small intestine	-	✓	**Ovary**	✓	✓
Bone	✓	✓			

Table 1. Tissue distribution of ER subtypes in humans.

1.3 Mechanism of estrogen action

Estrogens act on target tissues by binding to ERs. These proteins function as transcription factors when they are activated by a ligand. Biological action of ERs involves complex and broad mechanisms. For the ERs two main mechanisms of action have been described, including a genomic and a non-genomic pathway (Figure 2).

The *genomic action* of ERs occurs in the nucleus of the cell, when the receptor binds specific DNA sequences directly ("direct activation" or classical pathway) or indirectly ("indirect activation" or non-classical pathway). In the absence of ligand, ERs are associated with heat-shock proteins. The Hsp90 and Hsp70 associated chaperone machinery stabilizes the ligand binding domain (LBD) and makes it accessible to the ligand. Liganded ER dissociates from the heat-shock proteins, changes its conformation, dimerizes, and binds to specific DNA sequences called estrogen responsive elements (EREs) in order to regulate transcription (Nilsson *et al.*, 2001). In the presence of the natural ligand E2, ER induces chromatin remodelling and increases transcription of estrogen regulated genes (Berno *et al.*, 2008).

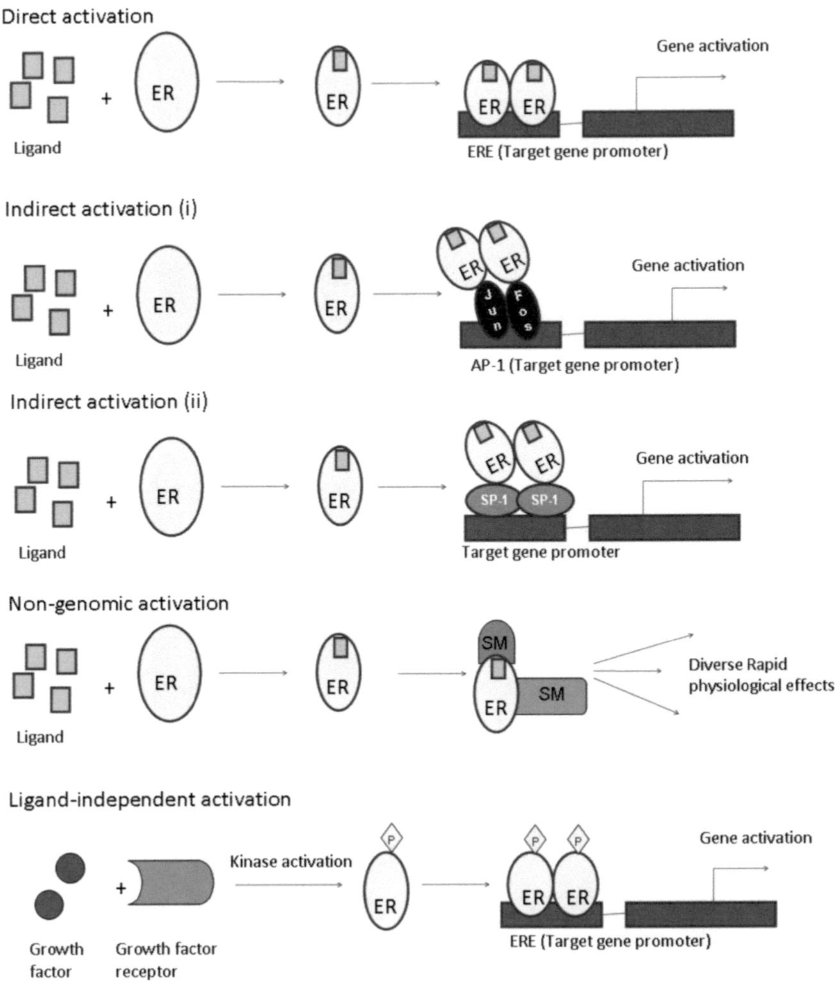

Fig. 2. Mechanisms of estrogen receptor (ER) action. In the *direct activation*, ERs dimerize after ligand binding and attach to the ERE in the promoter of target genes. In the *indirect activation* manner, ligand-bound ER dimers might activate transcription of non-ERE containing genes, by binding to other transcription factors (e.g. AP1 or SP1). In the *non-genomic pathway*, ligand-bound ERs interact directly with and change the function of proteins some of which function as 'second messengers' (SM). ERs can also be activated by phosphorylation in the absence of ER ligands (*ligand-independent activation*). Based on Morani *et al.* 2008 (Morani *et al.*, 2008).

In the non-classical pathway, AP-1 (DeNardo *et al.*, 2005) and SP-1 (Kim *et al.*, 2003) are alternative regulatory DNA sequences used by both isoforms of the receptor, ERα and ERβ, to modulate gene expression. In this case, ER does not interact directly with DNA but interacts with other DNA-bound transcription factors such as c-Jun or c-Fos, or with other proteins (Kushner *et al.*, 2003). Both AF-1 and AF-2 domains of ER are required for the interaction with Fos/Jun complex and both receptors differentially affect AP-1 dependent genes. In the presence of ERα, E2 works as AP-1 agonist by enhancing activity of the proteins at AP-1 sites (Brzozowski *et al.*, 1997), while in the presence of ERβ it antagonizes

AP-1 activity (Nilsson *et al.*, 2001). When both receptors are present, ERβ inhibits the action of ERα on AP-1 promoters (Matthews *et al.*, 2006). Interactions of ERs with other transcription factors might be also selectively modulated by different ligands, such as genistein and quercetin, which are not able to stimulate AP-1 dependent transcription (Figtree *et al.*, 2003; Schreihofer, 2005).

Even though ERs are considered transcription factors they can act through *non-genomic* mechanisms. Rapid ER effects were first observed in 1960s when administration of a physiological dose of E2 was reported to increase uterine cAMP levels in ovariectomized rats within 15 seconds (Szego and Davis, 1967), a time scale that is considered too fast for a genomic action. There is still no agreement if receptors responsible for rapid actions of estrogens are the same proteins as nuclear ERs or distinct G-protein coupled steroid receptors (Funakoshi *et al.*, 2006; Maggiolini *et al.*, 2004; Pedram & Levin, 2006; Warner & Gustafsson, 2006). However, a broad range of other rapid pathways induced by E2 has been identified so far. Some of these pathways include MAPK/ERK pathway, activation of endothelial nitric oxide synthase (eNOS), PLC stimulated IP_3 production, calcium influx and PI3K/Akt pathway activation (Stirone *et al.*, 2005; Virgili F, 2004; Ascenzi *et al.*, 2006). Similarly to non-classic mechanisms of activation, phytoestrogens might affect rapid pathways in a different way than E2. Quercetin for example has been shown to fail to phosphorylate ERK-2 kinase (opposite to E2) nor did it stimulate transcription of Cyclin D1, the transcription of which sometimes depends on rapid ER pathways (Virgili F, 2004). The stimulation of eNOS, which plays a role in cardiovascular health effects induced by E2 also seems to be regulated differently by phytoestrogens. Rapid activation of eNOS in the presence of E2 is dependent on ERα (Simoncini *et al.*, 2005), while both receptors are required for prolonged effects. However phytoestrogens do not activate eNOS in a rapid manner but seem to activate it through a prolonged, ERβ dependent transcriptional mechanism (Simoncini *et al.*, 2005).

In addition to ligand dependent mechanisms, ERα has ligand independent activity mediated through AF-1, which has been shown to be associated with stimulation of MAPK through growth factors such as Insulin like Growth Factor – 1 (IGF-1) and Epidermal Growth Factor (EGF). Activity of AF-1 is dependent on phosphorylation of Ser 118. A good example of the cross-talk between ER and growth factor signalling is phosphorylation of Ser 118 by MAPK in response to growth factors, such as IGF-1 and EGF (Kato *et al.*, 1995). The importance of growth factors in ER signalling is well illustrated by the fact that EGF can mimic effects of E2 in the mouse reproductive tract (Nilsson *et al.*, 2001).

1.4 Ligand dependent effects and cofactors

The overall biological effects of E2 and other estrogenic compounds are the result of complex interplay between various mechanisms, which largely depend on cellular context, ratio between ER subtypes, expression of coactivators in the cell, sequences of target EREs but also cross-talk with growth factor pathways and activity of kinases and phosphatases. All these factors together enable a precise and targeted response to the natural hormone. However a broad range of pathways involved in ER signaling provides many points of possible signal modulation by estrogens and estrogen-like compounds and small structural changes between different ligands might result in significantly different responses.

Structural differences in the LBD underlie differences in affinity and transcriptional activity of certain ER ligands and provide one of the mechanisms for selective modulation of ER responses. ERβ has an impaired AF-1 domain compared with ERα and the necessary synergy with AF-2 is dramatically reduced (Cowley & Parker, 1999). These differences suggest that it is possible to develop ligands with different affinities, potencies, and agonist vs antagonist behavior for the two ER subtypes.

It has been demonstrated that E2 has higher affinity towards ERα than to ERβ (Bovee *et al.*, 2004; Veld *et al.*, 2006), and certain selective estrogen receptor modulators (SERMs) might exhibit a preference towards one of the receptors (Escande *et al.*, 2006). Plant derived phytoestrogens, which are structurally similar to E2 (Figure 3) provide a good example of ligand selectivity (Kuiper *et al.*, 1998). Genistein is the major isoflavone present in soy and fava beans whereas quercetin is present in red onions, apples, cappers or red grapes among others (Kuiper *et al.*, 1998). *In vitro* studies with reporter gene assays proved that phytoestrogens are able to stimulate ERE-dependent genes at high concentrations. Therefore they are considered weak ER agonists with the majority of them preferentially binding to ERβ (Chrzan & Bradford, 2007; Harris *et al.*, 2005). The main hypothesis on the positive role of phytoestrogens in modulation of ER signaling is their higher affinity towards the ERβ subtype, which can silence ERα dependent signaling and decrease overall cell sensitivity to E2 (Hall & McDonnell, 1999), which is thought to be significant in cancer prevention.

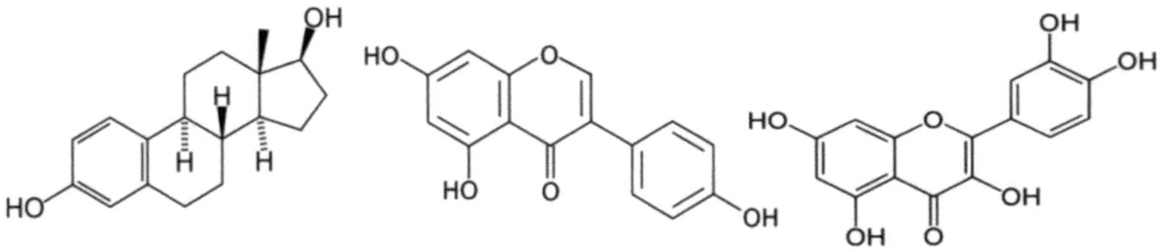

Fig. 3. Chemical structure of estradiol, genistein and quercetin.

ERs can associate with distinct subsets of coactivators and corepressors depending on binding affinities and relative abundance of these factors (Chen & Evans, 1995; Halachmi *et al.*, 1994). Several ER coactivators and corepressors have been described (Nilsson *et al.*, 2001). Differences between ERα and ERβ in coactivator and corepressor recruitment have also been reported (Cowley & Parker, 1999; Suen *et al.*, 1998), and therefore this preferential binding of certain coactivators and corepressors to one of the ERs may have consequences for specific ligand signalling and the ultimate biological effect elicited by ligand binding.

NCoR and SMRT corepressors and the p160 family coactivators are widely expressed (Horlein *et al.*, 1995; Misiti *et al.*, 1998; Oñate *et al.*, 1995). Low levels of SRC-3 have been demonstrated for human proliferating endometrium with increased expression in the late secretory phase (Gregory *et al.*, 2002) while overexpression of SRC-3 is frequently observed in breast, ovarian, and prostate cancers (Anzick *et al.*, 1997; Gnanapragasam *et al.*, 2001; McKenna *et al.*, 1999). Similar expression levels of CBP, p300, AIB1, GRIP1, p300, NCoR, and SMRT have been measured for Ishikawa uterine and MCF-7 breast cancer cells (Shang and Brown, 2002). High levels of SRC-1 expression are found in Ishikawa cells, and this might correlate with the agonist activity of tamoxifen in this cell line (Shang and Brown, 2002). We have seen in our studies (Sotoca *et al.*, 2011), that the T47D breast cancer cells express the ER coactivator PRMT1. Recruitment of this coactivator is accompanied by histone methylation (Huang *et al.*, 2005; Klinge *et al.*, 2004). Recently, PRMT1 gene expression has been used as a marker of unfavourable prognosis for colon cancer patients (Mathioudaki *et al.*, 2008).

Thus, other signalling events within the cell may affect nuclear receptor transcriptional responses via alteration in the expression of certain coregulators, and therefore it is predicted that significant differences in coactivator and corepressor expression found in various cell and tissue types would be important determinants of specific receptor modulator activity.

In addition, distribution of particular splicing variants of both ERs should be taken into account when considering tissue response to estrogens and cofactor recruitment as they have differential and sometimes antagonistic properties and their relative abundance might significantly influence biological responses to hormones. The main physiological role of ER splice variants in breast cancer development is however far from clear and might be a crucial determinant for clinical parameters.

2. ER Isoforms: ERα and ERβ

Full length ERα and ERβ proteins are approximately 66 and 59 kDa respectively (Ascenzi *et al.*, 2006; Fuqua *et al.*, 1999), although as a result of alternative splicing both receptors can form different isoforms. ERα has been shown to form over 20 alternative splice variants in breast cancer and other tumors (Poola *et al.*, 2000), three of them with proven functionality, while at least five ERβ variants have been reported in human (Lewandowski *et al.*, 2002).

The function and physiological significance of all isoforms have not been described so far, but some of them are powerful modulators of ER signaling pathways in normal tissues.

2.1 ERα splice variants

The two most referenced ERα isoforms that seem to be of particular significance are **ERα46** and **ERα36** as they were reported to oppose genomic actions of full length **ERα66** (figure 4).

The **ERα46** isoform has been identified in the MCF7 breast cancer cell line (Penot *et al.*, 2005) in which it is coexpressed with full length ERα66. The presence of ERα46 has also been confirmed in osteoblasts (Wang *et al.*, 2005) and endothelial cells (Figtree *et al.*, 2003). This isoform is formed by skipping exon 1 encoding the N-terminus (A/B) and it is devoid of AF-1 activity. In contrast with full length ERα66, the truncated isoform ERα46 does not mediate E2 dependent cell proliferation and high levels of this isoform have been shown to be associated with cell cycle arrest in the G0/G1 phase and a state of refraction to E2 stimulated growth, which is normally reached at hyperconfluency of the cells (Penot *et al.*, 2005). Similarly to ERβ, ERα46 is a potent ligand-dependent transcription factor containing AF-2 and a powerful inhibitor of ERα AF-1 dependent transcription (Figtree *et al.*, 2003). By inhibition of ERα66 dependent gene transcription, ERα46 isoform inhibits estrogenic induction of c-Fos and Cyclin D1 promoters, which are involved in cell cycle control. Coexpression of ERα46 with ERα66 in an SaOs osteoblast cell line results in concentration dependent inhibition of E2 stimulated cell proliferation (Ogawa *et al.*, 1998b), an effect similar to the consequence observed with coexpression of ERα with ERβ (Sotoca *et al.*, 2008; Ström *et al.*, 2004).

The second truncated ERα isoform ERα36 was first described recently (Wang *et al.*, 2005), and it has been shown to lack both the AF-1 and AF-2 transactivation functions of full length ERα. However it has functional DBD, partial dimerization and LBD domains. ERα36 contains an exon coding for myristoylation sites, hence predicting an interaction with the plasma membrane. Transcription of this ERα36 isoform is initiated from a previously unidentified promoter in the first intron of the ERα gene and the unique 27 amino acid C-terminal sequence is encoded by a novel ERα exon, localized downstream of exon 8 to replace the last 138 amino acids encoded by exon 7-8 (Wang *et al.*, 2005).

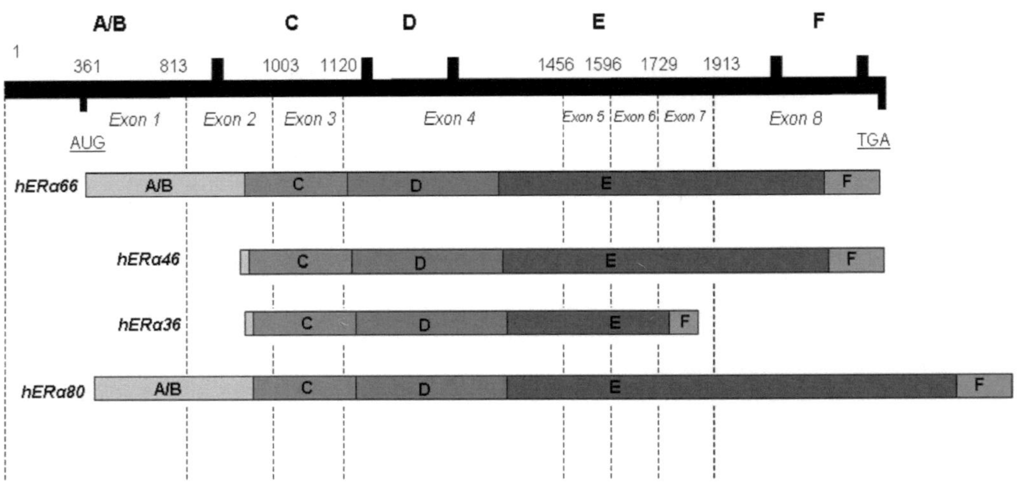

Fig. 4. Schematic comparison between full length ERα and its most referenced truncated isoforms.

This novel isoform has been cloned from a human placenta cDNA library, which indicates that it is a naturally occurring isoform of ERα. With no functional AF-1 and AF-2 ERα36 does not have any direct transcriptional activity. However, it is a robust inhibitor of full length ERα and ERβ dependent transactivation (ZhaoYi Wang *et al.*, 2006). It is mainly localized in the plasma membrane and works in a different way than full length protein. Even though it lacks transcriptional activity it can activate non genomic ER pathways such as MAPK/ERK signaling in response to E2 which is of particular significance in response to antiestrogens such as tamoxifen, 4OH-tamoxifen and ICI-182.780 (ZhaoYi Wang *et al.*, 2006). As a result of MAPK/ERK pathway activation by E2 and these antiestrogens a signal is transduced to the nucleus and consequently Elk1 transcription factor is activated. The effect of MAPK/ERK activation mediated by ERα36 is increased cell proliferation in response to E2 as well as antiestrogens in doses that shut down transcriptional activity of full length ERα and ERβ proteins (ZhaoYi Wang *et al.*, 2006).

The **ERα80** isoform was detected in the MCF7:2A cell line, which is a subclone MCF7 cell line derived from long term growth in the absence of E2. This ERα80 isoform was produced by duplication of exons 6 and 7 (Pink *et al.*, 1996). No evident function has been described so far.

Several other multiple splice variants (**ERαΔE2, ERαΔE3, ERαΔE4, ERαΔE5, ERαΔE6, ERαEΔ5,7, ERαEΔ7**…) as a result of exon splicing deletions have been confirmed in human (Poola *et al.*, 2000; Zhang *et al.*, 1996) showing a dominant inhibitory effect in normal ER function. A list of selected ERα splice variants and their expression in various breast tissues (normal and tumor) and breast cancer cell lines is given in Table 2.

Splice variant	Breast	MCF7	T47D	MDA-MB-231	MDA-MB-435	BT-474	BT20	ZR-75	References
ERα36	+			+					(Shi *et al.*, 2009; Lee *et al.*, 2008; ZhaoYi Wang *et al.*, 2006)
ERα46 (or ERαΔ1)		+							(Penot *et al.*, 2005)
ERαΔ2	+	+	+		+				(Wang and Miksicek, 1991; Zhang *et al.*, 1996; Bollig and Miksicek, 2000; Poola and Speirs, 2001; Miksicek *et al.*, 1993; Poola *et al.*, 2000)
ERαΔ3	+	+	+	+					(Wang and Miksicek, 1991; Poola and Speirs, 2001; Bollig and Miksicek, 2000; Zhang *et al.*, 1996; Koduri *et al.*, 2006; Erenburg *et al.*, 1997; Miksicek *et al.*, 1993; Fuqua *et al.*, 1993)
ERαΔ4		+			+			+	(Pfeffer *et al.*, 1993; Zhang *et al.*, 1996; Bollig and Miksicek, 2000; Poola *et al.*, 2000; Poola and Speirs, 2001)
ERαΔ5	+	+	+	+	+	+	+	+	(Zhang *et al.*, 1993; Zhang *et al.*, 1996; Bollig and Miksicek, 2000; Poola and Speirs, 2001; Zhang *et al.*, 1996; Fuqua *et al.*, 1991; Daffada *et al.*, 1994)
ERαΔ6	+								(Poola and Speirs, 2001; Bollig and Miksicek, 2000)
ERαΔ7	+	+	+	+					(Wang and Miksicek, 1991; Fuqua *et al.*, 1992; Poola and Speirs, 2001; Bollig and Miksicek, 2000; Fuqua *et al.*, 1992; Miksicek *et al.*, 1993)
ERαΔ5,7	+								(Zhang *et al.*, 1996)

Table 2. List of selected ERα splice variants and their expression in various breast tissues (normal and tumour) and breast cancer cell lines.

2.2 ERβ splice variants

The presence of ERβ isoforms has been confirmed in various human cell lines as well as in a broad range of tissues at different levels (Leung *et al.*, 2006; Moore *et al.*, 1998), which provides another possible mechanism of tissue-dependent modulation of the ER response. Therefore distribution of particular isoforms of both ERs should be taken into account when considering tissue response to estrogens as they have differential and sometimes antagonistic properties and their differential distribution might significantly influence biological response to hormone.

Different isoforms of ERβ have been described (figure 5) with a variable C-terminus, and which were cloned from a testis cDNA library (Moore *et al.*, 1998). At present their functional significance is poorly understood. The ERβ isoform whose function has been described in most detail of all ERβ isoforms studied is **ERβ1**, which is a full length protein with LBD and active AF-2 domain. **ERβ2, 4 and 5** have a shortened Helix 11 and a full length Helix 12 is present only in ERβ1 and β2. In ERβ2, Helix 12 has a different orientation than in ERβ1 due to the shorter Helix 11. It has been reported that the displaced Helix 12 in ERβ2 limits ligand access to the binding pocket. As a consequence of their altered structure, ERβ2, 4 and 5 cannot form homodimers and have no transcriptional activity on their own, although they have been shown to heterodimerize with ERβ1 upon E2 treatment and enhance its AF-2 mediated transcriptional activity (Leung *et al.*, 2006). Studies of interactions between different ERβ isoforms with ERα are very limited. However **ERβ2** (also named **ERβcx**) was shown to limit DNA binding of ERα66 and inhibit its transcriptional activity in similar manner to ERβ1 (Ogawa *et al.*, 1998b).

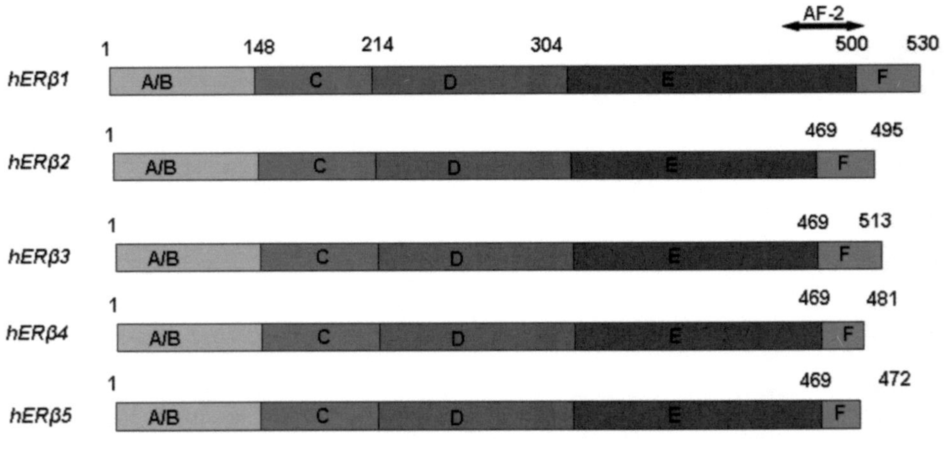

Fig. 5. Comparison between full length ERβ and it is most referenced truncated isoforms.

Two new exon-deleted variants were detected in the cancer cell line MDA-MB-231, **ERβΔ1,2,5** and **ERβΔ1,2,5,6** of approximately 35 and 28 kDa, respectively (Treeck *et al.*, 2008). Both proteins are predicted not to contain AF-1, and to have deletions in the DBD and LBD. Therefore, these two variants are expected to be devoid of or have significantly reduced ligand-dependent and ligand independent activities, and their expression did not affect growth of cancer cell lines tested. A list of selected ERβ splice variants and their expression in various breast tissues (normal and tumor) and breast cancer cell lines is given in Table 3.

Various studies reveal that physiological levels of ERα and ERβ may vary depending on the cell or tissue type (Enmmark *et al.*, 1997; Bonkhoff *et al.*, 1999; Makinen *et al.*, 2001; Pearce *et al.*, 2004) and as a consequence the biological response to endogenous or exogenous ligands can differ significantly.

Splice variant	Breast	MCF7	T47D	MDA-MB-231	MDA-MB-435	BT20	References
ERβ2	+	+	+	+		+	(Davies *et al.*, 2004; Zhao *et al.*, 2007; Girault *et al.*, 2004; Saji *et al.*, 2005; Cappelletti *et al.*, 2006; Leung *et al.*, 2006)
ERβ3	+						(Girault *et al.*, 2004)
ERβ4	+		+		+		(Moore *et al.*, 1998; Girault *et al.*, 2004; Poola *et al.*, 2005)
ERβ5	+	+	+	+	+	+	(Davies *et al.*, 2004; Girault *et al.*, 2004; Moore *et al.*, 1998; Fuqua *et al.*, 1999; Leung *et al.*, 2006; Cappelletti *et al.*, 2006)
ERβΔ2	+			+			(Poola *et al.*, 2002a)
ERβΔ3	+						(Poola *et al.*, 2002a; Poola *et al.*, 2002b)
ERβΔ4	+						(Poola *et al.*, 2002a; Poola *et al.*, 2002b)
ERβΔ5	+	+		+			(Poola *et al.*, 2002a; Speirs *et al.*, 2000; Leygue *et al.*, 1998)
ERβΔ6	+						(Poola *et al.*, 2002a; Leygue *et al.*, 1998)
ERβΔ1,2,5		+		+			(Treeck *et al.*, 2008)
ERβΔ1,2,5,6		+		+			(Treeck *et al.*, 2008)

Table 3. List of selected ERβ splice variants and their expression in various breast tissues (normal and tumour) and breast cancer cell lines.

3. Conclusion

Cell proliferation in normal developing breast tissue is stimulated by estrogens and estrogens may prevent osteoporosis by increasing bone mineral density (Douchi *et al.*, 2007). However, as cells can have their own set of ER splice variants that varies in time and abundance the estrogen receptor proteins can be expected to have a role in developmental regulation depending on splice variant and ligand present. ER splice variants are widely expressed in normal, premalignant and cancerous tissues and cell lines [reviewed in (Taylor *et al. 2010*)]. Co-expression of splice variants remains under investigation to understand its biological implications. Here, we briefly summarize ER expression and its role in positive or negative transcriptional activation in breast cancer.

Several studies have demonstrated that estrogens stimulate the growth of a large proportion of ERα positive breast cancers (Lazennec, 2006; Monroe *et al.*, 2005; Pedram *et al.*, 2006; Weitzmann & Pacifici, 2006). Furthermore, a decreased ERβ expression in cancer tissues as compared to benign tumours or normal tissues has been reported, whereas ERα expression seems to persist (Lazennec *et al.*, 2001, Bardin *et al.*, 2004). Recent progress in cellular experiments confirmed that ERβ opposes ERα actions in breast cancer cell lines (Sotoca *et al.*, 2011; Sotoca *et al.*, 2008; Ström *et al.*, 2004).

The main roles of ER splice variants in breast cancer development are, however, far from clear (Davies *et al.*, 2004; Saji *et al.*, 2005). ERα positivity in breast cancer in vivo is strongly associated with more favourable clinicopathological parameters. ERβ positive patients have been shown to have favourable prognosis and better survival due to better endocrine-treatment response compared with ERβ negative breast tumor patients (Davies *et al.*, 2004; Saji *et al.*, 2005).

When bound to estrogens as homodimers, each receptor activates transcription of certain target genes bearing a classical ERE in their promoter region. However, estrogen binding to ERβ can also inhibit gene transcription via AP-1 sites, while binding to ERα leads to activation. Furthermore, when heterodimers are formed, when the two receptors are co-expressed, ERβ can inhibit ERα function. Given that ER regulates cell proliferation by different mechanisms, we summarize (Table 4 and 5) by which molecular characteristics of ER this proliferation is driven.

Full activation of AF-1 in ERα induces cell proliferation in breast cancer cells (Fujita *et al.*, 2003). AF-1 activity of estrogen-ERβ is weaker compared with that of estrogen-ERα on ERE, whereas their AF-2 activities are similar (Cowley & Parker, 1999). In general ERβ has antiproliferative effects in breast cancer cells. All ERβ variants have negative effect on ERα by heterodimerization and reduce or abrogate both ligand-dependent and ligand-independent activities. Especially the ERβ2 isoform inhibits ERα-mediated estrogen action. In addition, several short ERα isoforms are able to oppose genomics actions of ERβ.

The most important point is that ERα expression induces significant cell proliferation in the absence of ERβ but not the other way around. Cell proliferation is triggered by classical genomic and non-genomic pathways. Only the wild type ERα isoform is able to induce hormone-dependent proliferation. It has been shown that most of the ER variants do not mediate ligand-dependent proliferation.

In conclusion, the overall biological effects of E2 and other estrogenic compounds in breast cancer cells are the result of complex interplay between various mechanisms, which depend on cellular context, balance between ER subtypes, coactivators and corepressors, sequences of target EREs but also cross-talk with growth factor pathways and activity of certain kinases and phosphatases. All these factors taken together enable response to estrogens or antiestrogens.

Isoform	Feature	AF-1	DBD	LBD	AF-2
ERα66	Wild type form Induces cell proliferation	+	+	+	+
ERα46	Does not mediate E2-dependent proliferation Opposes genomic action of ERα66 and ERβ Inhibitor of AF-1 dependent transactivation Potent AF-2 ligand dependent transcription activity	-	+	+	+
ERα36	Opposes genomic actions of ERα66 Can activate non-genomic ER pathways via MAPK/ERK No direct transcriptional activity Inhibitor of ERα and β dependent transactivation	-	+	+	-
ERα80	Not described	+	+	+*	+*
ERαΔ2	No transcriptional regulation	-	-	-	-
ERαΔ3	Binds ligand Dominant negative at ERE Interacts with AP-1 sites Suppresses E2-stimulated gene expression	+	-	+	+
ERαΔ4	Dominant negative transcriptional effect	+	-	-	+
ERαΔ5	Dominant positive transcriptional effect Dominant negative at ERE Coexpresses with ERα and enhances ERE-Luc	+	+	-	-
ERαΔ7	Dimerizes with ERα and hERβ Binds to ERE Dominant negative transcriptional effect	+	+	+	-

Table 4. Summary of ERα mechanism.

Isoform	Features	AF-1	DBD	LBD	AF-2
ERβ1	Wild type form	+	+	+	+
ERβ2	Dimerizes with ERβ1 and ERα Does not bind ligand	+	+	+	-
ERβ3	Dimerizes with ERβ1 Does not bind ligand	+	+	+	-
ERβ4	Dimerizes with ERβ1 and ERα Does not bind ligand	+	+	+	-
ERβ5	Dimerizes with ERβ1 and ERα Does not bind ligand	+	+	+	-
ERβΔ5	Negative effect on ERβ1 and ERα	+	+		
ERβΔ1,2,5	Reduced both ligand-dependent and ligand independent activities	-	+	-	+
ERβΔ1,2,5,6	Reduced both ligand-dependent and ligand independent activities	-	+	-	+

Table 5. Summary of ERβ mechanism.

4. Acknowledgements

This research was partially funded by the Graduate School of Voeding, Levensmiddelentechnologie, Agrobiotechnologie en Gezondheid (VLAG) (project number 61.61.100.040).

5. References

Anzick SL, Kononen J, Walker RL, Azorsa DO, Tanner MM, Guan X-Y *et al* (1997). AIB1, a Steroid Receptor Coactivator Amplified in Breast and Ovarian Cancer. *Science* 277: 965-968.

Ascenzi P, Bocedi A, Marino M (2006). Structure-function relationship of estrogen receptor [alpha] and [beta]: Impact on human health. *Molecular Aspects of Medicine* 27: 299.

Bardin A, Boulle N, Lazennec G, Vignon F, Pujol P (2004). Loss of ER{beta} expression as a common step in estrogen-dependent tumor progression. *Endocrine-Related Cancer* 11: 537-551.

Berno V, Amazit L, Hinojos C, Zhong J, Mancini MG, Dave Sharp Z *et al* (2008). Activation of Estrogen Receptor-alpha by E2 or EGF Induces Temporally Distinct Patterns of Large-Scale Chromatin Modification and mRNA Transcription. *PLoS ONE* 3: e2286.

Bollig A, Miksicek RJ (2000). An Estrogen Receptor-{alpha} Splicing Variant Mediates Both Positive and Negative Effects on Gene Transcription. *Mol Endocrinol* 14: 634-649.

Bonkhoff H, Fixemer T, Hunsicker I, Remberger K (1999). Estrogen Receptor Expression in Prostate Cancer and Premalignant Prostatic Lesions. *Am J Pathol* 155:641-647.

Bovee TFH, Helsdingen RJR, Rietjens IMCM, Keijer J, Hoogenboom RLAP (2004). Rapid yeast estrogen bioassays stably expressing human estrogen receptors [alpha] and [beta], and green fluorescent protein: a comparison of different compounds with both receptor types. *The Journal of Steroid Biochemistry and Molecular Biology* 91: 99.

Brzozowski AM, Pike ACW, Dauter Z, Hubbard RE, Bonn T, Engstrom O *et al* (1997). Molecular basis of agonism and antagonism in the oestrogen receptor. *Nature* 389: 753-758.

Cappelletti V, Miodini P, Fronzo GD, Daidone MG (2006). Modulation of estrogen receptor-beta isoforms by phytoestrogens in breast cancer cells. *International journal of Oncology* 28: 1185-91.

Chen JD, Evans RM (1995). A transcriptional co-repressor that interacts with nuclear hormone receptors. *Nature* 377: 454-457.

Chrzan BG, Bradford PG (2007). Phytoestrogens activate estrogen receptor beta1 and estrogenic responses in human breast and bone cancer cell lines. *Molecular Nutrition & Food Research* 51: 171-177.

Cowley SM, Hoare S, Mosselman S, Parker MG (1997). Estrogen Receptors alpha and beta Form Heterodimers on DNA. *Journal of Biological Chemistry* 272: 19858-19862.

Cowley SM, Parker MG (1999). A comparison of transcriptional activation by ER alpha and ER beta. *Journal of Steroid Biochemistry and Molecular Biology* 69: 165-175.

Daffada AAI, Johnston SRD, Nicholls J, Dowsett M (1994). Detection of wild type and exon 5-deleted splice variant oestrogen receptor (ER) mRNA in ER-positive and -negative breast cancer cell lines by reverse transcription/polymerase chain reaction. *Molecular Endocrinology* 13: 265-273.

Davies MPA, O'Neill PA, Innes H, Sibson DR, Prime W, Holcombe C *et al* (2004). Correlation of mRNA for oestrogen receptor beta splice variants ER{beta}1, ER{beta}2/ER{beta}cx and ER{beta}5 with outcome in endocrine-treated breast cancer. *J Mol Endocrinol* 33: 773-782.

DeNardo DG, Kim H-T, Hilsenbeck S, Cuba V, Tsimelzon A, Brown aPH (2005). Global Gene Expression Analysis of Estrogen Receptor Transcription Factor Cross Talk in Breast Cancer: Identification of Estrogen-Induced/Activator Protein-1-Dependent Genes. *Molecular Endocrinology* 19: 362-378.

Douchi T, Yonehara Y, Kosha S, Iwamoto I, Rai Y, Sagara Y *et al* (2007). Bone mineral density in breast cancer patients with positive estrogen receptor tumor status. *Maturitas* 57: 221-225.

Enmark E, Pelto-Huikko M, Grandien K, Lagercrantz S, Lagercrantz J, Fried G *et al* (1997). Human Estrogen Receptor {beta}-Gene Structure, Chromosomal Localization, and Expression Pattern. *J Clin Endocrinol Metab* 82: 4258-4265.

Erenburg I, Schachter B, Lopez RMy, Ossowski L (1997). Loss of an Estrogen Receptor Isoform (ER{alpha}{Delta}3) in Breast Cancer and the Consequences of Its Reexpression: Interference with Estrogen-Stimulated Properties of Malignant Transformation. *Mol Endocrinol* 11: 2004-2015.

Escande A, Pillon A, Servant N, Cravedi J-P, Larrea F, Muhnd P *et al* (2006). Evaluation of ligand selectivity using reporter cell lines stably expressing estrogen receptor alpha or beta. *Biochemical Pharmacology* 71: 1459-1469.

Figtree GA, McDonald D, Watkins H, Channon KM (2003). Truncated Estrogen Receptor {alpha} 46-kDa Isoform in Human Endothelial Cells: Relationship to Acute Activation of Nitric Oxide Synthase. *Circulation* 107: 120-126.

Fujita T, Kobayashi Y, Wada O, Tateishi Y, Kitada L, Yamamoto Y *et al* (2003). Full activation of estrogen receptor alpha (ER alpha) activation function-1 (AF-1) induces proliferation of breast cancer cells. *J. Biol. Chem.*: M301031200.

Funakoshi T, Yanai A, Shinoda K, Kawano MM, Mizukami Y (2006). G protein-coupled receptor 30 is an estrogen receptor in the plasma membrane. *Biochemical and Biophysical Research Communications* 346: 904.

Fuqua SAW, Allred DC, Elledge RM, Krieg SL, Benedix MG, Nawaz Z *et al* (1993). The ER-positive / PgR-negative breast cancer phenotype is not associated with mutations within the DNA binding domain. *Breast Cancer Research and Treatment* 26: 191-202.

Fuqua SAW, Fitzgerald SD, Allred DC, Elledge RM, Nawaz Z, McDonnell DP *et al* (1992). Inhibition of Estrogen Receptor Action by a Naturally Occurring Variant in Human Breast Tumors. *Cancer Research* 52: 483-486.

Fuqua SAW, Fitzgerald SD, Chamness GC, Tandon AK, McDonnell DP, Nawaz Z *et al* (1991). Variant Human Breast Tumor Estrogen Receptor with Constitutive Transcriptional Activity. *Cancer Research* 51: 105-109.

Fuqua SAW, Schiff R, Parra I, Friedrichs WE, Su J-L, McKee DD *et al* (1999). Expression of Wild-Type Estrogen Receptor {{beta}} and Variant Isoforms in Human Breast Cancer. *Cancer Res* 59: 5425-5428.

Girault I, Andrieu C, Tozlu S, Spyratos F, Bièche I, Lidereau R (2004). Altered expression pattern of alternatively spliced estrogen receptor [beta] transcripts in breast carcinoma. *Cancer Letters* 215: 101-112.

Gnanapragasam VJ, Leung HY, Pulimood AS, Neal DE, Robson CN (2001). Expression of RAC 3, a steroid hormone receptor co-activator in prostate cancer. *Br J Cancer* 85: 1928-1936.

Gosden J, Middleton P, Rout D (1986). Localization of the human oestrogen receptor gene to chromosome 6q24----q27 by in situ hybridization. *Cytogenetics and cell genetics* 43: 218-20.

Gregory CW, Wilson EM, Apparao KBC, Lininger RA, Meyer WR, Kowalik A *et al* (2002). Steroid Receptor Coactivator Expression throughout the Menstrual Cycle in Normal and Abnormal Endometrium. *J Clin Endocrinol Metab* 87: 2960-2966.

Halachmi S, Marden E, Martin G, MacKay H, Abbondanza C, Brown M (1994). Estrogen receptor-associated proteins: possible mediators of hormone-induced transcription. *Science* 264: 1455-1458.

Hall JM, McDonnell DP (1999). The Estrogen Receptor {beta}-Isoform (ER{beta}) of the Human Estrogen Receptor Modulates ER{alpha} Transcriptional Activity and Is a Key Regulator of the Cellular Response to Estrogens and Antiestrogens. *Endocrinology* 140: 5566-5578.

Harris DM, Besselink E, Henning SM, Go VLW, Heber aD (2005). Phytoestrogens induce differential estrogen receptor alpha- or Beta-mediated responses in transfected breast cancer cells. *Experimental Biology and Medicine* 230: 558-568.

Heldring N, Pike A, Andersson S, Matthews J, Cheng G, Hartman J *et al* (2007). Estrogen Receptors: How Do They Signal and What Are Their Targets. *Physiological Reviews* 87: 905-931.

Horlein AJ, Naar AM, Heinzel T, Torchia J, Gloss B, Kurokawa R *et al* (1995). Ligand-independent repression by the thyroid hormone receptor mediated by a nuclear receptor co-repressor. *Nature* 377: 397-404.

Huang S, Litt M, Felsenfeld G (2005). Methylation of histone H4 by arginine methyltransferase PRMT1 is essential in vivo for many subsequent histone modifications. *Genes & Development* 19: 1885-1893.

Kato S, Endoh H, Masuhiro Y, Kitamoto T, Uchiyama S, Sasaki H *et al* (1995). Activation of the Estrogen Receptor Through Phosphorylation by Mitogen-Activated Protein Kinase. *Science* 270: 1491-1494.

Katzenellenbogen BS, Katzenellenbogen JA (2000). Estrogen receptor transcription and transactivation: Estrogen receptor alpha and estrogen receptor beta - regulation by selective estrogen receptor modulators and importance in breast cancer. *Breast Cancer Res* 2: 335 - 344.

Kim K, Thu N, Saville B, Safe aS (2003). Domains of Estrogen Receptor alpha (ERalpha) Required for ERalpha/Sp1-Mediated Activation of GC-Rich Promoters by Estrogens and Antiestrogens in Breast Cancer Cells *Molecular endocrinology* 17: 804-817.

Klinge CM, Jernigan SC, Mattingly KA, Risinger KE, Zhang aJ (2004). Estrogen response element-dependent regulation of transcriptional activation of estrogen receptors alpha and beta by coactivators and corepressors. *Journal of Molecular Endocrinology* 33: 387-410.

Koduri S, Goldhar A, Vonderhaar B (2006). Activation of vascular endothelial growth factor (VEGF) by the ER-α variant, ERΔ3. *Breast Cancer Research and Treatment* 95: 37-43.

Kuiper GGJM, Enmark E, Pelto-Huikko M, Nilsson S, Gustafsson J-A (1996). Cloning of a novel estrogen receptor expressed in rat prostate and ovary. *PNAS* 93: 5925-5930.

Kuiper GGJM, Lemmen JG, Carlsson B, Corton JC, Safe SH, van der Saag PT *et al* (1998). Interaction of Estrogenic Chemicals and Phytoestrogens with Estrogen Receptor {beta}. *Endocrinology* 139: 4252-4263.

Kushner PJ, Webb P, Uht RM, Liu M-M, and Richard H. Price J (2003). Estrogen receptor action through target genes with classical and alternative response elements. *Pure Applied Chemistry* 75, Nos: 1757-1769.

Lazennec G (2006). Estrogen receptor beta, a possible tumor suppressor involved in ovarian carcinogenesis. *Cancer Letters* 231: 151-7.

Lazennec G, Bresson D, Lucas A, Chauveau C, Vignon F (2001). ER beta inhibits proliferation and invasion of breast cancer cells. *Endocrinology* 142: 4120-30.

Lee LMJ, Cao J, Deng H, Chen P, Gatalica Z, Wang Z-Y (2008). ER-α36, a Novel Variant of ER-α, is Expressed in ER-positive and -negative Human Breast Carcinomas. *Anticancer research* 28: 479-483.

Leung Y-K, Mak P, Hassan S, Ho S-M (2006). Estrogen receptor (ER)-beta isoforms: A key to understanding ER-beta signaling. *PNAS* 103: 13162-13167.

Lewandowski S, Kalita K, Kaczmarek L (2002). Estrogen receptor [beta]: Potential functional significance of a variety of mRNA isoforms. *FEBS Letters* 524: 1-5.

Leygue E, Dotzlaw H, Watson PH, Murphy LC (1998). Altered Estrogen Receptor {alpha} and {beta} Messenger RNA Expression during Human Breast Tumorigenesis. *Cancer Res* 58: 3197-3201.

Maggiolini M, Vivacqua A, Fasanella G, Recchia AG, Sisci D, Pezzi V *et al* (2004). The G Protein-coupled Receptor GPR30 Mediates c-fos Up-regulation by 17{beta}-Estradiol and Phytoestrogens in Breast Cancer Cells. *J. Biol. Chem.* 279: 27008-27016.

Makinen S, Makela S, Weihua Z, Warner M, Rosenlund B, Salmi S, Hovatta O, Gustafsson J-A (2001). Localization of oestrogen receptors alpha and beta in human testis. *Mol Hum Reprod* 7:497-503

Mathioudaki K, Papadokostopoulou A, Scorilas A, Xynopoulos D, Agnanti N, Talieri M (2008). The PRMT1 gene expression pattern in colon cancer. *Br J Cancer* 99: 2094-2099.

Matsuzaki S, Fukaya T, Suzuki T, Murakami T, Sasano H, Yajima A (1999). Oestrogen receptor {alpha} and ß mRNA expression in human endometrium throughout the menstrual cycle. *Mol. Hum. Reprod.* 5: 559-564.

Matthews J, Gustafsson J-A (2003). Estrogen Signaling: A Subtle Balance Between ER{alpha} and ER{beta}. *Mol. Interv.* 3: 281-292.

Matthews J, Wihlen B, Tujague M, Wan J, Strom A, Gustafsson J-A (2006). Estrogen Receptor (ER) {beta} Modulates ER{alpha}-Mediated Transcriptional Activation by Altering the Recruitment of c-Fos and c-Jun to Estrogen-Responsive Promoters. *Mol Endocrinol* 20: 534-543.

McKenna NJ, Lanz RB, O'Malley BW (1999). Nuclear Receptor Coregulators: Cellular and Molecular Biology. *Endocr Rev* 20: 321-344.

Miksicek RJ, Lei Y, Wang Y (1993). Exon skipping gives rise to alternatively spliced forms of the estrogen receptor in breast tumor cells. *Breast Cancer Research and Treatment* 26: 163-174.

Misiti S, Schomburg L, M. Yen P, Chin WW (1998). Expression and Hormonal Regulation of Coactivator and Corepressor Genes. *Endocrinology* 139: 2493-2500.

Monroe DG, Secreto FJ, Subramaniam M, Getz BJ, Khosla S, Spelsberg TC (2005). Estrogen Receptor {alpha} and {beta} Heterodimers Exert Unique Effects on Estrogen- and Tamoxifen-Dependent Gene Expression in Human U2OS Osteosarcoma Cells. *Mol Endocrinol* 19: 1555-1568.

Moore JT, McKee DD, Slentz-Kesler K, Moore LB, Jones SA, Horne EL *et al* (1998). Cloning and Characterization of Human Estrogen Receptor [beta] Isoforms. *Biochemical and Biophysical Research Communications* 247: 75-78.

Morani A, Warner M, Gustafsson JÅ (2008). Biological functions and clinical implications of oestrogen receptors alfa and beta in epithelial tissues. *Journal of Internal Medicine* 264: 128-142.

Nilsson S, Makela S, Treuter E, Tujague M, Thomsen J, Andersson G *et al* (2001). Mechanisms of Estrogen Action. *Physiol. Rev.* 81: 1535-1565.

Ogawa S, Inoue S, Watanabe T, Hiroi H, Orimo A, Hosoi T *et al* (1998a). The Complete Primary Structure of Human Estrogen Receptor [beta] (hER[beta]) and Its Heterodimerization with ER [alpha]in Vivoandin Vitro. *Biochemical and Biophysical Research Communications* 243: 122-126.

Ogawa S, Inoue S, Watanabe T, Orimo A, Hosoi T, Ouchi Y *et al* (1998b). Molecular cloning and characterization of human estrogen receptor betacx: a potential inhibitor ofestrogen action in human. *Nucl. Acids Res.* 26: 3505-3512.

Oñate SA, Tsai SY, Tsai M-J, O'Malley BW (1995). Sequence and Characterization of a Coactivator for the Steroid Hormone Receptor Superfamily. *Science* 270: 1354-1357.

Pearce ST, Jordan VC (2004). The biological role of estrogen receptors alpha and beta in cancer. *Critical Reviews in Oncology/Hematology* 50: 3-22.

Pedram A, Levin MRaER (2006). Nature of Functional Estrogen Receptors at the Plasma Membrane. *Molecular Endocrinology* 20: 1996-2009.

Pedram A, Razandi M, Wallace DC, Levin ER (2006). Functional Estrogen Receptors in the Mitochondria of Breast Cancer Cells. *Mol. Biol. Cell* 17: 2125-2137.

Penot G, Le Peron C, Merot Y, Grimaud-Fanouillere E, Ferriere F, Boujrad N *et al* (2005). The Human Estrogen Receptor-{alpha} Isoform hER{alpha}46 Antagonizes the Proliferative Influence of hER{alpha}66 in MCF7 Breast Cancer Cells. *Endocrinology* 146: 5474-5484.

Pfeffer U, Fecarotta E, Castagnetta L, Vidali G (1993). Estrogen Receptor Variant Messenger RNA Lacking Exon 4 in Estrogen-responsive Human Breast Cancer Cell Lines. *Cancer Research* 53: 741-743.

Pink JJ, Wu SQ, Wolf DM, Bilimoria MM, Jordan VC (1996). A novel 80 kDa human estrogen receptor containing a duplication of exons 6 and 7. *Nucl. Acids Res.* 24: 962-969.

Poola I, Abraham J, Baldwin K (2002a). Identification of ten exon deleted ER[beta] mRNAs in human ovary, breast, uterus and bone tissues: alternate splicing pattern of estrogen receptor [beta] mRNA is distinct from that of estrogen receptor [alpha]. *FEBS Letters* 516: 133-138.

Poola I, Abraham J, Baldwin K, Saunders A, Bhatnagar R (2005). Estrogen receptors beta4 and beta5 are full length functionally distinct ERβ isoforms. *Endocrine* 27: 227-238.

Poola I, Abraham J, Liu A (2002b). Estrogen receptor beta splice variant mRNAs are differentially altered during breast carcinogenesis. *The Journal of Steroid Biochemistry and Molecular Biology* 82: 169-179.

Poola I, Koduri S, Chatra S, Clarke R (2000). Identification of twenty alternatively spliced estrogen receptor alpha mRNAs in breast cancer cell lines and tumors using splice targeted primer approach. *The Journal of Steroid Biochemistry and Molecular Biology* 72: 249-258.

Poola I, Speirs V (2001). Expression of alternatively spliced estrogen receptor alpha mRNAs is increased in breast cancer tissues. *The Journal of Steroid Biochemistry and Molecular Biology* 78: 459-469.

Saji S, Hirose M, Toi M (2005). Clinical significance of estrogen receptor β in breast cancer. *Cancer Chemotherapy and Pharmacology* 56: 21-26.

Schreihofer DA (2005). Transcriptional regulation by phytoestrogens in neuronal cell lines. *Molecular and Cellular Endocrinology* 231: 13-22.

Sentis S, Le Romancer M, Bianchin C, Rostan M-C, Corbo L (2005). Sumoylation of the Estrogen Receptor {alpha} Hinge Region Regulates Its Transcriptional Activity. *Mol Endocrinol* 19: 2671-2684.

Shang Y, Brown M (2002). Molecular Determinants for the Tissue Specificity of SERMs. *Science* 295: 2465-2468.

Shi L, Dong B, Li Z, Lu Y, Ouyang T, Li J *et al* (2009). Expression of ER-{alpha}36, a Novel Variant of Estrogen Receptor {alpha}, and Resistance to Tamoxifen Treatment in Breast Cancer. *J Clin Oncol* 27: 3423-3429.

Simoncini T, Fornari L, Mannella P, Caruso A, Garibaldi S, Baldacci C *et al* (2005). Activation of nitric oxide synthesis in human endothelial cells by red clover extracts. *Menopause:The Journal of The North American Menopause Society* 12: 69-77.

Sotoca AM, Sollewijn Gelpke MD, Boeren S, Stöm A, Gustafsson J-Å, Murk AJ *et al* (2011). Quantitative proteomics and transcriptomics addressing the estrogen receptor subtype-mediated effects in T47D breast cancer cells exposed to the phytoestrogen genistein. *Molecular and cellular proteomics* 10: M110.002170.

Sotoca AM, van den Berg H, Vervoort J, van der Saag P, Ström A, Gustafsson J-Å *et al* (2008). Influence of Cellular ER{alpha}/ER{beta} Ratio on the ER{alpha}-Agonist Induced Proliferation of Human T47D Breast Cancer Cells. *Toxicological sciences* 105: 303-311.

Speirs V, Adams IP, Walton DS, Atkin SL (2000). Identification of Wild-Type and Exon 5 Deletion Variants of Estrogen Receptor {beta} in Normal Human Mammary Gland. *J Clin Endocrinol Metab* 85: 1601-1605.

Stirone C, Duckles SP, Krause DN, Procaccio V (2005). Estrogen Increases Mitochondrial Efficiency and Reduces Oxidative Stress in Cerebral Blood Vessels. *Mol Pharmacol* 68: 959-965.

Ström A, Hartman J, Foster JS, Kietz S, Wimalasena J, Gustafsson aJ-Å (2004). Estrogen receptor beta inhibits 17beta-estradiol-stimulated proliferation of the breast cancer cell line T47D. *Proc Natl Acad Sci U S A* 101: 1566-71.

Suen C-S, Berrodin TJ, Mastroeni R, Cheskis BJ, Lyttle CR, Frail DE (1998). A Transcriptional Coactivator, Steroid Receptor Coactivator-3, Selectively Augments Steroid Receptor Transcriptional Activity. *Journal of Biological Chemistry* 273: 27645-27653.

Szego CM, Davis JS (1967). Adenosine 3',5'-monophosphate in rat uterus: acute elevation by estrogen. *Proc Natl Acad Sci USA* 58: 1711–1718.

Taylor SE, Martin-Hirsch PL, Martin FL (2010). Oestrogen receptor splice variants in the pathogenesis of disease. *Cancer Letters* 288: 133-148

Treeck O, Juhasz-Boess I, Lattrich C, Horn F, Ortmann RGaO (2008). Effects of exon-deleted estrogen receptor b transcript variants on growth, apoptosis and gene expression of human breast cancer cell lines. *Breast Cancer Reserach Treatment* 110: 507–520.

Veld MGRt, Schouten B, Louisse J, Es DSv, Saag PTvd, Rietjens IMCM *et al* (2006). Estrogenic Potency of Food-Packaging-Associated Plasticizers and Antioxidants As Detected in ER and ER Reporter Gene Cell Lines. *Journal or Agriculture and Food Chemistry* 54 4407 -4416.

Virgili F AF, Ambra R, Rinna A, Totta P, Marino M. (2004). Nutritional flavonoids modulate estrogen receptor alpha signaling. *IUBMB Life* 56: 145-151.

Wang Y, Miksicek RJ (1991). Identification of a Dominant Negative Form of the Human Estrogen Receptor. *Molecular Endocrinology* 5: 1707-1715.

Wang Z, Zhang X, Shen P, Loggie BW, Chang Y, Deuel TF (2005). Identification, cloning, and expression of human estrogen receptor-[alpha]36, a novel variant of human estrogen receptor-[alpha]66. *Biochemical and Biophysical Research Communications* 336: 1023-1027.

Warner M, Gustafsson J-A (2006). Nongenomic effects of estrogen: Why all the uncertainty? *Steroids* 71: 91-95.

Weitzmann MN, Pacifici R (2006). Estrogen deficiency and bone loss: an inflammatory tale. *J. Clin. Invest.* 116: 1186-1194.

Zhang Q-X, Borg A, Fuqua SAW (1993). An Exon 5 Deletion Variant of the Estrogen Receptor Frequently Coexpressed with Wild-Type Estrogen Receptor in Human Breast Cancer. *Cancer Research* 53: 5882-5884.

Zhang Q-X, Hilsenbeck SG, Fuqua SAW, Borg Å (1996). Multiple splicing variants of the estrogen receptor are present in individual human breast tumors. *The Journal of Steroid Biochemistry and Molecular Biology* 59: 251-260.

Zhao C, Matthews J, Tujague M, Wan J, Strom A, Toresson G *et al* (2007). Estrogen Receptor {beta}2 Negatively Regulates the Transactivation of Estrogen Receptor {alpha} in Human Breast Cancer Cells. *Cancer Res* 67: 3955-3962.

ZhaoYi Wang, XinTian Zhang, Peng Shen, Brian W Loggie, YunChao Chang, Deuel. TF (2006). A variant of estrogen receptor-{alpha}, hER-{alpha}36: Transduction of estrogen- and antiestrogen-dependent membrane-initiated mitogenic signaling. *PNAS* 103: 9063-9068.

6

GPCRs and G Protein Activation

Waelbroeck Magali
Université Libre de Bruxelles
Belgium

1. Introduction

An efficient intercellular communication system is essential to allow the correct functioning of multicellular organisms. This necessitates extracellular messengers (hormones, or neurotransmitters) as well as receptors , that is, proteins capable of recognizing these extracellular messengers and transducing a signal inside the cell. Each cell expresses several different types of receptors: signal transduction is both temporally and spatially integrated in order to generate the appropriate cellular response to each physiological situation.

Hydrophobic ligands are able to penetrate inside the cell: they recognize intracellular receptors that migrate to the nucleus and regulate protein transcription. Hydrophilic ligands, in contrast, are unable to cross the plasma membrane: these extracellular ligands recognize transmembrane receptors that then produce intracellular messengers to affect the target cell function. Several families of transmembrane receptors are known: some are ligand-gated ion channels, and regulate the transmembrane voltage (depolarization or hyperpolarisation) or the intracellular Ca^{++} concentration; others are ligand-activated enzymes: some synthetize cGMP, others phosphorylate specific target proteins upon ligand recognition; and yet other receptors (known as "G Protein Coupled Receptors" or "GPCRs") activate intracellular trimeric G proteins in response to extracellular signals. These receptor-activated G proteins in turn activate enzymes responsible for "second messenger" synthesis (adenylate cyclase → cAMP, or phospholipase C → Inositol trisphosphate (Inositol(1,4,5)P_3 or "IP3") and diacylglycerol), regulate ion channels, or activate other ("small") G proteins.

2. G protein coupled receptors

2.1 A few examples

The human genome contains at least 800 GPCRs, grouped in five main families (Fredriksson *et al.*, 2003). One of the best characterized GPCR, rhodopsin, is responsible for vision in the dark: it captures photons thanks to its prosthetic group (11-cis retinal), and leads to phosphodiesterase activation in retina rod cells. It is extremely abundant in the rod cell disks, comparatively easy to purify, and therefore has been very extensively studied for many years by biochemists. Other GPCRs allow us to taste and smell, control our appetite, fertility, stress, heart rate and breathing, etc. Adrenaline (the stress hormone), histamine (allergic reactions), glucagon (glycemia control), but also taste and odorant receptors, luteotropic and follicular stimulating hormone receptors (ovule and spermatozoid development), etc. recognize GPCRs and induce G protein activation.

2.2 GPCR families

All G protein coupled receptors possess a glycosylated extracellular amino-terminal (N-term) and an intracellular carboxyl-terminal (C-term) domain, separated by 7 transmembrane helices (TM1 to TM7) joined by three intracellular (IC1 to IC3) and three extracellular (EC1 to EC3) loops (Figure 1).

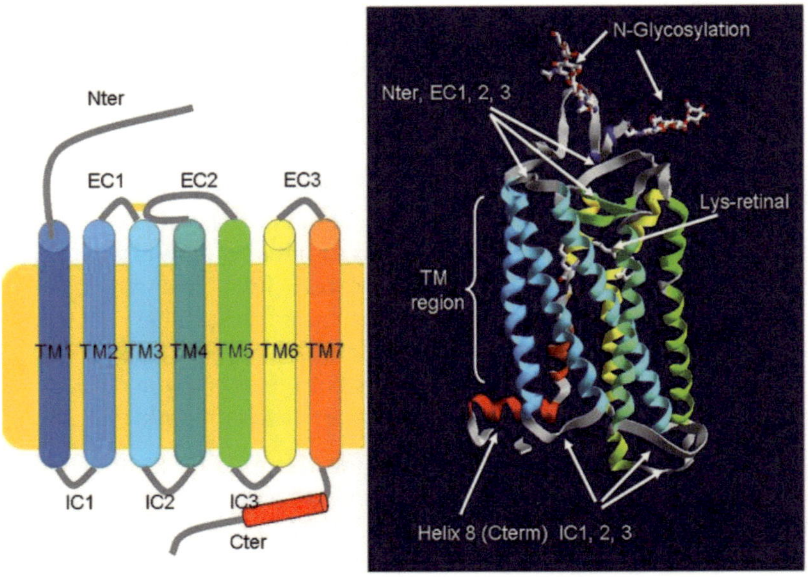

Fig. 1. Ribbon representation of the X-ray structure of rhodopsin. Left: schematic representation of a GPCR, showing the TM helices, intracellular (IC) and extracellular (EC) loops. Right: ribbon representation of the crystal structure of rhodopsin (1GZM). The prosthetic group, retinal, is covalently bound to a lysine side chain in TM7 (sticks).

The first and second EC loops are joined by a conserved disulfide bridge. The C-term region begins by an intracellular α-helix, H8, which lies horizontally on the plasma membrane: it forms an aromatic cluster with a tyrosine side chains from TM7 and interacts with the phospholipid head groups through lysine and arginine side chains. The 7 helices are arranged in a bundle (Figure 1). GPCRs have been identified in animals, yeast, plants. They probably arise from a common ancestor (Fredriksson *et al.*, 2003). Several hundred putative GPCRs have been identified in the human genome where they represent 1-3% of the genes (Fredriksson *et al.*, 2003). The vast majority of GPCRs (including most odorant receptors) share "signature" amino acids with rhodopsin (see below). They have been grouped in "Family A" (Kolakowski, Jr., 1994) or "rhodopsin-like receptor family" (Fredriksson *et al.*, 2003). Other GPCR families do not possess these highly conserved amino acids, but share other signature amino acids. For instance, all "family B" (secretin-receptor like) receptors possess a typical N-terminal "sushi" domain with three conserved disulfide bridges and have very strong sequence homologies in the transmembrane domain. Fifteen of these receptors with a comparatively short N-term domain ("sushi" domain only) are specialized in recognition of peptide hormones and neurotransmitters (glucagon, Growth Hormone Releasing Hormone (GHRH or GRF), parathyroid hormone (PTH) and others).

The majority of family B receptors possess several additional modules (EGF-like, Immunoglobulin-like, etc) before the sushi domain, suggesting that they might function as adhesion proteins (Mizuno and Itoh, 2011). Most of these are "orphan" receptors (that is: their ligand is unknown) and their ability to activate G proteins has not been proven yet. "Family C" receptors recognize amino acids (metabotropic receptors for glutamate and $GABA_B$ receptors for GABA) or calcium ions, through N-terminal "venus flytrap" domains (Jensen *et al.*, 2002; Wellendorph and Brauner-Osborne, 2009). Some of the taste receptors also are GPCRs: sweet and "umami" tasting molecules are recognized by "family C" GPCRs, and the "bitter" taste, by GPCRs that present very little homology with the other GPCRs, and form an additional receptor family (Fredriksson *et al.*, 2003).

2.3 Conserved residues in "family A" receptors

Rhodopsin and related receptors possess a few extremely conserved residues in each TM helix. In the Ballesteros and Weinstein nomenclature, the most conserved amino acid in each TM helix is numbered $X^{h.50}$ (where "h" is the helix number): for instance, $R^{3.50}$ is the most conserved amino acid in TM3; $D^{3.49}$ and $Y^{3.51}$ are the two conserved amino acids immediately preceding and following this arginine.

Some of these very conserved side chains are involved in structural features, like the prolines in helices 5, 6 and 7 that induce kinks in the TM helices. In the different family A receptors crystal structures (rhodopsin but also β-adrenergic, adenosine, histamine H3 receptors), the conserved asparagine of the TM7 $NPxxY(x)_{5-6}F$ motif is part of a hydrogen bond network involving TM1, TM2 ($D^{2.50}$) and TM7, while the tyrosine in this motif constrains TM7 in contact with aromatic side chains in the C term helix 8. Other conserved side chains play a role in the resting and/or active receptor conformation For instance, the arginine of the TM3 "DRY motif" ($E/DR^{3.50}Y$) at the intracellular end of the third transmembrane helix forms in rhodopsin a H bond network with $E^{3.49}$, $E^{6.30}$ and $T^{6.34}$. The "ionic lock" $R^{3.50}$-$E^{6.30}$ stabilizes the resting state: it is broken up in metarhodopsin II (the active rhodopsin conformation). In that structure, $R^{3.50}$ folds back inside the G protein to interact with $Y^{5.58}$, thereby creating an intracellular binding pocket, able to accommodate the G-protein. The ionic lock is less stable in the β-adrenergic receptors compared to rhodopsin, and this is perhaps responsible for their detectable constitutive activity (ability to activate G proteins in the absence of agonist) (Moukhametzianov *et al.*, 2011). $W^{6.48}$ of the $CWxP^{6.50}$ motif is in very close contact with the agonist ligands, and was thought to trip the switch of receptor activation by toggling between different rotamer conformations and thereby affecting the position of neighbouring aromatic side chains. Although this hypothesis is supported by computational mapping (Bhattacharya and Vaidehi, 2010), the toggle is not evident in the metarhodopsin II (Standfuss *et al.*, 2011; Choe *et al.*, 2011) or β_2-adrenergic receptor crystal structures (Rasmussen *et al.*, 2011b; Rasmussen *et al.*, 2011a).

3. Trimeric G proteins

3.1 G protein subtypes and GPCR effectors

The G proteins that transduce the signal from GPCRs are heterotrimeric (Gαβγ) proteins. Some of the mRNAs encoding the Gα subunits are subject to alternative splicing so that sixteen genes encode 23 known Gα proteins: (Birnbaumer, 2007). The Gα proteins are anchored to the plasma membrane by N-terminal myristoylation or palmitoylation. They can be grouped into four families based upon sequence homologies, and each GPCR has a preference for a single Gα or for a single family of Gα subunits. Each Gα subunit regulates one or a few effectors (Birnbaumer, 2007):

- G proteins in the G_s ($G_{s/olf}$) G protein family stimulate adenylate cyclase,
- G proteins in the G_i ($G_{i/o/t/gust/z}$) G protein family inhibit adenylate cyclase and/or regulate ion channels,
- G proteins in the $G_{q/11}$ ($G_{q/11/14/15/16}$) G protein family activate phospholipase C,
- G proteins in the G12/13 G protein family activate "Guanyl nucleotide Exchange Factors" (GEFs) that in turn activate another group of "small" (monomeric) G proteins, the Rho G proteins

The carboxyl-terminal Gα sequence is the major determinant for receptor recognition: exchanging this sequence allows the construction of promiscuous chimeric G proteins that can be used to drive GPCR coupling to a non-physiological effector (Kostenis *et al.*, 2005).

There are five known human Gβ and 12 Gγ genes (Birnbaumer, 2007). Most but not all of the Gβγ and Gα- Gβγ combinations are allowed. All Gγ subunits are C-terminally prenylated (some with geranyl-geranyl, others with farnesyl groups) and carboxymethylated: this helps to anchor the Gβγ subunits to the plasma membrane. The C-terminal sequence determines the nature of the prenyl group (farnesyl or geranyl-geranyl) modifying the Gγ subunit; both the C-terminal sequence and the prenyl group play an active role in the recognition of both rhodopsin and phospholipids (Katadae *et al.*, 2008). Although the literature on this subject is sparse, there is some evidence that other GPCRs also recognize preferentially specific Gβγ subunits (Jian *et al.*, 2001; Kisselev and Downs, 2003; Birnbaumer, 2007). The Gβγ subunits recognize and regulate a growing list of effectors, including ion channels, phospholipase C (PLC), phosphoinositide-3′ kinase-γ (PI3Kγ), various adenylate cyclase isoforms, etc. Different PLC isoforms respond differently to different Gβγ isoforms; and the cardiac ATP-inhibited inwardly rectifying K^+ channel (KirATP) is either inhibited or activated by Gβγ depending on the nature of the Gβ subunit (Birnbaumer, 2007).

3.2 G proteins as (inefficient) GTPases

G proteins are (poor) GTPases (Birnbaumer, 2007): they hydrolyze GTP slowly to GDP + inorganic phosphate, then release GDP extremely slowly. The GDP release and the GTP hydrolysis reactions are highly regulated, accompanied by conformation changes, and used as molecular clocks.

Trimeric G proteins are no exception to this rule: GTP binding is necessary to allow transient effectors activation (Oldham and Hamm, 2006; Birnbaumer, 2007). As summarized in Figure 2, the GDP release from trimeric G proteins is accelerated by G Protein Coupled Receptors (GPCRs) that function as "Guanyl nucleotides Exchange Factors" (GEFs): they allow GDP release, and this is rapidly followed by GTP recognition and dissociation of the two G protein subunits. Both subunits can then transiently recognize their respective effectors. The GTP hydrolysis reaction (leading to signal interruption) is accelerated by "Regulators of G protein Signaling" (RGS) proteins, that function as "GTPase Activator Proteins" (GAPs) and accelerate signal interruption.

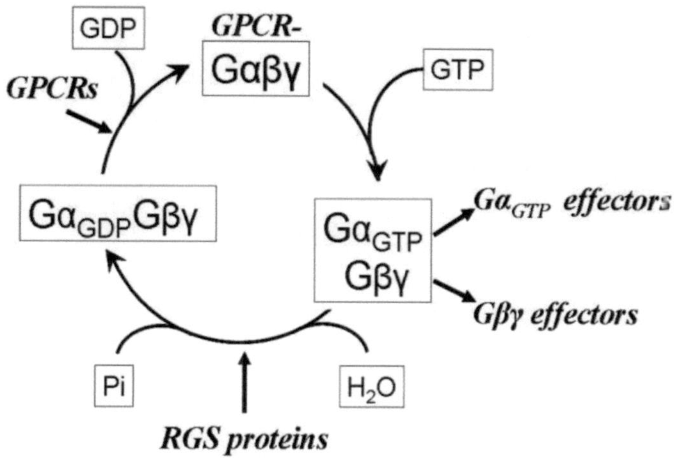

Fig. 2. the G protein activation cycle. In the resting state, the G protein is trimeric ($G\alpha_{GDP}G\beta\gamma$) and occupied by GDP. GPCRs interact with resting (GDP-bound) G proteins, facilitate the GDP release and stabilize an empty G protein conformation. GTP induces the dissociation of the two G protein subunits, $G\alpha_{GTP}$ and Gβγ: this allows both subunits to recognize and regulate their respective effectors (enzymes, channels or regulators of G protein signaling). $G\alpha_{GTP}$ hydrolyses GTP to GDP and the $G\alpha_{GDP}$ complex recognizes Gβγ with a very high affinity: the resting trimeric complex reforms spontaneously. GTP hydrolysis can be accelerated by "Regulators of G protein Signaling" (RGS) molecules.

3.3 G protein structures

The G proteins regulated by GPCRs are heterotrimeric (Birnbaumer, 2007). The three polypeptide chains form two independent subunits: the Gα and Gβγ subunits (Figure 3). The Gα protein structure can be divided into two domains held together by mutual interactions with the guanyl nucleotide: a N-terminal "Ras-like domain" (with strong structural homology with the small GTPases, Ras) and a C-terminal α-helical domain (Figure 3). In the agonist-receptor-G protein complex, the guanyl nucleotide has dissociated, and the helical domain "floats away" from the Ras -like domain (Rasmussen *et al.*, 2011b). The Gβ protein "WD repeats" (blue) forms a 7 blades beta-propeller domain, and the Gγ protein (green) wraps around Gβ, one of the two small α-helices forming a coiled-coil with the Gβ protein α-helix (Figure 3). Gβγ forms a stable complex that cannot be dissociated without denaturation but Gα can dissociate from Gβγ upon GTP binding.

As shown in Figure 4, the conformation of three segments ("switch regions") of Gα changes during the GTPase catalytic cycle (Oldham and Hamm, 2006; Rasmussen *et al.*, 2011b). This regulates the interaction of Gα with Gβγ and with its effectors. Switch 2 together with either switch 1 or 3 indeed forms part of the Gα - protein binding interface (Figure 5). It participates to the recognition of Gβγ, but also of effectors and "Regulators or G protein Signaling" (RGS) proteins (Figure 5): the Gα- Gβγ dissociation is essential to allow effectors activation by $G\alpha_{GTP}$.

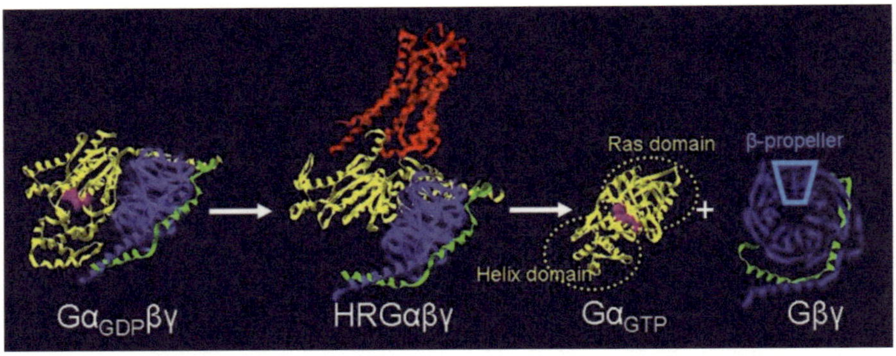

Fig. 3. Ribbon representations of G protein structures. The Gα subunits are presented in yellow, the β subunits in blue and the γ subunits in green and the β2-adrenergic receptor, in red. GDP and GTP are shown in fuchsia as space-filling structures. From left to right: the G_i protein (1GG2), the β2 adrenergic receptor-agonist-$G_s$complex (3SN6), the activated $G\alpha_{GTP}$ (1GIL) and Gβ1γ2 (1TBG) structures. The latter two structures have been rotated separately by approximately 90° compared to $G\alpha_{GDP}\beta\gamma$, to show the different domains (center right: the ras-like and α-helical domains of the Gα subunit; far right: a β-propeller structure in the Gβ subunit).

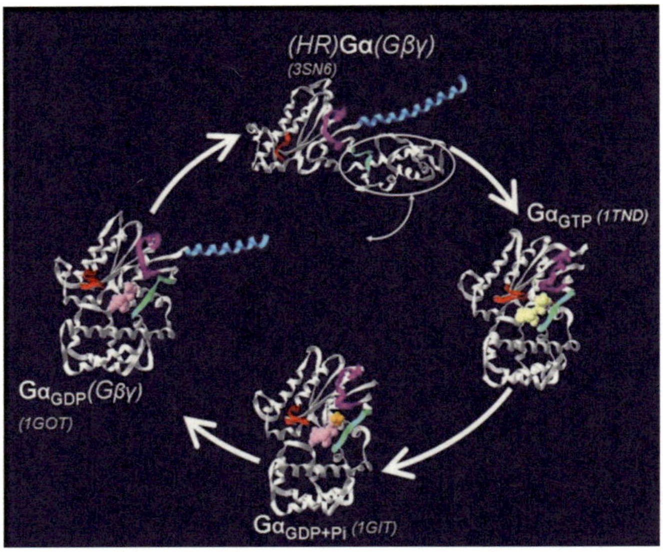

Fig. 4. Effect of guanyl nucleotide binding on the Gα subunit conformation. Ribbon representation of Gα in different crystallized complexes : the Gα subunit only is shown for simplicity; the Gβγ subunit when present would be in front and to the right of the Gα subunit. The N-terminal region, when structurally defined (stabilized by interactions with Gβγ) is represented by a blue ribbon. Three regions change conformation during the GTPase catalytic cycle: "switch 1" is shown in green, "switch 2" in gold, and "switch 3" in red. GDP (pink), GTP (yellow) and phosphate (pink) are shown as space filling structures. Left: the $G\alpha_t$ subunit in the $G\alpha_{GDP}\beta\gamma$ complex (PDB 1GOT); top center: the structure of $G\alpha_S$ in the ternary complex, HRG (PDB 3SN6), stabilized by a nanobody; right: the structure of the GTP analogue GTPγS-activated $G\alpha_t$ (PDB 1TND) and bottom center: the structure of $G\alpha_i$ bound to GDP and inorganic phosphate during GTP hydrolysis (PDB 1GIT).

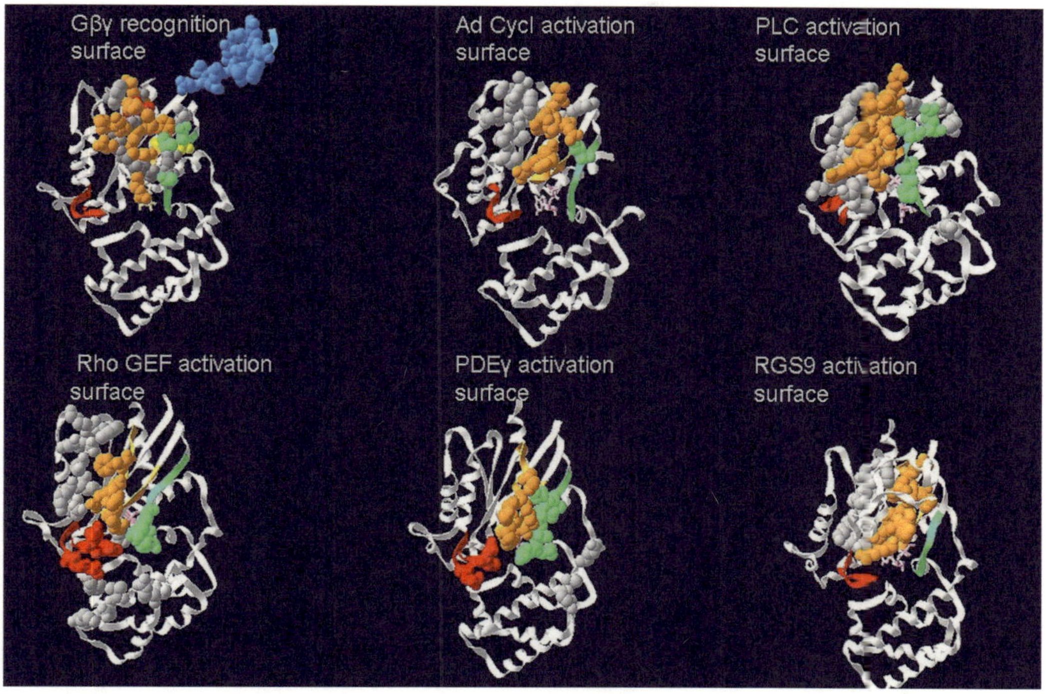

Fig. 5. Ribbon representation of Gα subunits in complex with Gβγ or with their effector and regulator proteins, showing the side chains that belong to the protein binding site. The Gα subunit only is shown for simplicity. The amino acids that belong to the protein binding sites of the different Gα structures are shown as space filling. When structurally defined, the ribbon (and side chains) that belong to the N-terminal helix are shown in blue, those that belong to switch 1 in green, to switch 2 in orange, to switch 3 in red, and to the rest of the protein in light grey. Top left: the GDP-bound transducin-G_i chimera showing the interaction surface with Gβγ (1GOT); top center: GTP*-bound $G\alpha_S$ showing the interaction surface with adenylate-cyclase (1CUL); top right: GTP-bound $G\alpha_q$ showing the interaction surface with phospholipase-C (3OHM); bottom left:GTP-bound $G\alpha_{13}$ showing the interaction surface with the Rho GEF, p115 (1SHZ); bottom center: GTP-bound transducin-G_i chimera showing the interaction surface with the phosphodiesterase inhibitor subunit, PDEγ (1FQJ) and bottom right, GTP-bound transducin-G_i chimera showing the interaction surface with RGS9 in a complex with PDEγ and RGS9 (1FQK).

Gβγ is also capable of activating certain effectors. Its binding site for Gα overlaps in part the Gβγ-effector and Gβγ- regulator binding sites: the dissociation of $G\alpha_{GTP}$ from Gβγ is essential to allow Gβγ to recognize its effectors (Figure 6).

GTP hydrolysis, rapidly followed by the release of the phosphate ion, modifies the switch regions conformation (Figure 4). The conformation change does not only inhibit the Gα-effector interaction (Figure 5) but also favors Gβγ recognition by $G\alpha_{GDP}$, thereby also inactivating Gβγ (Figure 6). Agonist-bound receptors interact with both G protein subunits (Figure 3): the formation of the "ternary complex" is an essential step for G protein activation, but the ternary complex Gα subunit is not in the right conformation to activate G protein effectors (Figure 3).

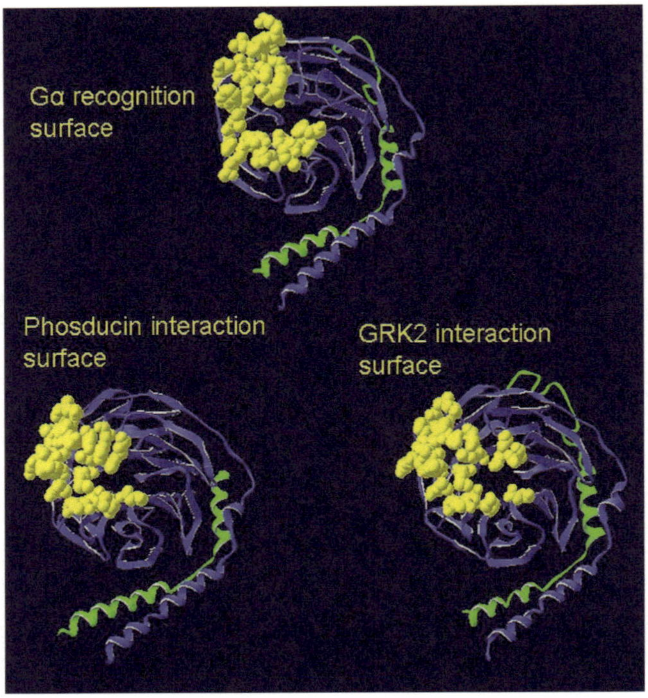

Fig. 6. Ribbon representation of Gβγ showing the interaction surface with Gα or with effectors. Ribbon representation of Gβγ in different crystallized complexes - the Gβ (blue) and Gγ (green) subunits only are shown for simplicity. The amino acids that belong to the Gα recognition surface (PDB ref 2BCG: top), to the phosducin binding site (1GP2: bottom left) or to the kinase, GRK2 binding surface (bottom right : 1OMW) are shown as space filling structures, in yellow.

4. G protein activation kinetics

Rhodopsin and related receptors **catalyze** G protein activation (Hamm, 1998): this means that each receptor sequentially activates several G proteins by facilitating the release of GDP, thereby allowing GTP binding. Rhodopsin does not enter in one of the Enzyme Commission "E.C." subclasses, as it does not catalyze the rupture or formation of covalent bond(s). The equations describing the reaction kinetics are nevertheless identical to those describing "ping pong" (double displacement) enzyme reaction (Waelbroeck *et al.*, 1997; Heck and Hofmann, 2001; Ernst *et al.*, 2007): G protein binding (substrate 1) is followed by GDP release (product 1), and GTP binding (substrate 2) is followed by the release of the activated G protein (product 2).

The kinetics of transducin activation by rhodopsin have been analyzed in detail (Heck and Hofmann, 2001; Ernst *et al.*, 2007). Rhodopsin recognizes transiently the GDP-bound trimeric G protein, transducin, and activates transducin at the diffusion limit (Ernst *et al.*, 2007). The physiological concentrations of the two "substrates" (GDP-bound transducin and GTP) are close to their respective Michaelis constants, K_M. In the case of double displacement reactions, it is unfortunately impossible to the individual rate constants of each reaction from the kinetic data (K_M, V_{max}). The [Substrate]/K_M ratios at physiological concentrations nevertheless strongly suggest that the concentrations of the four reaction intermediates, Rh*, Rh*-G_{GDP}, Rh*-G and Rh*-G_{GTP} (where Rh* is the light activated rhodopsin) are similar (Roberts and Waelbroeck, 2004): none of the reaction intermediates accumulates. These characteristics are reminiscent of the properties of triose phosphate isomerase and other "kinetically perfect enzymes" (Albery and Knowles, 1976): all the reaction intermediates have very similar free energies at physiological substrate concentrations and the energy barriers separating the different enzyme states are very low, thereby allowing the reaction to proceed at the diffusion limit.

Enzymes accelerate reactions by stabilizing the "transition state", that is, the state with the highest energy along the reaction coordinates. Trimeric G proteins cannot be purified in the absence of guanyl nucleotides: they are unstable when empty. As explained below, agonists stabilize the agonist-receptor-G protein ternary complex (that includes an empty G protein): like enzymes, active GPCRs catalyze G protein activation by decreasing the free energy of the transition state, that is, the empty G protein (Waelbroeck, 1999). GTP recognition by the G protein destabilizes the ternary complex: this induces activated G protein release - and allows the catalytic activation of several G proteins by a single receptor.

5. Ligand binding studies and the ternary complex model

Ligands that induce G protein activation are termed "agonists", and ligands that do not affect the receptor activity, "antagonists". Even at 100% receptor occupancy, some agonists have a larger effect than others on G protein activation: the more effective agonists are called "full agonists" and the less efficient compounds, "partial agonists". More recently, it has been demonstrated that most GPCRs have the ability to activate (inefficiently) their cognate G proteins in the absence of any ligand: this is called "constitutive activity". Compounds that counteract the receptors' constitutive activity are called "inverse agonists".

Agonist binding to GPCRs in the absence of either GDP or GTP facilitates the formation of the ternary complex involving the receptor, an agonist, and an empty G protein (Figure 3) (Lefkowitz *et al.*, 1976; De Lean *et al.*, 1980; Rasmussen *et al.*, 2011b). This is evident from the formation of a high molecular weight "ternary complex" (ligand-receptor-G protein, LRG) with a much higher affinity for agonists compared to isolated receptors. Guanyl nucleotides (GTP, GTP analogues or GDP) destabilize the G protein interaction with agonist-bound receptors by markedly decreasing the G-protein affinity for the receptor. When recognizing the high affinity ternary complex, guanyl nucleotides dramatically increase the agonists' dissociation rate, and decrease the receptor affinity for agonists while increasing their affinity for inverse agonists (see for instance (Lefkowitz *et al.*, 1976; Berrie *et al.*, 1979)); GDP is typically needed in larger concentrations than GTP or GTP analogues.

The effect of GTP on ligand binding can be used as a measure of the relative stability of the ternary complex compared to the binary ligand-receptor complex (Lefkowitz *et al.*, 1976). Full agonists have a higher affinity in the absence of GTP and inverse agonists have a higher affinity in its presence: the effect of GTP on ligand recognition is correlated with the ligands' ability to induce or inhibit G protein activation by the receptor (Lefkowitz *et al.*, 1976; De Lean *et al.*, 1980).

The original ternary complex model (Figure 7: top left) (Lefkowitz *et al.*, 1976) was designed to describe ligand binding to GPCRs. It describes the allosteric interactions between the ligand (L) and the G protein (G) recognizing different binding sites on the same receptor (R). Guanyl nucleotides were assumed to "prevent" G protein interaction with the receptor. The ternary complex model was later completed to the cubic ternary complex model (Figure 7: bottom right) (Weiss *et al.*, 1996a; Weiss *et al.*, 1996c; Weiss *et al.*, 1996b). Two receptor conformations (R and R*) without and with the ability to activate G proteins respectively are assumed to coexist at equilibrium (R←→R*) in the absence and presence of ligands or G proteins. Agonists and G proteins favor the active (R*) receptor conformation while inverse agonists stabilize the inactive (R) conformation. As in the ternary complex model, the cubic model describes the binding (as opposed to functional) properties of GPCRs; guanyl nucleotides are assumed to "prevent" the receptor-G protein interaction.

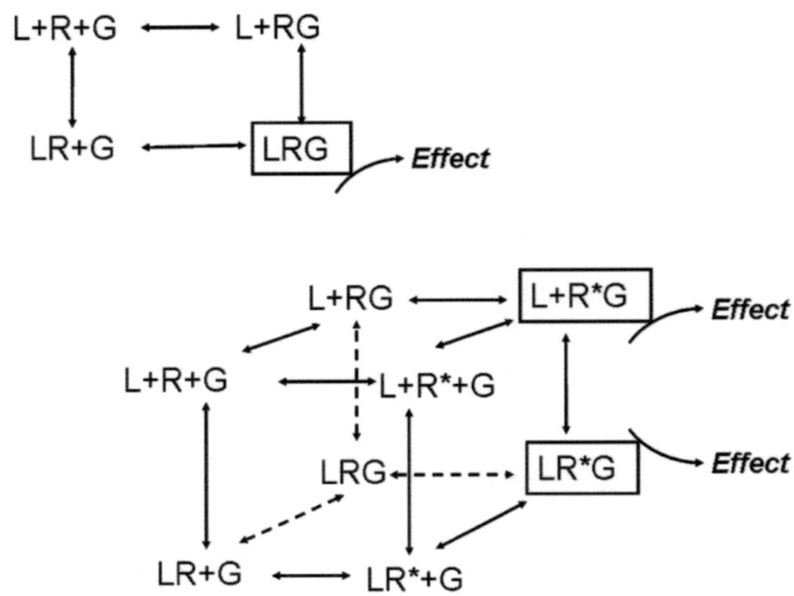

Fig. 7. The ternary complex model. Top left: the ternary complex model assumes that the receptor (R) can interact simultaneously with a ligand (L) and the G protein (G). Agonists facilitate and antagonists inhibit the receptor-G protein interaction; the ternary complex (LRG) is somehow responsible for transduction of the effect. Bottom right: the "cubic ternary complex model" assumes that the receptor can be found in a resting (R) or in an active (R*) conformation. G proteins (G) and agonist ligands stabilize R* while inverse agonists stabilize the R conformation. R*G and LR*G complexes are responsible for the biological effects of the active receptor.

Both ternary complex models have had a tremendous impact on our vision of GPCR function: the agonist-receptor-G protein complex is more and more often considered as "**the** active receptor". It should be remembered, however, that the ternary complex is certainly not "biologically active": it accumulates only under conditions where the G protein is unable to activate its' effectors (in the absence of GTP), and the G protein conformation in the β-adrenergic receptor-G protein complex (Figure 3) is not compatible with G_S-adenylate cyclase interaction (Figure 5)! The ternary complex model was designed to describe ligand binding to the receptors, as opposed to effectors activation.

6. "Resting" and "active" receptor structures

When rhodopsin is illuminated, its conformation passes through a number of intermediates (bathorhodopsin ($t_{1/2}$ 50ns), lumirhodopsin ($t_{1/2}$ 50µs), followed by metarhodopsin I and II) before releasing the all-trans retinal. Metarhodopsin II is biologically active: it catalyses G protein (transducin) activation.

The three dimensional structure of rhodopsin and several "family A" GPCRs has been elucidated by X-ray crystallography in the absence and presence of antagonists, agonists, G protein surrogates, or of the trimeric-G_s protein (Choe *et al.*, 2011; Lebon *et al.*, 2011; Rasmussen *et al.*, 2011a; Rasmussen *et al.*, 2011b; Rosenbaum *et al.*, 2011; Standfuss *et al.*, 2011; Warne *et al.*, 2011; Xu *et al.*, 2011). A conserved ionic bond between rhodopsin arginine 135 ($R^{3.50}$) (in the conserved E/DRY motif at the intracellular end of TM3) and glutamate 247 ($E^{6.30}$) , at the end of the third intracellular loop-TM6 junction tethers rhodopsin TM3 to TM6. In antagonist-bound β2-adrenergic receptors this hydrogen bond between the conserved arginine and glutamate is not visible in the crystal and β2-adrenergic receptors are known to activate slightly their cognate Gs G protein even in the absence of agonist; the ionic bridge is less stable than in rhodopsin (Moukhametzianov *et al.*, 2011).

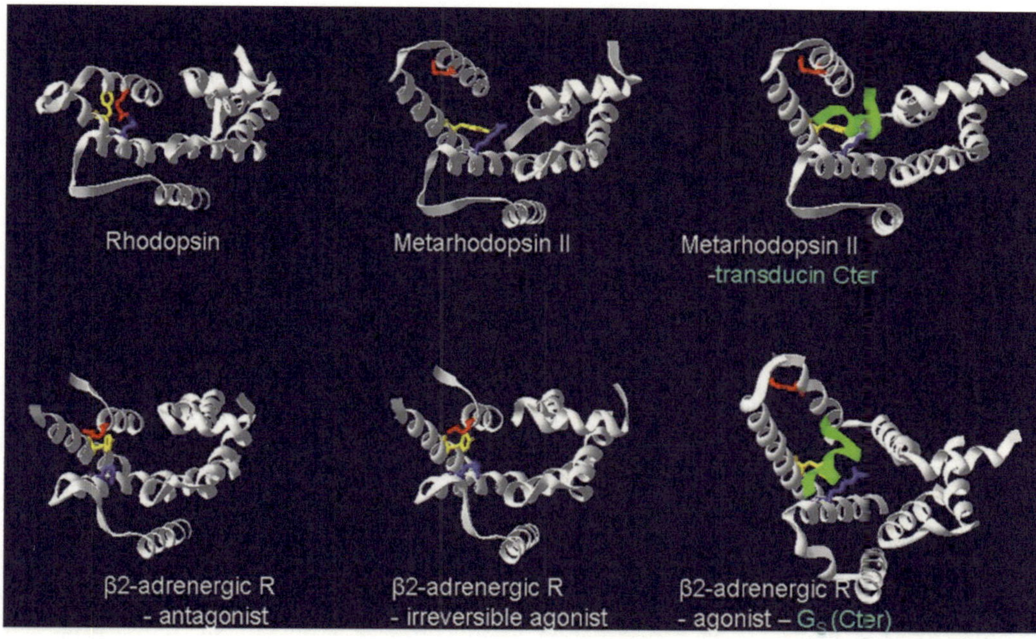

Fig. 8. Resting and activated GPCR conformations. Ribbon representation of the IC and TM regions of rhodopsin (1GZM), metarhodopsin II (3PXO), metarhodopsin II - Gt C-term complex (3PQR), antagonist-bound β_2-adrenergic receptor (2RH1), agonist-bound β_2-adrenergic receptor (3PDS) and agonist-Gs (Cterm) bound β_2-adrenergic receptor (3SN6), seen from the cytosol. The conserved TM3 arginine ($R^{3.50}$), TM5 tyrosine ($Y^{5.58}$) and TM6 glutamate ($E^{6.30}$) side chains are shown in red, yellow and blue respectively. The C-terminal transducin peptide (top right) and the C-terminal region of Gαs (bottom right) are shown as green ribbons.

In contrast with (dark adapted) rhodopsin, a large intracellular binding pocket is present between the TM helices of metarhodopsin II (Choe *et al.*, 2011): the distance between the conserved arginine ($R^{3.50}$) at the intracellular end of TM3 (E/DRY motif) and the conserved glutamate at the junction between the third intracellular loop and TM6 ($E^{6.30}$) increases from less than 3.3Å in rhodopsin (PDB 1GZM), bathorhodopsin (PDB 2G87) and lumirhodopsin (PDB 2HPY) to >15 Å in opsin (PDB 3CAP, 3DBQ) and metarhodopsin II (PDB 3PQR, 3PXO) (Figure 8). Arginine $R^{3.50}$ forms in opsin and metarhodopsin II a strong ion-dipole interaction with the conserved tyrosine, $Y^{5.58}$ in TM5. The G-protein binding pocket is created by the rotation of TM 5 and 6. It is large enough to accommodate the C-terminal helix of the transducin Gα subunit or of $G\alpha_S$, at almost 40° from the membrane surface (Park *et al.*, 2008; Scheerer *et al.*, 2008; Choe *et al.*, 2011) (Figure 8): this movement forces the opening of the GDP binding pocket and release GDP from the G protein (Rasmussen *et al.*, 2011b).

The sixth transmembrane helix (TM6) of the crystallized β1-, β2-adrenergic and adenosine A_{2A} receptors remains very close to TM3 even in the presence of agonists (Rasmussen *et al.*, 2011a; Rosenbaum *et al.*, 2011; Warne *et al.*, 2011); and the conserved E/DRY motif arginine folds towards the cytoplasm, in the direction of the conserved TM6 glutamate: the G protein binding pocket is unavailable (Figure 8). An open, "metarhodopsin II-like" structure is achieved by β2-adrenergic receptors only in the presence of a G protein surrogate or of G_S (Rasmussen *et al.*, 2011a; Rosenbaum *et al.*, 2011; Rasmussen *et al.*, 2011b): TM5 and TM6 rotate away from TM3, and the arginine side chain $R^{3.50}$ toggles away from the IC loop $E^{6.30}$ towards the conserved tyrosine, $Y^{5.38}$.

6.1 Is the "active receptor" mobile or rigid?

It is well known that crystallization rigidifies proteins and can lead to selection of an unusual conformation stabilized by "within the crystal" (non physiological) protein-protein interactions. For instance, opsin and metarhodopsin II conserve the same "opened" conformation in the crystal in the absence and presence of a transducin surrogate (Figure 8) (Altenbach *et al.*, 2008). Nevertheless, opsin – in contrast to metarhopsin II – has a very low ability to activate transducin: it is thus clear that crystallization of opsin "selected" a protein conformation with a low probability in intact membranes, stabilized through its contacts with other opsin molecules in the crystal.

X-ray diffraction studies have a tremendous impact on our perception of protein structure: they enhance the impression that proteins are rigid molecules with a well defined, stable conformation. In addition, the activity of most allosteric enzymes can explained in terms of two conformations with very different enzyme activities, stabilized by allosteric enhancers and inhibitors, respectively (Monod *et al.*, 1965). This is not an absolute rule, however: some enzymes change markedly in conformation upon substrate binding and dissociation – a phenomenon known as "induced fit" (Ma and Nussinov, 2010) and this can play an important role in enzyme regulation (Heredia *et al.*, 2006; Molnes *et al.*, 2011).

As suggested above, the ternary complex model had a tremendous impact on the way we understand GPCR activation, to the extent that the agonist-receptor-(empty) G protein complex is now described as "the" (one and only?) active receptor conformation – despite the fact that the Gα subunit conformation in the ternary complex (Figure 3) is not compatible with effectors activation. This interpretation was further supported by the initial computational mapping of conformational energy landscape of the β2-adrenergic receptor:

preliminary results (Bhattacharya and Vaidehi, 2010) indeed suggested that full agonists-bound receptors switch spontaneously to a more stable, active conformation very similar to metarhodopsin II. Detailed computational mapping in the presence of water and lipid molecules of rhodopsin (Provasi and Filizola, 2010) and of the agonist-bound β2-adrenergic receptor (Niesen *et al.*, 2011) however indicate that both proteins are flexible and able to sample a large number of conformations. In the case of the β2-adrenergic receptor, reversible "shearing" movements of TM5-6 relative to TM 1-4 and 7 and "breathing" movements (opening and closing of the ligand binding pocket) have been predicted.

In the case of traditional receptors (including the β2-adrenergic receptor), GTP has a tremendous effect on agonists' recognition: agonists have a significantly lower affinity and much greater dissociation rate in the presence of GTP. This suggests that the predominant receptor conformation under "functional" conditions (in the presence of GTP) is different from the ternary complex conformation (that accumulates in the absence of GTP). I should like to suggest that most GPCRs are able to recruit G proteins while in the "closed" (low agonist affinity) conformation (Hu *et al.*, 2010), open a G protein binding pocket and force GDP release to achieve the "high affinity" (ternary complex) conformation, then return to the "closed" conformation upon GTP recognition and activated G protein release. Agonists do not only facilitate the transition between the "closed" and "opened" conformations described by X-ray diffraction, but also decrease the free energy difference between the ternary complex and uncoupled receptors, thereby stabilizing the empty G protein conformation and facilitating GDP release, GTP binding. G protein dissociation from the receptor is then necessary to complete G protein activation.

6.2 Partial GPCR activation: Agonist efficacy

Some compounds seem less efficient than others to activate G protein coupled receptors: the rate of G protein activation by agonist-bound receptors varies depending on the ligand. Two explanations are usually put forward to account for this very common observation: partial agonists might stabilize the same "active" receptor conformation as full agonists but to a lesser extent; alternatively, they might stabilize an alternative receptor conformation, not quite as appropriate as the conformation induced by full agonists for G protein activation. These two explanations are non-exclusive and both explanations might in fact be correct at least where β2-adrenergic agonists are concerned (Bhattacharya and Vaidehi, 2010). Indeed, while dopamine was predicted to stabilize (less efficiently) the same "opened" receptor conformation as norepinephrine, salbutamol was predicted to stabilize a slightly different, less opened, receptor conformation. Yet a third explanation has been suggested for muscarinic receptors: agonists dissociate from muscarinic receptors with a rate constant comparable to the G protein exchange reaction rate. The efficacy of agonists activating M_3 muscarinic receptors was correlated with their dissociation rate constant, suggesting that the G protein activation reaction can be aborted prematurely if the agonist dissociates too early in the reaction cycle (Sykes *et al.*, 2009).

7. Do GPCRs function as monomers or dimers?

$GABA_B$ receptors (a "family C" GPCR) function as obligate dimers (Jones *et al.*, 1998; White *et al.*, 2002). One of the two subunits is trapped intracellularly by an endoplasmic reticulum retention signal; the second forms non-functional homodimers. Upon coexpression, formation of a heterodimer is driven by dimerization of the N-terminal region, and by formation of a coiled coil by the C-terminal regions α-helices. This masks the E.R. retention signal of the first subunit, and allows the expression of the functional heterodimer at the plasma membrane, (Jones *et al.*, 1998; White *et al.*, 2002). Likewise, all other family C receptors form heterodimers.

Rhodopsin, the best known "family A" GPCR, forms quasi crystalline arrays in rod cells disk membranes: this led to the suggestion that not only family C receptors but all GPCRs might function as homo- or heterodimers. Non-radiative energy transfer between two fluorophores ("FRET") or from a luminescent protein to a fluorophore ("BRET") can be easily demonstrated if the "donor" and "acceptor" molecules are close enough (typically less than 50 Å from each other). Chimeric constructs including "donor" and "acceptor" proteins (luciferase, fluorescent proteins from jellyfish, etc.) and the protein of interest can be built by molecular biology techniques; alternatively, the donor and acceptor fluorophores can be tagged chemically to the protein of interest, or to an antibody raised against this protein. BRET and FRET have been used to demonstrate not only protein-protein interaction, but also conformational changes of a single protein (by tagging for instance the N- and C-terminal of the protein of interest). In analogy with "family C" receptors, the vast majority of family A and several family B receptors have been shown by BRET or FRET experiments to either dimerize or oligomerize. This idea raised a lot of interest, because the potential consequences of dimerization are so multiple and important (Milligan, 2009; Milligan, 2010; Birdsall, 2010):

- Dimerization is essential for "family C" receptor expression at the plasma membrane (Temussi, 2009)) and might play a role in several other systems;
- Dimerization affects the "pharmacology" of some receptors. For instance: the sweet taste is sensed by a T1R2-T1R3 heterodimer, while "umami" is detected by a T1R1-T1R3 heterodimer (Temussi, 2009) (NB. The bitter taste is sensed by a non-family C GPCR, the T2R receptor; and salt and acid are recognized by "ligand gated channels" receptors (Temussi, 2009)). Likewise, dimerization of some GPCRs (i.e. dopamine, opiate or taste receptors) has been shown to alter their pharmacological properties - suggesting that their interaction is stable enough to affect the receptor conformation (Milligan, 2010; Milligan, 2009). Negative cooperativity has been observed between agonists binding to TSH and chemokine receptor dimers (Springael *et al.*, 2005; Urizar *et al.*, 2005)): binding of one agonist ligand to the dimer decreased the affinity of the second agonist by increasing its dissociation rate.
- Dimerization may have important functional consequences: a single agonist is sufficient for activation of G_S by the TSH receptor, but (low affinity) double occupancy of the dimer is necessary to support the activation of $G_{q/11}$ proteins and of phospholipase C (see below).
- Most if not all GPCRs do not only interact with G proteins, but also with other associated proteins, often in an agonist-modulated manner (Magalhaes *et al.*, 2011). Two receptors rather than one might be necessary to form optimal interactions with the receptor-associated proteins; alternatively, interaction of one subunit in the dimer with an associated protein might hinder or prevent the recognition of the second receptor subunit by steric hindrance, leading to "half of the sites reactivity".
- Several of the receptor-associated proteins act as scaffolds, recruiting in their turn other proteins in the vicinity of the receptor and of each other (Magalhaes *et al.*, 2011).

Dimerization of the receptors might be necessary to bring together some of the different accessory proteins recruited by each monomer.

It is unfortunately necessary to reassess the presence and consequences of dimerization for each receptor of interest: no generalization can be made in this respect. Indeed:

- Monomeric rhodopsin and β-adrenergic receptors, isolated and reconstituted in high density lipoprotein particles, function normally (Whorton *et al.*, 2007; Whorton *et al.*, 2008); and the receptor associated proteins arrestins, like GRKs, are able to recognize one receptor per protein (Hanson *et al.*, 2007; Bayburt *et al.*, 2011).
- Muscarinic M1 receptors dimerize only transiently: monomeric and dimeric forms are present at comparable concentrations in the plasma membrane at equilibrium and dimers dissociate rapidly ((Hern et al., 2010); see also (Johnston et al., 2011)). Cross-talk between two or more receptors is likely to necessitate strong interactions between the different monomers.
- Muscarinic M3 receptors (McMillin *et al.*, 2011) (and perhaps other receptors: (Johnston *et al.*, 2011)) are able to use several dimerization interfaces: this might explain why so many different dimerization interfaces have been observed when studying different receptors. This does not support the hypothesis that proteins like arrestin, G proteins or receptor kinases **need** a dimer for receptor recognition: the relative position of the two receptor monomers would be very important in that case.

8. Receptor promiscuity and biased signaling

GPCR "promiscuity" is defined as the ability of a given receptor to activate several different effectors (for review: (Hermans, 2003). While most receptors can probably induce parallel signaling by the $G\alpha_{GTP}$ and $G\beta\gamma$ subunits (see above), some are capable of activating different G proteins; and some use both G protein dependent and G protein independent signaling pathways.

"Biased signaling" refers to the observation that when receptors two or more signaling pathways, a few agonists preferentially use only one of the signaling pathways available to the other agonists - an observation that suggests that the activated receptor takes different conformations, depending on the agonist occupying its binding site.

8.1 Activation of several G protein subtypes by the same receptors

Each cell expresses several G proteins, belonging or not to the same family: all these G proteins will compete for recognition of each activated GPCR. Most G_i-coupled receptors activate several G_i isoforms with variable efficiency; some G_i and G_S coupled receptors activate in addition $G_{q/11}$ G proteins - less efficiently, or only at much higher agonist concentrations... Does this reflect a lower (but measurable) affinity of the non-cognate G protein, or less efficient activation?

GPCRs catalyze G protein activation: they should be considered like honorary enzymes. If several substrates compete for transformation by the same enzyme, the proportion of substrates transformed by the enzyme per minute, at steady state, is proportional to their relative substrate concentration over specificity constant ratios, $[S]/K_S$:

$$\frac{v^A}{v^B} = \frac{[A]/K_S^A}{[B]/K_S^B} \quad (1)$$

(where A and B represent the two substrates (G proteins), respectively, and K_S^A and K_S^B are their respective specificity constants : $K_S = \frac{k_{cat}}{K_M}$).

The equation is extremely similar to the equation describing the competition of several ligands for the same receptor: the proportion of receptor occupied by each ligand ([RA] and [RB]) is proportional to their relative ligand concentration over dissociation constant ratios:

$$\frac{[RA]}{[RB]} = \frac{[A]/K_D^A}{[B]/K_D^B} \quad (2)$$

The meaning of "K_D" and "K_S" is however very different: the dissociation constant, $K_D = 1/K_{affinity}$, is a concentration. It measures the ligand concentration necessary to occupy, at equilibrium and in the absence of competitors, 50% of the receptors. The specificity constant K_S, in contrast is a bimolecular reaction rate constant and measured in $M^{-1}sec^{-1}$. It measures the rate of formation of the "productive complex", ES† in the absence of alternative substrates of inhibitors.

Multiple G protein signaling has more often been observed in transfected systems, where it depends on the receptor expression level (for review: (Hermans, 2003)). Transiently expressed α2 adrenergic receptors inhibit adenylate cyclase at low agonist concentrations but activate the enzyme at high agonist concentrations (Fraser et al., 1989). Adenylate cyclase inhibition but not activation is prevented by Gi protein inactivation by pertussis toxin (Fraser et al., 1989): these results indicate that α2 adrenergic receptors are capable of activating both Gi and GS. The equations above predict that the relative activation rate of "Gi" and "GS" is proportional to their relative concentrations. Activation of GS by α2 adrenergic receptors is observed only at very high agonist concentrations: this suggests that, at very high agonist concentrations, Gi becomes unable to compete for receptor activation: in contrast with Gi-GDP, the activated Gi-GTP complex is probably unable to recognize agonist-bound α2 adrenergic receptors (Waelbroeck, 2001).

A few GPCRs are capable of activating several G proteins in physiological settings: the G protein specificity is not always "absolute". Although this is unusual in Family A, some G protein coupled receptors can be expressed as related isoforms due to alternative splicing of RNA expressed from a single gene or to RNA editing: this may lead to receptor isoforms with different abilities to activate G proteins (Hermans, 2003; Bresson-Bepoldin *et al.*, 1998). Alternatively, post-translational modifications such as phosphorylation of the receptor may alter its G protein specificity: β_2-adrenergic receptors activate G_S proteins, leading to adenylate cyclase and protein kinase A stimulation, then - after phosphorylation by protein kinase A - activate G_i proteins (Zamah *et al.*, 2002). The TSH receptor is able to activate G proteins from all four families (Allgeier *et al.*, 1997; Laugwitz *et al.*, 1996). Its binding properties are compatible with the hypothesis that it forms a stable dimer, and that occupancy of the dimer by one TSH molecule decreases the affinity of the second binding site ("negative cooperativity" (Urizar *et al.*, 2005)). While signaling through G_S is induced at very low TSH concentrations, low affinity occupancy of two binding sites per dimer appears to be necessary to drive receptor activation of G_i (Allen *et al.*, 2011).

8.2 GPCR phosphorylation and desensitization

Activated rhodopsin (metarhodopsin II) activates the rhodopsin kinase (GRK1), which in turn phosphorylates preferentially the activated rhodopsin (Premont and Gainetdinov, 2007). Both activation to metarhodopsin II and phosphorylation synergistically increase the rhodopsin affinity for an adaptor protein, arrestin. This protein competitively inhibits transducin recognition by steric hindrance - resulting in rhodopsin desensitization: light activated rhodopsin becomes unable to activate transducin and signaling is "arrested".

Likewise, ligand-activated GPCRs recognize and activate "GRKs" (G protein coupled Receptor Kinases), that in turn preferentially phosphorylate activated GPCRs (Premont and Gainetdinov, 2007; Huang and Tesmer, 2011). Most GPCRs are, in addition, targets for "second messenger activated kinases": they possess consensus sequences for protein kinase A that is activated in response to the increased cAMP, or for protein kinase C, activated by the phospholipase C signaling pathway (cytosolic Ca2+ and diacylglycerol). Receptor phosphorylation by these kinases will lead to "heterodesensitization", since a given agonist can induce the desensitization of receptors it does not activate.

Mammalian cells express seven GRKs: two of them (GRK1 and 7) are found only in rod and cone cells in the retina; GRK4 is found mainly in the testes and to a lesser extent in some brain regions and in the kidney, and the last four (GRK2, 3, 5 and 6) are ubiquitous (Yang and Xia, 2006). They can be subdivided in three subgroups: GRK1 and 7; GRK2 and 3, and GRK 4, 5 and 6. The C-terminal region of GRK2 and 3 is longer than in other GRKs and possesses a "Pleckstrin Homology" (PH) domain: these two GRKs are cytosolic and recruited by Gβγ in response to G protein activation (Yang and Xia, 2006). In contrast, GRK 1 and 7 are C-terminal farnesylated, GRK4 and 6 are palmitoylated on C-terminal cysteines and GRK4-6 have a highly conserved binding site for phosphatidyl inositol 4-phosphate: the PH domain and post-translational modifications facilitate the permanent localization of these GRKs at the plasma membrane (Yang and Xia, 2006). The N-terminal region of all GRKs is similar and important for receptor recognition - GRKs are highly specific in their receptor preference (Yang and Xia, 2006).

Three dimensional structures for at least one representative of the three GRK families have been determined by X-ray diffraction (Figure 9). In most structures, the N-terminal region (that is essential for receptor recognition) is undefined, and the active cleft is too "open" for substrate recognition, suggesting that the kinases usually crystallize in the resting conformation. Very recently, GRK6 was crystallized in a form very likely resembling its' active conformation, with a relatively "closed" active cleft (Figure 10) this structure probably resembles the active GRK (Boguth et al., 2010). It is characterized by a well defined extended N-terminal α-helix, that could easily be fitted - superimposed on the G protein C-terminal α-helix (Boguth et al., 2010) - in the intracellular pocket formed in the metarhodopsin II structure (Figure 10). This would bring the active cleft in close proximity to the receptors IC3 and Cter - the two regions that are phosphorylated by GRKs.

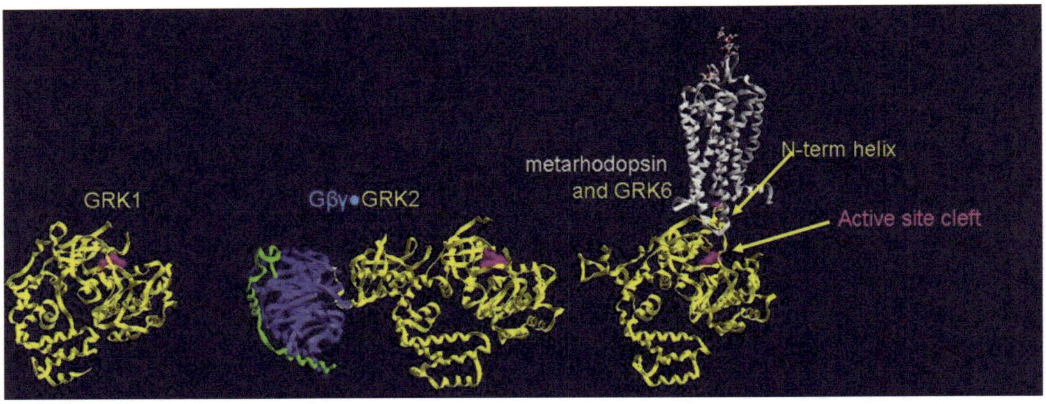

Fig. 9. Representative X-ray structures of GRKs from the three families. Ribbon structure of GRK1 (left: 3C4W), of the GRK2-Gβγ complex (center: 3KRW), and of the presumed "active conformation" of GRK6 (right: 3DQB) superimposed on the transducin C-term peptide in close apposition with the metarhodopsin II structure (3NYN), to form a hypothetical active GRK-receptor complex (according to (Boguth *et al.*, 2010)). The GRKs are shown by a yellow ribbon, co-crystallized ATP or ATP analogues in pink to identify the active site; Gβ and Gγ is shown in blue and green, respectively, and metarhodopsin II, in light grey.

Most GRKs are probably able to regulate GPCR signal transduction by phosphorylation-independent mechanisms. All GRKs have a "Regulator of G protein Signaling (RGS) homology" (RH) domain, and GRK2 and 3 have been shown to specifically interact through this domain with $G\alpha_q$ family members, thereby blocking their interaction with their effector, phospholipase C (see Figure 5). At least some GRKs are able to compete with G protein recognition by the activated receptor and/or compete with effector proteins for Gβγ recognition (Yang and Xia, 2006). By phosphorylating the receptor, they also increase markedly the receptor affinity for "arrestin" molecules that compete with G proteins for receptor recognition, facilitate receptor internalization in endosomes, and may serve as "scaffold", allowing "G protein independent signaling" (see below) (Premont and Gainetdinov, 2007; Huang and Tesmer, 2011).

Since GRKs and G proteins compete for the same (active) receptor conformation, the sequence of receptor recognition is important: GPCRs should recognize first the G proteins, then GRKs. "Sequential" recognition of two ligands is easily explained under the assumption that they have different *dissociation* rate constants (Motulsky and Mahan, 1984): the ligand with the faster dissociation rate constant will occupy the receptor rapidly, then progressively give place to the ligand with the slower dissociation rate constant (see Figure 10).

The most important factor under non equilibrium conditions is the relative **dissociation** (not association) rate constant of the two ligands. This might seem counterintuitive, but can easily be explained. Let us first examine the case of two ligands with different affinities due to different association rate constants. The lower affinity ligand will be needed in larger concentrations to significantly occupy the receptors at equilibrium: its lower association rate constant is then automatically compensated by the larger ligand concentration used. (The association rate is equal to k_{on}[L], where k_{on} is the association rate constant and [L], the ligand concentration). In contrast, if the two ligands have different affinities because of different dissociation rate constant: the larger dissociation rate constant of the low affinity ligand cannot be compensated by the larger ligand concentrations used to occupy the receptor at equilibrium: the dissociation rate, $k_{off}[LR]$ does not depend on the free ligand concentration. In order for the G protein, GRKs (and arrestin) to recognize sequentially the receptors, it is therefore necessary and sufficient that they have a different dissociation rate constants from the receptor. This is not a problem, as a very rapid G protein dissociation from the receptor is also necessary to allow receptor recycling and efficient catalytic activation of the G proteins…

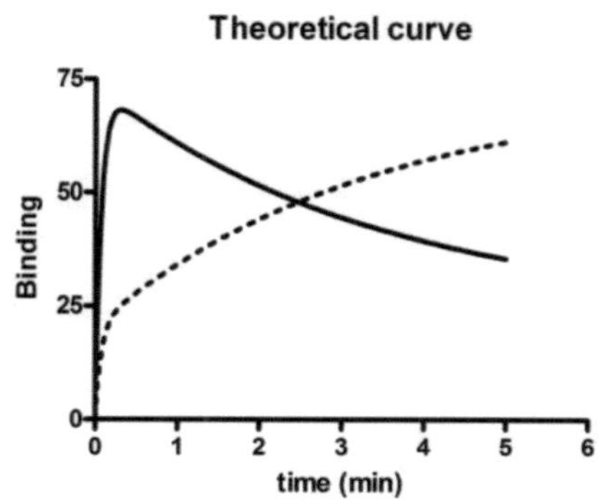

Fig. 10. Competitive binding of two ligands to the same receptor as a function of time. Ligand A (full line) has a k_{on}=$10^8M^{-1}min^{-1}$, k_{off} = 1 min^{-1} and is present at a concentration of 100 nM (10 K_D); it will occupy 24% of the receptors at equilibrium in the presence of ligand B. Ligand B (hatched line), has the same k_{on} = $10^8min^{-1}M^{-1}$, a lower dissociation rate (k_{off} = .0.1 min^{-1}) and is present at a concentration of 30 nM (30 K_D): it will occupy 73% of the receptors at equilibrium, in the presence of ligand A.

8.3 Arrestin recognition by GPCRs

All mammalian cells express at least one of the four "arrestins": rod and cone cells from the visual system express arrestins 1 and 4, respectively; arrestins 2 and 3 (also known as β-arrestin 1 and 2) are ubiquitously expressed. These proteins recognize and are activated by multi-phosphorylated, activated GPCRs: arrestin (arrestin 1) is specific for rhodopsin, arrestin 4, for the iodopsins, and arrestins 2 and 3 recognize most if not all GPCRs. Phosphorylation and receptor activation synergistically enhance rhodopsin-arrestin interactions: light activated rhodopsin and resting but phosphorylated rhodopsin have a 10-100 fold lower affinity for arrestin, and rhodopsin does not detectably interact with arrestin 1. β-arrestin binding to "traditional" GPCRs is affected more by phosphorylation than by agonist binding (Gurevich *et al.*, 1995).

All known arrestin 3D structures are rather similar to visual arrestin (Figure 11). They can be subdivided into two concave β-sheet domains held together by a hinge region, an ionic bridge network between two arginine and three aspartate side chains (center of the structure on Figure 11), and by interactions between the C-term tail, the first N-term β strand and the α helix (left of Figure 11).

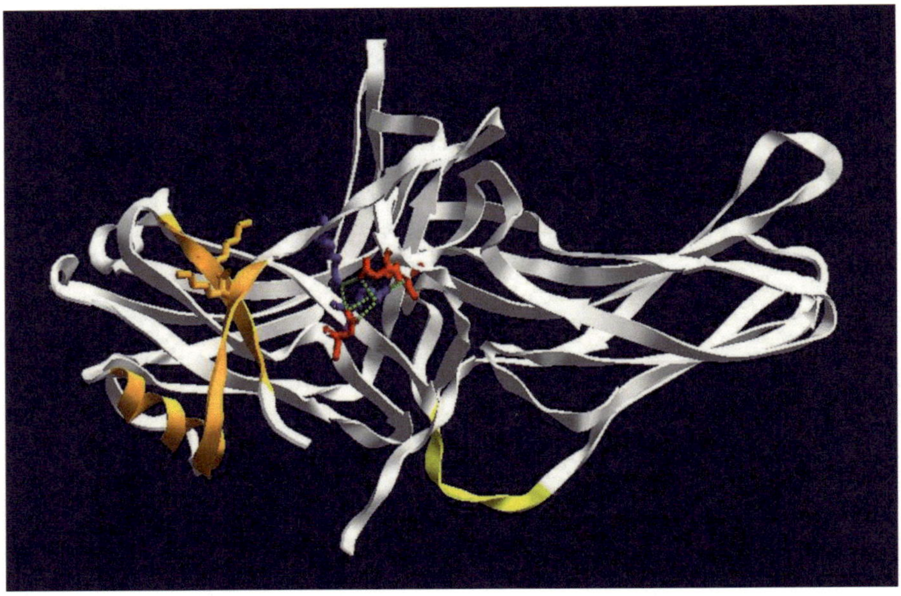

Fig. 11. The visual arrestin 1 crystal structure 1CF1. The N-terminal lobe is shown on the left, and the C-terminal lobe on the right. They are joined by a single random coil stretch (yellow) but held together through ionic interactions between buried arginine (blue) and aspartate (red) side chains, and by close contacts of the first N-terminal β-sheet stretch and α-helix with a C-terminal β-sheet stretch (orange). The two lysine side chains that are important for preferential phosphorylated>non-phosphorylated (light activated) rhodopsin recognition are shown in orange (left).

Each of the two arrestin domains is large enough to interact with a rhodopsin monomer (Figure 12). Even though visual arrestin forms a one to one complex with rhodopsin both in vitro (Bayburt *et al.*, 2011) and in vivo (Hanson *et al.*, 2007), several side chains covering both domains are implicated in rhodopsin recognition or rhodopsin - GPCR discrimination (Bayburt *et al.*, 2011; Vishnivetskiy *et al.*, 2011; Skegro *et al.*, 2007) (Figure 12). This indicates that arrestin undergoes a significant conformation change when it recognizes the phosphorylated receptors. This is confirmed by the observation that the arrestin sensitivity to proteolytic degradation increases upon GPCR recognition, and that the intramolecular BRET between the N- and C-terminal region of a luciferase–arrestin–Yellow Fluorescent Protein (YFP) construct is markedly affected by arrestin recognition of agonist-bound receptors (Shukla *et al.*, 2008).

At least two rhodopsin Ser/Thr must be phosphorylated to allow arrestin interaction with metarhodopsin; three phosphates support stronger arrestin binding, and heavier phosphorylation promotes arrestin binding , in addition, to neighbouring dark (inactive) rhodopsin and to phospho-opsin, two unpreferred rhodopsin forms (Vishnivetskiy *et al.*, 2007). Likewise, β-arrestin recognition increases mainly in response to multi-phosphorylation of the GPCRs C-terminal or IC3 sequence rather than in response to agonist binding (Gurevich *et al.*, 1995; Oakley *et al.*, 2000). "Phosphoserine/phosphothreonine rich" patches are necessary for stable, high affinity arrestin recognition (Oakley *et al.*, 2001). GPCRs that present patches of phosphorylated Ser/Thr residues (angiotensin II type 1A, neurotensin 1, vasopressin V2, thyrotropin-releasing hormone and substance P receptors) have a high affinity and do not discriminate the arrestin 1, 2 and 3 isoforms; they are rapidly internalized and recycle inefficiently or not at all. In contrast, β_2-adrenergic, μ opioid, endothelin type A, dopamine D_{1A}, and α_{1b} adrenergic receptors (with separate phosphorylated Ser/Thr residues) have a low affinity for β-arrestin 2 (arrestin 3), an even lower affinity for β-arrestin 1 (arrestin 2) and do not detectably recruit arrestin 1. Upon internalization, these receptors are rapidly dephosphorylated and recycled to the plasma membrane (Oakley *et al.*, 2000).

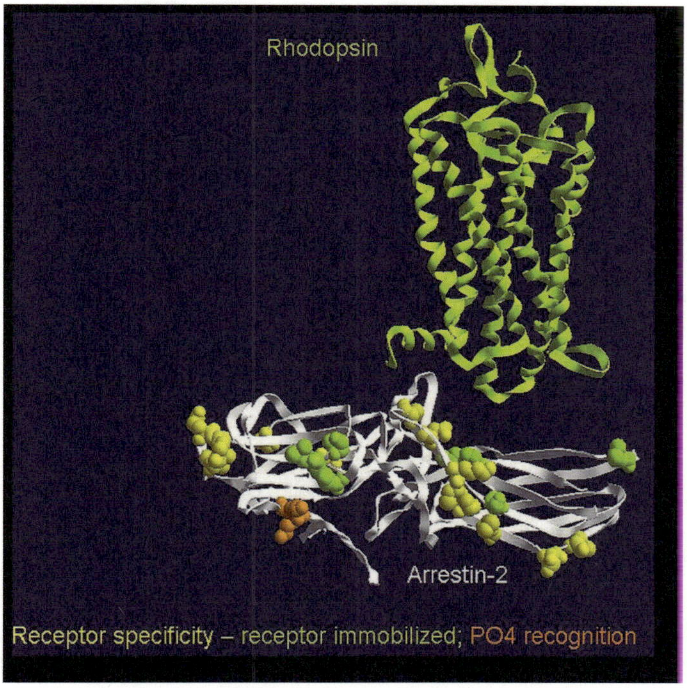

Fig. 12. Arrestin 3 (β-arrestin 2), close to opsin structure. Grey ribbon: arrestin 3 ribbon structure structure (1JSY), showing some of the side chains that are immobilized upon dark phosphorhodopsin recognition (light green), involved in the discrimination of light activated phosphorhodopsin from carbachol-activated phosphorylated M2 muscarinic receptor (yellow) or necessary for recognition of the phosphoserine/threonines (tan) (Vishnivetskiy *et al.*, 2011). Green ribbon: opsin structure 3CAP is shown for size comparison.

Arrestins change conformation upon receptor recognition (Shukla *et al.*, 2008) and behave as receptor-dependent "scaffold proteins" bringing together a number of other proteins (for review: (Premont and Gainetdinov, 2007; DeFea, 2011)). Some of their binding sites are shown in Figure 13.

Several β-arrestin scaffolds have been identified: this protein can recruit either MAP kinase partners, PI3Kinase or Akt, phosphodiesterase of actin assembly proteins scaffolds when bound to activated, phosphorylated receptors (DeFea, 2011). The different binding sites are very close (Figure 13): only some well-defined complexes can be formed or dissociated in response to agonist-receptor recognition by arrestin. The factors determining which complex is formed in response to a given receptor are still elusive.

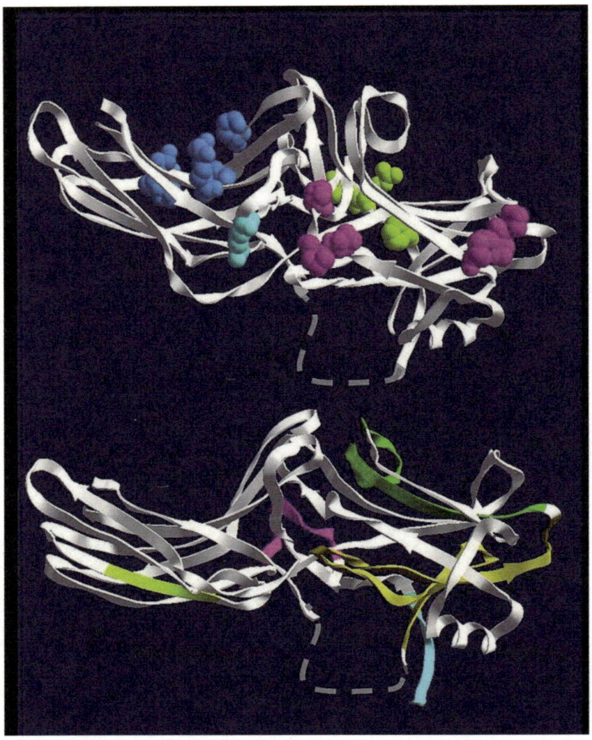

Fig. 13. Scaffolding sites on arrestin. Top: 3 arrestin 3 structure 3P2D The Proline-rich regions that allow SH3 domains recognition are shown in fuchsia, the PIP_2 binding site, in blue and side chains essential for β-adaptin binding, in green. The clathrin recognition site LφxφE is in the hatched (unstructured) region. R307, that is essential for cRaf1 recognition, is shown in light blue. Bottom : arrestin 2 structure (1JSY) showing the partially overlapping Ask1 and MEK binding regions (kaki), the MKK4 (dark green), PDED5 (light green), ERK2 (fuchsia) binding sites, and the partially overlapping Akt and clathrin binding sites (light blue) (according to (DeFea, 2011)).

9. References

Albery WJ and Knowles JR (1976) Free-Energy Profile of the Reaction Catalyzed by Triosephosphate Isomerase. *Biochemistry* 15: 5627-5631.

Allen MD, Neumann S and Gershengorn MC (2011) Occupancy of Both Sites on the Thyrotropin (TSH) Receptor Dimer Is Necessary for Phosphoinositide Signaling. *FASEB J*.

Allgeier A, Laugwitz KL, Van Sande J, Schultz G and Dumont JE (1997) Multiple G-Protein Coupling of the Dog Thyrotropin Receptor. *Mol Cell Endocrinol* 127: 81-90.

Altenbach C, Kusnetzow AK, Ernst OP, Hofmann KP and Hubbell WL (2008) High-Resolution Distance Mapping in Rhodopsin Reveals the Pattern of Helix Movement Due to Activation. *Proc Natl Acad Sci U S A* 105: 7439-7444.

Bayburt TH, Vishnivetskiy SA, McLean MA, Morizumi T, Huang CC, Tesmer JJ, Ernst OP, Sligar SG and Gurevich VV (2011) Monomeric Rhodopsin Is Sufficient for Normal Rhodopsin Kinase (GRK1) Phosphorylation and Arrestin-1 Binding. *J Biol Chem* 286: 1420-1428.

Berrie CP, Birdsall NJ, Burgen AS and Hulme EC (1979) Guanine Nucleotides Modulate Muscarinic Receptor Binding in the Heart. *Biochem Biophys Res Commun* 87: 1000-1005.

Bhattacharya S and Vaidehi N (2010) Computational Mapping of the Conformational Transitions in Agonist Selective Pathways of a G-Protein Coupled Receptor. *J Am Chem Soc* 132: 5205-5214.

Birdsall NJ (2010) Class A GPCR Heterodimers: Evidence From Binding Studies. *Trends Pharmacol Sci* 31: 499-508.

Birnbaumer L (2007) Expansion of Signal Transduction by G Proteins. The Second 15 Years or So: From 3 to 16 Alpha Subunits Plus Betagamma Dimers. *Biochim Biophys Acta* 1768: 772-793.

Boguth CA, Singh P, Huang CC and Tesmer JJ (2010) Molecular Basis for Activation of G Protein-Coupled Receptor Kinases. *EMBO J* 29: 3249-3259.

Bresson-Bepoldin L, Jacquot MC, Schlegel W and Rawlings SR (1998) Multiple Splice Variants of the Pituitary Adenylate Cyclase-Activating Polypeptide Type 1 Receptor Detected by RT-PCR in Single Rat Pituitary Cells. *J Mol Endocrinol* 21: 109-120.

Choe HW, Kim YJ, Park JH, Morizumi T, Pai EF, Krauss N, Hofmann KP, Scheerer P and Ernst OP (2011) Crystal Structure of Metarhodopsin II. *Nature* 471: 651-655.

De Lean A, Stadel JM and Lefkowitz RJ (1980) A Ternary Complex Model Explains the Agonist-Specific Binding Properties of the Adenylate Cyclase-Coupled Beta-Adrenergic Receptor. *J Biol Chem* 255: 7108-7117.

DeFea KA (2011) Beta-Arrestins As Regulators of Signal Termination and Transduction: How Do They Determine What to Scaffold? *Cell Signal* 23: 621-629.

Ernst OP, Gramse V, Kolbe M, Hofmann KP and Heck M (2007) Monomeric G Protein-Coupled Receptor Rhodopsin in Solution Activates Its G Protein Transducin at the Diffusion Limit. *Proc Natl Acad Sci U S A* 104: 10859-10864.

Fraser CM, Arakawa S, McCombie WR and Venter JC (1989) Cloning, Sequence Analysis, and Permanent Expression of a Human Alpha 2-Adrenergic Receptor in Chinese Hamster Ovary Cells. Evidence for Independent Pathways of Receptor Coupling to Adenylate Cyclase Attenuation and Activation. *J Biol Chem* 264: 11754-11761.

Fredriksson R, Lagerstrom MC, Lundin LG and Schioth HB (2003) The G-Protein-Coupled Receptors in the Human Genome Form Five Main Families. Phylogenetic Analysis, Paralogon Groups, and Fingerprints. *Mol Pharmacol* 63: 1256-1272.

Gurevich VV, Dion SB, Onorato JJ, Ptasienski J, Kim CM, Sterne-Marr R, Hosey MM and Benovic JL (1995) Arrestin Interactions With G Protein-Coupled Receptors. Direct Binding Studies of Wild Type and Mutant Arrestins With Rhodopsin, Beta 2-Adrenergic, and M2 Muscarinic Cholinergic Receptors. *J Biol Chem* 270: 720-731.

Hamm HE (1998) The Many Faces of G Protein Signaling. *J Biol Chem* 273: 669-672.

Hanson SM, Gurevich EV, Vishnivetskiy SA, Ahmed MR, Song X and Gurevich VV (2007) Each Rhodopsin Molecule Binds Its Own Arrestin. *Proc Natl Acad Sci U S A* 104: 3125-3128.

Heck M and Hofmann KP (2001) Maximal Rate and Nucleotide Dependence of Rhodopsin-Catalyzed Transducin Activation: Initial Rate Analysis Based on a Double Displacement Mechanism. *J Biol Chem* 276: 10000-10009.

Heredia VV, Thomson J, Nettleton D and Sun S (2006) Glucose-Induced Conformational Changes in Glucokinase Mediate Allosteric Regulation: Transient Kinetic Analysis. *Biochemistry* 45: 7553-7562.

Hermans E (2003) Biochemical and Pharmacological Control of the Multiplicity of Coupling at G-Protein-Coupled Receptors. *Pharmacology & Therapeutics* 99: 25-44.

Hern JA, Baig AH, Mashanov GI, Birdsall B, Corrie JE, Lazareno S, Molloy JE and Birdsall NJ (2010) Formation and Dissociation of M1 Muscarinic Receptor Dimers Seen by Total Internal Reflection Fluorescence Imaging of Single Molecules. *Proc Natl Acad Sci U S A* 107: 2693-2698.

Hu J, Wang Y, Zhang X, Lloyd JR, Li JH, Karpiak J, Costanzi S and Wess J (2010) Structural Basis of G Protein-Coupled Receptor-G Protein Interactions. *Nat Chem Biol* 6: 541-548.

Huang CC and Tesmer JJ (2011) Recognition in the Face of Diversity: Interactions of Heterotrimeric G Proteins and G Protein-Coupled Receptor (GPCR) Kinases With Activated GPCRs. *J Biol Chem* 286: 7715-7721.

Jensen AA, Greenwood JR and Brauner-Osborne H (2002) The Dance of the Clams: Twists and Turns in the Family C GPCR Homodimer. *Trends Pharmacol Sci* 23: 491-493.

Jian X, Clark WA, Kowalak J, Markey SP, Simonds WF and Northup JK (2001) Gbetagamma Affinity for Bovine Rhodopsin Is Determined by the Carboxyl-Terminal Sequences of the Gamma Subunit. *J Biol Chem* 276: 48518-48525.

Johnston JM, Aburi M, Provasi D, Bortolato A, Urizar E, Lambert NA, Javitch JA and Filizola M (2011) Making Structural Sense of Dimerization Interfaces of Delta Opioid Receptor Homodimers. *Biochemistry* 50: 1682-1690.

Jones KA, Borowsky B, Tamm JA, Craig DA, Durkin MM, Dai M, Yao WJ, Johnson M, Gunwaldsen C, Huang LY, Tang C, Shen Q, Salon JA, Morse K, Laz T, Smith KE, Nagarathnam D, Noble SA, Branchek TA and Gerald C (1998) GABA(B) Receptors Function As a Heteromeric Assembly of the Subunits GABA(B)R1 and GABA(B)R2. *Nature* 396: 674-679.

Katadae M, Hagiwara K, Wada A, Ito M, Umeda M, Casey PJ and Fukada Y (2008) Interacting Targets of the Farnesyl of Transducin Gamma-Subunit. *Biochemistry* 47: 8424-8433.

Kisselev OG and Downs MA (2003) Rhodopsin Controls a Conformational Switch on the Transducin Gamma Subunit. *Structure* 11: 367-373.

Kolakowski LF, Jr. (1994) GCRDb: a G-Protein-Coupled Receptor Database. *Receptors Channels* 2: 1-7.

Kostenis E, Waelbroeck M and Milligan G (2005) Techniques: Promiscuous Galpha Proteins in Basic Research and Drug Discovery. *Trends Pharmacol Sci* 26: 595-602.

Laugwitz KL, Allgeier A, Offermanns S, Spicher K, Van Sande J, Dumont JE and Schultz G (1996) The Human Thyrotropin Receptor: a Heptahelical Receptor Capable of Stimulating Members of All Four G Protein Families. *Proc Natl Acad Sci U S A* 93: 116-120.

Lebon G, Warne T, Edwards PC, Bennett K, Langmead CJ, Leslie AG and Tate CG (2011) Agonist-Bound Adenosine A2A Receptor Structures Reveal Common Features of GPCR Activation. *Nature* 474: 521-525.

Lefkowitz RJ, Mullikin D and Caron MG (1976) Regulation of Beta-Adrenergic Receptors by Guanyl-5'-Yl Imidodiphosphate and Other Purine Nucleotides. *J Biol Chem* 251: 4686-4692.

Ma B and Nussinov R (2010) Enzyme Dynamics Point to Stepwise Conformational Selection in Catalysis. *Curr Opin Chem Biol* 14: 652-659.

Magalhaes AC, Dunn H and Ferguson SS (2011) Regulation of G Protein-Coupled Receptor Activity, Trafficking and Localization by GPCR-Interacting Proteins. *Br J Pharmacol.*

McMillin SM, Heusel M, Liu T, Costanzi S and Wess J (2011) Structural Basis of M3 Muscarinic Receptor Dimer/Oligomer Formation. *J Biol Chem* 286: 28584-28598.

Milligan G (2009) G Protein-Coupled Receptor Hetero-Dimerization: Contribution to Pharmacology and Function. *Br J Pharmacol* 158: 5-14.

Milligan G (2010) The Role of Dimerisation in the Cellular Trafficking of G-Protein-Coupled Receptors. *Curr Opin Pharmacol* 10: 23-29.

Mizuno N and Itoh H (2011) Signal Transduction Mediated Through Adhesion-GPCRs. *Adv Exp Med Biol* 706: 157-166.

Molnes J, Teigen K, Aukrust I, Bjorkhaug L, Sovik O, Flatmark T and Njolstad PR (2011) Binding of ATP at the Active Site of Human Pancreatic Glucokinase--Nucleotide-Induced Conformational Changes With Possible Implications for Its Kinetic Cooperativity. *FEBS J* 278: 2372-2386.

Monod J, Wyman J And Changeux JP (1965) On the Nature of Allosteric Transitions: a Plausible Model. *J Mol Biol* 12: 88-118.

Motulsky HJ and Mahan LC (1984) The Kinetics of Competitive Radioligand Binding Predicted by the Law of Mass Action. *Mol Pharmacol* 25: 1-9.

Moukhametzianov R, Warne T, Edwards PC, Serrano-Vega MJ, Leslie AG, Tate CG and Schertler GF (2011) Two Distinct Conformations of Helix 6 Observed in Antagonist-Bound Structures of a Beta1-Adrenergic Receptor. *Proc Natl Acad Sci U S A* 108: 8228-8232.

Niesen MJ, Bhattacharya S and Vaidehi N (2011) The Role of Conformational Ensembles in Ligand Recognition in G-Protein Coupled Receptors. *J Am Chem Soc.*

Oakley RH, Laporte SA, Holt JA, Barak LS and Caron MG (2001) Molecular Determinants Underlying the Formation of Stable Intracellular G Protein-Coupled Receptor-Beta-Arrestin Complexes After Receptor Endocytosis*. *J Biol Chem* 276: 19452-19460.

Oakley RH, Laporte SA, Holt JA, Caron MG and Barak LS (2000) Differential Affinities of Visual Arrestin, Beta Arrestin1, and Beta Arrestin2 for G Protein-Coupled

Receptors Delineate Two Major Classes of Receptors. *J Biol Chem* 275: 17201-17210.

Oldham WM and Hamm HE (2006) Structural Basis of Function in Heterotrimeric G Proteins. *Q Rev Biophys* 39: 117-166.

Park JH, Scheerer P, Hofmann KP, Choe HW and Ernst OP (2008) Crystal Structure of the Ligand-Free G-Protein-Coupled Receptor Opsin. *Nature* 454: 183-187.

Premont RT and Gainetdinov RR (2007) Physiological Roles of G Protein-Coupled Receptor Kinases and Arrestins. *Annu Rev Physiol* 69: 511-534.

Provasi D and Filizola M (2010) Putative Active States of a Prototypic G-Protein-Coupled Receptor From Biased Molecular Dynamics. *Biophys J* 98: 2347-2355.

Rasmussen SG, Choi HJ, Fung JJ, Pardon E, Casarosa P, Chae PS, Devree BT, Rosenbaum DM, Thian FS, Kobilka TS, Schnapp A, Konetzki I, Sunahara RK, Gellman SH, Pautsch A, Steyaert J, Weis WI and Kobilka BK (2011a) Structure of a Nanobody-Stabilized Active State of the Beta(2) Adrenoceptor. *Nature* 469: 175-180.

Rasmussen SG, Devree BT, Zou Y, Kruse AC, Chung KY, Kobilka TS, Thian FS, Chae PS, Pardon E, Calinski D, Mathiesen JM, Shah ST, Lyons JA, Caffrey M, Gellman SH, Steyaert J, Skiniotis G, Weis WI, Sunahara RK and Kobilka BK (2011b) Crystal Structure of the Beta(2) Adrenergic Receptor-Gs Protein Complex. *Nature*.

Roberts DJ and Waelbroeck M (2004) G Protein Activation by G Protein Coupled Receptors: Ternary Complex Formation or Catalyzed Reaction? *Biochem Pharmacol* 68: 799-806.

Rosenbaum DM, Zhang C, Lyons JA, Holl R, Aragao D, Arlow DH, Rasmussen SG, Choi HJ, Devree BT, Sunahara RK, Chae PS, Gellman SH, Dror RO, Shaw DE, Weis WI, Caffrey M, Gmeiner P and Kobilka BK (2011) Structure and Function of an Irreversible Agonist-Beta(2) Adrenoceptor Complex. *Nature* 469: 236-240.

Scheerer P, Park JH, Hildebrand PW, Kim YJ, Krauss N, Choe HW, Hofmann KP and Ernst OP (2008) Crystal Structure of Opsin in Its G-Protein-Interacting Conformation. *Nature* 455: 497-502.

Shukla AK, Violin JD, Whalen EJ, Gesty-Palmer D, Shenoy SK and Lefkowitz RJ (2008) Distinct Conformational Changes in Beta-Arrestin Report Biased Agonism at Seven-Transmembrane Receptors. *Proc Natl Acad Sci U S A* 105: 9988-9993.

Skegro D, Pulvermuller A, Krafft B, Granzin J, Hofmann KP, Buldt G and Schlesinger R (2007) N-Terminal and C-Terminal Domains of Arrestin Both Contribute in Binding to Rhodopsin. *Photochem Photobiol* 83: 385-392.

Springael JY, Urizar E and Parmentier M (2005) Dimerization of Chemokine Receptors and Its Functional Consequences. *Cytokine Growth Factor Rev* 16: 611-623.

Standfuss J, Edwards PC, D'Antona A, Fransen M, Xie G, Oprian DD and Schertler GF (2011) The Structural Basis of Agonist-Induced Activation in Constitutively Active Rhodopsin. *Nature* 471: 656-660.

Sykes DA, Dowling MR and Charlton SJ (2009) Exploring the Mechanism of Agonist Efficacy: a Relationship Between Efficacy and Agonist Dissociation Rate at the Muscarinic M3 Receptor. *Mol Pharmacol* 76: 543-551.

Temussi PA (2009) Sweet, Bitter and Umami Receptors: a Complex Relationship. *Trends Biochem Sci* 34: 296-302.

Urizar E, Montanelli L, Loy T, Bonomi M, Swillens S, Gales C, Bouvier M, Smits G, Vassart G and Costagliola S (2005) Glycoprotein Hormone Receptors: Link Between Receptor Homodimerization and Negative Cooperativity. *EMBO J* 24: 1954-1964.

Vishnivetskiy SA, Gimenez LE, Francis DJ, Hanson SM, Hubbell WL, Klug CS and Gurevich VV (2011) Few Residues Within an Extensive Binding Interface Drive Receptor Interaction and Determine the Specificity of Arrestin Proteins. *J Biol Chem* 286: 24288-24299.

Vishnivetskiy SA, Raman D, Wei J, Kennedy MJ, Hurley JB and Gurevich VV (2007) Regulation of Arrestin Binding by Rhodopsin Phosphorylation Level. *J Biol Chem* 282: 32075-32083.

Waelbroeck M (1999) Kinetics Versus Equilibrium: the Importance of GTP in GPCR Activation. *Trends Pharmacol Sci* 20: 477-481.

Waelbroeck M (2001) Activation of Guanosine 5'-[Gamma-(35)S]Thio-Triphosphate Binding Through M(1) Muscarinic Receptors in Transfected Chinese Hamster Ovary Cell Membranes; 1. Mathematical Analysis of Catalytic G Protein Activation. *Mol Pharmacol* 59: 875-885.

Waelbroeck M, Boufrahi L and Swillens S (1997) Seven Helix Receptors Are Enzymes Catalysing G Protein Activation. What Is the Agonist Kact? *J Theor Biol* 187: 15-37.

Warne T, Moukhametzianov R, Baker JG, Nehme R, Edwards PC, Leslie AG, Schertler GF and Tate CG (2011) The Structural Basis for Agonist and Partial Agonist Action on a Beta(1)-Adrenergic Receptor. *Nature* 469: 241-244.

Weiss JM, Morgan PH, Lutz MW and Kenakin TP (1996a) The Cubic Ternary Complex Receptor-Occupancy Model. I. Model Description. *J Theor Biol* 178: 151-167.

Weiss JM, Morgan PH, Lutz MW and Kenakin TP (1996b) The Cubic Ternary Complex Receptor-Occupancy Model. II.Understanding Apparent Affinity. *J Theor Biol* 178: 169-182.

Weiss JM, Morgan PH, Lutz MW and Kenakin TP (1996c) The Cubic Ternary Complex Receptor-Occupancy Model. III. Resurrecting Efficacy. *J Theor Biol* 181: 381-397.

Wellendorph P and Brauner-Osborne H (2009) Molecular Basis for Amino Acid Sensing by Family C G-Protein-Coupled Receptors. *Br J Pharmacol* 156: 869-884.

White JH, Wise A and Marshall FH (2002) Heterodimerization of Gamma-Aminobutyric Acid B Receptor Subunits As Revealed by the Yeast Two-Hybrid System. *Methods* 27: 301-310.

Whorton MR, Bokoch MP, Rasmussen SG, Huang B, Zare RN, Kobilka B and Sunahara RK (2007) A Monomeric G Protein-Coupled Receptor Isolated in a High-Density Lipoprotein Particle Efficiently Activates Its G Protein. *Proc Natl Acad Sci U S A* 104: 7682-7687.

Whorton MR, Jastrzebska B, Park PS, Fotiadis D, Engel A, Palczewski K and Sunahara RK (2008) Efficient Coupling of Transducin to Monomeric Rhodopsin in a Phospholipid Bilayer. *J Biol Chem* 283: 4387-4394.

Xu F, Wu H, Katritch V, Han GW, Jacobson KA, Gao ZG, Cherezov V and Stevens RC (2011) Structure of an Agonist-Bound Human A2A Adenosine Receptor. *Science* 332: 322-327.

Yang W and Xia SH (2006) Mechanisms of Regulation and Function of G-Protein-Coupled Receptor Kinases. *World J Gastroenterol* 12: 7753-7757.

Zamah AM, Delahunty M, Luttrell LM and Lefkowitz RJ (2002) Protein Kinase A-Mediated Phosphorylation of the Beta 2-Adrenergic Receptor Regulates Its Coupling to Gs and Gi. Demonstration in a Reconstituted System. *J Biol Chem* 277: 31249-31256.

7

Application of Quantitative Immunogold Electron Microscopy to Determine the Distribution and Relative Expression of Homo- and Heteromeric Purinergic Adenosine A1 and P2Y Receptors

Kazunori Namba
National Institute of Sensory Organs, National Tokyo Medical Center, Tokyo Japan

1. Introduction

The idea that G-protein-coupled receptors (GPCRs) may generate or modify various functions as dimmers or higher-order oligomers is now generally accepted. Significant numbers of GPCRs exit as heteromeric assemblies (refered to as hetero-oligomerization), generating novel functions for ligand binding and second messengers, and in turn creating unique receptor trafficking systems for pharmacological profiles (Angers *et al.*, 2002, Bulenger *et al.*, 2005). This is also true of the purinergic receptor family. Over recent years, we have explored many biochemical and pharmacological aspects of this particular family via hetero-oligomerization between metabotropic (i.e. G protein-coupled) purinergic receptors (particularly between P1 and P2), in which the agonists are metabolites playing important role in the purinergic signaling cascade.

Purines such as adenosine triphosphate (ATP), via their specific P1 and P2 receptors, mediate a variety of physiological processes including pathophysiology, neurotransmission, neuromodulation, pain, cardiac function, immune responses and almost every aspect of development (Abbracchio *et al.*, 2009; Burnstock, 2007; Burnstock, 2008; Ralevic *et al.*, 1998). P1 receptors are further sub-classified into A_1, A_{2A}, A_{2B} and A_3 sub-types, all of which are G protein-coupled receptors (GPCRs). The adenosine A_1 receptor (A_1R) is known to regulate Ca^{2+}/K^+ channels, adenylate cyclase, and phospholipase C by coupling to $G_{i/o}$ proteins (Ralevic *et al.*, 1998). The P2 receptors can be further sub-classified into ligand-gated ion channel-type $P2X_{(1-7)}$ receptors, and G protein-coupled $P2Y_{(1, 2, 4, 6, 11, 12, 13, 14)}$ receptors. $P2Y_2R$-specific pharmacology (induction of Ca^{2+} release) has been analyzed in detail using CHO-K1 cells (Mehta *et al.*, 2008). In hippocampal astrocytes, $P2Y_1R$- and $P2Y_2R$-mediated Ca^{2+} responses differentially show two forms of activity-dependent negative feedback of synaptic transmission via the phospholipase C beta-IP^3 pathway (Fam *et al.*, 2003). $P2Y_2R$ modulation of pain responses has also been reported (Molliver *et al.*, 2002). Today, homo- or hetero-oligomers of many kinds of GPCRs have been reported (Bouvier, 2001) and the hetero-oligomerization of GPCRs affects various aspects of receptor function, including the alteration of ligand-binding specificity and cellular trafficking. We previously demonstrated that A_1R associates with $P2Y_1R$ in co-transfected HEK293T cells and in rat brain homogenates, whereby a $P2Y_1R$ agonist stimulates A_1R signaling via $G_{i/o}$ (Yoshioka *et al.*, 2001, Yoshioka *et al.*, 2002). Furthermore, in co-transfected HEK293T cells, hetero-oligomers display unique pharmacology whereby simultaneous activation of the two receptors attenuates A_1R signaling via $G_{i/o}$, but synergistically enhances $P2Y_2R$ signaling via $G_{q/11}$ (Suzuki *et al.*, 2006). Because A_1R are widely expressed in the brain (Yoshioka *et al.*, 2002), it is likely that these receptors also associate directly *in situ*; however, direct evidence of their oligomerization or precise co-localization in brain has yet to be demonstrated. In our laboratory, we are developing a new method, immunogold electron microscopic observation using different sized-immunogold particles enable visualize the oligomerization of A_1R and $P2Y_2R$ (Namba *et al.*, 2010). The aim of the study was to determine whether A_1R and $P2Y_2R$ associate with each other in the rat brain by looking for receptor complexes with immunogold electron microscopy (IEM). This method also provides information concerning the localization and density of GPCR monomers and oligomers expressed in transfected cells, that are also applicable to tissues such as brain.

In this chapter, we describe both pre- and post-embedding electronmicroscopic techniques to identify cells or tissues expressing GPCRs utilizing differently-sized immunogold particles, and review IEM quantification as an efficient approach to analyze two specific types of data. One data set represents the classification of receptor formations. A_1R and $P2Y_1R$ ($P2Y_2R$) produce five receptor formations which are made up of monomers (A_1R, P2YR), homo-oligomers (A_1R- A_1R, P2YR- P2YR) and hetero-oligomers (A_1R-P2YR). The second dataset describes the estimation of receptor expression levels by counting immunoreactive immunogold particles at the cell surface.

Establishing specific expression patterns of GPCRs at the ultrastructural level, and detecting homo- and hetero-oligomers of GPCRs in both co-transfected cultured cells and tissues, will enable us to visually understand some of the phenomena underlying signal transduction signalling pathways operating via GPCRs in a heteromeric dependent manner. It is widely accepted that drug discovery targets for rapid remedies are likely to be specific receptors expressed upon the cytoplasmic membrane. In order to establish the precise effects of new drugs, the expression patterns and expression level of A_1R and $P2Y_1R$ ($P2Y_2R$) represent significant factors to be considered, especially with regard to their association with cross-talk systems.

2. Immunostaining of GPCRs in transfected HEK293T cells and brain sections

Double immunofluororescence microscopic methods is now generally employed for studying the co-localization of GPCRs in transfected cells. Transient transfection using HEK293T cells with epitope tagged-receptors (Hemagglutinin: HA- or Myc-) in expression plasmids has been performed routinely in our laboratory (Yoshioka *et al.*, 2002; Nakata *et al.*, 2006). Before commencing immunogold electron microscopy, we routinely analyse, the subcelluar distribution of HA-A_1R and Myc-$P2Y_1R$ ($P2Y_2R$) in co-transfected cells by immunocytochemistry and confocal laser microscopy (Yoshioka *et al.*, 2001; Namba *et al.*, 2010). Confocal imaging of co-localized GPCRs provides highly detailed information regarding their co-localization upon cellular organelles, an important feature for the subsequent analysis of co-localization in ultrastructural images obtained by transmission electronmicroscopy. If two genes are co-localized at specific cellular organelles, then there is a much higher probability of hetero-oligomerization. Thus, confocal images of co-localization between A_1R and $P2Y_1R$ provide an important opportunity to determine whether immunoelectronmicroscopy is possible.

2.1 Results: Co-localization of A_1R and $P2Y_1R$ ($P2Y_2$) in transfected HEK293T cells

Confocal imaging for studying the GPCRs using transfected HEK293T cells is the most common method of co-localization of GPCRs. In our laboratory, the co-localization of A_1R and $P2Y_1R$ ($P2Y_2R$) in co-transfected HEK293T cells has been examined by the double immunostaining of HA-A_1R and Myc-$P2Y_1R$ (or HA-A_1R and Myc-$P2Y_2R$) in order to compare localization pattern. Confocal images of co-transfected HEK293T cells double labeled for HA-A_1R (red) and Myc-$P2Y_2R$ (green) are shown in Fig.1. As co-localization occurred upon the plasma membrane, this data supports the heteromeric association of A_1R and $P2Y_2R$. A similar pattern of co-localization for A_1R and $P2Y_2R$ has been demonstrated in rat brain sections as shown in Fig.2.

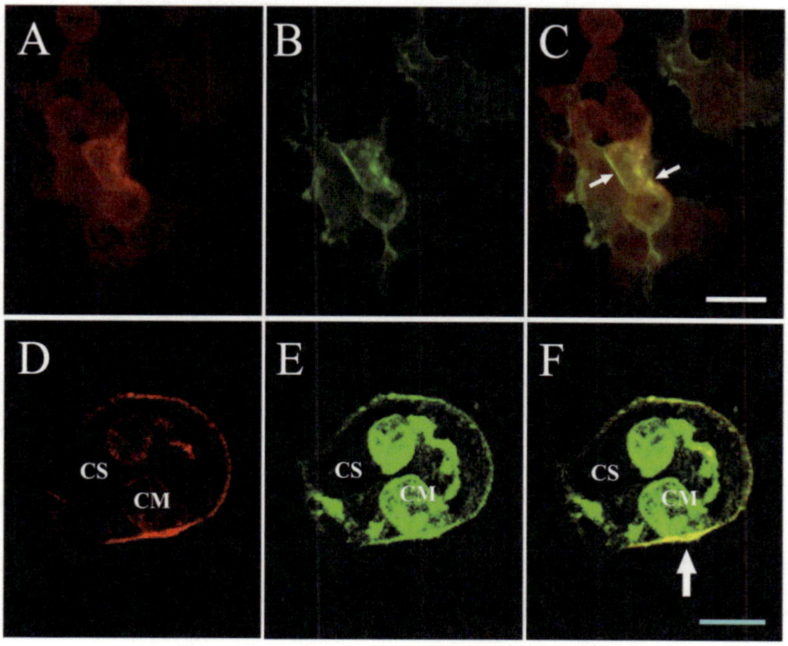

Fig. 1. Co-localization of A_1R and $P2Y_2R$. *A-C*. Confocal images of double immunostained HA-A_1R (A; red), Myc-$P2Y_2R$ (B; green), and their merged images (C; yellow) in co-transfected HEK293T cells. The co-localization of HA-A_1R and Myc-$P2Y_2R$ is evident at the cell surface membrane (C; small arrow). White bar = 50 µm (A-C). Confocal images of double immunofluorescence for HA-A_1R (D; red), Myc-$P2Y_2R$ (E; green), and their merged images (F; yellow) in co-transfected HEK293T cells are also obtained. Co-localizations of A_1R and $P2Y_1R$ (F) was detected upon the cell surface membrane, but was not as evident upon inner cellular membranes (F; arrow). Cyan bar = 10 µm (D-F). Fluorescent images were obtained via confocal laser scanning microscopy (Zeiss LSM410, Carl Zeiss, Oberkochen, Germany) at two levels: 30-µm(A-C), and 15-µm(D-F). At each level, serial images were collected at 1-µm intervals through a total sectional thickness of 40-µm. Serial optical sections were recorded using an air objective lens of (20 X and 40X, numerical aperture; 0.6).

Both receptors were localized predominantly upon the cell surface and cytosolic membranes (Fig. 1. A,B). Merged images showed co-localization mainly in cell membranes (Fig. 1. C.). Our negative controls showed no positive signals in non-transfected HEK293T cells, indicating that the immunoreactivity observed in Fig. 1 was specific to the expressed receptors (data not shown).

2.2 Results: Immunohistochemical studies in rat brain

We examined the expression of A_1R and P2YR in brain using using immunohistochemistry (Yoshioka *et al.*, 2002, Namba *et al.*, 2010). Prominent staining of A_1R and $P2Y_2R$ were observed, particularly in Purkinje cells (Fig. 2A-C). Expression was predominantly restricted to cell bodies and neuronal dendrites. Importantly, co-localization of A_1R and $P2Y_2R$ was observed in cell bodies within the cerebellum, but was detected within the nucleus of Purkinje cells.

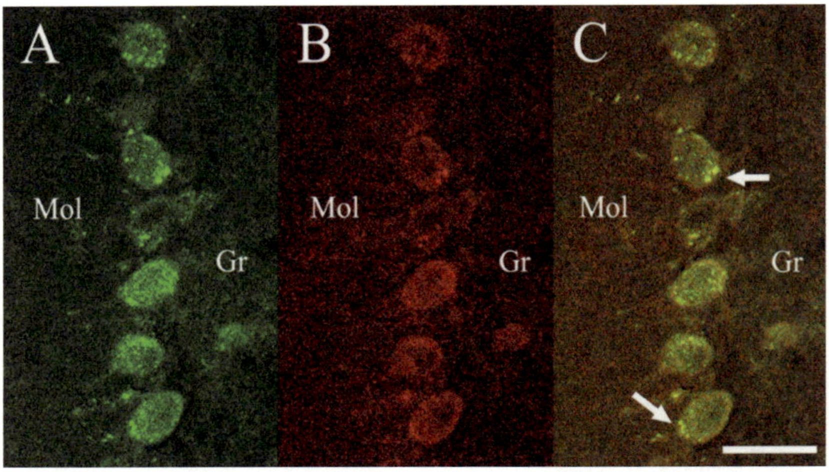

Fig. 2. Confocal images of double immunofluorescence stained A_1R (A; green), $P2Y_2R$ (B; red), and their merged images (C; yellow) in Purkinje cells. Mol: cerebellar molecular layer, Gr: cerebellar granule cell layer. Co-localization of A_1R and $P2Y_2R$ (C; yellow) were detected in the soma of the Purkinje cells (arrows). Bar = 50 μm. Fluorescent images were collected via confocal laser scanning microscopy (Zeiss LSM410, Carl Zeiss) and each 10-μm optical slice consisted of a stack of of 20 sections (0.5-μm thick). Serial optical sections were recorded using an air objective lens of (40X, numerical aperture; 0.6).

3. Pre and post-embedding immunogold electron microscopy of transfected HEK293T cells

The monomeric- or hetero-oligomerization of intrinsic GPCRs cannot be ascertained by immunoelectronmicroscopic examination of brain tissues alone. Data concerning the hetero-oligomerization of GPCRs in brain tissues is typically acquired from three experimental phases. The first phase involves immunoelectronmicroscopic data acquired from pre-embedding methods and gene transfected cells. This provides important information as to whether hetero-expressed GPCRs can oligomerize or not. In other words, this method compares the expression patterns and co-localization of differently-sized immunogold particles using transfected- or non transfected-cells, yielding data that can determine the occurrence of GPCR hetero-oligomerization when using the same antibodies and immunoreactive conditions. Additionally, a particular advantage of using pre-embedding methods is that native antigenicity is maintained. The second phase is to acquire data from gene transfected cells using post-embedding methods. The reason why immunoelectronmicroscopic observation of tissues is applied with post-embedding methods is because it is difficult for a specific epitope anti-body to penetrate into the cytoplasmic region of tissue cells. Before experimenting on tissues, it is important to confirm patterns of immune-reaction with transfected cells using post-embedding methods in order to form positive controls for specific tissues. The last phase is to acquire data from tissues using post-embedding methods. It is suggested that the hetero-oligomerization of GPCRs in tissues would be very precise as an oligomer of different-sized gold particles in a given case of comparative data from single transfected GPCRs and co-transfected GPCRs.

3.1 Results: Immunogold electron microscopic observations of HA-A_1R and Myc-P2Y_2R expressed in transfected cells

In our laboratory, we examined the cellular localization of HA-A_1R/Myc-P2Y_2R in co-transfected HEK293T cells using post-embedding methods, anti-HA or anti-Myc IEM (Figs. 3A-D) (Namba *et al.*, 2010). Immunogold particles were localized individually or in clusters, indicating that both HA-A_1R and Myc-P2Y_2R form monomers and homo-oligomers. Specificities of the gold-labeled anti-HA and anti-Myc antibodies were demonstrated by incubating A_1R-transfected HEK293T cells with a mixture of both antibodies. Data showed that only A_1R-labeled particles were present (Fig. 3A). No significant patterns were detected with either anti-HA and anti-Myc antibodies in mock-transfected HEK293T cells or with only secondary alone (i.e., no primary antibodies) in HA-A_1R-transfected HEK293T cells (data not shown). Also, when Myc-P2Y_2R-transfected HEK293T cells were incubated with both anti-HA and anti-Myc antibodies, we detected single particles (monomers) scattered all over the cells (Fig.3B). Another control for hetero-oligomerrization, HA-A_1R-transfected HEK293T cells incubated with both anti-HA and Myc-P2Y_2R, we observed all over the cells (Fig. 3D, inner cellular site). In HEK293T cells co-transfected with both HA-A_1R and Myc-P2Y_2R, clusters of different-sized particles were observed mainly at the cell surface (Fig. 3C) suggesting the formation of hetero-oligomers.

We would also like to introduce another means of investigating the cellular localization of anti-A_1R/anti-P2Y_1R in co-transfected HEK293T cells using post-embedding methods. Using mouse anti-A_1R or rabbit anti- P2Y_1R antibodies, hetero-oligomeric gold particles were clearly observed, predominantly at the cell surface (Fig.3C). This pattern concurred with patterns defined using pre-embedding methods and gold-labeled anti-HA and anti-Myc antibodies. The frequency of A_1R and P2Y_1R (P2Y_2R) hetero-oligomers detected using post- embedding methods was smaller than that detected with pre-embedding methods using fresh specimens. This was likely to be due to polymerization occuring during the embedding process (data not shown). We consider that the native antigenicities of GPCRs in transfected cells may be reduced by polymerization treatment with LR-white, though closely-related patterns of immunoreactivity were obtained in our laboratory across differeing methods (Fig. 4). HA-A_1R transfected HEK293T cells incubated with mouse anti-A_1R using post-embedding methods indicated patterns (Fig. 3A, large particles) identical to those arising from pre-embedding methods (Data not shown). This data indicates that the immunoreactivety of mouse anti-A_1R antibodies using LR-white post-embedding were effective in HA-A_1R transfected HEK293T cells. Hetero-oligomeric gold particles of the mouse anti-A_1R or rabbit anti-$P2Y_1R$ antibodies were observed at the cell surface (Fig.4B). HA-A_1R-transfected HEK293T cells incubated with either mouse anti-A_1R or rabbit anti-$P2Y_1R$ were also seen scattered all over the cells (Fig. 4C, cellular surface).

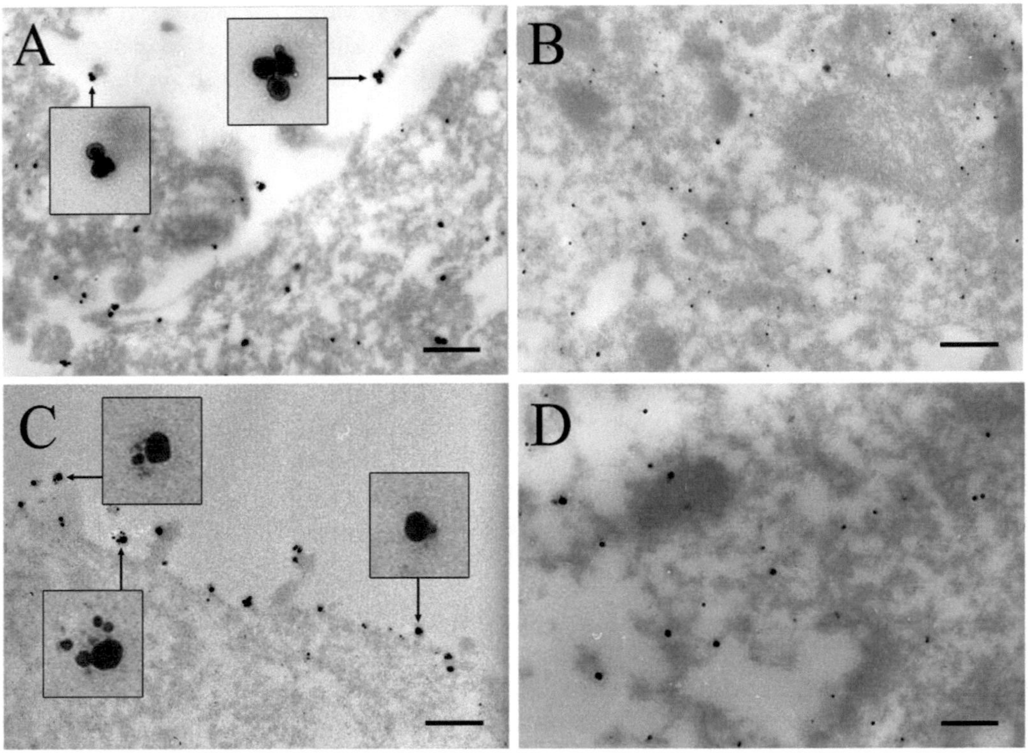

Fig. 3. Immunogold electron microscopy (post-embedding) method to visualise A_1R and $P2Y_2R$ in transfected HEK293T cells using nanogold particles. A: Localization of HA-A_1R (large particles) detected with anti-HA in HA-A_1R-transfected HEK293T cells. B: Localization of Myc-$P2Y_2R$ (small particles) detected with anti-Myc in Myc-$P2Y_2R$-transfected HEK293T cells. C: Anti-HA and anti-Myc immuno-localization of anti-A_1R and Myc-$P2Y_2R$ in co-transfected HEK293T cells. D: HA-A_1R-transfected HEK293T cells incubated with both anti-HA and anti-Myc. Bars represent 100 nm.

4. Post-embedding immunogold electron microscopy of brain tissues

In keeping with observations gained by post-embedding methods for the study of co-transfected HEK293T cells described in Section 3.1. of this chapter, we should highlight that it is also possible to apply immunogold staining using post-embedding methods for the study of brain tissues. Comparing the dose of immunoreactivity from gold particles reflecting hetero-oligomers using co-transfected culture cells and post-embedding methods is essential in acquiring immunogold pattern data from hetero-oligomers *in situ*.

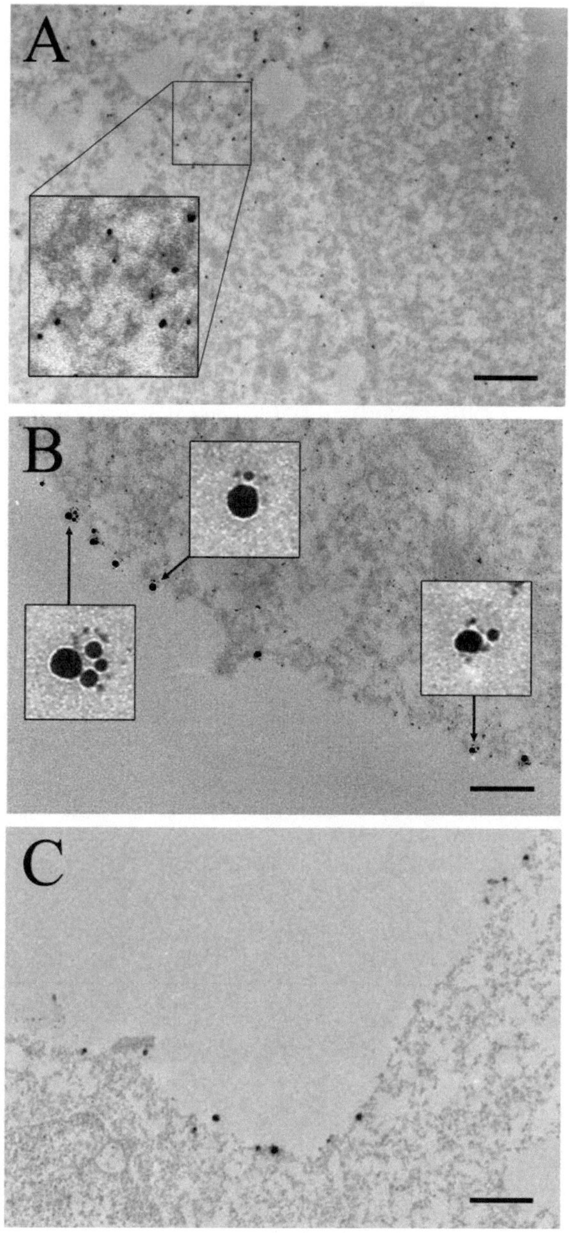

Fig. 4. Immunogold electron microscopy to visualise A_1R and $P2Y_1R$ in transfected HEK293T cells using nanogold particles. A: Localization of HA-A_1R (small particles) detected with mouse anti- A_1R in HA-A_1R -transfected HEK293T cells. B: Mouse anti-A_1R and rabbit anti-$P2Y_1R$ immuno-localization of HA-A_1R and Myc-$P2Y_1R$ in co-transfected HEK293T cells. C: HA-A_1R-transfected HEK293T cells incubated with both mouse anti-A_1R and rabbit anti-$P2Y_1R$. Bars represent 100 nm.

There are two reasons for this. Firstly, the antigenicity for the two receptor antibodies must accurately reflect hetero-oligomers or single expression. Secondly, the immunoreactivity of a particular antibody could be variable depending upon the methodology utilized, for example whether post- or pre-embedding methods were deployed. Usually, immunoreactive conditions during pre-embedding methods are much better than during post-embedding methods. However, immunoreactions involving inner tissues are technically difficult to perform. In the following section, we introduce how we can image the hetero-oligomerization of A_1R and $P2Y_1R$ in brain tissues using post-embedding immunogold electron microscopy.

4.1 Results: Immunogold electron microscopic observations of A_1R and P2YR expressed in brain tissues

We incubated post-embedded, primary antibody-stained rat brain tissues with two secondary antibodies labeled with gold particles (a 5-nm gold particle-conjugated goat anti-mouse IgG antibody for A_1R, and a 10-nm gold particle-conjugated goat anti-rabbit IgG antibody for $P2Y_1R$). As negative controls, brain tissues were stained with only secondary antibodies conjugated with different sized gold particles; no significant immunoreactivity was observed under the experimental conditions (data not shown). As found with transfected HEK293T cells (3.2-3.4), we observed clusters of different-sized gold particles at cytoplasmic membranes in cell bodies, indicating the presence of heteromeric complexes of endogenous A_1R and $P2Y_1R$ in the rat cerebellum (Fig. 5). Significant immunoreactivity was

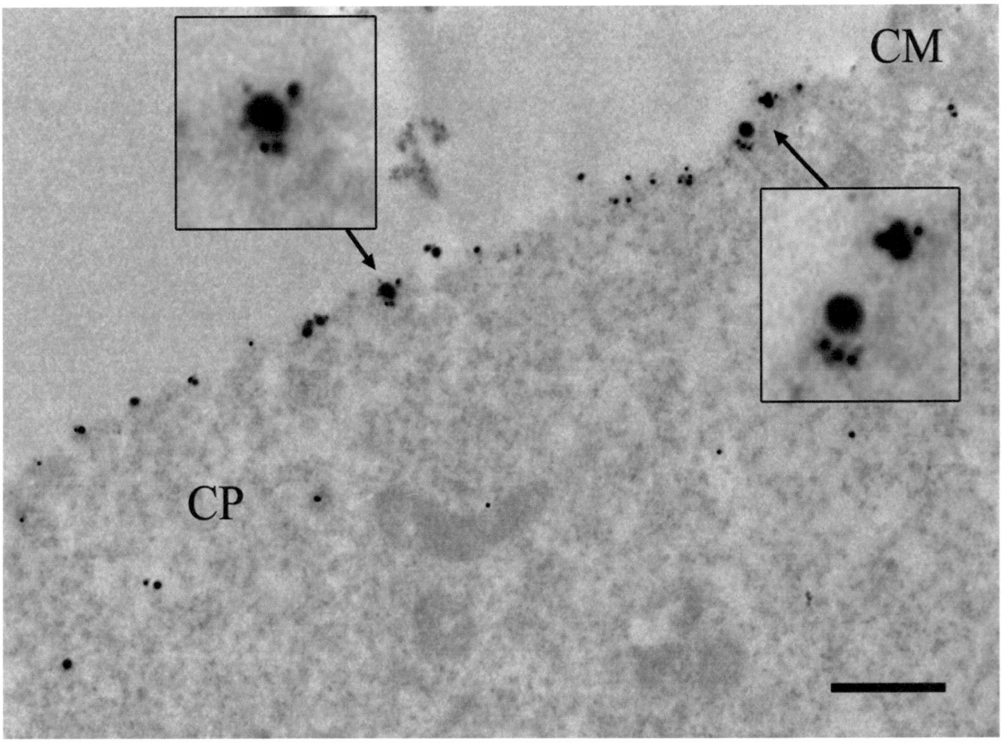

Fig. 5. Immunogold electron microscopy (post-embedding) method for the visualisation of A_1R and $P2Y_1R$ in rat brain using nanogold particles. Localization of A_1R (small particles) and $P2Y_1R$ (large particles) in cell surface of Purkinje cells detected with both anti-A_1R and anti-$P2Y_1R$ Arrows indicate two adjacent receptors on the cell membrane. Bars represent 100 nm. CM, cell membrane; CP, cytoplasm.

detected in the cell surface region (Fig. 3F). In our earlier experiments, oligomerization of A_1R and $P2Y_2R$ in rat brain tissues under the same experimental conditions involved hippocampal pyramidal cells, cerebellum, and pyramidal cells in the forebrain (Namba *et al.*, 2010). Hetero- and homo-oligomers of both $A_1R/P2Y_1R$ and $A_1R/P2Y_2R$ were detected in significant numbers at the cell surface in both transfected HEK293T cells and native brains.

5. Data analysis: Comparison of the frequencies of monomers, homo-oligomers, and hetero-oligomers between $P2Y_1R/A_1R$ and $P2Y_2R/A_1R$

Gold-staining was quantified in the following way. Firstly, gene-transfected HEK293T cells exhibiting the highest number of total immuno-reacted gold particles were defined as 100% labeling. Since co-transfected HEK293T cells that displayed unique pharmacology in our previous study (Suzuki *et al.*, 2006) exhibited more than 20% hetero-oligomeric gold particles, we used this number as a threshold in the current study. Thus, cells with more than 20% hetero-oligomeric particles were defined as being "significantly stained", and those with 20% or less were defined as "not significantly stained".

The proportions of relative distributions for A_1R and P2YR between cell surfaces and inner cytoplasmic membranes were clearly different (Fig. 6). The tendency for the proportional distribution of A_1R and P2YR at the surface of HEK293T cells concur with data from brain

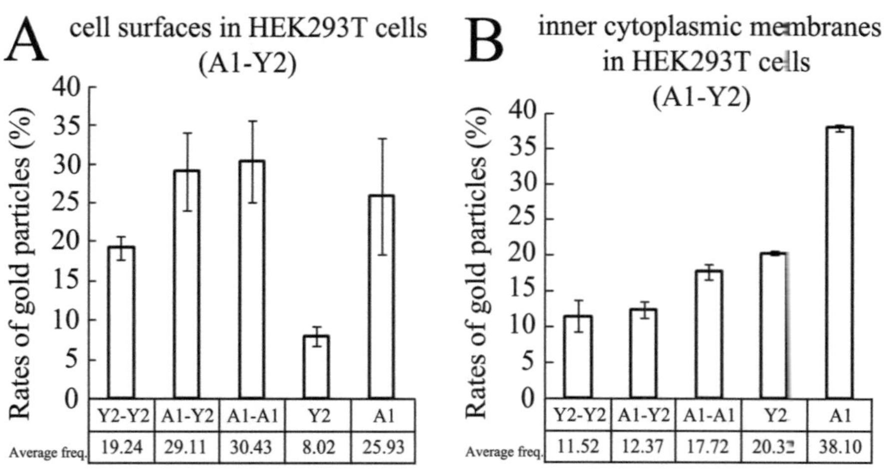

Fig. 6. Comparison of the relative distributions for A_1R- and $P2Y_2R$-conjugated gold particles between the cell surface and inner cytoplasmic membranes.

tissues. Based on this, the numbers of immunogold particles at the surface of each cell type in brain tissues were determined. We defined single particles located independently as monomers (A_1R and P2YR in Fig. 7), complexes composed of clusters of the same-sized gold particles as "homo-oligomers" (A_1R-A_1R or P2YR-P2YR in Fig. 7), and those of different sized gold particles as "hetero-oligomers" (A_1R-P2YR in Fig. 7). Separate calculations were carried out for particles in Purkinje cells (Fig. 7A, C), hippocampal pyramidal neurons (Fig. 7B, D), and cortical neurons (Fig. 7E). To do this, gold particles were counted the number of in three cells for each region. We previously counted immunogold particles in co-transfected HEK293T cells (Namba *et al.*, 2010). The total number of immunoreactive gold particles on

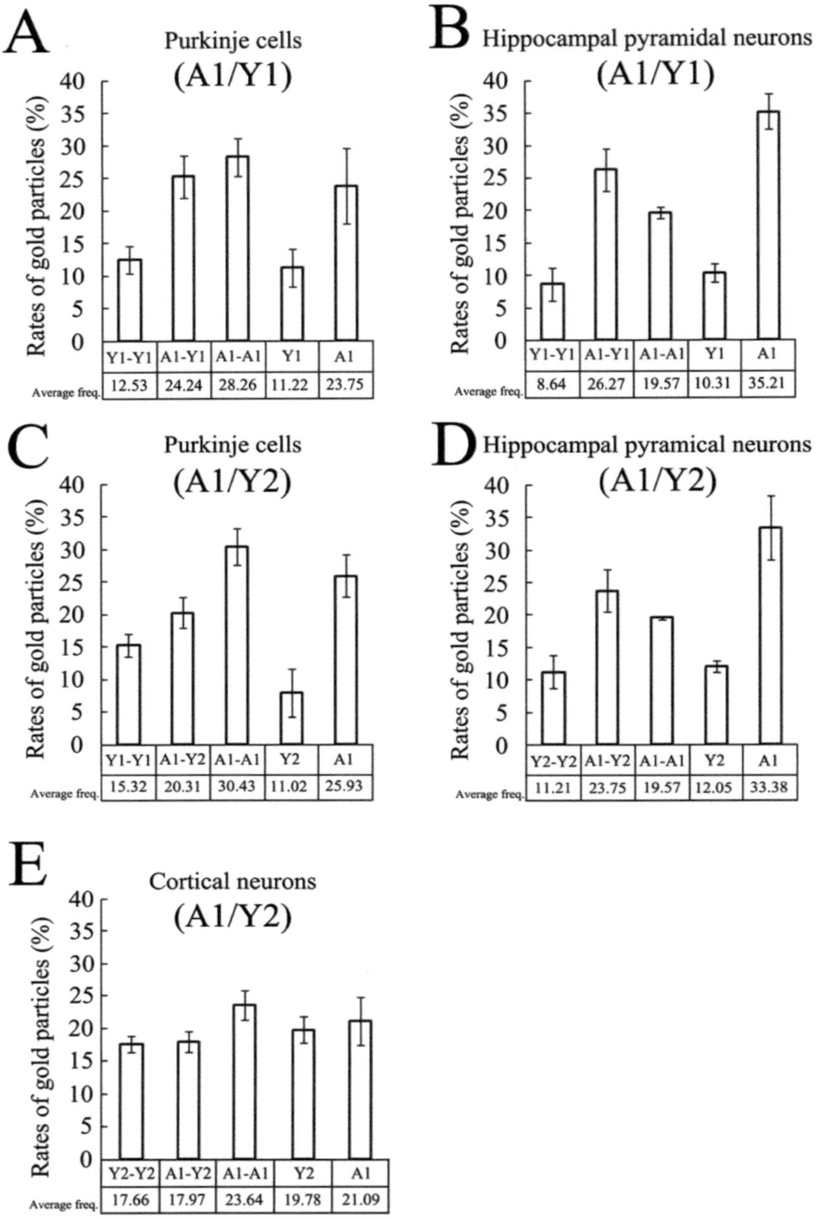

Fig. 7. Bar graphs comparing the relative distributions of A_1R(A1)/$P2Y_1R$(Y1) immunoreactive elements in Purkinje cells (A) Hippocampal pyramidal cells (B), and A_1R(A1)/$P2Y_2R$(Y2) immunoreactive elements in Purkinje cells (C), Hippocampal pyramidal cells (D) and Cortical neurons (E). P2YR-P2YR, A_1R-A_1R and A_1R-P2YR oligomers are indicated by Y1-Y1 (Y2-Y2), A1-A1 and A1-Y1 (A1-Y2), respectively. The total number of immunoreactive gold particles on the cell surface was defined as 100%. Each column represents the average frequency (± SD) from three cells. Data describing the average numbers of gold particles are shown in tables under the graphs. Data represent the mean of three independent experiments.

each cell surface was defined as 100%. From a total of 12 photos from each brain area (i.e., 36 photos) and from transfected cells that were reacted under the same conditions as the brain sections for each immunostaining, we selected three photos of each specimen containing whole cells for comparison.

We then counted gold particles on the surfaces of cells in the cerebellum, hippocampus and cortical neurons, and classified them as monomers, homo-oligomers, or hetero-oligomers. While the homo-oligomerization ratios (A_1R-A_1R/P2YR-P2YR) displayed different patterns between Purkinje cells and hippocampal pyramidal cells, the rates of hetero-oligomerization were particularly prominent in hippocampal pyramidal cells among them. Curiously, the frequency of A_1R or $P2Y_1R$ hetero-oligomerization was slightly higher than that of A_1R or $P2Y_2R$ in both tissues (Fig. 7A, B). This indicated that the hetero-oligomerization of A_1R or $P2Y_1R$ are the dominant form in both Purkinje cells and hippocampal pyramidal cells.

6. Discussion

Previous reports describe electron microscopic studies of plasma membranes for homo-oligomeric B_1 Bradykinin receptor complexes (Kang *et al.*, 2005), heteromeric-oligomerization of $GABA_B$ R1 and R2 receptors (Charara *et al.*, 2004), and the localization of A_1R with caveolin-3 in rat ventricular cardiomyocytes (Lasley *et al.*, 2000). An immunological study suggested that A_1R forms oligomers the cortex of the pig brain (Ciruela *et al.*, 1995), and a FRET study demonstrated the oligomerization of $P2Y_2R$ in transfected HEK293 cells (Kotevic *et al.*, 2005). The hetero-oligomerization of A_1R-$P2Y_1R$ on postsynaptic neurons was also analyzed by IEM (Tonazzini *et al.*, 2007). The present study provides the first detailed evidence of an interaction between endogenous A_1R and $P2Y_2R$ in brains using IEM.

The homo- oligomerization of A_1R and its structural profile were previously analyzed in our laboratory by computational prediction, co-immunoprecipitation, and BRET analysis with differently tagged A_1Rs (Suzuki *et al.*, 2009); homo-oligomers and monomers were easily distinguished by IEM. This particular study confirmed the existence of homo-oligomers (A_1R-A_1R and $P2Y_2R$-$P2Y_2R$) using IEM. Interestingly, the percentage of A_1R homo-oligomers was higher than that of $P2Y_2R$ in both rat brain and transfected HEK293T cells (Namba *et al.*, 2010). By contrast, the ratio of heteromeric gold-particle clusters were different in the cortex, hippocampus, and cerebellum. Importantly, both homo-oligomeric and hetero-oligomeric gold-particles were reduced in number at inner cytoplasmic membranes than at the cell surface (data not shown). In general, most GPCRs oligomers have been observed at the cell surface (Minneman, 2007; Bulenger *et al.*, 2005).

While the frequencies of A_1R and $P2Y_1R$ homo-oligomers and monomers were similar in the cerebellum (Fig. 5) and in transfected HEK293T cells (Fig. 4B), the ratio of the different receptor oligomers occurred in different patterns in each of the three brain areas (Fig. 7). Total numbers of hetero-oligomers observed on the cell surface and in the cytoplasm were clearly different (Fig. 6A, B) and may reflect the process of receptor maturation and association of the A_1R-P2YR complex. However, hetero-oligomers were unmistakably detected at the cell surface by IEM (Fig. 3C, Fig. 4B).

As a signaling pathway, $P2Y_1R$ and $P2Y_2R$ display different ligand specificities. As ligands, ATP and UTP fully activate $P2Y_2R$. However, UTP is not an agonist for $P2Y_1R$. In addition,

ADP is a strong agonist for $P2Y_1R$ but not $P2Y_2R$ (Abbracchio *et al.*, 2006). Many previous studies suggest that A_1R-$P2Y_1R$ and A_1R-$P2Y_2R$ hetero-oligomers exhibit general pharmacological profiles, possibly because of differences in the conformational changes induced by oligomerization (Nakata *et al.*, 2010). The hetero-oligomerization of A_1R-$P2Y_1R$ inhibits adenylyl cyclase activity via the $G_{i/o}$ protein linked effector. The hetero-oligomerization of A_1R-$P2Y_2R$ resulted in an increase in intracellular Ca^{2+} levels induced by $P2Y_2R$ activation of $G_{q/11}$ which was synergistically enhanced by the simultaneous addition of an A_1R agonist in the co-expressing cells (Suzuki *et al.*, 2006). Differences in the amounts of hetero-oligomerization between A_1R-$P2Y_1R$ and A_1R-$P2Y_2R$ were observed (Fig. 7). Assuming that the number of hetero-oligomers formed is functionally dominant, the dominancy of the signaling via A_1R-$P2Y_1R$ may be generated by competitive antagonism in pharmacology between $P2Y_1R$ and $P2Y_2R$ in order to oligomerize with A_1R. This hypothesis, however, requires further investigation.

In our previous study, the hippocampal hetero-oligomerization of A_1R and $P2Y_2R$ was far more pronounced than in other regions of the brain (Namba *et al.*, 2010). Another research group suggested that the hetero-oligomerization, or cross-talk between A_1R and $P2Y_1R$ is involved in regulation of glutamate release in the hippocampus (Tonazzini *et al.*, 2007). The relative distributions of immunoreactivity for $GABA_B$ R2 and $GABA_B$ R1 were also different in the basal ganglia and globus pallidus/substantia nigra, which suggests the possible co-existence and hetero-oligomerization of the two types of receptors at various pre-/postsynaptic sites (Charara *et al.*, 2004). From the present study, it can be speculated that the A_1R/$P2Y_2R$ hetero-oligomer might be responsible for down regulation, via hippocampal Ca^{2+} secretion, of synaptic functions (Safiulina *et al.*, 2006). Furthermore, the abundant formation of A_1R/$P2Y_1R$ or A_1R/$P2Y_2R$ hetero-oligomers in the cerebellum revealed in this present study supports the idea that the unique signal transduction generated by hetero-oligomerization, including the enhancement of Ca^{2+} signaling via $G_{q/11}$, observed in transfected cells, also occurs in the cerebellum.

There are many families of GPCRs expressed in whole brain, most of which remain a mystery. However, it is clear that GPCR hetero-oligomerization is common in the brain and exhibits unique pharmacology in this region, thus implying that associated signal transduction pathways can be anticipated in this region. The methodology described here using immunogolod particles is one of the most influential techniques available to elucidate the ingenious mechanism underlying GPCR hetero-oligomerization.

7. Summary

In summary, IEM provided direct evidence for the existence of homo- and hetero-oligomers of A_1R and $P2Y_2R$, not only in co-transfected cultured cells, but also *in situ* on the surface of neurons in various brain regions. The molecular mechanisms responsible for the control of A_1R and P2YR monomer/homo-oligomer/hetero-oligomer ratios remain to be elucidated. Future investigation of GPCR oligomer formation is indispensable for revealing the elaborate mechanisms of cellular function.

The importance of these novel experimental procedures using IEM is to provide information concerning crosstalk between small molecules with high angle views of whole cells, although these methods do require a high level of technical skill. The development of ingenious histochemical and immunoelectronmicroscopic methods has made it possible to visuallize crosstalk and provide specific insight into the nature of hetero-oligomers of not only GPCRs, but also various proteins expressed by cells.

8. Acknowledgments

Most of the procedures described in this work were performed in Dr. Hiroyasu Nakata's laboratory at the Tokyo Metropolitan Institute for Neuroscience, Tokyo, Japan. The author wishes to express sincere appreciation and gratitude to Dr. Nakata and his collaborators.

9. References

Abbracchio, M. P., Burnstock, G., Boeynaems, J. M., Barnard, E. A., Boyer, J. L., Kennedy, C., Knight, G. E., Fumagalli, M., Gachet, C., Jacobson, K. A., Weisman, G. A. (2006). International Union of Pharmacology LVIII: update on the P2Y G protein-coupled nucleotide receptors: from molecular mechanisms and pathophysiology to therapy. *Pharmacol Rev*. 58(3): 281–341.

Abbracchio, M. P., Burnstock, G., Verkhratsky, A. & Zimmermann, H. (2009). Purinergic signalling in the nervous system: an overview. *Trends Neurosci*. 32(1): 19–29.

Angers, S., Salahpour, A. & Bouvier M. (2002). Dimerization: an emerging concept for G protein-coupled receptor ontogeny and function. *Annu Rev Pharmacol Toxicol*. 42: 409-435.

Bouvier, M. (2001). Oligomerization of G-protein-coupled transmitter receptors. *Nat Rev Neurosci*. 2(4): 274-286.

Bulenger, S., Marullo, S. & Bouvier M. (2005). Emerging role of homo- and heterodimerization in G-protein-coupled receptor biosynthesis and maturation. *Tre*nds Pharmacol Sci. 26(3): 131-137.

Burnstock, G. (2007). Physiology and pathophysiology of purinergic neurotransmission. *Physiol Rev*. 87(2): 659–797.

Burnstock, G. (2008). Purinergic signalling and disorders of the central nervous system. *Nat Rev Drug Discov*. 7(7): 575–590.

Charara, A., Galvan, A., Kuwajima, M., Hall, R. A. & Smith, Y. (2004). An electron microscope immunocytochemical study of GABA(B) R2 receptors in the monkey basal ganglia: a comparative analysis with GABA(B) R1 receptor distribution. *J Comp Neurol*. 476(1): 65-79.

Ciruela, F., Casado, V., Mallol, J., Canela, E. I., Lluis, C. & Franco, R. (1995). Immunological identification of A_1 adenosine receptors in brain cortex. *J Neurosci Res*. 42(6): 818-828.

Fam, S. R., Gallagher, C. J., Kalia, L. V. & Salter, M. W. (2003). Differential frequency dependence of P2Y1- and P2Y2- mediated Ca 2+ signaling in astrocytes. *J Neurosci*. 23(11): 4437-4444.

Kang, D. S., Gustafsson, C., Mörgelin, M. & Leeb-Lundberg, L. M. (2005). B1 bradykinin receptor homo-oligomers in receptor cell surface expression and signaling: effects of receptor fragments. *Mol Pharmacol*. 67(1): 309-318.

Kotevic, I., Kirschner, K. M., Porzig, H. & Baltensperger, K. (2005). Constitutive interaction of the P2Y2 receptor with the hematopoietic cell-specific G protein G(alpha16) and evidence for receptor oligomers. *Cell Signal*. 17(7): 869-880.

Lasley, R. D., Narayan, P., Uittenbogaard, A. & Smart, E. J. (2000). Activated cardiac adenosine A(1) receptors translocate out of caveolae. *J Biol Chem*. 275(6): 4417-4421.

Mehta, B., Begum, G., Joshi, N. B. & Joshi, P. G. (2008). Nitric oxide-mediated modulation of synaptic activity by astrocytic P2Y receptors. *J Gen Physiol*. 132(3): 339-349.

Minneman, K. P. (2007). Heterodimerization and surface localization of G protein coupled receptors. *Biochem Pharmacol*. 73(8): 1043-1050.

Molliver, D. C., Cook, S. P., Carlsten, J. A., Wright, D. E. & McCleskey, E. W. (2002). ATP and UTP excite sensory neurons and induce CREB phosphorylation through the metabotropic receptor, P2Y2. *Eur J Neurosci*. 16(10): 1850-1860.

Nakata, H., Suzuki, T., Namba, K. & Oyanagi, K. (2010). Dimerization of G protein-coupled purinergic receptors: increasing the diversity of purinergic receptor signal responses and receptor functions. *J Recept Signal Transduct Res*. 30(5): 337-346.

Nakata, H., Yoshioka, K., Namba, K. & Kamiya, T. (2006). *Oligomerization of G protein-coupled Purinergic Receptors*, InTech, ISBN 978-0-8493-2771-1, Gunma University, Japan.

Namba, K., Suzuki, T. & Nakata, H. (2010). Immunogold electron microscopic evidence of in situ formation of homo- and heteromeric purinergic adenosine A1 and P2Y2 receptors in rat brain. *BMC Res Notes*. 3: 323.

Ochiishi, T., Chen, L., Yukawa, A., Saitoh, Y., Sekino, Y., Arai, T., Nakata, H. & Miyamoto, H. (1999). Cellular localization of adenosine A1 receptors in rat forebrain: immunohistochemical analysis using adenosine A1 receptor-specific monoclonal antibody. *J Comp Neurol*. 411(2): 301-316.

Ralevic, V., Burnstock, G. (1998). Receptors for purines and pyrimidines. *Pharmacol Rev*. 50(3): 413–492.

Safiulina, V. F., Afzalov, R., Khiroug, L., Cherubini, E. & Giniatullin, R. (2006). Reactive oxygen species mediate the potentiating effects of ATP on GABAergic synaptic transmission in the immature hippocampus. *J Biol Chem*. 281(33):23464-23470.

Suzuki, T., Namba, K., Tsuga, H. & Nakata, H. (2006). Regulation of pharmacology by hetero-oligomerization between A1 adenosine receptor and P2Y2 receptor. *Biochem Biophys Res Commun*. 351(2): 559–565.

Suzuki, T., Namba, K., Yamagishi, R., Kaneko, H., Haga, T. & Nakata, H. (2009). A highly conserved tryptophan residue in the fourth transmembrane domain of the A adenosine receptor is essential for ligand binding but not receptor homodimerization. *J Neurochem*. 110(4):1352-1362.

Tonazzini, I., Trincavelli, M. L., Storm-Mathisen, J., Martini, C. & Bergersen, L. H. (2007). Co-localization and functional cross-talk between A1 and P2Y1 purine receptors in rat hippocampus. *Eur J Neurosci*. 26(4): 890-902.

Yoshioka, K., Saitoh O. & Nakata, H. (2001). Heteromeric association creates a P2Y-like adenosine receptor. *Proc Natl Acad Sci USA*. 98(13): 7617-7622.

Yoshioka, K., Hosoda, R., Kuroda, Y. & Nakata, H. (2002). Heterooligomerization of adenosine A1 receptors with P2Y1 receptors in rat brains. *FEBS Lett*. 531(2): 299–303.

Yoshioka, K., Saitoh, O. & Nakata, H. (2002). Agonist-promoted heteromeric oligomerization between adenosine A(1) and P2Y(1) receptors in living cells. *FEBS Lett*. 523(1-3): 147-151.

Part 2

Enzymes

8

Carbonic Anhydrase and Heavy Metals

Maria Giulia Lionetto, Roberto Caricato, Maria Elena Giordano,
Elisa Erroi and Trifone Schettino
University of Salento - Dept. of Biological and Environmental Sciences and Technologies
Italy

1. Introduction

Carbonic anhydrase (CA; EC 4.2.1.1) is a zinc metalloenzyme catalysing the reversible hydration of CO_2 to produce H^+ and HCO_3^-. Its activity is virtually ubiquitous in nature. The fundamental role of this biochemical reaction in diverse biological systems has driven the evolution of several distinct and unrelated families of CAs. Five CA families, referred as α-, β-, γ-CA, δ, and ζ-CAs have been identified in animals, plants and bacteria (Hewett-Emmett and Tashian, 1996; Supuran, 2010). These are the α-CAs, present in vertebrates, bacteria, algae and plants; the β-CAs, predominantly in bacteria, algae and plants; the γ-CAs, mainly present in archaea and some bacteria; the δ-CAs and ζ-CAs only found in some marine diatoms (Supuran, 2010).

The monomeric α-carbonic anhydrases are by far the best studied, being found in animals. In mammals at least 16 different CA isoforms were isolated and several novel isozymes have also been identified in non-mammalian vertebrates. The α-CA isoenzymes differ in their kinetic properties, their tissue distribution and subcellular localization, and their susceptibility to various inhibitors. In general, there are three distinct groups of CA isozymes within the α-CA gene family. One of these groups contains the cytoplasmic CAs, which includes mammalian CA I, II, III, V, VII and XIII. These isozymes are found in the cytoplasm of various tissues, with the exception of the mitochondrial confined CA V. Another group of isozymes, termed the membrane-bound CAs, consists of mammalian CA IV, IX, XII, XIV and XV (Esbaugh and Tufts, 2006). These isozymes are associated with the plasma membranes of many different tissue types. The final group contains several very intriguing isozymes, CA VIII, X and XI, which are termed the CA-related proteins (CA-RP; Tashian et al., 2000). These isozymes have lost classical CA activity - the hydration/dehydration of CO_2 - and have no known physiological function; however, their highly conserved nature does suggest a very important role in vertebrates (Tashian et al., 2000).

The β-carbonic anhydrases are dimers, tetramers, or octamers and include the majority of the higher plant CA isoforms (Kimber and Pai, 2000). The γ-carbonic anhydrase is a homotrimer that has been reported for the bacterium *Methanosarcina thermophila* (Alber and Ferry, 1994). The δ class has its prototype in the monomeric CA TWCA1 from the marine diatom *Thalassiosira weissflogii* (Roberts et al., 1997; Tripp et al., 2001). The ζ-CAs are probably monomer with three slightly different active sites on the same protein backbone (Xu et al 2008).

All CAs are metalloenzymes but whereas α-, β-, and δ-CAs use Zn(II) ions at the active site, the γ-CAs are probably Fe(II) enzymes (Ferry et al., 2010), but they are active also with bound Zn(II) or Co(II) ions, and the ζ-class uses also Cd(II) to perform the physiologic reaction catalysis (Lane et al., 2000; Lane et al., 2005).

CA plays key roles in a wide variety of physiological processes involving CO_2 and HCO_3^-. In animals the various CA isozymes are found in many different tissues and are involved in a number of different physiological processes, including bone resorption, calcification, ion transport, acid–base transport, and a number of different metabolic processes such as biosynthetic reactions (gluconeogenesis, lipogenesis, and ureagenesis). In algae and plants they play an important role in photosynthesis (Ivanov et al, 2007; Zhang et al., 2010; Cannon Gordon et al., 2010).

Considerable advances towards a detailed understanding of the catalytic mechanism of the zinc enzyme carbonic anhydrase have been made during the past years as a result of the application of crystallographic and kinetic methods to wild-type and mutant enzymes. Moreover, a great amount of work has been performed on CA inhibitors, first of all sulfonamides, RSO_2NH_2, which represent the classical CA inhibitors (CAIs) and are in clinical use for more than 50 years as diuretics and systemically acting antiglaucoma drugs (Supuran, 2010).

The review focuses on one interesting but less investigated aspect of the biochemistry of this metalloenzyme, encompassing several areas of interest from human health to environmental science: the relationships between carbonic anhydrase and heavy metals. Heavy metals are chemical elements with a density higher than 5.0 g/cm^3, characterized by high reactivity, redox behaviour, and complex formation based on the characteristic of the outer *d* electron shell. In the scientific literature the following elements are normally ascribed to the heavy metal groups: aluminium, iron, silver, barium, beryllium, manganese, mercury, molybdenum, nickel, lead, copper, tin, titanium, tallium, vanadium, zinc. Some metalloids, such as arsenic, bismuthum, and selenium, are also included in the heavy metals groups.

Heavy metals generally regarded as essential for animals in trace amounts include zinc, the known cofactor of CAs, iron, copper, manganese, chromium, molybdenum and selenium. They are essential because they form an integral part of one or more enzymes involved in a metabolic or biochemical process. Besides essential metals, a number of other heavy metals, such as arsenic, lead, cadmium, mercury, have no known function in the body and are referred as toxic metals. However, also essential metals become toxic when their levels in the body exceed the homeostatic capacity of the organism. The intracellular levels of essential metals are regulated by transporters (which translocate metal across the plasma membrane) as well as by metallothionein and other metal binding proteins (Maret and Wolfgang, 2011). The toxicity of heavy metals is generally ascribed to their high affinity for nucleophilic groups like sulfhydryls. In fact they are soft donors and will therefore readily bind to soft acceptors such as sulphydryl groups.

Recently, a number body of evidence has emerged regarding the effect of several heavy metals on carbonic anhydrase catalytic activity and protein expression. These studies encompass a wide area of interest from human health to environmental sciences.

2. Heavy metals as carbonic anhydrase cofactors

CAs catalyze the reversible hydration of carbon dioxide to bicarbonate and protons by means of a metal-hydroxide ($Lig^3M^{2+}(OH)^-$) mechanism, although the α-CAs possess other catalytic activities such as esterase, phosphatase, cyanate/cyanamide hydrase, etc. (Supuran et al., 2003; Supuran and Scozzafava 2007; Innocenti et al., 2008). In the α-, γ, and δ-CA classes, Lig^3 is always constituted by three His residues. The metal (M) is ZnII for all classes. The zinc atom is in the +2 state and is located in a cleft near the center of the enzyme. The role of zinc in carbonic anhydrase is to facilitate the deprotonization of water with the formation of the nucleophilic hydroxide ion, which can attack carbonyl group of carbon dioxide to convert it into bicarbonate. This is obtained through the +2 charge of the zinc ion which attracts the oxygen of water, deprotonates water, thus converting it into a better nucleophile able to attack the carbon dioxide.

Water naturally deprotonates itself, but it is a rather slow process. Zinc deprotonates water by providing a positive charge for the hydroxide ion. The proton is donated temporarily to the surrounding amino acid residues, and then it is given to the environment, while allowing the reaction to continue. Zinc is able to help the deprotonation of water by lowering the pKa of water. Therefore, more water molecules are now able to deprotonate at a lower pH than normal, increasing the number of hydroxide ions available for the nucleophilic attack to carbon dioxide (Berg, 2007).

The affinity of carbonic anhydrase for zinc is in subpicomolar range, as assessed for studies on the α-class (Tripp et al., 2001). Cox et al. (2000) and Hunt et al. (1999) ascribed a role for hydrophobic core residues in human CA-II that are important for preorienting the histidine ligands in a geometry that favours zinc binding and destabilizes geometries that favour other metals. In particular, mutagenesis experiments demonstrated that substitutions of these amino acids at position 93, 95, and 97 decrease the affinity of zinc, thereby altering the metal binding specificity up to 10^4-fold. Furthermore, the free energy of the stability of native CAII, determined by solvent-induced denaturation, correlates positively with increased hydrophobicity of the amino acids at positions 93, 95, and 97 as well as with zinc affinity (Hunt et al., 1999).

β-CAs, present in green plants and cyanobacteria, contain also Zn^{2+} in the active site but are differentiated from α-CAs by virtue of the fact that the active site is coordinated by a pair of cysteine residues and a single histidine residue, whereas the fourth ligand may be either a water molecule/hydroxide ion, or a carboxylate from a conserved aspartate residue in some β-CAs (Type II β-CAs) [Trip et al., 2001; Xu et al., 2008]. The metal hydroxide catalytic mechanism seems to be also valid for these enzymes [Supuran, 2008].

Besides zinc, other metals have demonstrated to be physiologically relevant cofactor for some CAs. In fact, in the γ-CAs metal may also be FeII (Ferry et al., 2010). Cam, the prototypic γ-class carbonic anhydrase, from the anaerobic methane producing Archaea species *Methanosarcina thermophila,* contains zinc in the active site when overproduced in *Escherichia coli* and purified aerobically [Alber et al., 1996], while it has 3-fold greater carbonic anhydrase activity and contains Fe^{2+} in the active site (Fe-Cam) when purified anaerobically from *E. coli* or overproduced in the closely related species M. *acetivorans* and purified anaerobically. Soluble Fe^{2+} is abundant in oxygen free environments and available to anaerobic microbes. The different results obtained in aerobic and anaerobic conditions is explained by the fact that in aerobic conditions Fe^{3+} is oxidized and rapidly loss from CAM enzyme, substituted by Zn^{2+} contaminating buffers not treated with chelating agents. These results indicate Fe^{2+} as the physiologically relevant metal [MacAuley et al., 2009; Tripp et al., 2004] in the active site for CAM enzyme. Interestingly, evidence for the role of ferrous ion in CA has been obtained also for the α class. In fact carbonic anhydrase activity from duck erythrocytes is increased in the presence of iron in the incubation medium suggesting a role for iron in the active site (Wu et al., 2007).

The ζ-CA naturally uses Cd^{2+} as its catalytic metal in marine diatoms (Lane and Morel, 2000; Lane et al., 2005; Park et al., 2008). This cdmium-CA (CDCA1) consists of three tandem CA repeats (R1–R3), which share 85% identity in their primary sequences (Lane et al., 2005). Although CDCA1 was initially isolated as a Cd enzyme, it is actually a "cambialistic" enzyme since it can use either Zn or Cd for catalysis—and spontaneously exchanges the two metals (Xu et al., 2008). Kinetic data show that the replacement of Zn by Cd results nonetheless in a decrease in catalytic efficiency (Xu et al., 2008). In the active site, Cd is coordinated by three invariant residues in CDCA of all diatom species (Park et al., 2007): Cys 263, His 315 and Cys 325. The tetrahedral coordination of Cd is completed by a water molecule. The use of Cd in CDCA is thought to explain the nutrient-like concentration profile of Cd in the oceans, where the metal is impoverished at the surface by phytoplankton uptake and regenerated at depth by remineralization of sinking organic matter (Lane and Morel 2000). It is cycled in the water column like an algal nutrient. It is thought that the expression of a CDCA in diatoms, which are responsible for about 40% of net marine primary production, represents an adaptation to life in a medium containing vanishingly small concentrations of essential metals (Xu et al., 2008). As suggested by Xu et al. (2008) the remarkable ability to make use of cadmium, an element known for its toxicity, gave presumably a significant competitive advantage to diatoms in the oceans, poor in metals, with respect to other species, and could have contributed to the global ratiation of diatoms during the Cenozoic Era and to the parallel decrease in atmospheric CO_2.

Moreover, Co(II) has been shown to replace Zn(II) in α-, β and γ-CA (Hoffmann et al., 2011). Cobalt ionic radius and polarizability are very similar to those of Zn(II). In contrast to Zn(II) (d^{10}), the d^7electron configuration of Co(II) is accessible to electronic spectroscopic methods (, yielding information about the interactions protein-metal. As a result, spectroscopy of Co(II) substituted CA isozymes has been used to probe the environment of the metal ions in the active sites and get information on the nature of the first coordination sphere of the metal (Hoffmann et al., 2011). The Co-containing form of the enzyme generally shows a marked decrease in activity compared with the native Zn form (Tu and Silverman, 1985). The demonstration that Zn can be extracted from a protein and replaced with Co *in vitro* does not demonstrate that such metal substitution takes place in vivo. The evidence for in vivo Co substitution in a CA was for the first time provided by Morel et al (1994) and Yee and Morel (1996) in the diatoms *T. weissflogii,* who demonstrated ^{65}Zn and ^{57}Co bands to co-migrate with a single band of CA activity on a native gel of diatom proteins.

3. Heavy metals as inhibitors of carbonic anhydrase activity

Several heavy metals were demonstrated to *in vitro* inhibit CA activity in a variety of organisms, including fishes, crabs, bovines, and humans.

The early work of Christensen and Tucker (1976) demonstrated carbonic anhydrase inhibition by heavy metals for the first time in fish. The study was carried out on red blood cells CA of the teleost *Oncorhynchus mykiss*. Erythrocyte CA, which represents the most abundant pool of the enzyme in fish, appeared significantly in *vitro* inhibited by several heavy metals cations, such as Cd^{2+}, Cu^{2+}, Ag^{+}, and Zn^{2+} (Tab1).

In the intestine and gills of the European eel, *Anguilla anguilla*, Lionetto et al. (1998; 2000) found cadmium to significantly inhibit carbonic anhydrase activity. The inhibition appeared tissue specific (Lionetto et al., 1998; Lionetto et al., 2000). The gill CA was much more sensitive to the heavy metal as compared to the enzyme activity in the intestine, as observed by comparing the IC50 values (Tab1). In particular in the intestine the inhibitory effect of cadmium was more pronounced on the cytosolic than the membrane-bound CA, which revealed only a partial inhibition at high concentrations. Moreover CA activity inhibition showed a certain time-dependence, with a delay of at least 10 min and 30 min for the cytosolic isoform and the membrane bound isoform respectively. The authors attributed this behaviour to the time required by cadmium for displacing the metal (zinc) associated with the enzyme, giving an inactive Cd-substituted carbonic anhydrase. Cadmium is a bivalent metal, similar in many respects to zinc: both are in the same group of the periodic table, contain the same common oxidation state (+2), and when ionized have almost the same size. Due to these similarities, cadmium can replace zinc in many biological systems. Moreover, the delayed inhibition of membrane-bound CA with respect to the cytosolic isoform was explained by a more difficult access of cadmium to the active site of the enzyme bound to the membrane. In fact, it has to be considered that the membrane-bound CA is stabilised by disulfide bonds (Whitney and Briggle, 1982) which could contribute to a less sensitivity of the membrane bound CA to cadmium.

As suggested by Lionetto et al (2000), the observed *in vitro* inhibition of cadmium on CA activity could be useful in the understanding of the toxic effects that the heavy metal can elicits on fish physiology *in vivo*. The inhibitory effect on gill CA activity suggests that the heavy metal might interfere with a number of physiological functions in which gill CA is involved as gas exchanges (Randall and Daxbaeck, 1984), acid-base balance (Heisler, 1984), osmoregulation (Henry, 1984) and clearance of the waste products from nitrogenous metabolism (Evans and Cameron, 1986). Morgan et al (2004) directly demonstrated in *in vivo* expoxure experiments on rainbow trout that inhibition of branchial CA was able to induce an early decline in the gill Na^{+} and Cl^{-} uptake. With regards to the intestine, the physiological role of the cytosolic CA is that of generating HCO_3^- from metabolic CO_2 while the role of the CA enzyme associated to the brush-border membrane should be that of mediating the environmental HCO_3^- uptake (Maffia et al., 1996). Therefore, the inhibitory effect of cadmium on intestinal CA isoforms should interfere with bicarbonate balance and in turn with systemic acid-base balance and osmoregulation in fish. In fact, as previously shown (Schettino et al., 1992), the HCO_3^- entry via the membrane-bound CA in the cell across the luminal membrane of the enterocytes seems to be essential for maintaining a steady intracellular HCO_3^- concentration and/or pHi; as a consequence the salt transport in eel intestine occurs at a highest rate and the passive water loss is recovered so solving in part the osmoregulatory problem in marine fish. Therefore, inhibition of CA enzymes by cadmium could alter $[HCO_3^-]$i and/or pHi leading to a reduction of salt absorption and consequently impairing the osmoregulation of marine fish.

More recently, Soyut et al (2008) demonstrated Co^{2+}, Cu^{2+}, Zn^{2+}, Ag^{+}, and Cd^{2+} to be potent inhibitor for brain CA enzyme activity in Rainbow trout (*Oncorhynchus mykiss),* with the following sequence $Co^{2+} > Zn^{2+} > Cu^{2+} > Cd^{2+} > Ag^{+}$. They also demonstrated that Co^{2+}, Ag^{+}, and Cd^{2+} inhibit the enzyme with competitive manner, Cu^{2+} inhibits with noncompetitive manner, and Zn^{2+} with uncompetitive manner.

Ceyhun et al., 2011 *in vitro* demonstrated Al^{+3}, Cu^{+2}, Pb^{+2}, Co^{+3}, Ag^{+1}, Zn^{+2} and Hg^{+2} to exert inhibitory effects on fish liver CA. Metal ions inhibited the enzyme activity at low concentrations. Al^{+3} and Cu^{2+} resulted the most potent inhibitors of the CA enzyme. All the metals inhibited CA in competitive manner and aluminium showed to be the best inhibitor for fish liver CA. Concerning the mechanism of inhibition, the authors argued a possible interaction of the metal with the histidines exposed on the surface of the molecule and/or other aminoacids around the active site.

In invertebrates Vitale et al (1999)demonstrated cadmium, copper and zinc to *in vitro* inhibit CA activity in the gills of the estuarine crabs *Chasmagnathus granulate* (Tab.1*). The* inhibitory potentials of the three metals on CA was in the following sequence: $Cu^{2+} > Zn^{2+} > Cd^{2+}$. The observed inhibitory effect *in vitro* was confirmed by a corresponding inhibitory effect *in vivo.*

In the euryhaline crabs *Callinectes sapidus and Carcinus maenas* Skaggs et al (2002) also documented a significant *in vitro* inhibition of gill CA by Ag^{+}, Cd^{2+}, Cu^{2+} and Zn^{2+}. The binding affinities of the metals were one thousand times weaker for cytoplasmic CA from the gills of *C. maenas* than that from *C. sapidus*. The large differences in Ki values (Tab.1) suggests the presence of two different CA isoforms in the gills of these species, with *Callinectes sapidus* possessing a highly metal-sensitive CA isoform and *Carcinus maenas* having a metal-resistant isoform. Interestingly, heavy metal inhibition of CA from the gills of another euryhaline crab, *Chasmagnathus granulata,* (as reported by Vitale et al., 1999, see above) appears to be intermediate between that found in the other two species. Moreover, in *Callinectes sapidus* CA isolated from the cytoplasmic pool of gill homogenates was much more sensitive to heavy metal inhibition than was CA from the microsomal fraction, which is believed to be anchored to the basolateral membrane, and as such, it exists within a lipid-rich environment. The authors argued that metal could be sequestered in the lipid component of the microsomal fraction and, therefore, higher amounts of metals are required to achieve an effective concentration of free metals available for CA inhibition. However, the authors did not considered the time-dependence of the inhibition, which can be an important aspect to be taken into account (see Lionetto et al., 2000) in the analysis of membrane bound vs cytosolic isoform CA inhibition.

In humans Ekinci et al (2007) demonstrated the inhibition of two human carbonic anhydrase isozymes *in vitro*, the cytosolic HCA I and II by lead, cobalt and mercury. Lead was a noncompetitive inhibitor for HCA-I and competitive for HCA-II, cobalt was competitive for HCA-I and noncompetitive for HCA-II and mercury was uncompetitive for both HCA-I and HCA-II. Lead was the best inhibitor for both HCA-I and HCA-II.

In tab.1 the Ki, IC50 values and the type of inhibition for several heavy metals on CA from different vertebrate and invertebrate species is summarized. A great variability among species, tissues and metals can be observed. This suggests that the inhibitory mechanisms through which heavy metals exert their effect on carbonic anhydrase activity could be different for different isoenzymes and that also small structural differences between CA isoforms could result in different metal binding affinities.

Metal	Average value of Ki (M)	IC50 (M)	Type of inhibition	Tissue	Species	Ref
Cd^{2+}	n.d.	$9.979\ 10^{-6}$	n.d.	gills	*Anguilla anguilla*	Lionetto et al 2000
	n.d.	$3.64\ 10^{-5}$	n.d.	Intestine (cytosolic isoform)	*Anguilla anguilla*	Lionetto et al 2000
	n.d.	$2.15\ 10^{-5}$	n.d.	gills	*Chasmagnathus granulata*	Vitale et al., 1999
	n.d.	$9.00\ 10^{-4}$	n.d.	Red blood cells	*Ictalurus punctatus*	Christensen and Tucker, 1976
	$94.16\ 10^{-3}$M	8.25 ± 10^{-2}	Competitive	brain	*Oncorhynchus mykiss*	Soyut et al., 2008
	$5.0\ 10^{-7}$	n.d.	n.d.	Gills (cytoplasmic isofom)	*Callinectes sapidus*	Skaggs and Hery, 2002
	$6.0\ \text{-}25.0\ 10^{-4}$	n.d.	n.d.	Gills (cytoplasmic isofom)	*Carcinus maenas*	Skaggs and Hery, 2002
Ag^{+}	$193.8\ 10^{-3}$M	$1.59\ 10^{-1}$	Competitive	brain	*Oncorhynchus mykiss*	Soyut et al., 2008
	$6.40\ 10^{-4}$	$3.79\ 10^{-4}$	Competitive	liver	*Dicentrarchus labrax*	Ceyhun et al., 2011
	n.d.	$3.50\ 10^{-5}$	n.d.	Red blood cells	*Ictalurus punctatus*	Christensen and Tucker, 1976
	$5.0–0.10\ 10^{-8}$	n.d.	n.d.	Gills (cytoplasmic isofom)	*Callinectes sapidus*	Skaggs and Hery, 2002
	$6.0\ \text{-}25.0\ 10^{-4}$	n.d.	n.d.	Gills (cytoplasmic isofom)	*Carcinus maenas*	Skaggs and Hery, 2002
Zn^{2+}	$2.15\ 10^{-3}$M	$3.10\ 10^{-4}$	Uncompetitive	brain	*Oncorhynchus mykiss*	Soyut et al., 2008
	$7.21\ 10^{-4}$	$3.90\ 10^{-4}$	Competitive	liver	*Dicentrarchus labrax*	Ceyhun et al., 2011
	n.d.	$7.00\ 10^{-4}$	n.d.	Red blood cells	*Ictalurus punctatus*	Christensen and Tucker, 1976
	n.d.	$1.62\ 10^{-5}$	n.d.	gills	*Chasmagnathus granulata*	Vitale et al., 1999
	$6.0\ \text{-}25.0\ 10^{-4}$	n.d.	n.d.	Gills (cytoplasmic isofom)	*Carcinus maenas*	Skaggs and Hery, 2002

Metal	Average value of Ki (M)	IC50 (M)	Type of inhibition	Tissue	Species	Ref
Cu^{2+}	27.6 10^{-3}M	3.00 10^{-2}	Non competitive	brain	*Oncorhynchus mykiss*	Soyut et al., 2008
	1.75 10^{-5}	7.15 10-5	Competitive	liver	*Dicentrarchus labrax*	Ceyhun et al., 2011
	n.d.	6.50 10^{-5}	n.d.	Red blood cells	*Ictalurus punctatus*	Christensen and Tucker, 1976
	n.d.	3.75 10^{-6}	n.d.	gills	*Chasmagnathus granulata*	Vitale et al., 1999
	3.60 10^{-7}	n.d.	n.d.	Gills (cytoplasmic isofom)	*Callinectes sapidus*	Skaggs and Hery, 2002
	6.0 -25.0 10^{-4}	n.d.	n.d.	Gills (cytoplasmic isofom)	*Carcinus maenas*	Skaggs and Hery, 2002
Co^{2+}	5 10^{-5}M	1.40 10^{-5}	competitive	brain	*Oncorhynchus mykiss*	Soyut et al., 2008
	5.32 10^{-4}	3.16 10^{-4}	competitive	liver	*Dicentrarchus labrax*	Ceyhun et al., 2011
	3.91 10^{-3}	n.d.	competititve	Erytrocytes (CAI)	*Homo sapiens*	Ekinci et al., 2007
	1.7 10^{-3}	n.d.	non competitive	Erytrocytes (CAII)	*Homo sapiens*	Ekinci et al., 2007
Al^{3+}	1.48 10^{-4}	6.92 10^{-5}	competitive	liver	*Dicentrarchus labrax*	Ceyhun et al., 2011
Pb^{2+}	2.42 10^{-4}	1.13 10^{-4}	competitive	liver	*Dicentrarchus labrax*	Ceyhun et al., 2011
	9.90 10^{-4}	n.d.	Non competitive	Erytrocytes (CAI)	*Homo sapiens*	Ekinci et al., 2007
	5.6 10^{-5}	n.d.	uncompetitive	Erytrocytes (CAII)	*Homo sapiens*	Ekinci et al., 2007
Hg^{2+}	7.68 10^{-4}	4.48 10^{-4}	competitive	liver	*Dicentrarchus labrax*	Ceyhun et al., 2011
	1.42 10^{-3}	n.d.	uncompetitive	Erytrocytes (CAI)	*Homo sapiens*	Ekinci et al., 2007
	3.12 10^{-4}	n.d.	uncompetitive	Erytrocytes (CAII)	*Homo sapiens*	Ekinci et al., 2007

Table 1. Ki, IC50 and type of inhibition for several heavy metals in different species and tissues as assessed in *in vitro* studies.

Concerning the mechanisms of inhibition some heavy metals are believed to bind to CA not at the specific catalytic site of CO_2 hydration but nearby in a pocket, the so called 'proton shuttle' as demonstrated for human CAII (Tu et al., 1981). His-64 is a proton shuttle in catalysis, where it accepts the proton product (via the bridging solvent molecules) from zinc-bound water as zinc-bound hydroxide is regenerated; subsequently, the proton product is passed along to buffer (Liang et al, 1988; Tu et al., 1989; Vedani et al., 1989). The mechanism of inhibition of heavy metals on proton shuttle has been elucidated for copper on human CA II. Cu^{2+} is believed to competitively inhibit CAII by binding to the imidazole side chain of His-64, blocking its role in proton transfer from the zinc-bound water molecule to buffer molecules located outside of the active site region [Tu et al., 1981]. However, the knowledge of the mechanism of action of other metals on different CA isoforms is lacking. It cannot be excluded the CA binding to other different parts of the protein, possibly cysteine residues, as demonstrated in studies with other enzymes for silver and mercury.

4. Heavy metals as modulators of carbonic anhydrase activity and expression

If it has been widely demonstrated *in vitro* that heavy metals are able to inhibit CA activity in a variety of organisms, on the contrary little is known about the *in vivo* effects of trace metals on the activity and the expression of this metalloenzyme. The major information regards Zn^{2+}, while very few is known about other metals.

In humans early studies demonstrated that dietary zinc deficiency significantly reduces zinc concentrations of serum and in turn CA activity in erythrocytes (Hove,1940, Rahman et al., 1961; Kirchgessner et al., 1975) suggesting a possible influence of Zn^{2+} on CA protein expression. These early data have been more recently confirmed by Lukaski (2005) who demonstrated zinc concentration of serum and erythrocyte to be positively correlated to CA activity *in vivo*. Low dietary zinc decreases erythrocyte carbonic anhydrase activity and, in turn, impairs cardiorespiratory function in men during exercise (Lukaski et al., 2005). In ducks Zn^{2+} at a low level (up to 1.25 μM Zn) induced the rise of CA activity in erythrocytes (Wu et al., 2007). In parotid saliva of patients with CAVI deficiency Zn^{2+} treatment was able to stimulate synthesis/secretion of CAVI (Henkin et al., 1999), probably through stimulation of CAIV gene. In rats Zn^{2+} deficiency significantly reduced CAII protein expression in the submandibular gland (Goto et al., 2008).

As regards other metals Grimes et al (1997) reported the depression of CAIII mRNA and, in turn, CAIII protein in the mouse mutant 'toxic milk' (tx) liver following copper accumulation, Kuhara et al (2011) found CAIII suppression by copper accumulation during carcinogenesis, while Wu et al (2007) found iron at low levels to induce a rise in CA activity in duck erythrocytes.

Recently, Caricato et al (2010) demonstrated for the first time CA activity and protein expression to be enhanced by the exposure to the trace element cadmium in animals, opening new perspective in the comprehension of the functioning and regulation of this enzyme. Digestive gland CA activity showed a weak sensitivity to *in vitro* cadmium exposure since only high concentrations of $CdCl_2$ (from 10^{-5} to 10^{-3} M) were able to exert a significant inhibition. On the contrary digestive gland CA activity showed a significant increment in cadmium exposed animals (about 40% after two week of exposure). This was the first time that CA activity appears to be increased by cadmium in animals. Carbonic anhydrases from the microalgae *Chlamydomonas reinhardtii* (Wang at al., 2005) and *Thalassiosira weissflogii* (Morel et al., 1994; Lee et al., 1995) are the only other examples reported in nature of CA activity increase induced by cadmium exposure. Evidence of *in vivo* utilization of Cd in CA has been found in microalgae (Price and Morel, 1990; Morel et al., 1994; Lee et al. 1995, Xu et al., 2008). In these organisms the ability of Cd to substitute for Zn at the active site of the enzyme is reflected in the regulation of the enzyme expression. In *Thalassiosira weissflogii* a cadmium-containing CA was found to be expressed during zinc limitation (Lane and Morel, 2000; Lane et al., 2005). This cadmium CA (CDCA1) which naturally uses Cd as its catalytic metal (Trip et al., 2001; Lane et al., 2005) has been ascribed to a novel ζ-CA class (see above). Genes coding for similar proteins have been identified in other cultured diatoms (Park et al., 2007). In mussel digestive gland western blotting analysis clearly demonstrated the enhancement of CA protein expression following cadmium exposure, according to the enzymatic activity data (Caricato et al., 2010). Laboratory experimental results were confirmed by a field experiment. Mussels exposed for 30 days to an anthropogenic impacted site showed a significant increase in CA activity and protein expression with respect to animals exposed for 30 days in a control site. If the new synthesized enzyme is a Cd-CA is not possible to say at the moment. If it was the case, then the increase in CA would not be a direct adaptive response to Cd pollution; rather, Cd could remove any limitations placed on CA synthesis by the availability of Zn. However, future studies will be needed to clarify this intriguing aspect of the research.

5. Carbonic anhydrase and heavy metals interactions: Potential applications

In the last years the interactions between carbonic anhydrase and heavy metals have found a number of applications in environmental and health fields, including the development of biomarkers of pollution exposure, in vitro bioassays, and biosensors.

5.1 Carbonic anhydrase sensitivity to heavy metals and development of biomarkers of pollution exposure

Pollution by trace metals is a world-wide problem due to the persistency and continuing accumulation of metals in the environment (de Mora et al. 2004; Hwang et al 2006). Heavy metals may enter the organisms through food, water, air, or absorption through the skin. As a result of mining, waste disposal and fuel combustion the environment is becoming increasingly contaminated with heavy metals.

In recent years the increasing sensibility to pollution problems has promoted the development of environmental "diagnostic" tools for early warning detection of pollution. Pollution monitoring has been increasingly concerned with the use of biological responses to pollutants at molecular and cellular level for evaluating biological hazard of toxic chemicals. Methods based on biological effects and their underlying mechanisms can complement the use of analytical chemistry in environmental monitoring. The major advantages of such biological, mechanism-based methods are their toxicological specificity, rapidity, and low cost. Toxicological specificity refers to the relationship between the assay response and the toxic potential rather than simply the contaminant concentrations (provided by chemical analysis) of the sample being analyzed. Moreover, biological assays provide rapid, sensitive, easily learnt and readily interpretable new useful tools for environmental biomonitoring and risk assessment. They include biomarkers, and *in vivo* and *in vitro* bioassays. It is known that the harmful effects of pollutants are typically first manifested at lower levels of biological organization before disturbances are realized at population, community and ecosystem levels (Adams, 1990). This is the reason why in recent years the study of molecular and cellular effects of pollutants has given important advancement in the developing of biologically-based methodologies useful for environmental biomonitoring and risk assessment. Enzymatic inhibition studies have been a very fruitful field for environmental monitoring application as biomarker of exposure/effect. Biomarkers are defined as pollutant induced variation in cellular or biochemical components occurring in organisms as a result of natural exposure to contaminants in their environment (Depledge, 1994). As reported by several authors, the evaluation of biomarkers in bioindicator organisms sampled in one or more areas suspected of chemical contamination and their comparison with organisms sampled in a control area can allow the evaluation of the potential risk of toxicological exposure of the studied community (Lionetto et al., 2003; Lionetto et al., 2004).

Carbonic anhydrase sensitivity to heavy metal exposure has been recently explored for its possible applications as biomarker of exposure to heavy metal pollution (Lionetto et al. 2006; Caricato et al, 2010b.) in "sentinel" organisms. Lionetto et al., (2006) investigated CA activity inhibition by heavy metals in the filter feeding *Mytilus galloprovincialis*, widely used in pollution monitoring programs as sentinel organism (Jernelov et al., 1996). Following *in vitro* and *in vivo* exposure to cadmium, mantle CA activity was significant inhibited. The inhibitory effect of cadmium on mantle CA activity can explain results previously obtained by Soto et al. (2000), who observed a significant decreased in shell growth in *M. galloprovincialis* exposed to heavy metals. The sensitivity of CA to heavy metals in mussels appears to be tissue-specific. In fact, as reported above, in mussel's digestive gland CA activity and expression was found to increase following Cd exposure (Caricato et al., 2010). Because of the widely application of *M. galloprovincialis* in environmental quality monitoring and assessment, data on tissue specific sensitivity of carbonic anhydrase to heavy metals represent a starting point for future potential application of CA activity changes as biomarker of exposure to heavy metals in the sentinel organism *M. galloprovincialis*.

Other studies carried out on corals have suggested alteration in CA activity as potential biomarker of exposure to environmental chemical stress. CA activity has been demonstrated to be inhibited by heavy metal exposure in anemones and corals (Gilbert and Guzman, 2001), where the enzyme plays a key role in the calcification process. Coral growth has been shown to be an effective indicator of the overall health of a coral reef ecosystem and reduced growth can reflect impaired photosynthetic output of the zooxanthellae and/or changes in enzyme activity (Moya et al., 2008). In an era of climate change and ocean acidification, where factors impacting growth and resilience factors are becoming important, understanding the biological effects of metal exposure to these keystone tropical organisms may be critical (Bielmyer et al., 2010).

5.2 Carbonic anhydrase based bioassay

Bioassays use biological systems to detect the presence of toxic chemicals in the environmental matrices (water, sediment, sewage, soil, etc.). In recent years, *in vitro* bioassays, employing cultured cells or cellular extracts, are increasingly being developed and used to detect the presence of contaminants. Examples include assays that measure enzyme inhibition, receptor-binding, or changes in gene expression in *in vitro* systems. Although *in vitro* assay is not a substitute for biomarker approach, it can be used as an adjunct model to whole-animal *in vivo* exposure and to ecotoxicological evaluation of the potential risk of trace pollutants in aquatic environments. They are rapid, low cost and simple tools to be utilized in combination with chemical analysis, for the pre-screening of the environmental samples that should be analyzed. Lionetto et al (2005; 2006) explored the possible application of heavy metal CA inhibition for the development of an *in vitro* bioassay applicable to the determination of the toxicity of environmental aqueous samples. They developed rapid and sensitive chemical hazard detection system for standardizing rapid, sensitive, and low cost CA based *in vitro* bioassay (Schettino et al., 2008).

6. Carbonic anhydrase-based biosensing of metal ions

In the last years the affinity of carbonic anhydrase for metal ions has been applied for the development of fluorescence based biosensors for determination of free metal ions in solution using variants of human carbonic anhydrase (apoCA). In particular, Cu^{2+}, Co^{2+}, Zn^{2+}, Cd^{2+}, and Ni^{2+} have been determined at concentration down the picomolar range (Fierke and Thompson, 2001; Thompson and Jones, 1993; Mey et al., 2011) by changes in fluorescence emission (Thompson et al., 2000) and excitation wavelength ratios (Thompson et al., 2002), lifetimes (Thompson and Patchan, 1995), and anisotropy (polarization) (Elbaum et al., 1996; Thompson et al., 2000). The sensitivity, selectivity, analyte binding, kinetics and stability of the biosensors have been improved by subtle modification of the protein structure by directed mutagenesis (Kiefer et al., 1995; Hunt et al., 1999; DiTusa et al., 2001; McCall et al., 2004; Burton et al, 2000). These studies have hallowed the development of highly selective and sensitive fluorescence-based biosensors for Zn^{2+} e Cu^{2+}, which have been shown to be viable approach in some important applications. In fact, the CA-based Cu^{2+} biosensor has been used to obtain real-time measurement of free Cu(II) at picomolar concentrations in seawater (Zeng et al., 2003), while the CA-base Zn^{2+} biosensor has been used for measurement of free Zn ion at picomolar levels in cultured cells (Bozym et al, 2004).

7. Conclusions

Although carbonic anhydrase represents one of the most investigated metalloenzyme in nature, its interaction with heavy metals has been only partially elucidated to date and some issues still remains to be explored. An intriguing aspect that needs more investigation is the *in vivo* effect of heavy metals on CA expression. From the few studies available in literature some metals appear to be important modulator of the expression of this protein. The understanding of the underlying mechanisms could open new perspective in the comprehension of the functioning and regulation of this enzyme. Another intriguing aspect of the biochemistry of CA is the inhibition by heavy metals. It has been documented for some species and some metals, but the mechanisms behind the inhibition, its metal specificity and isoform specificity remains still unknown. These aspects merits in depth examination and open new perspective for drug design and biomarkers development.

8. References

Adams, S.M.; Crumby, W.D.; Greeley, M.S.; Ryon, M.G. & Schilling, E.M. (1990). Relationship between Physiological and Fish Population Responses in a Contaminated Stream. Environmental Toxicology and Chemistry, Vol. 11, Issue 11, (November 1992), pp. 1549-1557, ISSN: 0730-7268

Alber, B.E. & Ferry, J.G. (1994). A carbonic anhydrase from the archaeon *Methanosarcina thermophila. Proceedings of the National Academy of Sciences of the United States of America*, Vol.91, (July 1994), pp. 6909-6913, ISSN 0027-8424

Alber, B.E. & Ferry, J.G. (1996). Characterization of heterologously produced carbonic anhydrase from *Methanosarcina thermophila,* The Journal of Bacteriology, Vol.178, No.11, (June 1996), pp.3270–3274, ISSN 0021-9193

Alber, B.E.; Colangelo, C.M.; Dong, J.; Staalhandske, C.M.V.; Baird, T.T.; Tu, C.; Fierke, C.A.; Silverman, D.N.; Scott, R.A. & Ferry, J.G. (1999). Kinetic and Spectroscopic Characterization of the Gamma-Carbonic Anhydrase from the Methanoarchaeon *Methanosarcina thermophil. Biochemistry,* Vol.38, Issue 40, (October 1999), pp. 13119–13128, ISSN 0006-2960

Berg, J.M. (2007). Biochemistry, 6th Ed., Sara Tenney. ISBN0-7167-8724-5

Bertini, I. & Luchinat, C. (1984). High spin cobalt(II) as a probe for the investigation of metalloproteins. *Advances in inorganic biochemistry,* Vol. 6, pp. 71-111, ISSN 0190-0218

Bielmyer, G.K.; Grosell, M.; Bhagooli, R.; Baker, A.C.; Langdon, C.; Gillette, P. & Capo, T.R. (2010). Differential effects of copper on three species of scleractinian corals and their algal symbionts (Symbiodinium spp.). *Aquatic Toxicology,* Vol. 97, No. 2, (April 2010), pp. 125–133, ISSN 0166-445X

Bozym, R.A.; Zeng, H.H.; Cramer, M.; Stoddard, A.; Fierke, C.A. & Thompson, R.B. (2004). In vivo and intracellular sensing and imaging of free zinc ion. In: *Proceedings of the SPIE Conference on Advanced Biomedical and Clinical Diagnostic Systems II*, Cohn, G.E., Grundfest, W.S., Benaron, D.A.; and Vo-Dinh, T., eds. Bellingham, WA: SPIE, 2004.

Burton, R.E.; Hunt, J.A.; Fierke, C.A. & Oas, T.G. (2000). Novel disulfide engineering in human carbonic anhydrase II using the PAIRWISE side-chain geometry database. *Protein Science*, Vol. 9, Issue 4, (April 2000), pp. 776-785, ISSN 0961-8368

Cannon, G.C.; Heinhorst, S. & Kerfeld C.A. (2010). Carboxysomal carbonic anhydrases: Structure and role in microbial CO(2) fixation. *Biochimica et Biophysica Acta-Proteins and Proteomics*, Vol. 1804, Issue 2, (February 2010), pp. 382-392, ISSN 1570-9639

Caricato, R.; Lionetto, M.G.; Dondero, F.; Viarengo, A. & Schettino, T. (2010a) Carbonic anhydrase activity in *Mytilus galloprovincialis* digestive gland: sensitivity to heavy metal exposure. *Comparative Biochemistry and Phyiology Part C: Toxicology & Pharmacology,* Vol.152C, Issue 3, (September 2010), pp.241-247, ISSN 1532-0456

Caricato, R.; Lionetto, M.G. & Schettino, T. (2010b). Seasonal variation of biomarkers in *Mytilus galloprovincialis* sampled inside and outside Mar Piccolo of Taranto (Italy). *Chemistry and Ecology,* Vol.26, supplement 1, (June 2010), pp.143-153, ISSN 0275-7540

Ceyhun, S.B.; Şentürk, M.; Yerlikaya, E.; Erdoğan, O.; Küfrevioğlu, Ö.I. & Ekinci, D. (2011). Purification and characterization of carbonic anhydrase from the teleost fish *Dicentrarchus labrax* (European Seabass) liver and toxicological effects of metals on enzyme activity. *Environmental Toxicology and Pharmacology.* Vol.32, Issue 1, (July 2011), pp. 69–74, ISSN 1382-6689

Christensen, G.M. & Tucker, J.H. (1976). Effects of selected water toxicants on the in vitro activity of fish carbonic anhydrase. *Chemico Biological Interactions,* Vol.13, Issue 2, (May 1976), pp.181–92, ISSN 0009-2797

Cox, J.D.; Hunt, J.A.; Compher, K.M.; Fierke, C.A. & Christianson, D.W. (2000). Structural Influence of Hydrophobic Core Residues on Metal Binding and Specificity in Carbonic Anhydrase II. *Biochemistry,* Vol.39, No.45, (November 2000), pp.13687–13694, ISSN 0006-2960

De Mora, S.; Fowler, S.W.; Wyse, E. & Azemard, S. (2004). Distribution of heavy metals in marine bivalves, fish and coastal sediments in the Gulf and Gulf of Oman, *Marine Pollution Bulletin,* Vol. 49, Issue 5-6, (September 2004), pp.410-424, ISSN 0025-326X

Depledge, M.H. (1994). The rational basis for the use of biomarkers as ecotoxicological tools. In: *Nondestructive Biomarkers in Vertebrates,* M.C., Fossi; C., Leonzio (Eds.), pp. 271–295, Lewis Publisher, ISBN 978-0873716482, Boca Raton, USA

DiTusa, C.A.; McCall, K.A.; Chritensen, T.; Mahapatro, M.; Fierke, C.A. & Toone, E.J. (2001). Thermodynamics of metal ion binding. 2. Metal ion binding by carbonic anhydrase variants. *Biochemistry,* Vol.40, Issue 18, (May 2001), pp.5345-5351, ISSN 0006-2960

Ekinci, D.; Beydemir, Ş. & Küfrevioğlu Ö.İ. (2007). In vitro inhibitory effects of some heavy metals on human erytrocyte carbonic anhydrases. *Journal of Enzyme Inhibition and Medicinal Chemistry,* Vol.22, Issue 6, pp.745–750, ISSN 1475-6366

Elbaum, D.; Nair, S.K.; Patchan, M.W.; Thompson, R.B. & Christianson, D.W. (1996). Structure-based design of a sulfonamide probe for fluorescence anisotropy detection of zinc with a carbonic anhydrase-based biosensor. *Journal of the American Chemical Society,* Vol. 118, Issue 35, (September 1996), pp. 8381-8387, ISSN 0002-7863

Esbaugh, A.J. & Tufts, B.L. (2006). The structure and function of carbonic anhydrase isozymes in the respiratory system of vertebrates. *Respiratory Physiology & Neurobiology,* Vol.154, Issue 1-2, (November 2006), pp. 185–198, ISSN 1569-9048

Evans, D.H. & Cameron, J.N. (1986). Gill ammonia transport. *Journal of Experimental Zoology,* Vol.239, Issue 1, (July 1986), pp. 17–23, ISSN 0022-104X

Ferry, J.F. (2010). The gamma class of carbonic anhydrases. *Biochimica et Biophysica Acta (BBA) - Proteins & Proteomics,* Vol.1804, Issue 2, (February 2010), pp. 374-38, ISSN 1570-9639

Fierke, C.A. & Thompson, R.B. (2001). Fluorescence-based biosensing of zinc using carbonic anhydrase. *Biometals,* Vol. 14, Issue 3-4, (September 2001), pp. 205-222, ISSN 0966-0844

Gilbert, A.L. & Guzman, H.M. (2001). Bioindication potential of carbonic anhydrase activity in anemones and corals. *Marine Pollution Bulletin,* Vol. 42, Issue 9, (September 2001), pp. 742–744, ISSN 0025-326X

Goto, T.; Shirakawa, H.; Furukawa Y. & Komai, M. (2008). Decreased expression of carbonic anhydrase isozyme II, rather than of isozyme VI, in submandibular glands in long-term zinc-deficient rats. British Journal of Nutrition, Vol.99, Issue 2, (February 2008), pp. 248-53, ISSN 0007-1145

Grimes, A.; Paynter, J.; Walker, I.D.; Bhave, M. & Mercer, J.F.B. (1997). Decreased carbonic anhydrase III levels in the liver of the mouse mutant "toxic milk" (tx) due to copper accumulation, *Biochemical Journal,* Vol. 321, Part. 2, (January 1997), pp. 341–346, ISSN 0264-6021

Heisler, N. (1984). Acid–base regulation in fishes. In: *Hoar,* W.S., Randall, D.J. (Eds.), Fish Physiology, vol. 10A. Academic Press, New York, pp. 315–401.

Henkin, R.; Martin, B.M. & Agarwal, R. (1999). Efficacy of exogenous oral zinc in treatment of patients with carbonic anhydrase VI deficiency. *American Journal of the Medical Science, Vol.* 318, Issue 6, (December 1999), pp. 392-405, ISSN 0002-9629

Henry, R.P. (1984). The role of carbonic anhydrase in blood ion and acid–base regulation. *American Zoologist*. Vol. 24(1), pp. 241–253, 0003-1569.

Hewett-Emmett, D. & Tashian, R.E. (1996). Functional diversity, conservation, and convergence in the evolution of the α-, ß-, and γ-carbonic anhydrase gene families. *Molecular Phylogenetics and Evolution, Vol.* 5, Issue 1, (February 1996), pp. 50–77, ISSN 1055-7903

Hoffmann, K.M.; Samardzic, D.; van den Heever, K. & Rowlett, R.S. (2011). Co(II)-substituted Haemophilus influenzae b-carbonic anhydrase: Spectral evidence for allosteric regulation by pH and bicarbonate ion. *Archives of Biochemistry and Biophysics,* Vol. 511, Issue 1-2, (July 2011), pp. 80–87, ISSN 0003-9861

Hove, C.; Elvehjem, C.A. & Hart, E.B. (1940). The relation of zinc to carbonic anhydrase. J Biol Chem 136:425–434

Hunt, J.A. & Fierke, C.A. (1997). Selection of carbonic anhydrase variants displayed on phage. Aromatic residues in zinc binding site enhance metal affinity and equilibration kinetics. *Journal of Biological Chemistry,* Vol.272, Issue 33, (August 1997), pp. 20364−20372, ISSN 0021-9258

Hunt, J.A.; Ahmed, M. & Fierke, C.A. (1999). Metal Binding Specificity in Carbonic Anhydrase is Influenced by Conserved Hydrophobic Amino Acids. *Biochemistry,* Vol. 38, Issue 28, (July 1999), pp. 9054-9060, ISSN 0006-2960

Hwang, H.; Green, P.G.; Higashi, R.M. & Young, T.M. (2006). Tidal salt marsh sediment in California, USA. Part 2: Occurrence and anthropogenic input of trace metals, *Chemosphere,* Vol. 64, Issue 11, (September 2006), pp.1899-1909, ISSN 0045-6535

Innocenti, A.; Scozzafava, S.; Parkkila, L.; Puccetti, G.; De Simone, G. & Supuran, C.T. (2008). Investigations of the esterase, phosphatase, and sulfatase activities of the cytosolic mammalian carbonic anhydrase isoforms I, II, and XIII with 4-nitrophenyl esters as substrates. *Bioorganic & Medicinal Chemistry Letters,* Vol. 18, Issue 7, (April), pp.2267–2271, ISSN 0960-894X

Ivanov, B.N.; Ignatova, L.K. & Romanova, A.K. (2007). Diversity in Forms and Functions of Carbonic Anhydrase in Terrestrial Higher Plants. *Russian Journal of Plant Physiology,* Vol. 54, No. 2, (March-April 2007), pp. 143–162, ISSN 1021-4437

Jernelov, A. (1996). The international mussel watch: a global assessment of environmental levels of chemical contaminats. *Science of the Total Environment,* Vol.188, Supplement 1, (September 1996), pp. 37-44, ISSN: 0048-9697

Kiefer L.L.; Paterno S.A. & Fierke, C.A. (1995). Hydrogen-Bond Network in the Metal-Binding Site of Carbonic Anhydrase Enhances Zinc Affinity and Catalytic Efficency. *Journal of the American Chemical Society,* Vol. 117, Issue 26, (July 1995), pp. 6831-6837, ISSN 0002-7863

Kimber, M.S. & Pai, E.F. (2000). The active site architecture of Pisum sativum β-carbonic anhydrase is a mirror image of that of α-carbonic anhydrases. *Embo Journal, Vol.* 19, Issue 7, (April 2000), pp.1407–1418, ISSN 0261-4189

Kirchgessner M, Stadler AE, Roth HP (1975) Carbonic anhydrase activity and erythrocyte count in the blood of zinc-deficient rats. *Bioinorganic chemistry,* Vol. 5, Issue 1, pp. 33-38, ISSN: 0006-3061

Kuhara, M.; Wang, J.; Flores, M.J.; Qiao, Z.; Koizumi, Y.; Koyota, S.; Taniguchi, M. & Sugiyama, T. (2011). Sexual dimorphism in LEC rat liver: suppression of carbonic anhydrase III, by copper accumulation during hepatocarcinogenesis. Biomedical Research, Vol. 32 (2), (April 2011), pp. 111-117, ISSN 0388-6107

Lane, T.W. & Morel, F.M.M. (2000). A biological function for cadmium in marine diatoms. *Proceedings of the National Academy of Sciences of the United States of America*, Vol.97, Issue 9, (April 2000), pp. 4627–4631, ISSN 0027-8424

Lane, T.W.; Saito, M.A.; George, G.N.; Pickering, I.J.; Prince, R.C. & Morel, F.M.M. (2005). A cadmium enzyme from a marine diatom. *Nature,* Vol. 435, Issue 7038, (May 2005), pp.42–42, - ISSN 0028-0836

Lee, J.G.; Roberts, S.B. & Morel, F.M.M. (1995). Cadmium a nutrient for the marine diatom. *Limnology and Oceanography*, Vol. 40, Issue 6, (September 1995), pp. 1056-1063, ISSN 0024-3590

Liang, J.Y. & Lipscomb, W. N (1988). Hydration of CO_2 By Carbonic-Anhydrase - Intramolecular Proton Transfer between Zn^{2+}-Bound H_2O and Histidine-64 in Human Carbonic Anhydrase-II. *Biochemistry*, Vol. 27, Issue 23, (November 1988), pp. 8676-8682, ISSN 0006-2960

Lionetto, M.G.; Maffia, M.; Cappello, M.S.; Giordano, M.E.; Storelli, C. & Schettino, T. (1998). Effect of cadmium on carbonic anhydrase and Na+-K+-ATPase in eel, Anguilla anguilla, intestine and gills. *Comparative Biochemistry and Physiology A- Molecular and Integrative Physiology*, Vol.120, Issue 1, (May 1998), pp.89-91, ISSN 1095-6433

Lionetto, M.G.; Giordano, M.E.; Vilella, S. & Schettino, T. (2000). Inhibition of eel enzymatic activities by cadmium. *Aquatic Toxicology*, Vol. 48,Issue 4, (April 2000), pp. 561-571, ISSN: 0166-445X

Lionetto, M.G.; Caricato, R.; Giordano, M.E.; Pascariello, M.F.; Marinosci, L. & Schettino, T. (2003). Integrated use of biomarkers (acetylcholinesterase and antioxidant enzymatic activities) in Mytilus galloprovincialis and Mullus barbatus in an Italian coastal marine area. *Marine Pollution Bulletin*, Vol. 46, Issue 3, (March 2003), pp. 324-330, ISSN 0025-326X

Lionetto, M.G.; Caricato, R.; Erroi, E.; Giordano, M.E. & Schettino, T. (2005). Carbonic anhydrase based environmental bioassay. *International Journal of Environmental Analytical Chemistry*, Vol.85, Issue 12-13, (October-November 2005), pp. 895-903, ISSN: 0306-7319

Lionetto, M.G.; Caricato, R.; Erroi, E.; Giordano, M.E. & Schettino, T. (2006). Potential application of carbonic anhydrase activity in bioassay and biomarker studies. *Chemistry and Ecology*, Vol. 22, Supplement 1, pp. 119-125, ISSN: 0275-7540

Lukaski, H.C. (2005). Low dietary zinc decreases erythrocyte carbonic anhydrase activities and impairs cardiorespiratory function in men during exercise. *American Journal of Clinical Nutrition*, Vol 81, Issue 5, (May 2005), pp.1045–1051, ISSN 0002-9165

MacAuley, S.R.; Zimmerman, S.A.; Apolinario, E.E.; Evilia, C.; Hou, Y.; Ferry, J.G. & Sowers, K.R. (2009). The archetype γ-class carbonic anhydrase (Cam) contains iron when synthesized in vivo, *Biochemistry*, Vol. 48, Issue 5, (February 2009) pp 817–819, ISSN: 0006-2960

Maffia, M.; Trischitta, F.; Lionetto, M.G.; Storelli, C. & Schettino, T. (1996). Bicarbonate absorption in eel intestine: Evidence for the presence of membrane-bound carbonic anhydrase on the brush border membranes of the enterocyte. *Journal of Experimental Zoology*, Vol: 275, Issue 5, (August 1996), pp. 365-373, ISSN 0022-104X

Maret, W. (2011). Metals on the move: zinc ions in cellular regulation and in the coordination dynamics of zinc proteins. *Biometals*,Vol. 24, Issue 3, (June 2011), pp. 411-418, ISSN: 0966-0844

Marouan, R.; Cecchi, A; Montero, J.L.; Innocenti, A.; Vullo, D; Scozzafava, A.; Winum, J.Y. &. Supuran, C.T. (2008) Carbonic Anhydrase Inhibitors: Design of Membrane-Impermeant Copper(II) Complexes of DTPA-, DOTA-, and TETA-Tailed Sulfonamides Targeting the Tumor-Associated Transmembrane Isoform IX. ChemMedChem, Vol. 3, Issue 11, (November 2008), pp. 1780 – 1788, ISSN 1860-7179

McCall, K.A. & Fierke, C.A. (2004). Probing Determinants of the Metal Ion Selectivity in Carbonic Anhydrase Using Mutagenesis. Biochemistry, Vol. 43, Issue 13, (April 2004), pp. 3979-3986, ISSN 0006-2960

Mei, Y.J.; Frederickson, C.J.; Giblin, L.J.; Weiss, J.H.; Medvedeva, Y. & Bentley, PA (2011). Sensitive and selective detection of zinc ions in neuronal vesicles using PYDPY1, a simple turn-on dipyrrin. Chemical Communications, Vol. 47, Issue 25, pp. 7107-7109, ISSN 1359-7345

Morel, F.M.M.; Reinfelder, J.R.; Roberts, S.B.; Chamberlain, C.P.; Lee, J.G. & Yee, D. (1994). Zinc and carbon co-limitation of marine phytoplankton. *Nature*, Vol. 369, Issue 6483, (June 1994), pp. 740–742, ISSN: 0028-0836

Morgan, I.J.; Henry, R.P. & Wood, C.M. (1997). The mechanism of acute silver nitrate toxicity in freshwater rainbow trout (Oncorhynchus mykiss) is inhibition of gill Na^+ and Cl^- transport. *Aquatic Toxicology*,Vol.38, Issue 1-3, (May 1997), pp. 145–63, ISSN: 0166-445X

Morgan, T.P.; Grosell, M.; Gilmour, K.M.; Playle, R.C. & Wood, C.M. (2004). Time course analysis of the mechanism by which silver inhibits active Na+ and Cl- uptake in gills of rainbow trout. *American Journal of Physiology-Regulatory Integrative and Comparative Physiology*, Vol. 287, Issue 1, (July 2004), ISSN 0363-6119

Moya, A.; Ferrier-Pages, C.; Furla, P.; Richier, S.; Tambutte, E.; Allemand, D. & Tambutte, S. (2008). Calcification and associated physiological parameters during a stress event in the scleractinian coral Stylophora pistillata. *Comparative Biochemistry and Physiology A – Molecular & Integrative Physiology*, Vol. 151, Issue 1, (September 2008), pp. 29–36, ISSN 1095-6433

Park, H.; McGinn, P.J. & Morel, F.M.M. (2008). Expression of cadmium carbonic anhydrase of diatoms in seawater. Aquatic Microbial. Ecology, Vol.51, Issue2, (May 2008), pp. 183–193, ISSN: 0948-3055

Park, H.; Song, B. & Morel, F.M.M. (2007). Diversity of the cadmium-containing carbonic anhydrase in marine diatoms and natural waters. *Environmental Microbioogyl*, Vol. 9, Issue 2, (Feb 2007), pp. 403–413, ISSN: 1462-2912

Price, N.M. & Morel, F.M.M. (1990). Cadmium and cobalt substitution for zinc in a marine diatom. *Nature*, Vol. 344, Issue 6267, (April 1990), pp. 658–660, ISSN 0028-0836

Rahman, M.W.; Davies, R.E.; Deyoe, C.W.; Reid, B.L. &. Couch, J.R. (1961). Role of zinc in the nutrition of growing pullets. *Poultry Science*, Vol. 40, pp. 195-200, ISSN 0032-5791.

Randall, D.J. & Daxbaeck, C. (1984). Oxygen and carbon dioxide transfer across fish gills. In: Hoar, W.S., Randall, D.J. (Eds.), Fish Physiology, vol. 10A. Academic Press, New York, pp. 263–314, ISBN 0-12-350430-9

Roberts, S.B.; Lane, T.W. & Morel, F.M.M. (1997). Carbonic anhydrase in the marine diatom Thalassiosira weissflogii (Bacillariophyceae). Journal of Phycology, Vol. 33, Issue 5, (October 1997), pp. 845–850, ISSN 0022-3646

Rowlett, R.S. (2010). Structure and catalytic mechanism of the β-carbonic anhydrases. Biochimica et Biophysica Acta (BBA) - Proteins & Proteomics, Vol.1804, Issue 2, (February 2010), pp. 362-373, ISSN: 1570-9639

Schettino, T.; Trischitta, F; Denaro, M.G.; Faggio, C. & Fucile, I. (1992). Requirement of HCO_3^- For Cl^--Absorption in Seawater-Adapted Eel Intestine. *Pflugers Archiv-European Journal of Physiology*, Vol. 421, Issue 2-3, (June 1992), pp. 146-154, ISSN 0031-6768

Schettino, T.; Lionetto, M.G. & Erroi, E. (2008). Enzymatic method for the dectetion of the tozicity of aqueous environmental matrices. Patent n. MI2008A008813, PCT/EP2008/064703

Skaggs, H.S. & Henry, R.P. (2002). Inhibition of carbonic anhydrase in the gills of two euryhaline crabs, Callinectes sapidus and Carcinus maenas, by heavy metals. *Comparative Biochemistry and Physiology C-Toxicology & Pharmacology*, Vol. 133, Issue 4, (December 2002), pp. 605-612, ISSN 1532-0456

Soto, M.; Ireland, M.P. & Marigómez, I. (2000). Changes in mussel biometry on exposure to metals: implications in estimation of metal bioavailability in "Mussel-Watch" programmes. *The Sciences of the Total Environment*, Vol. 247, Issue 2-3, (March 2000), pp. 175-187, ISSN 0048-9697

Soyut, H.; Beydemir, Ş. & Hisar, O. (2008). Effects of Some Metals on Carbonic Anhydrase from Brains of Rainbow Trout. *Biological Trace Element Research*, Vol.123, Issue 1-3, (June 2008), pp.179–190, ISSN 0163-4984

Supuran, C.T.; Scozzafava, A. & Casini, A. (2003). Carbonic anhydrase inhibitors. *Medicinal Research Reviews*, Vol.23, Issue 2, (March 2003), pp. 146– 189, ISSN: 0198-6325

Supuran, C.T. & Scozzafava, A. (2007). Carbonic anhydrases as targets for medicinal chemistry. *Bioorganic & Medicinal Chemistry*, Vol. 15, Issue 13, (July 2007), pp. 4336–4350, ISSN: 0968-0896

Supuran C.T., (2008). Carbonic anhydrases: novel therapeutic applications for inhibitors and activators. Nature Reviews Drug Discovery, Vol.7, Issue 2, (February), pp.168–181, ISSN 1474-1776

Supuran, C.T. (2010). Carbonic anhydrase inhibitor. *Bioorganic & Medicinal Chemistry Letters*, Vol. 20, Issue 12, (June), pp. 3467-3474, ISSN: 0960-894X

Tashian, R.E.; Hewett-Emmett, D.; Carter, N.D. & Bergenhem, N.C.H. (2000). Carbonic anhydrase (CA)-related proteins (CA-RPs) and transmembrane proteins with CA or CA-RP domains. In: Chegwidden, W.R.; Carter, N.D. & Edwards, Y.H. (Eds.), Carbonic anhydrase (CA)-related proteins (CA-RPs) and transmembrane proteins with CA or CA-RP domains. *The Carbonic Anhydases: New Horizons*. Birkhauser, Basel, pp. 105–120, ISBN 3-7643-5670-7

Thompson, R.B. & Jones, E.R. (1993). Enzyme-based fiber optic zinc biosensor. Analitical Chemistry, Vol.65, Issue 6, March 1993), pp. 730-734, ISSN 0003-2700

Thompson, R.B. & Patchan, M.W. (1995). Lifetime-Based Fluorescence Energy-Transfer Biosensing of Zinc. *Analytical Biochemistry*, Vol. 227, Issue 1, (May 1995), pp. 123-128, ISSN 0003-2697

Thompson, R.B.; Maliwal, B.P. & Zeng H.H. (2000). Zinc biosensing with multiphoton excitation using carbonic anhydrase and impoved fluorofores. Journal of Biomedical Optics, Vol. 5, Issue 1, (January 2000), pp. 17-22, ISSN 1083-3668

Thompson, R.B.; Thompson, R.B. & Meisinger, J.J. (2002). Fluorescent zinc indicators for neurobiology. Journal of Neuroscience Methods, Vol. 118, Issue 1, (July 2002), pp. 63-75, ISSN 0047-2425

Tripp, B.C.; Bell, C.B.; Cruz, F.; Krebs, C.& Ferry, J.G. (2004). A role for iron in an ancient carbonic anhydrase, Journal of Biological Chemistry, Vol. 279, Issue 20, (May), pp. 21677-21677, ISSN 0021-9258

Tripp, B.C.; Smith K.S. & Ferry J.G. (2001)., Journal of Biological Chemistry, Vol.276, Issue 52, (December), pp. 48615– 48618, ISSN: 0021-9258

Tu, C.; Silverman, D. N.; Forsman, C.; Jonsson, B.-H. & Lindskog, S. (1989). Role of Histidine-64 in the Catalytic Mechanism of Human Carbonic Anhydrase-II Studied with a Site-Specific Mutation. Biochemistry, Vol.28, Issue 19, (September 1989), pp. 7913-7918, ISSN 0006-2960

Tu, C.K. & Silverman, D.N. (1985). Catalysis by cobalt(II)- substituted carbonic anhydrase II of the exchange of oxygen-18 between CO2 and H2O. *Biochemistry*, Vol.24, Issue 21, pp. 5881–5887, ISSN: 0006-2960

Tu, C; Wynns, G.C. & Silverman, D.N. (1981). Inhibition by cupric ions of ^{18}O exchange catalyzed by human carbonic anhydrase II. Relation to the Interaction Between Carbonic Anhydrase and Hemoglobin. *The Journal of Biological Chemistry*, Vol. 256, No.18, (September 1981), pp. 9466-9470, ISSN 0021-9258

Vedani, A.; Huhta, D. W. & Jacober, S. P. J. (1989). Metal Coordination, H-Bound Network Formation, and Protein-Solvent Interactions in Native and Complexed Human Carbonic Anhydrase-I - a Molecular Mechanism Study. *Journal of the American Chemical Society*, Vol.111, Issue 11, (May 1989), pp. 4075-4081, ISSN: 0002-7863

Viarengo, A.; Ponzano, E.; Dondero, F. & Fabbri, R. (1997). A simple spectrophotometric method for metallothionein evaluation in marine organisms: an application to Mediterranean and Antartic molluscs. *Marine Environmental Research*, Vol. 44, Issue 1, (July 1997), pp. 69–84, ISSN 0141-1136

Vitale, A.M.; Monserrat, J.M.; Casthilo, P. & Rodriguez, E.M. (1999). Inhibitory effects of cadmium on carbonic anhydrase activity and ionic regulation of the estuarine crab, Chasmagnathus granulata (Decapoda, Grapsidae). *Comparative Biochemistry and physiology C-Toxicology & Pharmacology*, Vol. 122, Issue 1, (January 1999), pp. 121-129, ISSN 1532-0456

Wang, B.; Liu, C.Q. & Wu, Y. (2005). Effect of Heavy Metals on the Activity of External Carbonic Anhydrase of Microalga Chlamydomonas reinhardtii and Microalgae from Karst Lakes. *Bulletin of Environmental Contamination and Toxicology*, Vol. 74, Issue 2, (February 2005), pp. 227-233, ISSN 0007-4861

Whitney, P.L. & Briggle, T.V. (1982). Membrane-associated carbonic anhydrase purified from bovine lung. *The Journal of Biological Chemistry*, Vol.257, (October 1982), pp. 12056-12059, ISSN 0021-9258

Wu, Y.; Zhao, X.; Li, P. & Huang, H. (2007). Impact of Zn, Cu, and Fe on the Activity of Carbonic Anhydrase of Erythrocytes in Ducks. *Biological Trace Element Research*, Vol. 118, No. 3, (September 2007), pp. 227-232, ISSN 0163-4984

Xu, Y.; Feng, L.; Jeffrey, P.D.; Shi, Y. & Morel, F.M. (2008). Structure and metal exchange in the cadmium carbonic anhydrase of marine diatoms. *Nature*, Vol. 452, Issue 7183, (March 2008), pp.56-U3, ISSN 0028-0836

Yee, D. & Morel, F.M.M. (1996). In vivo substitution of zinc by cobalt in carbonic anhydrase of a marine diatom. *Limnology and Oceanography*, Vol.41, Issue 3, (May 1996), pp. 573–577, ISSN 0024-3590

Zeng, H.H.; Thompson, R.B.; Maliwal, B.P.; Fones, G.R.; Moffet, J.W. & Fierke, C.A. (2003). Real-time determination of picomolar free Cu(II) in seawater using a fluorescence-based fiber optic biosensor. *Analytical Chemitry*, Vol.75, Issue 24, (December 2003), pp. 6807-6812, ISSN 0003-2700

Zhang, B.Y.; Yang, F.; Wang, G.C. & Peng, G. (2010) Cloning and Quantitative Analysis of the Carbonic Anhydrase Gene from Porphyra yezoensis. *Journal of Phycology*, Vol. 46, Issue 2, (April 2010), pp.290-296, ISSN 0022-3646

9

Enzymology of Bacterial Lysine Biosynthesis

Con Dogovski[1*] et al.
[1]*Department of Biochemistry and Molecular Biology, Bio21 Molecular Science and Biotechnology Institute, University of Melbourne, Parkville, Victoria Australia*

1. Introduction

Lysine is an essential amino acid in the mammalian diet, but can be synthesised *de novo* in bacteria, plants and some fungi (Dogovski et al., 2009; Hutton et al., 2007). In bacteria, the lysine biosynthesis pathway, also known as the diaminopimelate (DAP) pathway (Fig. 1), yields the important metabolites *meso*-2,6-diaminopimelate (*meso*-DAP) and lysine. Lysine is utilised for protein synthesis in bacteria and forms part of the peptidoglycan cross-link structure in the cell wall of most Gram-positive species; whilst *meso*-DAP is the peptidoglycan cross-linking moiety in the cell wall of Gram-negative bacteria and also Gram-positive *Bacillus* species (Burgess et al., 2008; Mitsakos et al., 2008; Voss et al., 2010) (Fig. 1).

The synthesis of *meso*-DAP and lysine begins with the condensation of pyruvate (PYR) and *L*-aspartate-semialdehyde (ASA) by the enzyme *dihydrodipicolinate synthase* (DHDPS, EC 4.2.1.52) (Blickling et al., 1997a; Mirwaldt et al., 1995; Voss et al., 2010; Yugari & Gilvarg, 1965). The product of the DHDPS-catalysed reaction is an unstable heterocycle, 4-hydroxy-2,3,4,5-tetrahydro-*L,L*-dipicolinic acid (HTPA) (Fig. 1). HTPA is non-enzymatically dehydrated to produce dihydrodipicolinate (DHDP), which is subsequently reduced by the NAD(P)H-dependent enzyme, *dihydrodipicolinate reductase* (DHDPR, EC 1.3.1.26), to form *L*-2,3,4,5,-tetrahydrodipicolinate (THDP) (Dommaraju et al., 2011; Girish et al., 2011; Reddy et al., 1995, 1996) (Fig. 1). The metabolic pathway then diverges into four sub-pathways depending on the species, namely the succinylase, acetylase, dehydrogenase and aminotransferase pathways (Dogovski et al., 2009; Hutton et al., 2007) (Fig. 1)

The most common of the alternative metabolic routes is the succinylase pathway, which is inherent to many bacterial species including *Escherichia coli*. This sub-pathway begins with the conversion of THDP to N-succinyl-*L*-2-amino-6-ketopimelate (NSAKP) catalysed by *2,3,4,5-tetrahydropyridine-2-carboxylate N-succinyltransferase* (THPC-NST, EC 2.3.1.117).

* Sarah. C. Atkinson[1], Sudhir R. Dommaraju[1], Matthew Downton[2], Lilian Hor[1], Stephen Moore[2], Jason J. Paxman[1], Martin G. Peverelli[1], Theresa W. Qiu[1], Matthias Reumann[2], Tanzeela Siddiqui[1], Nicole L. Taylor[1], John Wagner[2], Jacinta M. Wubben[1] and Matthew A. Perugini[1,3]
[1]*Department of Biochemistry and Molecular Biology, Bio21 Molecular Science and Biotechnology Institute, University of Melbourne, Parkville, Victoria, Australia*
[2]*IBM Research Collaboratory for Life Sciences-Melbourne, Victorian Life Sciences Computation Initiative, University of Melbourne, Parkville, Victoria, Australia*
[3]*Department of Biochemistry, La Trobe Institute for Molecular Science, La Trobe University, Melbourne, Australia*

NSAKP is then converted to N-succinyl-*L,L*-2,6,-diaminopimelate (NSDAP) by *N-succinyldiaminopimelate aminotransferase* (NSDAP-AT, EC 2.6.1.17), which is subsequently desuccinylated by *succinyldiaminopimelate desuccinylase* (SDAP-DS, EC 3.5.1.18) to form *L,L*-2,6-diaminopimelate (*LL*-DAP) (Kindler & Gilvarg., 1960; Ledwidge & Blanchard., 1999; Simms et al., 1984) (Fig. 1). *LL*-DAP is then converted to *meso*-DAP by the enzyme diaminopimelate epimerase (DAPE, EC 5.1.1.7) (Wiseman, & Nichols, 1984) (Fig. 1).

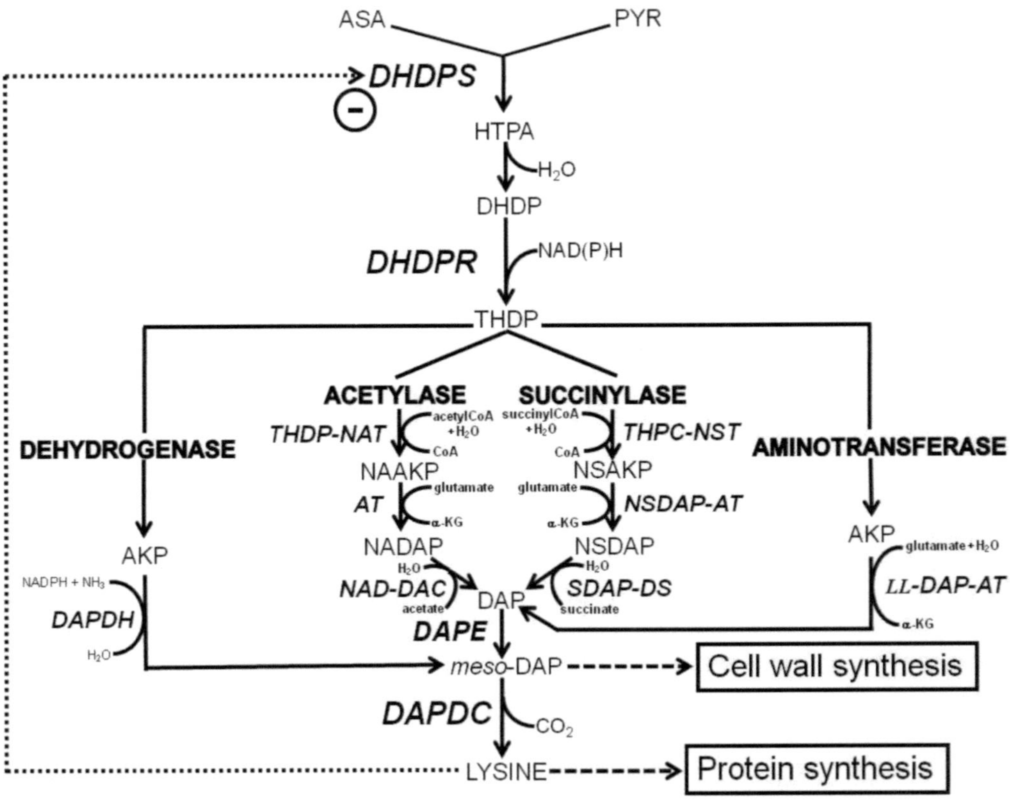

Fig. 1. Diaminopimelate pathway in bacteria.

As for the succinylase pathway, the acetylase pathway involves four enzymatic steps, but incorporates N-acetyl groups rather than N-succinyl moieties. This pathway is common to several *Bacillus* species, including *B. subtilis* and the anthrax-causing pathogen *B. anthracis* (Chatterjee & White., 1982; Peterkofsky & Gilvarg., 1961; Sundharadas & Gilvarg., 1967). The sub-pathway begins with the conversion of THDP to N-acetyl-(*S*)-2-amino-6-ketopimelate (NAAKP) catalysed by *tetrahydrodipicolinate N-acetyltransferase* (THDP-NAT, EC 2.3.1.89), followed by conversion to N-acetyl-(2*S*)-2,6,-diaminopimelate (NADAP) by *aminotransferase A* (ATA, EC 2.6.1). NADAP is subsequently deacetylated to form DAP by the enzyme *N-acetyldiaminopimelate deacetylase* (NAD-DAC, EC 3.5.1.47) (Fig. 1). As in the succinylase pathway, *LL*-DAP is then converted to *meso*-DAP by DAPE (Fig. 1).

There are also two additional sub-pathways that are less common to bacteria. The aminotransferase pathway, catalysed by the enzyme *diaminopimelate aminotransferase* (*LL*-DAP-AT, EC 2.6.1.83), is found in plant, eubacterial and archaeal species (Hudson et al., 2006). This sub-pathway involves the conversion of the acyclic form of THDP, 1,2-amino-6-ketopimelate (AKP), to *meso*-DAP in a single step. *LL*-DAP is then converted in the second step of the sub-pathway to *meso*-DAP by DAPE, as for the acetylase and succinylase pathways (Fig. 1). The dehydrogenase pathway, which is common to *Corynebacterium* and some *Bacillus* species, converts THDP to *meso*-DAP, also in a single step (Misono et al., 1976). This sub-pathway employs the NADPH-dependent enzyme, *diaminopimelate dehydrogenase* (DAPDH, EC 1.4.1.16), which also employs AKP as the substrate (Fig. 1).

All four alternative pathways then converge to utilise the same enzyme for the final step of lysine biosynthesis, namely diaminopimelate decarboxylase (DAPDC, EC 4.1.1.20) (Ray et al., 2002). DAPDC catalyses the decarboxylation of *meso*-DAP to yield lysine and carbon dioxide. This step is important for the overall regulation of the lysine biosynthesis pathway since the downstream product, lysine, has been shown to allosterically inhibit DHDPS from plants and Gram-negative bacteria (Section 2.1.1, Fig. 1). DHDPS is therefore considered the rate-limiting enzyme of the pathway.

This book chapter will describe the function, structure, and regulation of the key enzymes functioning in the lysine biosynthesis pathway. Furthermore, given that several of these enzymes are the products of essential bacterial genes that are not expressed in humans, the pathway is of interest to antibiotic discovery research (Dogovski et al., 2009; Hutton et al., 2007). Accordingly, the chapter will also review the current status of rational drug design initiatives targeting essential enzymes of the lysine biosynthesis pathway in pathogenic bacteria.

2. Dihydrodipicolinate synthase

2.1 Function of DHDPS

Dihydrodipicolinate synthase (DHDPS, EC 4.2.1.52) was first purified in 1965 from *E. coli* extracts (Yugari & Gilvarg, 1965). The enzyme is the product of the *dap*A gene, which has been shown to be essential in several bacterial species (Dogovski et al., 2009; Hutton et al., 2007). The *dap*A product, DHDPS, catalyses the condensation of pyruvate (PYR) and aspartate semialdehyde (ASA) to form 4-hydroxy-2,3,4,5-tetrahydro-*L,L*-dipicolinic acid (HTPA) (Fig. 1). It was first suggested that the product released by DHDPS was dihydrodipicolinate (DHDP), but studies using ^{13}C-labelled pyruvate support the view that the product is the unstable heterocycle HTPA (Blickling et al. 1997a). Rapid decomposition of the ^{13}C-NMR signals of HTPA following its production indicate that formation of DHDP occurs via a nonenzymatic step.

In all cases examined, the DHDPS-catalysed reaction proceeds via a ping-pong kinetic mechanism in which pyruvate binds the active site, resulting in the release of a protonated water molecule. ASA then binds and is condensed with pyruvate to form the heterocyclic product, HTPA (Blickling et al. 1997a).

In the first step of the mechanism, the active site lysine, (Lys161 in *E. coli* DHDPS) forms a Schiff base with pyruvate (Laber et al., 1992) (Fig 2). Formation of the Schiff base proceeds through a tetrahedral intermediate. It is proposed that a catalytic triad of three residues - Tyr133, Thr44 and Tyr107 (*E. coli* numbering) - act as a proton relay to transfer protons to and from the active site via a water-filled channel leading to bulk solvent (Dobson et al., 2004a). The Schiff base (imine) is converted to its enamine form, which then adds to the aldehyde group of ASA (Blickling et al., 1997a; Dobson et al., 2008). In aqueous solution, ASA is known to exist in the hydrated form rather than the aldehyde, but the biologically-relevant form of the substrate remains to be determined. HTPA is then formed by nucleophilic attack of the amino group of ASA onto the intermediate imine, leading to cyclisation and detachment of the product from the enzyme, with release of the active site lysine residue (Fig 2).

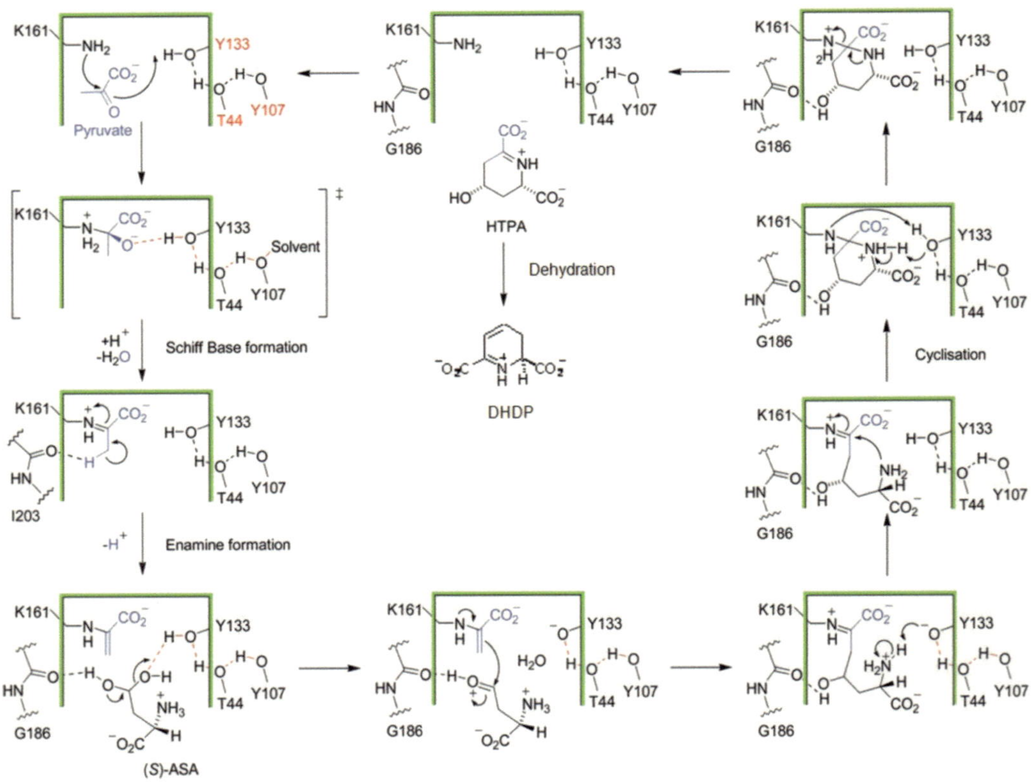

Fig. 2. The catalytic mechanism of DHDPS.

2.1.1 Regulation of DHDPS activity

In some organisms, the activity of DHDPS is regulated allosterically by lysine via a classical feedback inhibition process. Lysine feedback inhibition of DHDPS has been investigated in several plant, Gram-negative and Gram-positive bacterial species to date. Studies involving *Daucus carota sativa* (Matthews et al., 1979), *Pivus sativum* (Dereppe et al., 1992), *Spinacia aloeracea* (Wallsgrove et al., 1980), *Triticum aestivium* (Kumpaisal et al., 1987), and *Zea mays* (Frisch et al., 1991) show that DHDPS from plant species are generally strongly inhibited by lysine (IC_{50} = 0.01-0.05 mM). In contrast, DHDPS from bacteria are significantly less sensitive to lysine inhibition than their plant counterparts. For example, DHDPS from Gram-negative bacteria, such as *E. coli* (Dobson et al., 2005a; Yugari and Gilvarg, 1965), *Niesseria meningitidis* (Devenish et al., 2009), and *Sinorhizobium. meliloti* (Phenix & Palmer, 2008), display IC_{50} values that range from 0.25 mM to 1.0 mM. Whereas, the enzyme from Gram-positive bacteria such as *Bacillus anthracis* (Domigan et al., 2009). *Bacillus cereus* (Hoganson & Stahly, 1975), *Corynebacterium glutamicum* (Cremer et al., 1988), *Lactobacillus plantarum* (Cahyanto et al., 2006) and *Staphylococcus aureus* (Burgess et al., 2008) show little or no inhibition by lysine.

The crystal structure of DHDPS in complex with lysine from *E. coli* shows that the lysine allosteric binding site is situated in a crevice at the interface of the tight dimer, distal from the active site, but connected to the active site via a water channel (Blickling et al., 1997a). Two inhibitory lysine molecules are bound in close proximity within van der Waals contact to each other. Seven residues located within the allosteric site bind lysine, namely Ala49, His53, His56, Gly78, Asp80, Glu84, and Tyr106 (Blickling et al., 1997a).

Studies show that lysine inhibition is cooperative with the second lysine molecule binding 10^5 times more tightly than the first (Blickling et al., 1997a). The mechanism by which lysine exerts regulatory control over bacterial DHDPS is not well understood, although kinetic and structural studies suggest that it is an allosteric inhibitor, causing partial inhibition (approximately 90%) at saturating concentrations (Blickling et al., 1997a). It has recently been suggested that lysine exerts some effect on the first half reaction by attenuating proton-relay and also the function of Arg138, thought to be crucial for ASA binding (Dobson et al., 2004b). The crystal structure of the *E. coli* DHDPS-lysine complex was solved in the absence of substrate; however, thermodynamic studies have illustrated that the substrate pyruvate has a substantial effect on the nature of enzyme-inhibitor association (Blickling et al., 1997a).

2.2 Structure of DHDPS

2.2.1 Subunit and quaternary structure of DHDPS

DHDPS from *B. anthracis* (Blagova et al., 2006; Voss et al., 2010), *E. coli* (Mirwaldt et al., 1995), *Mycobacterium tuberculosis* (Kefala et al., 2008), *Thermoanaerobacter tengcongensis* (Wolterink-van Loo et al., 2008), *Thermotoga maritima* (Pearce et al., 2006), and several other species is a homotetramer in both crystal structure and solution (Fig. 3). In *E. coli*, the monomer is 292 amino acids in length and is composed of two domains (Mirwaldt et al., 1995). The N-terminal domain is a $(\beta/\alpha)_8$ TIM-barrel (residues 1-224) with the active site located within the centre of the barrel (Fig. 3). The C-terminal domain (residues 225-292) consists of three α-helices and contains several key residues that mediate tetramerisation (Dobson et al., 2005a). The association of the four monomers leaves a large water-filled cavity in the centre of the tetramer, such that each monomer has contacts with two neighbouring monomers only. The tetramer can also be described as a dimer of dimers, with strong interactions between the monomers A & B and C & D at the so-called tight dimer interface, and weaker interactions between the dimers A-B and C-D at the weak dimer interface (Dobson et al., 2005a) (Fig. 3).

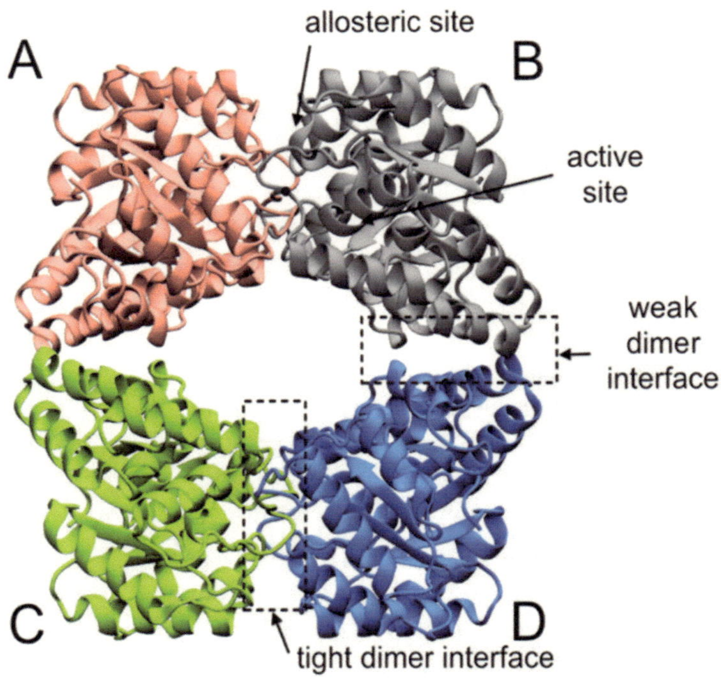

Fig. 3. *E. coli* DHDPS structure . The active sites, allosteric sites, dimerisation interface (tight dimer interface) and tetramerisation interface (weak dimer interface) are shown (PDB: 1YXC).

2.2.2 Active site

The active site is located in cavities formed by the two monomers of the dimer. A long solvent-accessible catalytic crevice with a depth of 10 Å is formed between β-strands 4 and 5 of the barrel (Mirwaldt et al., 1995). Lys161, involved in Schiff-base formation is situated in the β-barrel near the catalytic triad of three residues, namely Tyr133, Thr44 and Tyr107, which act as a proton shuttle (Blickling et al., 1997a) (Fig. 4). Thr44 is hydrogen bonded to both Tyr133 and Tyr107 and its position in the hydrogen-bonding network may play a role in Schiff base formation and cyclisation (Dobson et al., 2005a). The dihedral angles of Tyr107 fall in the disallowed region of the Ramachandran plot, suggesting an important role in the enzyme's function (Mirwaldt et al., 1995). It is believed to be involved in shuttling protons between the active site and solvent (Dobson et al., 2005a). In contrast, Tyr133 plays an important role in substrate binding, donating a proton to the Schiff base hydroxyl. It is also thought to coordinate the attacking amino group of ASA, which requires the loss of a proton subsequent to cyclisation (Fig. 2). A marked reduction in activity is observed in single substitution mutants, highlighting the importance of this catalytic triad (Dobson et al., 2004a).

Situated at the entrance to the active site, Arg138 is essential for ASA binding (Dobson et al., 2005b). In the *E. coli* DHDPS structure, a hydrogen bond is formed between Arg138 and Tyr107 (Dobson et al., 2004a) and a water mediated hydrogen bond is formed between Arg138 and Tyr133 (Dobson et al., 2005a). Arg138 is thus also important for stabilisation of the catalytic triad, both of which are highly conserved in all DHDPS enzymes (Dobson et al., 2005a).

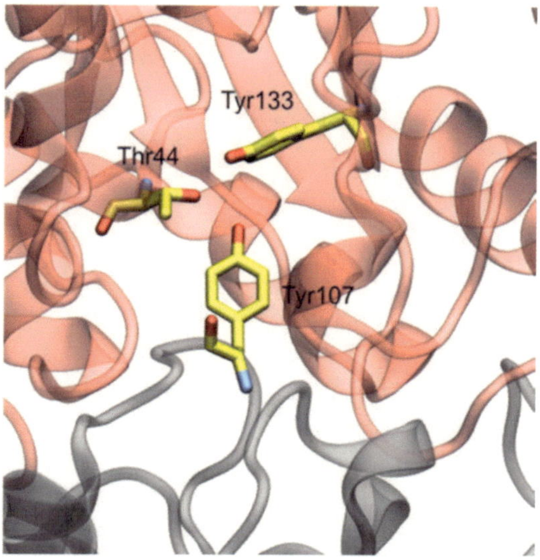

Fig. 4. *E. coli* DHDPS active site, illustrating the catalytic triad Thr44, Tyr133 and Tyr107 interdigitating from the opposing monomer (PDB: 1YXC).

2.2.3 Tight dimer interface

In *E. coli* DHDPS, 1400 Å^2 of surface area from one monomer in each dimer is buried at the tight dimer interface (Dobson et al., 2005a) (Fig. 3). This interface is made up of 25 residues from each monomer, with hydrogen bonds formed between Ser111 and Cys141, and hydrophobic interactions between Leu51 and Ala81, among others. In addition, Tyr107 of one monomer is coordinated with Tyr106 from the adjacent subunit, interdigitating across the monomer-monomer interface and thus forming a stabilising hydrophobic sandwich-like stacking of aromatic rings.

2.2.4 Weak dimer interface

The tight dimer units of the *E. coli* DHDPS tetramer associate via two isologous interfaces formed between corresponding monomers (Fig. 3). This interface buries approximately 538 Å^2 of surface area. Nine residues from each monomer are involved in contacts at the weak dimer interface (Mirwaldt et al., 1995), situated within the α6, α7 and α9-helices. The interface is stabilised by hydrophobic contacts between Leu167, Thr168 and Leu197 (Dobson et al., 2004a). The importance of Leu197 at the interface has been demonstrated with mutations resulting in a dimeric species, unable to form a tetramer (Griffin et al., 2008, 2010). This interface is not conserved in other DHDPS structures. A greater number of contacts are observed at the weak dimer interface in DHDPS from *B. anthracis* (Blagova et al., 2006; Voss et al, 2010), *M. tuberculosis* (Kefala et al., 2008), and most strikingly, *T. maritima* (Pearce et al., 2006) with 20 residues involved in many interactions.

2.2.5 Allosteric site

As described in Section 2.1.1, lysine is an allosteric modulator of DHDPS function, partially inhibiting DHDPS activity. The lysine binding site is situated in a crevice at the interface of the tight dimer, distal from the active site, but connected via a water filled channel (Fig. 3). The crystal structure of lysine-bound *E. coli* DHDPS shows two lysine molecules bound per dimer (four per tetramer) with each molecule interacting with both monomers and the adjacent lysine molecule (Blickling et al., 1997c) (Fig. 5).

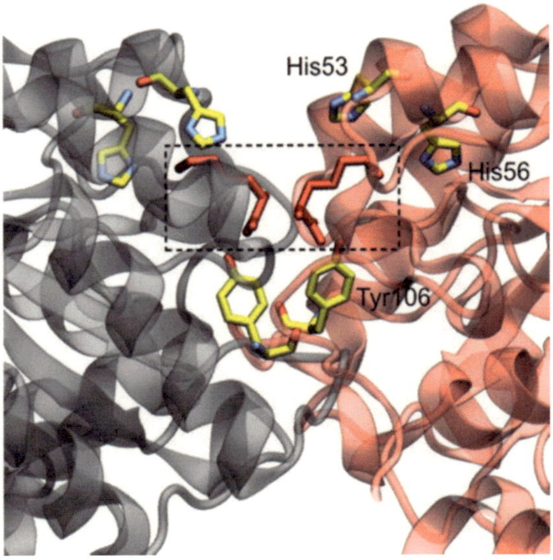

Fig. 5. *E. coli* DHDPS allosteric site with two lysine (boxed - red) molecules bound (PDB: 2ATS).

As stated earlier, seven residues are believed to be involved in binding lysine; Ser48, Ala49, His53, His56, Asn80, Glu84 and Tyr106. All these residues show slightly altered conformations in the presence of lysine, moving to accommodate the molecule (Blickling et al., 1997a). Importantly, Tyr106 moves towards the carboxyl group of lysine, which alters the aromatic stacking of Tyr106 and Tyr107. Otherwise, very few changes are observed upon lysine binding, with no significant secondary structure or quaternary structure change occurring (Dobson et al., 2005a). Most of the residues identified as important in the lysine allosteric binding site are not conserved in those DHDPS enzymes that are not inhibited by lysine (Burgess et al., 2008; Kefala et al., 2008; Voss et al., 2010; Wolterink-van Loo et al., 2008).

2.2.6 Alternative quaternary architecture

Whilst the DHDPS monomer from most bacteria has a molecular mass of approximately 31 kDa, the plant enzymes are larger. For example, DHDPS from *Nicotiana sylvestris* (Blickling et al., 1997b) has a relative molecular mass of 36 kDa, whilst DHDPS from *Pivus sativum* (Dereppe et al., 1992) has been reported to be a homotrimer of 43 kDa monomers based on gel filtration liquid chromatography studies, although this result is uncorroborated. The only plant DHDPS structure solved to date is from *N. sylvestris* (Blickling et al., 1997b). As for the bacterial enzymes, it is a homotetramer, described as a dimer of dimers. The contact areas within the tight dimer are similar within the plant and bacterial enzyme, with 13 of the 19 residues contributing to the interface conserved in both bacteria and plants. However, as Figure 6 shows, the plant dimer of dimers has an alternative architecture, namely the residues involved at the weak dimer interface are located on the opposite face of the monomer. The plant enzyme can thus be described as a "back-to-back" arrangement of dimers

(Fig. 6) compared to the "head-to-head" arrangement observed for bacterial DHDPS (Fig. 3). Compared to the bacterial interface, the weak dimer interface of *N. sylvestris* DHDPS is larger than its bacterial counterpart, burying 810 Å^2 surface area, which is reflected in the greater number of residues contributing to inter-subunit contacts. The additional residues of the C-terminus, as well as the novel quaternary structure of *N. sylvestris* DHDPS, reduces the central water filled cavity and results in a tetramer where all subunits are in contact with each other (Fig. 6). Despite the significant structural differences between the plant and bacterial enzymes, the position and orientation of all active site residues are conserved. Most strikingly, considering the rearrangement of dimers, lysine binds at an equivalent binding pocket at the interfaces of the two monomers of a dimer in both the *E. coli* (Blickling et al., 1997a) and *N. sylvestris* (Blickling et al., 1997b) enzymes. The lysine molecules also bind in the same orientation, with coordination of the α-amino and α-carboxyl groups almost identical.

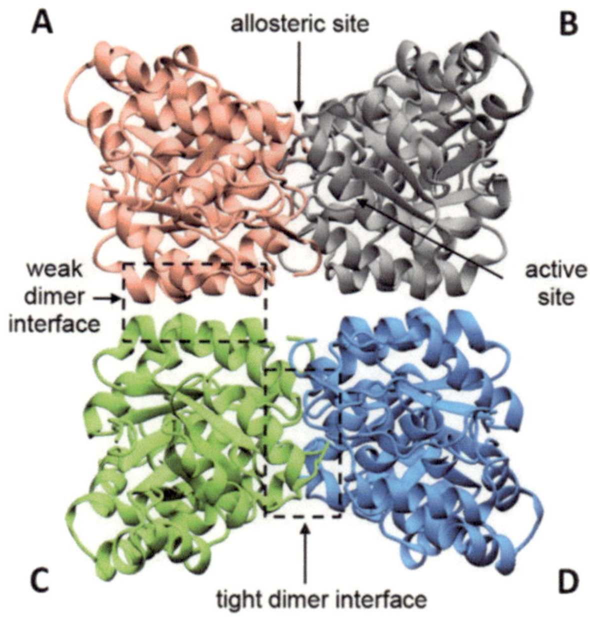

Fig. 6. *N. sylvestris* DHDPS, showing an alternate architecture ("back-to-back" arrangement of dimers) to that observed for bacterial DHDPS enzymes.

In addition, other quaternary structures of bacterial DHDPS enzymes have been reported. For example, DHDPS from methicillin-resistant *S. aureus* (MRSA) has recently been shown to be a dimer in solution (Burgess et al., 2008), with enzymatic activity similar to that of the wild-type *E. coli* tetramer. Several DHDPS enzymes have also been identified in *Agrobacterium tumefaciens,* with two forms crystallising as dimers (3B4U, 2R8W) and one as a hexamer (2HMC), although the function of this enzyme has yet to be confirmed.

2.3 Inhibition of DHDPS

A number of potential DHDPS inhibitors have been synthesised and characterised. A variety of heterocyclic analogues of DHDP and HTPA have been shown to act as moderate inhibitors of DHDPS (Hutton et al., 2007). Analogues of the cyclic lactol form of ASA, including homoserine lactone and 2-aminocyclopentanone, show non-competitive moderate inhibition with K_i = 12-24 mM (Hutton et al., 2007). Analogues of the straight chain hydrate form of ASA have also been investigated, with aspartic acid showing mixed type inhibition with K_i = 90-140 μM (Hutton et al., 2007). Product analogues have also been investigated, exhibiting moderate DHDPS inhibition. More success was achieved with inhibitors based on the acyclic enzyme-bound DHDPS intermediates, such as diethyl (*E,E*)-4-oxo-2,5-heptadienedioate (Turner et al., 2005) and a bis-oxime ester (Boughton et al., 2008), which irreversibly inhibit DHDPS. Interestingly, several of these compounds have displayed clear differentiation in inhibition of DHDPS enzymes from different species (Mitsakos et al., 2008), suggesting the potential for targeting compounds to specific pathogens.

3. Dihydrodipicolinate reductase

3.1 Function of DHDPR

Dihydrodipicolinate reductase (DHDPR, EC 1.3.1.26) was first isolated from *E. coli* in 1965 (Farkas & Gilvarg, 1965). Since then, the enzyme has been characterised from several species including *B. cereus* (Kimura & Goto, 1977), *Bacillus megaterium* (Kimura & Goto, 1977), *Bacillus subtilis* (Kimura, 1975), *C. glutamicum* (Cremer et al., 1988), *Methylophilus methylotrophus* (Gunji et al., 2004), *M. tuberculosis* (Cirilli et al., 2003), *S. aureus* (Dommaraju et al., 2011; Girish et al., 2011), and *T. maritima* (Pearce et al., 2008). DHDPR catalyses the second step in the lysine biosynthesis pathway (Fig. 1), the pyridine nucleotide-dependent reduction of dihydrodipicolinate (DHDP) to form *L*-2,3,4,5,-tetrahydrodipicolinate (THDP) (Dogovski et al., 2009; Hutton et al., 2007).

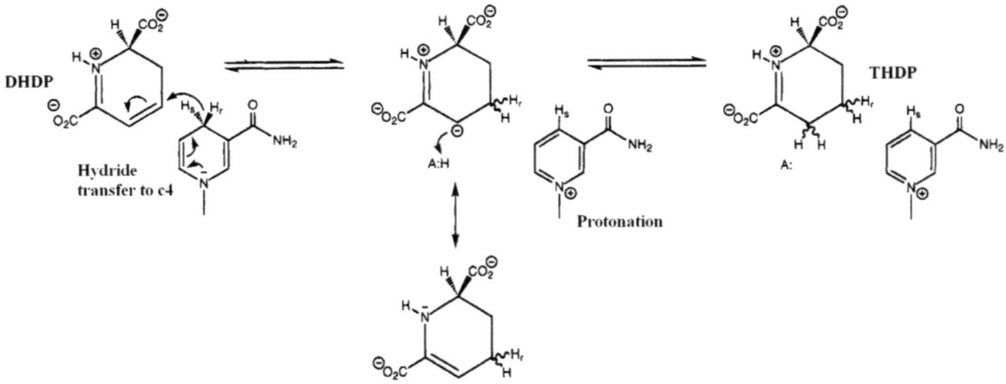

Fig. 7. Schematic representation of the catalytic mechanism of DHDPR.

In *E. coli*, DHDPR is encoded by the *dap*B gene, which is also an essential bacterial gene (Dogovski et al., 2009; Hutton et al., 2007). The open reading frame encodes a 273 amino acid polypeptide with a monomeric molecular weight of 28,758 Da. The enzyme functions by utilising either phosphorylated or non-phosphorylated pyridine nucleotides, NAD(P)H, as hydrogen donors to carry out its reaction. The kinetic mechanism of *E. coli* DHDPR is ordered and sequential (Reddy et al., 1995), involving binding of NAD(P)H followed by DHDP. The reaction is initiated by hydride transfer from the 4-pro-*R* position of NAD(P)H to the C4-position of DHDP, with the resultant enamine then undergoing tautomerisation to form THDP. Upon completion of the reaction, the release of the product THDP is followed by $NAD(P)^+$ release (Reddy et al., 1995) (Fig. 7).

3.1.2 Nucleotide preference of bacterial DHDPR

Pyridine nucleotide-dependent dehydrogenases typically have a strong preference for either NADPH or NADH as co-factors (Cirilli et al., 2003; Pearce et al., 2008; Reddy et al., 1996). In most cases dual-cofactor enzymes preferentially utilise NADPH over NADH. In light of this observation, there has been significant interest in studying the molecular basis of nucleotide preference. All NAD-dependent dehydrogenases contain the consensus sequence G<u>X</u>GXXG or G<u>X</u>XGXXG and conserved acidic amino acids 20-30 residues downstream of this glycine rich region (Dommaraju et al., 2011). The main chain nitrogen of the second residue (<u>X</u>) in the consensus sequence interacts with this conserved acidic residue. *E. coli* DHDPR has an unusual pyridine nucleotide specificity, exhibiting only a modest selectivity for its nucleotides. Kinetic studies show that *E. coli* DHDPR utilises NADH only slightly more efficiently than NADPH (Reddy et al., 1996). This is consistent with the observation that the binding affinity of *E. coli* DHDPR to NADH (K_D = 0.26 μM) is stronger than that of NADPH (K_D = 1.8 μM) (Reddy et al., 1996). Structural studies of *E. coli* DHDPR show the existence of hydrogen bonds between the side-chain of the acidic residue Glu38 and that of the O3′ of the adenine ribose of NADH. It is hypothesised that the basic residue Arg39, also found in the nucleotide binding pocket, can interact with the negatively charged 2′ phosphate of NADPH, thus enabling the enzyme to utilise both NADH and NADPH. Kinetic analysis of DHDPR from *M. tuberculosis* also shows that the enzyme exhibits only a moderate preference for NADH. The crystal structures of *M. tuberculosis* DHDPR in two ternary complexes (DHDPR-2,6-PDC-NADH and DHDPR-2,6-PDC-NADPH) demonstrate that the number of hydrogen bonds between DHDPR and the nucleotides NADH and NADPH are very similar (Cirilli et al., 2003; Reddy et al., 1996; Scapin et al., 1997).

3.2 Structure of DHDPR

3.2.1 Subunit and quaternary structure of DHDPR

The three-dimensional structure of DHDPR has been elucidated by X-ray crystallography from five diverse bacterial species, namely, *Bartonella henselae*, (PDB: 3IJP), *E. coli* (Scapin et al., 1995, 1997), *M. tuberculosis* (Cirilli et al., 2003), *S. aureus* (Girish et al., 2011), and *T. maritima* (Pearce et al., 2008). DHDPR from *E. coli* (Fig. 8) was the first DHDPR enzyme to be extensively studied in terms of structure and function (Farkas & Gilvarg, 1965; Reddy et al., 1995; Scapin et al., 1995, 1997).

DHDPR is a tetrameric enzyme consisting of four identical monomers (Fig. 8). Each monomer is comprised of an N-terminal nucleotide binding domain and a C-terminal substrate binding domain (Fig. 9). In *E. coli* DHDPR, the nucleotide binding domain is formed by the first 130 and last 36 residues of the polypeptide chain, whereas the substrate binding domain is formed by residues 130-240. The nucleotide binding domain consists of four α-helices and seven β-strands, which are arranged to form a Rossmann (dinucleotide binding) fold. The substrate binding domain contains two α-helices and four β-strands, which form an open mixed β-sandwich (Scapin et al., 1995). Interactions between the four subunits of the tetramer occur exclusively between residues of the substrate binding domain. A long loop (Leu182 to Gly204) also extends from the substrate binding domain and plays an important role in maintaining the quaternary structure of the enzyme. The four monomers interact by pairing the four β-strands on the substrate binding domain to form a 16-stranded, mixed, flattened β-barrel (Fig. 8). This central barrel is anchored by the four long loops (Leu182 to Gly204) that extend from the body of the substrate binding domain of each monomer and wrap around the mixed β-sheet of the neighboring monomer. Residues 65-74 and 127-130 form flexible hinge regions between the nucleotide and substrate binding domains (Scapin et al., 1995).

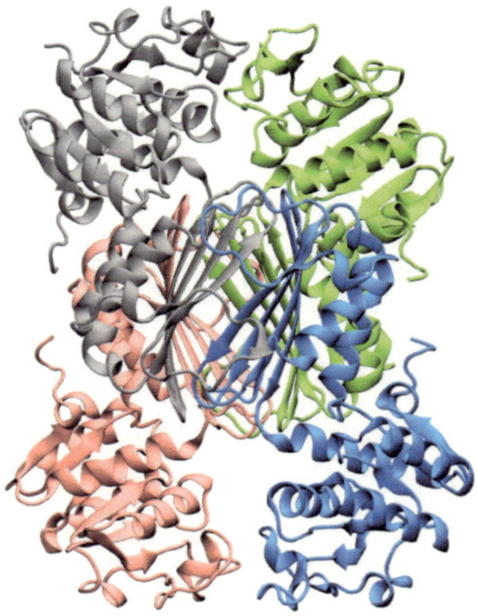

Fig. 8. Structure of *E. coli* DHDPR (PDB: 1ARZ).

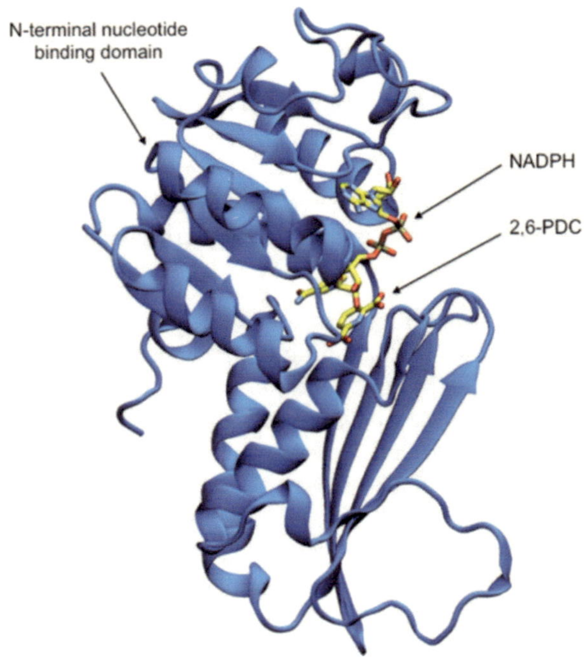

Fig. 9. Structure of the *E. coli* DHDPR monomer bound to NADH and the substrate analogue, 2,6-PDC (PDB: 1ARZ).

3.2.2 Substrate binding site

The consensus sequence, E(L/A)HHXXKXDAPSGTA is found in the substrate binding domain of all known bacterial DHDPR enzymes (Pavelka et al., 1997). This sequence is thought to contain residues involved in binding of substrate and/or catalysis. Molecular modelling studies, using the apo form (enzyme in the absence of substrate) of *E. coli* DHDPR as a structural template, suggest a cluster of five basic residues are the key catalytic site residues (Scapin et al., 1997), namely His159, His160, Arg161, His162 and Lys163 (all contained within the consensus sequence). These residues are located in the loop connecting β-strand B7 to α-helix A5. Structural studies of *E. coli* DHDPR in complex with NADH and the substrate analogue and inhibitor, 2,6-pyridinedicarboxylate (2,6-PDC), show that 2,6-PDC is bound to the substrate binding domain of DHDPR, in a spherical cavity bordered by residues from both the nucleotide binding (Gly102-Phe106 and Ala126-Ser130) and substrate binding domains (Ile155-Gly175 and Val217-His220) (Scapin et al., 1997). The bound inhibitor makes several hydrogen bonding interactions with the atoms of the conserved E(L/A)HHXXKXDAPSGTA motif. Similar interactions are observed between 2,6-PDC and DHDPR from *M. tuberculosis* (Cirilli et al., 2003).

3.2.3 Nucleotide binding site

The nucleotide binding domain of DHDPR adopts a Rossmann fold, which is typical of nucleotide-dependent dehydrogenases (Fig. 9). The consensus sequence (V/I)(A/G)(V/I)-XGXXGXXG located within this domain, is conserved in all NAD(P)H-dependent dehydrogenases, including DHDPR (Pavelka et al., 1997). Structural analyses of *E. coli* DHDPR show that this motif extends from the C-terminal end of β-strand B1 to the loop that connects B1 to α-helix A1. An acidic residue (Glu38 in *E. coli* DHDPR) is located approximately 20 amino acids downstream of the conserved consensus sequence. The two hydroxyl groups from the adenine ribose are known to interact with the side-chain of Glu38 and also the backbone atoms of the glycine rich motif GXXGXXG. Several hydrophobic interactions exist between the adenine ring of NADH and the residues Arg39, Gly84 and His88. The pyrophosphate group of NADH is located over the α-helix A1 and interacts with residues contained within the loop connecting β-strand B1 and α-helix A1 (Reddy et al., 1996; Scapin et al., 1997).

3.3 Inhibition of DHDPR

The substrate analogue, 2,6-PDC, is a competitive inhibitor (K_i = 26 μM) of DHDPR (Scapin et al., 1995) (Fig. 10A). Other substrate analogues such as picolinic acid (Fig. 10B), isopthalic acid (Fig. 10C), pipecolic acid (Fig. 10D) and dimethyl chelidamate (Fig. 10E), are much weaker inhibitors, each displaying an IC_{50} > 10 mM (Hutton et al., 2003). A vinylogous amide that acts as a competitive inhibitor of DHDPR (K_i = 32 μM) has been described and is one of the most potent inhibitors of DHDPR reported to date (Caplan et al., 2000). Molecular modeling in tandem with conventional drug screening strategies has identified novel inhibitors, including sulfones and sulfonamides, with K_i values ranging from 7-90 μM (Caplan et al., 2000). However, a sub-micromolar inhibitor of DHDPR has not been discovered to date.

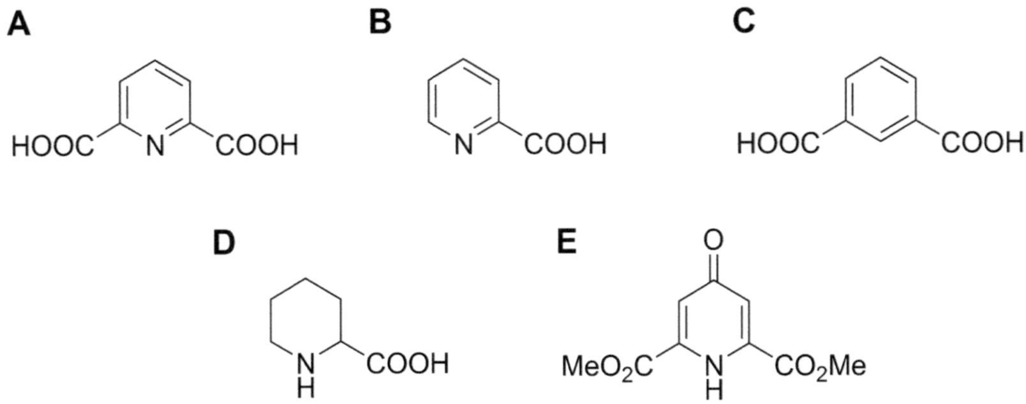

Fig. 10. Inhibitors of DHDPR.

4. Succinylase pathway

4.1 Tetrahydrodipicolinate N-succinyltransferase

Tetrahydrodipicolinate N-succinyltransferase (THPC-NST, EC 2.3.1.117) is a succinyl-coenzyme A (SCoA) dependant enzyme that catalyses the conversion of cyclic *L*-2,3,4,5,-tetrahydrodipicolinate (THDP) to acyclic N-succinyl-*L*-2-amino-6-ketopimelate (NSAKP) (Simms et al., 1984) (Fig. 1). The reaction occurs via a *L*-2-amino-6-ketopimelate (AKP) intermediate. The transfer of an acyl group functions to maintain a linear conformation of the product of the reaction (NSKAP) and exposes the 6-keto group for subsequent transamination (Beaman et al., 2002). Substrate and cofactor kinetic parameters for *E. coli* THPC-NST have been determined. Studies show that the K_M^{app} for THDP and succinyl-CoA are 20 μM and 15 μM, respectively (Berges et al., 1986b; Simms et al., 1984).

The *dap*D gene encoding THPC-NST is found in a large number of bacterial species including *E. coli* and *Mycobacterium* species (Beaman et al., 1997; Richaud et al., 1984; Schuldt et al., 2009). Expression of this gene in *E. coli* is weakly inhibited by lysine (Ou et al., 2008; Richaud et al., 1984). THPC-NST enzymes characterised to date are comprised of approximately 290 residues and show greater than 18% sequence identity (Beaman et al., 1997; Richaud et al., 1984; Schuldt et al., 2009).

The crystal structure of THPC-NST from *Mycobacterium bovis* (Fig. 11) shows that the enzyme forms a homotrimer. The monomer consists of three domains, namely, the (i) N-terminal, (ii) left handed parallel β-helix (LβH), and (iii) C-terminal domains (Beaman et al., 1997). The N-terminal domain is comprised of four α-helices and two hairpin loops. The LβH domain, comprising 50% of the subunit, contains the hexapeptide repeat motif ([LIV]-[GAED]-X_2-[STAV]-X) within each turn of the β-helix. The LβH domain is interrupted by two loops, including a flexible loop (residues 166-175) that is involved in binding substrate. The C-terminal domain consists of a β-stranded structure. All three domains contribute to inter-subunit contacts. The structure of THPC-NST from other bacterial species have since been determined and show a high degree of similarity to that of *M. bovis* THPC-NST (Nguyen et al., 2008; Schuldt et al., 2009).

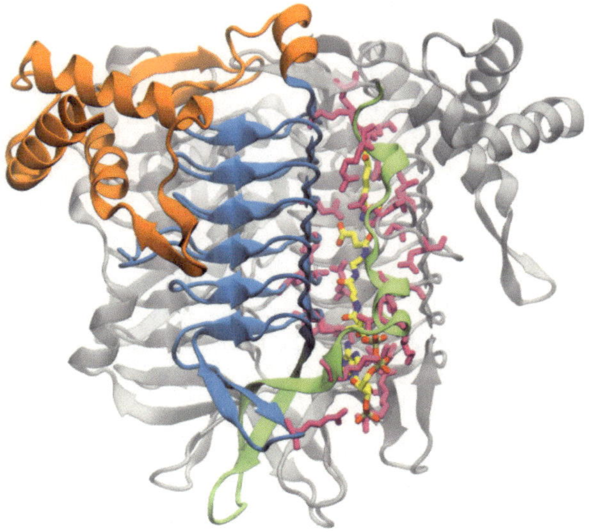

Fig. 11. Structure of trimeric *M. bovis* THPC-NST in complex with *L*-2-aminopimelate and succinamide-CoA. The N-terminal (orange), LβH (blue) and C-terminal (green) domains are indicated. The substrate *L*-2-aminopimelate (yellow) and cofactor succinamide-CoA (yellow) are bound via the THPC-NST active site residues (pink) (PDB: 1KGQ).

Crystal structures of *M. bovis* THPC-NST in complex with substrate analogs and several forms of coenzyme A have resulted in a model describing substrate binding and catalysis (Beaman et al., 1998, 2002). Self-association of the monomer subunit results in a homotrimer complex containing three active sites. The AKP and SCoA binding sites are located at the LβH domain interfaces. Binding of SCoA and possibly AKP is thought to promote a large conformational change that encloses the bound substrate and cofactor within the active site. In this state, the 2-amino group of AKP is placed in close proximity to the SCoA thioester, allowing nucleophilic attack and transfer of the succinyl group (Beaman et al., 2002).

Studies have shown that *L*-2-aminopimelic acid, an analog of AKP, is an inhibitor of THPC-NST, although it does not display antibacterial activity (Berges et al., 1986a). However, peptide derivatives of 2-aminopimelic acid show significant antibacterial activity against a range of Gram-negative bacteria (Berges et al., 1986a).

4.2 N-succinyldiaminopimelate aminotransferase

N-succinyldiaminopimelate aminotransferase (NSDAP-AT, EC 2.6.1.17) catalyses the conversion of NSKAP to N-succinyl-*L,L*-2,6,-diaminopimelate (NSDAP) (Fig. 1). The reaction begins by the formation of a Schiff base linkage between an active site lysine and the cofactor pyridoxal-5′-phosphate (PLP). An amino group, donated by glutamate, is transferred to PLP, to form pyridoxamine phosphate (PMP). The enzyme subsequently transfers the amino group from PMP to NSAKP to yield N-succinyl-*L,L*-2,6,-diaminopimelate (NSDAP) and α-ketoglutarate (Peterkofsky & Gilvarg., 1961; Ledwidge & Blanchard., 1999). Studies of *E. coli* NSDAP-AT report K_M values for the substrates NSKAP and glutamate of 0.5 mM and 0.52 mM, respectively (Peterkofsky & Gilvarg., 1961).

The gene encoding NSDAP-AT (*dapC*), is found in a large number of bacterial species including *Bordetella pertussis* (Fuchs et al., 2000), *C. glutamicum* (Hartmann et al., 2003), *E. coli,* (Peterkofsky & Gilvarg., 1961) and *M. tuberculosis* (Weyand et al., 2006). In *E. coli,* the gene encoding NSDAP-AT is annotated *argD* (Ledwidge & Blanchard., 1999). This enzyme also functions as a N-acetylornithine aminotransferase, a component of the arginine biosynthesis pathway. The *dapC* gene in *B. pertussis* (Fuchs et al., 2000), *C. glutamicum,* (Hartmann et al., 2003), and *E. coli* (Bukari & Taylor., 1971) has been found to map in close proximity to the *dapD* gene on the chromosome. Sequence analyses have shown that NSDAP-AT consists of approximately 400 residues and shares greater than 26% identity across species (Fuchs et al., 2000; Hartmann et al., 2003; Peterkofsky & Gilvarg., 1961; Weyand et al., 2006). The NSDAP-AT sequence is characterised by the presence of the PLP binding sequence motif, SLSKXSNVXGXRAG, that includes an active site lysine residue (underlined) (Fuchs et al., 2000).

Structure studies of *M. tuberculosis* NSDAP-AT in complex with PLP shows that the enzyme forms a homodimer (Fig. 12). The structure is characteristic of the aminotransferase family of class I PLP-binding proteins (Weyand et al., 2007). The monomer subunit is comprised of (i) an α-helical N-terminal extension, (ii) a central domain comprising an 8-stranded β-sheet surrounded by 8 α-helices, and (iii) a C-terminal domain consisting of a four stranded β-sheet flanked by 4 α-helices. The active site of each subunit is located at the dimer interface with residues from both subunits contributing to the architecture of the active sites. PLP is bound to the active site Lys232, presumably via a Schiff base, and makes a number of noncovalent contacts with other residues within the active site via a hydrogen bond network.

A number of hydrazino-dipeptide analogs of NSDAP inhibit NSDAP-AT with K_i values ranging from 22-556 nM and show significant antibacterial activity against *E. coli* (Cox et al., 1998).

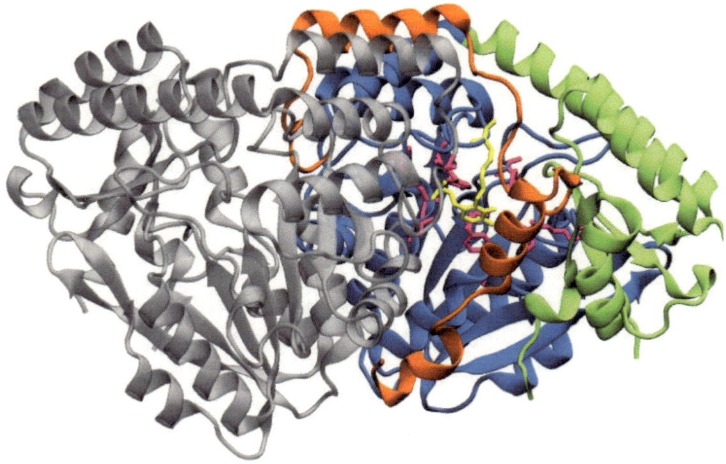

Fig. 12. Structure of dimeric *M. tuberculosis* NSDAP-AT in complex with PLP. The α-helical N-terminal extension (orange), central (blue) and C-terminal (green) domains are indicated. The cofactor PLP (yellow) is bound by the NSDAP-AT active site residues (pink) (PDB: 2O0R).

4.3 Succinyldiaminopimelate desuccinylase

Succinyldiaminopimelate desuccinylase (SDAP-DS, EC 3.5.1.18) catalyses the hydrolysis of N-succinyl-*L,L*-2,6,-diaminopimelate (NSDAP) to yield *L,L*-2,6-diaminopimelate (DAP) and succinate (Kindler & Gilvarg., 1960) (Fig. 1). Kinetic parameters for SDAP-DS from several bacterial species have been reported, with substrate K_M and k_{cat} values ranging from 0.73 - 1.3 mM and 140 - 200 s^{-1}, respectively (Bienvenue et al., 2003; Born et al., 1998; Lin et al., 1988).

The gene encoding SDAP-DS, *dap*E, is present in a large number of bacterial species including, *C. glutamicum* (Wehrmann et al., 1994), *E. coli* (Bouvier et al., 1992), *Haemophilus influenzae,* (Born et al., 1998) and *Salmonella enterica* (Broder & Miller., 2003). In general, SDAP-DS contains approximately 375 residues and shares greater than 22% sequence identity across bacterial species. Alignment of SDAP-DS amino acid sequences show conservation of histidine and glutamate metal binding residues that are characteristic of metal-dependent amidases (Born et al., 1998).

Consistent with the conservation of metal binding residues, the activity of SDAP-DS enzymes are dependent on Zn^{2+} ions (Born et al., 1998; Lin *et al.*, 1988). Futhermore, studies involving Zn K-edge extended X-ray absorption fine structure (EXAFS) analyses of *H. influenzae* SDAP-DS indicate that the enzyme contains dinuclear Zn^{2+} active sites (Cosper et al., 2003). Studies of *H. influenzae* SDAP-DS mutants by kinetics, electronic absorption spectroscopy and electron paramagnetic resonance spectroscopy showed that His67 and His349 coordinate Zn^{2+} ions, with His67 functioning in catalysis (Gillner et al., 2009). A similar study showed that residue Glu134 is also involved in catalysis, possibly functioning as an acid/base (Davis et al., 2006).

The crystal structure of zinc bound SDAP-DS has been determined (Fig. 13). Studies have shown that the enzyme forms a homodimer, with each monomer subunit containing a catalytic domain and a dimerisation domain (Nocek et al., 2010). The core of the catalytic domain is composed of an eight-stranded twisted β-sheet that is sandwiched between seven α-helices. The dimerisation domain adopts a two layer α+β sandwich fold and is comprised of a four stranded antiparallel β-sheet and two α-helices.

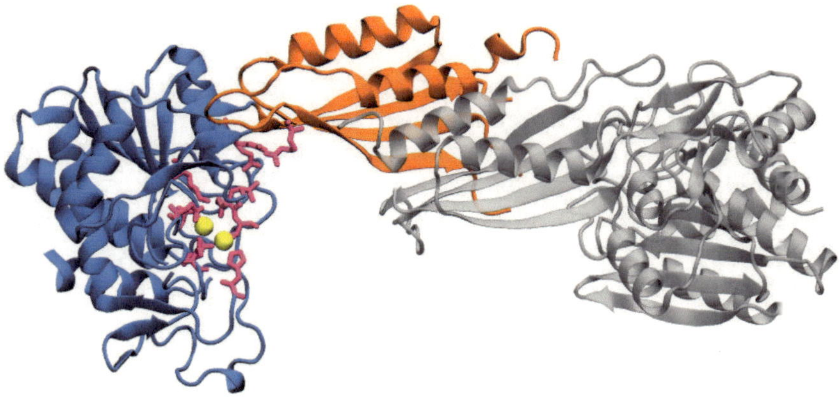

Fig. 13. Structure of dimeric *H. influenzae* SDAP-DS in complex with two zinc ions. the dimerisation (orange) and catalytic (blue) domains are indicated. Zinc ions (yellow) are bound by SDAP-DS active site residues (pink) (PDB: 3IC1).

The catalytic domain incorporates a negatively charged active site cleft, containing two zinc ions. One zinc ion is coordinated by the imidazole group and sidechain oxygens of His67 and Glu163, respectively, whilst another zinc ion is coordinated in a similar manner by His349 and Glu135. The zinc ions are bridged together by interaction with Asp100 and a water/hydroxide.

The availability of a structural model has resulted in a proposed mechanism for hydrolysis of NSDAP by SDAP-DS (Born et al., 1998; Nocek et al., 2010). It is hypothesised that NSDAP adopts an extended conformation when bound to the active site of the enzyme. The NSDAP amide carbonyl coordinates to an active site Zn^{2+} ion and becomes avaliable for nucleophilic attack. This binding event displaces a bridging water molecule, resulting in its hydrolysis by Glu134 and the generation of a zinc bound nucleophilic hydroxide. The hydroxide then attacks the target carbonyl carbon to form a η-1-μ-transition-state complex, which then resolves to release DAP and succinate.

The DAP isomers *L,L*-DAP and *D,L*-DAP are competitive inhibitors of *H. influenzae* SDAP-DS, exhibiting K_i values of 8 and 12 mM, respectively (Born et al., 1998). Studies employing Zn K-edge EXAFS suggest that the *H. influenzae* SDAP-DS inhibitor, 5-mercaptopentanoic acid, may exert its effect through binding to active site Zn^{2+} ions (Cosper et al., 2003).

5. Acetylase pathway

5.1 Tetrahydrodipicolinate N-acetyltransferase

Tetrahydrodipicolinate N-acetyltransferase (THDP-NAT, EC 2.3.1.89) is an acetyl-coenzyme A (ACoA) dependant enzyme that catalyses the conversion of cyclic THDP to acyclic N-acetyl-(*S*)-2-amino-6-ketopimelate (NAAKP) (Chatterjee & White., 1982) (Fig. 1). The transferred acyl group maintains the linear conformation of the product and exposes the 6-keto group for subsequent transamination (Beaman et al., 2002). Crude cell extracts from *B. megaterium* were found to contain active THDP-NAT (Chatterjee & White., 1982). THDP-NAT enzymes are thought to be largely confined to *Bacillus sp.* (Weinberger & Gilvarg., 1970).

5.2 Aminotransferase A

Aminotransferase A (ATA, EC 2.6.1) is a PLP-dependant enzyme that catalyses the conversion of NAAKP and glutamate to N-acetyl-(2*S*)-2,6,-diaminopimelate (NADAP) and α-ketoglutarate (Ledwidge & Blanchard., 1999; Peterkofsky & Gilvarg., 1961) (Fig. 1). It has been speculated that the ATA reaction mechanism resembles that of NSDAP-AT (Section 4.2) (Ledwidge & Blanchard., 1999; Peterkofsky & Gilvarg., 1961). Crude cell extracts from *B. megaterium* were found to contain active ATA (Chatterjee & White., 1982), with ATA activity identified by monitoring enzyme activity in the reverse direction utilising an acid ninhydrin assay (Chatterjee & White., 1982; Sundharadas & Gilvarg., 1967).

5.3 N-acetyldiaminopimelate deacetylase

N-acetyldiaminopimelate deacetylase (NAD-DAC, EC 3.5.1.47) catalyses the hydrolysis of NADAP to form DAP and acetate (Fig. 1). NAD-DAC was first identified from studies involving the isolation of a *B. megaterium* DAP auxotroph (Saleh & White., 1979; Sundharadas & Gilvarg., 1967). The mutant strain possesses a non-functional form of NAD-

DAC and consequently accumulates NADAP. Early studies of this enzyme centred on Gram-positive species, with NAD-DAC activity identified by utilising an acid ninhydrin assay to detect NADAP formation in crude cell extracts (Chatterjee & White., 1982; Weinberger & Gilvarg., 1970). The distribution of NAD-DAC has since been investigated in large number of Gram-negative and Gram-positive bacteria. Interestingly, the enzyme appears to be restricted to *Bacillus sp.* (Weinberger & Gilvarg., 1970).

6. Aminotransferase pathway

6.1 Function of diaminopimelate aminotransferase

Diaminopimelate aminotransferase (*LL*-DAP-AT, EC 2.6.1.83) is a PLP-dependant enzyme that catalyses the conversion of *L*-2,3,4,5,-tetrahydrodipicolinate (THDP) to *L,L*-2,6-diaminopimelate (*LL*-DAP) (Fig. 1). This transamination reaction utilises glutamate as an amino donor to yield α-ketoglutarate. (Hudson et al., 2006, 2008; Liu et al., 2010; McCoy et al., 2006)

The enzyme was first isolated from plant and cyanobacterial species and thus demonstrated a new branch of the lysine biosynthesis pathway existed (Hudson et al., 2006). Although plants are known to synthesise lysine *de novo*, components of the pathway required for conversion of THDP to *meso*-DAP had not been identified previously despite years of investigation. Studies of crude cell extracts had shown that plants do not catalyse reactions specific to the succinylase, acetylase or dehydrogenase branches of the pathway. This was subsequently confirmed with the observation that annotated plant genomes, including that from *Arabidopsis thaliana*, lack some or all genes associated with the three classical branches (Chatterjee et al., 1994; Hudson et al., 2005). The identification and characterisation of *LL*-DAP-AT from *A. thaliana* demonstrated for the first time the means by which plant species catalyse the conversion of THDP to *meso*-DAP via the aminotransferase sub-pathway (Hudson et al., 2006).

More recently *LL*-DAP-AT has been identified in algal, archaeal and bacterial species including, *Chlamydia trachomatis* (McCoy et al., 2006), *Chlamydomonas reihardtii* (Hudson et al., 2011), *Methanocaldococcus jannaschii* (Liu et al., 2010), and *Protochlamydia amoebophila* (McCoy et al., 2006). Comparative genomic analyses shows that *LL*-DAP-AT is restricted to the eubacterial lineages, *Bacteroidetes, Chlamydiae, Chloroflexi, Cyanobacteria, Desulfuromonadales, Firmicutes*, and *Spirochaeta*; and the archaea, *Archaeoglobaceae* and *Methanobacteriaceae* (Hudson et al., 2008). The phylogeny of *LL*-DAP-AT from these species has established the existence of two classes of *LL*-DAP-AT orthologues, namely, DapL1 and DapL2, which differ significantly in primary amino acid sequence. DapL1 and DapL2 are found predominantly in eubacteria and archaea, respectively (Hudson et al., 2008).

LL-DAP-AT enzymes are classified as members of the PLP-dependant protein superfamily of class I/II aminotransferases (Hudson et al., 2008; Jensen et al., 1996; Sung et al., 1991). Orthologues are in general 410 amino acids in length and can share as little as 29% sequence identity. Kinetic parameters for the *LL*-DAP-AT reaction have been determined for enzymes from a number of species, including *A. thaliana, C. trachomatis, Desulfitobacterium hafniense, Leptospira interrogans, Methanobacterium thermoautotrophicus, Morella thermoacetica*, and *P. amoebophila*. (Hudson et al., 2006, 2008; McCoy et al., 2006). In the human pathogen *C.*

trachomatis, the K_M values for the substrates THDP and glutamate have been reported as 19 μM and 2.1 μM, respectively (Hudson et al., 2008).

6.2 Structure of *LL*-DAP-AT

At present, the PDB reports twelve *LL*-DAP-AT X-ray crystal structures from three species, namely, *A. thaliana, C. trachomatis and C. reihardtii* (Watanabe et al., 2007, 2008, 2011; Dobson et al., 2011). The tertiary and quaternary structure of all three proteins are very similar with *LL*-DAP-AT existing as a homodimer (Fig. 14).

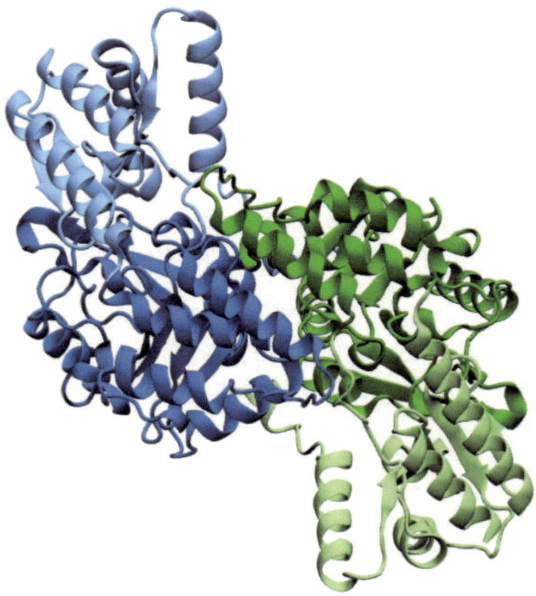

Fig. 14. Structure of dimeric *C. trachomatis LL*-DAP-AT. Monomers, indicated in blue and green, associate to form a functional dimer (PDB: 3ASA).

The subunit structure of *C. trachomatis LL*-DAP-AT is described as containing two domains, a large domain (LD) (residues 48-294) and a small domain (SD) (residues 1-47 and 295-394) (Watanabe et al., 2011; Watanabe & James, 2011). The LD is composed of α-β-α sandwich, whilst the SD assumes an α-β complex (Fig. 14). The LD is involved in binding PLP and also dimer formation, whereas the SD forms an N-terminal arm and also the C-terminal region. The active site is situated in a groove between the two domains of the monomer (Fig. 14). Importantly, the dimer structure is proposed to be essential for function as both subunits participate in substrate binding. Study of the structures of apo and ligand-bound forms of *C. trachomatis LL*-DAP-AT have revealed that the enzyme adopts an open and closed conformation (Watanabe et al., 2011). In the absence of ligand, the enzyme assumes an open state, whereby the active site is exposed to solvent. Upon PLP binding, the enzyme adopts a closed conformation. Within the active site, PLP is covalently linked to Lys236 via a Schiff base and is stabilised through an aromatic stacking interaction with Tyr128. PLP also forms a network of hydrogen bonding interactions with residues within the enzyme active site (Watanabe et al., 2011) (Fig 15).

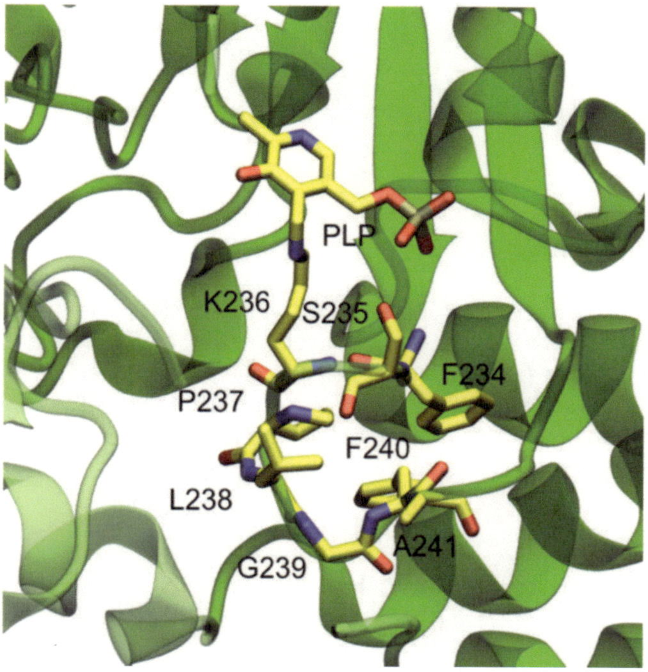

Fig. 15. Catalytic site of *LL*-DAP-AT from *C. trachomatis* (PDB: 3ASA). Ligand binding induces a closed conformation. PLP is covalently linked to Lys236 *via* a Schiff base.

6.3 Inhibition of *LL*-DAP-AT

A number of potential *LL*-DAP-AT inhibitors have been synthesised and characterised. In a screen involving 29,201 molecules, 15 compounds displayed IC_{50} values ranging from 20 μM to 60 μM, with the best hit being an aryl hydrazide showing an IC_{50} of 5 μM (Fan et al., 2010). However, the best hit appears to be an uncompetitive inhibitor and probably reacts irreversibly with PLP. Analogues of this compound have been synthesised and studies show that they fail to effectively inhibit *LL*-DAP-AT. In addition, there are two rhodanine-based molecules reported that show IC_{50} values of 41 μM and 46 μM (Fan et al., 2010).

7. Diaminopimelate epimerase

7.1 Function of DAPE

Diaminopimelate epimerase (DAPE, EC 5.1.1.7) catalyses the penultimate step in the lysine biosynthetic pathway whereby *L,L*-2,6-diaminopimelate (*LL*-DAP) is converted to *meso*-DAP (Fig. 1, Fig. 16). In *E. coli*, the enzyme is encoded by the *dap*F gene and is constitutively expressed (Neidhardt & Curtiss, 1996). DAPE was first characterised in 1957 using enzyme derived from crude extracts of *E. coli* (Work, 1957). The enzyme specifically recognises the *LL*-DAP isomer (Anita et al., 1957), whereas the *DD*-DAP isomer is not a substrate or inhibitor of the enzyme. Early studies noted that DAPE was inhibited by low concentrations of thiol-binding reagents and could be reactivated by reducing agents, suggesting the presence of an essential sulfhydryl group (Work, 1957). This finding was subsequently confirmed upon purification of DAPE to homogeneity (Wiseman, & Nichols, 1984).

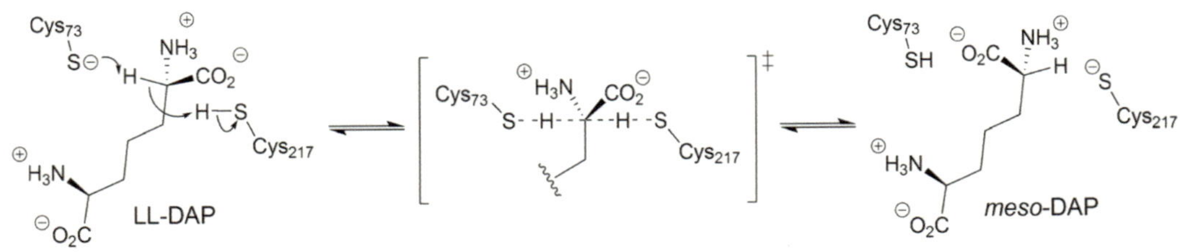

Fig. 16. DAPE catalysed reaction.

DAPE catalyses the conversion of *LL*-DAP to *meso*-DAP by employing a "two-base" mechanism (Wiseman, & Nichols, 1984). The reaction involves two active site Cys residues, where the first Cys residue (73 in *H. influenzae*) acts as base abstracting proton from *LL*-DAP, while the second Cys residue (217 in *H. influenzae*) re-protonates the molecule to generate *meso*-DAP. The enzyme is also capable of catalysing the reverse reaction, with the two Cys residues reversing their roles (Wiseman, & Nichols, 1984).

7.2 Structure of DAPE

The structures of DAPE from four species have been described. These include DAPE from *B. anthracis* (PDB:2OTN), *H. influenza* (Cirilli et al., 1998; Lloyd et al., 2004), and *M. tuberculosis* (Usha et al., 2009); and also the plant species *A. thaliana* (Pillai et al., 2009). The enzyme is a symmetrical monomer comprised of two domains containing eight β-strands and two α-helices (Cirilli et al., 1998) (Fig. 17).

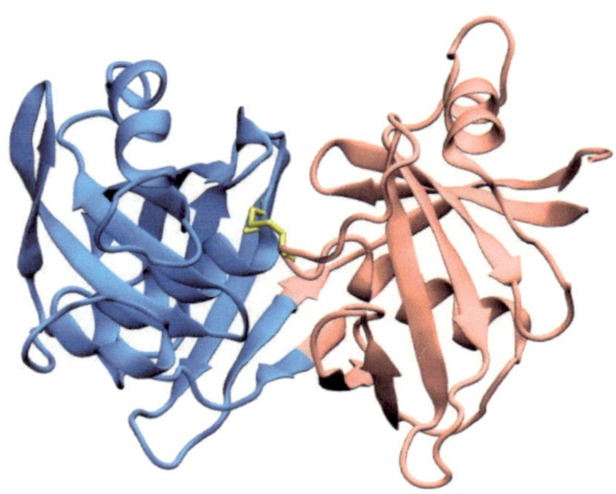

Fig. 17. Structure of DAPE from *H. influenzae*. Domains are coloured pink and blue, active site cysteines (disulfide linked) are shown in yellow (PDB: 1BWZ).

This fold, first observed in *H. influenzae* DAPE, is now referred to as the DAP epimerase-like fold. The structure of DAPE from *H. influenzae* shows that each domain of the enzyme contributes one active site Cys (residues 73 and 217). The distal, non-reacting end of the substrate interacts via a number of hydrogen bonds to residues Asn157, Asp190, Arg209, Asn64, and Glu208 (Fig. 18). The nature of this interaction ensures that only the *LL*-DAP stereoisomer is recognised. Interestingly, DAPE adopts two distinct conformational states. In the absence of substrate, the enzyme exists in an open conformation, and upon binding substrate adopts a closed conformation (Pillai et al., 2007).

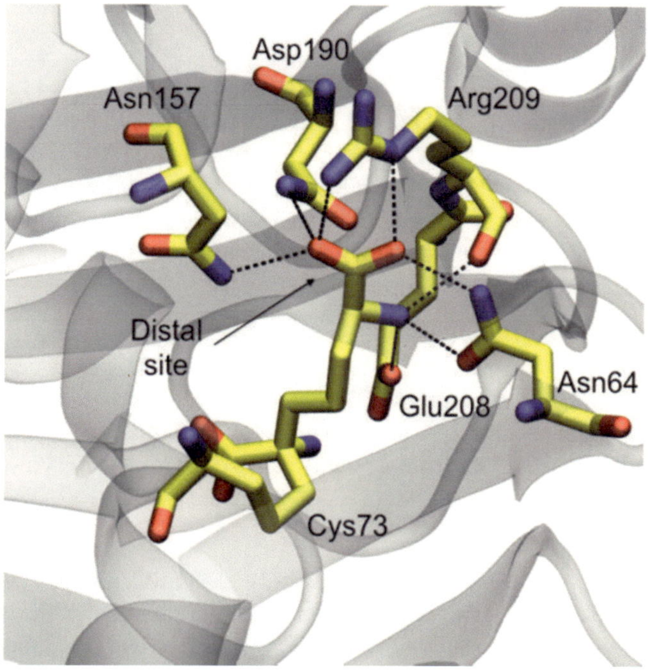

Fig. 18. Catalytic site of DAPE from *H. influenzae*. Hydrogen bond interactions (black dotted lines) at the distal site of the substrate analogue *LL*-AziDAP (arrow indicating position of the analogue) (PDB: 2GKE).

7.3 Inhibition of DAPE

Substrate analogues of DAP have been used as the basis for the generation of inhibitors of DAPE. These inhibitors take advantage of the anionic character at the α-carbon during the reaction or mimic the planar transition state. The most potent inhibitors are shown in Fig. 19 (Williams et al., 1996).

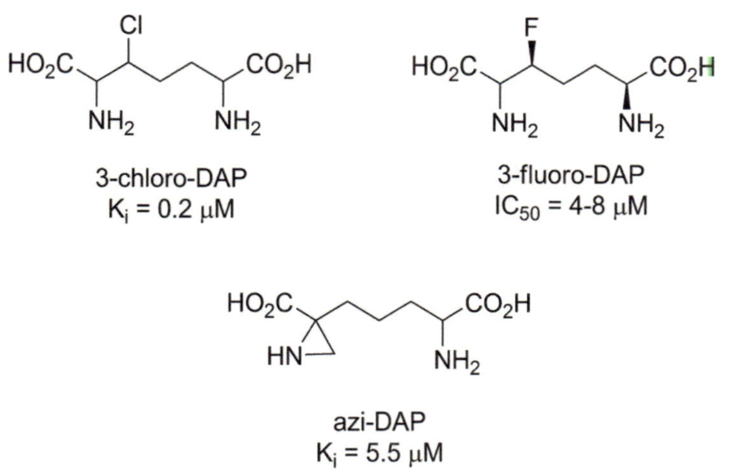

Fig. 19. Inhibitors of DAPE.

8. Dehydrogenase pathway

8.1 *Diaminopimelate dehydrogenase*

Diaminopimelate dehydrogenase (DAPDH EC 1.4.1.16) is a NADPH dependant enzyme that catalyses the reductive amination of *L*-2-amino-6-ketopimelate (AKP), the acyclic form of *L*-2,3,4,5,-tetrahydrodipicolinate (THDP), to produce *meso*-DAP (Misono et al., 1976; Misono & Soda., 1980) (Fig. 1). It is assumed that the reaction occurs via an imine intermediate as a result of amination of *L*-2-amino-6-ketopimelate. Reduction of the imine by hydride transfer from NADPH generates *meso*-DAP (Scapin et al., 1998).

Only a small group of Gram-positive and Gram-negative bacteria posses DAPDH activity. These include *Bacillus sphaericus, Brevibacterium sp., C. glutamicum* and *Proteus vulgaris* (Misono et al., 1979). Characterised DAPDH enzymes are comprised of approximately 320 residues and share greater than 27% sequence identity (Ishino et al., 1987; Hudson et al., 2011b). Kinetic studies of DAPDH from *C. glutamicum* has yielded K_M values for NADPH, *L*-2-amino-6-ketopimelate and ammonia of 0.13 mM, 0.28 mM and 36 mM, respectively (Misono *et al.*, 1986).

Some bacterial species possessing DAPDH activity use multiple pathways to synthesise lysine. For example, *C. glutamicium* (Schrumpf et al., 1991) can synthesise lysine by either the dehydrogenase or succinylase pathway, whilst *Bacillus macerans* (Hudson et al., 2011b) can employ enzymes of the dehydrogenase or acetylase pathways.

DAPDH from *C. glutamicum* forms a homodimer (Scapin et al., 1996) (Fig. 20). The DAPDH monomer subunit is comprised of (i) a dinucleotide binding domain, that is similar to but not identical to a classical Rossman fold, (ii) a dimerisation domain, and (iii) a C-terminal domain (Fig. 20). Monomer subunits interact via two α-helices and a three-stranded antiparallel β-sheet to form the dimer.

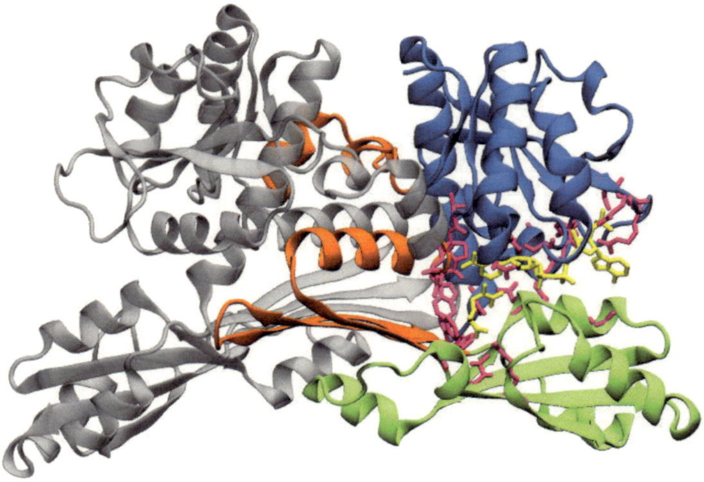

Fig. 20. Structure of dimeric *C. glutamicum* DAPDH in complex with NADPH and *L*-2-amino-6-methylene-pimelate. The dimerisation (orange), dinucleotide binding (blue), and C-terminal (green) domains are indicated. The cofactor NADPH (yellow) and inhibitor *L*-2-amino-6-methylene-pimelate (yellow) are bound by active site residues (pink) (PDB:1F06).

The crystal structure of the *C. glutamicum* DAPDH in complex with ligand shows that the oxidised cofactor, NADP+, is bound within each of the dinucleotide binding domains (Scapin et al., 1996). The domains exhibit open and closed conformations thought to represent the binding and active states of DAPDH, respectively (Scapin et al., 1996). In the closed conformation the $NADP^+$ pyrophosphate forms seven additional noncovalent contacts. Subsequent studies demonstrate the product, *meso*-DAP, binds within an elongated cavity formed at the interface of the dimerisation and dinucleotide binding domains (Scapin et al., 1998).

Crystal structures of *C. glutamicum* complexed with the inhibitors (2*S*,5*S*)-2-amino-3-(3-carboxy-2-isoxazolin-5-yl)-propanoic acid (K_i = 4.2 μM) and *L*-2-amino-6-methylene-pimelate (K_i = 5 μM) show that they form similar interactions with DAPDH as the product *meso*-DAP (Scapin et al., 1998). An additional hydrogen bond between the α-amino group of the *L*-2-amino-6-methylene-pimelate and the indole ring of DAPDH Trp144 is thought to account for the strong competitive inhibition observed (Scapin et al., 1998).

9. Diaminopimelate decarboxylase

9.1 Function of DAPDC

Diaminopimelate decarboxylase (DAPDC, EC 4. 1. 1. 20) is a PLP-dependant enzyme that is responsible for catalysing the final reaction of the lysine biosynthesis pathway (Fig. 1). In this non-reversible reaction, DAPDC converts the substrate *meso*-DAP to lysine and carbon dioxide (Fig. 21). Unlike other PLP-dependant decarboxylases that decarboxylate an *L*-stereocentre, DAPDC specifically cleaves the *D*-stereocentre carboxyl group. Thus, the enzyme possesses a means to differentiate between two stereocentres (Gokulan et al., 2003; Ray et al., 2002). DAPDC is classified as a type III class PLP enzyme, from the alanine racemase family.

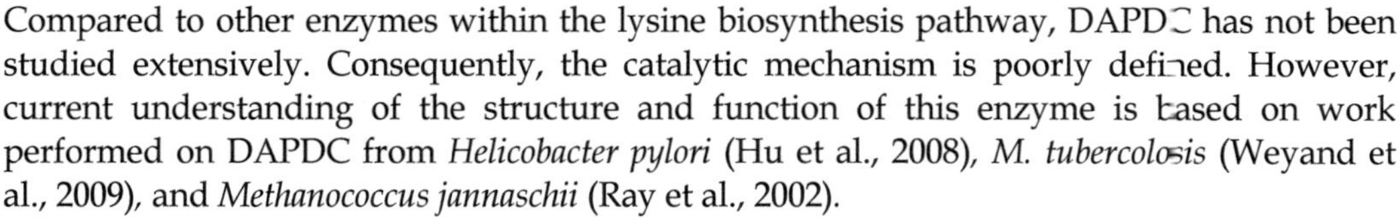

Fig. 21. DAPDC catalysed reaction.

Compared to other enzymes within the lysine biosynthesis pathway, DAPDC has not been studied extensively. Consequently, the catalytic mechanism is poorly defined. However, current understanding of the structure and function of this enzyme is based on work performed on DAPDC from *Helicobacter pylori* (Hu et al., 2008), *M. tubercolosis* (Weyand et al., 2009), and *Methanococcus jannaschii* (Ray et al., 2002).

9.2 Structure of DAPDC

The crystal structures of DAPDC from seven species have been determined. There appears to be no consensus in quaternary structure of the enzyme as monomeric, dimeric, and tetrameric forms of DAPDC have been described. This is unusual, and possibly not a true reflection of what occurs in nature. Studies have shown that the active site of DAPDC is located at the dimer interface (Hu et al., 2008; Ray et al., 2002; Weyand et al., 2009). This implies that the dimer is the minimal catalytic unit. Therefore, monomeric forms of DAPDC are likely to be non-functional; however, this does not rule out the existence of active tetrameric forms of DAPDC.

In species such as *M. jannaschii* (Ray et al., 2002) and *M. tubercolosis,* (Gokulan et al., 2003; Weyand et al., 2009) DAPDC is composed of a homodimer, whereby subunits associate to form a head-to-tail quaternary architecture (Fig. 22).

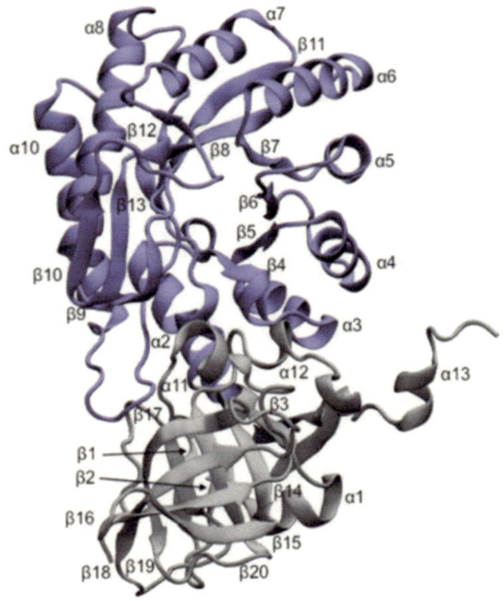

(A)

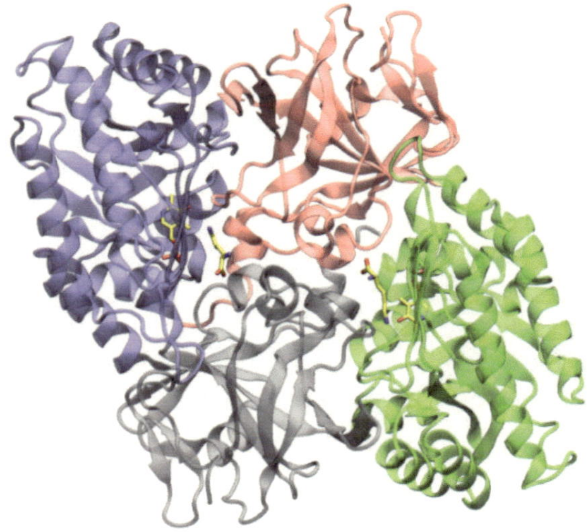

(B)

Fig. 22. Structure of *M. tuberculosis* DAPDC. (A) *M. tuberculosis* DAPDC monomer - The N-terminal (purple) and C-terminal (grey) domains are indicated. (B) *M. tuberculosis* DAPDC dimer - The active site is situated at the homodimer interface. PLP (yellow) and lysine (yellow) are located within the active site cavity (PDB: 1HKV).

The DAPDC monomer is composed of two domains, consisting of an N-terminal 8-fold α/β-barrel domain and a C-terminal β-sheet domain (Gokulan et al., 2003; Ray et al., 2002) (Fig. 22A). In *M. tuberculosis* DAPDC, the N-terminal α/β-barrel domain (residues 48-308) is comprised of β-strands β4–β13 and helices α2-α10 (Fig. 22A). The C-terminal domain (residues 2-47 and 309-446) is comprised of β-strands β1-β3, β14-β21 and helices α1, α11-α13 (Gokulan et al., 2003) (Fig. 22A). The active site is located at the interface between the α/β-barrel domain of one subunit and β-sheet domain of both subunits (Gokulan et al., 2003) (Fig. 22B).

The X-ray structure of *H. pylori* DAPDC has allowed identification of key residues involved in substrate and cofactor recognition. The enzyme was crystallised in the presence of PLP and lysine. The *H. pylori* structure is very similar to that of *M. tuberculosis* DAPDC, forming a homodimer in a head-to-tail conformation. In this enzyme, PLP forms Schiff base linkages with Lys46 and lysine to produce a lysine-PLP external aldimine. This aldimine is believed to mimic the catalytic intermediate formed between *meso*-DAP and PLP (Hu et al. 2008).

9.3 Inhibition of DAPDC

Diaminopimelic acid analogues (Fig. 23) have been synthesised to study the inhibition of DAPDC from *B. sphaericus* (Kelland et al., 1986). Mixtures of isomers of N-hydroxydiaminopimelate and N-aminodiaminopimelate are potent competitive inhibitors of DAPDC, with K_i values of 0.91 mM and 0.1 mM, respectively. Lanthionine sulfoxides (Fig. 23) are good competitive inhibitors, providing about 50% inhibition at 1 mM. Weaker competitive inhibitors include the *meso* and *LL*-isomers of lanthionine sulfone and lanthionine, whereas the *DD*-isomers (Fig. 23) were less effective.

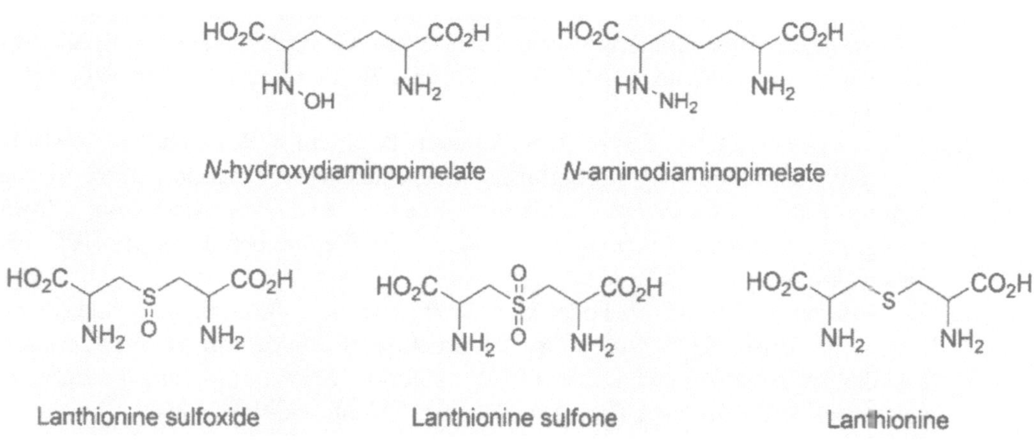

Fig. 23. Inhibitors of DAPDC.

10. Conclusions

Significant advances in our understanding of the enzymes of the lysine biosynthetic pathway have occurred in recent years, particularly through detailed kinetic and structural studies of wild-type and mutant enzymes. While advances in inhibitor design have not been as dramatic, our increased structural knowledge augurs well for the design of potent enzyme inhibitors in the near future, and subtle differences between the structures of the enzymes from different pathogenic species offers great potential of designing pathogen-specific antibiotics. The improvements in our understanding of the lysine biosynthetic pathway in recent years will no doubt advance our efforts toward the ultimate goal of developing novel antibiotics that target this essential bacterial pathway.

11. References

Anita, M., Hoare, D. S., & Work, E. (1957). The stereoisomers of diaminopimelic acid. *Biochem. J.*, Vol.65, No.3, (March 1957), pp. 448-459, ISSN 0264-6021

Beaman, T. W., Binder, D. A., Blanchard, J. S. & Roderick, S. L. (1997). Three-dimensional structure of tetrahydrodipicolinate N-succinyltransferase. *Biochemistry,* Vol.36, No.3, (January 1997), pp. 489-494, ISSN 0006-2960

Beaman, T. W., Blanchard, J. S., & Roderick, S. L. (1998). The conformational change and active site structure of tetrahydrodipicolinate N-succinyltransferase. *Biochemistry,* Vol.37, No.29, (July 1998), pp. 10363-10369, ISSN 0006-2960

Beaman, T. W., Vogel, K. W., Drueckhammer, D. G., Blanchard, J. S., & Roderick, S. L. (2002). Acyl group specificity at the active site of tetrahydridipicolinate N-succinyltransferase. *Protein Sci.*, Vol.11, No.4, (April 2002), pp. 974-9, ISSN 0961-8368

Berges, D. A., DeWolf, W. E. Jr., Dunn, G. L., Grappel, S. F., Newman, D. J., Taggart, J. J., & Gilvarg, C. (1986a). Peptides of 2-aminopimelic acid: antibacterial agents that inhibit diaminopimelic acid biosynthesis. *J. Med. Chem.*, Vol.29, No.1, (January 1986), pp. 89-95, ISSN 0022-2623

Berges, D. A., DeWolf, W. E. Jr., Dunn, G. L., Newman, D. J., Schmidt, S. J., Taggart, J. J., & Gilvarg, C. (1986b). Studies on the active site of succinyl-CoA:tetrahydrodipicolinate N-succinyltransferase. Characterization using analogs of tetrahydrodipicolinate. *J. Biol. Chem.*, Vol.261, No.14, (May 1986), pp. 6160-6167, ISSN 0021-9258

Bienvenue, D. L., Gilner, D. M., Davis, R. S., Bennett, B., & Holz, R. C. (2003). Substrate specificity, metal binding properties, and spectroscopic characterization of the DapE-encoded N-succinyl-L,L-diaminopimelic acid desuccinylase from *Haemophilus influenzae. Biochemistry,* Vol.42, No.36, (September 2003), pp. 10756-63, ISSN 0006-2960

Blagova, E., Levdikov, V., Milioti, N., Fogg, M. J., Kalliomaa, A. K., Brannigan, J. A., Wilson, K. S., & Wilkinson, A. J. (2006). Crystal structure of dihydrodipicolinate synthase (BA3935) from *Bacillus anthracis* at 1.94 Å resolution. *Proteins: Structure, Function, & Bioinformatics,* Vol.62, No.1, (January 2006), pp. 297-301, ISSN 0887-3585

Blickling, S., Renner, C., Laber, B., Pohlenz, H., Holak, T. A., & Huber, R. (1997a). Reaction mechanism of *Escherichia coli* dihydrodipicolinate synthase investigated by X-ray crystallography and NMR spectroscopy. *Biochemistry,* Vol.36, No.1, (January 1997), pp. 24-33, ISSN 0006-2960

Blickling, S., Beisel, H. G., Bozic, D., Knäblein, J., Laber, B., & Huber, R. (1997b). Structure of dihydrodipicolinate synthase of *Nicotiana sylvestris* reveals novel quaternary structure. *J. Mol. Biol.*, Vol.274, No.4, (December 1997), pp. 608-621, ISSN 0022-2836

Blickling, S., & Knablein, J. (1997c). Feedback inhibition of dihydrodipicolinate synthase enzymes by L-lysine. *Biol. Chem.*, Vol.378, No.3-4, (March-April 1997), pp. 207-210, ISSN 1431-6730

Born, T. L., Zheng, R., & Blanchard, J. S. (1998). Hydrolysis of N-succinyl-L,L-diaminopimelic acid by the *Haemophilus influenzae* dapE-encoded desuccinylase: metal activation, solvent isotope effects, and kinetic mechanism. *Biochemistry*, Vol.37, No.29, (July 1998), pp. 10478-87, ISSN 0006-2960

Boughton, B. A., Dobson, R. C., Gerrard, J. A., & Hutton, C. A. (2008). Conformationally constrained diketopimelic acid analogues as inhibitors of dihydrodipicolinate synthase. *Bioorg. Med. Chem. Lett.*, Vol.18, No.2, (January 2008), pp. 460-463, ISSN 0960-894X

Bouvier, J., Richaud, C., Higgins, W., Bögler, O., & Stragier, P. (1992). Cloning, characterization, and expression of the dapE gene of *Escherichia coli*. *J. Bacteriol.*, Vol.174, No.16, (August 1992), pp. 5265-5271, ISSN 0021-9193

Broder, D. H., & Miller, C. G. (2003). DapE can function as an aspartyl peptidase in the presence of Mn2+. *J. Bacteriol.*, Vol.185, No.16, (August 2003), pp. 4748-54, ISSN 0021-9193

Bukhari, A. I., & Taylor, A. L. (1971). Genetic analysis of diaminopimelic acid- and lysine-requiring mutants of *Escherichia coli*. *J. Bacteriol.*, Vol.105, No.3, (March 1971), pp. 844-854, ISSN 0021-9193

Burgess, B. R., Dobson, R. C. J., Bailey, M. F., Atkinson, S. C., Griffin, M. D. W., Jameson, G. B., Parker, M. W., Gerrard, J. A., & Perugini, M. A. (2008). Structure and evolution of a novel dimeric enzyme from a clinically-important bacterial pathogen. *J. Biol. Chem.*, Vol.283, No.41, (October 2008), pp. 27598-27603, ISSN 0021-9258

Cahyanto, M. N., Kawasaki, H., Fujiyama, K., & Seki, T. (2006). Regulation of aspartokinase, aspartate semialdehyde dehydrogenase, dihydrodipicolinate synthase and dihydrodipicolinate reductase in *Lactobacillus plantarum*. *Microbiology*, Vol.152, No.Pt 1, (January 2006), pp. 105-112, ISSN 1350-0872

Caplan, J. F., Zheng, R., Blanchard, J. S., & Vederas, J. C. (2000). Vinylogous amide analogues of diaminopimelic acid (DAP) as inhibitors of enzymes involved in bacterial lysine biosynthesis. *Org. Lett.*, Vol.2, No.24, (November 2000), pp. 3857-60, ISSN 1523-7060

Cremer, J., Treptow, C., Eggeling, L., & Sahm, H. (1988). Regulation of enzymes of lysine biosynthesis in *Corynebacterium glutamicum*. *J. Gen. Microbiol.*, Vol.134, No.12, (December 1988), pp. 3221-3229, ISSN 0022-1287

Chatterjee, S. P., & White, P. J. (1982). Activities and regulation of the enzymes of lysine biosynthesis in a lysine-excreting strain of *Bacillus megaterium*. *J. Gen. Microbiol.*, Vol.128, (October 1982), pp. 1073-1081, ISSN 0022-1287

Chatterjee S. P., Singh B. K., & Gilvarg C. (1994). Biosynthesis of lysine in plants: the putative role of meso-diaminopimelate dehydrogenase. *Plant Mol. Biol.*, Vol.26, No.1, (October 1994), pp. 285-290, ISSN 0167-4412

Cirilli, M., Zheng, R., Scapin, G., & Blanchard, J. S. (1998). Structural symmetry: the three-dimensional structure of *Haemophilus Influenzae* diaminopimelate epimerase. *Biochemistry*, Vol.37, No.47, (November 1998), pp. 16452-16458, ISSN 0006-2960

Cirilli, M., Scapin, G., Sutherland, A., Vederas, J. C., & Blanchard, J. S. (2000). The three-dimensional structure of the ternary complex of *Corynebacterium glutamicum* diaminopimelate dehydrogenase-NADPH-L-2-amino-6-methylene-pimelate. *Protein Sci.*, Vol.9, No.10, (October 2000), pp. 2034-2037, ISSN 0961-8368

Cirilli, M., Zheng, R., Scapin, G., & Blanchard, J. S. (2003). The three-dimensional structures of the *Mycobacterium tuberculosis* dihydrodipicolinate reductase-NADH-2,6-PDC and -NADPH-2,6-PDC complexes. Structural and mutagenic analysis of relaxed nucleotide specificity. *Biochemistry*, Vol.42, No.36, (August 2003), pp. 10644-50, ISSN 0006-2960

Cosper, N. J., Bienvenue, D. L., Shokes, J. E., Gilner, D. M., Tsukamoto, T., Scott, R. A., & Holz, R. C. (2003). The dapE-encoded N-succinyl-l,l-diaminopimelic acid desuccinylase from *Haemophilus influenzae* is a dinuclear metallohydrolase. *J. Am. Chem. Soc.*, Vol.125, No.48, (December 2003), pp. 14654-14655, ISSN 0002-7863

Cox, R. J., Schouten, J. A., Stentiford, R. A., & Wareing, K. J. (1998). Peptide inhibitors of N-succinyl diaminopimelic acid aminotransferase (DAP-AT): a novel class of antimicrobial compounds. *Bioorg. Med. Chem. Lett.*, Vol.8, No.8, (April 1998), pp. 945-950, ISSN 1464-3405

Davis, R., Bienvenue, D., Swierczek, S. I., Gilner, D. M., Rajagopal, L., Bennett, B., & Holz, R. C. (2006). Kinetic and spectroscopic characterization of the E134A- and E134D-altered dapE-encoded N-succinyl-L,L-diaminopimelic acid desuccinylase from *Haemophilus influenzae*. *J. Biol. Inorg. Chem.*, Vol.11, No.2, (March 2006), pp. 206-216, ISSN 0949-8257

Dereppe, C., Bold, G., Ghisalba, O., Ebert, E., & Schar, H. (1992). Purification and Characterization of Dihydrodipicolinate Synthase from Pea. *Plant Physiol.*, Vol.98, No.3, (March 1992), pp. 813-821, ISSN 0032-0889

Devenish, S. R. A., Huisman, F. H., Parker, E. J., Hadfield, A. D., & Gerrard, J. A. (2009). Cloning and characterisation of dihydrodipicolinate synthase from the pathogen *Neisseria meningitidis*. *Biochimica et Biophysica Acta*, Vol.1794, No.8, (August 2009), pp. 1168-1174, ISSN 0006-3002

Dobson, R. C., Valegård, K., & Gerrard, J. A. (2004a). The crystal structure of three site-directed mutants of *Escherichia coli* dihydrodipicolinate synthase: further evidence for a catalytic triad. *J. Mol. Biol.*, Vol.338, No.2, (April 2004), pp. 329-339, ISSN 0022-2836

Dobson, R. C. J., Griffin, M. D. W., Roberts, S. J., & Gerrard, J. A. (2004b). Dihydrodipicolinate synthase (DHDPS) from *Escherichia coli* displays partial mixed inhibition with respect to its first substrate, pyruvate. *Biochimie*, Vol.86, No.4-5, (April-May 2004), pp. 311-315, ISSN 0300-9084

Dobson, R. C., Griffin, M. D., Jameson, G. B., Gerrard, J. A. (2005a). The crystal structures of native and (S)-lysine-bound dihydrodipicolinate synthase from *Escherichia coli* with improved resolution show new features of biological significance. *Acta Crystallogr. D Biol. Crystallogr.*, Vol.61, No.Pt8, (August 2005), pp. 1116-1124, ISSN 0907-4449

Dobson, R. C., Devenish, S. R. A., Turner, L. A., Clifford, V. R., Pearce, F. G., Jameson, G. B., & Gerrard, J. A. (2005b). Role of arginine 138 in the catalysis and regulation of

Escherichia coli dihydrodipicolinate synthase. *Biochemistry*, Vol.44, No.39, (October 2005), pp. 13007-13013, ISSN 0006-2960

Dobson, R. C. J., Griffin, M. D. W., Devenish, S. R. A., Pearce, F. G., Hutton, C. A., Gerrard, J. A., Jameson, G. B., & Perugini, M. A. (2008) Conserved main-chain peptide distortions: a proposed role for Ile203 in catalysis by dihydrodipicolinate synthase. *Protein Sci.*, Vol.17, No.12, (December 2008), pp. 2080-2090, ISSN 0961-8368

Dobson, R. C., Girón, I., & Hudson, A. O. (2011), L,L-diaminopimelate aminotransferase from *Chlamydomonas reinhardtii*: a target for algaecide development. *PLoS one.*, Vol.6 No.5, (May 2011), In press, ISSN 1932-6203

Dogovski, C., Atkinson, S. C., Dommaraju, S. R., Hor, L., Dobson, R. C. J., Hutton C. A., Gerrard, J. A., & Perugini, M. A. (2009). Lysine biosynthesis in bacteria: an unchartered pathway for novel antibiotic design. In: *Encyclopedia Of Life Support Systems*, Volume 11 (Biotechnology Part I), pp116-136, edited by H.W. Doelle and S. Rokem, Eolss Publishers, Oxford ,UK < http://www.eolss.net>

Domigan, L. J., Scally, S. W., Fogg. M. J., Hutton, C. A., Perugini, M. A., Dobson, R. C. J., Muscroft-Taylor, A., Gerrard, J. A. & Devenish, S. R. A. (2009) Characterisation of dihydrodipicolinate synthase from *Bacillus anthracis*. *BBA Proteins* Vol.1794, No.10, (October 2009) pp. 1510-1516, ISSN 1570-9639

Dommaraju, S. R., Dogovski, C., Czabotar, P. E., Hor, L., Smith, B. J., & Perugini, M. A. (2011). Catalytic Mechanism and Cofactor Preference of Dihydrodipicolinate Reductase from Methicillin-Resistant *Staphylococcus aureus*. *Arch. Biochem. Biophys.*, Vol.512, No.2, (August 2011), pp. 167-74, ISSN 0003-9861

Fan, C., Clay, M. D., Deyholos, M. K., & Vederas, J. C. (2010). Exploration of inhibitors for diaminopimelate aminotransferase. *Bioorg. Med. Chem.*, Vol.18, No.6, (March 2010), pp. 2141-2151, ISSN 0968-0896

Farkas, W., & Gilvarg, C. (1965). The reduction step in diaminopimelic acid biosynthesis. *J. Biol. Chem.*, Vol.240, No.12, (December 1965), pp. 4717-22, ISSN 0021-9258

Frisch, D. A., Gengenbach, B. G., Tommey, A. M., Seliner, J. M., Somers, D. A., & Myers, D. E. (1991). Isolation and Characterization of Dihydrodipicolinate Synthase from Maize. *Plant Physiol.*, Vol.96, No.2, (June 1991), pp. 444-452, ISSN 0032-0889

Fuchs, T. M., Schneider, B., Krumbach, K., Eggeling, L., & Gross, R. (2000). Characterization of a *Bordetella pertussis* diaminopimelate (DAP) biosynthesis locus identifies dapC, a novel gene coding for an N-succinyl-L,L-DAP aminotransferase. *J. Bacteriol.*, Vol.182, No.13, (July 2000), pp. 3626-3631, ISSN 0021-9193

Gillner, D. M., Bienvenue, D. L., Nocek, B. P., Joachimiak, A., Zachary, V., Bennett, B., & Holz, R. C. (2009). The dapE-encoded N-succinyl-L,L-diaminopimelic acid desuccinylase from *Haemophilus influenzae* contains two active-site histidine residues. *J. Biol. Inorg. Chem.*, Vol.14, No.1, (January 2009), pp. 1-10, ISSN 0949-8257

Girish, T. S., Navratna, V., & Gopal, B. (2011). Structure and nucleotide specificity of *Staphylococcus aureus* dihydrodipicolinate reductase (DapB). *FEBS Lett.*, Vol.585, No.16, (August 2011), pp. 2561-7, ISSN 0014-5793

Gokulan, K., Rupp, B., Pavelka, M. S., Jr., Jacobs, W. R., Jr., & Sacchettini, J. C. (2003). Crystal structure of *Mycobacterium tuberculosis* diaminopimelate decarboxylase, an essential enzyme in bacterial lysine biosynthesis. *J. Biol. Chem.*, Vol.278, No.20, (March 2003), pp. 18588-18596, ISSN 0021-9258

Griffin, M. D. W., Dobson, R. C. J., Pearce, F. G., Antonio, L., Whitten, A. E., Liew, C. K., Mackay, J. P., Trewhella, J., Jameson, G. B., Perugini, M. A., & Gerrard, J. A. (2008). Evolution of quaternary structure in a homotetrameric protein. *J. Mol. Biol.*, Vol.380, No.4, (July 2008), pp. 691-703, ISSN 0022-2836

Griffin, M. D. W., Dobson, R. C. J., Gerrard, J. A., & Perugini, M. A. (2010) Exploring the dimerdimer interface of the dihydrodipicolinate synthase tetramer: how resilient is the interface? *Arch. Biochem. Biophys.*, Vol.494, No.1, (February 2010), pp. 58-63, ISSN 0003-9861

Gunji, Y., Tsujimoto, N., Shimaoka, M., Ogawa-Miyata, Y., Sugimoto, S., & Yasueda, H. (2004). Characterization of the L-lysine biosynthetic pathway in the obligate methylotroph *Methylophilus methylotrophus*. *Biosci. Biotechnol. Biochem.*, Vol.68, No.7, (July 2004), pp. 1449-60, ISSN 0916-8451

Hartmann, M., Tauch, A., Eggeling, L., Bathe, B., Möckel, B., Pühler, A., & Kalinowski, J. (2003). Identification and characterization of the last two unknown genes, dapC and dapF, in the succinylase branch of the L-lysine biosynthesis of *Corynebacterium glutamicum*. *J. Biotechnol.*, Vol.104, No.1-3, (September 2003), pp. 199-211, ISSN 0168-1656

Hoganson, D.A., & Stahly, D.P. (1975). Regulation of dihydrodipicolinate synthase during growth and sporulation of *Bacillus cereus*. *J. Bacteriol.*, Vol.124, No.3, (December 1975), pp. 1344-1350, ISSN 0021-9193

Hu, T., Wu, D., Chen, J., Ding, J., & Jiang, H, Shen, X. (2008). The catalytic intermediate stabilized by a "down" active site loop for diaminopimelate decarboxylase from *Helicobacter pylori*. Enzymatic characterization with crystal structure analysis. *J. Biol. Chem.*, Vol.283, No.30 (May 2008), pp. 21284-21293, ISSN 0021-9258

Hudson, A. O., Bless, C., Macedo, P., Chatterjee, S. P., Singh, B. K., Gilvarg, C., & Leustek, T. (2005). Biosynthesis of lysine in plants: evidence for a variant of the known bacterial pathways. *Biochim. Biophys. Acta.*, Vol.1721, No.1-3, (January 2005), pp. 27-36, ISSN 0006-3002

Hudson, A. O., Singh, B. K., Leustek, T., & Gilvarg, C. (2006). An LL-Diaminopimelate Aminotransferase Defines a Novel Variant of the Lysine Biosynthesis Pathway in Plants. *Plant Physiol.*, Vol.140, No.1, (January 2006), pp. 292-301, ISSN 0032-0889

Hudson, A. O., Gilvarg, C., & Leustek, T. (2008). Biochemical and Phylogenetic Characterization of a novel diaminopimelate biosynthesis pathway in prokaryotes identifies a diverged form of LL-diaminopimelate aminotransferase. *J. Bact.*, Vol.190. No.9, (May 2008), pp. 3256–3263, ISSN 0021-9193

Hudson, A. O., Girón, I., & Dobson, R. C. (2011a). Crystallization and preliminary X-ray diffraction analysis of L,L-diaminopimelate aminotransferase (DapL) from *Chlamydomonas reinhardtii*. *Acta Cryst. F*, Vol.67, No.Pt1, (January 2011), pp. 140-3, ISSN 1744-3091

Hudson, A. O., Klartag, A., Gilvarg, C., Dobson, R. C., Marques, F. G., & Leustek, T. (2011b). Dual diaminopimelate biosynthesis pathways in *Bacteroides fragilis* and *Clostridium thermocellum*. *Biochim. Biophys. Acta.*, Vol.1814, No.9, (September 2011), pp. 1162-1168, ISSN 0006-3002

Hutton, C. A., Southwood, T. J., & Turner, J. J. (2003). Inhibitors of lysine biosynthesis as antibacterial agents. *Mini Rev. Med. Chem.*, Vol.3, No.2, (March 2003), pp. 115-27, ISSN 1389-5575

Hutton, C. A., Perugini, M. A., & Gerrard, J. A. (2007). Inhibition of lysine biosynthesis: an emerging antibiotic strategy. *Molecular BioSystems*, Vol.3, No.7, (July 2007), pp. 458-465, ISSN 1742-2051

Ishino, S., Mizukami, T., Yamaguchi, K., Katsumata, R., & Araki, K. (1987). Nucleotide sequence of the meso-diaminopimelate D-dehydrogenase gene from *Corynebacterium glutamicum*. *Nucleic Acids Res.*, Vol.15, No.9, (May 1987), pp. 3917, ISSN 0305-1048

Jensen, R. A., & Gu, W. (1996). Evolutionary recruitment of biochemically specialized subdivisions of Family I within the protein superfamily of aminotransferases. *J. Bact.*, Vol.178, No.8, (April 1996), pp. 2161-2171., ISSN 0021-9193

Karsten, W. E. (1997). Dihydrodipicolinate synthase from *Escherichia coli*: pH dependent changes in the kinetic mechanism and kinetic mechanism of allosteric inhibition by lysine. *Biochemistry*, Vol.36, No.7 (February 1997), pp. 1730–1739, ISSN 0006-2960

Kefala, G., Evans, G. L., Griffin, M. D. W., Devenish, S. R. A., Pearce, F. G., Perugini, M. A., Gerrard, J. A., Weiss, M. S., & Dobson, R. C. J. (2008). Crystal structure and kinetic study of dihydrodipicolinate synthase from *Mycobacterium tuberculosis*. *Biochem. J.*, Vol.411, No.2, (April 2008), pp. 351-360, ISSN 0264-6021

Kelland, J. G., Arnold, L. D., Palcic, M. M., Pickard, M. A., & Vederas, J. C. (1986). Analogs of diaminopimelic acid as inhibitors of meso-diaminopimelate decarboxylase from *Bacillus sphaerius* and wheat germ. *J. Biol. Chem.*, Vol.261, No.28, (October 1986), pp. 13216-13223, ISSN 0021-9258

Kimura, K. (1975). A new flavin enzyme catalyzing the reduction of dihydrodipicolinate in sporulating *Bacillus subtilis*: I. Purification and properties. *J. Biochem.*, Vol.77, No.2, (July 1975), pp. 405-13, ISSN 0021-924X

Kimura, K., & Goto, T. (1977). Dihydrodipicolinate reductases from *Bacillus cereus* and *Bacillus megaterium*. *J. Biochem.*, Vol.81, No.5, (May 1977), pp. 1367-73, ISSN 0021-924X

Kindler, S. H., & Gilvarg, C. (1960). N-Succinyl-L-2,6-diaminopimelic acid deacylase. *J. Biol. Chem.*, Vol.235, (December 1960), pp. 3532-3535, ISSN 0021-9258

Kumpaisal, R., Hashimoto, T., & Yamada, Y. (1987). Purification and Characterization of Dihydrodipicolinate Synthase from Wheat Suspension Cultures. *Plant Physiol.*, Vol.85, No.1, (September 1987), pp. 145-151, ISSN 0032-0889

Laber, B., Gomis-Rüth, F., & Romão, M. J., & Huber, R. (1992). *Escherichia coli* dihydrodipicolinate synthase. Identification of the active site and crystallization. *Biochem. J.*, Vol.288, No.Pt2, (December 1992), pp. 691-695, ISSN 0264-6021

Ledwidge, R., & Blanchard, J. S. (1999). The dual biosynthetic capability of N-acetylornithine aminotransferase in arginine and lysine biosynthesis. *Biochemistry*, Vol.38, No.10, (March 1999), pp. 3019-3024, ISSN 0006-2960

Lin, Y. K., Myhrman, R., Schrag, M. L., & Gelb, M. H. (1988). Bacterial N-succinyl-L-diaminopimelic acid desuccinylase. Purification, partial characterization, and substrate specificity. *J. Biol. Chem.*, Vol.263, No.4, (February 1988), pp. 1622-7, ISSN 0021-9258

Liu, Y., White, R. H., & Whitman, W. B. (2010). Methanococci use the diaminopimelate aminotransferase (DapL) pathway for lysine biosynthesis. *J. Bact.*, Vol.192, No.13, (July 2010), pp. 3304–3310, ISSN 0021-9193

Lloyd, A. J., Huyton, T., Turkenburg, J., & Roper, D. I. (2004). Refinement of *Haemophilus influenzae* diaminopimelic acid epimerase (DapF) at 1.75 A resolution suggests a mechanism for stereocontrol during catalysis. *Acta Cryst. Sect. D,* Vol.60, No.2, (November 1998), pp. 397-400, ISSN 1399-0047

Matthews, B. F., & Widholm, J. M. (1979). Expression of aspartokinase, dihydrodipicolinic acid synthase and homoserine dehydrogenase during growth of carrot cell suspension cultures on lysine- and threonine-supplemented media. *Z Naturforsch [C]*, Vol.34, No.12, (December 1979), pp. 1177-1185, ISSN 0939-5075

McCoy, A. J., Adams, N. E., Hudson, A. O., Gilvarg, C., Leustek, T., & Maurelli, A. T. (2006). L,L-diaminopimelate aminotransferase, a trans-kingdom enzyme shared by Chlamydia and plants for synthesis of diaminopimelate/lysine. *PNAS*, Vol.103, No.47, (November 2006), pp. 17909-17914, ISSN 0027-8424

Mirwaldt, C., Korndorfer, I., & Huber, R. (1995). The crystal structure of dihydrodipicolinate synthase from *Escherichia coli* at 2.5 Å resolution. *J. Mol. Biol.*, Vol.246, No.1, (February 1995), pp. 227-239, ISSN 0022-2836

Misono, H., Togawa, H., Yamamoto, T., & Soda, K. (1976). Occurrence of meso-alpha, epsilon-diaminopimelate dehydrogenase in *Bacillus sphaericus*. *Biochem. Biophys. Res. Commun.*, Vol.72, No.1, (September 1976), pp. 89-93, ISSN 0006-291X

Misono, H., Togawa, H., Yamamoto, T., & Soda, K. (1979). Meso-alpha,epsilon-diaminopimelate D-dehydrogenase: distribution and the reaction product. *J. Bacteriol.*, Vol.137, No.1, (January 1979), pp. 22-27, ISSN 0021-9193

Misono, H., & Soda, K. (1980). Properties of meso-alpha,epsilon-diaminopimelate D-dehydrogenase from *Bacillus sphaericus*. *J. Biol. Chem.*, Vol.255, No.22, (November 1980), pp. 10599-10605, ISSN 0021-9258

Misono, H., Ogasawara, M., & Nagasaki, S. (1986). Characterization of meso-diaminopimelate dehydrogenase from *Corynebacterium glutamicum* and its distribution in bacteria. *Agric. Biol. Chem.*, Vol.50, No.11, (March 1986), pp. 2729-2734, ISSN 0002-1369

Mitsakos, V., Dobson, R. C. J., Pearce, F. G., Devenish, S. R., Evans, G. L., Burgess, B. R., Perugini, M. A., Gerrard, J. A., & Hutton, C. A. (2008). Inhibiting dihydrodipicolinate synthase across species: towards specificity for pathogens? *Bioorg. Med. Chem. Lett.*, Vol.18, No.2, (January 2008), pp. 842-844, ISSN 0960-894X

Neidhardt, F. C., & Curtiss, R. (1996). *Escherichia coli and Salmonella Cellular and Molecular Biology* (2nd), ASM Press, ISBN 9781555810849, Washington D.C.

Nguyen, L., Kozlov, G., & Gehring, K. (2008). Structure of *Escherichia coli* tetrahydrodipicolinate N-succinyltransferase reveals the role of a conserved C-terminal helix in cooperative substrate binding. *FEBS Lett.*, Vol.582, No.5, (March 2008), pp. 623-626, ISSN 0014-5793

Nocek, B. P., Gillner, D. M., Fan, Y., Holz, R. C., & Joachimiak, A. (2010). Structural basis for catalysis by the mono- and dimetalated forms of the dapE-encoded N-succinyl-L,L-diaminopimelic acid desuccinylase. *J. Mol. Biol.*, Vol.397, No.3, (April 2010), pp. 617-626, ISSN 0022-2836

Ou, J., Yamada, T., Nagahisa, K., Hirasawa, T., Furusawa, C., Yomo, T., & Shimizu, H. (2008). Dynamic change in promoter activation during lysine biosynthesis in *Escherichia coli* cells. *Mol. Biosyst.,* Vol.4, No.2, (February 2008), pp. 128-34, ISSN 1742-2051

Pavelka Jr., M. S., Weisbrod, T. R., & Jacobs Jr., W. R. (1997). Cloning of the dapB gene, encoding dihydrodipicolinate reductase, from *Mycobacterium tuberculosis. J. Bacteriol.,* Vol.179, No.8, (April 1997), pp. 2777-82, ISSN 0021-9193

Pearce, F. G., Perugini, M. A., Mckerchar, H. J., & Gerrard, J. A. (2006). Dihydrodipicolinate synthase from *Thermotoga maritima. Biochem J.,* Vol.400, No.2, (December 2006), pp. 359-366, ISSN 0264-6021

Pearce, F. G., Sprissler, C., & Gerrard, J. A. (2008). Characterization of dihydrodipicolinate reductase from *Thermotoga maritima* reveals evolution of substrate binding kinetics. *J. Biochem.,* Vol.143, No.5, (May 2008), pp. 617-23, ISSN 0021-924X

Peterkofsky, B., & Gilvarg, C. (1961). N-Succinyl-L-diaminopimelic-glutamic transaminase. *J. Biol. Chem.,* Vol.236, (May 1961), pp. 1432-1438, ISSN 0021-9258

Phenix, C. P., & Palmer, D. R. (2008). Isothermal titration microcalorimetry reveals the cooperative and noncompetitive nature of inhibition of *Sinorhizobium meliloti* L5-30 dihydrodipicolinate synthase by (S)-lysine. *Biochemistry,* Vol.47, No.30, (July 2008), pp. 7779-7781, ISSN 0006-2960

Pillai, B., Cherney, M., Diaper, C. M., Sutherland, A., Blanchard, J. S., Vederas, J. C., & James, M. N. G. (2007). Dynamics of catalysis revealed from the crystal structures of mutants of diaminopimelate epimerase. *Biochem. Bioph. Res. Co.,* Vol.363, No.3, (November 2007), pp. 547-553, ISSN 0006-291X

Pillai, B., Moorthie, V. A., Van Belkum, M. J., Marcus, S. L., Cherney, M. M. Diaper, C. M., Vederas, J. C., & James, M. N. G. (2009). Crystal Structure of Diaminopimelate Epimerase from *Arabidopsis thaliana,* an Amino Acid Racemase Critical for L-Lysine Biosynthesis. *J. Mol. Biol.,* Vol.385, No.2, (January 2009), pp. 580-594, ISSN 0022-2836

Ray, S. S., Bonanno, J. B., Rajashankar, K. R., Pinho, M. G., He, G., & De Lencastre, H. (2002). Cocrystal structures of diaminopimelate decarboxylase: mechanism, evolution, and inhibition of an antibiotic resistance accessory factor. *Structure,* Vol.10, No.11, (November 2002), pp. 1499-1508, ISSN 0969-2126

Reddy, S. G., Sacchettini, J. C., & Blanchard, J. S. (1995). Expression, purification, and characterization of *Escherichia coli* dihydrodipicolinate reductase. *Biochemistry,* Vol.34, No.11, (March 1995), pp. 3492-501, ISSN 0006-2960

Reddy, S. G., Scapin, G., & Blanchard, J. S. (1996). Interaction of pyridine nucleotide substrates with *Escherichia coli* dihydrodipicolinate reductase: thermodynamic and structural analysis of binary complexes. *Biochemistry,* Vol.35, No.41, (October 1996), pp. 13294-302, ISSN 0006-2960

Richaud, C., Richaud, F., Martin, C., Haziza, C., & Patte, J. C. (1984). Regulation of expression and nucleotide sequence of the *Escherichia coli* dapD gene. *J. Biol. Chem.,* Vol.259, No.23, (December 1984), pp. 14824-14828, ISSN 0021-9258

Saleh, F., & White, P. J. (1979). Metabolism of DD-2,6-diaminopimelic acid by a diaminopimelate-requiring mutant of *Bacillus megaterium. J. Gen. Microbiol.,* Vol.115, (February 1979), pp. 95-100, ISSN 0022-1287

Scapin, G., Blanchard, J. S., & Sacchettini, J. C. (1995). Three-Dimensional Structure of *Escherichia coli* Dihydrodipicolinate Reductase. *Biochemistry*, Vol.34, No.11, (March 1995), pp. 3502–3512, ISSN 0006-2960

Scapin, G., Reddy, S. G., & Blanchard, J. S. (1996). Three-dimensional structure of meso-diaminopimelic acid dehydrogenase from *Corynebacterium glutamicum*. *Biochemistry*, Vol.35, No.42, (October 1996), pp. 13540-13551, ISSN 0006-2960

Scapin, G., Reddy, S. G., Zheng, R., & Blanchard, J. S. (1997). Three-Dimensional Structure of *Escherichia coli* Dihydrodipicolinate Reductase in Complex with NADH and the Inhibitor 2,6-Pyridinedicarboxylate. *Biochemistry*, Vol.36, No.49, (December 1997), pp. 15081–15088, ISSN 0006-2960

Scapin, G., Cirilli, M., Reddy, S. G., Gao, Y., Vederas, J. C., & Blanchard, J. S. (1998). Substrate and inhibitor binding sites in *Corynebacterium glutamicum* diaminopimelate dehydrogenase. *Biochemistry*, Vol.37, No.10, (March 1998), pp. 3278-3285, ISSN 0006-2960

Schrumpf, B., Schwarzer, A., Kalinowski, J., Pühler, A., Eggeling, L., & Sahm, H. (1991). A functionally split pathway for lysine synthesis in *Corynebacterium glutamicium*. *J. Bacteriol.*, Vol.173, No.14, (July 1991), pp. 4510-4516, ISSN 0021-9193

Schuldt, L., Weyand, S., Kefala, G., & Weiss, M. S. (2009). The three-dimensional Structure of a mycobacterial DapD provides insights into DapD diversity and reveals unexpected particulars about the enzymatic mechanism. *J. Mol. Biol.*, Vol.389, No.5, (April 2009), pp. 863-879, ISSN 0022-2836

Simms, S. A., Voige, W. H., & Gilvarg, C. (1984). Purification and characterization of succinyl-CoA: tetrahydrodipicolinate N-succinyltransferase from *Escherichia coli*. *J. Biol. Chem.*, Vol.259, No.5, (March 1984), pp. 2734-2741, ISSN 0021-9258

Sundharadas, G., & Gilvarg, C. (1967). Biosynthesis of alpha,epsilon-diaminopimelic acid in *Bacillus megaterium*. *J. Biol. Chem.*, Vol.242, No.17, (September 1967), pp. 3983-3984, ISSN 0021-9258

Sung, M. H., Tanizawa, K., Tanaka, H., Kuramitsu, S., Kagamiyama, H., Hirotsu, K., Okamoto, A., Higuchi, T., & Soda, K. (1991). Thermostable aspartate aminotransferase from a thermophilic Bacillus species. Gene cloning, sequence determination, and preliminary x-ray characterization. *J. Biol. Chem.*, Vol.266, No.4, (February 1991), pp. 2567-2572, ISSN 0021-9258

Turner, J. J., Healy, J. P., Dobson, R. C. J., Gerrard, J. A., & Hutton, C. A. (2005). Two new irreversible inhibitors of dihydrodipicolinate synthase: diethyl (E,E)-4-oxo-2,5-heptadienedioate and diethyl (E)-4-oxo-2-heptenedioate. *Bioorg. Med. Chem. Lett.*, Vol.15, No.4, (February 2005), pp. 995-998, ISSN 0960-894X

Usha, V., Dover, L. G., Roper, D. I., Futterer, K., & Besra, G. S. (2009). Structure of the diaminopimelate epimerase DapF from *Mycobacterium tuberculosis*. *Acta Cryst. Sect. D*, Vol.65, No 4, (April 2009), pp. 383-387, ISSN 0907-4449

Voss, J. E., Scally, S. W., Taylor, N. L., Atkinson, S. C., Griffin, M. D. W., Hutton, C. A., Parker, M. W., Alderton, M. R., Gerrard, J. A., Dobson, R. C. J., Dogovski, C., & Perugini, M. A. (2010). Substrate-mediated stabilization of a tetrameric drug target reveals achilles heel in anthrax. *J. Biol. Chem.*, Vol.285, No.8, (February 2010), pp. 5188-5195, ISSN 0021-9258

Wallsgrove, R. M., & Mazelis, M. (1980). The enzymology of lysine biosynthesis in higher plants: complete localization of the regulatory enzyme dihydrodipicolinate synthase in the chloroplasts of spinach leaves. *FEBS Lett.*, Vol.116, No.2, (July 2008), pp. 189-192, ISSN 0014-5793

Watanabe, N., Cherney, M. M., van Belkum, M. J., Marcus, S. L., Flegel, M. D., Clay, M. D., Deyholos, M. K., Vederas, J. C., & James, M. N. (2007). Crystal structure of LL-diaminopimelate aminotransferase from *Arabidopsis thaliana*: a recently discovered enzyme in the biosynthesis of L-lysine by plants and Chlamydia. *J. Mol. Biol.*, Vol.371, No.3, (August 2007), pp. 685-702, ISSN 0022-2836

Watanabe, N., Clay, M. D., van Belkum M. J., Cherney, M. M., Vederas, J. C., & James, M. N. (2008). Mechanism of substrate recognition and PLP-induced conformational changes in LL-diaminopimelate aminotransferase from *Arabidopsis thaliana*. *J. Mol. Biol.*, Vol.384, No.5, (December 2008), pp. 1314-1329, ISSN 0022-2836

Watanabe, N., Clay, M. D., van Belkum, M. J., Fan, C., Vederas, J. C., & James, M. N. (2011). The Structure of LL-Diaminopimelate Aminotransferase from *Chlamydia trachomatis*: Implications for Its Broad Substrate Specificity. *J. Mol. Biol.*, Vol.411, No.3, (August 2011), pp. 649-660, ISSN 0022-2836

Watanabe, N., & James, M. N. (2011). Structural insights for the substrate recognition mechanism of LL-diaminopimelate aminotransferase. *Biochim. Biophys. Acta.*, (March 2011), In press, ISSN 0006-3002

Wehrmann, A., Eggeling, L., & Sahm, H. (1994). Analysis of different DNA fragments of Corynebacterium glutamicum complementing dapE of *Escherichia coli*. *Microbiology*, Vol.140, No.12, (December 1994), pp. 3349-3356, ISSN 1350-0872

Weinberger, S., & Gilvarg, C. (1970). Bacterial distribution of the use of succinyl and acetyl blocking groups in diaminopimelic acid biosynthesis. *J. Bacteriol.*, Vol.101, No.1, (January 1970), pp. 323-324, ISSN 0021-9193

Weyand, S., Kefala, G. & Weiss, M. S. (2006). Cloning, expression, purification, crystallization and preliminary X-ray diffraction analysis of DapC (Rv0858c) from *Mycobacterium tuberculosis*. *Acta Crystallogr. Sect. F Struc.t Biol. Cryst. Commun.*, Vol.62, No.8, (August 2006), pp. 794-797, ISSN 1744-3091

Weyand, S., Kefala, G., & Weiss, M. S. (2007). The three-dimensional structure of N-succinyldiaminopimelate aminotransferase from *Mycobacterium tuberculosis*. *J. Mol. Biol.*, Vol.367, No.3, (March 2007), pp. 825-838, ISSN 0022-2836

Weyand, S., Kefala, G., Svergun, D. I., & Weiss, M. S. (2009). The three-dimensional structure of diaminopimelate decarboxylase from *Mycobacterium tuberculosis* reveals a tetrameric enzyme organisation. *J. Struct. Funct. Genomics*, Vol.10, No.3, (September 2009), pp. 209-217, ISSN 1345-711X

Williams, R. M., Fegley, G. J., Gallegos, R., Schaefer, F., & Pruess, D. L. (1996). Asymmetric Syntheses of (2S,3S,6S)-, (2S,3S,6R)-, and (2R,3R,6S)-2,3-Methano-2,6-diaminopimelic Acids. Studies Directed to the Design of Novel Substrate-based Inhibitors of L,L-Diaminopimelate Epimerase. *Tetrahedron*, Vol.52, No.4, (January 1996), pp. 1149-1164, ISSN 0040-4020

Wiseman, J. S., & Nichols, J. S. (1984). Purification and properties of diaminopimelic acid epimerase from *Escherichia coli*. *J. Biol. Chem.*, Vol.259, No.14, (July 1984), pp. 8907-8914, ISSN 0021-9258

Wolterink-van Loo, S., Levisson, M., Cabrières, M.C., Franssen, M.C.R., & van der Oost, J. (2008). Characterization of a thermostable dihydrodipicolinate synthase from *Thermoanaerobacter tengcongensis. Extremophiles*, Vol.12, No.3, (May 2008), pp. 461-469, ISSN 1431-0651

Work, E. (1962). Diaminopimelic racemase. In: *Methods in Enzymology,* Colowick, S. P. & Kaplan, N. O., pp. 858-864, Academic Press, ISBN 0076-6879, New York

Yugari, Y., & Gilvarg C. (1965). The condensation step in diaminopimelate synthesis. *J. Biol. Chem.*, Vol.240, No.12, (December 1965), pp. 4710-4716, ISSN 0021-9258

10

Enzyme-Mediated Preparation of Flavonoid Esters and Their Applications

Jana Viskupicova[1,2], Miroslav Ondrejovic[3,4] and Tibor Maliar[3]
[1]*Institute of Experimental Pharmacology and Toxicology, Slovak Academy of Sciences*
[2]*Department of Biochemistry and Microbiology*
Slovak University of Technology in Bratislava
[3]*Department of Biotechnology, University of SS. Cyril and Methodius in Trnava*
[4]*Department of Biocentrum, Food Research Institute in Bratislava*
Slovakia

1. Introduction

Flavonoids comprise a group of plant polyphenols with a broad spectrum of biological activities. They have been shown to exert beneficial effects on human health and play an important role in prevention and/or treatment of several serious diseases, such as cancer, inflammation and cardiovascular disease (Middleton et al., 2000; Rice-Evans, 2001). Flavonoids are important beneficial components of food, pharmaceuticals, cosmetics and various commodity preparations due to their antimutagenic, hepatoprotective (Stefani et al., 1999), antiallergic (Berg & Daniel, 1988), antiviral (Middleton & Chithan, 1993) and antibacterial activity (Tarle & Dvorzak, 1990; Tereschuk et al., 1997; Singh & Nath, 1999; Quarenghi et al., 2000; Rauha et al., 2000). They are known to inhibit nucleic acid synthesis (Plaper et al., 2003; Cushnie & Lamb, 2006), cause disturbance in membranes (Stepanovic et al., 2003; Stapleton et al., 2004; Cushnie & Lamb, 2005) and affect energy metabolism (Haraguchi et al., 1998). But the most studied activity is their antioxidant action since they can readily eliminate reactive oxygen and nitrogen species or degradation products of lipid peroxidation and are thus effective inhibitors of oxidation (Ross & Kasum, 2002).

However, their commercial applications are limited due to low solubility in lipophilic environment and low availability for a living organism. Although aglycons, prenylated and methoxylated flavonoid derivatives may be implemented into such systems, they are rarely found in nature and are often unstable. In some plant species, the last step in the flavonoid biosynthesis is terminated by acylation which is known to increase solubility and stability of glycosylated flavonoids in lipophilic systems. Selectively acylated flavonoids with different aliphatic or aromatic acids may not only improve physicochemical properties of these molecules (Ishihara & Nakajima, 2003) but also introduce various beneficial properqties to the maternal compound. These include penetration through the cell membrane (Suda et al., 2002; Kodelia et al., 1994) enhanced antioxidant activity (Viskupicova et al., 2010; Katsoura et al., 2006; Mellou et al., 2005), antimicrobial (Mellou et al., 2005), anti-proliferative (Mellou et al., 2006) and cytogenic (Kodelia et al., 1994) effect and improvement of thermostability and light-resistivity of certain flavonoids.

In nature, flavonoid acylation is catalyzed by various acyltransferases which are responsible for the transfer of aromatic or aliphatic acyl groups from a CoA-donor molecule to hydroxyl residues of flavonoid sugar moieties (Davies & Schwinn, 2006). Acylation is widespread especially among anthocyanins; more than 65% are reported to be acylated (Andersen & Jordheim, 2006). While the exact role of plant acylation is not yet fully understood, it is known that these modifications modulate the physiological activity of the resulting flavonoid ester by altering solubility, stability, reactivity and interaction with cellular targets (Ferrer et al., 2008). Acylation might be a prerequisite molecular tag for efficient vacuolar uptake of flavonoids (Kitamura, 2006; Nakayama et al., 2003). Some acylated flavonoids have been found to be involved in plant-insect interactions; they act as phytoalexins, oviposition stimulants, pollinator attractants (Iwashina, 2003), and insect antifeedants (Harborne & Williams, 1998). With respect to novel biological activities, acylation of flavonoids can result in changes in pigmentation (Bloor, 2001), insect antifeedant activity (Harborne & Williams, 1998) and antioxidant properties (Alluis & Dangles, 1999).

Over the past 15 years, there has been a substantial effort to take advantage of this naturally occurring phenomenon and to implement acylation methods into laboratories. However, the use of acyltransferases as modifying agents is rather inconvenient, as they require corresponding acylcoenzyme A, which must be either in stoichiometric amounts or regenerated *in situ*. Natural acyltransferases and cell extracts from *Ipomoea batatas* and *Perilla frutescens* containing acyltransferases were applied for selective flavonoid modification with aromatic acids (Tab.1) (Nakajima et al., 2000; Fujiwara et al., 1998).

Acyltransferase	**Plant source**	**References**
hydroxycinnamoyl-CoA:anthocyanin 3-*O*-glucosid-6''-*O*-acyltransferase	*Perilla frutescens*	Yonekura-Sakakibara et al., 2000
malonyl-CoA:anthocyanin 3-*O*-glucosid-6''-*O*-malonyltransferase	*Dahlia variabilis*	Wimmer et al., 1998
hydroxycinnamoyl-CoA:anthocyanin 5-*O*-glucosid-6''-*O*-acyltransferase	*Gentiana triflora*	Tanaka et al., 1996
hydroxycinnamoyl-CoA:anthocyanidin 3-rutinosid acyltransferase	*Petunia hybrida*	Brugliera & Koes, 2003
malonyl-CoA:anthocyanidin 5-*O*-glucosid-6''-*O*-malonyltransferase	*Salvia splendens*	Suzuki et al., 2001

Table 1. Acyltransferase catalysis of flavonoid acylation and their nature sources.

To solve this problem, the chemical approach was first investigated. It possessed a low degree of regioselectivity of esterification and drastic reaction conditions had to be applied (Patti et al., 2000). Later on, hydrolytic enzymes (lipases, esterases and proteases) have been recognized as useful agents due to their large availability, low cost, chemo-, regio- and enantioselectivity, mild condition processing and no need of cofactors (Collins & Kennedy, 1999; Nagasawa & Yamada, 1995).

Since the enzymatic preparation of flavonoid derivatives is a matter of several years, commercial applications have just been emerging. There are several patented inventions available to date, oriented on the flavonoid ester production and their use for the manufacture of pharmaceutical, dermopharmaceutical, cosmetic, nutritional or agri-foodstuff compositions

(Fukami et al., 2007; Moussou et al., 2004, 2007; Ghoul et al., 2006; Bok et al., 2001; Perrier et al., 2001; Otto et al., 2001; Nicolosi et al., 1999; Sakai et al., 1994).

This review presents available information on enzyme-mediated flavonoid acylation *in vitro*, emphasizing reaction parameters which influence performance and regioselectivity of the enzymatic reaction. In the second part, the paper focuses on biological effects of synthesized flavonoid esters as well as of those isolated from nature. Finally, the paper ends with application prospects of acylated flavonoids in the food, pharmaceutical and cosmetic industry.

2. Flavonoid esterification

Presently, the enzyme-catalyzed flavonoid esterification in organic media is a well-mastered technique for synthesis of selectively modified flavonoids. Results in this field suggest that a high degree of conversion to desired esters can be achieved when optimal reaction conditions are applied. The key factors, which influence regioselectivity and the performance of the enzymatic acylation of flavonoids, include type and concentration of enzyme, structure and concentration of the substrates (acyl donor, acyl acceptor and their ratio), nature of the reaction media, water content in the media, reaction temperature and nature of the reaction as reviewed in Chebil et al., 2006, 2007.

2.1 Enzymes

To date, the use of proteases, esterases, acyltransferases and lipases has been investigated in order to find the most potent biocatalyst for selective flavonoid acylation. These enzymes are often in the immobilized form which improves enzyme stability, facilitates product isolation, and enables enzyme reuse (Adamczak & Krishna, 2004).

2.1.1 Proteases

Proteases represent a class of enzymes which occupy a pivotal position with respect to their physiological roles as well as their commercial applications. They represent the first group of hydrolytic enzymes used for flavonoid modification. They perform both hydrolytic and synthetic functions. Since they are physiologically necessary for living organisms, proteases occur ubiquitously in diverse sources, such as plants, animals, and microorganisms. They are also classified as serine proteases, aspartic proteases, cysteine proteases, threonine proteases and metalloproteases, depending on the nature of the functional group at the active site.

Proteases have a large variety of applications, mainly in the detergent and food industries. In view of the recent trend of developing environmentally friendly technologies, proteases are envisaged to have extensive applications in leather treatment and in several bioremediation processes. Proteases are also extensively used in the pharmaceutical industry (Rao et al., 1998). Protease subtilisin was the first enzyme used for flavonoid ester synthesis conducted by Danieli et al. (1989, 1990). Later on, subtilisin was used for selective rutin acylation in organic solvents (Xiao et al., 2005; Kodelia et al., 1994). However, it has been reported that reactions catalyzed by subtilisin led to low conversion yields and a low degree of regioselectivity was observed (Danieli et al., 1990). These authors reported that the structure of the sugar moiety affected the regioselectivity. For flavonoid acylation, especially serine proteases (subtilisin) have been used in ester synthesis (Danieli et al., 1989, 1990; Kodelia et al., 1994).

2.1.2 Esterases

Esterases (carboxyl esterases, EC 3.1.1.1) represent a diverse group of hydrolases catalyzing the cleavage and formation of ester bonds with wide distribution in animals, plants and microorganisms. A classification scheme for esterases is based on the specificity of the enzymes for the acid moiety of the substrate, such as the carboxyl esterases, aryl esterases, acetyl esterases, cholin esterases, cholesterol esterases, etc. (Jeager et al., 1999). Esterases show high regio- and stereospecificity, which makes them attractive biocatalysts for the production of optically pure compounds in fine-chemicals synthesis (reviewed in Bornscheuer, 2002).

They have the same reaction mechanism as lipases, but differ from them by their substrate specificity, since they prefer short-chain fatty acids, whereas lipases usually prefer long-chain fatty acids. Another difference lies in the interfacial activation (Hidalgo & Bornscheuer, 2006). In contrast to lipases, only a few esterases have commercial applications in organic synthesis because lipases are generally more entantioselective and resistant to organic solvents. The most widely used esterase is the preparation isolated from pig liver (Hidalgo & Bornscheuer, 2006). The practical applications of esterases in enzymatic transformation of flavonoids are not very attractive as it enables the implementation only of the molecule of a short aliphatic chain length, such as acetate, propionate and butyrate (Sakai et al., 1994).

2.1.3 Lipases

Today lipases stand amongst the most important biocatalysts in industry. Among them, microbial lipases find the biggest application use. They can be classified according to sequence alignment into three major groups: mammalian lipases (e.g. porcine pancreatic lipase), fungal lipases (*Candida rugosa* and *Rhizomucor* family) and bacterial lipases (*Staphylococcus* and *Pseudomonas* family) (Hidalgo & Bornscheuer, 2006). More than 50% of the reported lipases are produced by yeast in the forms of various isozymes (Vakhlu & Kour, 2006).

Lipases (triacylglycerol acylhydrolases, EC 3.1.1.3) belong to the class of serine hydrolases. They catalyze a wide range of reactions, including hydrolysis, inter-esterification, alcoholysis, acidolysis, esterification and aminolysis (Vakhlu & Kour, 2006). Under natural conditions, they catalyze the hydrolysis of ester bonds at the hydrophilic-hydrophobic interface. At this interface, lipases exhibit a phenomenon termed interfacial activation, which causes a remarkable increase in activity upon contact with a hydrophobic surface. The catalytic process involves a series of differentiated stages: contact with the interface, conformational change, penetration in the interface, and finally the catalysis itself (Hidalgo & Bornscheuer, 2006). Under certain experimental conditions, such as in the absence of water, they are capable of reversing the reaction. The reverse reaction leads to esterification and formation of glycerides from fatty acids and glycerol (Saxena et al., 1999). This synthetic activity of lipases is being successfully utilized also in flavonoid ester production.

Candida antarctica lipase B (CALB) is one of the most widely used biocatalysts in organic synthesis on both the laboratory and the commercial scale (Anderson et al., 1998; Uppenberg et al., 1995) due to its ability to accept a wide range of substrates, its non-aqueous medium tolerance and thermal deactivation resistance (Degn et al., 1999; Anderson et al., 1998; Cordova et al., 1998; Drouin et al., 1997). CALB belongs to the α/β hydrolase-fold superfamily with a conserved catalytic triad consisting of Ser105-His224-Asp187 (Uppenberg et al., 1995). It comprises 317 amino acid residues. The active site contains an oxyanion hole which stabilizes the transition state and the oxyanion in the reaction intermediate (Haeffner et al., 1998). Reaction mechanism of CALB follows the bi-bi ping-pong mechanism, illustrated in Fig.1 (Kwon et al., 2007). The substrate molecule reacts with serine of the active site forming a tetrahedral intermediate which is stabilized by catalytic residues of His and Asp. In the next step alcohol is released and the acyl-enzyme complex is created. A nucleophilic attack (water in hydrolysis, alcohol in transesterification) causes another tetrahydral intermediate formation. In the last step, the intermediate is split into product and enzyme and is recovered for the next catalytic cycle (Patel, 2006).

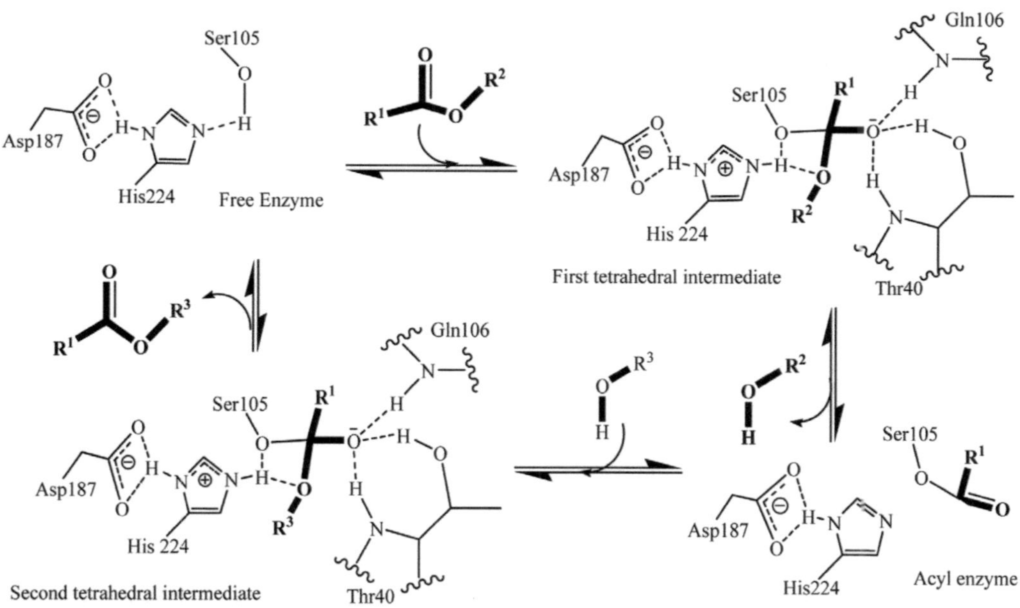

Fig. 1. Reaction mechanism catalyzed by *Candida antarctica* lipase (Kwon et al., 2007).

The active site of CALB consists of a substrate-nonspecific acyl-binding site and a substrate specific alcohol-binding site (Cygler & Schrag, 1997; Uppenberg et al., 1995). It is selective for secondary alcohols (Uppenberg et al., 1995), as reflected by the geometry of the alcohol-binding site (Lutz, 2004). In contrast to most lipases, CALB has no lid covering the entrance to the active site and shows no interfacial activation (Martinelle et al., 1995). CALB is being frequently used in acylation of various natural compounds such as saccharides, steroids and natural glycosides, including flavonoids (Riva, 2002; Davis & Boyer, 2001). The proper enzyme selection plays multiple roles in flavonoid acylation. The biocatalyst significantly influences the regioselectivity of the reaction. Information is available mainly on the use of lipases for flavonoid ester synthesis; especially the use of lipase B from *Candida antarctica*, which is preferred due to its acceptance of a wide range of substrates, good catalytic activity and a high degree of regioselectivity (Viskupicova et al., 2010; Katsoura et al., 2006, 2007; Ghoul et al., 2006; Mellou et al., 2005, 2006; Stevenson et al., 2006; Ardhaoui et al., 2004a, 2004b, 2004c; Passicos et al., 2004; Moussou et al., 2004; Gayot et al., 2003; Ishihara & Nakajima, 2003; Ishihara et al., 2002; Kontogianni et al., 2001, 2003; Nakajima et al., 1999, 2003; Gao et al., 2001; Otto et al., 2001; Danieli et al., 1997).

As for flavonoid aglycons, only two enzymes have been reported to be capable of acylating this skeleton - lipase from *Pseudomonas cepacia* and carboxyl esterase. Lambusta et al. (1993) investigated the use of *P. cepacia* lipase for catechin modification. They discovered that the acylation took place on the C5 and C7 hydroxyls. Sakai et al. (1994) observed that carboxyl esterase showed regioselectivity towards C3-OH of catechin. Sakai et al. (1994) explored the use of carboxyl esterase from *Streptomyces rochei* and *Aspergillus niger* for the 3-*O*-acylated catechin production.

2.2 Reaction conditions

The performance and regioselectivity of the enzyme-catalyzed flavonoid transformation is affected by several factors, including the type of enzyme, the nature of medium, reaction conditions, water content in the media, structure and concentration of substrates and their molar ratio. By varying these factors, significant changes in ester production and regioselectivity can be achieved.

2.2.1 Reaction media

Reaction media play an important role in enzymatic transformations. Methodologies for enzymatic flavonoid acylation have focused on searching a reaction medium which allows appropriate solubility of polar acyl acceptor (flavonoid glycoside) and nonpolar acyl donor as well as the highest possible enzymatic activity. Moreover, the medium has often been required to be nontoxic and harmless to biocatalyst. In order to meet the above-mentioned requirements, several scientific teams have dealt with proper medium selection (Viskupicova et al., 2006; Mellou et al., 2005; Kontogianni et al., 2001, 2003; Gao et al., 2001; Nakajima et al., 1999; Danieli et al., 1997).

Non-aqueous biocatalysis has several advantages over conventional aqueous catalysis: the suppression of hydrolytic activity of the biocatalyst which is carried out in water (Fossati & Riva, 2006), the enhanced solubility of hydrophobic substrates, the improvement of enzyme enantioselectivity, the exclusion of unwanted side reactions, the easy removal of some products, the enhanced enzyme thermostability and the elimination of microbial contamination (Rubin-Pitel & Zhao, 2006; Torres & Castro, 2004). Laane (1987) pointed out that log P, as a solvent parameter, correlated best with enzyme activity. Zaks & Klibanov (1988) reported that the activity of lipases was higher in hydrophobic solvents than in hydrophilic ones. Narayan & Klibanov (1993) claimed that it was hydrophobicity and not polarity or water miscibility which was important, whereas the log P parameter could be called a measure of solvent hydrophobicity. Trodler & Pleiss (2008), using multiple molecular dynamics simulations, showed that the structure of CALB possessed a high stability in solvents. In contrast to structure, flexibility is solvent-dependent; a lower dielectric constant led to decreased protein flexibility. This reduced flexibility of CALB in non-polar solvents is not only a consequence of the interaction between organic solvent molecules and the protein, but it is also due to the interaction with the enzyme-bound water and its exchange on the surface (Trodler & Pleiss, 2008). In organic solvents, the surface area has been suggested to be reduced, leading to improved packing and increased stability of the enzyme (Toba & Merz, 1997).

Polar aprotic solvents such as dimethyl sulfoxid (DMSO), dimethylformamide (DMF), tetrahydrofuran (THF) and pyridine were first investigated (Nakajima et al., 1999; Danieli et al., 1997). However, it was observed that enzyme activity was readily deactivated in these solvents. To date enzymatic acylation of flavonoids has been successfully carried out in various organic solvents (Tab.2), while the most frequently used are 2-methylbutan-2-ol and acetone because of their low toxicity, their polarity allowing proper solubilization of substrates and high conversion yields.

Solvent	Reference
2-Methylbutan-2-ol	Ghoul et al., 2006; Ardhaoui et al., 2004a, 2004b, 2004c; Passicos et al., 2004; Gayot et al., 2003
Acetone	Ghoul et al., 2006; Mellou et al., 2005, 2006; Kontogianni et al., 2001, 2003; Ishihara et al., 2002, Ishihara & Nakajima, 2003; Nakajima et al., 1999, 2003; Danieli et al., 1997
Acetonitrile	Ghoul et al., 2006; Ishihara & Nakajima, 2003; Nakajima et al., 1997, 1999
2-Methylpropan-2-ol	Ghoul et al., 2006; Stevenson et al., 2006; Mellou et al., 2005; Moussou et al., 2004; Kontogianni et al., 2001, 2003; Otto et al., 2001
Dioxane	Ghoul et al., 2006; Danieli et al., 1997
Pyridine	Danieli et al., 1990, 1997
THF, DMSO, DMF	Kontogianni et al., 2001, 2003; Danieli et al., 1997
Binaric mixtures of solvents	Ghoul et al., 2006; Gao et al., 2001; Nakajima et al., 1999; Danieli et al., 1997

Table 2. Organic solvents used in flavonoid acylation.

The effect of the solvent on conversion yield depends on the nature of both the acyl donor and the flavonoid (Chebil et al., 2006). Although much has been done in this area, it is quite difficult to deduce any general conclusion on solvent choice because the available data are controversial and sometimes even contrary.

Recently, ionic liquids have received growing attention as an alternative to organic solvents used for the enzymatic transformation of various compounds (Katsoura et al., 2006; Kragl et al., 2006; Jain et al., 2005; Lozano et al., 2004; Reetz et al., 2003; Van Rantwick et al., 2003). The potential of these "green solvents" lies in their unique physicochemical properties, such as non-volatility, nonflammability, thermal stability and good solubility for many polar and less polar organic compounds (Jain et al., 2005; Wilkes, 2004; Itoh et al., 2003; Van Rantwick et al., 2003). Probably the most promising advantage of the use of ionic liquids is their potential application in food, pharmaceutical and cosmetic preparations due to their reduced toxicity (Jarstoff et al., 2003). Due to the many above-mentioned advantages of ionic liquids for enzyme-mediated transformations, several flavonoid esters have been recently prepared in such media (Katsoura et al., 2006, 2007; Kragl et al., 2006). The biocatalytic process showed significantly higher reaction rates, regioselectivity and yield conversions compared to those achieved in organic solvents. Thus ionic liquid use seems to be a challenging approach to conventional solvent catalysis.

The solvent-free approach for elimination of the co-solvent of the reaction has been recently introduced as an alternative for conventional solvents (Enaud et al., 2004; Kontogianni et al., 2001, 2003). It is based on the use of one reactant in the role of the solvent. The authors reported rapid reaction rates; however, the conversion yields were slightly decreased. In spite of the attractiveness, the use of solvent-free systems is characterized by a serious drawback due to the necessity to eliminate the excess of the acyl donor for the recovery of the synthesized products (Chebil et al., 2006).

2.2.2 Water content

Water content in reaction media is a crucial parameter in lipase-catalyzed synthesis as it alters the thermodynamic equilibrium of the reaction towards hydrolysis or synthesis. Moreover, it is involved in noncovalent interactions which keep the right conformation of an enzyme catalytic site (Foresti et al., 2007). The amount of water required for the catalytic process depends on the enzyme, its form (native or immobilized), the enzyme support, and on the solvent nature (Arroyo et al., 1999; Zaks & Klibanov, 1988). The influence of water content in the reaction system on enzyme activity is variable with various enzymes (lipase from *Rhizomucor miehei, Rhizomucor niveus, Humicola lanuginosa, Candida rugosa, Pseudomonas cepacia*).

In general, the water amount which is considered to be optimal for esterifications in organic solvents is 0.2 – 3% (Rocha et al., 1999; Yadav & Piyush, 2003; Iso et al., 2001). The enzymatic esterification of flavonoids in non-aqueous media is greatly influenced by the water content of the reaction system (Ardhaoui et al., 2004b; Gayot et al., 2003; Kontogianni et al., 2003). Ardhaoui et al. (2004b) observed the best enzyme activity when water content was maintained at 200 ppm. Gayot et al. (2003) found that the optimal value of water in an organic reaction medium equaled 0.05% (v/v). Kontogianni et al. (2003) reported that highest flavonoid conversion was reached when initial water activity was 0.11 or less.

2.2.3 Temperature

Temperature represents a significant physical factor in enzyme-catalyzed reactions. It affects viscosity of the reaction medium, enzyme stability, and substrate and product solubility.

Since lipase from *C. antarctica* belongs to thermostable enzymes, improved catalytic activity was observed at higher temperatures (Arroyo et al., 1999). To date, flavonoid transformation has been carried out in the temperature range 30 – 100°C (Ghoul et al., 2006; Katsoura et al., 2006; Stevenson et al., 2006; Mellou et al., 2005; Ardhaoui et al., 2004a, 2004b, 2004c; Moussou et al., 2004; Passicos et al., 2004; Enaud et al., 2004; Gayot et al., 2003; Kontogianni et al., 2003; Ishihara et al., 2002; Gao et al., 2001; Otto et al., 2001; Nakajima et al., 1999; Danieli et al., 1990). The choice of temperature depends on the enzyme and solvent used. The majority of authors performed flavonoid acylation at 60°C due to the best enzyme activity, good solubility of substrates and highest yields of resulting esters reached (Viskupicova et al., 2006, 2010; Ghoul et al., 2006; Katsoura et al., 2006; Stevenson et al., 2006;

Ardhaoui et al., 2004a, 2004b, 2004c; Moussou et al., 2004; Passicos et al., 2004; Enaud et al., 2004; Gayot et al., 2003; Otto et al., 2001). Our results on the effect of temperature on naringin conversion are presented in Fig.2 and are in accordance with other authors (Viskupicova et al., 2006).

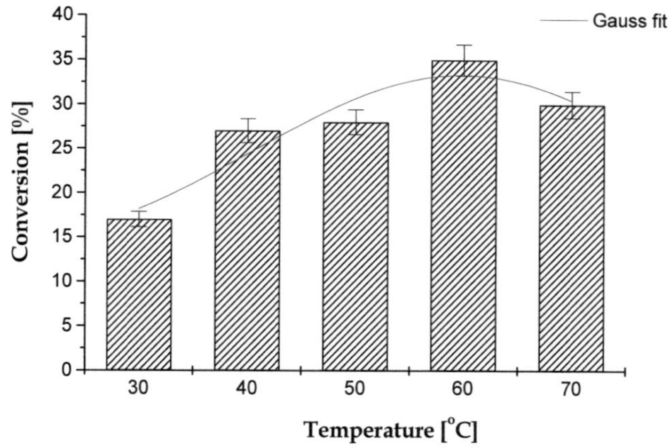

Fig. 2. Effect of temperature on naringin conversion to naringinpalmitate in 2-methylbutan-2-ol catalyzed by *C. antarctica* lipase after 24 h (Viskupicova et al., 2006).

2.3 Acyl donors and acceptors

2.3.1 Acyl donor

Since lipase-catalyzed acylation takes place through the formation of an acyl-enzyme intermediate, the nature of the acyl donor has a notable effect on reactivity. The ideal acyl donor should be inexpensive, fast acylating, and completely non-reactive in the absence of the enzyme (Ballesteros et al., 2006). Many acylating agents have been tested in flavonoid esterification, such as aromatic or aliphatic organic acids, substituted or not (Tab.3). Special attention was attributed to fatty acid ester production (Katsoura et al., 2006; Mellou et al., 2005, 2006; Ardhaoui et al., 2004a, 2004b, 2004c; Enaud et al., 2004; Gayot et al., 2003; Kontogianni et al., 2003). This approach enables to improve flavonoid solubility and stability in lipophilic systems. The proper acyl donor selection may significantly influence not only the physicochemical but also biological properties of the resulting esters.

A simple way to increase the reaction rate and conversion yield in acylation is to use an excess of acyl donor (Patti et al., 2000). Many authors have tried to determine the optimal molar ratio of flavonoid/acyl donor in order to achieve the highest possible yields. The molar ratios 1:1 to 1:15 (acyl acceptor/acyl donor) have been investigated, whereas the majority agreed on the ratio 1:5 to be the most suitable for the best reaction performance (Mellou et al., 2006; Gayot et al., 2003; Ishihara & Nakajima, 2003; Ishihara et al., 2002; Kontogianni et al., 2001). A better solution is offered by the use of special acyl donors which ensure a more or less irreversible reaction. This can be achieved by the introduction of electron-withdrawing substituents (esters), resulting in higher conversion yields and reaction rates. The use of vinyl esters allows a several times faster reaction progress than do other activated esters (Ballesteros et al., 2006). Enzymatic synthesis of flavonoid esters can be realized by two basic approaches, i.e. esterification and transesterification (Fig.3).

Aliphatic acids	Aromatic acids
Acetic*	Caffeic*
Malic*	p-Coumaric*
Malonic*	Ferulic*
Succinic*	Gallic*
Tartaric*	p-Hydroxybenzoic*
Butyric	Sinapic*
Crotonic	Benzoic
n-Butanoic	Cinnamic
Isobutyric	Isoferulic
Isovaleric	Methylsinapic
Lactic	
3-methylbutyric	
Quinic	
Vinylpropionic	
Tiglic	

*acyl donors found in anthocyanins

Table 3. Acyl donors found in flavonols, flavones (Williams, 2006) and anthocyanins (Andersen & Jordheim, 2006).

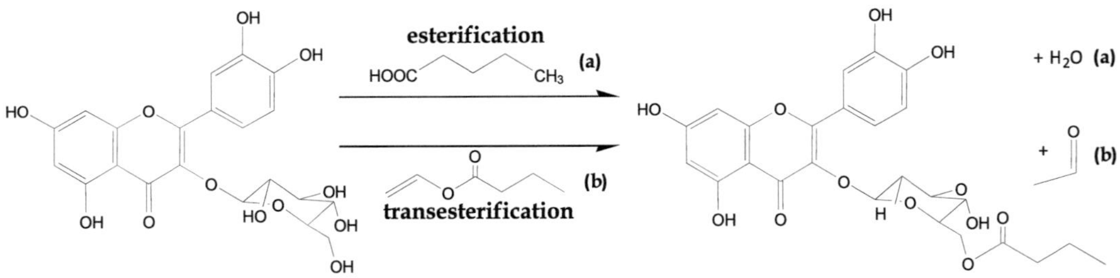

Fig. 3. Mechanism of isoquercitrin esterification and transesterification (Chebil et al., 2006).

Pleiss et al. (1998) studied the acyl binding site of CALB and found the enzyme to be selective for short and medium fatty acid chain length. This fact may be attributed to the structure of the lipase acyl binding pocket, which is an elliptical, narrow cleft of 9.5 × 4.5 Å. With increasing carbon number of a fatty acid or molecule size, the steric hindrance is involved resulting in low efficiency of the enzymatic reaction (Riva et al., 1988; Wang et al., 1988; Carrea et al., 1989). This fact was experimentally confirmed by Katsoura et al. (2006) and by Viskupicova & Ondrejovic (2007) whose results showed higher performance of the naringin and rutin esterification when fatty acids up to C10 were introduced. On the other hand, Ardhaoui et al. (2004b) and Kontogianni et al. (2003) reported that the fatty acid chain length had no significant effect on conversion yield when fatty acids of a medium and high chain length were used.

Thus, the effect of fatty acid chain length on flavonoid acylation still remains a matter of discussion. Our team conducted a series of experiments with both saturated and unsaturated fatty acids and found a correlation between log P of the acids tested and conversion yields (Viskupicova et al., 2010). It would be interesting to take this parameter into consideration when assessing the influence of an acyl donor on the reaction progress.

Only little progress has been achieved in flavonoid esterification with aromatic acids (Stevenson et al., 2006; Enaud et al., 2004; Gao et al., 2001; Nakajima et al., 2000). It has been observed that the performance of the process depends mainly on the nature of the substitutions, the position of the hydroxyls and the length of the spacers.

2.3.2 Acyl acceptor

The structure of acyl acceptor (flavonoid), especially stereochemistry of glycosidic bonds, plays an important role in flavonoid acylation. The structural differences, such as the number and position of hydroxyl groups, the nature of saccharidic moiety, as well as the position of glycosidic bonds, influence the flavonoid solubility, and thus affect the overall conversion yield.

Available studies are concerned mainly with acylation on flavonoid glycosides. Among polyphenolic compounds, naringin and rutin are the most widely used substrates. For the naringin molecule, which possesses a primary hydroxyl group on glucose, the acylation takes place on the 6''-OH (Katsoura et al., 2006; Konntogianni et al., 2001, 2003; Ishihara et al., 2002; Gao et al., 2001; Otto et al., 2001; Danieli et al., 1990) since the primary hydroxyl is favored by CALB (Fig.4). However, in rutin, which has no primary hydroxyl available, either the 3''-OH of glucose (Ishihara et al., 2002; Danieli & Riva, 1994) or the 4'''-OH of rhamnose (Fig.4) (Viskupicova et al., 2010; Mellou et al., 2006; Ardhaoui et al., 2004a, 2004b, 2004c) can be acylated. Danieli et al. (1997) observed the rutin-3'',4'''-*O*-diester formation. When subtilisin was used as biocatalyst, naringin-3''-*O*-ester and rutin-3''-*O*-ester were synthesized (Danieli et al., 1990).

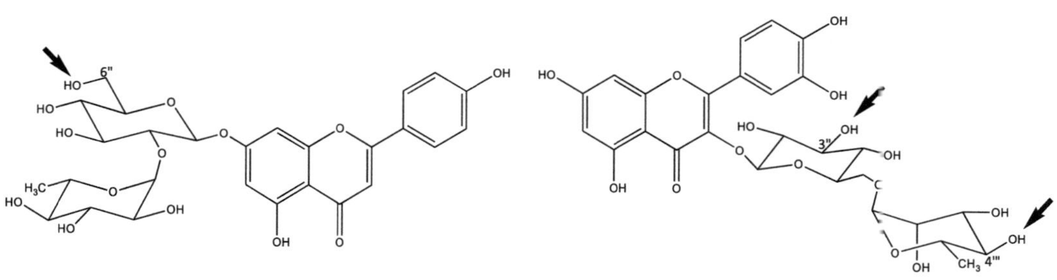

Fig. 4. Acylation sites of naringin (left) and rutin (right) molecule.

The concentration of the flavonoid also affects the performance of the acylation reaction. The conversion yield and the initial rate rise with increasing flavonoid concentration. However, the amount of flavonoid is limited by its solubility in a reaction medium (Chebil et al., 2006, 2007).

3. Influence of flavonoid derivatization on biological activities

3.1 Esters with aromatic acids

Aromatic acids, along with flavonoids, belong to the group of phenols of secondary metabolism of living organisms. The described secondary metabolites represent a store of biologically active compounds, displaying various biological activities. We can therefore assume that physicochemical and biological properties of the initial flavonoids may be improved by acylation of flavonoids with aromatic acids. However, by this reaction a new compound can also gain novel activities provided by the aromatic acids.

Flavonoid acylation with aromatic acids was reported to improve physiological activities, such as UV-absorbing capacity, radical scavenging ability (Delazar et al., 2005; Ishihara & Nakajima, 2003; Harborne & Williams, 2000; Alluis & Dangles 1999; Jungblut et al., 1995) pigment stabilization (especially anthocyanins) (Ishihara & Nakajima 2003), and interaction with cellular targets (Ferrer et al., 2008).

Flavonoid esters acylated with *p*-coumaric acid were found to increase antioxidant (Pajero et al., 2005) and anti-inflammatory activities (Harborne & Williams, 2000), as well as antiproliferative and cytotoxic effects on various cancer cell lines (Mitrokotsa et al., 1993). Moreover, *p*-coumaroyl esters of quercetin and kaempferol were reported to have positive effects on cerebrovascular disorders (Calis et al., 1995). Similarly, flavonoid esters esterified with cinnamic acid were shown to exhibit antiproliferative activity against several human cancer cell lines (Duarte-Almeida et al., 2007). Flavonoid acylation with caffeic acid contributes to the enhancement of antioxidant properties (Pajero et al., 2005). Flavonolignans acylated with truxinic acid were shown to possess hepatoprotective as well as anticancer activity (Sharma et al., 2003).

3.2 Esters with aliphatic acids

Biological activities of aliphatic acids are not of a big importance in comparison with aromatic acids. These compounds are mainly accepted as energy storage and components of several compartments of cells, such as membranes, enzymes, surfactants, etc. In the literature, more studies can be found describing changes in biological activities of flavonoids after their acylation with aliphatic acids.

The aliphatic acylation of anthocyanins with malonic acid is important for enhancing the pigment solubility in water, protecting glycosides from enzymatic degradation and stabilizing anthocyanin structures (Nakayama et al., 2003). Several *in vitro* observations suggest that acylation with malonic acid or sinapic acid is crucial for efficient flavonoid accumulation in plants.

Fatty acid esters of catechins were reported to display antitumor, antibacterial and 5-α reductase inhibiting activity (Fukami et al., 2007) as well as antioxidant properties (Sakai et al., 1994). Lee et al. (2003) reported anti-atherogenic activity of two naringenin derivatives, 7-*O*-oleic ester and 7-*O*-cetyl ether.

Acylation of the flavonoid molecule with polyunsaturated fatty acids introduces potential antitumor and antiangiogenic properties (Mellou et al., 2006). Anticarcinogenic effects were observed also in silybin esters acylated with butyric and lauric acid (Xanthakis et al., 2010). Recently, we found that acylation of rutin with unsaturated fatty acids, such as oleic, α-linoleic and linolenic, increased the antioxidant potential of the initial compound (Viskupicova et al., 2010). This observation is in accordance with the results of Mellou et al. (2006) and Katsoura et al. (2006).

In the field of fatty acid ester synthesis, information on the photoprotective effectiveness of new quercetin derivatives acylated with acetic, propionic and palmitic acids, has been reported. The authors found that esterification with a short side-chain (such as acetate or propionate) may improve migration through the aqueous environment and interaction with or penetration into phospholipid membranes (Saija et al., 2003).

Recent experimental findings indicate that acylation of flavonoid may increase enzyme inhibitory activity. Lin et al. (2010) observed increased 5α-reductase inhibition after acylation of (-)-epigallocatechins. Salem et al. (2011) showed that the acylation of isorhamnetin-3-*O*-glucoside with different aliphatic acids enhanced its capacity to inhibit xanthine oxidase. Our recent investigations showed that lipophilic rutin and naringin esters were strong inhibitors of transport enzymes such as sarcoplasmic reticulum Ca^{2+}-ATPase and plasma membrane Ca^{2+}-ATPase (Augustyniak et al., 2010; Viskupicova et al., 2009), and thus might be useful in calcium regulation. We presume that there might be a general mechanism involved in the enhanced inhibitory activity of the acylated flavonoids on structurally diverse classes of enzymes which seems to be donated by the medium to long fatty acid chains.

4. Application perspectives

The following section provides a summary of patented inventions available in the commercial sphere. These include practical applications in food, pharmaceuticals and cosmetics.

4.1 Food

The major contribution of acylated flavonoids in the food industry lies in the improvement of stability and solubility of initial molecules, e.g. by reducing lipid oxidation in oil/fat based food systems, desirable modification of unwanted sensory properties of certain flavonoids, taking advantage of pigment stabilization by the means of flavonoid acylation, or other food characteristics. Furthermore, selectively acylated flavonoids may cause significant changes in their bioavailability and bioactivity, and when consumed, may thus play a role in preventing diseases.

Flavonoid acylation is a useful tool for modification of sensory properties of food. While flavonoids provide a variety of health benefits, flavonoid-containing food often suffers from bitter and astringent taste. Degenhardt et al. (2007) found that certain glycosylation and acylation patterns can effectively modulate these negative taste factors in edible preparations, pharmaceutical preparations and cosmetics with mouth contact (i.e. tooth paste, mouth wash). Both the taste intensity and the taste profile perception are improved by the novel compounds. Ghoul et al. (2006) introduced a process for the selective preparation of acylated flavonoid glycosides with improved stability and solubility in various preparations with their antioxidant effect remaining intact or being improved.

Another particular advantage obtained by these modified flavonoids is the bifunctional character of their molecule with higher biological activity. Free unsaturated fatty acids represent a potential risk because they are highly reactive and by creating free radicals they cause undesirable damage in food. Enzymatic synthesis of flavonoids with unsaturated fatty acids was found to be a useful solution for the stabilization of these highly oxidizable acids (Viskupicova et al., 2010; Mellou et al., 2006).

Another important benefit of acylated anthocyanins lies in the use as food colorants which can serve as a useful alternative to synthetic additives (Giusti & Wrolstad, 2003; Fox, 2000; Asen et al., 1979). The discovery of acylated anthocyanins with increased stability has shown that these pigments may provide food products with the desirable color and stability at a wide pH range. Examples of suitable acylated anthocyanin sources may be radishes, red potatoes, red cabbage, black carrots, and purple sweet potatoes (reviewed in Giusti & Wrolstad, 2003). The invention of Asen et al. (1979) refers to a stable food colorant from a natural source. It relates to an anthocyanin isolated from the Heavenly Blue Morning Glory (*Ipomoea tricolor* Cav cv), peonidin 3-(dicaffeylsophoroside)-5-glucoside, which is characterized by the stability of colors ranging from purplish-red to blue produced in food and beverage products at pH values from about 2.0 to about 8.0. Fox (2000) reported the invention referring to a stable, ruby red natural colorant (anthocyanins acylated with chlorogenic acid) derived from purple sunflower hulls, useful as a coloring agent in food products, cosmetics, pharmaceuticals and other materials.

4.2 Pharmaceuticals

In recent years, coronary artery diseases, such as atherosclerosis and hypercholesterolemia, represent a major cause of death, exceeding even oncological causes or infectious diseases. Novel acylated flavanone derivatives are effective in the treatment or prevention of elevated blood lipid level-related diseases, e.g. hyperlipidemia, arteriosclerosis, angina pectoris, stroke and hepatic diseases since they exert inhibitory effects on acylcolicholesterol acyl transferase activity and HMG-CoA reductase activity. In spite of their potent efficacies, the flavanone derivatives exhibited no toxicity or mitogenicity in tests using mice (Bok et al., 2001).

Mellou et al. (2005) carried out enzymatic acylation on Greek endemic plants and reported that this modification increased both their antioxidant activity towards isolated low-density lipoproteins (LDL) and serum model and antimicrobial activity against two Gram-positive bacteria, *Staphylococcus aureus* and *Bacillus cereus*. Katsoura et al. (2006) also found that biocatalytic acylation of rutin with various acyl donors affected its antioxidant potential towards both isolated LDL and total serum model *in vitro*. A significant increase in antioxidant activity was observed for rutin-4'''-oleate.

The 6''-*O*-esterification of kaempferol-3-*O*-glucoside (astragalin) with *p*-coumaric acid was found to increase its anti-inflammatory activity eight times compared to the initial flavonoid, while addition of another *p*-coumaroyl group at 2'' position gave an activity 30 times greater than that of astragalin (Harborne & Williams, 2000). Another kaempferol derivative, kaempferol 3-(2'',3''-di-E-*p*-coumaroylrhamnoside), was found to possess a cytotoxic effect. It significantly modulated the proliferation of promyelocytic cell line HL60 and MOLT3 (a T-ALL with phenotypic characteristics of cortical thymocytes) (Mitrokotsa et al., 1993). Also Demetzos et al. (1997) synthesized novel flavonoid esters with cytotoxic activity. These acetylated esters of tiliroside exhibited a strong cytotoxic effect against four leukemic cell lines (HL60, DAUDI, HUT78 and MOLT3), whilst the maternal compound had no effect (Demetzos et al., 1997). Tricin-7-*O*-β-(6''-methoxycinnamic)-glucoside, a flavone from sugarcane, was found to exhibit antiproliferative activity against several human cancer cell lines, with higher selectivity toward cells of the breast resistant NIC/ADR line (Duarte-Almeida et al., 2007). Mellou et al. (2006) provided evidence that flavonoid derivatives esterified with polyunsaturated fatty acids were able to decrease the production of vascular endothelial growth factor by K562 human leukemia cells unlike the initial flavonoids, indicating that these novel compounds might possess improved anti-angiogenic and anti-tumor properties. Anticancer acitivity was established also in two *O*-acylated flavonoids, daglesiosides I and II, which were isolated from the leaves of *Pseudotsuga menziesii* (Sharma et al., 2003).

Parejo et al. (2005) examined quercetagetin glycosides acylated with caffeic and *p*-coumaric acid for antioxidant activity. They found that these compounds exhibited a high radical scavenging activity in comparison with reference compounds. Fatty acid derivatives of catechins are described as having antitumorigenesis promoting activity or 5-α reductase inhibiting activity, as well as antibacterial activity (Fukami et al., 2007). Since these acylated catechin compounds have a greatly superior solubility in fats and oils than any catechins previously known, they may be used as a highly effective antioxidative agents (Sakai et al., 1994).

A different catechin derivative, 3-*O*-octanoyl-(+)-catechin, was synthesized by Aoshima et al. (2005) by incorporation of an octanoyl chain into (+)-catechin. This ester was found to be more efficient than catechin in inhibiting the response of ionotropic gamma-aminobutyric acid receptors and Na+/glucose cotransporters expressed in *Xenopus oocytes* in a noncompetitive manner. Moreover, it induced a nonspecific membrane current and decreased the membrane potential of the oocyte. This newly synthesized catechin derivative possibly binds to the lipid membrane more strongly than do catechin, (-)-epicatechin gallate, or (-)-epigallocatechin-3-gallate, and as a result it perturbs the membrane structure (Aoshima et al., 2005).

4.3 Cosmetics

The majority of cosmetic or dermopharmaceutical compositions consist of a fatty phase, the oily products of which have a certain tendency to oxidize, even at room temperature. The consequence of this oxidation is to profoundly modify the properties, which makes them unusable after a variable time period. In order to protect the compositions with respect to these oxidation phenomena, it is common practice to incorporate protective agents which act as antioxidizing agents (N'guyen, 1995). By virtue of the skin-protecting and skin-cleansing properties of flavonoids and their effects against aging, against skin discoloration and on the appearance of the skin, they have been used as constituents of cosmetic or dermopharmaceutical compositions. They also act on the mechanical properties of hair (Ghoul et al., 2006).

Moussou et al. (2007) found that the esters of flavonoids with omega-substituted C6 to C22 fatty acids have the property to protect the skin cells against damage caused by UV radiation. According to the invention, these esters of flavonoids protect skin cells against UVA and UVB radiation in a more effective manner than flavonoids alone. Moreover, these esters demonstrated their property to stimulate glutathione metabolism of human skin cells after UVA irradiation, i.e. to stimulate their cellular defenses. They have also anti-inflammatory and soothing properties, as demonstrated by the inhibition of released protein kinase PGE2 after UVB irradiation. Thus these flavonoid esters may be used to protect the skin and scalp and/or to fight against UV and sun damage, erythema, sunburn, mitochondrial or nuclear DNA damage, to prevent or fight photo-aging, providing improvement for signs of aging as skin wrinkles, elasticity lost and decrease in skin thickness (Moussou et al., 2007).

Perrier et al. (2001) discovered that specific flavonoid esters can be stabilized while preserving their initial properties, particularly free radical inhibition and enzyme inhibition, and for applications associated with these properties: venous tonics, agents for increasing the strength of blood capillaries, inhibitors of blotchiness, inhibitors of chemical, physical or actinic erythema, agents for treating sensitive skin, decongestants, draining agents, slimming agents, anti-wrinkle agents, stimulators of the synthesis of the components of the extracellular matrix, toners for making the skin more elastic and anti-ageing agents (Perrier et al., 2001).

5. Conclusions

Flavonoids, having a wide spectrum of health-beneficial activities, seem to be applicable in various areas of national management from food additivization to pharmaceutical preparations with the purpose of prevention and/or treatment of important civilization diseases. Their chemical structure determines not only biological effects on human health but also their solubility, stability and bioavailability. Recently, selective enzyme-mediated acylation of flavonoids has been introduced to confer improved biological properties to the novel compounds including both biological activity of initial flavonoid and other parameters determined by the chemical structure of an acyl donor. In the past, proteases, esterases and acyltransferases were used for the preparation of acylated flavonoids. In light of our review, immobilized lipases, especially *Candida antarctica* B lipase, are suitable for this purpose. Not only the given enzyme but also the reaction conditions have a distinct influence on the performance of acylation. This aspect must be considered when producing acylated flavonoids in technology scale for potential uses in the food, pharmaceutical and cosmetic industry.

6. Acknowledgment

The work was supported by The Slovak Research and Development Agency in the frame of the Project APVV-VMSP-II-0021-09 and by The Agency of the Ministry of Education, Science, Research and Sport of the Slovak Republic for the Structural Funds of EU, OP R&D of ERDF in the frame of the Project „Evaluation of natural substances and their selection for prevention and treatment of lifestyle diseases (ITMS 26240220040).

7. References

Adamczak, M. & Krishna, S.H. (2004). Strategies for improving enzymes for efficient biocatalysis. *Food Technology and Biotechnology*, Vol.42, pp. 251-264.

Alluis, B. & Dangles, O. (1999). Acylated flavone glucosides: synthesis, conformational investigation, and complexation properties. *Helvetica Chimica Acta*, Vol.82, pp. 2201-2212.

Andersen, Ø.M. & Jordheim, M. (2006). The anthocyanins. In: Andersen, Ø.M.; Markham K.R.: Flavonoids: chemistry, biochemistry and applications. New York: Taylor & Francis Group, CRC Press, 471 - 551. ISBN: 0-8493-2021-6.

Anderson, E.M.; Larsson, K.M. & Kirk, O. (1998). One biocatalyst – many applications: the use of *Candida antarctica* B-lipase in organic synthesis. *Biocatalysis and Biotransformation*, Vol.16, pp. 181-204.

Aoshima, H.; Okita, Y.; Hossain, S.J.; Fukue, K.; Mito, M.; Orihara, Y.; Yokoyama, T.; Yamada, M.; Kumagai, A.; Nagaoka, Y.; Uesatos, S. & Hara, Y. (2005). Effect of 3-*O*-octanoyl-(+)-catechin on the responses of GABAa receptors and Na+/glucose cotransporters expressed in *Xenopus oocytes* and on the oocyte membrane potential. *Journal of Agricultural and Food Chemistry*, Vol.53, pp. 1955-1959.

Ardhaoui, M.; Falcimaigne, A.; Engasser, J.-M.; Moussou, P.; Pauly, G. & Ghoul, M. (2004a). Acylation of natural flavonoids using lipase of *Candida antarctica* as biocatalyst. *Journal of Molecular Catalysis B: Enzymatic*, Vol.29, pp. 63-67.

Ardhaoui, M.; Falcimaigne, A.; Ognier, S.; Engasser, J.-M.; Moussou, P.; Pauly, G. & Ghoul, M. (2004b). Effect of acyl donor chain length and substitutions pattern on the enzymatic acylation of flavonoids. *Journal of Biotechnology*, Vol.110, pp. 265-272.

Ardhaoui, M.; Falcimaigne, A.; Engasser, J.-M.; Moussou, P.; Pauly, G. & Ghoul, M. (2004c). Enzymatic synthesis of new aromatic and aliphatic esters of flavonoids using *Candida antarctica* as biocatalyst. *Biocatalysis and Biotransformation*, Vol.22, pp. 253-259.

Arroyo, M.; Sánchez-Montero, J.M. & Sinisterra, J.V. (1999). Thermal stabilization of immobilized lipase B from *Candida antarctica* on different supports. Effect of water activity on enzymatic activity in organic media. *Enzyme & Microbial Technology*, Vol.24, pp. 3-12.

Asen, S.; Stewart, R.N. & Norris, K.H. (1979). Stable foods and beverages containing the anthocyanin, peonidin-3-(dicaffeylsophoroside)-5-glucosid. Patent US 4172902.

Augustyniak, A.; Bartosz, G.; Cipak, A.; Duburs, G.; Horakova, L.; Luczaj, W.; Majekova, M.; Odysseos, A.D.; Rackova, L.; Skrzydlewska, E.; Stefek, M.; Strosova, M.; Tirzitis, G.; Viskupicova, J.; Vraka, P.S. & Zarkovic, N. (2010). Natural and synthetic antioxidants: an updated overview. *Free Radical Research*, Vol.44, pp. 1216-1262.

Ballesteros, A.; Plou, F.J.; Alcade, M.; Ferrer, M.; Garcia-Arellano, H.; Reyes-Duarte, D. & Ghazi, I. (2006). Enzymatic synthesis of sugar esters and oligosaccharides from renewable resources. In: Patel, R.N.: Biocatalysis in the pharmaceutical and biotechnology industries. New York: CRC Press, Taylor & Francis Group, 463-488. ISBN: 0-8493-3732-1.

Berg, P.A. & Daniel, P.T. (1988). Plant flavonoids in biology and medicine II. *Progress in Clinical and Biological Research*, Vol.280, pp. 157-171.

Bloor, S.J. (2001). Deep blue anthocyanins from blue *Dianella* berries. *Phytochemistry*, Vol.58, pp. 923-927.

Bok, S.-H.; Jeong, T.-S.; Lee, S.-K.; Kim, J.-R.; Moon, S.-S.; Choi, M.-S.; Hyun, B.-H.; Lee, C.-H. & Choi, Y.-K. (2001). Flavanone derivatives and composition for preventing or treating blood lipid level-related diseases comprising same. Patent US 6455577.

Bornscheuer, U.T. (2002). Microbial carboxyl esterases: classification, properties and application in biocatalysis. *FEMS Microbiology Reviews*, Vol.26, pp. 73-81.

Calis, I.; Ozipek, M. & Ruedi, P. (1995). Enzyme-mediated regioselective acylation of flavonoid glycosides. FABAD Journal of Pharmaceutical Sciences, Vol.20, pp. 55-59.

Carrea, G.; Riva, S.; Secundo, F. & Danieli, B. (1989). Enzymatic synthesis of various 1-*O*-sucrose and 1-*O*-fructose esters. *Journal of the Chemical Society Perkin Transactions*, Vol.1, pp. 1057-1061.

Chebil, L.; Humeau, C.; Falcimaigne, A.; Engasser, J.-M. & Ghoul, M. (2006). Enzymatic acylation of flavonoids. *Process Biochemistry*, Vol.41, pp. 2237-2251.

Chebil, L.; Anthoni, J.; Humeau, C.; Gerardin, C.; Engasser, J.-M. & Ghoul, M. (2007). Enzymatic acylation of flavonoids: Effect of the nature of the substrate, origin of lipase, and operating conditions on conversion yield and regioselectivity. *Journal of Agricultural and Food Chemistry*, Vol.55, pp. 9496-9502.

Collins, A.M. & Kennedy, M.J. (1999). Biotransformations and bioconversions in New Zealand: Past endeavours and future potential. *Australasian Biotechnology*, Vol.9, pp. 86-94.

Cordova A.; Iverson T. & Hult K. (1998). Lipase catalysed formation of macrocycles by the ring opening polymerisation of ε-caprolactone. *Polymer*, Vol.39, pp. 6519-6524.

Cushnie, T.P.T. & Lamb, A.J. (2005). Detection of galangin-induced cytoplasmic membrane damage in *Staphylococcus aureus* by measuring potassium loss. *Journal of Ethnopharmacology*, Vol.101, pp. 243-248.

Cushnie, T.P.T. & Lamb, A.J. (2006). Assessment of the antibacterial activity of galangin against 4-quinolone resistant strains of *Staphylococcus aureus*. *Phytomedicine*, Vol.13, pp. 191-197.

Cygler, M., & Schrag, J. (1997). Structure as a basis for understanding interfacial properties of lipases. *Methods in Enzymology*, Vol.284, pp. 3-27.

Danieli, B.; De Bellis; P.; Carrea, G. & Riva, S. (1989). Enzyme-mediated acylation of flavonoid monoglycosides. *Heterocycles*, Vol.29, pp. 2061-2064.

Danieli, B.; De Bellis, P.; Carrea, G. & Riva, S. (1990). Enzyme-mediated regioselective acylations of flavonoid disaccharide monoglycosides. *Helvetica Chimica Acta*, Vol.73, pp. 1837-1844.

Danieli, B.; Luisetti, M.; Sampognaro, G.; Carrea, G. & Riva, S. (1997). Regioselective acylation of polyhydroxylated natural compounds catalyzed by *Candida antarctica* lipase B (Novozym 435) in organic solvents. *Journal of Molecular Catalysis B: Enzymatic*, Vol.3, pp. 193-201.

Davis, B. G. & Boyer, V. (2001). Biocatalysis and enzymes in organic synthesis. *Natural Product Reports*, Vol.18, pp. 618-640.

Davies, K.M. & Schwinn, K.E. (2006). Molecular biology and biotechnology of flavonoid biosynthesis. In: Andersen, Ø.M.; Markham, K.R.: Flavonoids: chemistry, biochemistry and applications. New York: Taylor & Francis Group, CRC Press, 143-218. ISBN: 0-8493-2021-6.

Degenhardt, A.; Ullrich, F.; Hofmann, T. & Stark, T. (2007). Flavonoid sugar addition products, method for manufacture and use thereof. Patent US 20070269570.

Degn P.; Pedersen L.H.; Duus J. & Zimmerman, W. (1999). Lipase catalysed synthesis of glucose fatty acid esters in *t*-butanol. *Biotechnology Letters*, Vol.21, pp. 275-280.

Delazar, A.; Celik, S.; Gokturk, R.S.; Nahar, L. & Sarker, S.D. (2005). Two acylated flavonoid glycosides from *Stachys bombycina*, and their free radical scavenging activity. *Pharmazie*, Vol.60, pp. 878-880.

Demetzos, C.; Magiatis, P.; Typas, M.A.; Dimas, K.; Sotiriadou, R.; Perez, S. & Kokkinopoulos, D. (1997). Biotransformation of the flavonoid tiliroside to 7-methylether tiliroside: bioactivity of this metabolite and of its acetylated derivative. *Cellular and Molecular Life Sciences*, Vol.53, pp. 587-592.

Drouin, J.; Costante, J. & Guibé-Jampel, E. (1997). A thermostable microbial enzyme for fast preparative organic chemistry: Preparation of R-(+)-1-phenylethanol from (±)-1-phenylethyl pentanoateand *n*-butanol. *Journal of Chemical Education*, Vol.74, pp. 992-995.

Duarte-Almeida, J.M.; Negri, G.; Salatino, A.; De Carvalho, J.E. & Lajolo, F.M. (2007). Antiproliferative and antioxidant activities of a tricin acylated glycoside from sugarcane (*Saccharum officinarum*) juice. *Phytochemistry*, Vol.68, pp. 1165-1171.

Enaud, E.; Humeau, C.; Piffaut, B. & Girardin, M. (2004). Enzymatic synthesis of new aromatic esters of phloridzin. *Journal of Molecular Catalysis B: Enzymatic*, Vol.27, pp. 1-6.

Ferrer, J.-L.; Austin, M.B.; Stewart, C. & Noel, J.P. (2008). Structure and function of enzymes involved in the biosynthesis of phenylpropanoids. *Plant Physiology and Biochemistry*, Vol.46, pp. 356-370.

Foresti, M.L.; Pedernera, M.; Bucalá, V. & Ferreira, M.L. (2007). Multiple effects of water on solvent-free enzymatic esterifications. *Enzyme & Microbial Technology*, Vol.41, pp. 62-70.

Fossati, E. & Riva, S. (2006). Stereoselective modifications of polyhydroxylated steroids. In: Patel, R.N.: Biocatalysis in the pharmaceutical and biotechnology industries. New York: CRC Press, Taylor & Francis Group, 591-604. ISBN: 0-8493-3732-1.

Fox, G.J. (2000). Natural red sunflower anthocyanin colorant with naturally stabilized color qualities, and the process of making. Patent US 6132791.

Fujiwara, H.; Tanaka, Y.; Yonekura-Sakakibara, K.; Fukuchi-Mizutani, M.; Nakao, M.; Fukui, Y.; Yamaguchi, M.; Ashikari, T. & Kusumi, T. (1998). cDNA cloning, gene expression and subcellular localization of anthocyanin 5-aromatic acyltransferase from *Gentiana triflora*. *The Plant Journal*, Vol.16, pp. 421-431.

Fukami, H.; Nakao, M.; Namikawa, K. & Maeda, M. (2007). Esterified catechin, process for producing the same, food and drink or cosmetic containing the same. Patent EP 1849779.

Gao, C.; Mayon, P.; Macmanus, D.A. & Vulfson, E. N. (2001). Novel enzymatic approach to the synthesis of flavonoid glycosides and their esters. *Biotechnology & Bioengineering*, Vol.71, pp. 235-243.

Gayot, S.; Santarelli, X. & Coulon, D. (2003). Modification of flavonoid using lipase in non-conventional media: effect of the water content. *Journal of Biotechnology*, Vol.101, pp. 29-36.

Ghoul, M.; Engasser, J.-M.; Moussou, P.; Pauly, G.; Ardhaoui, M. & Falcimaigne, A. (2006). Enzymatic production of acyl flavonoid derivatives. Patent US 20060115880.

Giusti, M.M. & Wrolstad, R.E. (2003). Acylated anthocyanins from edible sources and their applications in food systems. *Biochemical Engineering Journal*, Vol.14, pp. 217-225.

Haeffner, F.; Norin, T. & Hult, K. (1998). Molecular modeling of the enantioselectivity in lipase-catalyzed transesterification reactions. *Biophysical Journal*, Vol.74, pp. 1251-1262.

Haraguchi, H.; Tanimoto, K.; Tamura, Y.; Mizutani, K. & Kinoshita, T. (1998). Mode of antibacterial action of retrochalcones from *Glycyrrhiza inflata*. *Phytochemistry*, Vol.48, pp. 125-129.

Harborne, J.B. & Williams, C.A. (1998). Anthocyanins and other flavonoids. *Natural Product Reports*, Vol.15, 631-652.

Harborne, J.B. & Williams, C.A. (2000). Advances in flavonoid research since 1992. *Phytochemistry*, Vol.55, pp. 481-504.

Hidalgo, A. & Bornscheuer, U.T. (2006). Direct evolution of lipases and esterases for organic synthesis. In: Patel, R.N.: Biocatalysis in the pharmaceutical and biotechnology industries. New York: CRC Press, Taylor & Francis Group, 159-179. ISBN: 0-8493-3732-1.

Ishihara, K. & Nakajima, N. (2003). Structural aspects of acylated plant pigments: stabilization of flavonoid glucosides and interpretation of their functions. *Journal of Molecular Catalysis B: Enzymatic*, Vol.23, No.2-6, pp. 411-417.

Ishihara, K.; Nishimura, Y.; Kubo, T.; Okada, C.; Hamada, H. & Nakajima, N. (2002). Enzyme-catalyzed acylation of plant polyphenols for interpretation of their functions. *Plant Biotechnology*, Vol.19, pp. 211-214.

Iso, M.; Chen, B.; Eguchi, M.; Kudo, T. & Shrestha, S. (2001). Production of biodiesel fuel from triglycerides and alcohol using immobilized lipase. *Journal of Molecular Catalysis B: Enzymatic*, Vol.16, pp. 53-58.

Itoh, T.; Nishimura, Y.; Ouchi, N. & Hayase, S. (2003). 1-Butyl-2,3-dimethylimidazolium tetrafluoroborate: the most desirable ionic liquid solvent for recycling use of enzyme in lipase-catalyzed transesterification using vinyl acetate as acyl donor. *Journal of Molecular Catalysis B: Enzymatic*, Vol.26, pp. 41-45.

Iwashina, T. (2003). Flavonoid function and activity to plants and other organisms. *Biological Science in Space*, Vol.17, pp. 24-44.

Jain, N.; Kumar, A.; Chauhan, S. & Chauhan, S.M.S. (2005). Chemical and biochemical transformations in ionic liquids. *Tetrahedron*, Vol.61, pp. 1015-1060.

Jarstoff, B.; Störmann, R.; Ranke, J.; Mölter, K.; Stock, F.; Oberheitmann, B.; Hoffmann, J.; Nüchter, M.; Ondruschka, B. & Filser, J. (2003). How hazardous are ionic liquids? Structure–activity relationships and biological testing as important elements for sustainability evaluation. *Green Chemistry*, Vol.5, pp. 136-142.

Jeager, K.E.; Dijkstra, B.W. & Reetz, M.T. (1999). Bacterial biocatalysts: molecular biology, three-dimensional structures, and biotechnological applications. *Annual Review of Microbiology*, Vol.53, pp. 315-351.

Jungblut, T.P.; Schnitzler, J.P.; Hertkorn, N.; Metzger, J.W.; Heller, W. & Sandermann, H. (1995). Acylated flavonoids as plant defense compounds against environmentally relevant ultraviolet-B radiation in *Scots pine* seedlings. Phytochemicals and health. Proceedings, tenth annual Penn State Symposium in Plant Physiology, Rockville, Md.: American Society of Plant Physiologists, 266 – 267.

Katsoura, M.H.; Polydera, A.C.; Tsironis, L.; Tselepis, A.D. & Stamatis, H. (2006). Use of ionic liquids as media for the biocatalytic preparation of flavonoid derivatives with antioxidant potency. *Journal of Biotechnology*, Vol.123, pp. 491-503.

Katsoura, M.H.; Polydera, A.C.; Katapodis, P.; Kolisis, F.N. & Stamatis, H. (2007). Effect of different reaction parameters on the lipase-catalyzed selective acylation of polyhydroxylated natural compounds in ionic liquids. *Process Biochemistry*, Vol.42, pp. 1326-1334.

Kitamura, S. (2006). Transport of flavonoids: from cytosolic synthesis to vacuolar accumulation. In Grotewold, E.: The science of flavonoids. Springer Science Business Media, New York, 123-146. ISBN: 0-387-28821-X

Kodelia, G.; Athanasiou, K. & Kolisis, F.N. (1994). Enzymatic synthesis of butyryl-rutin ester in organic solvents and its cytogenetic effects in mammalian cells in culture. *Applied Biochemistry and Biotechnology*, Vol.44, pp. 205-212.

Kontogianni, A.; Skouridou, V.; Sereti, V.; Stamatis, H. & Kolisis, F.N. (2003). Lipase-catalyzed esterification of rutin and naringin with fatty acids of medium carbon chain. *Journal of Molecular Catalysis B: Enzymatic*, Vol.21, pp. 59-62.

Kontogianni, A.; Skouridou, V.; Sereti, V.; Stamatis, H. & Kolisis, F.N. (2001). Regioselective acylation of flavonoids catalyzed by lipase in low toxicity media. *European Journal of Lipid Science and Technology*, Vol.103, pp. 655-660.

Kragl, U.; Kaftzik, N.; Schofer, S. & Wasserscheid, P. (2006). U.S. Patent No. 20060211096.

Kwon, C.H.; Dae, Y.S.; Jong, H.L. & Seung, W.K. (2007). Molecular modeling and its experimental verification for the catalytic mechanism of *Candida antarctica* lipase B. *Journal of Microbiology and Biotechnology*, Vol.17, No.7, pp. 1098-1105.

Laane, C. (1987). Medium engineering for bioorganic synthesis. *Biocatalysis*, Vol.30, pp. 80-87.

Lambusta, D.; Nicolosi, G.; Patti, A. & Piattelli, M. (1993). Enzyme-mediated regioprotection-deprotection of hydroxyl groups in (+)-catechin. *Synthesis*, Vol.11, pp. 1155-1158.

Lee, S.; Lee, C.-H.; Moon, S.-S.; Kim, E.; Kim, C.-T.; Kim, B.-H.; Bok, S.-H.; Jeong, T.-S. (2003). Naringenin derivatives as anti-atherogenic agents. *Bioorganic & Medicinal Chemistry Letters*, Vol.13, pp. 3901-3903.

Lin, S.F.; Lin, Y.-H.; Lin, M.; Kao, Y.-F.; Wang, R.-W.; Teng, L.-W.; Chuang, S.-H.; Chang, J.-M. Yuan, T.-T.; Fu, K.C.; Huang, K.P.; Lee, Y.-S.; Chiang, C.-C.; Yang, S.-C.; Lai, C.-L.; Liao, C.-B.; Chen, P.; Lin, Y.-S.; Lai, K.-T.; Huang, H.-J.; Yang, J.-Y.; Liu, C.-W.; Wei, W.-Y.; Chen, C.-K.; Hiipakka, R.A.; Liao, S. & Huang, J.-J. (2010). Synthesis and structure-activity relationship of 3-*O*-acylated (-)-epigallocatechins as 5α-reductase inhibitors. *European Journal of Medicinal Chemistry*, Vol.45, pp. 6068-6076.

Lozano, P.; De Diego, T.; Carrie, D.; Vaultier, M. & Iborra, J. L. (2004). Synthesis of glycidyl esters catalyzed by lipases in ionic liquids and supercritical carbon dioxide. *Journal of Molecular Catalalysis A: Chemical*, Vol.214, pp. 113-119.

Lutz, S. (2004). Engineering lipase B from *Candida antarctica*. *Tetrahedron: Asymmetry*, Vol.15, pp. 2743-2748.

Martinelle, M.; Holmquist, M. & Hult, K. (1995). On the interfacial activation of *Candida antarctica* lipase A and B as compared with *Humicola lanuginosa* lipase. *Biochimica et Biophysica Acta*, Vol.1258, pp. 272-276.

Mellou, F.; Lazari, D.; Skaltsa, H.; Tselepis, A.D.; Kolisis, F.N. & Stamatis, H. (2005). Biocatalytic preparation of acylated derivatives of flavonoid glycosides enhances their antioxidant and antimicrobial activity. *Journal of Biotechnology*, Vol.116, pp. 295-304.

Mellou, F.; Loutrari, H.; Stamatis, H.; Roussos, C. & Kolisis, F.N. (2006). Enzymatic esterification of flavonoids with unsaturated fatty acids: Effect of novel esters on vascular endothelial growth factor release from K562 cells. *Process Biochemistry*, Vol.41, pp. 2029-2034.

Middleton, E.Jr. & Chithan, K. (1993). The impact of plant flavonoids on mammalian biology: implications for immunity, inflammation and cancer. In.: Harborne, J.B.: The flavonoids: advances in research since 1986. London, Chapman and Hall, 619-652, ISBN 0-412-48070-0.

Middleton, E.Jr.; Kandaswami, C. & Theoharides, T.C. (2000). The effects of plant flavonoids on mammalian cells: Implications for inflammation, heart disease, and cancer. *Pharmacological Reviews*, Vol.52, pp. 673-751.

Mitrokotsa, D.; Mitaku, S.; Demetzos, C.; Harvala, C.; Mentis, A.; Perez, S. & Kokkinopoulos, D. (1993). Bioactive compounds from the buds of *Platinus orientalis* and isolation of a new kaempferol glycoside. *Planta Medica*, Vol.59, pp. 517-520.

Moussou, P.; Falcimaigne, A.; Ghoul, M.; Danoux, L. & Pauly, G. (2007). Esters of flavonoids with ω-substituted C6-C22 fatty acids. Patent US 20070184098.

Moussou, P.; Falcimaigne. A.; Pauly, G.; Ghoul, M.; Engasser, J.-M. & Ardhaoui, M. (2004). Preparation of flavonoid derivatives. Patent EP1426445.

Nagasawa, T. & Yamada, H. (1995). Microbial production of commodity chemicals. *Pure & Applied Chemistry*, Vol.67, pp. 1241-1256.

Nakajima, N.; Ishihara, K.; Hamada, H.; Kawabe, S.-I. & Furuya, T. (2000). Regioselective acylation of flavonoid glucoside with aromatic acid by an enzymatic reaction system from cultured cells of *Ipomoea batatas*. *Journal of Bioscience & Bioengineering*, Vol.90, pp. 347-349.

Nakajima, N.; Ishihara, K.; Itoh, T.; Furuya, T. & Hamada, H. (1999). Lipase catalysed direct and regioselective acylation of flavonoid glucoside for mechanistic investigation of stable plant pigments. *Journal of Bioscience and Bioengineering*, Vol.61, pp. 1926-1928.

Nakajima, N.; Sugimoto, M.; Yokoi, H.; Tsuji, H. & Ishihara, K. (2003). Comparison of acylated plant pigments: Light-resistance and radical-scavenging ability. *Bioscience, Biotechnology, and Biochemistry*, Vol.67, pp. 1828-1831.

Nakayama, T.; Suzuki, H. & Nishino, T. (2003). Anthocyanin acyltransferases: specificities, mechanism, phylogenetics, and applications. *Journal of Molecular Catalysis B: Enzymatic*, Vol.23, pp. 117-132.

Narayan, V.S. & Klibanov, A.M. (1993). Are water-immiscibility and apolarity of the solvent relevant to enzyme efficiency? *Biotechnology & Bioengineering*, Vol.41, pp. 390-393.

N'guyen, Q.L. (1995). Cosmetic or dermopharmaceutical composition containing, in combination, a lauroylmethionate of a basic amino acid and at least one polyphenol. Patent US 5431912.

Nicolosi, G.; Piattelli, M.; Lambusta, D. & Patti, A. (1999). Biocatalytic process for the preparation of 3-*O*-acyl-flavonoids. Patent WO/1999/066062.

Otto, R.; Geers, B.; Weiss, A.; Petersohn, D.; Schlotmann, K. & Schroeder, K.R. (2001). Novel flavone glycoside derivatives for use in cosmetics, pharmaceuticals and nutrition. Patent EP 1274712.

Parejo, I.; Bastida, J.; Viladomat, F. & Codina, C. (2005). Acylated quercetagetin glycosides with antioxidant activity from *Tagetes maxima*. *Phytochemistry*, Vol.66, pp. 2356-2362.

Patel, R.N. (2006). Biocatalysis for synthesis for chiral pharmaceutical intermediates. In: Patel, R.N.: Biocatalysis in the pharmaceutical and biotechnology industries. New York: CRC Press, Taylor & Francis Group, 103-158. ISBN: 0-8493-3732-1.

Passicos, E.; Santarelli, X. & Coulon, D. (2004). Regioselective acylation of flavonoids catalyzed by immobilized *Candida antarctica* lipase under reduced pressure. *Biotechnology Letters*, Vol.26, pp. 1073-1076.

Patti, A.; Piattelli, M. & Nicolosi, G. (2000). Use of *Mucor miehei* lipase in the preparation of long chain 3-*O*-acylcatechins. *Journal of Molecular Catalysis B: Enzymatic*, Vol.10, pp. 577-582.

Perrier, E.; Mariotte, A.-M.; Boumendjel, A. & Bresson-Rival, D. (2001). Flavonoide esters and their use notably in cosmetics. Patent US 6235294.

Plaper, A.; Golob, M.; Hafner, I.; Oblak, M.; Solmajer, T. & Jerala, R. (2003). Characterization of quercetin binding site on DNA gyrase. *Biochemical and Biophysical Research Communications*, Vol.306, pp. 530-536.

Pleiss, J.; Fischer, M. & Schmid, R. D. (1998). Anatomy of lipase binding sites: the scissile fatty acid binding site. *Chemistry and Physics of Lipids*, Vol.93, pp. 67-80.

Quarenghi, M.V.; Tereschuk, M.L.; Baigori, M.D. & Abdala, L.R. (2000). Antimicrobial activity of flowers from *Anthemis cotula*. *Fitoterapia*, Vol.71, pp. 710-712.

Rao, M.B.; Tanksale, A.M.; Ghatge, M.S. & Deshpande, V.V. (1998). Molecular and biotechnological aspects of microbial proteases. *Microbiology and Molecular Biology Reviews*, Vol.62, pp. 597-635.

Rauha, J.P.; Remes, S.; Heinonen, M.; Hopia, A.; Kähkönen, M.; Kujala, T.; Pihlaja, K.; Vuorela, H. & Vuorela, P. (2000). Antimicrobial effect of Finnish plant extracts containing flavonoids and other phenolic compounds. *International Journal of Food Microbiology*, Vol.56, pp. 3-12.

Reetz, M.T.; Wiesenhöffer, W.; Francio, G. & Leitner, W. (2003). Continuous flow enzymatic kinetic resolution and enantiomer separation using ionic liquid/supercritical carbon dioxide media. *Advanced Synthesis & Catalysis*, Vol.345, pp. 01221-01228.

Rice-Evans, C.A. (2001). Flavonoid Antioxidants. *Current Medicinal Chemistry*, Vol.8, pp. 797-807.

Riva, S. (2002). Enzymatic modification of the sugar moieties of natural glycosides. *Journal of Molecular Catalysis B: Enzymatic*, Vol.19-20, pp. 43-54.

Riva, S.; Chopineau, J.; Kieboom, A.P.G. & Klibanov, A.M. (1988). Protease-catalyzed regioselective esterification of sugars and related compounds in anhydrous dimethylformamide. *Journal of the American Chemical Society*, Vol.110, pp. 584-589.

Rocha, J.; Gil, M. & Garcia, F. (1999). Optimisation of the enzymatic synthesis of *n*-octyl oleate with immobilized lipase in the absence of solvents. *Journal of Chemical Technology & Biotechnology*, Vol.74, pp. 607-612.

Ross, J.A. & Kasum, C.M. (2002). Dietary flavonoids: bioavailability, metabolic effects, and safety. *Annul Reviews in Nutrition*, Vol.22, pp. 19-34.

Rubin-Pitel, S.B. & Zhao, H. (2006). Recent Advances in biocatalysis by directed enzyme evolution. *Combinatorial Chemistry & High Throughput Screening*, Vol 9, pp. 247-257.

Saija, A.; Tomaino, A.; Trombetta, D.; Pellegrino, M.L.; Tita, B.; Messina, C.; Bonina, F.P.; Rocco, C.; Nicolosi, G. & Castelli, F. (2003). '*In vitro*' antioxidant and photoprotective properties and interaction with model membranes of three new quercetin esters. *European Journal of Pharmaceutics and Biopharmaceutics*, Vol.56, pp. 167-174.

Sakai, M.; Suzuki, M.; Nanjo, F. & Hara, Y. (1994). 3-*O*-acylated catechins and methods of producing same. Patent EP 0618203.

Salem, J.H.; Chevalot, I.; Harscoat-Schiavo, C.; Paris, C.; Fick, M. & Humeau, C. (2011). Biological activities of flavonoids from *N. retusa* and their acylated derivatives. *Food Chemistry*, Vol.124, pp. 486-494.

Saxena, R.K.; Ghosh, P.K.; Gupta, R.; Davidson, W.S.; Bradoo, S. & Gulati, R. (1999). Microbial lipases: potential biocatalysts for the future industry. *Current Science*, Vol.77, pp. 101-115.

Sharma, G.; Singh, R.P.; Chan, D.C. & Agarwal, R. (2003) Silibinin induces growth inhibition and apoptotic cell death in human lung carcinoma cells. *Anticancer Research*, Vol.23, pp. 2649-2655.

Singh, R.K. & Nath, G. (1999). Antimicrobial activity of *Elaeocarpus sphaericus*. *Phytotherapy Research*, Vol.13, pp. 448-450.

Stapleton, P.D.; Shah, S.; Anderson, J.C.; Hara, Y.; Hamilton-Miller, J.M.T. & Taylor, P.W. (2004). Modulation of beta-lactam resistance in *Staphylococcus aureus* by catechins and gallates. *International Journal of Antimicrobial Agents*, Vol.23, pp. 462-467.

Stefani, E.D.; Boffetta, P.; Deneo-Pellegrini, H.; Mendilaharsu, M.; Carzoglio, J.C.; Ronco, A. & Olivera, L. (1999). Dietary antioxidants and lung cancer risk: a case-control study in Uruguay. *Nutrition and Cancer*, Vol.34, pp. 100–110.

Stepanovic, S.; Antic, N.; Dakic, I. & Svabic-Vlahovic, M. (2003). *In vitro* antimicrobial activity of propolis and synergism between propolis and antimicrobial drugs. *Research in Microbiology*, Vol.158, pp. 353-357.

Stevenson, D.E.; Wibisono, R.; Jensen, D.J.; Stanley, R.A. & Cooney, J.M (2006). Direct acylation of flavonoid glycosides with phenolic acids catalyzed by *Candida antarctica* lipase B (Novozym 435®). *Enzyme and Microbial Technology*, Vol.39, pp. 1236-1241.

Suda, I.; Oki, T.; Masuda, M.; Nishiba, Y.; Furuta, S.; Matsugano, K.; Sugita, K. & Terahara, N. (2002). Direct absorption of acylated anthocyanin in purple-fleshed sweet potato into rats. *Journal of Agricultural and Food Chemistry*, Vol.50, pp. 1672-1676.

Tarle, D. & Dvorzak, I. (1990). Antimicrobial activity of the plant *Cirsium oleraceum* (L.). Scop. *Acta Pharmaceutica Jugoslavica*, Vol.40, pp. 567-571.

Tereschuk, M.L.; Riera, M.V.; Castro, G.R. & Abdala, L.R. (1997). Antimicrobial activity of flavonoids from leaves of *Tagetes minuta*. *Journal of Ethnopharmacology*, Vol.56, pp. 227-232.

Toba, S. & Merz, K. M. (1997). The concept of solvent compatibility and its impact on protein stability and activity enhancement in non-aqueous solvents. *Journal of the American Chemical Society*, Vol.119, pp. 9939-9948.

Torres, S. & Castro, G.R. (2004). Non-aqueous biocatalysis in homogeneous solvent systems. *Food Technology & Biotechnology*, Vol.42, pp. 271-277.

Trodler, P. & Pleiss, J. (2008). Modeling structure and flexibility of *Candida antarctica* lipase B in organic solvents. *BMC Structural Biology* [online].

Uppenberg, J.; Öhrner, N.; Norin, M.; Hult, K.; Kleywegt, G.J.; Patkar, S.; Waagen, V.; Anthonsen, T. & Jones, T.A. (1995). Crystallographic and molecular-modelling studies of lipase B from *Candida antarctica* reveal a stereospecificity pocket for secondary alcohols. *Biochemistry*, Vol.33, pp. 16838-16851.

Van Rantwick, F.; Lau, R.M. & Sheldon, R.A. (2003). Biocatalytic transformations in ionic liquids. *Trends in Biotechnology*, Vol.21, pp. 131-138.

Vakhlu, J. & Kour, A. (2006). Yeast lipases: enzyme purification, biochemical properties and gene cloning. *Electronic Journal of Biotechnology* [online].

Viskupicova, J.; Danihelova, M.; Ondrejovic, M.; Liptaj, T. & Sturdik, E. (2010). Lipophilic rutin derivatives for antioxidant protection of oil-based foods. *Food Chemistry*, Vol.123, pp. 45-50.

Viskupicova, J.; Maliar, T.; Psenakova, I. & Sturdik, E. (2006). Eznymatic acylation of naringin. *Nova Biotechnologica*, Vol.6, pp. 149-159 [In Slovak].

Viskupicova, J. & Ondrejovic, M. (2007). Effect of fatty acid chain length on enzymatic esterification of rutin. In: Book of abstracts of the 1st International Conference of Applied Natural Sciences, November 7 – 9, 2007 (pp. 59), Trnava: UCM.

Viskupicova, J.; Strosova, M.; Sturdik, E. & Horakova, L. (2009). Modulating effect of flavonoids and their derivatives on sarcoplasmic reticulum Ca^{2+}-ATPase oxidized by hypochloric acid and peroxynitrite. *Neuroendocrinology Letters*, Vol.30, pp. 148-151.

Wang, Y.F.; Lalonade, J.J.; Momongan, M.; Bergbreiter, D.E. & Wong, C.H. (1988). Lipase-catalyzed irreversible transesterifications using enol esters as acylating reagents: preparative enantio and regioselective synthesis of alcohols, glycerol derivatives, sugars, and organometallics. Journal of the American Chemical Society, Vol.110, pp. 7200-7205.

Williams, C.A. (2006). Flavone and flavonol O-glycosides. In: Andersen, Ø.M.; Markham, K.R.: Flavonoids: chemistry, biochemistry and applications. New York: Taylor & Francis Group, CRC Press, 397-441. ISBN: 0-8493-2021-6.

Wilkes, J.S. (2004). Properties of ionic liquids solvents for catalysis. *Journal of Molecular Catalysis A: Chemical*, Vol.214, pp. 11-17.

Xanthakis, E.; Theodosiou, E.; Magkouta, S.; Stamatis, H. & Loutrari, H. (2010). Enzymatic transformation of flavonoids and terpenoids: structural and functional diversity of the novel derivatives. *Pure & Applied Chemistry*, Vol.82, pp. 1-16.

Xiao, Y.-M.; Wu, Q.; Wu, W.-B.; Zhang, Q.-Y. & Lin, X.-F. (2005). Controllable regioselective acylation of rutin catalyzed by enzymes in non-aqueous solvents. *Biotechnology Letters*, Vol.27, pp. 1591-1595.

Yadav, G.D. & Piyush, S.L. (2003). Kinetics and mechanism of synthesis of butyl isobutyrate over immobilised lipases. *Biochemical Engineering Journal*, Vol.16, pp. 245-252.

Zaks, A. & Klibanov, A.M. (1988). Enzyme catalysis in nonaqueous solvents. *Journal of Biological Chemistry*, Vol.263, pp. 3194-3201.

Part 3

Metabolism and Mechanism

11

Glucose Metabolism and Cancer

Lei Zheng[1,2], Jiangtao Li[2] and Yan Luo[3]
[1]*The Sidney Kimmel Comprehensive Cancer Center, The Skip Viragh Center for Pancreatic Cancer, The Sol Goldman Pancreatic Cancer Center Department of Oncology, and Department of Surgery, The Johns Hopkins University School of Medicine, Baltimore, Maryland*
[2]*Department of Surgery, The Second Affiliated Hospital Zhejiang University College of Medicine, Hangzhou*
[3]*Section of Biochemistry and Genetics, School of Basic Medical Sciences and Cancer Institute, the Second Affiliated Hospital Zhejiang University College of Medicine, Hangzhou*
[1]*USA*
[2,3]*China*

1. Introduction

An outstanding biochemical characteristic of neoplastic tissues is that despite ample oxygen supply, glycolysis is the dominant pathway for adenosine 5'-triphosphate (ATP) production, a phenomenon termed "the Warburg effect" (Warburg et al., 1927; Warburg, 1956). This aerobic glycolysis seems unexpected as one would imagine that cancer cells should have, given sufficient oxygen, adapted to utilize the complete oxidative phosphorylation to maximize ATP production. In addition, for producing same level of ATP, glycolysis would consume >15-fold more glucose, resulting in an addiction for glucose during active tumor growth. There must be a biological logic for cancer cells to prefer utilization of glycolysis for ATP production: they obtain and/or sustain growth advantage at the cost of an "addiction" to glycolysis. Mechanistically, there have been many hypotheses proposed to explain this phenomenon and two are more prominent: one is that by consuming excessive glucose, cancer cells may change the tumor's microenvironment to gain growth advantage over normal cells; the second is that, to sustain a dominant glycolysis process, cancer cells alter the expression or functions of glucose metabolic enzymes, which may have resulted in additional changes that promote cancer development (Vander Heiden et al., 2009). A growing body of evidence has supported both hypotheses and will be reviewed here.

2. Altered glycolytic enzymes in cancer cells

In cancer cells, the activities or expression levels of many enzymes participating in glucose metabolism are altered, those involved in glycolysis in particular. The glycolysis commonly refers to the reactions that covert glucose into pyruvate or lactate (Figure 1). Usually, one or more isoforms of glycolytic enzymes have altered expression patterns or activities in cancer cells, which was previously reviewed (Herling et al., 2011; Porporato et al., 2011).

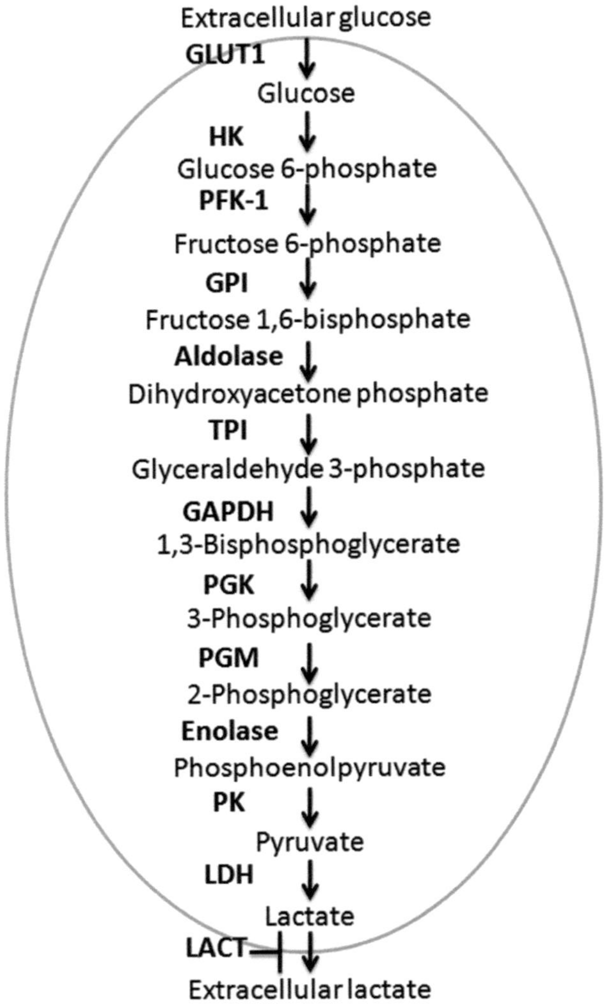

Fig. 1. Glycolysis in cancer cells

- GLUT1, an isoform of **glucose transporters** (GLUTs) is overexpressed in many types of human malignancies, whereas the insulin-sensitive GLUT4 is downregulated in cancer cells. The imbalanced expression of GLUT1 versus GLUT4 in cancer cells may have contributed to the insulin-independent glucose uptake in cancer cells. In addition, GLUT3 was also reported to be overexpressed in cancer cells (Smith, 1999; Medina and Owen, 2002; Noguchi et al., 1998).
- HK-2, an isoform of **hexokinase** that is one of the three rate limiting glycolytic enzymes, is overexpressed in cancer cells and was shown to contribute to the Warburg effect. The high glycolytic rate characteristic of hypoxic solid tumors is attributed to the overexpression of HK-2 (Mathupala et al., 2009; Wolf et al.).
- The expression level of the **glucose 6-phosphate isomerase** (GPI) has also been reported to be elevated in its mRNA level in different human cancer cell lines (Funasaka et al., 2005).

- The alteration of **phosphofructokinase-1** (PFK-1) in cancer cells is not at expression level, but at the level of the enzyme activity. PFK in cancer cells is less sensitive to the inhibition by its allosteric regulators such as citrate and ATP (Meldolesi et al., 1976). Additionally, the expression of all four genes of the **Phosphofructokinase-2** (PFK-2/FBPase/PFKFB) family is inducible by hypoxia, among which the gene encoding *PFKFB3* is highly expressed in several types of human neoplasm (Minchenko et al., 2005; Atsumi et al., 2002; Kessler et al., 2008).
- The expression of **aldolase** isoenzymes is downregulated in some cancer types such as hepatocellular carcinoma (Song et al., 2004) and upregulated in some other tumor types such as pancreatic ductal adenocarcinoma (Cui et al., 2009).
- **Triosephosphate isomerase** (TPI) was detected in the plasma of cancer patients (Robert et al., 1961) and autoantibodies against TPI was also detected in sera from breast cancer patients (Tamesa et al., 2009).
- **Phosphoglycerate kinase-1** (PGK-1) is overexpressed in majority of the pancreatic ductal adenocarcinomas and can also be detected in sera of patients with these tumors (Hwang et al., 2006).
- **Phosphoglycerate mutase** (PGM/PGAM) M type subunit (PGM-M) is overexpressed in many cancers including lung, colon, liver and breast (Durany et al., 2000; Durany et al., 1997).
- The expression of **Enolase 1** (ENOA) is upregulated at the transcriptional and / or translational level in multiple types of tumor including brain, breast, cervix, colon, eye, gastric, head and neck, kidney, leukemia, liver, lung, muscle, ovary, pancreas, prostate, skin and testis(Capello et al.). It is also differentially regulated at the post-translational level in cancer cells as compared with normal cells. ENOA in tumor cells is subjected to more acetylation, methylation and phoshorylation than in normal tissues (Capello et al.). Specific acetylated residues of ENOA were found in cervix, pancreatic, and colon cancers. Five aspartate and five glutamate residues were found to be specifically methylated in pancreatic cancer. Several serine and threonine residues were found to be specifically phosphorylated in leukemia, cervix, and lung cancers. Although ENOA is phosphorylated at Serine 419 in both normal and malignant pancreatic tissues, this phosphorylated form of ENOA is overexpressed in pancreatic cancer(Zhou et al., 2010).
- **Pyruvate kinase** has two isoforms, M and L, which have tissue-specific expression. Normal proliferating cells including embryonic cells and adult stem cells selectively express the M2 isoform (PKM2) (Reinacher and Eigenbrodt, 1981; Yamada and Noguchi, 1999). During tissue differentiation in development, embryonic PKM2 is replaced by tissue-specific isoforms. However, the tissue-specific expression pattern of PK is disrupted during tumorigenesis (Hacker et al., 1998). PKM2 is re-expressed and becomes the only predominant isoform in cancer cells (Mazurek et al., 2005).
- The **lactate dehydrogenase** (LDH) family of tetrameric enzymes catalyzes the pyruvate reduction into lactate. LDHs are formed by four subunits of two different isozymes of either LDH-H or LDH-M. LDH-H is encoded by the *LDH-B* gene and is ubiquitously expressed, and LDH-M is encoded by *LDH-A*. LDH-M has a higher *K*m for pyruvate and a higher *V*max for pyruvate reduction than LDH-H (Markert et al., 1975). Consequently, LDHs predominantly comprising of the LDH-M subunits drive the reduction of pyruvate to lactate; LDHs predominantly comprising of the LDH-H subunits drive the oxidation of lactate to pyruvate. Many types of tumor cells manifest a high expression of the *LDH-A* gene. Elevated expression of LDH5, which comprises of

four LDH-M subunits, is an unfavorable prognostic factor for many human malignancies (Koukourakis et al., 2003; Koukourakis et al., 2005; Koukourakis et al., 2009). In addition, the glycolysis process in cancer cells is reportedly to be associated with hypermethylation of the *LDH-B* gene promoter, which is linked to gene silencing (Leiblich et al., 2006; Thangaraju et al., 2009).

- Plasma membrane **lactate transport** (LACT) is facilitated by the family of proton-linked moncarboxylate transporters (MCTs) or by SMCT1, a sodium coupled lactate transporter. The MCT4 isoform is upregulated in many cancer types. The expression of MCT2, an isoform mainly implicated in lactate import, is decreased in tumor cell lines. SMCT1, also implicated in lactate import, is downregulated in a number of cancer types including colon, thyroid, and stomach(Herling et al., 2011; Porporato et al., 2011).

3. Glucose metabolic pathways switched by oncogenes and tumor suppressors

It has been long proposed that the altered expression or enzyme activities of glycolytic enzymes are regulated by oncogenes and tumor suppressor genes.

- The expression of GLUT1 has been demonstrated to be controlled by the hypoxia-inducible transcription factor HIF-1, c-Myc, and Akt (Chen et al., 2001; Osthus et al., 2000; Rathmell et al., 2003).
- The gene encoding HK-2, but not HK-1, is known to be a transcriptional target of HIF-1(Rempel et al., 1996). It was later shown that HIF-1 cooperates with c-Myc to transactivate HK-2 under hypoxia (Kim et al., 2007). The phosphorylated form of HK-2 interacts with the voltage-dependent anion channel (VDAC) at the outer mitochondrial membrane (Bustamante and Pedersen, 1977; Nakashima et al., 1986; Gottlob et al., 2001). The interaction of HK-2 with VDAC interferes with the binding of the pro-apoptotic protein Bax to VDAC thus preventing the formation of the channel through which cytochrome c can escape from mitochondria to trigger apoptosis (Pastorino et al., 2002). Therefore, overexpression of HK-2 in cancer cells leads to a switch from HK-1 to HK-2 and offers a metabolic advantage by protecting cancer cells against apoptosis(Porporato et al., 2011).
- Overexpression of GPI can be induced by HIF-1 and VEGF(Funasaka et al., 2005).
- The expression of PFK-2 genes in tumor cells is shown to be regulated by Ras and src (Yalcin et al., 2009) and that of PFKFB3 is demonstrated to be induced by HIF-1, c-myc, ras, src, and loss of function of p53 (Minchenko et al., 2002). Among the four PFK-2 genes, PFKFB3 is the most significantly induced in response to hypoxia. Hypoxia-induced PFK-2 activity of human PFKFB3 is further enhanced through phosphorylation of the serine 462 residue (Marsin et al., 2002). This phosphorylation process may involve AMP-activated protein kinase (AMPK) and Akt(Shaw and Cantley, 2006; Yun et al., 2005).
- Although no evidence has suggested that the expression level of **Glyceraldehyde-3-phosphate dehydrogenase** (GAPDH) is altered in cancer cells, its expression has been demonstrated to be highly dependent on the proliferative state of the cells and can by regulated by HIF-1, p53, and c-jun(Colell et al., 2009; Colell et al., 2007).
- The gene encoding ENOA is a target of, and its expression is upregulated by, c-Myc (Sedoris et al. 2010). Notably, a growing body of evidence has suggested that ENOA

is a tumor-associated antigen(Capello et al., 2011). In patients of many different cancer types, including pancreatic, leukemia, melanoma, head and neck, breast and lung, anti-ENOA autoantibodies have been detected(Capello et al., 2011). In pancreatic cancer patients, the anti-ENOA autoantibodies are directed against phosphorylated Serine 419(Tomaino et al., 2011). One study has shown that, in pancreatic cancer, ENOA elicits a CD4+ and CD8+ T cell response both in vitro and in vivo (Cappello et al., 2009). In pancreatic cancer patients, production of anti-ENOA IgG is correlated with the ability of T cells to be activated in response to ENOA(Cappello et al., 2009). In patients with oral squamous cell carcinoma, an HLA-DR8-restricted peptide (amino acid residues 321–336) of human ENOA recognized by CD4+ T cell has been identified (Kondo et al., 2002).

- As described above, PKM2 is a predominant isoform of PKM in cancer cells. PKM2 is less active than other PKs but is however the only PK subject to regulation by the allosteric activator fructose-1,6-bi-phosphate (FBP) and possessing the capability to bind phosphotyrosine proteins (Christofk et al., 2008). Binding of phosphotyrosine peptides to PKM2 leads to dissociation of FBP hence lowering the PKM2 enzyme activity, which may provide a link between cell growth signals and the Warburg effect given that many growth signals and oncogenic pathways involve tyrosine kinases. The unique feature of the PKM2 isoform provides a mechanism by which oncogenes regulate glycolysis and the Warburg effect, and offers an advantage in metabolic plasticity by equipping cancer cells with an exquisitely regulated switch between promoting ATP production and cell proliferation; and lowering the PKM2 activity upon growth signaling was proposed to allow efficient biomass building (anabolic) pathways branched from glycolysis (Christofk et al., 2008). It was suggested that PKM2 is regulated by HIF-1 at the transcriptional level (Discher et al., 1998). Recently, it was demonstrated that, after nuclear translocation, PKM2 cooperates with HIF-1 to transactivate genes, the products of which are involved in promoting the glycolysis and tumor angiogenesis (Luo et al., 2011). Despite these advancements, the regulatory mechanism of PKM2 in cancer cells is not fully understood.
- The expression of PGM/PGAM was found to be downregulated by p53 (Kondoh et al., 2005). Thus, loss of p53 function is anticipated to induce the expression of PGM/PGAM. Phosphoenolpyruvatue, the substrate for PK in cells, transfers the phosphate to the histidine residue located in the catalytic center of the human PGAM1; however, this reaction occurs only in those PKM2-expressing cells but is independent of the PKM2 enzyme activity. Thus, histidine phosphorylation of PGAM1 may provide an alternate glycolytic step in proliferating cancer cells where low PK activity accompanies the expression of PKM2 (Vander Heiden et al., 2010b).
- LDH-A is a target gene of c-Myc and HIF-1 (Dang et al., 2009). Loss of LDH-A functions results in diminished cellular transformation, anchorage independent tumor growth under hypoxic conditions, or xenograft tumor growth (Shim et al., 1997).

Therefore, activation of oncogenes and loss of tumor suppressors are believed to underlie the metabolic switch in cancer cells. Many cancer associated gene products are involved, including c-Myc, NF-kB, Akt, and multiple types of tyrosine kinase including epidermal growth factors (EGFs) and insulin-like growth factor 1 (IGF-1) receptor (Levine and Puzio-Kuter, 2010). Several pathways have been shown to be important for the regulation of glucose metabolism including the PI3K-AKT-mTOR and c-Myc pathways and both have

HIF-1 as their downstream effector. Consistently, among the HIF-1 regulated genes, most are those encoding glycolytic enzymes (Semenza, 2003). The loss of PTEN and concurrent increase of Akt and mTOR lead to the HIF-1 activation and the Warburg effect (Arsham et al., 2002; Zundel et al., 2000).

Another oncogene, K-Ras, can alter glucose metabolism so as to provide tumor cells with a selective advantage. In cells with mutated K-Ras, GLUT1 is upregulated, leading to an augmented glucose uptake, glycolysis and lactate production. Interestingly, in these cells mitochondrial functions and oxidative phosphorylation are not compromised, which allows increased survival rate of the K-Ras mutant cells during glucose deprivation (Annibaldi and Widmann, 2010).

Loss of p53 functions also leads to the Warburg effect. As described above, p53 represses transcription of the genes encoding GLUT1 and 4, and induces transcription of the TIGAR gene, which in turn lowers the intracellular level of PFK/FBPase (Bensaad et al., 2006). In addition, p53 inhibits the PI3K-Akt-mTOR pathways. This appears to be mediated by the transcriptional targets of p53 including PTEN, IGF-binding protein 3, tuberous sclerosis protein TSC-2, and the beta subunit of AMPK (Feng et al., 2007). Conceivably, loss of p53 functions and subsequent loss of expression of these target genes lead to a high HIF level and establishment of the Warburg effect.

4. Tumor "friendly" microenvironment attributed to altered glucose metabolism

One would ask: what is the advantage for cancer cells to use energy-inefficient glycolysis under adequate oxygen supply. A potential advantage, as a result of Warburg effect, is high production of lactic acid due to enhanced glycolysis. Positive correlation between lactate serum levels and tumor burden in cancer patients has been well documented, implicating a role of acidic microenvironment in promoting tumor growth and development (McCarty and Whitaker, 2011).

First, accumulated evidence has suggested that acidic environment amplifies the capacity of invasion and metastasis of cancer cells. For instance, acid pretreatment of tumor cells enhance their ability to form metastases in tumor-transplanted mice (Rofstad et al., 2006). Consistently, it has been shown that increasing tumor pH via bicarbonate therapy significantly reduces the number and the size of metastases in a mouse model of breast cancer (Robey et al., 2009).

Second, the extracellular pH of solid tumors is significantly more acidic than that of normal tissues, thus impairing the uptake of weakly basic chemotherapeutic drugs (Raghunand et al., 1999). Several anticancer drugs such as doxorubicin, mitoxantrone and vincristine are weak bases that are protonated in slightly acid tumor microenvironments. The protonated forms of the drugs cannot easily diffuse across the plasma membrane and therefore their cellular uptake is suppressed. It has been demonstrated that the addition of sodium bicarbonate in the drinking water enhanced the anti-tumor effect of doxorubicin on xenotransplanted tumors presumably by enhancing the intracellular drug delivery through raising the pH of the extracellular milieu in mice (Raghunand et al., 1999). The reverse situation was also demonstrated in another study showing that glucose administration to mice led to a lower efficacy of doxorubicin on tumors presumably due to a decrease in the extracellular pH (Gerweck et al., 2006).

Third, acidic microenvironment inhibits anti-tumor immune response. For instance, lactic acid suppressed the proliferation and cytokine production of human cytotoxic T lymphocytes (CTLs) up to 95% and led to a 50% decrease in cytotoxic activity (Fischer et al., 2007). Activated lymphocytes themselves use glycolysis, which relies on the efficient secretion of lactic acid. Export of lactic acid from lymphocytes depends on a gradient between intracellular and extracellular lactic acid concentration. High extracelullar acidity would diminish this gradient and block the secretion of lactic acid from lymphocytes. The accumulation of intracellular lactic acid eventually disturbs the glycolysis process hence affecting the activity of lymphocytes. Acidification similarly inhibits the activity of other immune cells such as dendritic cells.

5. Coupled biological and metabolic processes and the logic of a mammalian metabolic cycle

Glycolytic enzymes have multiple cellular functions. For instance, GAPDH has been implicated in numerous non-glycolytic functions (Colell et al., 2007; McKnight, 2003). In 2003, we published a paper that describes the isolation and characterization of OCA-S, which is a transcription cofactor complex that directly stimulates the transcription of the histone H2B gene in an S-phase-specific manner (Zheng et al., 2003). Surprisingly, a key component of the OCA-S complex represents a nuclear form of GAPDH, which regulates H2B transcription in a redox dependent manner. LDH was later shown to be an essential OCA-S component as well and can exercise the enzyme activity to reverse in vitro inhibition of H2B transcription by converting NADH to NAD^+ in the presence of substrate pyruvate (Dai et al., 2008). Conceivably, the participation of these glycolytic enzymes in such a cell cycle event would subject cell cycle regulation to altered glucose metabolism in cancer cells, providing yet another mechanistic explanation of cancer growth and development. The "moonlighting" participation of the glycolytic enzymes in a cellular process would in theory impose a dynamic modulation of the redox status in the cellular compartment where this process is executed and subsequently affect the functions of other redox-sensitive proteins in the same intracellular compartment. Thus, in addition to histone expression, our study has suggested that other cellular processes including cell cycle regulation, DNA replication and damage repair are potentially all coupled through the redox signals (Yu et al., 2009). The coupling of these cellular processes is apparently crucial for the maintenance of chromatin integrity during cell cycle, and thus altered glucose metabolism in cancer cells potentially would disrupt the coupling of these processes and make cancer genomes more error-prone.

It is well delineated that in yeast, quite a few biological and metabolic processes are known to be compartmentalized in time, termed the yeast metabolic cycle (YMC) that is in sync with the cell cycle progression (Tu et al., 2005). The YMC has oxidative, reductive/building and reductive/charging phases, and the S-phase of yeast cell is synchronized with the most reductive stage of YMC. We subsequently found that the oxidative and reductive phases in mammalian cells are also synchronized with the cell cycle, and our study demonstrated that the free NAD^+/NADH ratio fluctuated in a defined manner during cell cycle(Yu et al., 2009). At G1 phase, the intracellular NAD^+/NADH ratio is high, suggesting that G1 cells maintain an oxidative status. Upon entering S phase, the ratio becomes lower, corresponding to a reductive status. When the cells exit S phase and enter G2 phase, the NAD^+/NADH ratio becomes higher again. This oxidative status appears to be maintained until cells enter the next S phase. This phenomenon has been dubbed mammalian metabolic cycle (MMC) for its similarity with YMC. The fluctuating NAD^+/NADH ratios in a mammalian cell cycle must reflect overall oscillatory cellular metabolism; whether and how glucose metabolism is synchronized with the cell cycle remains to be explored. Nonetheless, this synchronization must have been very precisely regulated. Conceivably, if glucose metabolism is altered, the cell cycle must be coordinately modulated, and vice versa. Therefore, cancer cells may have acquired the growth and proliferative advantage over normal cells through alteration in glucose metabolism.

6. Targeting glycolysis for cancer treatment

The aberrant metabolic pathways underlying the Warburg effect are being considered as novel targets for cancer therapy. Several strategies have been employed to target glucose metabolic pathways for cancer treatment.

First, inhibitors of glycolytic enzymes or glycolytic pathways are being searched to identify therapeutic agents that can inhibit cancer growth and development.

A number of small molecules have been reported to target glycolysis although none to date has been shown to have specific molecular targets. For example, 3-bromopyruvate, a highly active alkylating agent, was reported to target HK-2 and/or GAPDH (Dang et al., 2009). 2-deoxyglucose (2-DG) can be phosphorylated by HK-2, which in turn inhibits HK-2 (Ralser et al., 2008). It is also shown to be a GLUT inhibitor.

Many efforts have been made to identify specific inhibitors. Lonidamine has been described as a specific inhibitor of mitochondria-bound HK (Floridi et al., 1981) and has been tested in multiple clinical trials including a phase II study in combination with diazepam for the treatment of glioblastoma patients (Porporato et al., 2011). Unfortunately, none of these clinical trials have successfully shown its therapeutic benefit in terms of time-to-progression and overall survival (Oudard et al., 2003); one study showed its severe hepatic adverse effects.

Drugs that target more specific metabolic control points of glycolysis in cancer cells, such as PKM2 or LDH-A, warrant investigation as potential cancer therapies. gossypol/AT-101, a natural product and a non-specific LDH inhibitor that has more preferential inhibitory activity on malarial and spermoctye LDHs, has already been tested in human clinical trials for its anti-cancer effect(Porporato et al., 2011). A selective competitive inhibitor of LDH5, 3-dihydroxy-6-methyl-7-(phenylmethyl)-4-propylnaphthalene-1-carboxylicacid (FX11), has been identified through screening a library of compounds derived from gossypol (Yu et al., 2001). FX11 has been shown to suppress in vivo xenograft tumor growth of human B lymphoid tumor and pancreatic cancer cells (Le et al. 2010), providing a strong rationale for clinical development of therapeutic agents targeting LDH-A. Recently, *N*-Hydroxy-2-carboxy-substituted indole compounds have been identified as LDH5-specific inhibitors (Granchi et al., 2011).

Several clinical trials with the PKM2 inhibitor TLN-232/CAP-232, a seven amino-acid peptide, have been initiated. Encouraging preliminary results demonstrated that it is safe, well tolerated, and may offer disease control (Porporato et al., 2011). New small molecule inhibitors of PKM2 were also screened. Among them, the most potent one resulted in decreased glycolysis and increased cell death in respond to loss of growth factor signaling, supporting the feasibility and viability of targeting glucose metabolism as a novel strategy to treating human cancers (Vander Heiden et al., 2010a).

Conceivably, ENOA, as a tumor-associated antigen with an ability of eliciting both B cell and T cell immune response (Capello et al., 2011), is an ideal target of cancer vaccine and immunotherapy. Comparing to small molecule inhibitors, immunotherapy offers superior target specificity and may be used as an alternative approach to target other glycolysis enzymes.

Second, inhibition of glycolytic enzyme or glycolysis pathways serves as a strategy to enhance the sensitivity of tumor cells to conventional cytotoxic chemotherapy agents.

Inhibition of LDH-A has been shown to re-sensitize Taxol-resistant cancer cells to Taxol (Zhou et al.). 2-DG is another example. The safety of using it as an anti-cancer agent has been questioned notably because of brain toxicity (Tennant et al., 2010). However, it has a proven efficacy in sensitizing human osteosarcoma and non-small cell lung cancers to adriamycin and paclitaxel (Maschek et al., 2004). Recently, a Phase I clinical study for prostate cancer has defined a maximum tolerance dose of 45mg/kg for Phase II trials (Stein et al., 2010). It will be interesting to test whether 2-DG can enhance the efficacy of chemotherapy agents even if it cannot offer anti-cancer activity by itself at this dose level.

As aforementioned, chemosensitivity is enhanced by counteracting the acidification of tumor's microenvironment. Inhibitors of glycolytic enzymes may impose an alkalizing effect in tumor's microenvironment particularly at the tumor tissue level and thus may have a more specific and powerful role in enhancing the sensitivity of tumor cells to basic chemotherapy drugs. Major targets for counteracting the acidification of tumor's microenvironment include carbonic anhydrasesis (CA)-9 and -12, sodium–proton exchanger 1 (NHE1), sodium bicarbonate cotransporter (NBC), vacuolar ATPase (V-ATPase), sodium–potassium (NaK) ATPase, and MCT4 (Porporato et al., 2011). Indisulam is a leading compound for CA9 inhibition. It is a sulfonamide derivative and shown to inhibit CA9 at nanomolar concentrations (Abbate et al., 2004; Supuran, 2008; Owa et al., 2002). It has been tested in multiple clinical trials for the treatment of melanoma, lung, pancreatic and metastatic breast cancers and has not been found in the completed clinical trials to have antitumor efficacy as a single agent (Talbot et al., 2007). Girentuximab, a specific antibody targeting CA9, is now being tested in Phase III clinical trials for the treatment of clear-cell renal cell carcinoma (Reichert, 2011). Several inhibitors of membrane-bound V-ATPase have been reported to have antitumor activity in preclinical studies (Perez-Sayans et al., 2009). Other enzymes involved in the cellular export of protons are also studied as targets of anti-cancer therapeutic development. However, future studies should emphasize on combining anti-acidification therapies with cytotoxic chemotherapy or immunotherapy to achieve the effective anti-cancer treatment.

Third, combination of inhibitors of glucose metabolic enzymes with inhibitors of oncogenic pathways may result in synergistic anti-tumor effects.

Inhibitors of oncogenic pathways have been extensively tested for cancer therapy, with only moderate success in a few types of human cancers. Among them, inhibitors of the Ras pathway are essentially not effective. As K-ras mutated cancer cells have an enhanced survival when glucose is deprived, a combinatorial treatment with both glycolysis inhibitors and Ras pathway inhibitors may target Ras-mutated cancer cells more effectively and more specifically. Supporting this hypothesis, the hexokinase inhibitor 3-bromopyruvate was demonstrated to be highly toxic specifically to cancer cells with K-ras mutation, but not to cancer cells with wild-type K-ras (Yun et al., 2009). BAY87-2243, which is a small molecule inhibitor of HIF-1 activity and of HIF-1α stability, and EZN-2968, which is an antisense oligonucleotide targeting HIF-1α, had been tested in clinical trials (Greenberger et al., 2008). Metformin is an AMPK-activating drug and is currently used for type-2 diabetes treatment. Epidemiological studies have shown reduced incidence of cancer in diabetic patients treated with metformin (Evans et al., 2005; Jalving et al., 2010; Libby et al., 2009). It is highly intriguing to test whether this clinically safe and known glucose metabolism modulating drug can enhance anti-cancer activity of cytotoxic chemotherapy and/or further lower the cancer recurrence following adjuvant chemotherapy. Similarly, other combination treatments with inhibitors of both glucose metabolisms and oncogenic pathways such as Akt, mTOR, etc. also warrant investigation.

7. References

Abbate, F., Casini, A., Owa, T., Scozzafava, A., and Supuran, C.T. (2004). Carbonic anhydrase inhibitors: E7070, a sulfonamide anticancer agent, potently inhibits cytosolic isozymes I and II, and transmembrane, tumor-associated isozyme IX. *Bioorg Med Chem Lett 14*:217-223.

Annibaldi, A., and Widmann, C. (2011). Glucose metabolism in cancer cells. *Current opinion in clinical nutrition and metabolic care 13*:466-470.

Arsham, A.M., Plas, D.R., Thompson, C.B., and Simon, M.C. (2002). Phosphatidylinositol 3-kinase/Akt signaling is neither required for hypoxic stabilization of HIF-1 alpha nor sufficient for HIF-1-dependent target gene transcription. *The Journal of biological chemistry 277*:15162-15170.

Atsumi, T., Chesney, J., Metz, C., Leng, L., Donnelly, S., Makita, Z., Mitchell, R., and Bucala, R. (2002). High expression of inducible 6-phosphofructo-2-kinase/fructose-2,6-bisphosphatase (iPFK-2; PFKFB3) in human cancers. *Cancer research 62*:5881-5887.

Bensaad, K., Tsuruta, A., Selak, M.A., Vidal, M.N., Nakano, K., Bartrons, R., Gottlieb, E., and Vousden, K.H. (2006). TIGAR, a p53-inducible regulator of glycolysis and apoptosis. *Cell 126*:107-120.

Bustamante, E., and Pedersen, P.L. (1977). High aerobic glycolysis of rat hepatoma cells in culture: role of mitochondrial hexokinase. *Proceedings of the National Academy of Sciences of the United States of America 74*:3735-3739.

Capello, M., Ferri-Borgogno, S., Cappello, P., and Novelli, F. (2011). alpha-Enolase: a promising therapeutic and diagnostic tumor target. *The FEBS journal 278*:1064-1074.

Cappello, P., Tomaino, B., Chiarle, R., Ceruti, P., Novarino, A., Castagnoli, C., Migliorini, P., Perconti, G., Giallongo, A., Milella, M., *et al.* (2009). An integrated humoral and cellular response is elicited in pancreatic cancer by alpha-enolase, a novel pancreatic ductal adenocarcinoma-associated antigen. *Int J Cancer 125*:639-648.

Chen, C., Pore, N., Behrooz, A., Ismail-Beigi, F., and Maity, A. (2001). Regulation of glut1 mRNA by hypoxia-inducible factor-1. Interaction between H-ras and hypoxia. *The Journal of biological chemistry 276:* 9519-9525.

Christofk, H.R., Vander Heiden, M.G., Harris, M.H., Ramanathan, A., Gerszten, R.E., Wei, R., Fleming, M.D., Schreiber, S.L., and Cantley, L.C. (2008). The M2 splice isoform of pyruvate kinase is important for cancer metabolism and tumour growth. *Nature 452*:230-233.

Colell, A., Green, D.R., and Ricci, J.E. (2009). Novel roles for GAPDH in cell death and carcinogenesis. *Cell death and differentiation 16*:1573-1581.

Colell, A., Ricci, J.E., Tait, S., Milasta, S., Maurer, U., Bouchier-Hayes, L., Fitzgerald, P., Guio-Carrion, A., Waterhouse, N.J., Li, C.W., *et al.* (2007). GAPDH and autophagy preserve survival after apoptotic cytochrome c release in the absence of caspase activation. *Cell 129*:983-997.

Cui, Y., Tian, M., Zong, M., Teng, M., Chen, Y., Lu, J., Jiang, J., Liu, X., and Han, J. (2009). Proteomic analysis of pancreatic ductal adenocarcinoma compared with normal adjacent pancreatic tissue and pancreatic benign cystadenoma. *Pancreatology 9*:89-98.

Dai, R.P., Yu, F.X., Goh, S.R., Chng, H.W., Tan, Y.L., Fu, J.L., Zheng, L., and Luo, Y. (2008). Histone 2B (H2B) expression is confined to a proper NAD+/NADH redox status. *The Journal of biological chemistry 283*:26894-26901.

Dang, C.V., Le, A., and Gao, P. (2009). MYC-induced cancer cell energy metabolism and therapeutic opportunities. *Clin Cancer Res 15*:6479-6483.

Discher, D.J., Bishopric, N.H., Wu, X., Peterson, C.A., and Webster, K.A. (1998). Hypoxia regulates beta-enolase and pyruvate kinase-M promoters by modulating Sp1/Sp3 binding to a conserved GC element. *The Journal of biological chemistry 273*:26087-26093.

Durany, N., Joseph, J., Campo, E., Molina, R., and Carreras, J. (1997). Phosphoglycerate mutase, 2,3-bisphosphoglycerate phosphatase and enolase activity and isoenzymes in lung, colon and liver carcinomas. *British journal of cancer 75*:969-977.

Durany, N., Joseph, J., Jimenez, O.M., Climent, F., Fernandez, P.L., Rivera, F., and Carreras, J. (2000). Phosphoglycerate mutase, 2,3-bisphosphoglycerate phosphatase, creatine kinase and enolase activity and isoenzymes in breast carcinoma. *British journal of cancer 82*:20-27.

Evans, J.M., Donnelly, L.A., Emslie-Smith, A.M., Alessi, D.R., and Morris, A.D. (2005). Metformin and reduced risk of cancer in diabetic patients. *BMJ 330*:1304-1305.

Feng, Z., Hu, W., de Stanchina, E., Teresky, A.K., Jin, S., Lowe, S., and Levine, A.J. (2007). The regulation of AMPK beta1, TSC2, and PTEN expression by p53: stress, cell and tissue specificity, and the role of these gene products in modulating the IGF-1-AKT-mTOR pathways. *Cancer research 67*:3043-3053.

Fischer, K., Hoffmann, P., Voelkl, S., Meidenbauer, N., Ammer, J., Edinger, M., Gottfried, E., Schwarz, S., Rothe, G., Hoves, S., *et al.* (2007). Inhibitory effect of tumor cell-derived lactic acid on human T cells. *Blood 109*:3812-3819.

Floridi, A., Paggi, M.G., Marcante, M.L., Silvestrini, B., Caputo, A., and De Martino, C. (1981). Lonidamine, a selective inhibitor of aerobic glycolysis of murine tumor cells. *J Natl Cancer Inst 66*:497-499.

Funasaka, T., Yanagawa, T., Hogan, V., and Raz, A. (2005). Regulation of phosphoglucose isomerase/autocrine motility factor expression by hypoxia. *Faseb J 19*:1422-1430.

Gerweck, L.E., Vijayappa, S., and Kozin, S. (2006). Tumor pH controls the in vivo efficacy of weak acid and base chemotherapeutics. *Molecular cancer therapeutics 5*:1275-1279.

Gottlob, K., Majewski, N., Kennedy, S., Kandel, E., Robey, R.B., and Hay, N. (2001). Inhibition of early apoptotic events by Akt/PKB is dependent on the first committed step of glycolysis and mitochondrial hexokinase. *Genes & development 15*:1406-1418.

Granchi, C., Roy, S., Giacomelli, C., Macchia, M., Tuccinardi, T., Martinelli, A., Lanza, M., Betti, L., Giannaccini, G., Lucacchini, A., *et al.* (2011). Discovery of Nhydroxyindole-based inhibitors of human lactate dehydrogenase isoform A (LDH-A) as starvation agents against cancer cells. *J Med Chem 54*:1599-1612.

Greenberger, L.M., Horak, I.D., Filpula, D., Sapra, P., Westergaard, M., Frydenlund, H.F., Albaek, C., Schroder, H., and Orum, H. (2008). A RNA antagonist of hypoxia-inducible factor-1alpha, EZN-2968, inhibits tumor cell growth. *Molecular cancer therapeutics 7*:3598-3608.

Hacker, H.J., Steinberg, P., and Bannasch, P. (1998). Pyruvate kinase isoenzyme shift from L-type to M2-type is a late event in hepatocarcinogenesis induced in rats by a choline-deficient/DL-ethionine-supplemented diet. *Carcinogenesis 19*:99-107.

Herling, A., Konig, M., Bulik, S., and Holzhutter, H.G. (2011). Enzymatic features of the glucose metabolism in tumor cells. *The FEBS journal 278*:2436-2459.

Hwang, T.L., Liang, Y., Chien, K.Y., and Yu, J.S. (2006). Overexpression and elevated serum levels of phosphoglycerate kinase 1 in pancreatic ductal adenocarcinoma. *Proteomics 6*:2259-2272.

Jalving, M., Gietema, J.A., Lefrandt, J.D., de Jong, S., Reyners, A.K., Gans, R.O., and de Vries, E.G. (2010). Metformin: taking away the candy for cancer? *Eur J Cancer 46*:2369-2380.

Kessler, R., Bleichert, F., Warnke, J.P., and Eschrich, K. (2008). 6-Phosphofructo-2-kinase/fructose-2,6-bisphosphatase (PFKFB3) is up-regulated in high-grade astrocytomas. *J Neurooncol 86*:257-264.

Kim, J.W., Gao, P., Liu, Y.C., Semenza, G.L., and Dang, C.V. (2007). Hypoxia-inducible factor 1 and dysregulated c-Myc cooperatively induce vascular endothelial growth factor and metabolic switches hexokinase 2 and pyruvate dehydrogenase kinase 1. *Molecular and cellular biology 27*:7381-7393.

Kondo, H., Sahara, H., Miyazaki, A., Nabeta, Y., Hirohashi, Y., Kanaseki, T., Yamaguchi, A., Yamada, N., Hirayama, K., Suzuki, M., *et al.* (2002). Natural antigenic peptides from squamous cell carcinoma recognized by autologous HLA-DR8-restricted CD4+ T cells. *Jpn J Cancer Res 93*:917-924.

Kondoh, H., Lleonart, M.E., Gil, J., Wang, J., Degan, P., Peters, G., Martinez, D., Carnero, A., and Beach, D. (2005). Glycolytic enzymes can modulate cellular life span. *Cancer research 65*:177-185.

Koukourakis, M.I., Giatromanolaki, A., Simopoulos, C., Polychronidis, A., and Sivridis, E. (2005). Lactate dehydrogenase 5 (LDH5) relates to up-regulated hypoxia inducible factor pathway and metastasis in colorectal cancer. *Clin Exp Metastasis 22*:25-30.

Koukourakis, M.I., Giatromanolaki, A., Sivridis, E., Bougioukas, G., Didilis, V., Gatter, K.C., and Harris, A.L. (2003). Lactate dehydrogenase-5 (LDH-5) overexpression in non-small-cell lung cancer tissues is linked to tumour hypoxia, angiogenic factor production and poor prognosis. *British journal of cancer 89*: 877-885.

Koukourakis, M.I., Giatromanolaki, A., Winter, S., Leek, R., Sivridis, E., and Harris, A.L. (2009). Lactate dehydrogenase 5 expression in squamous cell head and neck cancer relates to prognosis following radical or postoperative radiotherapy. *Oncology 77*:285-292.

Le, A., Cooper, C.R., Gouw, A.M., Dinavahi, R., Maitra, A., Deck, L.M., Royer, R.E., Vander Jagt, D.L., Semenza, G.L., and Dang, C.V. (2010). Inhibition of lactate dehydrogenase A induces oxidative stress and inhibits tumor progression. *Proceedings of the National Academy of Sciences of the United States of America 107*:2037-2042.

Leiblich, A., Cross, S.S., Catto, J.W., Phillips, J.T., Leung, H.Y., Hamdy, F.C., and Rehman, I. (2006). Lactate dehydrogenase-B is silenced by promoter hypermethylation in human prostate cancer. *Oncogene 25*:2953-2960.

Levine, A.J., and Puzio-Kuter, A.M (2010). The control of the metabolic switch in cancers by oncogenes and tumor suppressor genes. *Science 330*:1340-1344.

Libby, G., Donnelly, L.A., Donnan, P.T., Alessi, D.R., Morris, A.D., and Evans, J.M. (2009). New users of metformin are at low risk of incident cancer: a cohort study among people with type 2 diabetes. *Diabetes Care 32*:1620-1625.

Luo, W., Hu, H., Chang, R., Zhong, J., Knabel, M., O'Meally, R., Cole, R.N., Pandey, A., and Semenza, G.L. (2011). Pyruvate kinase M2 is a PHD3-stimulated coactivator for hypoxia-inducible factor 1. *Cell 145*:732-744.

Markert, C.L., Shaklee, J.B., and Whitt, G.S. (1975). Evolution of a gene. Multiple genes for LDH isozymes provide a model of the evolution of gene structure, function and regulation. *Science 189*:102-114.

Marsin, A.S., Bouzin, C., Bertrand, L., and Hue, L. (2002). The stimulation of glycolysis by hypoxia in activated monocytes is mediated by AMP-activated protein kinase and inducible 6-phosphofructo-2-kinase. *The Journal of biological chemistry 277*:30778-30783.

Maschek, G., Savaraj, N., Priebe, W., Braunschweiger, P., Hamilton, K., Tidmarsh, G.F., De Young, L.R., and Lampidis, T.J. (2004). 2-deoxy-D-glucose increases the efficacy of adriamycin and paclitaxel in human osteosarcoma and non-small cell lung cancers in vivo. *Cancer research 64*:31-34.

Mathupala, S.P., Ko, Y.H., and Pedersen, P.L. (2009). Hexokinase-2 bound to mitochondria: cancer's stygian link to the "Warburg Effect" and a pivotal target for effective therapy. *Seminars in cancer biology 19*:17-24.

Mazurek, S., Boschek, C.B., Hugo, F., and Eigenbrodt, E. (2005). Pyruvate kinase type M2 and its role in tumor growth and spreading. Seminars in cancer biology *15*, 300-308.

McCarty, M.F., and Whitaker, J. Manipulating tumor acidification as a cancer treatment strategy. *Altern Med Rev 15*:264-272.

McKnight, S. (2003). Gene switching by metabolic enzymes--how did you get on the invitation list? *Cell 114*:150-152.

Medina, R.A., and Owen, G.I. (2002). Glucose transporters: expression, regulation and cancer. *Biological research 35*: 9-26.

Meldolesi, M.F., Macchia, V., and Laccetti, P. (1976). Differences in phosphofructokinase regulation in normal and tumor rat thyroid cells. *The Journal of biological chemistry 251*: 6244-6251.

Minchenko, A., Leshchinsky, I., Opentanova, I., Sang, N., Srinivas, V., Armstead, V., and Caro, J. (2002). Hypoxia-inducible factor-1-mediated expression of the 6-phosphofructo-2-kinase/fructose-2,6-bisphosphatase-3 (PFKFB3) gene. Its possible role in the Warburg effect. *The Journal of biological chemistry 277*: 6183-6187.

Minchenko, O.H., Ochiai, A., Opentanova, I.L., Ogura, T., Minchenko, D.O., Caro, J., Komisarenko, S.V., and Esumi, H. (2005). Overexpression of 6-phosphofructo-2-kinase/fructose-2,6-bisphosphatase-4 in the human breast and colon malignant tumors. *Biochimie 87*: 1005-1010.

Nakashima, R.A., Mangan, P.S., Colombini, M., and Pedersen, P.L. (1986). Hexokinase receptor complex in hepatoma mitochondria: evidence from N,N'-

dicyclohexylcarbodiimide-labeling studies for the involvement of the pore-forming protein VDAC. *Biochemistry 25*: 1015-1021.

Noguchi, Y., Yoshikawa, T., Marat, D., Doi, C., Makino, T., Fukuzawa, K., Tsuburaya, A., Satoh, S., Ito, T., and Mitsuse, S. (1998). Insulin resistance in cancer patients is associated with enhanced tumor necrosis factor-alpha expression in skeletal muscle. *Biochemical and biophysical research communications 253*:887-892.

Osthus, R.C., Shim, H., Kim, S., Li, Q., Reddy, R., Mukherjee, M., Xu, Y., Wonsey, D., Lee, L.A., and Dang, C.V. (2000). Deregulation of glucose transporter 1 and glycolytic gene expression by c-Myc. *The Journal of biological chemistry 275*:21797-21800.

Oudard, S., Carpentier, A., Banu, E., Fauchon, F., Celerier, D., Poupon, M.F., Dutrillaux, B., Andrieu, J.M., and Delattre, J.Y. (2003). Phase II study of lonidamine and diazepam in the treatment of recurrent glioblastoma multiforme. *J Neurooncol 63*:81-86.

Owa, T., Yokoi, A., Yamazaki, K., Yoshimatsu, K., Yamori, T., and Nagasu, T. (2002). Array-based structure and gene expression relationship study of antitumor sulfonamides including N-[2-[(4-hydroxyphenyl)amino]-3-pyridinyl]-4-methoxy-benzenesulfonamide and N-(3-chloro-7-indolyl)-1,4-benzenedisulfo-namide. *J Med Chem 45*:4913-4922.

Pastorino, J.G., Shulga, N., and Hoek, J.B. (2002). Mitochondrial binding of hexokinase II inhibits Bax-induced cytochrome c release and apoptosis. *The Journal of biological chemistry 277*:7610-7618.

Perez-Sayans, M., Somoza-Martin, J.M., Barros-Angueira, F., Rey, J.M., and Garcia-Garcia, A. (2009). V-ATPase inhibitors and implication in cancer treatment. *Cancer Treat Rev 35*:707-713.

Porporato, P.E., Dhup, S., Dadhich, R.K., Copetti, T., and Sonveaux, P. (2011). Anticancer targets in the glycolytic metabolism of tumors: a comprehensive review. *Front Pharmacol 2*:49.

Raghunand, N., He, X., van Sluis, R., Mahoney, B., Baggett, B., Taylor, C.W., Paine-Murrieta, G., Roe, D., Bhujwalla, Z.M., and Gillies, R.J. (1999). Enhancement of chemotherapy by manipulation of tumour pH. *British journal of cancer 80*:1005-1011.

Ralser, M., Wamelink, M.M., Struys, E.A., Joppich, C., Krobitsch, S., Jakobs, C., and Lehrach, H. (2008). A catabolic block does not sufficiently explain how 2-deoxy-D-glucose inhibits cell growth. *Proceedings of the National Academy of Sciences of the United States of America 105*:17807-17811.

Rathmell, J.C., Fox, C.J., Plas, D.R., Hammerman, P.S., Cinalli, R.M., and Thompson, C.B. (2003). Akt-directed glucose metabolism can prevent Bax conformation change and promote growth factor-independent survival. *Molecular and cellular biology 23*:7315-7328.

Reichert, J.M. (2011). Antibody-based therapeutics to watch in 2011. MAbs 3, 76-99.

Reinacher, M., and Eigenbrodt, E. (1981). Immunohistological demonstration of the same type of pyruvate kinase isoenzyme (M2-Pk) in tumors of chicken and rat. *Virchows Arch B Cell Pathol Incl Mol Pathol 37*:79-88.

Rempel, A., Mathupala, S.P., Griffin, C.A., Hawkins, A.L., and Pedersen, P.L. (1996). Glucose catabolism in cancer cells: amplification of the gene encoding type II hexokinase. *Cancer research 56*:2468-2471.

Robert, J., Van Rymenant, M., and Lagae, F. (1961). Enzymes in cancer. III. Triosephosphate isomerase activity of human blood serum in normal individuals and in individuals with various pathological conditions. *Cancer 14*:1166-1174.

Robey, I.F., Baggett, B.K., Kirkpatrick, N.D., Roe, D.J., Dosescu, J., Sloane, B.F., Hashim, A.I., Morse, D.L., Raghunand, N., Gatenby, R.A., *et al.* (2009). Bicarbonate increases tumor pH and inhibits spontaneous metastases. *Cancer research 69*:2260-2268.

Rofstad, E.K., Mathiesen, B., Kindem, K., and Galappathi, K. (2006). Acidic extracellular pH promotes experimental metastasis of human melanoma cells in athymic nude mice. *Cancer research 66*:6699-6707.

Sedoris, K.C., Thomas, S.D., and Miller, D.M. (2010). Hypoxia induces differential translation of enolase/MBP-1. *BMC cancer 10*:157.

Semenza, G.L. (2003). Targeting HIF-1 for cancer therapy. Nature reviews *3*, 721-732.

Shaw, R.J., and Cantley, L.C. (2006). Ras, PI(3)K and mTOR signalling controls tumour cell growth. *Nature 441*:424-430.

Shim, H., Dolde, C., Lewis, B.C., Wu, C.S., Dang, G., Jungmann, R.A., Dalla-Favera, R., and Dang, C.V. (1997). c-Myc transactivation of LDH-A: implications for tumor metabolism and growth. *Proceedings of the National Academy of Sciences of the United States of America 94*:6658-6663.

Smith, T.A. (1999). Facilitative glucose transporter expression in human cancer tissue. *British journal of biomedical science 56*:285-292.

Song, H., Xia, S.L., Liao, C., Li, Y.L., Wang, Y.F., Li, T.P., and Zhao, M.J. (2004). Genes encoding Pir51, Beclin 1, RbAp48 and aldolase b are up or down-regulated in human primary hepatocellular carcinoma. *World J Gastroenterol 10*:509-513.

Stein, M., Lin, H., Jeyamohan, C., Dvorzhinski, D., Gounder, M., Bray, K., Eddy, S., Goodin, S., White, E., and Dipaola, R.S. (2010). Targeting tumor metabolism with 2-deoxyglucose in patients with castrate-resistant prostate cancer and advanced malignancies. *Prostate 70*:1388-1394.

Supuran, C.T. (2008). Development of small molecule carbonic anhydrase IX inhibitors. *BJU Int 101 Suppl 4*:39-40.

Talbot, D.C., von Pawel, J., Cattell, E., Yule, S.M., Johnston, C., Zandvliet, A.S., Huitema, A.D., Norbury, C.J., Ellis, P., Bosquee, L., *et al.* (2007). A randomized phase II pharmacokinetic and pharmacodynamic study of indisulam as second-line therapy in patients with advanced non-small cell lung cancer. *Clin Cancer Res 13*:1816-1822.

Tamesa, M.S., Kuramitsu, Y., Fujimoto, M., Maeda, N., Nagashima, Y., Tanaka, T., Yamamoto, S., Oka, M., and Nakamura, K. (2009). Detection of autoantibodies against cyclophilin A and triosephosphate isomerase in sera from breast cancer patients by proteomic analysis. *Electrophoresis 30*:2168-2181.

Tennant, D.A., Duran, R.V., and Gottlieb, E. (2010). Targeting metabolic transformation for cancer therapy. *Nature reviews 10*:267-277.

Thangaraju, M., Carswell, K.N., Prasad, P.D., and Ganapathy, V. (2009). Colon cancer cells maintain low levels of pyruvate to avoid cell death caused by inhibition of HDAC1/HDAC3. *Biochem J 417*:379-389.

Tomaino, B., Cappello, P., Capello, M., Fredolini, C., Sperduti, I., Migliorini, P., Salacone, P., Novarino, A., Giacobino, A., Ciuffreda, L., *et al.* (2011). Circulating autoantibodies to phosphorylated alpha-enolase are a hallmark of pancreatic cancer. *J Proteome Res 10*:105-112.

Tu, B.P., Kudlicki, A., Rowicka, M., and McKnight, S.L. (2005). Logic of the yeast metabolic cycle: temporal compartmentalization of cellular processes. *Science 310*:1152-1158.

Vander Heiden, M.G., Cantley, L.C., and Thompson, C.B. (2009). Understanding the Warburg effect: the metabolic requirements of cell proliferation. *Science 324*:1029-1033.

Vander Heiden, M.G., Christofk, H.R., Schuman, E., Subtelny, A.O., Sharfi, H., Harlow, E.E., Xian, J., and Cantley, L.C. (2010a). Identification of small molecule inhibitors of pyruvate kinase M2. *Biochem Pharmacol 79*:1118-1124.

Vander Heiden, M.G., Locasale, J.W., Swanson, K.D., Sharfi, H., Heffron, G.J., Amador-Noguez, D., Christofk, H.R., Wagner, G., Rabinowitz, J.D., Asara, J.M., *et al.* (2010b). Evidence for an alternative glycolytic pathway in rapidly proliferating cells. *Science 329*:1492-1499.

Warburg, O. (1956). On the origin of cancer cells. *Science 123*:309-314.

Warburg, O., Wind, F., and Negelein, E. (1927). The Metabolism of Tumors in the Body. *J Gen Physiol 8*:519-530.

Wolf, A., Agnihotri, S., Micallef, J., Mukherjee, J., Sabha, N., Cairns, R., Hawkins, C., and Guha, A. Hexokinase 2 is a key mediator of aerobic glycolysis and promotes tumor growth in human glioblastoma multiforme. *The Journal of experimental medicine 208*:313-326.

Yalcin, A., Telang, S., Clem, B., and Chesney, J. (2009). Regulation of glucose metabolism by 6-phosphofructo-2-kinase/fructose-2,6-bisphosphatases in cancer. *Experimental and molecular pathology 86*:174-179.

Yamada, K., and Noguchi, T. (1999). Regulation of pyruvate kinase M gene expression. Biochemical and biophysical research communications *256*:257-262.

Yu, F.X., Dai, R.P., Goh, S.R., Zheng, L., and Luo, Y. (2009). Logic of a mammalian metabolic cycle: an oscillated NAD+/NADH redox signaling regulates coordinated histone expression and S-phase progression. *Cell cycle 8*: 773-779.

Yu, Y., Deck, J.A., Hunsaker, L.A., Deck, L.M., Royer, R.E., Goldberg, E., and Vander Jagt, D.L. (2001). Selective active site inhibitors of human lactate dehydrogenases A4, B4, and C4. *Biochem Pharmacol 62*:81-89.

Yun, H., Lee, M., Kim, S.S., and Ha, J. (2005). Glucose deprivation increases mRNA stability of vascular endothelial growth factor through activation of AMP-activated protein kinase in DU145 prostate carcinoma. *The Journal of biological chemistry 280*:9963-9972.

Yun, J., Rago, C., Cheong, I., Pagliarini, R., Angenendt, P., Rajagopalan, H., Schmidt, K., Willson, J.K., Markowitz, S., Zhou, S., *et al.* (2009). Glucose deprivation contributes to the development of KRAS pathway mutations in tumor cells. *Science 325*:1555-1559.

Zheng, L., Roeder, R.G., and Luo, Y. (2003). S phase activation of the histone H2B promoter by OCA-S, a coactivator complex that contains GAPDH as a key component. *Cell 114*:255-266.

Zhou, M., Zhao, Y., Ding, Y., Liu, H., Liu, Z., Fodstad, O., Riker, A.I., Kamarajugadda, S., Lu, J., Owen, L.B., *et al.* Warburg effect in chemosensitivity: targeting lactate dehydrogenase-A re-sensitizes taxol-resistant cancer cells to taxol. *Molecular cancer 9*: 33.

Zhou, W., Capello, M., Fredolini, C., Piemonti, L., Liotta, L.A., Novelli, F., and Petricoin, E.F. (2010). Mass spectrometry analysis of the post-translational modifications of alpha-enolase from pancreatic ductal adenocarcinoma cells. *J Proteome Res 9*:2929-2936.

Zundel, W., Schindler, C., Haas-Kogan, D., Koong, A., Kaper, F., Chen, E., Gottschalk, A.R., Ryan, H.E., Johnson, R.S., Jefferson, A.B., *et al.* (2000). Loss of PTEN facilitates HIF-1-mediated gene expression. *Genes & development 14*:391-396.

12

HIV-1 Selectively Integrates Into Host DNA *In Vitro*

Tatsuaki Tsuruyama
Department of Molecular Pathology, Graduate School of Medicine, Kyoto University
Kyoto, Kyoto Prefecture
Japan

1. Introduction

1.1 Summary

The biochemistry of retroviral integration selectivity is not fully understood. We modified the previously reported *in vitro* integration reaction protocol and developed a novel reaction system with higher efficiency. We used a DNA target composed of a repeat sequence DNA, 5′-(GTCCCTTCCCAGT)$_6$(ACTGGGAAGGGAC)$_6$-3′, that was ligated into a circular plasmid. Target DNA was reacted with a pre-integration (PI) complex that was formed by incubation of the end cDNA of the HIV-1 genome and recombinant integrase. It was confirmed that integration selectively occurred in the middle segment of the repeat sequence. On the other hand, both frequency and selectivity of integration markedly decreased when target sequences were used in which CAGT bases in the middle position of the original target sequence were deleted. Moreover, upon incubation with a combination of these deleted DNAs and the original sequence, the integration efficiency and selectivity towards the original target sequence were significantly reduced, which indicated interference effects by the deleted sequence DNAs. Efficiency and selectivity were also found to vary with changes in the manganese dichloride concentration of the reaction buffer, probably due to induction of fluctuation in the secondary structure of the substrate DNA. Such fluctuation may generate structural isomers that are favorable for selective integration into the target sequence DNA. In conclusion, there is considerable selectivity in HIV-integration into the specified target sequence. The present *in vitro* integration system will therefore be useful for monitoring viral integration activity or for testing of integrase inhibitors.

1.2 Background

Retroviral integration into host DNA is a critical step in the viral life cycle. Once integrated, the proviral genome will be stably duplicated along with the host cellular DNA duplication and will be transmitted to the daughter cells. Retroviruses can thus serve as powerful tools for the integration of foreign genes into a host genome (Fig. 1), and an MLV vector can be used to examine the function of introduced genes and the development of induced pluripotent stem (iPS) cells [1]. However, integrated retroviral genomes also have the potential to cause unexpected transformation through up-regulation of target genes by retroviral promoter elements located at the long terminal repeat (LTR) following integration [2]. Although

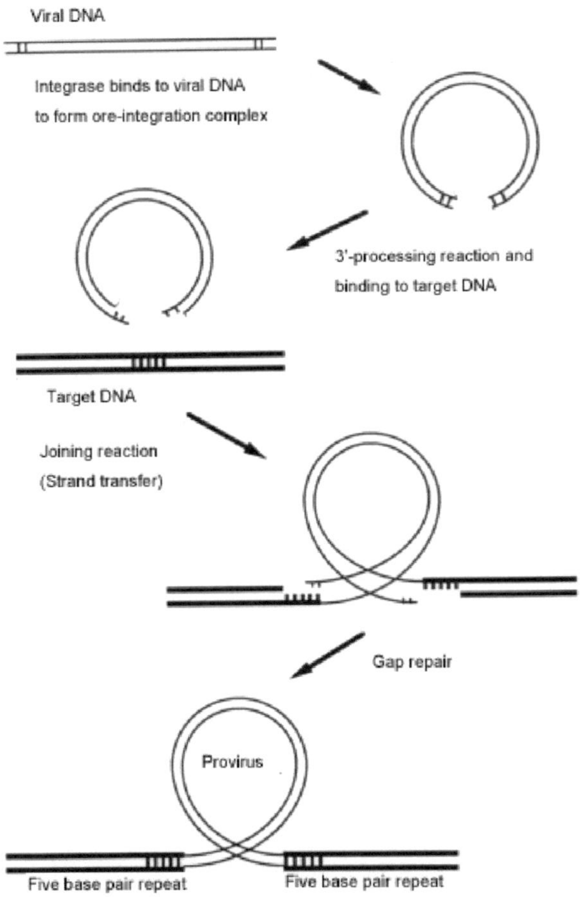

Fig. 1. Scheme of viral integration into target DNA: 3′-processing, join reaction, and gap repair. After the HIV-1 RNA genome has been transcribed into double-stranded DNA, the viral protein integrase binds to the termini of the viral DNA ends in a tetrameric fashion and the integrase creates overlapping 5′-ends by removing two nucleotides from the 3′-ends (3′end-processing). The HIV-1 DNA and the host cell DNA are ligated by synthesis of phosphodiester bonds between the terminal nucleotides of the viral 3′-ends and overlapping 5′-ends of the host chromosome. The non-homologous 5′-ends from the viral DNA are removed by integrase. Finally, the gaps are filled up by host cellular repair proteins, which recognize single strand breaks. A five base-repeat is observed in the flanking sequence after the gap repair reaction.

integration events have long been considered to be random, several recent findings have shown that integration of the murine leukemia retrovirus (MLV) and of HIV-1 is detected more frequently in actively transcribed genes [3, 4] or in promoter regions [5]. Previous statistical studies have also demonstrated that weak palindromic sequences are a common feature of the sites targeted for retroviral integration [6, 7]. Similar target preferences have also been reported for human T cell leukemia retrovirus type I (HTLV-I) integration sites [8]. Because of these findings, we investigated the biophysical mechanisms underlying *in vitro* integration. In the present study, we aimed to establish an *in vitro* integration assay using retroviral cDNA and integrase. Yoshinaga et al. previously reported the development of an *in vitro* integration assay using recombinant HIV-1 integrase, and short viral and target DNA sequences [9]. Using this method, they successfully detected retroviral cDNA-target DNA complexes *in vitro* and reported that the dinucleotide motif 5′-CA that is located at the proviral genome termini was essential for HIV-1 integration. Yoshinaga et al. called these dinucleotides the integration signal sequence. We modified Yoshinaga's method of *in vitro* integration in order to identify the precise HIV-1 integration sites using a target DNA that corresponds to an actual gene sequence.

In our previous study, we identified a common MLV integration site within the *signal transducer and activators of transcription* 5a (*Stat5a*) gene in MLV-induced spontaneous murine lymphoma in an inbred strain of mice, SL/Kh [10]. This is the first report of MLV integration target sequence. It has also been previously demonstrated that the *Stat5a* gene represents one of the common integration sites of MLV (the Mouse Retrovirus Tagged Cancer Gene Database (MRTCGD) (http://rtcgd.ncifcrf.gov/cgi-bin/mm7/easy_search.cgi) [11]. The encoded STAT5A protein is a transcription factor that is known to play an essential role in the development of myelo- and lympho-proliferative disease [12, 13]. In the current study, we modified this *Stat5a* gene sequence for use as a target for HIV-1 integration *in vitro*. This target gene consists of a 5′-CA-rich sequence, which may provide a useful clue for preparation of target DNA sequences, because terminal CA dinucleotide motifs are shared by MLV and HIV-1, as well as by HTLV-I proviruses.

2. *In vitro* integration assay used in previous studies

In many previous *in vitro* integration protocols, the double stranded DNA of the HIV-1 3'-LTR proviral end alone and the substrate DNA are mixed in an appropriate buffer containing $MnCl_2$ [14] (Fig. 2, 3). Subsequently, a PCR reaction using a primer set targeted to the proviral

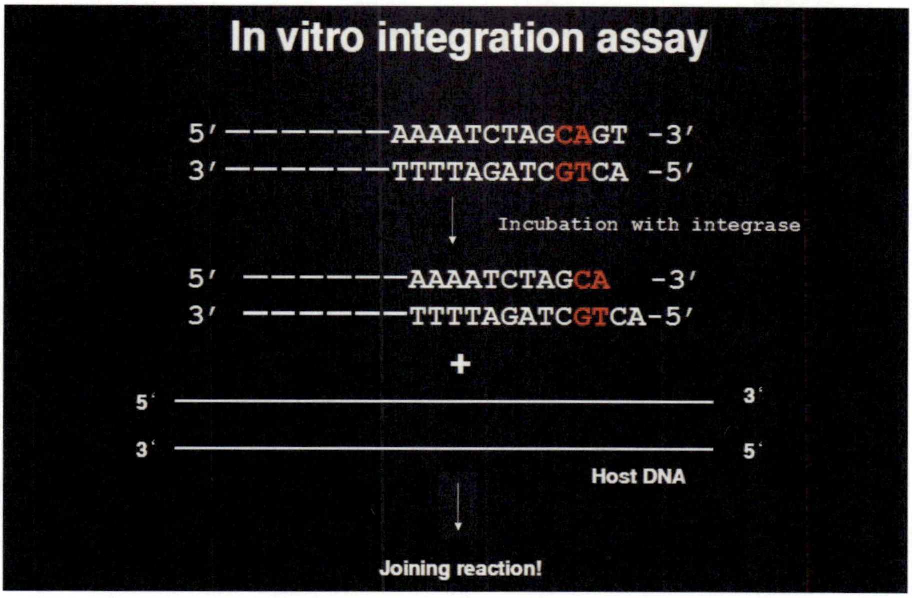

Fig. 2. Scheme of the previously described incubation with recombinant integrase

DNA and the target DNA amplifies a DNA segment that includes the viral-host DNA junction. In some protocols, even the viral DNA itself is used as the target DNA. Thus, previous studies paid little attention to the target sequence. It is commonly known that integrase binds to the proviral DNA in a regular tetrameric fashion. Indeed, some sequence motifs should be favored by an integrase oligomer, because a dimeric transcriptional factor protein has the ability to bind to palindromic sequence motifs such as E-box and GAS elements.

The 3′ end of the HIV-1 LTR sequence is shown. Red letters indicate the conserved dinucleotide motif. Incubation of the 3′ end of the HIV-1 LTR sequence DNA with integrase results in processing of the end of the 5′-GT dinucleotide. The resulting exposed hydroxyl group then attacks the target DNA.

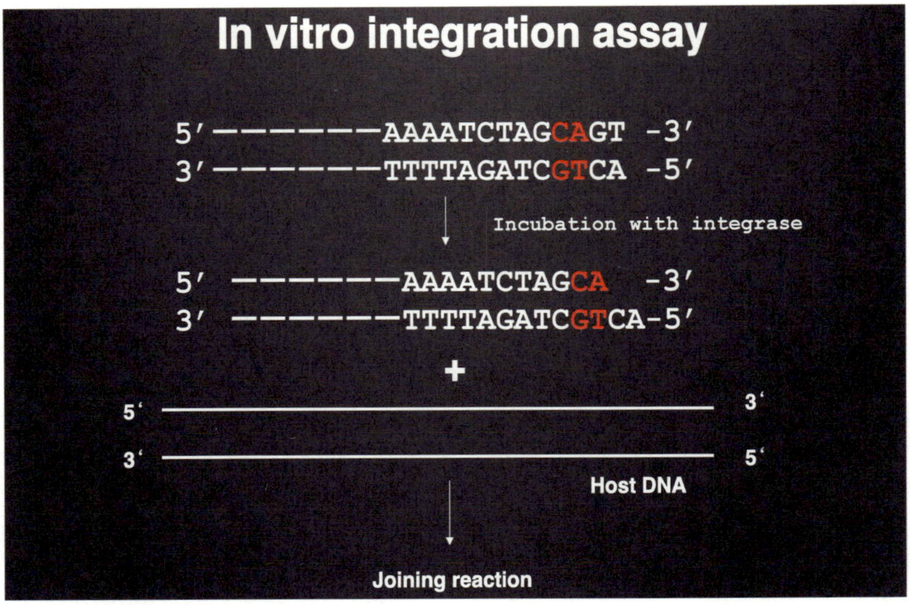

Fig. 3. Protocol of the previously described *in vitro* integration assay

A nick is introduced into the host DNA that is attached to a CA dinucleotide in the HIV-1 DNA end. Red letters indicate the conserved dinucleotide motif. Arrows indicate the PCR primer set and the direction of DNA polymerization.

2.1 Preparation of target sequences for *in vitro* integration

We recently reported the target sequence of MLV integration. We developed an inbred strain of mice suffering from spontaneous B cell lymphoma by MLV integration. MLV integration into *Stat5a* was identified in 25% of the lymphoma genome [15, 16] (Fig. 4). As depicted in Fig. 4, the hot spot of integration included a 5′-CA-rich sequence as well as a palindromic motif. Downward facing arrows in the figure indicated the MLV integration sites. The abundance of 5′-CA dinucleotides in the integration hot spots provided us with a hint for the preparation of target DNA for HIV integration, because these motifs are shared by the genome ends of MLV and HIV-1 (Fig. 5). We hypothesized that HIV-1 DNA also favors such a 5′-CA-rich sequence motif.

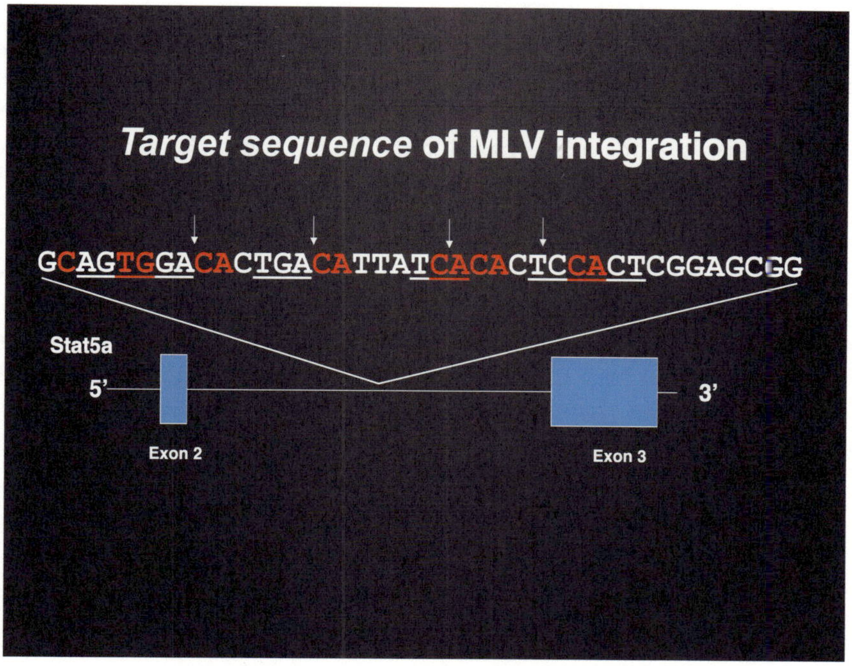

Fig. 4. Target sequence of MLV integration within the *Stat5a* gene.

Downward-facing arrows indicate the MLV integration sites (15, 16). Red letters highlight the CA-rich sequence. Blue boxes represent exons of the *Stat5a* gene. Underlines indicates the palindromic motif.

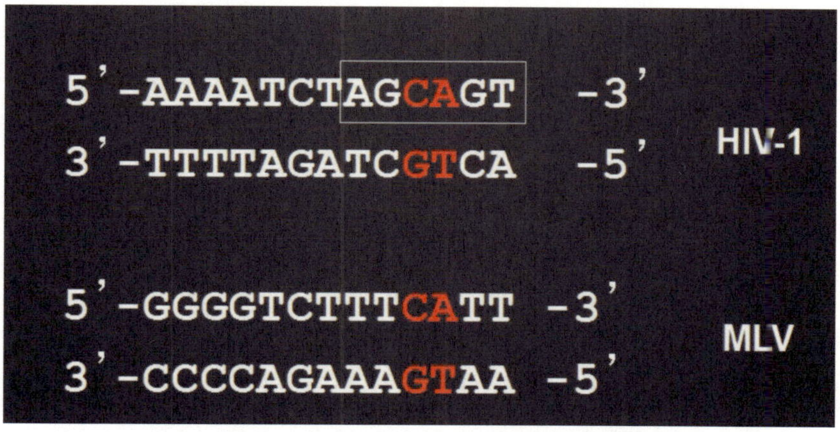

Fig. 5. Conserved 5′-CA dinucleotide in the proviral ends of HIV-1 and MLV.

A box indicate the integration signal sequence reported by Yoshinaga et al (9).

2.2 Modified *In vitro* integration assay

We then prepared a target sequence for HIV-1 integration. A repeat sequence was prepared in order to enhance integration efficiency. We used the repeat sequence, 5′-

(GTCCCTTCCC*AGT*)$_6$(*ACTG*GGAAGGGAC)$_6$-3′, or a modification of this sequence, which was ligated into a circular plasmid. The sequence within parenthesis is the unit of the repeat. This target sequence includes the 5′-CA dinucleotide motif, and includes 5′-AC at the HIV-1 DNA termini (Fig. 5). 5′-CAGT and 5′-ACTG (shown in *italics* in the above sequence) in the repeat units are also present in the HIV-1 proviral genome ends. This target DNA was reacted with recombinant integrase and formed a pre-integration (PI) complex. Figures 6 and 7 show our scheme of *in vitro* integration as well as the sequences of the HIV-1 proviral 5′- and 3′-ends. Following incubation of the proviral LTR sequence DNAs with recombinant integrase, the resultant pre-integration complexes were reacted with the target DNA. PCR amplification was performed and the integration sites were analyzed by direct sequencing. Unlike previously reported protocols, we used both 5′- and 3′-LTR sequences in our protocol. Such a target sequence unit was expected to directly interact with complementary HIV-1 DNA end sequences present in the target DNA. Complementarity between HIV-1 DNA and host DNA is shown in Fig. 8.

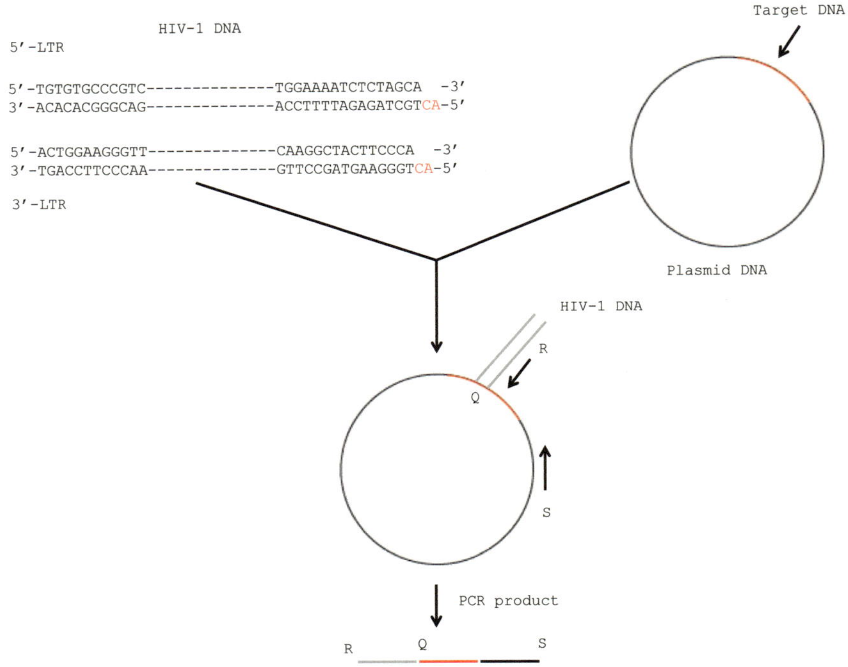

Fig. 6. Scheme of *in vitro* viral integration into the target sequence.

The red segment in the target sequence DNA that includes a circular plasmid represents the 144-bp target DNA, and the black line represents the remainder of the circular plasmid DNA used for ligation. Following incubation of the proviral LTR sequence DNAs with integrase, the resultant pre-integration complexes were reacted with the substrate DNA. Red letters in the HIV-1 cDNA represent the LTR termini. PCR amplification was performed using primers corresponding to regions in the proviral ends and a plasmid region. The integration sites were analyzed by direct sequencing. "Q" in the PCR product indicates the junction between the provirus and the target DNA and R&S represents the 3′-ends of the primer within the HIV-1 DNA and the plasmid DNA, respectively [14].

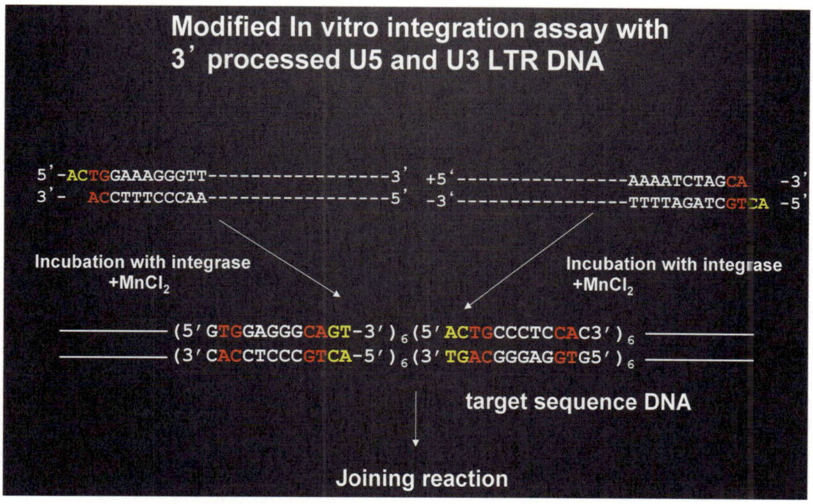

Fig. 7. Target sequence for integration.

The top sequences show the termini of the HIV-1 provirus. The bottom sequence indicates the target sequence and highlights the dinucleotide motif CA/TG (red), and the AC bases (yellow) that are also present at the HIV-1 DNA termini. We prepared a repeat sequence in order to enhance integration efficiency. The sequence shown in parenthesis is the unit of the repeat. Our protocol differs from previous protocols in that we used both 5′- and 3′-LTR sequences rather than a single 3′-LTR DNA.

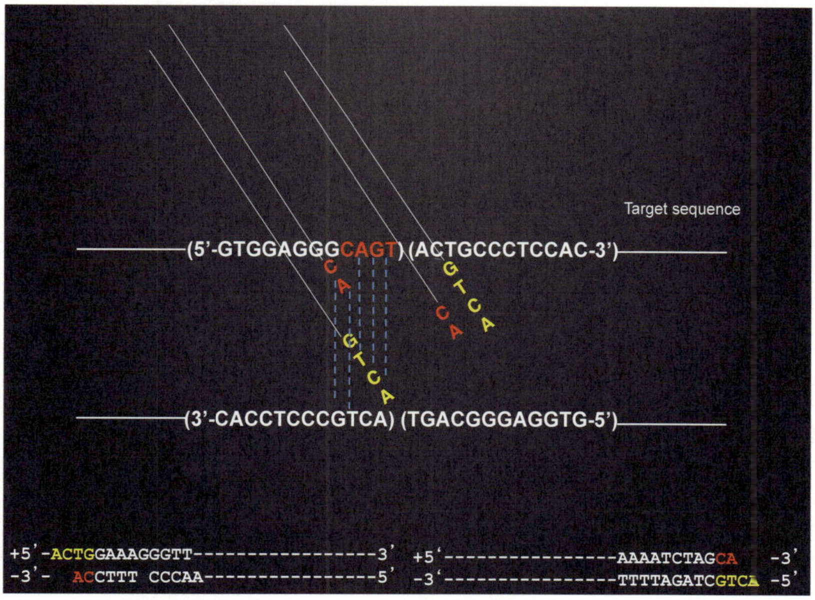

Fig. 8. Integration scheme.

The repeat sequence unit in the target sequence was expected to directly interact with HIV-1 DNA. The dotted lines indicate complementarity between target and viral DNA. The dsDNA sequence at the bottom indicates the 5′- and 3′-ends of the proviral HIV-1 DNA.

2.3 Reaction protocol

The detailed reaction protocol that we used is as follows. First, 75 ng of the U5′-LTR cDNA sequence of HIV-1.; (+) 5′-TGT GTG CCC GTC TGT TGT GTG ACT CTG GTA ACT AGA GAT CCT CAG ACC TTT TTG GTA GTG TGG AAA ATC TCT AGC A-3′ and (-) 5′-ACT GCT AGA GAT TTT CCA CAC TAC CAA AAA GGG TCT GAG GGA TCT CTA GTT ACC AGA GTC ACA CAA CAG ACG GGC ACA CA-3′, was incubated with 50 ng recombinant HIV-1 integrase in 10 μl of binding buffer for 1 h at 30°C. The binding buffer consisted of 1-0.1 mM $MnCl_2$, 80 mM glutamate potassium glutamate, 10 mM mercaptoethanol, 10% DMSO, and 35 mM MOPS (pH 7.2).

Similarly, the 3′-LTR cDNA sequence (75 ng) of HIV-1.; (+) 5′-ACT GGA AGG GTT AAT TTA CTC CAA GCA AAG GCA AGA TAT CC TTG ATT TGT GGG TCT ATA ACA CAC AAG GCT ACT TCC CA-3′ and (-) 5′-ACTG GGA AGT AGC CTT GTG TGT TAT AGA CCC ACA AAT CAA GGA TAT CTT GCC TTT GCT TGG AGT AAA TTA ACC CTT CCAGT-3′, was incubated with the recombinant retroviral integrase. After incubation, the double-stranded (ds) 5′-LTR DNA was combined with the ds 3′-LTR DNA for 1 h at 30 °C, and the LTR DNA was then further incubated with the target DNA for 1 h at 30°C. As controls, ds 5′-LTR DNA and ds 3′-LTR DNA were also individually incubated with the target DNA. For control target DNAs we synthesized four random 144-bp sequences, which were designed by a random number generator, and we ligated these sequences into circular DNA in the same manner as described below for the target DNA

In order to prevent non-specific reactions at the target DNA sequence, we ligated the target sequence DNA into circular plasmid DNA (Invitrogen pCR2.1 TOPO vector) and used this entire DNA as the target DNA for the assay (Fig. 6). The proportion of LTRs and target DNAs was optimized to prevent both non-specific reactions and integration due to an excess of LTRs. The DNA reacted in the buffer was purified using a QIA quick column (QIAGEN, GmbH, Germany). PCR amplification was then performed using retroviral primers: the HIV-1 U5′-LTR primer, 5′-GTG TGC CCG TCT GTT GTG TGA CTCTGG-3′, or the HIV-1 U3′-LTR primer, 5′-CTG GGA AGT AGC CTT GTG TGT TAT AG-3′, and a TOPO vector primer 5′-TCA CTC ATG GTT ATG GCA GC -3′ whose first nucleotide corresponds to nucleotide position 2222 in the TOPO-pCR2.1 plasmid (Invitrogen, Carlsberg, CA). Amplicon copy number was quantified following identification of the HIV-1-substrate DNA junction [14].

3. Results

3.1 Selective viral DNA integration into the target sequence

Figures 9 and 10 show the percentage of viral DNA integration into the target sequence or into the same length random sequences. Four types of the same length random sequences were used as controls. The horizontal blue line shows the percentage integration when uniform integration into the substrate DNA, including into the target DNA plus the circular plasmid, was thought to occur. These data indicate that the percentage of integration into the target sequence was significantly higher than that into random sequences. Also, when a target sequence was used in which the middle 5′-CA and 5′-GT nucleotides were deleted, the integration efficiency was significantly decreased. Thus, local nucleotide motifs within the target sequence affect integration efficiency.

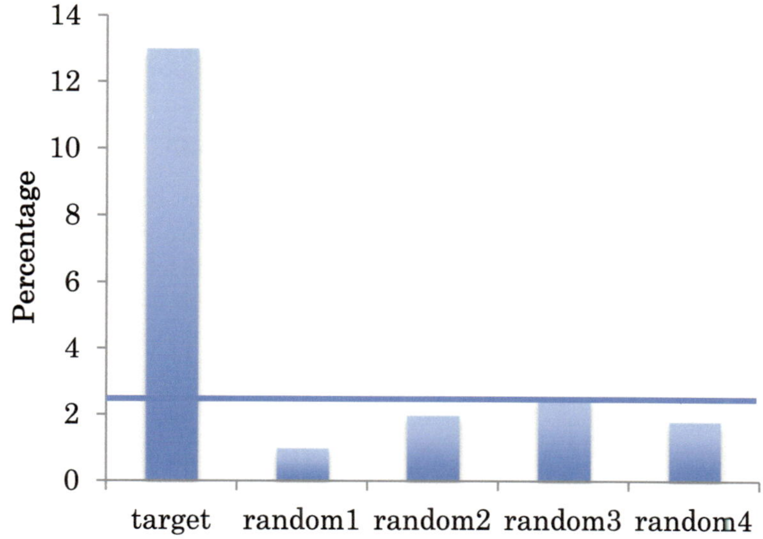

Fig. 9. Integration into target sequence DNA vs. random sequence DNAs.

The percentage of PCR product copies derived from viral DNA that had integrated into the target sequence or into random sequences is plotted vs. the total number of PCR product copies, including the PCR products that were integrated into the remainder of the DNA sequence of the plasmid. The horizontal line shows the ratio of these PCR products when integration was thought to occur in a uniform manner in the 4-kb substrate DNA.

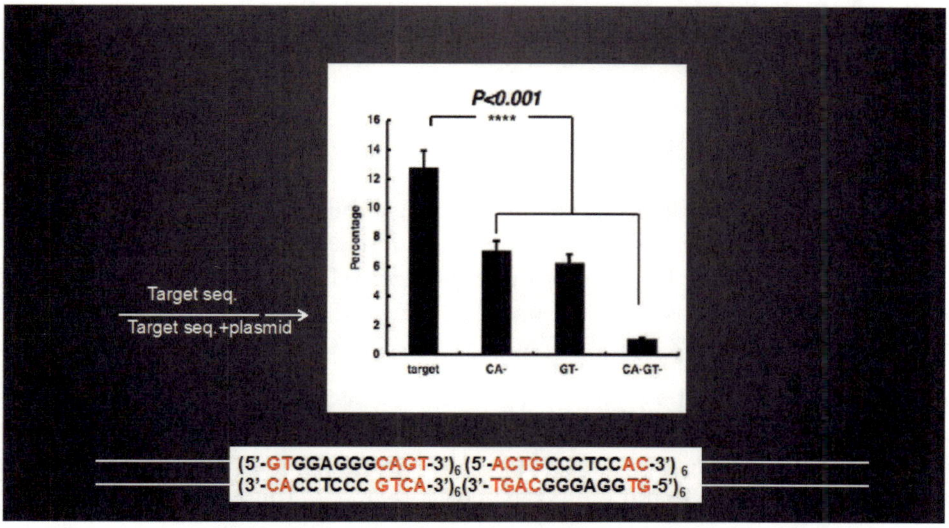

Fig. 10. A graph of the percentage integration into the target sequence or into random sequences.

The left arrows represents the percentage (~2.3%) of integration into target DNA to integration into control when integration was thought to occur in a uniform manner in the substrate DNA, the target sequence DNA plus the circular plasmid with which it was ligated. The integration efficiency was significantly decreased when CA and GT were removed from the middle region of the repeat. Thus, local nucleotide motifs affect integration efficiency (****$P < 0.001$).

3.2 *In vitro* integration site in the target repeat sequence DNA

Figure 11 shows the *in vitro* integration site in the target repeat sequence DNA. The entire target sequence is shown in this figure. The vertical axis indicates the percentage of PCR amplicons derived from the integration of individual LTR units. Integration efficiency was significantly higher when both the 5′- and the 3′-LTR DNA were used than when either LTR DNA was used alone. The use of both 5′- and 3′-LTR DNA is one of the unique points of our protocol, since previous protocols used a single 3′-LTR DNA (Figs. 7 and 11).

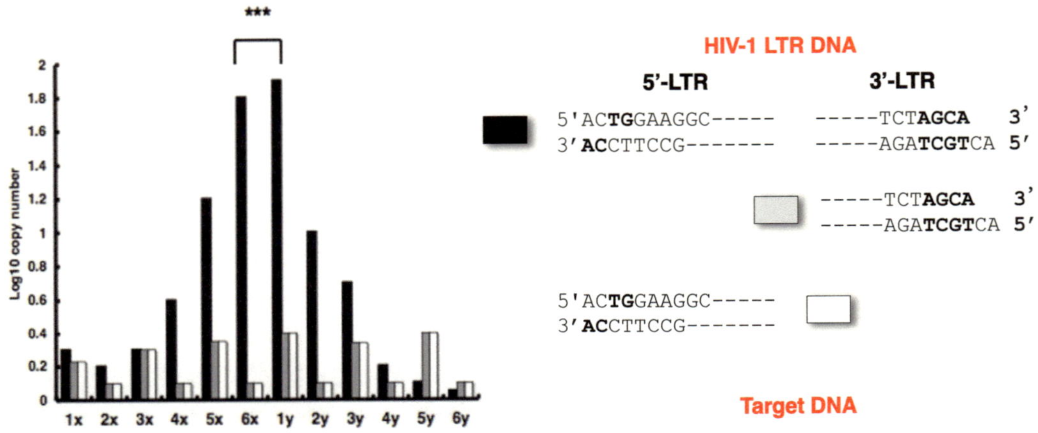

Fig. 11. *In vitro* integration site in the target repeat sequence DNA.

The vertical axis indicates the percentage of the PCR amplicons derived from proviral DNA integrating into individual units. The entire target sequence is shown. The integration efficiency was significantly higher when both 5′- and 3′-LTR DNA were used rather than when a single LTR DNA was used. The use of both 5′- and 3′-LTR DNA is one of the unique points of our protocol. x, GTGGAGGGCAGT; y, ACTGCCCCCAC. (*** $P < 0.001$)

Interestingly, we found that the middle segment of the target sequence was more favorable for integration, even though the same sequence units were repeated in the target sequence. To explain this observation, we considered the possibility that a structural factor may contribute to selective integration into the middle segment. Thus, if a single strand of the target DNA focally appeared by rewinding of the target double strand DNA, a long hairpin or cruciform structure may form in the target sequence site. It is probable that, if the target sequence DNA is open, or rewound, then the top of such a secondary structure would be favorable for integration. DNA folding thermodynamic analysis was performed to determine secondary structure in the target DNA and a hairpin structure was indeed predicted by this analysis (Fig. 12).

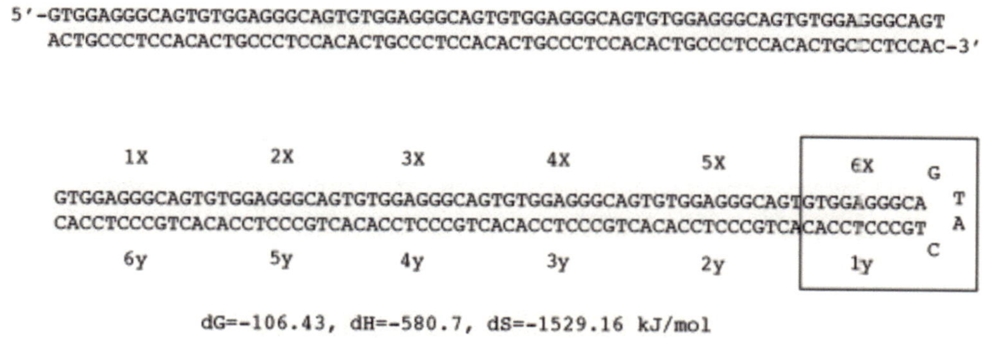

Fig. 12. A presumed secondary structure in the target sequence DNA.

The blue lines in the target DNA sequence shown at the top indicate the most frequent integration site. The presumed hairpin like structure shown at the bottom was constructed based on calculation of the target DNA sequence using the *m*-fold program (http://mfold.rna.albany.edu/?q=mfold/DNA-Folding-Form). dG, dH, and dS represent Gibbs' free energy, enthalpy, and entropy in ssDNA, respectively. A box represents the most frequent integration site.

3.2 Decoy effect of modified target sequences

We prepared two modified DNA target sequences in which the 5′-CA and 5′-GT were removed from the repeat unit at the middle site, termed modified sequence I and II, respectively (Fig. 13). PCR analysis of in vitro integration into modified sequence I or II revealed significant reductions in the number of copies of the PCR products compared to integration into the unmodified sequence. In addition, integration selectivity was not evident when using the modified DNA sequences ($P < 0.05$). We next mixed substrate DNA

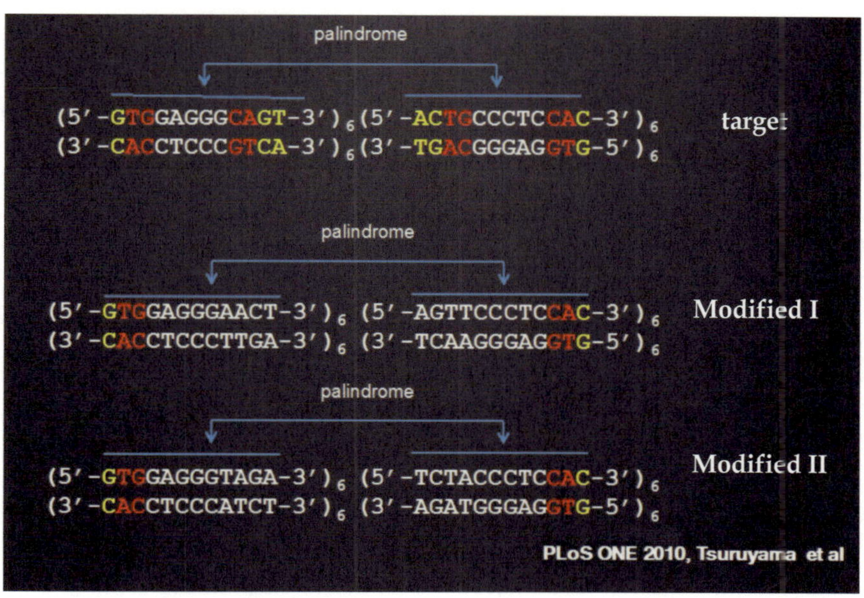

Fig. 13. Modified target DNA sequences.

containing the target sequence with substrate DNA containing modified sequence I or II in equal amounts, and examined the number of PCR product copies that originated from integration into the non-modified target sequence. Integration into the original, non-modified target sequence of the substrate DNA was significantly reduced when this DNA was mixed with the modified sequences (Fig. 14).

Two modified DNA sequences were prepared in which CA and GT were removed from the repeat unit at the middle site, termed modified sequence I and modified sequence II respectively. Red letters represent the TG/CA motifs. Yellow letters represent the GT/AC motifs that are observed in the HIV-1 proviral genome.

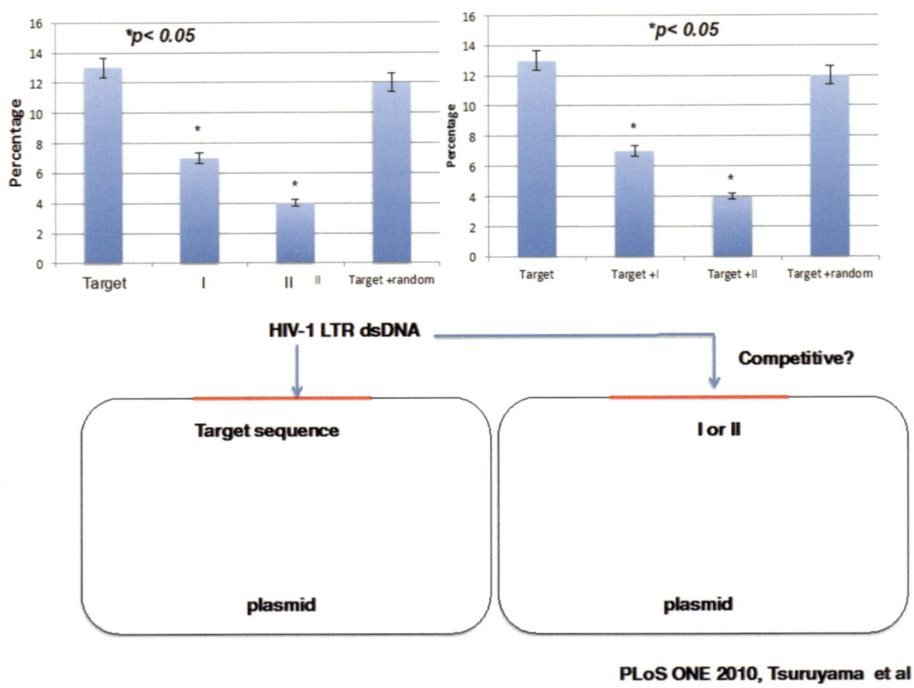

Fig. 14. *In vitro* integration using modified sequence I or II

The result showed significant reductions in the number of copies of PCR products derived from integrated DNA. In addition, integration selectivity was evidently suppressed when using the modified DNA targets (left graph, $*P < 0.05$). Substrate DNA containing the target sequence was then mixed with substrate DNA containing modified sequence I or II in equal amounts, and the percentage of PCR product copies originating from integration into the original target sequence were determined. Integration into the original target sequence was significantly reduced when this target was mixed with the modified sequences (right graph, $*P < 0.05$).

3.3 Biochemistry of the integrase: DNA structure fluctuation enhances selective integration

We digested circular DNA with HIV-1 integrase in a buffer containing various concentrations of manganese dichloride and measured the band intensity of linearized DNA following electrophoresis. The relative band intensity increased when the concentration of $MnCl_2$ in the reaction buffer exceeded 40 mM (Fig. 15). This result raised the question of how such fluctuation in DNA structure influences the selectivity of *in vitro* HIV-1 integration. Furthermore, the percentage of integration into the target sequence DNA was found to increase significantly when the concentration of $MnCl_2$ exceeded 40 mM. The ratio of the PCR product number derived from integration into the target sequence DNA was also found to increase significantly when the $MnCl_2$ concentration exceeded 40 mM (Fig. 16). In conclusion, such fluctuation may generate a favorable conformation of target DNA for integration of the HIV-1 LTR [16, 17].

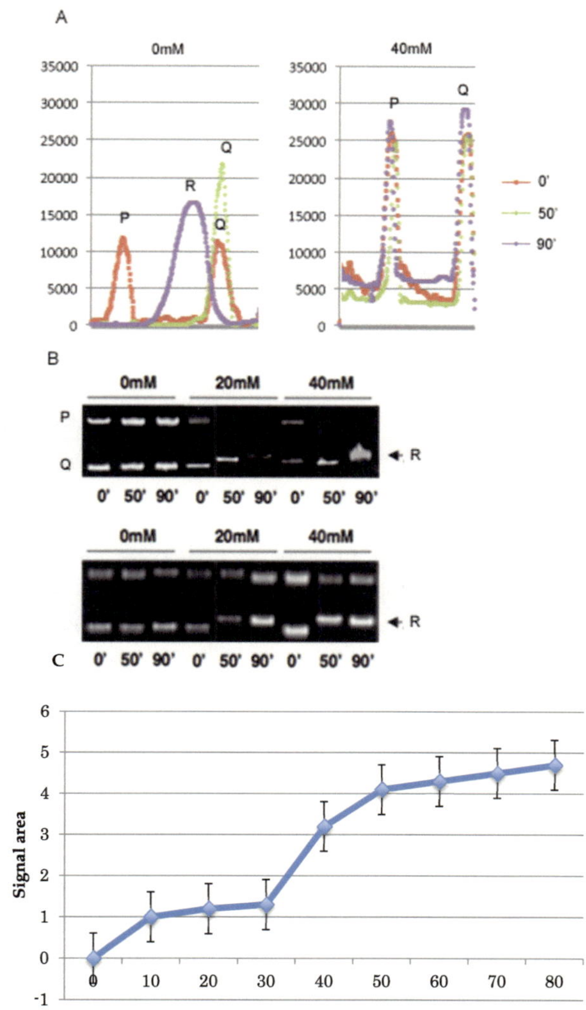

Fig. 15. Effect of $MnCl_2$ concentration and structural-fluctuation of target DNA on integration

(A) Electrophoretogram of plasmid plus target DNA (left) and plasmid DNA (right). P and Q indicate structural isomers of the circular DNA. R, a single fragment, indicates the digested DNA fragment after 90 min incubation. The vertical axis indicates relative signal intensity and the horizontal axis indicates electrophoretic mobility distance. (B) Electrophoresis of the DNA following incubation of recombinant integrase in buffer containing 0, 20, or 40 mM $MnCl_2$ following incubation for 0, 50, and 90 min. P, Q, and R are as described in (A). Upper and lower photographs display electrophoresis of plasmid plus target DNA and plasmid DNA, respectively. Significant fluctuation in R was observed in the electrophoresed DNA following 90 min incubation. (C) Relative signal area of the digested 4.0-kb substrate DNA corresponding to R in (B). Error bars represent standard deviation (S.D.). Fragment R area was calculated by integration of the individual curves shown in the electrophoretogram (A) with respect to electrophoretic mobility distance. Signal area significantly increased when 40 mM $MnCl_2$ was included in the buffer.

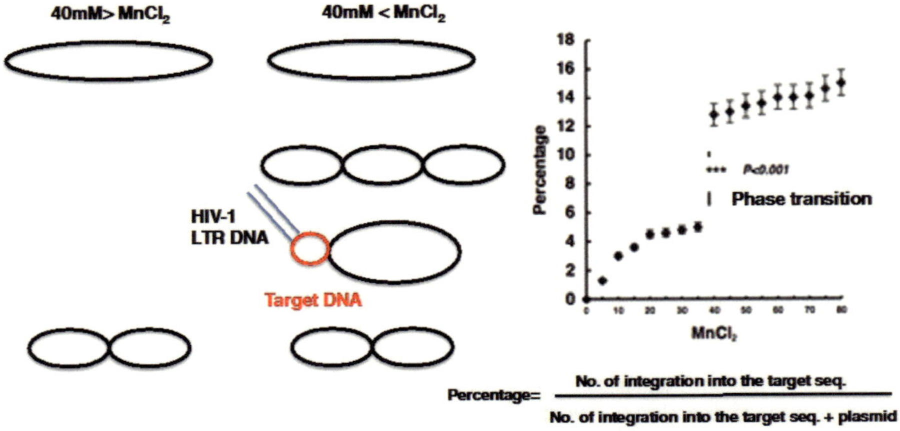

Fig. 16. DNA structural fluctuation and integration

A $MnCl_2$ concentration greater than 40 mM significantly increased the percentage of integration into the target sequence DNA. The percentage of copy number of PCR products derived from integration into the target sequence DNA was found to increase significantly when the concentration of $MnCl_2$ exceeded 40 mM. It is concluded that such fluctuation may generate a favorable conformation of target DNA (in red) for integration of HIV-1 LTR DNA.

4. Conclusion

In conclusion, selective HIV-1 integration was proved at an *in vitro* level in this study. The factors that determine this selectivity are (i) a sequence motif, including CAGT, and (ii) a structural factor that can be induced by fluctuation of a high concentration of $MnCl_2$ [16, 17]. The findings shown in Figs. 9 and 10 indicate that the percentage of integration into the target sequence was significantly greater than the integration rate into the random and deleted sequences. Moreover, the entire repeat sequence or secondary structure may be a target of integration.

In particular, our findings that sequences similar to the target DNA sequence interfere with integration (Fig. 14). Thus, a modified DNA can act as a decoy for the target DNA. In the present study, integration efficiency and selectivity were highly sensitive to $MnCl_2$ concentration in the reaction buffer. In particular, the integration efficiency and selectivity increased significantly when the $MnCl_2$ concentration was increased from 30 mM to 40 mM.

Fluctuations in the electrophoretic mobility of the substrate DNA also increased. These results suggest that there is a threshold concentration of $MnCl_2$ for *in vitro* integration, probably because $MnCl_2$ induces instability of secondary structure and therefore phase transition of the host DNA strand may occur. Target DNA can probably not generate the specified stable conformation under 40mM of $MnCl_2$. Based on these data as well as the data shown in Fig. 15, we propose that there are close correlations between structural changes in substrate DNA and integration selectivity and efficiency (Fig. 16, 17). We have used $MnCl_2$ for studies of *in vitro* integration because this salt is more appropriate than other salts for the generation of *in vivo* integration. However, during *in vivo* integration into the host genome, numerous DNA binding proteins and metal ions regulate the reaction in a complex manner. Therefore, the present data cannot be immediately applied to *in vivo* systems and further investigation using cell culture systems are necessary. Nevertheless, this *in vitro* integration assay is expected to facilitate understanding of the pathogenicity of HIV-1.

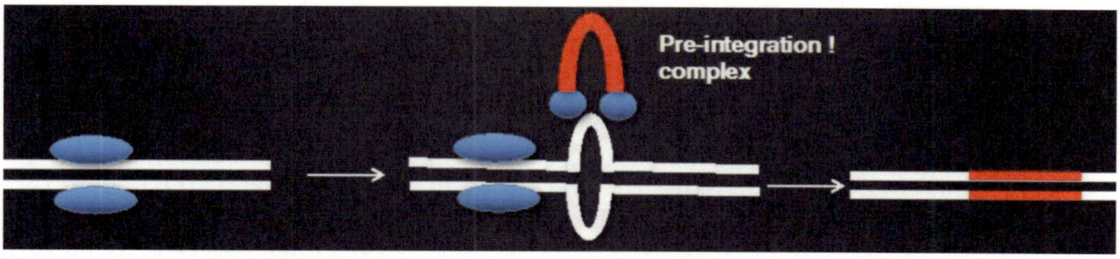

Fig. 17. A model of integration

The top of the secondary structure may be favorable for integration when the target DNA sequence is open or rewound by protein binding to the upstream of the target DNA sequence.

5. Acknowledgments

This work was supported by a Grant-in-Aid for Cancer Research from the Ministry of Education, Culture, Sports, Science, and Technology, Japan, and a Grant for Strategic Research on Cancer from the Ministry of Health, Labor, and Welfare, Japan (No. 72602-010-A03-0001) (http://www.jsps.go.jp/j-grantsinaid/index.html). The funders had no role in study design, data collection or analysis, in the decision to publish, or in preparation of the manuscript. The method described in this manuscript has been registered as "Nucleic acid having retroviral integration target-activity," patent number 4631084 in Japan (GenBank, DD323298). In relation to this patent, we also declare no conflict of interest. We are grateful to Masakazu Hatanaka and Tomokazu Yoshinaga for their helpful advice and insightful comments regarding this manuscript. In particular, we are also grateful to Dr. Tasuku Honjo and Dr. Hiroshi Hiai (Kyoto University) for their review of this study and for providing critical advice, and to Miss Hiroko Saito for her excellent technical support.

6. References

[1] Coffin, J.M.; Hughes, S.H. &Varmus, H.E. (1997). Retroviruses. New York: Cold Spring Harbor Laboratory Press.

[2] Wu, X.;Li Y.; Crise, B. & Burgess, S.M. (2003). Transcription start regions in the human genome are favored targets for MLV integration. Science 300:1749-1751.

[3] Schröder, A.R.; Shinn, P.; Chen, H.; Berry, C.; Ecker, J.R.; et al. (2002). HIV-1 integration in the human genome favors active genes and local hotspots. Cell 110: 521-529.

[4] Tsukahara, T.; Agawa, H.; Matsumoto, S.; Matsuda, M.; Ueno, S.; et al. (2006). Murine leukemia virus vector integration favors promoter regions & regional hot spots in a human T-cell line. Biochem Biophys Res Commun. 345:1099-1107.

[5] Holman, A.G. & Coffin, J.M. (2005). Symmetrical base preferences surrounding HIV-1,avian sarcoma/leukosis virus, and murine leukemia virus integration sites. Proc Natl Acad Sci U S A 102: 6103-6107.

[6] Wu, X.; Li, Y.; Crise, B.; Burgess, S.M. & Munroe, D.J. (2005). Weak palindromic consensus sequences are a common feature found at the integration target sites of many retroviruses. J Virol. 79: 5211-5214.

[7] Derse, D.; Crise, B.; Li, Y.; Princler, G.; Lum N.; et al. (2007). Human T-cell leukemia virus type 1 integration target sites in the human genome, comparison with those of other retroviruses. J Virol. 81: 6731-6741.

[8] Yoshinaga, T. & Fujiwara, T. (1995). Different roles of bases within the integration signal sequence of human immunodeficiency virus type 1 *in vitro*. J Virol. 69: 3233-3226.

[9] Tsuruyama, T.; Nakamura, T.; Jin, G.; Ozeki, M.; Yamada, Y.; et al. (2002). Constitutive activation of Stat5a by retrovirus integration in early pre-B lymphomas of SL/Kh strain mice. Proc Natl Acad. Sci. U. S. A. 99: 8253-8258.

[10] Akagi, K.; Suzuki, T.; Stephens, RM.; Jenkins, NA. & Copeland, NG (2004). RTCGD, retroviral tagged cancer gene database. Nucleic Acids Res. 32 (Database issue),D523-527.

[11] Nosaka, T.; Kawashima, T.; Misawa, K.; Ikuta, K.; Mui, AL.; et al. (1999). STAT5 as a molecular regulator of proliferation, differentiation, & apoptosis in hematopoieticcells. EMBO J 17: 4754-4765.

[12] Schwaller, J.; Parganas, E.; Wang, D.; Cain, D.; Aster, JC. et al. (2000). Stat5 is essential for the myelo- & lymphoproliferative disease induced by TEL/JAK2. Mol Cell 6:693-704.

[13] Tsuruyama,T.; Nakai, T.; Hiratsuka, T.; Jin, G.; Nakamura, T. & Yoshikawa, K. (2010).In vitro HIV-1 selective integration into the target sequence and decoy-effect of the modified sequence. PLoS One 5 : e13841.

[14] Tsuruyama, T.; Nakamura, T.; Jin G.; Ozeki, M.; Yamada, Y. &Hiai, H. Constitutive activation of Stat5a by retrovirus integration in early pre-B lymphomas of SL/Kh strain mice. Proc Natl Acad Sci U S A. 2002 Jun 11.;99(12):8253-825.

[15] Iwaki, T.; Makita, N. &Yoshikawa, K (2008). Folding transition of a single semiflexible polyelectrolyte chain through toroidal bundling of loop structures. Journal of Chemical Physics 129: 065103.

[16] Ueda, M.; & Yoshikawa, K. (1996). Phase transition and phase segregation in a single double-stranded DNA molecule. Physical Review Letters 77: 2133–2136.

13

Distinct Role for ARNT/HIF-1β in Pancreatic Beta-Cell Function, Insulin Secretion and Type 2 Diabetes

Renjitha Pillai and Jamie W. Joseph
School of Pharmacy, University of Waterloo, Waterloo
Canada

1. Introduction

Diabetes mellitus is a common metabolic syndrome that has become an epidemic in modern society and is characterized by either a near-complete lack of insulin production due to autoimmune destruction of pancreatic beta-cells as in type 1 diabetes or abnormal insulin secretion, beta-cell dysfunction and insulin resistance as in type 2 diabetes (T2D). T2D is a complex heterogeneous disease that is characterized by elevated fasting and postprandial blood glucose levels that can result in severe complications including renal failure, cardiovascular disease, blindness and slow wound healing (Lin and Sun, 2010). Pancreatic islet beta-cells play a critical role in maintaining blood glucose levels by secreting the hormone insulin following a meal. Insulin maintains blood glucose levels in the normal physiological range by promoting glucose uptake in muscles, liver and adipose tissue, and by inhibiting hepatic glucose production. Therefore, any defect in insulin secretion in response to a meal or defects in insulin action in peripheral tissues can lead to increased blood glucose levels (Tripathy and Chavez, 2010; Muoio & Newgard, 2008).

Abnormal insulin secretion is a hallmark of T2D. Despite the central role of insulin in maintaining glucose homeostasis, the fundamental biochemical mechanism regulating nutrient-stimulated insulin secretion from pancreatic beta-cells is still incompletely understood. Insulin secretion from the pancreatic beta-cells is regulated by nutrients, neurotransmitters and hormones. Among these three factors, nutrients, particularly glucose is the most dominant stimulatory signal for insulin secretion. Insulin secretion is biphasic with a first acute phase occurring within 10 minutes after a glucose load and a second more sustained phase that reaches a plateau very quickly as seen in mice or more gradually as seen in rats and humans (Gerich, 2002). Numerous models have been proposed over the last several decades to explain the mechanism governing glucose-stimulated insulin secretion (GSIS) from pancreatic beta-cells. The current model of GSIS holds that glucose enters beta-cells via the low affinity, high capacity glucose transporter 2 (GLUT2) and becomes phosphorylated by glucokinase (GK or hexokinase IV), which is the rate-limiting step in glycolysis. The glycolytic end product pyruvate then enters the tricarboxylic acid cycle (TCA), where oxidative phosphorylation occurs, leading to increased ATP production. The subsequent rise in the cytosolic ATP/ADP levels promotes closure of ATP-sensitive potassium channels (K_{ATP} channels) causing beta-cell membrane depolarization and activation of voltage-dependent Ca^{2+} channels (VDCC). The opening of VDCCs facilitates influx of extracellular Ca^{2+}, leading to a rise in the beta-cell cytosolic Ca^{2+} levels, which triggers exocytosis of the insulin-containing secretory granules (Figure 1) (Jensen et al., 2008; Prentki & Matchinsky, 1987; Ashcroft & Rorsman, 1989; Newgard & Matchinsky, 2001; Newgard & McGarry, 1995). This so-called "K_{ATP} channel-dependent" mechanism appears to be particularly important for the first, acute phase of insulin release. However, in the second and more sustained phase of insulin secretion a "K_{ATP} channel-independent" pathway also appears to play a key role in the regulation of GSIS in conjunction with the K_{ATP} channel-dependent pathway (Henquin et al., 2003, Ravier et al., 2009). Important support for "K_{ATP} channel-independent" pathway of GSIS comes from studies showing that glucose still causes a significant increase in insulin secretion in conditions where K_{ATP} channels are held open by application of diazoxide followed by membrane depolarization with high K^+, or in animals lacking functional K_{ATP} channels (Nenquin et al., 2004; Shiota et al., 2002; Szollozi et al., 2007; Ravier et al., 2009). These and more recent studies suggest that mitochondrial metabolism of glucose generates signals other than changes in the ATP/ADP ratio that are important for normal insulin secretion. Several molecules, including glutamate, malonyl-CoA/LC-CoA and NADPH, have been proposed as candidate coupling factors in GSIS (Maechler & Wollheim, 1999; Ivarsson et al., 2005; Corkey et al., 1989; Prentki et al., 1992).

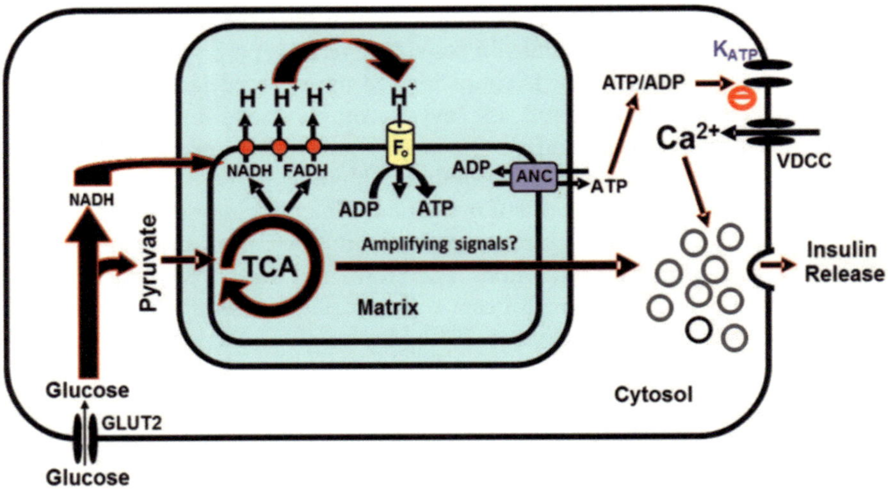

Fig. 1. Current model of glucose stimulated insulin secretion (GSIS) from pancreatic beta-cells. Glucose equilibrates across the plasma membrane through glucose transporter GLUT2, which initiates glycolysis. Pyruvate produced by glycolysis preferentially enters the mitochondria and is metabolized in the TCA cycle, producing reducing equivalents in the form of NADH and $FADH_2$. The transfer of electrons from these reducing equivalents through the mitochondrial electron transport chain is coupled with the pumping of protons from the mitochondrial matrix to the inter membrane space, leading to the generation of ATP. ATP is transferred to the cytosol through adenine nucleotide carrier (ANC), raising the ATP/ADP ratio. This results in the closure of the ATP sensitive K^+ channels (K_{ATP}), which in turn leads to membrane depolarization, opening of the voltage-sensitive Ca^{2+} channels, promoting calcium entry and increase in cytoplasmic Ca^{2+} leading to exocytosis of insulin granules. Glucose also generates amplifying signals other than ATP, which plays a significant role in the secretion of insulin from pancreatic beta-cells.

Maintenance of a functional mature beta-cell phenotype requires optimal expression of key transcription factors. Transcription factors regulate a variety of pancreatic beta-cell processes including cell differentiation, proliferation, cell signaling and apoptosis. By regulating the expression of specific sets of genes, transcription factors determine the spatio-temporal specificity of gene expression in most organisms, including mammals. Numerous studies have shown that transcription factors act synergistically to achieve normal beta-cell development and function (Cerf, 2006; Mitchell & Frayling, 2002; Lyttle et al., 2008). Development of the endocrine pancreas is initiated from multipotent precursor cells, which differentiate to form five different cell types in the pancreatic islet namely the α-cells (glucagon), β-cells (insulin), δ-cells (somatostatin), PP (pancreatic polypeptide) cells and ε-cells (ghrelin) (Steiner et al., 2010). The development of the islet architecture is regulated by an ordered system of transcriptional events activated by a hierarchy of transcription factors. Some of the major transcription factors represented in islets include several homeodomain factors like pancreatic and duodenal homeobox-1 (Pdx-1), paired box gene (Pax) Pax 4, Pax 6, Nkx 2.1 and Nkx 6.1 which are expressed in both progenitor as well as differentiated beta-cells. Pdx-1 and Nkx 2.2 are required for both early beta-cell differentiation and maintenance of a mature beta-cell phenotype (Habner et al., 2005). In addition, other transcription factors are important for maintenance of a mature beta-cell phenotype and their impairment may account for various pathophysiological abnormalities observed in type 2 diabetics. Among these, Pdx-1, neurogenin differentiation (NeuroD/BETA-2), foxhead box protein (FoxO-1), sterol regulatory element binding protein (SREBP-1c), and musculoaponeurotic fibrosarcoma oncogene homolog A (MafA) are the most studied (Johnson et al., 1994; Diraison et al., 2004; Kitamura et al., 2005).

In the context of T2D, it is a well-known fact that abnormal gene expression contributes to a myriad of beta-cell abnormalities. Support for this comes from studies of maturity-onset diabetes of the young (MODY), a monogenic form of T2D characterized by an early onset and defects in insulin secretion leading to hyperglycemia. With the exception of MODY-2, which is caused by a mutation in GK, MODY-1, 3, 4, 5 and 6 result from mutations in genes encoding transcription factors, hepatocyte nuclear factor (HNF) HNF-4α, HNF-1α, HNF-1β, Pdx-1, and NeuroD/BETA-2 respectively. These transcription factors regulate the expression of key genes involved in various aspects of beta-cell function (Stoffer & Zinkin, 1997; Habener et al., 1998; Fajans et al., 2001; Yamagata et al, 2003). Although there have been significant advancements in understanding the basic transcriptional network that exists in beta-cells, the exact mechanism of action of many of these factors still remains to be further defined. In this chapter we provide an overview of one of the recently described transcription factor in the context of impaired insulin secretion and beta-cell dysfunction, Aryl hydrocarbon receptor nuclear translocator (ARNT)/ hypoxia inducible factor 1β (HIF-1β), which is a master regulator of pancreatic beta-cell transcriptional network that regulates glucose metabolism and insulin secretion.

2. ARNT/HIF-1β

2.1 ARNT/HIF-1β structure and function

ARNT/HIF-1β belongs to a group of transcription factors, known as the basic helix loop helix - PER/ARNT/Sim (bHLH-PAS) family, which has a characteristic N-terminal bHLH motif for DNA binding, a central PAS domain which facilitates heterodimerization and a C-terminal transactivation domain for the recruitment of transcriptional coactivators such as CBP/p300 (Jain et al., 1994; Kobayashi et al., 1997). Recent evidence suggest that the PAS domain may also provide an additional binding site for coactivators and thereby recruiting them in a step necessary for transcriptional responses to hypoxia (Partch & Gardener, 2011). ARNT/HIF-1β acts as a common binding partner for most of the bHLH-PAS family of transcription factors and bind specific DNA sequences in the regulatory regions of the responsive genes. The half-site for ARNT/HIF-1β is on the 3' side of the 5'-GTG-3' recognition sequence. The sequence of the other half of the binding site depends upon the identity of the ARNT/HIF-1β dimerization partner (Swanson et al., 1995). DNA binding of ARNT/HIF-1β is mediated by its bHLH region and may also involve the PAS region. Dimerization between ARNT/HIF-1β and other bHLH-PAS proteins is mediated by their bHLH and PAS regions (Jiang et al., 1996, Lindebro et al., 1995). The human ARNT/HIF-1β gene is about 65 Kb in size, has 22 exons and is well conserved on an evolutionary scale (Scheel & Schrenk, 2000).

ARNT/HIF-1β was originally cloned as a factor required for the activity of the aryl hydrocarbon receptor (AhR). AhR induces a transcriptional response to various environmental pollutants, such as polycyclic aromatic hydrocarbons, heterocyclic amines, and polychlorinated aromatic compounds (Reyes et al., 1992). ARNT/HIF-1β was also identified as the β-subunit of a heterodimeric transcription factor, hypoxia-inducible factor 1α (HIF-1α) (Wang et al., 1995 (a)). Similar to HIF-1α, ARNT/HIF-1β gene expression and protein levels are significantly increased under hypoxic conditions suggesting that this gene plays an important role in the transcriptional response to low oxygen tension (Wang et al., 1995 (b)). Consistent with this idea, it has been shown that ARNT/HIF-1β is essential for the hypoxic induction of vascular endothelial growth factor (VEGF) and the glycolytic enzymes aldolase A (ALDO) and phosphoglycerate kinase (PGK) in a mouse hepatoma (Hepa 1c1c7) cell line (Li et al., 1996; Salceda et al., 1996). Unlike HIF-1α, which is exclusively expressed under hypoxic conditions, ARNT/HIF-1β is constitutively expressed in a number of tissues, such as the brain, heart, kidney, muscles, thymus, retina, olfactory epithelium and beta-cells of pancreas (Hirose et al., 1996).

2.2 ARNT/HIF-1β localization, binding partners, mechanism of action and lessons from knockout animals

ARNT/HIF-1β is a nuclear protein in most cell types, although it may also be located in the cytosol, particularly during embryogenesis. Studies conducted by Holmes and Pollenz (1997) in hepatic and non-hepatic cell lines derived from rat, mouse, human, and canine tissues confirm ARNT/HIF-1β as a nuclear transcription factor and showed that its physical interaction with DNA requires entry into the nucleus.

ARNT/HIF-1β serves as an obligatory binding partner for a number of other bHLH-PAS proteins, whose activity is modulated either by exogenous chemicals (AhR), hypoxia (HIF-1α, HIF-2α and HIF-3α), or which show tissue-specific expression pattern (e.g. SIM-1) (Salceda et al., 1996; Swanson et., 1995; Woods & Whitelaw, 2002). In addition to forming heterodimers, ARNT/HIF-1β appears to be capable of forming homodimers and bind to an E-box sequence 5'-CACGTG-3' (Antonsson et al., 1995). It was also shown that ARNT/HIF-1β homodimer regulates the transcription of murine cytochrome P450 (Cyp) 2a5 gene through a palindromic E-box element in the 5' regulatory region of Cyp2a5 gene in primary hepatocytes (Arpiainen et al., 2007). Two ARNT-related genes, ARNT-2 and ARNT-3 (also called BMAL-1 or MOP3) have been identified. ARNT-2 is more restricted in expression than ARNT/HIF-1β, but appears to dimerize with the same partner proteins as ARNT/HIF-1β (Hirose et al., 1996). ARNT-3 appears to have different dimerization potential than ARNT/HIF-1β (Ikeda & Nomura, 1997). The transactivation potential of ARNT/HIF-1β is not only determined through the recruitment of transcriptional cofactors, but also by signaling input from several protein kinases, such as PKC (Long et al., 1999).

The ARNT/HIF-1β/AhR heterodimer activates transcription of several genes involved in metabolism of foreign chemicals, including CYP1A1, CYP1B1, and NADP(H):oxidoreductase (NQO1) (Sogawa & Kuriayama, 1997; Beischlag et al., 2008). Transcriptional activation of these genes depends upon prior binding of AhR to xenobiotic ligands, including 2,3,7,8-tetrachlorodibenzo-p-dioxin (dioxin) and benzopyrene. The ARNT/HIF-1β/AhR heterodimer and ARNT/HIF-1β can have an impact on estrogen receptor (ER) activity. ARNT/HIF-1β interacts and functions as a potent coactivator of both ER-α and ER-β dependent transcription and it is believed that the C-terminal domain of ARNT/HIF-1β is essential for the transcriptional enhancement of ER activity (Lim et al., 2011;Rüegg et al., 2008).

ARNT/HIF-1β/HIF-1α heterodimer activity is primarily regulated by HIF-1α protein stability. Under normoxia, HIF-1α is hydroxylated by an oxygen requiring enzyme, prolyl hydroxylase (PHD), which is then targeted for ubiquitination by the E3 ubiquitin ligase, followed by binding to von Hippel-Lindau tumor suppressor (VHL) which leads to degradation of HIF-1α by the proteasome pathway. Conversely, under hypoxic conditions, a lack of oxygen inhibits hydroxylation, leading to stabilization of the HIF-1α protein and translocation of HIF-1α from the cytoplasm to the nucleus. In the nucleus, heterodimerization of HIF-1α with ARNT/HIF-1β is followed by binding to hypoxia response elements (HRE) in the promoter region of the target genes (Fedele et al., 2002) (Figure 2). Like HIF-1α, HIF-2α and 3α are stabilized by hypoxia and hypoglycemia, and activate transcription of genes involved in adapting to these adverse conditions, including the genes for erythropoietin (EPO), VEGF, and a number of enzymes of glycolysis including ALDO, phosphofructokinase (PFK) and lactate dehydrogenase (LDH) (Maltepe et al., 1997; Fraisl et al., 2009; Fedele et al., 2002). These studies suggest ARNT/HIF-1β is a central player in a number of signaling pathways and alterations in its activity can have serious impact on cellular responses to hypoxia, dioxin response and estrogen signaling in mammalian cells.

Observations from the ARNT/HIF-1β conditional knockout mice and whole body knockout mice have provided a wealth of information regarding the functional significance of this transcription factor in mammalian cells. Results obtained from the ARNT/HIF-1β null mice suggest that it plays a central role in embryonic development and physiological homeostasis as these mice are embryonic lethal due to severe defects in angiogenesis and placental development (Maltepe et al., 1997; Kozak et al., 1997). Data obtained from tissue specific ARNT/HIF-1β knockout mice demonstrates that disruption of ARNT/HIF-1β expression in liver and heart results in loss of AhR-stimulated gene transcription and that ARNT/HIF-1β is key to AhR function in these two mammalian tissues. It was also observed that ARNT/HIF-1β affects HIF-1α mediated target gene expression as several key genes including the expression of heme-oxygenase and glucose transporter-1 mRNA was abolished after treatment with $CoCl_2$, an agent that is thought to mimic hypoxia (Tomita et al., 2000).

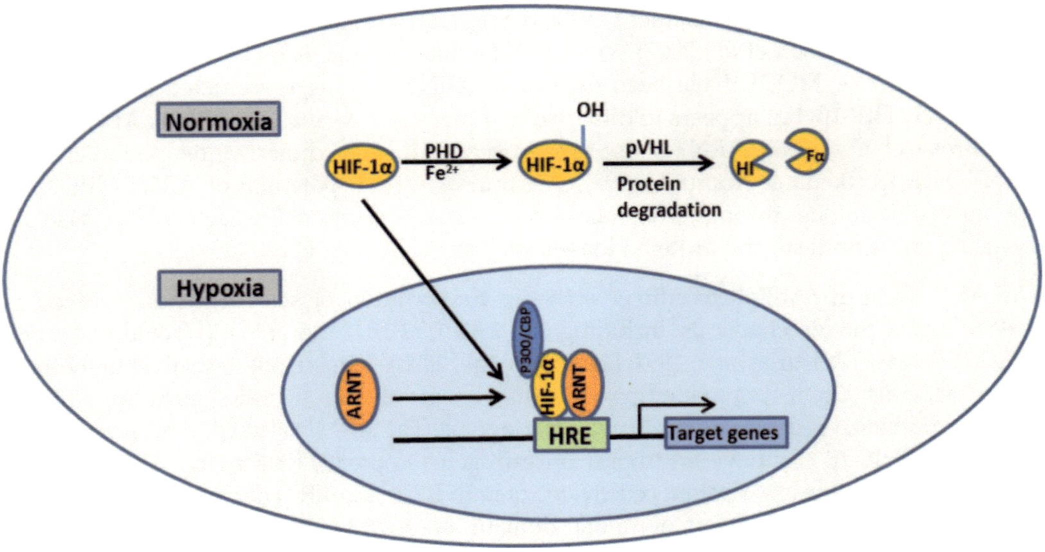

Fig. 2. Overview of gene regulation by ARNT/HIF-1β/HIF-1α complex in mammalian cells under normoxic and hypoxic conditions. In normoxic conditions, HIF-1α protein undergoes oxygen dependent hydroxylation by prolyl hydroxylases (PHD) and the hydroxylation site is recognized by pVHL, which targets the protein for ubiquitination by ubiquitin ligase, followed by degradation through ubiquitin proteasome pathway. During hypoxia, HIF-1α protein is not targeted for degradation and can translocate to the nucleus, where it heterodimerizes with ARNT/HIF-1β to form a stable transcriptional complex. The ARNT/HIF-1β/HIF-1α heterodimer then binds to the hypoxia response element (HRE) of target genes.

3. ARNT/HIF-1β and type 2 diabetes

3.1 ARNT/HIF-1β is reduced in human diabetic islets

In 2005, a study published in *Cell* (Gunton et al., 2005) suggested that ARNT/HIF-1β, a transcription factor with previously unknown functions in beta-cells, plays a significant role in mediating human beta-cell dysfunction in type 2 diabetics. Genome-wide gene expression profiling of islets obtained from human non-diabetics and type 2 diabetics revealed that the expression levels of ARNT/HIF-1β was reduced by 90% under the diabetic conditions. This was associated with reduced expression levels of several ARNT/HIF-1β target genes involved in glycolysis and insulin signaling. Several enzymes in glycolysis, including phosphoglucomutase (PGM), phosphoglucose isomerase (G6PI), PFK and ALDO were expressed at significantly lower levels as compared to those observed in normal islets. The low ARNT/HIF-1β expression levels observed under diabetic conditions was also associated with low gene expression levels of several key regulators in insulin signaling, such as the insulin receptor (IR), insulin receptor substrate 2 (IRS2), and protein kinase B (Akt2). Another interesting observation made in this study was that MODY genes, HNF-1α and HNF- 4α, were poorly expressed in human islets obtained from type 2 diabetics. HNF-4α, the gene mutated in MODY1, has been shown to interact with ARNT/HIF-1β possibly providing a connection between the two transcription factors (Tsuchiya et al., 2002).

In order to rule out the possibility that the profound ARNT/HIF-1β down regulation in pancreatic beta-cells is not caused by the diabetic environment, Gunton and co-workers demonstrated that an identical gene profile was observed in a beta-cell-specific ARNT/HIF-1β knockout mouse (β-ARNT KO). β-ARNT KO mice exhibited impaired GSIS and glucose intolerance, with no significant change in beta-cell insulin content and islet mass. The finding that ARNT/HIF-1β knockout mice have normal islet mass suggests that this transcription factor does not play a role in beta-cell differentiation.

Overall, a combination of *in vivo* and *in vitro* studies in humans and rodents have provided us with convincing evidence that reduction in ARNT/HIF-1β expression in human pancreatic beta-cells has negative consequences in terms of beta-cell function and insulin secretion. However, the extent of ARNT/HIF-1β mediated regulation of gene transcription is complex since it has the potential to bind with multiple partners affecting a multitude of signaling pathways.

3.2 ARNT/HIF-1β is reduced in human diabetic hepatic cells

In both rodents and humans, the liver plays a critical role in maintaining glucose and lipid homeostasis. During fasting, hepatic glucose production is critical for providing glucose for the brain, the kidneys and red blood cells. In liver, glucose is produced by glycogenolysis during the initial stages of fasting, however, after several hours of fasting, glucose production is primarily from gluconeogenesis, a process by which the liver produces glucose from precursors such as lactate and pyruvate (Michael et al., 2000; Saltiel & Kahn, 2001). Wang et al showed that ARNT/HIF-1β was severely reduced in the livers of human type 2 diabetics (Wang et al., 2009). Gene expression profiling of liver specimens from normal, obese and obese diabetic patients revealed a 30% reduction in the expression of ARNT/HIF-1β gene in obese diabetic individuals. The study demonstrated that the reduced expression of ARNT/HIF-1β in the livers of humans with T2D was associated with high glucose levels, high insulin levels, and insulin resistance. This study also suggested that insulin, not glucose regulates the expression of ARNT/HIF-1β gene and that ARNT/HIF-1β expression is reduced in both insulin-deficient and insulin-resistant states.

Wang et al (2009) also looked at the effects of liver-specific deletion of ARNT/HIF-1β gene in mice (L-ARNT KO) and demonstrated that there was an increase in gluconeogenesis, lipogenesis and increased serum insulin levels, all characteristic of human type 2 diabetics. The increase in hepatic gluconeogenesis and lipogenesis in L-ARNT KO mice was associated with the upregulation of several important gluconeogenic and lipogenic genes including PEPCK, G6Pase, SCD1 and FAS. Expression of C/EBPα and SREBP-1C, was also induced by 2-folds in L-ARNT KO mice. C/EBPα plays a major role in kick-starting hepatic glucose production at birth, and disruption of the C/EBPα gene in mice is known to cause hypoglycemia associated with the impaired expression of the gluconeogenic enzymes PEPCK and G6Pase (Pedersen et al., 2007; Qiao et al., 2006). SREBP-1C, on the other hand is a major player in lipogenesis (Horton et al., 2002). ARNT/HIF-1β may act as an upstream regulator of these transcription factors and play a key role in maintaining whole body glucose and lipid homeostasis. However, as seen in pancreatic beta-cells, the exact pathways targeted by ARNT/HIF-1β in liver cells are not clearly understood and is complicated by the fact that ARNT/HIF-1β has multiple binding partners.

3.3 ARNT/HIF-1β regulates glucose metabolism and insulin secretion in beta-cells

The central role played by ARNT/HIF-1β/HIF-1α heterodimer in the regulation of glucose homeostasis, particularly glycolysis has been well studied (Dery et al., 2005; Semenza et al., 1994) with a focus in cancer cell metabolism (Song et al., 2009; Semenza et al., 2000; Semenza, 2003). It is widely accepted that ARNT/HIF-1β/HIF-1α heterodimer plays a role in the Warburg effect, where cancer cells undergo a high rate of anaerobic glycolysis compared to normal cells. It has been suggested that the observed increase in glycolytic enzymes in these cancer cells is associated with increased HIF-1 activity, thus aiding in tumor formation and progression. Studies conducted in ARNT/HIF-1β mutant clonal cells indicate that it is an essential component of the HIF-1α complex and that absence of ARNT/HIF-1β leads to reduced cellular responses to stimuli such as hypoxia (Woods et al., 1996).

In pancreatic beta-cells, metabolism of glucose through aerobic glycolysis and oxidative phosphorylation plays a significant role in maintaining a normal secretory capacity. Beta-cells sense glucose and secrete appropriate amounts of insulin to promote glucose uptake by muscles and adipose tissue. Insulin also inhibits hepatic glucose production. Abnormal insulin secretion is one of the earliest detectable defects at the onset of T2D and despite its relevance, the mechanisms underlying GSIS are not completely understood. The generally accepted model of GSIS holds that metabolism of glucose in the beta-cells leads to a rise in the cytosolic ATP/ADP levels, which promotes closure of the K_{ATP} channel, increased cytosolic Ca^{2+} and triggers exocytosis of insulin-containing secretory granules (Henquin et al., 2003; Jensen et al., 2008). In beta-cells glucose derived pyruvate is directed mostly towards TCA for the production of ATP, since both the pentose phosphate pathway and anaerobic glycolysis is relatively inactive (Schuit et al., 1997). This exceptionally high dependence of beta-cells on the TCA cycle suggests that hypoxia or mechanisms reducing the aerobic capacity of beta-cells would probably have profound effects on GSIS.

It has been shown that down regulation of ARNT/HIF-1β in pancreatic beta-cells leads to loss of GSIS (Gunton et al., 2005; Pillai et al 2011). Our group has demonstrated that beta-cells with reduced ARNT/HIF-1β expression levels exhibit a 31% reduction in glycolytic flux without significant changes in glucose oxidation or the ATP/ADP ratio. Metabolomics analysis revealed that clonal beta-cells (832/13) treated with siRNAs against the ARNT/HIF-1β gene have lower levels of glycolytic, TCA cycle and fatty acid intermediates (Figure 3). It was also shown that the reduced levels of glycolysis, TCA and fatty acid intermediates were associated with a corresponding decrease in the expression of key genes in all three metabolic pathways including GLUT2, GK, PC, PDH, MEc, CIC DIC, CPT1a and FAS (Figure 4). The novel finding that reducing ARNT/HIF-1β levels leads to a profound reduction in PC, DIC, and OGC expression levels and a reduction in glycoslysis and TCA metabolites, even though glucose oxidation and ATP production were unaltered, is an unexpected result. These collective changes in metabolite levels suggest that the oxidative entry of pyruvate into the TCA cycle is preserved in the absence of ARNT/HIF-1β at the expense of a loss of anaplerosis. A key role for anaplerosis in insulin secretion is supported by the finding that pyruvate flows into mitochondrial metabolic pathways, in roughly equal proportions, through the anaplerotic (PC) and oxidative (PDH) entry points. Glucose carbon entering through the PC reaction leads to an increase in TCA intermediates (called anaplerosis) (Schuit et al., 1997; Khan et al., 1996). In addition, beta-cells contain enzymes that allow "cycling" of pyruvate via its PC-catalyzed conversion to oxaloacetate (OAA), metabolism of OAA to malate, citrate, or isocitrate in the TCA cycle, and subsequent recycling of these metabolites to pyruvate via several possible

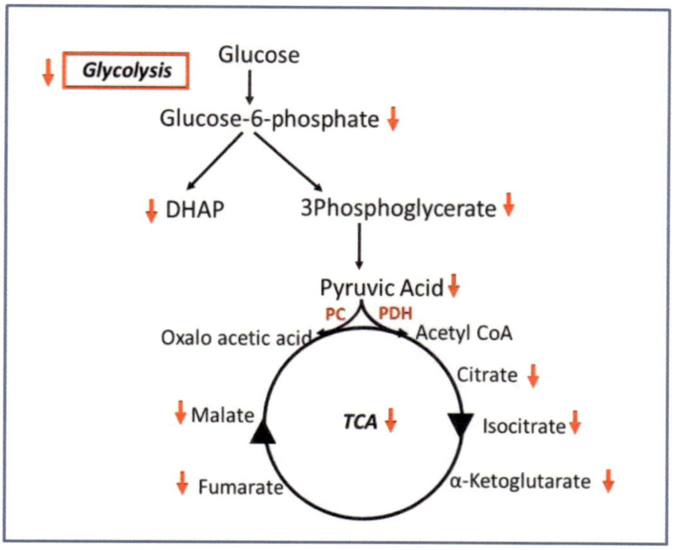

Fig. 3. Summary of the effects of siRNA mediated suppression of ARNT/HIF-1β in beta-cells. Several key metabolites in both glycolysis and TCA cycle were negatively affected by the knockdown of ARNT/HIF-1β. TCA, tricarboxylic acid cycle; DHAP, dihydroxyacetone phosphate; PC, pyruvate carboxylase; PDH, pyruvate dehydrogenase.

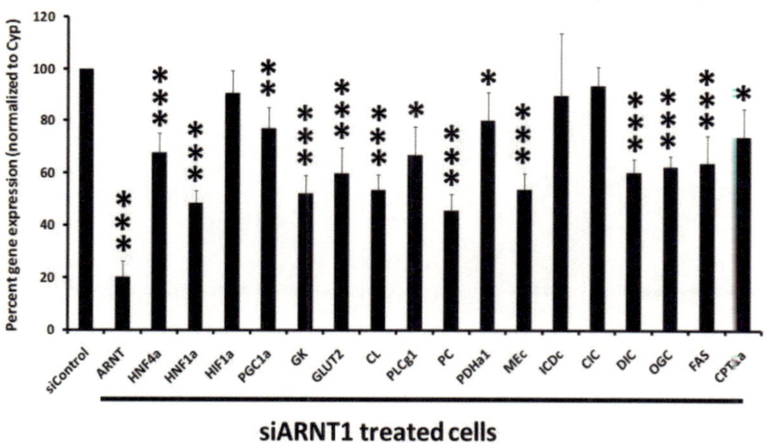

Fig. 4. Effects of siRNA mediated suppression of ARNT/HIF-1β on key genes involved in the metabolic regulation of β-cell function in 832/13 cells. Gene expression is expressed as a percentage of the target gene from siControl treated cells and corrected for by an internal control gene cyclophilin E (Cyp). There was no significant difference seen between treatment groups for Cyp (n=5). HNF4a, hepatocyte nuclear factor-4α; HNF1a, hepatocyte nuclear factor-1α; HIF1a, hypoxia inducible factor 1α; GK, glucokinase; GLUT2, glucose transporter-2; PLGg1, phospholipase γ-1; PC, pyruvate carboxylase; PDHa1, pyruvate dehydrogenase (α1 subunit); MEc, cytosolic malic enzyme; ICDc, cytosolic isocitrate dehydrogenase; CIC, citrate carrier; DIC, dicarboxylate carrier; OGC, α-ketoglutarate carrier; FAS, fatty acid synthase; CPT1a, Carnitine palmitoyl transferase 1α. * $P<0.05$, ** $P<0.01$, *** $P<0.001$ siControl vs siARNT1.

combinations of cytosolic and mitochondrial pathways (MacDonald et al., 1995). Numerous groups have shown that both pyruvate cycling and anaplerosis are important to maintain normal secretory capacity of beta-cells (Lu et al., 2002; Joseph et al., 2006; MacDonald et al., 2005; Ronnebaum et al., 2006). Since the amount of pyruvate is substantially lower in ARNT/HIF-1β depleted beta-cells, our data also suggests that this gene may play an important role in maintaining pyruvate cycling. However a direct link between ARNT/HIF-1β and pyruvate cycling has not yet been established. Figure 5 shows the diagrammatic representation of the transcriptional network regulated by ARNT/HIF-1β and its involvement glucose-stimulated anaplerosis and insulin release.

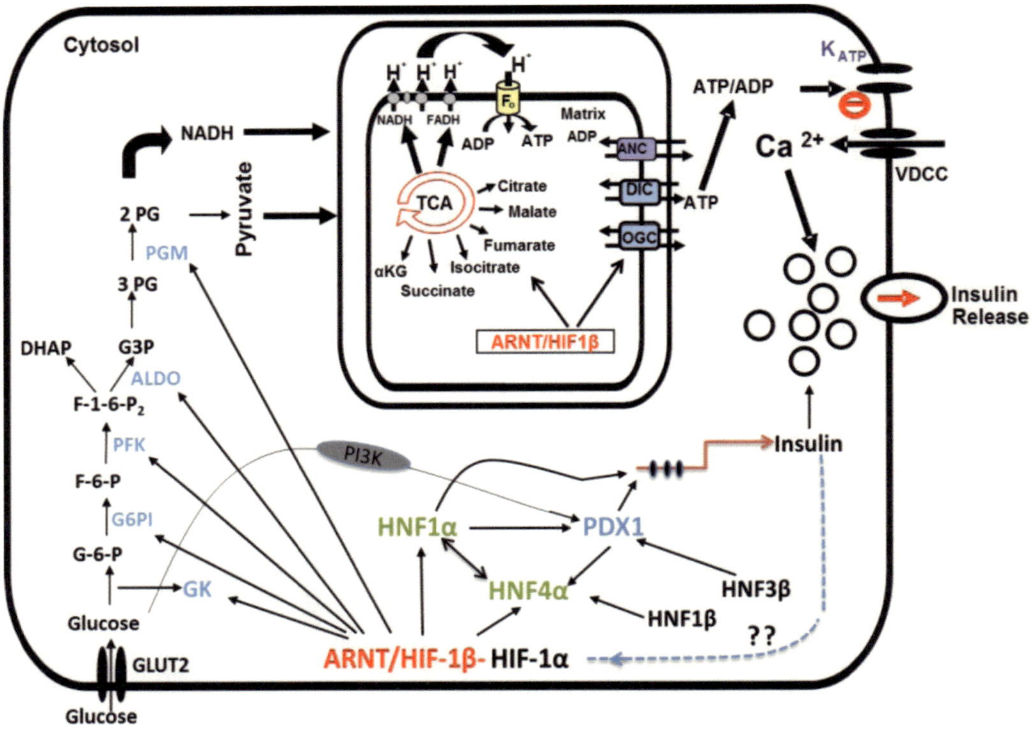

Fig. 5. Schematic of the transcriptional network regulated by ARNT/HIF-1β and its involvement in glucose-stimulated anaplerosis and insulin release from beta-cells. ARNT/HIF-1β regulates key genes in glycolysis and TCA cycle (shown in blue), including key metabolite carriers such as DIC and OGC. MODY genes regulated by ARNT/HIF-1β are shown in green. Interestingly, ARNT/HIF-1β does not seem to play a significant role in the regulation ATP production in beta-cells, however, it seems to be very important for glucose-induced anaplerosis, which provides crucial signals for GSIS.

3.4 ARNT/HIF-1β regulates beta-cell and hepatic transcriptional networks

Studies in ARNT/HIF-1β deficient beta-cells suggest that it plays a crucial role in the regulation of key genes involved in glucose metabolism and insulin secretion (Gunton et al., 2005, Pillai et al., 2011). ARNT/HIF-1β target genes in beta-cells include the MODY1 and MODY3 genes HNF4α and HNF1α, glucose metabolism genes GK, G6PI, PFK, aldolase, PC, PDH, MEc, DIC, OGC and insulin signaling genes IR, IRS2 and AKT2. In non-beta-cells it has been shown that ARNT/HIF-1β is essential for the normal function of HIF-1α, HIF2α, and AhR. These heterodimeric complexes are required for cellular responses to hypoxia (HIF proteins) and environmental toxins (AhR), respectively (Kozak et al., 1997; Kewley et al., 2004). It has been estimated that there are more than 13,000 putative ARNT/HIF-1β binding sites in promoters in the human genome (Gunton et al., 2005). Many of the target gene promoters have multiple potential binding sites. Thus it is reasonable to estimate that a substantial decrease in ARNT/HIF-1β would affect the expression of a large number of genes in humans. Although there is a lack of direct biochemical evidence, many of the genes found to be altered in association with decreased ARNT/HIF-1β gene expression have putative ARNT/HIF-1β-dimer consensus binding sites in their promoters (including HNF4α, HNF1α, Akt2, G6PI, PFK, and aldolase), suggesting a direct role for ARNT/HIF-1β containing dimers in the regulation of their expression (Figure 6).

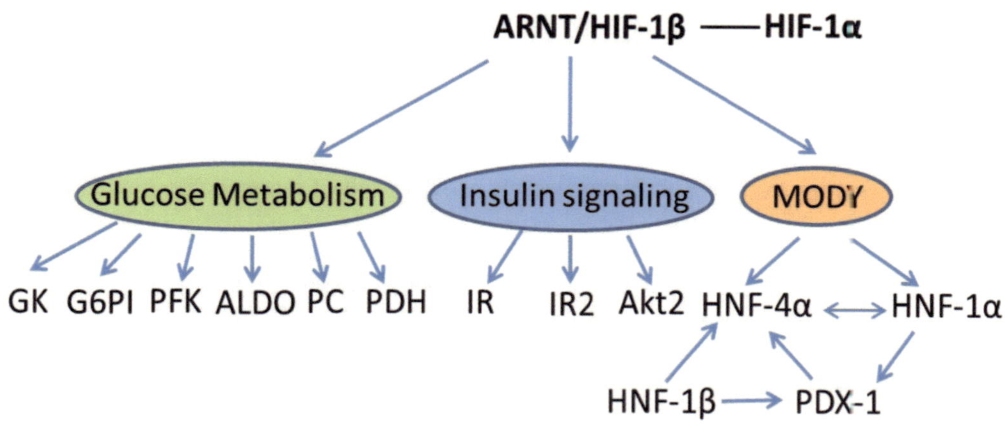

Fig. 6. Transcriptional network in pancreatic beta-cells regulated by ARNT/HIF-1β. ARNT/HIF-1β regulates several key genes involved in glucose metabolism, insulin signaling and MODY. Beta-cell specific knockout of ARNT/HIF-1β in mice leads to reduced expression of a number of important beta-cell genes including HNF-4α, HNF-1α, insulin receptor (IR), insulin receptor substrate-2 (IRS2), protein kinase b (Akt2), glucokinase (GK), glucose-6-phosphoisomerase (G6PI), phosphofructokinase (PFK), aldolase (ALDO), pyruvate carboxylase (PC) and pyruvate dehydrogenase (PDH).

In liver cells ARNT/HIF-1β has been shown to regulate the expression of several genes involved in glucose and lipid homeostasis. Support for the involvement of ARNT/HIF-1β in liver glucose homeostasis was provided by experiments showing that basal and insulin-induced expression of GLUT1, GLUT3, ALDO, PGK and VEGF were significantly reduced in ARNT/HIF-1β-defective HepG2 cells (Salceda et al., 1996). Wang *et al* (2009) demonstrated that a reduction of ARNT/HIF-1β in liver cells was associated with an increase in the expression of several important gluconeogenic and lipogenic genes including PEPCK, G6Pase, SCD1, FXR, C/EBPα, SREBP-1C, FBP-1 and FAS. The discovery that ARNT/HIF-1β may contribute to the regulation of beta-cell and hepatic genes suggests an essential role for this transcription factor in the regulation of glucose and lipid homeostasis (Figure 7).

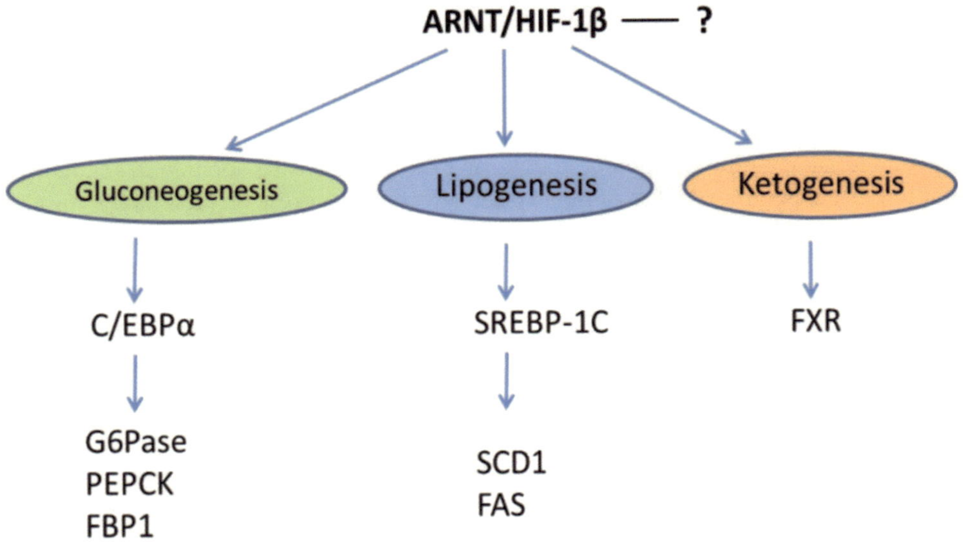

Fig. 7. ARNT/HIF-1β regulates the expression of several key genes involved in gluconeogenesis, lipogenesis and ketogenesis in the liver cells. Liver-specific knockout of ARNT/HIF-1β in mice leads to increased hepatic gluconeogensis and lipogenesis with a corresponding increase in the expression of phosphoenolpyruvate carboxykinase (PEPCK), Glucose-6-phosphatase (G6Pase), Fructose-1,6-biphosphatase (FBP1), Steroyl-CoA-desaturase (SCD1), Fatty acid synthase (FAS), CCAAT enhancer binding protein α (C/EBPα), sterol regulatory element binding protein (SREBP-1C) and Farsenoid X receptor (FXR). Adapted from Wang *et al.* (2009).

3.5 Regulation of ARNT/HIF-1β in beta-cells

Since ARNT/HIF-1β appears to be a major player in the pathogenesis of T2D, several attempts have been made to identify the upstream regulators of the gene in beta-cells. In 2008, Dror *et al.*, showed that glucose and endoplasmic reticulum Ca^{2+} channels regulate the expression of ARNT/HIF-1β in beta-cells via presinilin. Presinilin is a protein that has been implicated in the cellular response to reduced metabolic activity (Koo and Koppen, 2004). Overexpression of presenilin-1 in clonal Min6 beta-cells increased ARNT/HIF-1β suggesting that ARNT/HIF-1β may be a downstream target of presenilin (Dror et al., 2008). They demonstrated that this pathway is controlled by Ca^{2+} flux through intracellular channels. ARNT/HIF-1β has also recently been shown to be regulated by the carbohydrate-responsive element-binding protein (ChREBP), which is a transcription factor shown to regulate carbohydrate metabolism in the liver and pancreatic beta-cells in response to elevated glucose concentrations (Noordeen et al., 2009). In a genome-wide approach using high-density oligonucleotide arrays, the study showed that ChREBP binds directly to ARNT/HIF-1β promoter in Min6 clonal beta-cells. Accordingly, knockdown of ChREBP using siRNA resulted in an increase ARNT/HIF-1β mRNA levels whereas overexpression of ChREBP resulted in a decrease in ARNT/HIF-1β mRNA levels in rat. They also showed that incubating INS-1 (832/13) cells with glucose led to a substantial decrease in ARNT/HIF-1β mRNA levels.

Interestingly, it has also been shown that HIF-1α, the highly regulated binding partner for ARNT/HIF-1β, may be active under normoxic conditions in mouse and human beta-cells (Cheng et al., 2010). Coimmunoprecipitation studies demonstrated that HIF-1α was bound to ARNT/HIF-1β at the promoter region providing evidence for an interaction between HIF-1α and ARNT/HIF-1β in beta-cells. Treatment of diabetic mice with deferasirox (DFS), an agent that increases HIF-1α protein levels, improved glucose tolerance, normalized the expression of ARNT/HIF-1β and its target genes in human T2D islets. The same study also showed that HIF-2α, but not AhR, is another possible binding partner for ARNT/HIF-1β in pancreatic beta-cells. These studies provide a novel mechanism to regulate ARNT/HIF-1β gene expression in beta-cells.

Three studies published from independent laboratories studying the impact of increasing HIF-1α levels in beta-cells indicate that one has to be extremely cautious when using pharmacological agents, such as DFS, to activate HIF-1α in the islets (Puri et al., 2008; Zehetner et al., 2008; Cantley et al., 2009). These studies used the Cre-loxP system to conditionally delete VHL gene in beta-cells and showed that there were adverse effects associated with an increase in HIF-1α levels on beta-cell function. In all the three studies, increased HIF-1α levels were accompanied by severely impaired GSIS and increased lactate production, indicating a switch from aerobic to anaerobic glycolysis. Thus there appears to be a dose-response curve for the affects HIF-1α protein levels on beta-cell function (Cheng *et al.*, 2010). Although complete lack of HIF-1α seems deleterious to GSIS in mice and Min6 cells, milder increases are beneficial for beta-cell function. As seen in VHL knockout mice, very high levels of HIF-1α are detrimental for normal beta-cell function. Therefore, before we begin to develop a novel treatment regime that enhances HIF-1α or ARNT/HIF-1β activity in human diabetic islets, it is imperative that we understand the expected outcomes of such changes to avoid any detrimental effects.

4. Concluding remarks

It is well known that ARNT/HIF-1β plays a role in the cellular responses to hypoxia, however recent research has demonstrated a broader role for this transcription factor in maintaining glucose and lipid homeostasis in type 2 diabetics. It is now clear that a significant decrease in ARNT/HIF-1β gene expression in both the pancreatic beta-cells and the liver cells is deleterious and can result in T2D. Conversely, targeted disruption of ARNT/HIF-1β gene expression in the adipocytes followed by treatment of mice with a high fat diet improves insulin sensitivity and decreases adiposity (Jiang *et al.*, 2011). A central role for ARNT/HIF-1β in the regulation of key genes involved in glucose sensing, GSIS and insulin signaling in rodents as well in human islets suggest it plays an important role in maintaining normal beta-cell function. Current studies support the idea that ARNT/HIF-1β could act as an upstream regulator of many of the key genes involved in glucose and lipid homeostasis. Clearly, the transcriptional network regulated by ARNT/HIF-1β and genes that are under direct or indirect control of this transcription factor is very broad and hence any change in the regulation of ARNT/HIF-1β may have an impact on many signaling pathways. The fact that ARNT/HIF-1β is a binding partner for several other Per/ARNT/Sim transcription factor family members like HIF-1α, HIF-2α, HIF-3α and AhR makes it a significant member of this family of transcription factors. Improving our understanding of the beta-cell transcription factors, establishing their mechanism of action and hierarchy and finding ways to regulate their expression could prove beneficial in developing novel tools to prevent or correct beta-cell dysfunction in T2D.

5. References

Antonsson C, Arulampalam V, Whitelaw ML, Pettersson S & Poellinger L (1995): Constitutive function of the basic helix-loop-helix/PAS factor Arnt. Regulation of target promoters via the E box motif. *Journal of Biological Chemistry*, Vol. 270, pp 13968-13972.

Arpiainen S, Lämsä V, Pelkonen O, Yim SH, Gonzalez FJ & Hakkola J (2007). Aryl hydrocarbon receptor nuclear translocator and upstream stimulatory factor regulate Cytochrome P450 2a5 transcription through a common E-box site. *Journal of Molecular Biology*, Vol. 369, No. 3, pp 640-652.

Ashcroft FM & Rorsman P (1989). Electrophysiology of the pancreatic beta-cell. *Progress in Biophysics and Molecular Biology*, Vol. 54, pp 87-143.

Beischlag TV, Luis Morales J, Hollingshead BD & Perdew GH (2008). The aryl hydrocarbon receptor complex and the control of gene expression. *Critical Reviews in Eukaryotic Gene Expression*, Vol. 18, No. 3, pp 207-250.

Cantley J, Selman C, Shukla D, Abramov AY, Forstreuter F, Esteban MA, Claret M, Lingard SJ, Clements M, Harten SK, Asare-Anane H, Batterham RL, Herrera PL, Persaud SJ, Duchen MR, Maxwell PH, Withers DJ (2009). Deletion of the von Hippel-Lindau gene in pancreatic beta cells impairs glucose homeostasis in mice. *Journal of Clinical Investigation*, Vol. 119, pp 125–135.

Cerf ME (2006).Transcription factors regulating β-cell function. *European Journal of Endocrinology*, Vol. 155, pp 671-679.

Cheng K, Ho K, Stokes R, Scott C, Lau SM, Hawthorne WJ, O'Connell PJ, Loudovaris T, Kay TW, Kulkarni RN, Okada T, Wang XL, Yim SH, Shah Y, Grey ST, Biankin AV, Kench JG, Laybutt DR, Gonzalez FJ, Kahn CR & Gunton JE (2010). Hypoxia-inducible factor-1alpha regulates beta-cell function in mouse and human islets. *Journal of clinical investigation*, Vol. 120, No.6, pp 2171-2183.

Corkey BE, Glennon MC, Chen KS, Deeney JT, Matschinsky FM & Prentki M (1989). A role for malonyl-CoA in glucose-stimulated insulin secretion from clonal pancreatic beta-cells. *Journal of Biological Chemistry*, Vol. 264, pp 21608-21612.

Déry MA, Michaud MD & Richard DE (2005). Hypoxia-inducible factor 1: regulation by hypoxic and non-hypoxic activators. *International Journal of Biochemistry & Cell Biology*, Vol. 37, No. 3, pp 535-540.

Diraison F, Parton L, Ferré P, Foufelle F, Briscoe CP, Leclerc L & Rutter GA (2004). Over-expression of sterol-regulatory-element-binding protein-1c (SREBP1c) in rat pancreatic islets induces lipogenesis and decreases glucose-stimulated insulin release: modulation by 5-aminoimidazole-4-carboxamide ribonucleoside (AICAR), *Biochemical Journal*, Vol. 378, pp 769–778.

Dror V, Kalynyak TB, Bychkivska Y, Frey MH, Tee M, Jeffrey KD, Nguyen V, Luciani DS & Johnson JD (2008). Glucose and ER-calcium channels regulate HIF-1 via presenilin in pancreatic β–cells. *Journal of Biological Chemistry*, Vol. 283, No. 15, pp 9909–9916.

Fajans SS, Bell GI & Polonsky KS (2001). Molecular mechanisms and clinical pathophysiology of maturity-onset diabetes of the young. *New England Journal of Medicine*, Vol. 345, pp 971–980.

Fedele AO, Whitelaw ML & Peet DJ (2002). Regulation of gene expression by the hypoxia-inducible factors. *Molecular Interventions*, Vol. 2, pp 229–243.

Fraisl P, Mazzone M, Schmidt T & Carmeliet P (2009). Regulation of angiogenesis by oxygen and metabolism. *Developmental Cell*, Vol. 16, No. 2, pp 167-179.

Gerich JE (2002) Is reduced first-phase insulin release the earliest detectable abnormality in individuals destined to develop type 2 diabetes? *Diabetes*, Vol. 51, No.1, pp S117-S121.

Gunton JE, Kulkarni RN, Yim S, Okada T, Hawthorne WJ, Tseng YH, Roberson RS, Ricordi C, O'Connell PJ, Gonzalez FJ & Kahn CR (2005). Loss of ARNT/HIF-1beta mediates altered gene expression and pancreatic-islet dysfunction in human type 2 diabetes. *Cell*, Vol.122, pp 337-349.

Habener JF, Stoffers DA, Stanojevic & Clarke VW (1998). A newly discovered role of transcription factors involved in pancreas development and the pathogenesis of diabetes mellitus. *Proceedings of the associations of American physicians*, Vol. 110, pp 12–21.

Habener JF, Kemp DM & Thomas MK (2005). Minireview: transcriptional regulation in pancreatic development. *Endocrinology* Vol. 146, pp 1025 - 1034.

Henquin JC, Ravier MA, Nenquin M, Jonas JC & Gilon P (2003). Hierarchy of the beta-cell signals controlling insulin secretion. *European Journal of Clinical Investigation*, Vol. 33, pp 742-750.

Hirose K, Morita M, Ema M, Mimura J, Hamada H, Fujii H, Saijo Y, Gotoh O, Sogawa K & Fujii-Kuriyama Y (1996). cDNA cloning and tissue-specific expression of a novel basic helix-loop-helix/PAS factor (Arnt2) with close sequence similarity to the aryl hydrocarbon receptor nuclear translocator (Arnt). *Molecular Cell Biology* Vol. 16, pp 1706-1713.

Holmes JL & Pollenz RS (1997). Determination of aryl hydrocarbon receptor nuclear translocator protein concentration and subcellular localization in hepatic and nonhepatic cell culture lines: development of quantitative Western blotting protocols for calculation of aryl hydrocarbon receptor and aryl hydrocarbon receptor nuclear translocator protein in total cell lysates. *Molecular Pharmacology*, Vol. 52, No. 2, pp 202-11.

Horton, JD, Goldstein, JL & Brown M.S. (2002). SREBPs: activators of the complete program of cholesterol and fatty acid synthesis in the liver. *Journal of Clinical Investigation*, Vol. 109, pp 1125–1131.

Hussain K (2010). Mutations in pancreatic ß-cell Glucokinase as a cause of hyperinsulinaemic hypoglycaemia and neonatal diabetes mellitus. *Reviews in Endocrine & Metabolic Disorders*, Vol. 11, No. 3, pp 179-183.

Ikeda M & Nomura M (1997). cDNA cloning and tissue-specific expression of a novel basic helix-loop-helix/PAS protein (BMAL1) and identification of alternatively spliced variants with alternative translation initiation site usage. Biochemical & Biophysical Research Communications, Vol. 233, No. 1, pp 258-264.

Ivarsson R, Quintens R, Dejonghe S Tsukamoto K, in 't Veld P, Renström E & Schuit FC (2005). Redox control of exocytosis: regulatory role of NADPH, thioredoxin, and glutaredoxin. Diabetes Vol. 54, pp 2132-2142.

Jain S, Dolwick KM, Schmidt JV & Bradfield CA (1994). Potent transactivation domains of the Ah receptor and the Ah receptor nuclear translocator map to their carboxyl termini. Journal of Biological Chemistry, Vol. 269, pp 31518-31524.

Jensen MV, Joseph JW, Ronnebaum SM, Burgess SC, Sherry AD & Newgard CB (2008). Metabolic cycling in control of glucose-stimulated insulin secretion. American journal of physiology endocrinology & metabolism, Vol. 295, pp E1287-E1297.

Jiang BH, Rue E, Wang GL, Roe R & Semenza GL (1996). Dimerization, DNA binding, and transactivation properties of hypoxia-inducible factor 1. Journal of Biological Chemistry, Vol. 271, pp 17771–17778.

Jiang C, Qu A, Matsubara T, Chanturiya T, Jou W, Gavrilova O, Shah YM, & Gonzalez FJ (2011). Disruption of Hypoxia-Inducible Factor 1 in Adipocytes Improves Insulin Sensitivity and Decreases Adiposity in High-Fat Diet-Fed Mice. *Diabetes,* (Epub ahead of print).

Johnson J, Carlsson L, Edlund T & Edlund H (1994). Insulin-promoter-factor 1 is required for pancreas development in mice. *Nature,* Vol. 371, pp 606–609.

Joseph JW, Jensen MV, Ilkayeva O, Palmieri F, Alárcon C, Rhodes CJ & Newgard CB (2006). The mitochondrial citrate/isocitrate carrier plays a regulatory role in glucose-stimulated insulin secretion. *Journal of Biological Chemistry,* Vol. 281, pp 35624-35632.

Kazuya Y, Yoshida K, Murao K, Imachi H, Cao WM, Yu X, Li J, Ahmed RA, Kitanaka N, Wong NC, Unterman TG, Magnuson MA & Ishida T (2007). Pancreatic glucokinase is activated by insulin-like growth factor-1, *Endocrinology,* Vol.148, pp 2904–2913.

Kewley RJ, Whitelaw ML & Chapman-Smith A (2004). The mammalian basic helix-loop-helix/PAS family of transcriptional regulators. *International Journal of Biochemistry & Cell Biology,* Vol. 36, No. 2, pp 189-204.

Khan A, Ling Z & Landau BR (1996). Quantifying the carboxylation of pyruvate in pancreatic islets. *Journal of Biological Chemistry,* Vol. 271, 2539-2542.

Kitamura YI, Kitamura T, Kruse JP,. Raum JC, Stein R, Gu W & Accili D (2005). FoxO1 protects against pancreatic beta cell failure through NeuroD and MafA induction. *Cell Metabolism,* Vol. 2, pp 153–163.

Kobayashi A, Numayama-Tsuruta K, Sogawa K & Fujii-Kuriyama Y (1997). CBP/p300 functions as a possible transcriptional coactivator of Ah receptor nuclear translocator (Arnt). *Journal of Biological Chemistry,* Vol. 122, pp 703-710.

Koo EH, Kopan R (2004). Potential role of presenilin-regulated signaling pathways in sporadic neurodegeneration. Nature Medicine, Vol. 10 pp S26-33.

Kozak KR, Abbott B & Hankinson O (1997). ARNT-deficient mice and placental differentiation. *Developmental Biology,.* Vol. 191, No. 2, pp 297-305.

Li H, Dong L & Whitlock JP Jr (1994).Transcriptional activation function of the mouse Ah receptor nuclear translocator. *Journal of Biological Chemistry,* Vol. 269, pp 28098-28105.

Li H, Ko HP & Whitlock JP (1996). Induction of phosphoglycerate kinase 1 gene expression by hypoxia. Roles of Arnt and HIF-1alpha. *Journal of Biological Chemistry* Vol. 271, pp 21262-21267.

Lim W, Park Y, Cho J, Park C, Park J, Park Y-K, Park H & Lee Y (2011).Estrogen receptor beta inhibits transcriptional activity of hypoxia inducible factor-1 through the downregulation of arylhydrocarbon receptor nuclear translocator. *Breast Cancer Research*, Vol. 13, R32.

Lin Y & Sun Z. Current views on type 2 diabetes (2010). *Journal of Endocrinology, Vol.* 204, No. 1, pp 1–11.

Lindebro MC, Poellinger L & Whitelaw ML (1995). Protein-protein interaction via PAS domains: role of the PAS domain in positive and negative regulation of the bHLH/PAS dioxin receptor-Arnt transcription factor complex. *EMBO Journal,* Vol. 14, No. 14, pp 3528–3539.

Long WP, Chen X & Perdew GH (1999). Protein kinase C modulates aryl hydrocarbon receptor nuclear translocator protein-mediated transactivation potential in a dimer context. *Journal of Biological Chemistry,* Vol. 274, pp 12391-12400.

Lu D, Mulder H, Zhao P, Burgess SC, Jensen MV, Kamzolova S, Newgard CB, & Sherry AD (2002). 13C NMR isotopomer analysis reveals a connection between pyruvate cycling and glucose-stimulated insulin secretion (GSIS). *Proceedings of National Academy of Sciences,* Vol. 99, pp 2708-2713.

Lyttle BM, Li J, Krishnamurthy M, Fellows F, Wheeler MB, Goodyer CG & Wang R (2008). Transcription factor expression in the developing human fetal endocrine pancreas. *Diabetologia,* Vol. 51, No. 7, pp 1169-1180.

MacDonald MJ (1995) Influence of glucose on pyruvate carboxylase expression in pancreatic islets. *Archives of Biochemistry &.Biophysics*. Vol. 319, pp 128-132.

MacDonald MJ, Fahien LA, Brown LJ, Hasan NM, Buss JD & Kendrick MA (2005) Perspective: emerging evidence for signaling roles of mitochondrial anaplerotic products in insulin secretion. *American journal of physiology endocrinology & metabolism,* Vol. 288, pp E1-15.

Maechler P & Wollheim CB (1999) Mitochondrial glutamate acts as a messenger in glucose-induced insulin exocytosis. *Nature,* Vol. 402, pp 685-689.

Maltepe E, Schmidt JV, Baunoch D, Bradfield CA & Simon, MC (1997). Abnormal angiogenesis and responses to glucose and oxygen deprivation in mice lacking the protein ARNT. *Nature,* Vol. 386, pp 403-407.

Michael MD, Kulkarni RN, Postic C, Previs SF, Shulman GI, Magnuson MA & Kahn CR (2000). Loss of insulin signaling in hepatocytes leads to severe insulin resistance and progressive hepatic dysfunction. *Molecular Cell,* Vol. 6, pp 87–97

Mitchell SM & Frayling TM (2002). The role of transcription factors in maturity-onset diabetes of the young. *Molecular Genetics & Metabolism,* Vol. 77 No. 1-2, pp 35-43.

Muoio DM & Newgard CB (2008). Mechanisms of disease: molecular and metabolic mechanisms of insulin resistance and beta-cell failure in type 2 diabetes. Nature *Review Molecular Cell Biology,* Vol. 9, No. 3, pp 193-205.

Nenquin M, Szollosi A, guilar-Bryan L, Bryan J & Henquin JC (2004) Both triggering and amplifying pathways contribute to fuel-induced insulin secretion in the absence of sulfonylurea receptor-1 in pancreatic beta-cells. *Journal of Biological Chemistry,* Vol. 279, pp 32316-32324.

Newgard CB & Matchinsky FM (2001). Substrate control of insulin release. In: *Handbook of physiology,* section 7, The endocrine system volume II: The endocrine pancreas and regulation of metabolism, Jefferson AC, Cherrington A (eds), pp 125-151 Oxford University Press.

Newgard CB & McGarry JD (1995). Metabolic coupling factors in pancreatic b-cell signal transduction. *Annual Review of Biochemistry*, Vol. 64, pp 689-719.

Noordeen N, Khera TK, Sun G, Longbottom ER, Pullen TJ, Xavier GD, Rutter GA & Leclerc I (2009). ChREBP is a Negative Regulator of ARNT/HIF-1β Gene Expression in Pancreatic Islet β-Cells. *Diabetes*, Vol. 59, pp 153–160.

Partch CL & Gardner KH (2011). Coactivators necessary for transcriptional output of the hypoxia inducible factor, HIF, are directly recruited by ARNT PAS-B. *Proceedings of the National Academy of Sciences*, Vol. 108, No. 19, pp 7739-7744.

Pillai R, Huypens P, Huang M, Schaefer S, Sheinin T, Wettig SW & Joseph JW (2011). Aryl Hydrocarbon Receptor Nuclear Translocator/Hypoxia-inducible Factor-1β Plays a Critical Role in Maintaining Glucose-stimulated Anaplerosis and Insulin Release from Pancreatic β-Cells. *Journal of Biological Chemistry*, Vol. 286, No. 2, pp 1014-1024.

Pedersen TA, Bereshchenko O, Garcia-Silva, Ermakova, O, Kurz, E., Mandrup, S, Porse BT & Nerlov C (2007). Distinct C/EBPalpha motifs regulate lipogenic and gluconeogenic gene expression in vivo. *EMBO Journal*, Vol. *26*, pp 1081–1093.

Prentki M & Matschinsky FM (1987). Ca2+, cAMP, and phospholipid-derived messengers in coupling mechanisms of insulin secretion. *Physiology Review*, Vol. 67, pp 1185-1248.

Prentki M, Vischer S, Glennon MC, Regazzi R, Deeney JT & Corkey BE (1992) Malonyl-CoA and long chain acyl-CoA esters as metabolic coupling factors in nutrient-induced insulin secretion. *Journal of Biological Chemistry*, Vol. 267, pp 5802-5810.

Puri S, Cano DA, Hebrok M (2009). A role for von Hip- pel-Lindau protein in pancreatic beta-cell function. *Diabetes*, Vol. 58, No. *2*, pp 433–441.

Qiao L, MacLean PS, You H, Schaack J & Shao J. (2006). Knocking down liver CCAAT/enhancer-binding protein alpha by adenovirus-transduced silent interfering ribonucleic acid improves hepatic gluconeogenesis and lipid homeostasis in db/db mice. *Endocrinology*, Vol. 147, pp 3060–3069.

Ravier MA, Nenquin M, Miki T, Seino S, Henquin JC (2009). Glucose controls cytosolic Ca2+ and insulin secretion in mouse islets lacking adenosine triphosphate-sensitive K+ channels owing to a knockout of the pore-forming subunit Kir6.2. *Endocrinology* , Vol. 150, pp 33-45.

Reyes H, Reisz-Porszasz S & Hankinson O (1992). Identification of the Ah receptor nuclear translocator protein (Arnt) as a component of the DNA binding form of the Ah receptor. *Science*, Vol. 256, pp 1193-1195.

Ronnebaum SM, Ilkayeva O, Burgess SC, Joseph JW, Lu D, Stevens RD, Becker TC, Sherry AD, Newgard CB & Jensen MV (2006). A pyruvate cycling pathway involving cytosolic NADP-dependent isocitrate dehydrogenase regulates glucose-stimulated insulin secretion. *Journal of Biological Chemistry*, Vol. 281, pp 30593-30602.

Rüegg J, Swedenborg E, Wahlström D, Escande A, Balaguer P, Pettersson K, & Pongratz I (2008). The transcription factor aryl hydrocarbon receptor nuclear translocator functions as an estrogen receptor beta-selective coactivator, and its recruitment to alternative pathways mediates antiestrogenic effects of dioxin. *Molecular Endocrinology*, Vol. 22, pp 304-316.

Salceda S, Beck I & Caro J (1996). Absolute requirement of aryl hydrocarbon receptor nuclear translocator protein for gene activation by hypoxia. *Archives of Biochemistry & Biophysics*, Vol. 334, pp 389-394.

Saltiel AR & Kahn CR (2001). Insulin signalling and the regulation of glucose and lipid metabolism. *Nature,* Vol. 414, pp 799–806.

Schuit F, De Vos A, Farfari S, Moens K, Pipeleers D, Brun T & Prentki M (1997). Metabolic fate of glucose in purified islet cells. Glucose-regulated anaplerosis in beta cells. *Journal of Biological Chemistry,* Vol. 272, No. 30, pp 18572-18579.

Semenza GL, Roth PH, Fang HM & Wang GL (1994). Transcriptional regulation of genes encoding glycolytic enzymes by hypoxia-inducible factor 1. *Journal of Biological Chemistry* , Vol. 269, No. 38, pp 23757-23763.

Semenza GL, Agani F, Feldser D, Iyer N, Kotch L, Laughner E & Yu A (2000). Hypoxia, HIF-1, and the pathophysiology of common human diseases. *Advances in Experimental Medicine and Biology,* Vol. 475, pp 123–130.

Semenza, GL (2003). Targeting HIF-1 for cancer therapy. *Nature Review Cancer,* Vol. 3, pp 721-732.

Shiota C, Larsson O, Shelton KD, Shiota M, Efanov AM, Hoy M, Lindner J, Kooptiwut S, Juntti-Berggren L, Gromada J, Berggren PO & Magnuson MA (2002). Sulfonylurea receptor type 1 knock-out mice have intact feeding-stimulated insulin secretion despite marked impairment in their response to glucose. *Journal of Biological Chemistry,* Vol. 277: 37176-37183.

Sogawa K, Fujii-Kuriyama Y (1997). Ah receptor, a novel ligand-activated transcription factor. *J Biochem.* Vol. 122 No. 6, pp 1075-9.

Song IS, Wang AG, Yoon SY, Kim JM, Kim JH, Lee DS & Kim NS (2009). Regulation of glucose metabolism-related genes and VEGF by HIF-1α and HIF-1β, but not HIF-2α, in gastric cancer. *Experimental & Molecular Medicine,* Vol. 41, No. 1, pp 51-58.

Steiner DJ, Kim A, Miller M & Hara M (2010). Pancreatic islet plasticity Interspecies comparison of islet architecture and composition. *Islets* Vol. 2, No. 3, pp 135-145.

Stoffers DA & Zinkin NT (1997), Pancreatic agenesis attributable to a single nucleotide deletion in the human IPF1 gene coding sequence, Nature Genetics, Vol. 15, pp 106- 110.

Swanson HI, Chan WK & Bradfield CA (1995). DNA binding specificities and pairing rules of the Ah receptor, ARNT, and SIM proteins. *Journal of Biological Chemistry,* Vol. 270, pp 26292-302.

Szollosi A, Nenquin M & Henquin JC (2007) Overnight culture unmasks glucose-induced insulin secretion in mouse islets lacking ATP-sensitive K+ channels by improving the triggering Ca2+ signal. J.Biol.Chem. 282: 14768-14776

Tomita S, Sinal CJ, Yim SH & Gonzalez FJ (2000). Conditional disruption of the aryl hydrocarbon receptor nuclear translocator (Arnt) gene leads to loss of target gene induction by the aryl hydrocarbon receptor and hypoxia-inducible factor 1alpha. *Molecular Endocrinology,* Vol. 14, pp 1674-1681.

Tripathy D, Chavez AO (2010). Defects in insulin secretion and action in the pathogenesis of type 2 diabetes mellitus. *Current Diabetes Reports.* Vol. 10, No. 3, pp 184-91.

Tsuchiya T, Kominato K & Ueda M(2002). Human Hypoxic Signal Transduction through a Signature Motif in Hepatocyte Nuclear Factor 4. *Journal of Biochemistry,* Vol. 132, pp 37-44.

Wang GL & Semenza GL (1995 a). Purification and characterization of hypoxia-inducible factor 1. *Journal of Biological Chemistry,* Vol. 270, pp 1230-1237.

Wang GL, Jiang BH, Rue EA & Semenza GL (1995 b). Hypoxia-inducible factor 1 is a basic-helix-loop-helix-PAS heterodimer regulated by cellular O2 tension. *Proceedings of the national academy of sciences,* Vol. 92, No. 12, pp 5510-5514.

Wang XL, Suzuki R, Lee K, Tran T, Gunton JE, Saha AK, Patti ME, Goldfine A, Ruderman NB, Gonzalez FJ & Kahn CR (2009). Ablation of ARNT/HIF-1beta in liver alters gluconeogenesis, lipogenic gene expression, and serum ketones. *Cell Metabolism,* Vol. 9, pp 428-439.

Wood SM, Gleadle JM, Pugh CW, Hankinson O & Ratcliffe PJ (1996). The Role of the Aryl Hydrocarbon Receptor Nuclear Translocator (ARNT) in Hypoxic Induction of Gene Expression. Studies in ARNT-Deficient Cells. *Journal of Biological Chemistry,* Vol. 271, No. 25, pp 15117–15123.

Woods SL & Whitelaw ML (2002). Differential activities of murine single minded 1 (SIM1) and SIM2 on a hypoxic response element. Cross-talk between basic helix-loop-helix/per-Arnt-Sim homology transcription factors. *Journal of Biological Chemistry,* Vol. 277, No. 12, pp 10236-10243.

Yamagata K (2003). Regulation of pancreatic beta-cell function by the HNF transcription network: lessons from maturity-onset diabetes of the young (MODY). *Endocrinology Journal,* Vol. 50, No. 5, pp 491-499.

Zehetner J, Danzer C, Collins S, Eckhardt K, Gerber PA, Ballschmieter P, Galvanovskis J, Shimomura K, Ashcroft FM, Thorens B, Rorsman P & Krek W (2008). pVHL is a regulator of glucose metabolism and insulin secretion in pancreatic beta-cells. *Genes & Development,* Vol. 22, No. 22, pp 3135–3146.

14

Modulation of EAAC1-Mediated Glutamate Uptake by Addicsin

Mitsushi J. Ikemoto and Taku Arano
Biomedical Research Institute
National Institute of Advanced Industrial Science and Technology (AIST)
Graduate School of Science, Toho University
Japan

1. Introduction

Glutamate is the major excitatory neurotransmitter in the mammalian central nervous system (CNS). In addition to functioning as a neurotransmitter at the majority of brain synapses, it is the substrate for synthesis of the major inhibitory transmitter γ-aminobutyric acid (GABA). However, glutamate is also a neurotoxin, and a number of molecular control mechanisms are responsible for maintaining extracellular glutamate below excitotoxic levels. Na^+-dependent excitatory amino acid transporters (EAATs) are crucial regulators of extracellular glutamate and also act to control the dynamics of excitatory transmission in the CNS (Danbolt, 2001). The Na^+-dependent excitatory amino acid carrier 1 (EAAC1) is expressed in the somata and dendrites of many neuronal types, including pyramidal cells of the hippocampal formation and cortex, and many subtypes of GABAergic inhibitory neurons (Rothstein et al., 1994). The physiological significance of EAAC1 is unclear because the subcellular distribution and kinetic properties of this transporter would not allow for a substantial contribution to glutamate clearance from the synaptic cleft; rather, these functions are mediated by glial EAATs (EAAT1 and EAAT2) located in the perisynaptic region. Recent studies have demonstrated multiple functions for EAAC1 distinct from clearance of glutamate from CNS synapses (Kiryu-Seo et al., 2006; Levenson et al., 2002; Peghini et al., 1997; Sepkuty et al., 2002). For example, decreased EAAC1 expression in the CNS impairs neuronal glutathione (GSH) synthesis, leading to oxidative stress and age-dependent neurodegeneration (Aoyama et al., 2006), suggesting that aberrant EAAC1 expression contributes to the pathogenesis of neurodegenerative diseases.

Studies conducted over the past decade on the kinetics of EAAC1 and regulation of transporter expression and function have lead to a greater appreciation of the physiological and pathophysiological relevance of EAAC1 (Aoyama et al., 2008b; Danbolt, 2001; Kanai & Hediger, 2004; Nieoullon et al., 2006), but there are many issues to be resolved for a thorough understanding of the significance of EAAC1 in normal brain function and disease. In particular, the regulatory mechanisms of EAAC1-mediated glutamate uptake are largely unknown. The recent discovery of addicsin (glutamate transporter-associated protein 3-18, GTRAP3-18) as an EAAC1 binding protein has contributed greatly to our understanding of the regulatory mechanisms of EAAC1 activity (Lin et al., 2001). Furthermore, we recently proposed a regulatory model of EAAC1-mediated glutamate uptake by addicsin complexes (Akiduki & Ikemoto, 2008). In this chapter, we describe the regulation of EAAC1-mediated glutamate uptake based on our recent results. To better understand this regulatory mechanism, we first explain three key molecules involved in this regulatory pathway—EAAC1, addicsin, and ADP-ribosylation factor-like 6 interacting protein 1 (Arl6ip1).

1.1 EAAC1

The EAAC1 protein was first identified as a Na^+-dependent high-affinity glutamate transporter by expression cloning in *Xenopus* oocytes (Kanai & Hediger, 1992). Stoichiometric analysis demonstrates that EAAC1 transports L-glutamate, L-aspartate, and D-aspartate, accompanied by the cotransport of 3 Na^+ and 1 H^+, and the countertransport of 1 K^+ (Kanai & Hediger, 2003). In mammalian tissues, there are five different subtypes of EAATs—EAAT1 (glutamate/aspartate transporter, GLAST), EAAT2 (glutamate transporter 1, GLT-1), EAAT3 (EAAC1), EAAT4, and EAAT5 (Danbolt, 2001). These EAATs are structurally similar; all have eight transmembrane domains and a pore loop between the seventh and eighth domain. Most EAATs play an important role in removing extracellular glutamate from the synaptic and extrasynaptic space (Kanai & Hediger, 2003), particularly GLAST and GLT-1. These two isoforms are primarily expressed in glial cells and play a major role in protecting neurons from glutamate-induced toxicity (Rothstein et al., 1994) as well as terminating glutamatergic transmission (Rothstein et al., 1993; Tong & Jahr, 1994). In contrast, EAAC1 is diffusely localized to the cell bodies and dendrites of neurons and is enriched in cortical and hippocampal pyramidal cells as well as in some inhibitory neurons (Conti et al., 1998; Rothstein et al., 1994). This subcellular localization and restricted distribution indicate that EAAC1 does not play a major role in glutamate clearance from the synaptic cleft (Rothstein et al., 1996). Recent studies suggest that EAAC1 contributes to multiple physiological functions distinct from glutamate clearance. Indeed, EAAC1 transport provides cysteine as a substrate of GSH synthesis (Y. Chen & Swanson, 2003; Himi et al., 2003; Watabe et al., 2008; Zerangue & Kavanaugh, 1996). Neurons cannot transport extracellular GSH and therefore must transport cysteine from the extracellular space for *de novo* GSH synthesis from cysteine (Aoyama et al., 2008b). In the CNS, the depletion of GSH is associated with neurodegenerative disorders, including Alzheimer's and Parkinson's diseases (Ramassamy et al., 2000; Sian et al., 1994). Consistent with these results, EAAC1 knockout mice show oxidative stress in neurons and age-dependent neurodegeneration, pathologies that are rescued by *N*-acetylcysteine, a membrane-permeable cysteine precursor (Aoyama et al., 2006). These mice also show alteration of zinc homeostasis and increased neural damage after transient cerebral ischemia (Won et al., 2010). Furthermore, in a knockin mouse model of Huntington's disease, in which human *huntingtin* exon 1 with 140 CAG repeats was inserted into the wild-type low CGA repeat mouse *huntingtin* gene, oxidative stress and cell death were caused by abnormal Rab11-dependent EAAC1 trafficking to the cell surface (X. Li et al., 2010). In addition, 1-methyl-4-phenyl-1,2,3,6-tetrahydropyridine-treated mice, an animal model of Parkinson's disease, show reduced EAAC1-mediated neuronal cysteine uptake, impaired GSH synthesis, and motor dysfunction (Aoyama et al., 2008a). These results indicate that dysfunctional EAAC1-mediated cysteine transport increases neural vulnerability to oxidative stress and could contribute to the pathogenesis of neurodegenerative diseases.

In addition to cysteine transport, EAAC1 has several other functions unrelated to removal of extracellular glutamate. For instance, EAAC1 promotes GABA synthesis by supplying the substrate glutamate (Mathews & Diamond, 2003; Sepkuty et al., 2002). Therefore, EAAC1 can strengthen inhibitory synapses in response to elevations in extracellular glutamate and contribute indirectly to GABA release (Mathews & Diamond, 2003). Indeed, a loss of EAAC1 function leads to epilepsy (Sepkuty et al., 2002), underscoring the importance of EAAC1 in GABAergic transmission. Furthermore, EAAC1 plays a crucial role in preventing neuronal death by suppressing glutamate excitotoxicity (Kiryu et al., 1995; Murphy et al., 1989) and has a mitochondria-mediated anti-apoptotic function in injured motor neurons (Kiryu-Seo et al., 2006). These studies and those discussed in Section 3.4 strongly suggest that EAAC1 contributes to multiple functions in the CNS distinct from glutamate clearance.

The regulatory mechanisms of EAAC1 have been widely investigated *in vitro*. Cumulative evidence demonstrates that glutamate uptake by EAAC1 is facilitated by cell signaling molecules and accessory proteins that promote the redistribution of EAAC1 from the endoplasmic reticulum (ER) to the plasma membrane. First, several reports demonstrate that several kinase signaling cascades regulate EAAC1 activity. In C6BU-1 glioma cells and primary neuronal cultures, phorbol 12-myristate 13-acetate (PMA), a protein kinase C (PKC) activator, rapidly increases EAAC1-mediated glutamate uptake (Dowd & Robinson, 1996). This effect is regulated by mechanisms that are independent of *de novo* synthesis of new transporters but is related to the redistribution of EAAC1 from subcellular compartments to the plasma membrane (Davis et al., 1998; Fournier et al., 2004; Sims et al., 2000). Pharmacological analyses demonstrate that PKCα regulates EAAC1 translocation from intracellular compartments to the cell surface, and that PKCε increases EAAC1 functional activity (Gonzalez et al., 2002). PKCα interacts with EAAC1 in a PKC-dependent manner and phosphorylates EAAC1 (Gonzalez et al., 2003). Platelet-derived growth factor (PDGF) increases the delivery of EAAC1 to the cell surface through phosphatidylinositol 3-kinase (PI3K) activity (Fournier et al., 2004; Sheldon et al., 2006; Sims et al., 2000). Consistent with this result, wortmannin, a PI3K inhibitor, decreases cell surface expression of EAAC1 and inhibits EAAC1-mediated glutamate uptake (Davis et al., 1998). In addition, PKC and PDGF have different effects on trafficking and internalization of EAAC1; PMA, but not PDGF, reduces internalization of EAAC1 (Fournier et al., 2004). Thus, EAAC1 trafficking is regulated by two independent signaling pathways. In contrast, PKC negatively regulates EAAC1-mediated glutamate uptake in *Xenopus* oocytes (Trotti et al., 2001) and in Madin–Darby canine kidney (MDCK) cells (Padovano et al., 2009) by inhibiting cell surface expression through calcineurin-mediated internalization (Padovano et al., 2009; Trotti et al., 2001), suggesting that the regulatory mechanisms of EAAC1 surface expression and function by PKC are specific to cell type and depend on specific PKC isozymes. Second, accessory proteins regulate EAAC1 activity. For instance, δ opiod receptor interacts with EAAC1 and inhibits EAAC1-mediated glutamate uptake in *Xenopus* oocytes and rat hippocampal neurons (Xia et al., 2006). In addition, *N*-methyl-*D*-aspartate receptors containing NR1, NR2A, and/or NR2B interact with EAAC1 and facilitate the cell surface expression of EAAC1 in C6BU-1 cells and rat hippocampal neurons (Waxman et al., 2007). Moreover, the cell surface expression of EAAC1 is controlled by interactions with Na^+/H^+-exchanger regulatory factor 3 (NHERF-3, also called PDZK1) and adaptor protein 2 (AP-2). While NHERF-3 promotes the delivery of EAAC1 to the plasma membrane, AP-2 regulates constitutive endocytosis of EAAC1 in MDCK cells (D'Amico et al., 2010). Furthermore, reticulon 2B (RTN2B) interacts with EAAC1 and addicsin/GTRAP3-18, and promotes intracellular trafficking of EAAC1 in HEK293 cells and cultured cortical neurons (Liu et al., 2008). Addicsin/GTRAP3-18 interacts with EAAC1 and inhibits EAAC1 trafficking in HEK293 cells (Ruggiero et al., 2008). Thus, multiple regulatory mechanisms control EAAC1 trafficking and membrane expression, but the molecular details are generally unclear. In this study, we focus on the regulation of EAAC1 trafficking by addicsin.

1.2 Addicsin

In many papers, human addicsin and rat addicsin are called JWA and GTRAP3-18, respectively. Addicsin, GTRAP3-18, and JWA have been independently identified by several research groups (Ikemoto et al., 2002; Lin et al., 2001; Zhou et al., GeneBank, AF070523, unpublished observations). We first identified *addicsin* as a novel mRNA encoding a 22-kDa hydrophobic protein that is highly expressed in the basomedial nucleus of the mouse amygdala following repeated morphine administration (Ikemoto et al., 2002). Meanwhile, *GTRAP3-18* cDNA was identified as encoding an EAAC1 binding protein by yeast two-hybrid screening of a rat brain cDNA library using the C-terminal intracellular domain of EAAC1 as bait (Lin et al., 2001). The *JWA* gene was identified as an all-*trans* retinoic acid (RA)-responsive factor from human tracheobronchial epithelial cells (Zhou et al., GeneBank, AF070523, unpublished observations). Bioinformatic analysis demonstrates that JWA has a prenylated Rab acceptor 1 (PRA1) domain and 62% similarity with Jena-Muenchen 4 (JM4), a protein recently identified as PRA1 domain family member 2 (PRAF2) (Schweneker et al., 2005). Proteins containing a large PRA1 domain form a new family of PRA1 domain family proteins (PRAFs) that regulate intracellular protein trafficking. Thus, addicsin is a new member of the PRAF family, PRAF3.

The *addicsin* cDNA is approximately 1.4 kbp and consists of a 564-bp single open reading frame (Ikemoto et al., 2002). The *addicsin* gene contains three exons separated by two introns, and the sequence is highly conserved among vertebrates (Butchbach et al., 2002). Furthermore, *addicsin* is located on mouse chromosome 6, a location corresponding to human chromosome 3p (Butchbach et al., 2002; Ikemoto et al., 2002).

Mouse addicsin is a 22-kDa protein of 188 amino acids with putative transmembrane segments (Butchbach et al., 2002; Ikemoto et al., 2002). Mouse addicsin is 98% identical to rat GTRAP3-18 and 95% similar to human JWA (Butchbach et al., 2002; Ikemoto et al., 2002). Moreover, addicsin has two putative PKC phosphorylation motifs (amino acids 18–20 and 138–140) as well as two putative cAMP-dependent protein kinase and calcium/calmodulin-dependent protein kinase II phosphorylation motifs (amino acids 27–31 and 35–39) (Butchbach et al., 2002; Ikemoto et al., 2002) (Fig. 1). However, there is no evidence that these phosphorylation sites are phosphorylated by protein kinases *in vitro* and *in vivo*.

Expression profiles of addicsin and *addicsin* mRNA were investigated in the developing and mature brain. In the developing rat brain, the expression levels of addicsin decrease significantly from embryonic day 17 to post-natal day 0 (Maier et al., 2009). Meanwhile, *addicsin* mRNA levels increase gradually during early maturation, peaking around post-natal day 5, and then declining by about 50% by post-natal day 14 (Inoue et al., 2005). This developmental expression pattern corresponds to periods of elevated synaptogenesis, suggesting that addicsin is involved in synapse formation. Indeed, later in this chapter, we discuss evidence that addicsin participates in intracellular protein trafficking of neurotransmitter receptors. Addicsin is widely distributed in the brain (Akiduki et al., 2007; Butchbach et al., 2002). In the mature CNS, addicsin is expressed in the cerebral cortex, amygdala, striatum, hippocampus (CA1–3 fields), dentate gyrus, and cerebellum. Addicsin is expressed in the somata of glutamatergic and GABAergic neurons and exhibits presynaptic localization in restricted regions such as CA3 stratum lucidum (Akiduki et al., 2007). *In situ* hybridization analysis reveals that *addicsin* mRNA is widely distributed in the brain, predominantly expressed in principal neurons, including glutamatergic and GABAergic neurons in the mature CNS (Inoue et al., 2005). However, the precise subcellular localization of addicsin remains controversial. Recent reports found that addicsin is an integral ER membrane protein that prevents EAAC1 maturation and function by inhibiting ER trafficking (Ruggiero et al., 2008). However, our protein fractionation analysis using mouse whole brain lysates prepared in PBS, NaCl, or Na_2CO_3 buffer, all indicate that addicsin is predominantly present in the S1 soluble fraction, while the ER transmembrane protein calnexin is present in the P2 pellet fraction (Ikemoto et al., 2002). Our subcellular fractionation analysis with highly purified synaptic fractions prepared from mouse forebrain also support the notion that addicsin is present in the cytoplasmic and presynaptic membrane fractions (Akiduki et al., 2007). Furthermore, immunocytochemical studies reveal that addicsin is present in both the plasma membrane and the intracellular compartments, including the ER (Ikemoto et al., 2002; Watabe et al., 2007, 2008). Consistent with these findings, bioinformatic analysis demonstrates that the α-helix is not long enough for a transmembrane domain; nevertheless, addicsin is predicted to be a hydrophobic protein composed of 62% α-helix and 8% β-sheet (Butchbach et al., 2002), suggesting that it is membrane-associated. Further investigations are needed to clarify the subcellular localization of addicsin, but it is apparent that this protein can exist in both soluble and membrane-associated forms.

Addicsin easily forms homo- and heteromultimers (Ikemoto et al., 2002; Lin et al., 2001) and many reports demonstrate that addicsin can associate with a multitude of proteins (Akiduki & Ikemoto, 2008), including Arl6ip1 (Akiduki & Ikemoto, 2008), ARL6 (Ingley et al., 1999), δ opioid receptor (Wu et al., 2011), EAAC1 (Lin et al., 2001), Rab1 (Maier et al., 2009), and RTN2B (Liu et al., 2008). Moreover, recent studies using the yeast two-hybrid system revealed many potential addicsin-binding proteins (M.J. Ikemoto et al., unpublished data), strongly suggesting that addicsin exerts multiple physiological functions by forming various molecular complexes. It is vital to catalog these interacting proteins and to determine the presence and location of these molecular complexes.

These potential functions remain largely speculative, but molecular studies have provided several intriguing candidates (Fig. 2). First, addicsin is involved in apoptosis induced by 12-*O*-tetradecanoylphorbol-13-acetate, all-*trans* RA, *N*-(4-hydroxyphenyl) retinamide, arsenic trioxide, and cadmium (Mao et al., 2006; Zhou et al., 2008). Knockdown of addicsin attenuates all-*trans* RA-induced and arsenic trioxide-induced apoptosis (Mao et al., 2006; Zhou et al., 2008). Therefore, addicsin serves as a pro-apoptotic molecule. Second, addicsin acts as an environmental stress sensor to protect cells from oxidative stress and subsequent genomic damage. Addicsin is also involved in cellular responses to environmental stresses, including oxidative stress and heat shock, and in the differentiation of leukemia cells under nonphysiological conditions (Cao et al., 2007; Huang et al., 2006a, 2006b; T. Zhu et al., 2005). Addicsin is upregulated after exposure to the pro-oxidants benzo[α]pyrene and hydrogen peroxide through activation of the nuclear transcription factor I (NFI) (R. Chen et al., 2007). Addicsin facilitates DNA repair by interacting with X-ray cross-complementing group 1 protein, a regulator of the DNA base excision repair processes that translocates to the nucleus in response to oxidative stress (R. Chen et al., 2007; Wang et al., 2009). Thus, NFI-mediated addicsin upregulation protects against DNA damage induced by benzo[α]pyrene and hydrogen peroxide. Third, addicsin also inhibits cancer cell migration as was observed in HeLa, B16, and HCCLM3 cancer cells. (H. Chen et al., 2007). Addicsin has an important role in maintaining the stability of F-actin and in the initiation of actin cytoskeletal rearrangements. Moreover, knockdown of addicsin results in the inactivation of the MEK–ERK signaling cascade. Thus, addicsin inhibits cell migration by activating the mitogen-activated protein kinase (MAPK) cascade and regulating the rearrangement of the F-actin cytoskeleton (H. Chen et al., 2007). Fourth, addicsin participates in the regulation of GSH synthesis; the association of addicsin with EAAC1 at the plasma membrane inhibits the uptake of cysteine for GSH synthesis and thus determines the intracellular GSH content *in vitro* and *in vivo* (Watabe et al., 2007, 2008). This suggests that addicsin is a therapeutic target for enhancing GSH levels in patients with neurodegenerative disorders, such as Alzheimer's and Parkinson's diseases, associated with oxidative stress. Fifth, addicsin significantly inhibits neurite growth in differentiated CAD cells by inactivating Rab1, a positive regulator of ER-to-Golgi trafficking (Maier et al., 2009). Finally, addicsin participates in the regulation of EAAC1-mediated glutamate uptake (Akiduki & Ikemoto, 2008) and ER protein trafficking (Liu et al., 2008; Ruggiero et al., 2008). We discuss these latter two physiological functions in detail (Section 2).

1.3 Arl6ip1

The "ADP-ribosylation factor-like 6 interacting protein 1 (Arl6ip1)" is the new name assigned to three independently described factors: the original Arl6ip, apoptotic regulator in the membrane of the ER (ARMER), and protein KIAA0069. The Arl6ip1 protein was first identified by yeast two-hybrid screening using mouse ARL6 as bait (Ingley et al., 1999) and as a negative regulatory factor during myeloid differentiation by differential display (Pettersson et al., 2000). Moreover, a novel protein, designated ARMER, initially discovered as a false-positive clone by yeast two-hybrid screening using Bcl-xL as bait, is also Arl6ip1 (Lui et al., 2003). In addition, Arl6ip1 has more than 96% homology with the human protein KIAA0069, the product of a cDNA isolated from the human myeloblast cell line KG-1 during a systematic effort to characterize complete cDNAs (Nomura et al., 1994). Amino acid analysis of Arl6ip1 demonstrates that it is composed of 203 amino acids and encodes a 23-kDa protein with four putative transmembrane segments (Pettersson et al., 2000). Several studies indicate that Arl6ip1 is an integral membrane protein localized to the ER (Lui et al., 2003; Pettersson et al., 2000). Furthermore, computational analysis of the topology of Arl6ip1 demonstrates that the N- and C-terminal ends are both exposed to the cytoplasm (Lui et al., 2003). Consistent with these results, Arl6ip1 has two putative casein kinase II phosphorylation motifs (amino acids 18–21 and 128–131), three putative PKC phosphorylation motifs (amino acids 94–96, 115–117, and 128–130), a *N*-glycosylation motif (amino acids 6–9), a prenyl group-binding motif (amino acids 72–75), and an ER retention signal in the C-terminal cytoplasmic region (amino acids 200–203) (Akiduki &

Ikemoto, 2008; Lui et al., 2003) (Fig. 1). Thus, Arl6ip1 function may be controlled by diverse intracellular cell signals, but it is unknown whether these motifs are physiologically functional.

The functions of Arl6ip1 remain largely unknown, but culture studies have provided several intriguing possibilities. For example, Arl6ip1 protects HT1080 fibrosarcoma cells from apoptosis induced by serum starvation, doxorubicin, UV irradiation, tumor necrosis factor α, and ER stressors by inhibiting caspase-9 activity (Lui et al., 2003). In addition, Arl6ip1 suppresses cisplatin-induced apoptosis in CaSki human cervical cancer cells by regulating the expression of apoptosis-related proteins caspase-3, caspase-9, p53, NF-κB, MAPK, Bcl-2, Bcl-xL, and Bax (Guo et al., 2010a). Furthermore, Arl6ip1 is involved in cell growth, cell cycle progression, and invasion of cancer cells. Downregulation of Arl6ip1 suppresses cell proliferation and colony formation, arrests cell cycling at the G0/G1 phase, and inhibits migration of CaSki human cervical cancer cells (Guo et al., 2010b). Most relevant to the present discussion, Arl6ip1 is involved in the regulation of EAAC1. Recently, we demonstrated that Arl6ip1 is a novel addicsin-associating factor that indirectly promotes PKC-dependent EAAC1-mediated glutamate uptake by decreasing the number of addicsin molecules available for suppression of EAAC1 (Akiduki & Ikemoto, 2008).

2. Regulation of EAAC1 function by addicsin

The mechanisms by which addicsin regulates EAAC1 activity have not been definitively established. However, the discovery of addicsin/GTRAP3-18 has contributed greatly to our understanding of EAAC1 function. Recent evidence demonstrates two major mechanisms of addicsin-mediated regulation of EAAC1 activity. One regulatory pathway is dependent on the dynamic competition for free addicsin molecules by other addicsin molecules to form the homocomplex and by Arl6ip1 to form a heterocomplex. This addicsin–Arl6ip1 complex sequesters addicsin molecules and blocks the interaction of addicsin with EAAC1 in the plasma membrane, thereby reducing the inhibitory effect of addicsin on EAAC1-mediated glutamate uptake (Akiduki & Ikemoto, 2008; Lin et al., 2001) (Fig. 3). Second, addicsin functions as a negative regulator of EAAC1 trafficking through the ER and inhibits the cell surface expression of EAAC1 (Liu et al., 2008; Ruggiero et al., 2008). In this section, we discuss these two mechanisms in detail.

2.1 Modulation of EAAC1-mediated glutamate uptake by addicsin

As an introduction to addicsin/GTRAP3-18-mediated regulation of EAAC1 activity, we discuss two early papers in detail. Lin et al. demonstrated that addicsin/GTRAP3-18 binds to EAAC1 and inhibits EAAC1-mediated glutamate uptake by this direct interaction (Lin et al., 2001). The second is our study showing that addicsin inhibits EAAC1-mediated glutamate uptake in a PKC activity-dependent manner while Arl6ip1 promotes glutamate uptake (also in a PKC activity-dependent manner) by inhibiting the interaction of addicsin with EAAC1 (Akiduki & Ikemoto, 2008). Lin et al. first identified addicsin/GTRAP3-18 as an EAAC1-interacting protein by yeast two-hybrid screening of a rat brain cDNA library. To evaluate whether addicsin/GTRAP3-18 modulates EAAC1 function, they examined the effect of increasing addicsin/GTRAP3-18 expression on EAAC1-mediated glutamate uptake *in vitro* and *in vivo*. First, they showed that glutamate uptake decreased progressively with increasing expression of addicsin/GTRAP3-18 in HEK293 cells. Subsequent kinetic analyses in HEK293, C6BU-1, and COS7 cells revealed that elevated expression of addicsin decreased the glutamate affinity of EAAC1 without altering the maximal transport velocity (correlated with expression). Furthermore, HEK293 cells coexpressing addicsin/GTRAP3-18 and a truncated EAAC1 missing the addicsin/GTRAP3-18 association region showed higher glutamate uptake than cells expressing wild-type EAAC1. In addition, this truncated EAAC1 had a higher affinity for glutamate, suggesting that addicsin/GTRAP3-18 normally reduces EAAC1-mediated glutamate uptake by binding to this association region and reducing transporter glutamate affinity. Next, they evaluated the effect of intraventricular injection of an addicsin/GTRAP3-18 antisense mRNA on EAAC1-mediated glutamate uptake *in vivo*. The antisense treatment resulted in reduced addicsin/GTRAP3-18 expression, a significant increase in cortical EAAC1-mediated glutamate uptake, and an increase in glutamate affinity compared to saline-treated or sense mRNA-treated control animals. In conclusion, addicsin/GTRAP3-18 can negatively modulate EAAC1-mediated glutamate uptake by a direct interaction with EAAC1.

We first isolated addicsin as a novel protein richly expressed in the amygdala of mice under chronic morphine treatment. Addicsin has a tendency to form the multimeric complex *in vitro* (Ikemoto et al., 2002; Lin et al., 2001). The initial discovery of addicsin prompted us to perform yeast two-hybrid screening of an amygdala cDNA library constructed from chronic morphine-administered mice. From this screen, we identified Arl6ip1 as a candidate addicsin-interacting protein. As described in section 1.3, Arl6ip1 is an anti-apoptotic protein located in the ER. As previously described, addicsin inhibits EAAC1-mediated glutamate uptake by direct association at the plasma membrane (Lin et al., 2001), so we speculated that Arl6ip1 upregulates EAAC1-mediated glutamate transport by inhibiting the interaction between addicsin and EAAC1 (Fig. 3).

As a first step to verify this hypothesis, we investigated whether addicsin could bind Arl6ip1 *in vitro* and *in vivo*. To eliminate the possibility of false-positive clones, reconfirmation tests using a full length mouse Arl6ip1 as prey or bait were performed. This tests revealed the specific interaction with addicsin in the yeast AH109 strain. We next examined the reproducibility of this screening result by yeast two-hybrid screening using a different cDNA library prepared from whole brains of 7-week-old mice. We obtained 20 positive clones that clearly displayed α-galactosidase activity (the gene driven by the protein–protein interaction in the two-hybrid screen). Among these positive clones, 11 were identical to *Arl6ip1* cDNA (M.J. Ikemoto et al., unpublished data), confirming the interaction with addicsin and Arl6ip1 in the yeast AH109 strain. We then performed immunoprecipitation analysis, glycerol gradient analysis, and immunocytochemical analysis to directly test the interaction between Arl6ip1 and addicsin *in vitro*. For this purpose, we prepared cell lysates from NG108-15 cells expressing FLAG-tagged Arl6ip1 (Arl6ip1-FLAG), Myc-tagged addicsin (addicsin-myc), or both. Immunoprecipitation analysis of these cell lysates demonstrated that Arl6ip1-FLAG specifically interacted with addicsin-myc in the cell lysates prepared from coexpressing cells, but not from cells expressing Arl6ip1-FLAG or addicsin-myc alone. Glycerol gradient analysis revealed that the elution profile of Arl6ip1-FLAG was similar to that of addicsin-myc. The elution peaks of both proteins were observed in the fraction with a deduced molecular mass of 24 kDa. Moreover, the elution peak of the addicsin homodimer was present in the 44-kDa fraction, suggesting that addicsin forms Arl6ip1–addicsin heterodimers and addicsin–addicsin homodimers *in vitro*. Immunocytochemical analysis in NG108-15 cells overexpressing Arl6ip1-FLAG and addicsin-myc demonstrated subcellular colocalization (M.J. Ikemoto et al., unpublished data). To examine the interaction of both proteins *in vivo*, we performed *in vivo* immunoprecipitation assays of whole brain lysates using an anti-Arl6ip1 polyclonal antibody (generated from a synthetic peptide spanning amino acids 185–199 of mouse Arl6ip1) that again revealed a specific interaction between Arl6ip1 and addicsin. Western blot analysis demonstrated that Arl6ip1 was widely expressed in the mature brain and showed substantial regional overlap with addicsin. In addition, immunohistochemical staining confirmed that Arl6ip1 was widely expressed in the mature brain and localized in neuron-like cells. The neural expression pattern of Arl6ip1 was the same as addicsin, suggesting that Arl6ip1 is colocalized with addicsin in the mature CNS. We concluded that addicsin specifically interacted with Arl6ip1 *in vitro* and *in vivo*.

As a second step, we then determined the Arl6ip1- and addicsin-binding regions on addicsin. If Arl6ip1 does regulate EAAC1 activity by competitively binding to addicsin molecules and thus preventing the formation of addicsin homodimers that downregulate EAAC1 activity, the Arl6ip1- and addicsin-binding regions on addicsin should be located close enough for such a competitive interaction. Immunoprecipitation assays using several addicsin truncation mutants indicated that Arl6ip1 associated with full length addicsin (wt), a truncation lacking the C-terminal region at amino acids 145–188 (d1), a deletion mutant of the N-terminal domain at amino acids 1–102 (d2), and a mutant missing the region containing the C-terminal phosphorylation motif at amino acids 136–144 (d3). However, Arl6ip1 could not interact with a mutant lacking a portion of the hydrophobic region at amino acids 103–117 (d4). As expected, addicsin was able to associate with the wt, d1, d2, or d3 mutant, but not the d4 truncation mutant, indicating that the hydrophobic region at amino acids 103–117 of addicsin is a crucial domain for the formation of addicsin–addicsin homodimers and addicsin-Arl6ip1 heterodimers (Fig. 1). These results strongly support our hypothesis that Arl6ip1 antagonizes addicsin-mediated downregulation of EAAC1 activity by sequestering free addicsin.

As a third step, we investigated whether Arl6ip1 had a positive effect on EAAC1-mediated glutamate uptake. For this purpose, we selected C6BU-1 glioma cells that expressed EAAC1 as the principal or only EAAT (Palos et al., 1996). We created two stably expressing C6BU-1 cell lines, designated C6BU-1-pSw-addicsin and C6BU-1-pSw-Arl6ip1. In these cell lines, we could strictly control the expression levels of V5-tagged addicsin (addicsin-V5) or V5-tagged Arl6ip1 (Arl6ip1-V5) by exposure to 10 nM mifepristone (11β-[4-dimethylamino]phenyl-17β-hydroxy-17-[1-propynyl]estra-4,9-dien-3-one), a synthetic 19-norsteroid. In addition, a cell viability assay demonstrated that upregulation of Arl6ip1-V5 or addicsin-V5 by exposure to 10 nM mifepristone was not cytotoxic, making these cell lines excellent models to evaluate the effects of changing Arl6ip1 and addicsin expression on the functional activity of EAAC1. Compared to control cells untreated with mifepristone or the PKC agonist PMA, the upregulation of Arl6ip-V5 or addicsin-V5 by 10 nM mifepristone alone did not change EAAC1-mediated glutamate uptake. When these cells were stimulated with 100 nM PMA alone, the glutamate uptake activity in C6BU-1-pSw-addicsin cells and C6BU-1-pSw-Arl6ip1 cells increased about two-fold compared to untreated controls. EAAC1-mediated glutamate uptake was significantly lower in C6BU-1-pSw-addicsin cells stimulated with both mifepristone and PMA compared to C6BU-1-pSw-addicsin cells treated with PMA alone, indicating that activation of addicsin expression inhibited PKC-dependent EAAC1 activity. In contrast, C6BU-1-pSw-Arl6ip1 cells treated with PMA and mifepristone exhibited a three-fold increase in glutamate uptake compared to the same line treated with PMA alone, indicating that Arl6ip1 overexpression enhanced PKC-dependent EAAC1 activity. On the other hand, the nonstimulating PMA analog 4α phorbol did not increase glutamate uptake relative to controls.

To further support these conclusions, we performed a knockdown experiment by transient transfection of double-stranded siRNAs into C6BU-1-pSw-Arl6ip1 cells to investigate the effect of decreased addicsin expression on EAAC1-mediated glutamate uptake. As expected, cells transfected with either of two alternative addicsin siRNAs showed about a two-fold increase in glutamate uptake in response to PMA exposure compared to cells treated with control scrambled siRNA. The elevated glutamate uptake concomitant with addicsin knockdown strongly supported the proposed mechanism for EAAC1 regulation by addicsin and Arl6ip1.

To investigate the molecular mechanisms for altered EAAC1-mediated glutamate uptake in C6BU-1-pSw-Arl6ip1 cells, we performed kinetic analysis of glutamate flux across C6BU-1-pSw-Arl6ip1 cell membranes. When Arl6ip1 was conditionally overexpressed using mifepristone, PMA treatment increased the glutamate affinity but not the maximal velocity compared to vehicle-treated controls (PMA: K_m = 647 μM, V_{max} = 1.5 × 10^3 pmole/mg/min; vehicle: K_m = 824 μM, V_{max} = 1.5 × 10^3 pmole/mg/min) with no change of addicsin expression levels. Thus, Arl6ip1 promoted EAAC1-mediated glutamate uptake by increasing the catalytic efficacy of EAAC1. Specifically, Arl6ip1 blocked the addicsin-mediated reduction in EAAC1 glutamate affinity.

As a fourth step, we then examined the subcellular localization of Arl6ip1 in C6BU-1-pSw-Arl6ip1 cells. Western blot analysis revealed that Arl6ip1-V5 expression levels were unaffected by 100 nM PMA exposure. Immunocytochemical analysis demonstrated that Arl6ip1-V5 was predominantly localized to cytoplasmic structures such as the ER and that this subcellular expression pattern was not changed by PMA. Furthermore, cell biotinylation analysis indicated that Arl6ip1 did not interact with the plasma membrane, consistent with our previous result that Arl6ip1 failed to interact with EAAC1 by immunoprecipitation. Therefore, Arl6ip1 was localized to the ER under all conditions tested and acted to "trap" addicsin molecules in Arl6ip1–addicsin heterodimers, thus preventing the direct interaction of addicsin with EAAC1. To confirm our hypothesis, we produced an addicsin mutant that lacked interaction with Arl6ip1 but not with other addicsin molecules. Fine mutational analysis was used to separate the Arl6ip1- and addicsin-binding regions within the addicsin d4 region. We compared addicsin sequences among various species and noted that two amino acids at positions 110 and 112 of mouse addicsin were completely conserved from fruit fly to human. We created a double-mutated form of addicsin that substituted both the native tyrosine at amino acid 110 and the leucine at amino acid 112 with alanine. The mutant, designated addicsin Y110A/L112A (or addicsinYL), showed markedly less binding to Arl6ip1 (40% of wild-type addicsin) but normal wild-type binding to addicsin, as revealed by immunoprecipitation. In addition, a cell biotinylation assay indicated that addicsinYL was unable to localize to the plasma membrane, suggesting that addicsinYL lost EAAC1-binding activity. To evaluate the effect of addicsinYL on EAAC1-mediated glutamate uptake, we created a conditional C6BU-1 cell line, designated C6BU-1-pSw-addicsinYL. This cell line exhibited mifepristone-dependent upregulation of V5-tagged addicsinYL and increased glutamate uptake in response to PMA that was unchanged by mifepristone-induced upregulation of addicsinYL. That is, glutamate uptake was not reduced by induced addicsinYL expression. These data strongly suggest that addicsin is a key negative regulator of EAAC1 in the plasma membrane and that Arl6ip1 is a negative regulator of addicsin.

As a final step, we examined the effect of addicsin PKC phosphorylation sites on EAAC1-mediated glutamate uptake in C6BU-1 cells. Addicsin has putative PKC phosphorylation motifs at amino acids 18-20 and 138-140, and PKC activation increases EAAC1-mediated glutamate uptake. We established conditional C6BU-1 cell lines, designated C6BU-1-pSw-addicsinS18A and C6BU-1-pSw-addicsinS138A. C6BU-1-pSw-addicsinS18A cells expressed a V5-tagged addicsin point mutant that substituted native serine 18 for alanine in the N-terminal motif in response to mifepristone, while C6BU-1-pSw-addicsinS138A cells expressed a V5-tagged addicsin point mutant that substituted native serine 138 for alanine in the C-terminal motif. These cells showed no cytotoxicity in response to 10 nM mifepristone. In contrast to cells expressing wild-type addicsin, expression of addicsinS18A did not suppress the PMA-induced increase in EAAC1-mediated glutamate uptake. Moreover, increased expression of addicsinS18A caused a significant increase in glutamate uptake even without PMA stimulation by a dominant negative effect. Similarly, addicsinS138A expression did not suppress the PMA-induced increase in EAAC1-mediated glutamate uptake. Thus, these mutations abolished the inhibitory effect of addicsin. However, in contrast to addicsinS18A, addicsinS138A expression had no influence on EAAC1-mediated glutamate uptake activity in the absence of PMA stimulation. Both serine 18 and serine 138 within the putative PKC phosphorylation motifs are critical for the negative regulation of EAAC1-mediated glutamate uptake and suggest that the PKC phosphorylation site at serine 138 is functional under physiological conditions.

Based on these data, we proposed the regulatory model of EAAC1-mediated glutamate uptake illustrated in Fig. 3. If addicsin expression is high enough relative to Arl6ip1 to form many more addicsin homodimers than addicsin–Arl6ip1 heterodimers, EAAC1-mediated glutamate uptake is reduced. Furthermore, activation of the PKC isozyme that phosphorylates addicsin at S18 or S138 may further potentiate this negative regulation. On the other hand, if addicsin expression is low enough or Arl6ip1 expression high enough that formation of heterodimers predominates, fewer addicsin homodimers are available to suppress EAAC1 activity. The resulting decrease in addicsin–EAAC1 binding will enhance the catalytic efficacy of EAAC1, in a PKC-activity dependent manner. In sum, Arl6ip1 acts as a positive regulator of EAAC1-mediated glutamate uptake (Fig. 3) and may therefore possess significant neuroprotective efficacy against neurodegenerative diseases linked to excitotoxicity and oxidative stress.

2.2 Modulation of ER protein trafficking by addicsin

Addicsin is a member of the PRAF protein family with homology to PRA1 and PRAF2 (JM4) (Schweneker et al., 2005). PRA1 is associated with the Golgi membrane and interacts with Rab, a member of the Ras superfamily of small GTP-binding proteins, which regulates intracellular protein trafficking (Bucci et al., 1999; Liang & Li, 2000; Martincic et al., 1997).

Immunocytochemical studies reveal that mature addicsin is present in both the plasma membrane and the intracellular compartment, including the ER (Ikemoto et al., 2002; Watabe et al., 2007, 2008). Thus, addicsin may also be involved in intracellular protein trafficking. To investigate this possibility, we examined EAAC1 oligosaccharide residues under conditions of varying addicsin expression. The oligosaccharide residues on EAAC1 are an excellent indicator of the extent of ER-to-Golgi trafficking and plasma membrane localization because the newly synthesized EAAC1 is *N*-glycosylated with high mannose oligosaccharide chains that are subsequently processed into more complex sugar chains by resident Golgi enzymes (Yang & Kilberg, 2002). In HEK293T cells coexpressing addicsin, EAAC1 is predominantly modified by high mannose oligosaccharides, suggesting that EAAC1 proteins are largely confined to the ER. Furthermore, addicsin delays oligosaccharide maturation of EAAC1 but does not induce EAAC1 degradation (Ruggiero et al., 2008). These data suggest that addicsin delays ER-to-Golgi trafficking of EAAC1. Moreover, addicsin inhibits ER-to-Golgi trafficking of dopamine transporter, GABA transporter 1, and several G-protein-coupled receptors, including β_2-adrenergic receptor, α_1-β receptor, and D_2 receptor (Ruggiero et al., 2008). Furthermore, addicsin inhibits the function of RTN2B, a member of the reticulon protein family localized in the ER, which enhances ER-to-Golgi trafficking of EAAC1 (Liu et al., 2008). As addicsin, RTN2B, and EAAC1 are coexpressed in neurons, they may interact in one complex. Indeed, addicsin and EAAC1 can interact with RTN2B by binding to different regions of the protein. In addition, coexpression of RTN2B and EAAC1 in HEK293 cells increases EAAC1 cell surface expression, while increasing addicsin expression blocks this effect. Thus, EAAC1 trafficking is inhibited by addicsin and facilitated by RTN2B (Liu et al., 2008). Based on these data, Liu et al. proposed a model in which the regulation of ER trafficking governs the activity and density of EAAC1 at the plasma membrane. Under normal conditions, RTN2B facilitates EAAC1 trafficking from the ER because basal expression of addicsin is too low to have an inhibitory effect. Under stressful conditions, such as oxidative and chemical stress, addicsin expression is upregulated and the inhibitory effect on EAAC1 trafficking predominates over the facilitating effect of RTN2B (Liu et al., 2008). Addicsin can delay ER-to-Golgi trafficking of structurally and functionally distinct proteins in addition to EAAC1. Thus, addicsin is a stress-induced multifunctional protein that participates in various physiological and pathological functions by regulating ER trafficking of many membrane effector proteins, including receptors and transporters.

3. Addicsin & neurological disorders

Recent studies have also linked addicsin to the pathophysiology of several neurological diseases, including drug addiction, schizophrenia, and epilepsy. In this section, we focus on these diseases and review the putative pathophysiological functions of addicsin in the mammalian CNS.

3.1 Drug abuse

Several studies demonstrate that addicsin is involved in drug abuse, the development of morphine dependence (Ikemoto et al., 2002; Wu et al., 2011), and ethanol tolerance (C. Li et al., 2008). In an effort to clarify the molecular mechanism of opiate addiction, we performed subtractive hybridization of mRNA expressed in the amygdala of mice treated with repeated doses of morphine and identified *addicsin* mRNA as a factor selectively upregulated relative to drug-naïve mice (Ikemoto et al., 2000, 2002). Upregulation of *addicsin* mRNA was specifically induced by chronic, but not acute, morphine administration and was completely inhibited by coadministration of naloxone, an opiate receptor antagonist (Ikemoto et al., 2002). In that study, we used a morphine administration protocol that had been previously shown to induce morphine dependence and tolerance (Kaneto et al., 1973). Thus, our data strongly suggested that addicsin was involved in the development of morphine dependence in this animal model. Later reports have confirmed our findings by directly demonstrating that addicsin is directly involved in the development of morphine dependence (Wu et al., 2011). Chronic morphine treatment upregulated addicsin in prefrontal cortex, nucleus accumbens, and amygdala, which are regions known to be critical for the development of morphine dependence and other addictive behaviors. Furthermore, addicsin knockdown by infusion of addicsin antisense nucleotides into the cerebral ventricles significantly decreased withdrawal behaviors following chronic morphine treatment in rats (Wu et al., 2011). Addicsin knockdown suppressed the upregulation of δ opioid receptors, the activation of the dopamine- and cAMP-regulated phosphoprotein of 32 kDa (DARPP-32), and MAPK activation normally induced by chronic morphine treatment. Furthermore, addicsin knockdown enhanced the degradation of δ opioid receptors through the ubiquitin–proteasome pathway (Wu et al., 2011). These data suggest that addicsin directly contributes to the regulation of δ opioid receptor stability and the development of morphine dependence by suppressing δ opioid receptor expression and the activation of DARPP-32 and MAPK. The δ opioid receptor knockout mice do not develop analgesic tolerance to morphine without affecting the development of physical dependence (Kieffer & Gaveriaux-Ruff, 2002; Nitsche et al., 2002; Y. Zhu et al., 1999). Thus, further investigations are needed to clarify whether addicsin is involved in analgesic tolerance.

Ethanol-induced cellular responses are analogous to those elicited by heat shock stresses (Piper, 1995; Wilke et al., 1994). Similarly, addicsin expression is enhanced in response to various environmental stressors, such as oxidative stress and heat shock stress (R. Chen et al., 2007). Furthermore, our study demonstrated that addicsin plays an important role in the development of morphine dependence and tolerance (Ikemoto et al., 2002) In the light of these observations, addicsin is considered to be essential for the development of ethanol tolerance. To address this issue, addicsin knockdown flies were generated. To estimate ethanol tolerance objectively, the inebriation test was performed (Bellen, 1998). Flies were exposed to ethanol vapor, and the mean elution time (MET) was measured three times after inebriation. The addicsin knockdown flies showed no difference between the first MET and third MET, while wild-type flies exhibited a significant higher third MET (C. Li, et al., 2008), indicating that addicsin knockdown flies failed to acquire ethanol tolerance.

3.2 Schizophrenia

Glutamatergic neurotransmission and plasticity are disrupted in patients with schizophrenia (Javitt, 2010; Kantrowitz & Javitt, 2010; Paz et al., 2008). This has led some researchers to speculate that EAATs and EAAT-interacting proteins that regulate glutamate transport efficacy or transporter expression may be abnormal in patients with schizophrenia (Bauer et al., 2008; Huerta et al., 2006). Indeed, addicsin/JWA transcripts were overexpressed in the thalamus (Huerta et al., 2006) and the anterior cingulate cortex of schizophrenics as shown by *in situ* hybridization (Bauer et al., 2008). In these studies, the protein expression levels of addicsin/JWA were not determined. In addition, expression of EAAT3, the human homolog of EAAC1, was also upregulated in the anterior cingulate cortex of schizophrenic patients (Bauer et al., 2008). Furthermore, a microarray study of multiple human brain regions demonstrates that the anterior cingulate cortex is more vulnerable to these aberrant gene expression patterns (Katsel et al., 2005), and hypofrontality is a key feature of schizophrenia. Addicsin is thus a promising target for further research focusing on the role of glutamate transporters in schizophrenia. Moreover, addicsin regulates trafficking of a plethora of other membrane proteins, including dopamine receptors, suggesting another pathway through which addicsin participates in the pathogenesis of schizophrenia.

3.3 Epilepsy

Anatomical analysis of EAAT expression reveals that EAAC1 is enriched in neurons and particularly localized to inhibitory GABAergic neurons (Conti et al., 1998; He et al., 2000; Rothstein et al., 1994). Cerebroventricular injection of EAAC1 antisense oligonucleotides caused no elevation of extracellular glutamate in the rat striatum but did produce mild neurotoxicity and epileptiform activity (Rothstein et al., 1996). Furthermore, epilepsy in EAAC1 knockdown rats is caused by decreased GABA synthesis (Sepkuty et al., 2002). Glutamate is a precursor for GABA synthesis, so molecules that alter the intracellular availability of glutamate in GABAergic interneurons, including addicsin/GTRAP3-18, may have an important role in epileptogenesis or ictogenesis. In a recent study of the antiepileptic drug levetiracetam (LEV), changes in the expression of addicsin/GTRAP3-18, glutamate transporters, and GABA transporters were examined in a rat post-traumatic epilepsy model induced by $FeCl_3$ injection into the amygdala. Administration of LEV increased expression of EAAC1 and GABA transporter 3 (GAT-3) but decreased expression of addicsin/GTRAP3-18 in the rat hippocampal formation (Ueda et al., 2007). These results suggest that both the suppression of glutamatergic excitation and the enhancement of GABAergic inhibition induced by chronic LEV administration are due to the upregulation of EAAC1 and GAT-3 subsequent to downregulation of addicsin/GTRAP3-18. A long-lasting suppression of addicsin/GTRAP3-18 expression was observed in the rat pentylenetetrazole (PTZ)-induced kindling model of epilepsy (Ueda et al., 2006). Similarly, antisense-mediated knockdown of addicsin/GTRAP3-18 decreases seizure threshold and promotes PTZ kindling. In addition, addicsin/GTRAP3-18 knockdown increases basal release of glutamate and GABA in the rat hippocampal formation, indicating that knockdown of addicsin/GTRAP3-18 promotes GABA synthesis (Ueda et al., 2006). These studies, demonstrating that addicsin can increase GABA synthesis by increasing the substrate (i.e., glutamate) supply, define addicsin as a novel therapeutic target in epilepsy.

3.4 Other neurological disorders

Addicsin directly modulates glutamate and cysteine uptake by EAAC1, suggesting that addicsin participates in the pathogenesis of neurological disorders associated with excitotoxicity and oxidative stress. Here we briefly discuss some representative EAAC1 functions relevant to CNS pathology. A recent study demonstrated that EAAC1-deficient mice developed age-dependent brain atrophy and behavioral abnormalities in the cognitive and motivational domains. In addition, EAAC1 knockout mice displayed impaired GSH homeostasis and age-dependent neurodegeneration, and these pathologies were rescued by treatment with the membrane permeable cysteine precursor *N*-acetylcysteine (Aoyama et al., 2006). These EAAC1 knockout mice also display dicarboxylic aminoaciduria and significant motor impairments (Peghini et al., 1997). These results indicate that EAAC1 functions as a cysteine transporter in neurons and sustains intracellular GSH to ameliorate oxidative stress *in vivo*. Furthermore, neuronal glutamate uptake can also regulate memory formation (Levenson et al., 2000; Maleszka et al., 2000). The increase of EAAC1-mediated neuronal glutamate uptake is associated with the induction and expression of early phase long-term potentiation (LTP) in the CA1 area of the hippocampal formation and with contextual fear conditioning, a form of hippocampus-dependent memory thought to depend on induction of LTP (Levenson et al., 2002). These results suggest that regulation of glutamate uptake by EAAC1 is a physiologically important mechanism for the modulation of synaptic strength during long-term changes in synaptic efficacy (plasticity). Thus, dysfunction of EAAC1 induced by aberrant addicsin expression may lead to neurodegeneration and cognitive decline. Of particular interest is the role of addicsin in the pathogenesis of neurodegenerative diseases such as Alzheimer's and Parkinson's diseases. These questions warrant further research.

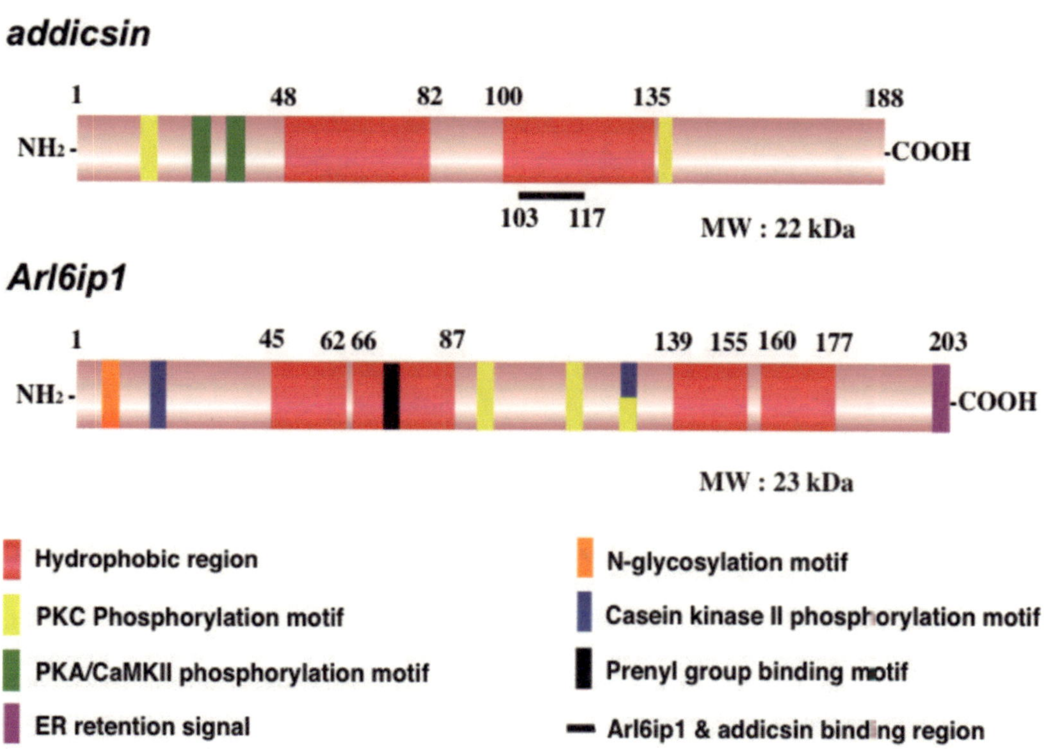

Fig. 1. A schematic presentation of addicsin and Arl6ip1

4. Future research perspective

Despite these advances, our understanding of the regulatory mechanisms of addicsin expression and the range of addicsin functions is far from complete. The elucidation of the regulatory mechanism of addicsin expression under basal and pathological conditions is essential for understanding the physiological and pathological roles of addicsin. For instance, while addicsin has consensus PKC phosphorylation sequences, it is unclear whether PKC actually phosphorylates addicsin and controls addicsin functions *in vivo*. It is also unknown whether or how PKC phosphorylation affects the interaction between addicsin and Arl6ip1. To overcome these challenges, it is crucial to clarify whether PKC phosphorylation sites of addicsin are physiologically controlled by PKC signaling and by which PKC isoforms. Furthermore, it remains controversial whether addicsin is an integral membrane protein. Our results strongly support the notion that addicsin is a membrane-associating protein with a soluble and membrane-localized form. Thus, it is important to clarify the different molecular features and functions of the soluble and membrane-localized forms of addicsin.

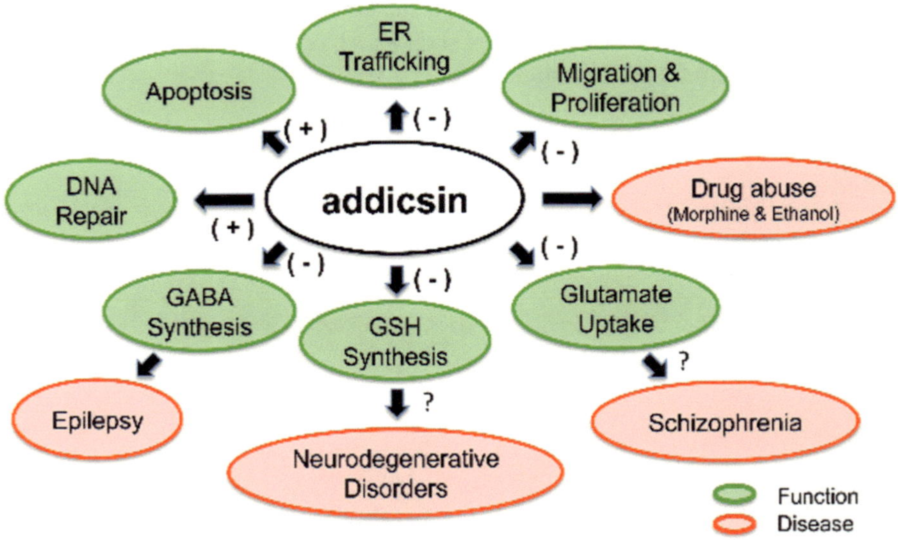

Fig. 2. A scheme of the proposed physiological functions of addicsin

Second, *in vivo* functional studies are still needed to clarify the physiological and pathological functions of addicsin. Accumulating evidence suggests that addicsin participates in various physiological and pathological processes *in vivo*, but the molecular mechanisms controlling the selective interaction of addicsin with multiple targets, including receptors and transporters, are unknown. Furthermore, many reports demonstrate that the physiological and pathological roles of addicsin are observed when expression of addicsin is increased by various stresses, including oxidative and chemical stress. Thus, the production of animal models that overexpressed addicsin in a tissue- or region-specific manner may be useful to analyze addicsin functions in various tissues, including the brain. At present, no studies have been undertaken in tissues outside the brain, although addicsin is ubiquitously expressed in kidney, heart, and liver (Butchbach et al., 2002; Ikemoto et al., 2002).

We believe that studies using transgenic or conditional knockin/knockout animal models will lead to novel insights into addicsin function. Of particular interest is whether dysfunctional addicsin expression or function can lead to neurodegenerative diseases through dysregulation of EAAC1 or other proteins. Finally, we hope that studies on addicsin will continue to advance our understanding of the role of addicsin in the pathogenesis of diseases, such as drug abuse, and lead to the development of curative therapies.

5. Conclusion

In this chapter, we argued that Arl6ip1 is a novel addicsin-interacting protein that indirectly promotes PKC-dependent, EAAC1-mediated glutamate uptake by inhibiting the interaction of addicsin with EAAC1 at the plasma membrane. Based on these findings, we proposed the regulatory model of EAAC1-mediated glutamate uptake illustrated in Fig. 3. In this model, EAAC1-mediated glutamate uptake activity can be negatively and positively regulated by PKC activity depending on dynamic modulation by addicsin complexes. Thus, the cellular dynamics of addicsin is a key element regulating EAAC1-mediated glutamate uptake. The study of addicsin is still in its infancy, but future findings on the physiological and pathophysiological functions of addicsin could greatly clarify the role of EAAC1 (and other proteins regulated by addicsin) in health and disease.

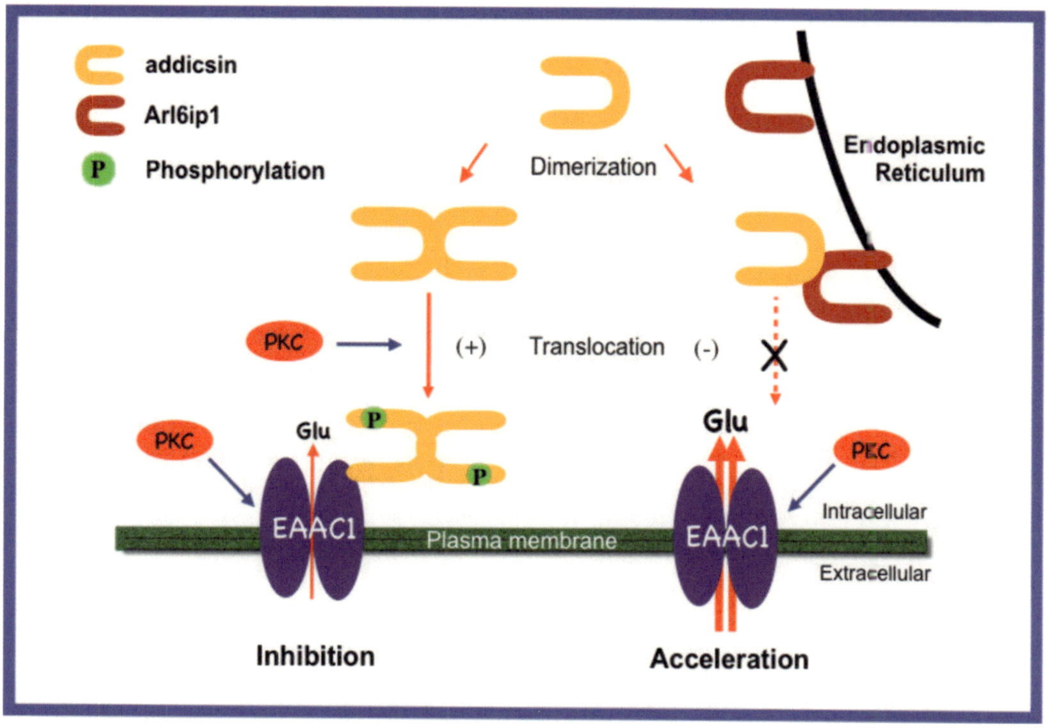

Fig. 3. A regulatory model of EAAC1-mediated glutamate uptake in C6BU-1 cells

6. Acknowledgment

This work was supported by a research grant from the National Institute of Advanced Industrial Science and Technology (AIST) of Japan. We thank Dr. S. Akiduki, M.Sc. M. Takumori, Dr. M. Ohtomi, Dr. K. Inoue, and Dr. T. Ochiishi for helpful discussions, and Ms. K. Nemoto for her excellent technical assistance.

7. References

Akiduki, S. & Ikemoto, M. J. (2008). Modulation of the Neural Glutamate Transporter EAAC1 by the Addicsin-Interacting Protein Arl6ip1, *The Journal of Biological Chemistry* Vol.283, No.46, (November 2008), pp. 31323-31332, ISSN 0021-9258

Akiduki, S., Ochiishi, T. & Ikemoto, M. J. (2007). Neural Localization of Addicsin in Mouse Brain, *Neuroscience Letters* Vol.426, No.3, (Octbor 2007), pp. 149-154

Aoyama, K., Suh, S. W., Hamby, A. M., Liu, J., Chan, W. Y., Chen, Y. & Swanson, R. A. (2006). Neuronal Glutathione Deficiency and Age-Dependent Neurodegeneration in the EAAC1 Deficient Mouse, *Nature Neuroscience* Vol.9, No.1, (January 2006), pp. 119-126

Aoyama, K., Matsumura, N., Watabe, M. & Nakaki, T. (2008a). Oxidative Stress on EAAC1 Is Involved in MPTP-Induced Glutathione Depletion and Motor Dysfunction, *The European Journal of Neuroscience* Vol.27, No.1, (January 2008), pp. 20-30, ISSN 1460-9568 (Electronic), 0953-816X (Linking)

Aoyama, K., Watabe, M. & Nakaki, T. (2008b). Regulation of Neuronal Glutathione Synthesis, *Journal of Pharmacological Sciences* Vol.108, No.3, (November 2008), pp. 227-238, ISSN 1347-8613

Bauer, D., Gupta, D., Harotunian, V., Meador-Woodruff, J. H. & McCullumsmith, R. E. (2008). Abnormal Expression of Glutamate Transporter and Transporter Interacting Molecules in Prefrontal Cortex in Elderly Patients with Schizophrenia, *Schizophrenia Research* Vol.104, No.1-3, (September 2008), pp. 108-120, ISSN 1573-2509

Bellen, H. J. (1998). The Fruit Fly: A Model Organism to Study the Genetics of Alcohol Abuse and Addiction?, *Cell* Vol.93, No.6, (June 1998), pp. 909-912, ISSN 0092-8674

Bucci, C., Chiariello, M., Lattero, D., Maiorano, M. & Bruni, C. B. (1999). Interaction Cloning and Characterization of the cDNA Encoding the Human Prenylated Rab Acceptor (PRA1), *Biochemical and Biophysical Research Communications* Vol.258, No.3, (May 1999), pp. 657-662, ISSN 0006-291X

Butchbach, M. E., Lai, L. & Lin, C. L. (2002). Molecular Cloning, Gene Structure, Expression Profile and Functional Characterization of the Mouse Glutamate Transporter (EAAT3) Interacting Protein GTRAP3-18, *Gene* Vol.292, No.1-2, (June 2002), pp. 81-90

Cao, X. J., Chen, R., Li, A. P. & Zhou, J. W. (2007). JWA Gene Is Involved in Cadmium-Induced Growth Inhibition and Apoptosis in HEK-293T Cells, *Journal of Toxicology and Environmental Health A* Vol.70, No.11, (June 2007), pp. 931-937

Chen, H., Bai, J., Ye, J., Liu, Z., Chen, R., Mao, W., Li, A. & Zhou, J. (2007). JWA as a Functional Molecule to Regulate Cancer Cells Migration Via MAPK Cascades and F-Actin Cytoskeleton, *Cellular Signalling* Vol.19, No.6, (June 2007), pp. 1315-1327

Chen, R., Qiu, W., Liu, Z., Cao, X., Zhu, T., Li, A., Wei, Q. & Zhou, J. (2007). Identification of JWA as a Novel Functional Gene Responsive to Environmental Oxidative Stress Induced by Benzo[α]Pyrene and Hydrogen Peroxide, *Free Radical Biology and Medicine* Vol.42, No.11, (June 2007), pp. 1704-1714

Chen, Y. & Swanson, R. A. (2003). The Glutamate Transporters EAAT2 and EAAT3 Mediate Cysteine Uptake in Cortical Neuron Cultures, *Journal of Neurochemistry* Vol.84, No.6, (March 2003), pp. 1332-1339, ISSN 0022-3042

Conti, F., DeBiasi, S., Minelli, A., Rothstein, J. D. & Melone, M. (1998). EAAC1, a High-Affinity Glutamate Tranporter, Is Localized to Astrocytes and GABAergic Neurons Besides Pyramidal Cells in the Rat Cerebral Cortex, *Cerebral Cortex* Vol.8, No.2, (March 1998), pp. 108-116, ISSN 1047-3211

D'Amico, A., Soragna, A., Di Cairano, E., Panzeri, N., Anzai, N., Vellea Sacchi, F. & Perego, C. (2010). The Surface Density of the Glutamate Transporter EAAC1 Is Controlled by Interactions with PDZK1 and AP2 Adaptor Complexes, *Traffic* Vol.11, No.11, (November 2010), pp. 1455-1470, ISSN 1600-0854 (Electronic), 1398-9219 (Linking)

Danbolt, N. C. (2001). Glutamate Uptake, *Progress in Neurobiology* Vol.65, No.1, (September 2001), pp. 1-105

Davis, K. E., Straff, D. J., Weinstein, E. A., Bannerman, P. G., Correale, D. M., Rothstein, J. D. & Robinson, M. B. (1998). Multiple Signaling Pathways Regulate Cell Surface Expression and Activity of the Excitatory Amino Acid Carrier 1 Subtype of Glu Transporter in C6 Glioma, *The Journal of Neuroscience* Vol.18, No.7, (April 1998), pp. 2475-2485, ISSN 0270-6474

Dowd, L. A. & Robinson, M. B. (1996). Rapid Stimulation of EAAC1-Mediated Na+-Dependent L-Glutamate Transport Activity in C6 Glioma Cells by Phorbol Ester, *Journal of Neurochemistry* Vol.67, No.2, (August 1996), pp. 508-516, ISSN 0022-3042

Fournier, K. M., Gonzalez, M. I. & Robinson, M. B. (2004). Rapid Trafficking of the Neuronal Glutamate Transporter, EAAC1: Evidence for Distinct Trafficking Pathways Differentially Regulated by Protein Kinase C and Platelet-Derived Growth Factor, *The Journal of Biological Chemistry* Vol.279, No.33, (August 2004), pp. 34505-34513, 0021-9258

Gonzalez, M. I., Kazanietz, M. G. & Robinson, M. B. (2002). Regulation of the Neuronal Glutamate Transporter Excitatory Amino Acid Carrier-1 (EAAC1) by Different Protein Kinase C Subtypes, *Molecular Pharmacology* Vol.62, No.4, (October 2002), pp. 901-910, ISSN 0026-895X

Gonzalez, M. I., Bannerman, P. G. & Robinson, M. B. (2003). Phorbol Myristate Acetate-Dependent Interaction of Protein Kinase C alpha and the Neuronal Glutamate Transporter EAAC1, *The Journal of Neuroscience* Vol.23, No.13, (July 2003), pp. 5589-5593, ISSN 1529-2401 (Electronic), 0270-6474 (Linking)

Guo, F., Li, Y., Liu, Y., Wang, J. & Li, G. (2010a). Arl6ip1 Mediates Cisplatin-Induced Apoptosis in Caski Cervical Cancer Cells, *Oncology Reports* Vol.23, No.5, (May 2010), pp. 1449-1455, ISSN 1791-2431 (Electronic), 1021-335X (Linking)

Guo, F., Liu, Y., Li, Y. & Li, G. (2010b). Inhibition of ADP-Ribosylation Factor-Like 6 Interacting Protein 1 Suppresses Proliferation and Reduces Tumor Cell Invasion in Caski Human Cervical Cancer Cells, *Molecular Biology Reports* Vol.37, No.8, (December 2010), pp. 3819-3825, ISSN 1573-4978 (Electronic), 0301-4851 (Linking)

He, Y., Janssen, W. G., Rothstein, J. D. & Morrison, J. H. (2000). Differential Synaptic Localization of the Glutamate Transporter EAAC1 and Glutamate Receptor Subunit GluR2 in the Rat Hippocampus, *The Journal of Comparative Neurology* Vol.418, No.3, (March 2000), pp. 255-269, ISSN 0021-9967

Himi, T., Ikeda, M., Yasuhara, T., Nishida, M. & Morita, I. (2003). Role of Neuronal Glutamate Transporter in the Cysteine Uptake and Intracellular Glutathione Levels in Cultured Cortical Neurons, *Journal of Neural Transmission* Vol.110, No.12, (December 2003), pp. 1337-1348, ISSN 0300-9564

Huang, S., Shen, Q., Mao, W. G., Li, A. P., Ye, J., Liu, Q. Z., Zou, C. P. & Zhou, J. W. (2006a). JWA, a Novel Signaling Molecule, Involved in All-*Trans* Retinoic Acid Induced Differentiation of HL-60 Cells, *Journal of Biomedical Science* Vol.13, No.3, (May 2006), pp. 357-371

Huang, S., Shen, Q., Mao, W. G., Li, A. P., Ye, J., Liu, Q. Z., Zou, C. P. & Zhou, J. W. (2006b). JWA, a Novel Signaling Molecule, Involved in the Induction of Differentiation of Human Myeloid Leukemia Cells, *Biochemical and Biophysical Research Communications* Vol.341, No.2, (March 2006), pp. 440-450

Huerta, I., McCullumsmith, R. E., Haroutunian, V., Gimenez-Amaya, J. M. & Meador-Woodruff, J. H. (2006). Expression of Excitatory Amino Acid Transporter Interacting Protein Transcripts in the Thalamus in Schizophrenia, *Synapse* Vol.59, No.7, (June 2006), pp. 394-402, ISSN 0887-4476

Ikemoto, M., Takita, M., Imamura, T. & Inoue, K. (2000). Increased Sensitivity to the Stimulant Effects of Morphine Conferred by Anti-Adhesive Glycoprotein SPARC in Amygdala, *Nature Medicine* Vol.6, No.8, (August 2000), pp. 910-915, ISSN 1078-8956

Ikemoto, M. J., Inoue, K., Akiduki, S., Osugi, T., Imamura, T., Ishida, N. & Ohtomi, M. (2002). Identification of Addicsin/GTRAP3-18 as a Chronic Morphine-Augmented Gene in Amygdala, *Neuroreport* Vol.13, No.16, (November 2002), pp. 2079-2084

Ingley, E., Williams, J. H., Walker, C. E., Tsai, S., Colley, S., Sayer, M. S., Tilbrook, P. A., Sarna, M., Beaumont, J. G. & Klinken, S. P. (1999). A Novel ADP-Ribosylation Like Factor (ARL-6), Interacts with the Protein-Conducting Channel SEC61beta Subunit, *FEBS Letters* Vol.459, No.1, (October 1999), pp. 69-74, ISSN 0014-5793

Inoue, K., Akiduki, S. & Ikemoto, M. J. (2005). Expression Profile of Addicsin/GTRAP3-18 Mrna in Mouse Brain, *Neuroscience Letters* Vol.386, No.3, (October 2005), pp. 184-188

Javitt, D. C. (2010). Glutamatergic Theories of Schizophrenia, *The Israel Journal of Psychiatry and Related Sciences* Vol.47, No.1, pp. 4-16, ISSN 0333-7308

Kanai, Y. & Hediger, M. A. (1992). Primary Structure and Functional Characterization of a High-Affinity Glutamate Transporter, *Nature* Vol.360, No.6403, (December 1992), pp. 467-471

Kanai, Y. & Hediger, M. A. (2003). The Glutamate and Neutral Amino Acid Transporter Family: Physiological and Pharmacological Implications, *European Journal of Pharmacology* Vol.479, No.1-3, (October 2003), pp. 237-247, ISSN 0014-2999

Kanai, Y. & Hediger, M. A. (2004). The Glutamate/Neutral Amino Acid Transporter Family SLC1: Molecular, Physiological and Pharmacological Aspects, *Pflugers Archiv European Journal of Physiology* Vol.447, No.5, (Feb 2004), pp. 469-479, ISSN 0031-6768 (Print),0031-6768 (Linking)

Kaneto, H., Koida, M., Nakanishi, H. & Sasano, H. (1973). A Scoring System for Abstinence Syndrome in Morphine Dependent Mice and Application to Evaluate Morphine Type Dependence Liability of Drugs, *The Japanese Journal of Pharmacology* Vol.23, No.5, (October 1973), pp. 701-707, ISSN 0021-5198

Kantrowitz, J. T. & Javitt, D. C. (2010). Thinking Glutamatergically: Changing Concepts of Schizophrenia Based Upon Changing Neurochemical Models, *Clinical Schizophrenia & Related Psychoses* Vol.4, No.3, (October 2010), pp. 189-200, ISSN 1935-1232

Katsel, P., Davis, K. L., Gorman, J. M. & Haroutunian, V. (2005). Variations in Differential Gene Expression Patterns across Multiple Brain Regions in Schizophrenia, *Schizophrenia Research* Vol.77, No.2-3, (September 2005), pp. 241-252, ISSN 0920-9964

Kieffer, B. L. & Gaveriaux-Ruff, C. (2002). Exploring the Opioid System by Gene Knockout, *Progress in Neurobiology* Vol.66, No.5, (Apr 2002), pp. 285-306, ISSN 0301-0082 (Print), 0301-0082 (Linking)

Kiryu, S., Yao, G. L., Morita, N., Kato, H. & Kiyama, H. (1995). Nerve Injury Enhances Rat Neuronal Glutamate Transporter Expression: Identification by Differential Display PCR, *The Journal of Neuroscience* Vol.15, No.12, (December 1995), pp. 7872-7878, ISSN 0270-6474

Kiryu-Seo, S., Gamo, K., Tachibana, T., Tanaka, K. & Kiyama, H. (2006). Unique Anti-Apoptotic Activity of EAAC1 in Injured Motor Neurons, *The EMBO Journal* Vol.25, No.14, (July 2006), pp. 3411-3421

Levenson, J., Endo, S., Kategaya, L. S., Fernandez, R. I., Brabham, D. G., Chin, J., Byrne, J. H. & Eskin, A. (2000). Long-Term Regulation of Neuronal High-Affinity Glutamate and Glutamine Uptake in *Aplysia*, *Proceedings of the National Academy of Sciences of the United States of America* Vol.97, No.23, (November 2000), pp. 12858-12863, ISSN 0027-8424

Levenson, J., Weeber, E., Selcher, J. C., Kategaya, L. S., Sweatt, J. D. & Eskin, A. (2002). Long-Term Potentiation and Contextual Fear Conditioning Increase Neuronal Glutamate Uptake, *Nature Neuroscience* Vol.5, No.2, (February 2002), pp. 155-161

Li, C., Zhao, X., Cao, X., Chu, D., Chen, J. & Zhou, J. (2008). The Drosophila Homolog of JWA Is Required for Ethanol Tolerance, *Alcohol and Alcoholism* Vol.43, No.5, (October 2008), pp. 529-536, ISSN 1464-3502 (Electronic), 0735-0414 (Linking)

Li, X., Valencia, A., Sapp, E., Masso, N., Alexander, J., Reeves, P., Kegel, K. B., Aronin, N. & Difiglia, M. (2010). Aberrant Rab11-Dependent Trafficking of the Neuronal Glutamate Transporter EAAC1 Causes Oxidative Stress and Cell Death in Huntington's Disease, *The Journal of Neuroscience* Vol.30, No.13, (March 2010), pp. 4552-4561, ISSN 1529-2401 (Electronic), 0270-6474 (Linking)

Liang, Z. & Li, G. (2000). Mouse Prenylated Rab Acceptor Is a Novel Golgi Membrane Protein, *Biochemical and Biophysical Research Communications* Vol.275, No.2, (August 2000), pp. 509-516, ISSN 0006-291X

Lin, C. I., Orlov, I., Ruggiero, A. M., Dykes-Hoberg, M., Lee, A., Jackson, M. & Rothstein, J. D. (2001). Modulation of the Neuronal Glutamate Transporter EAAC1 by the Interacting Protein GTRAP3-18, *Nature* Vol.410, No.6824, (March 2001), pp. 84-88

Liu, Y., Vidensky, S., Ruggiero, A. M., Maier, S., Sitte, H. H. & Rothstein, J. D. (2008). Reticulon RTN2b Regulates Trafficking and Function of Neuronal Glutamate Transporter EAAC1, *The Journal of Biological Chemistry* Vol.283, No.10, (March 2008), pp. 6561-6571, ISSN 0021-9258

Lui, H. M., Chen, J., Wang, L. & Naumovski, L. (2003). ARMER, Apoptotic Regulator in the Membrane of the Endoplasmic Reticulum, a Novel Inhibitor of Apoptosis, *Molecular Cancer Research* Vol.1, No.7, (May 2003), pp. 508-518, ISSN 1541-7786

Maier, S., Reiterer, V., Ruggiero, A. M., Rothstein, J. D., Thomas, S., Dahm, R., Sitte, H. H. & Farhan, H. (2009). GTRAP3-18 Serves as a Negative Regulator of Rab1 in Protein Transport and Neuronal Differentiation, *Journal of Cellular and Molecular Medicine* Vol.13, No.1, (January 2009), pp. 114-124, ISSN 1582-4934 (Electronic), 1582-1838 (Linking)

Maleszka, R., Helliwell, P. & Kucharski, R. (2000). Pharmacological Interference with Glutamate Re-Uptake Impairs Long-Term Memory in the Honeybee, *Apis Mellifera*, *Behavioural Brain Research* Vol.115, No.1, (October 2000), pp. 49-53, ISSN 0166-4328

Mao, W. G., Liu, Z. L., Chen, R., Li, A. P. & Zhou, J. W. (2006). JWA Is Required for the Antiproliferative and Pro-Apoptotic Effects of All-*Trans* Retinoic Acid in HeLa Cells, *Clinical and Experimental Pharmacology & Physiology* Vol.33, No.9, (September 2006), pp. 816-824, ISSN 0305-1870

Martincic, I., Peralta, M. E. & Ngsee, J. K. (1997). Isolation and Characterization of a Dual Prenylated Rab and VAMP2 Receptor, *The Journal of Biological Chemistry* Vol.272, No.43, (October 1997), pp. 26991-26998, ISSN 0021-9258

Mathews, G. C. & Diamond, J. S. (2003). Neuronal Glutamate Uptake Contributes to GABA Synthesis and Inhibitory Synaptic Strength, *The Journal of Neuroscience* Vol.23, No.6, (March 2003), pp. 2040-2048, ISSN 1529-2401 (Electronic), 0270-6474 (Linking)

Murphy, T. H., Miyamoto, M., Sastre, A., Schnaar, R. L. & Coyle, J. T. (1989). Glutamate Toxicity in a Neuronal Cell Line Involves Inhibition of Cystine Transport Leading to Oxidative Stress, *Neuron* Vol.2, No.6, (Jun 1989), pp. 1547-1558, ISSN 0896-6273 (Print), 0896-6273 (Linking)

Nieoullon, A., Canolle, B., Masmejean, F., Guillet, B., Pisano, P. & Lortet, S. (2006). The Neuronal Excitatory Amino Acid Transporter EAAC1/EAAT3: Does It Represent a Major Actor at the Brain Excitatory Synapse?, *Journal of Neurochemistry* Vol.98, No.4, (Aug 2006), pp. 1007-1018, ISSN 0022-3042 (Print), 0022-3042 (Linking)

Nitsche, J. F., Schuller, A. G., King, M. A., Zengh, M., Pasternak, G. W. & Pintar, J. E. (2002). Genetic Dissociation of Opiate Tolerance and Physical Dependence in Delta-Opioid Receptor-1 and Preproenkephalin Knock-out Mice, *The Journal of Neuroscience* Vol.22, No.24, (Dec 2002), pp. 10906-10913, ISSN 1529-2401 (Electronic), 0270-6474 (Linking)

Nomura, N., Nagase, T., Miyajima, N., Sazuka, T., Tanaka, A., Sato, S., Seki, N., Kawarabayasi, Y., Ishikawa, K. & Tabata, S. (1994). Prediction of the Coding Sequences of Unidentified Human Genes. II. The Coding Sequences of 40 New Genes (KIAA0041-KIAA0080) Deduced by Analysis of cDNA Clones from Human Cell Line KG-1, *DNA Research* Vol.1, No.5, pp. 223-229, ISSN 1340-2838

Padovano, V., Massari, S., Mazzucchelli, S. & Pietrini, G. (2009). PKC Induces Internalization and Retention of the EAAC1 Glutamate Transporter in Recycling Endosomes of MDCK Cells, *American Journal of Physiology. Cell Physiology* Vol.297, No.4, (October 2009), pp. C835-844, ISSN 1522-1563 (Electronic), 0363-6143 (Linking)

Palos, T. P., Ramachandran, B., Boado, R. & Howard, B. D. (1996). Rat C6 and Human Astrocytic Tumor Cells Express a Neuronal Type of Glutamate Transporter, *Brain Research. Molecular Brain Research* Vol.37, No.1-2, (April 1996), pp. 297-303, ISSN 0169-328X

Paz, R. D., Tardito, S., Atzori, M. & Tseng, K. Y. (2008). Glutamatergic Dysfunction in Schizophrenia: From Basic Neuroscience to Clinical Psychopharmacology, *European Neuropsychopharmacology* Vol.18, No.11, (November 2008), pp. 773-786, ISSN 0924-977X

Peghini, P., Janzen, J. & Stoffel, W. (1997). Glutamate Transporter EAAC-1-Deficient Mice Develop Dicarboxylic Aminoaciduria and Behavioral Abnormalities but No Neurodegeneration, *The EMBO Journal* Vol.16, No.13, (July 1997), pp. 3822-3832

Pettersson, M., Bessonova, M., Gu, H. F., Groop, L. C. & Jonsson, J. I. (2000). Characterization, Chromosomal Localization, and Expression During Hematopoietic Differentiation of the Gene Encoding Arl6ip, ADP-Ribosylation-Like Factor-6 Interacting Protein (ARL6), *Genomics* Vol.68, No.3, (September 2000), pp. 351-354, ISSN 0888-7543

Piper, P. W. (1995). The Heat Shock and Ethanol Stress Responses of Yeast Exhibit Extensive Similarity and Functional Overlap, *FEMS Microbiology Letters* Vol.134, No.2-3, (December 1995), pp. 121-127, ISSN 0378-1097

Ramassamy, C., Averill, D., Beffert, U., Theroux, L., Lussier-Cacan, S., Cohn, J. S., Christen, Y., Schoofs, A., Davignon, J. & Poirier, J. (2000). Oxidative Insults Are Associated with Apolipoprotein E Genotype in Alzheimer's Disease Brain, *Neurobiology of Disease* Vol.7, No.1, (February 2000), pp. 23-37, ISSN 0969-9961

Rothstein, J. D., Jin, L., Dykes-Hoberg, M. & Kuncl, R. W. (1993). Chronic Inhibition of Glutamate Uptake Produces a Model of Slow Neurotoxicity, *Proceedings of the National Academy of Sciences of the United States of America* Vol.90, No.14, (July 1993), pp. 6591-6595

Rothstein, J. D., Martin, L., Levey, A. I., Dykes-Hoberg, M., Jin, L., Wu, D., Nash, N. & Kuncl, R. W. (1994). Localization of Neuronal and Glial Glutamate Transporters, *Neuron* Vol.13, No.3, (September 1994), pp. 713-725, ISSN 0896-6273

Rothstein, J. D., Dykes-Hoberg, M., Pardo, C. A., Bristol, L. A., Jin, L., Kuncl, R. W., Kanai, Y., Hediger, M. A., Wang, Y., Schielke, J. P. & Welty, D. F. (1996). Knockout of Glutamate Transporters Reveals a Major Role for Astroglial Transport in Excitotoxicity and Clearance of Glutamate, *Neuron* Vol.16, No.3, (March 1996), pp. 675-686, ISSN 0896-6273

Ruggiero, A. M., Liu, Y., Vidensky, S., Maier, S., Jung, E., Farhan, H., Robinson, M. B., Sitte, H. H. & Rothstein, J. D. (2008). The Endoplasmic Reticulum Exit of Glutamate Transporter Is Regulated by the Inducible Mammalian Yip6b/GTRAP3-18 Protein, *The Journal of Biological Chemistry* Vol.283, No.10, (March 2008), pp. 6175-6183

Schweneker, M., Bachmann, A. S. & Moelling, K. (2005). JM4 Is a Four-Transmembrane Protein Binding to the CCR5 Receptor, *FEBS Letters* Vol.579, No.7, (March 2005), pp. 1751-1758, ISSN 0014-5793

Sepkuty, J. P., Cohen, A. S., Eccles, C., Rafiq, A., Behar, K., Ganel, R., Coulter, D. A. & Rothstein, J. D. (2002). A Neuronal Glutamate Transporter Contributes to Neurotransmitter GABA Synthesis and Epilepsy, *The Journal of Neuroscience* Vol.22, No.15, (August 2002), pp. 6372-6379

Sheldon, A. L., Gonzalez, M. I. & Robinson, M. B. (2006). A Carboxyl-Terminal Determinant of the Neuronal Glutamate Transporter, EAAC1, Is Required for Platelet-Derived Growth Factor-Dependent Trafficking, *The Journal of Biological Chemistry* Vol.281, No.8, (February 2006), pp. 4876-4886, ISSN 0021-9258

Sian, J., Dexter, D. T., Lees, A. J., Daniel, S., Agid, Y., Javoy-Agid, F., Jenner, P. & Marsden, C. D. (1994). Alterations in Glutathione Levels in Parkinson's Disease and Other Neurodegenerative Disorders Affecting Basal Ganglia, *Annals of Neurology* Vol.36, No.3, (September 1994), pp. 348-355, ISSN 0364-5134

Sims, K. D., Straff, D. J. & Robinson, M. B. (2000). Platelet-Derived Growth Factor Rapidly Increases Activity and Cell Surface Expression of the EAAC1 Subtype of Glutamate Transporter through Activation of Phosphatidylinositol 3-Kinase, *The Journal of Biological Chemistry* Vol.275, No.7, (February 2000), pp. 5228-5237, ISSN 0021-9258

Tong, G. & Jahr, C. E. (1994). Block of Glutamate Transporters Potentiates Postsynaptic Excitation, *Neuron* Vol.13, No.5, (November 1994), pp. 1195-1203

Trotti, D., Peng, J. B., Dunlop, J. & Hediger, M. A. (2001). Inhibition of the Glutamate Transporter EAAC1 Expressed in *Xenopus* Oocytes by Phorbol Esters, *Brain Research* Vol.914, No.1-2, (September 2001), pp. 196-203, ISSN 0006-8993

Ueda, Y., Doi, T., Nakajima, A., Tokumaru, J., Tsuru, N. & Ishida, Y. (2006). The Functional Role of Glutamate Transporter Associated Protein (GTRAP3-18) in the Epileptogenesis Induced by PTZ-Kindling, *The Annual Report of The Japan Epilepsy Research Foundation* Vol.17, pp. 33-40, Available from: http:hdl.handle.net/10458/2116

Ueda, Y., Doi, T., Nagatomo, K., Tokumaru, J., Takaki, M. & Willmore, L. J. (2007). Effect of Levetiracetam on Molecular Regulation of Hippocampal Glutamate and GABA Transporters in Rats with Chronic Seizures Induced by Amygdalar $FeCl_3$ Injection, *Brain Research* Vol.1151, (June 2007), pp. 55-61

Wang, S., Gong, Z., Chen, R., Liu, Y., Li, A., Li, G. & Zhou, J. (2009). JWA Regulates XRCC1 and Functions as a Novel Base Excision Repair Protein in Oxidative-Stress-Induced DNA Single-Strand Breaks, *Nucleic Acids Research* Vol.37, No.6, (April 2009), pp. 1936-1950

Watabe, M., Aoyama, K. & Nakaki, T. (2007). Regulation of Glutathione Synthesis Via Interaction between Glutamate Transport-Associated Protein 3-18 (GTRAP3-18) and Excitatory Amino Acid Carrier-1 (EAAC1) at Plasma Membrane, *Molecular Pharmacology* Vol.72, No.5, (November 2007), pp. 1103-1110

Watabe, M., Aoyama, K. & Nakaki, T. (2008). A Dominant Role of GTRAP3-18 in Neuronal Glutathione Synthesis, *The Journal of Neuroscience* Vol.28, No.38, (September 2008), pp. 9404-9413

Waxman, E. A., Baconguis, I., Lynch, D. R. & Robinson, M. B. (2007). *N*-Methyl-*D*-Aspartate Receptor-Dependent Regulation of the Glutamate Transporter Excitatory Amino Acid Carrier 1, *The Journal of Biological Chemistry* Vol.282, No.24, (June 2007), pp. 17594-17607, ISSN 0021-9258

Wilke, N., Sganga, M., Barhite, S. & Miles, M. F. (1994). Effects of Alcohol on Gene Expression in Neural Cells, *Experientia. Supplementum* Vol.71, pp. 49-59, ISSN 1023-294X

Won, S. J., Yoo, B. H., Brennan, A. M., Shin, B. S., Kauppinen, T. M., Berman, A. E., Swanson, R. A. & Suh, S. W. (2010). EAAC1 Gene Deletion Alters Zinc Homeostasis and Exacerbates Neuronal Injury after Transient Cerebral Ischemia, *The Journal of Neuroscience* Vol.30, No.46, (November 2010), pp. 15409-15418, ISSN 1529-2401 (Electronic), 0270-6474 (Linking)

Wu, Y., Chen, R., Zhao, X., Li, A., Li, G. & Zhou, J. (2011). JWA Regulates Chronic Morphine Dependence Via the Delta Opioid Receptor, *Biochemical and Biophysical Research Communications* Vol.409, No.3, (June 2011), pp. 520-525, ISSN 1090-2104 (Electronic), 0006-291X (Linking)

Xia, P., Pei, G. & Schwarz, W. (2006). Regulation of the Glutamate Transporter EAAC1 by Expression and Activation of Delta-Opioid Receptor, *The European Journal of Neuroscience* Vol.24, No.1, (July 2006), pp. 87-93, ISSN 0953-816X

Yang, W. & Kilberg, M. S. (2002). Biosynthesis, Intracellular Targeting, and Degradation of the EAAC1 Glutamate/Aspartate Transporter in C6 Glioma Cells, *The Journal of Biological Chemistry* Vol.277, No.41, (October 2002), pp. 38350-38357, ISSN 0021-9258

Zerangue, N. & Kavanaugh, M. P. (1996). Interaction of L-Cysteine with a Human Excitatory Amino Acid Transporter, *The Journal of Physiology* Vol.493 (Pt 2), (June 1996), pp. 419-423, ISSN 0022-3751

Zhou, J., Ye, J., Zhao, X. & Li, A. (2008). JWA Is Required for Arsenic Trioxide Induced Apoptosis in HeLa and MCF-7 Cells Via Reactive Oxygen Species and Mitochondria Linked Signal Pathway, *Toxicology and Applied Pharmacology* Vol.230, No.1, (July 2008), pp. 33-40, ISSN 0041-008X

Zhu, T., Chen, R., Li, A. P., Liu, J., Liu, Q. Z., Chang, H. C. & Zhou, J. W. (2005). Regulation of a Novel Cell Differentiation-Associated Gene, JWA During Oxidative Damage in K562 and MCF-7 Cells, *Journal of Biomedical Science* Vol.12, No.1, pp. 219-227

Zhu, Y., King, M. A., Schuller, A. G., Nitsche, J. F., Reidl, M., Elde, R. P., Unterwald, E., Pasternak, G. W. & Pintar, J. E. (1999). Retention of Supraspinal Delta-Like Analgesia and Loss of Morphine Tolerance in Delta Opioid Receptor Knockout Mice, *Neuron* Vol.24, No.1, (Sep 1999), pp. 243-252, ISSN 0896-6273 (Print), 0896-6273 (Linking)

15

Functional Genomics of Anoxygenic Green Bacteria *Chloroflexi* Species and Evolution of Photosynthesis

Kuo-Hsiang Tang
Carlson School of Chemistry and Biochemistry, and Department of Biology
Clark University, Worcester
USA

1. Introduction

In addition to the most recently reported aerobic anoxygenic phototrophic bacterium *Chloroacidobacterium thermophilium* [1], five phyla of phototrophic bacteria have been reported, including four phyla anoxygenic phototrophic bacteria (anaerobic and aerobic anoxygenic phototrophic Proteobacteria, filamentous anoxygenic phototrophs (FAPs), green sulfur bacteria and heliobacteria) and oxygenic phototrophic bacteria (cyanobacteria). According to 16S rRNA analysis, *Chloroflexi* species in FAPs are the earliest branching bacteria capable of photosynthesis [2,3] (**Fig. 1**), and the thermophilic bacterium *Chloroflexus* [Cfl.] *aurantiacus* among the *Chloroflexi* species has been long regarded as a key organism to resolve the obscurity of the origin and early evolution of photosynthesis. *Cfl. aurantiacus* can grow phototrophically under anaerobic conditions or chemotrophically under aerobic and dark conditions [4]. During phototrophic growth of *Cfl. aurantiacus*, the light energy is first absorbed by the peripheral light-harvesting complex chlorosomes, then transferred to the integral membrane B808-866 core antenna complex and finally to the reaction center (RC). *Cfl. aurantiacus* contains a chimeric photosystem that comprises some characters of green sulfur bacteria (chlorosomes) and anoxygenic phototrophic Proteobacteria (the B808-866 core antenna complex), and also has some unique electron transport proteins compared to other photosynthetic bacteria. The complete genomic sequence of *Cfl. aurantiacus* has been recently determined, analyzed and compared to the genomes of other photosynthetic bacteria [5].

Significant contributions of horizontal/lateral gene transfer among uni-cellular [6] and multi-cellular [7] organisms during the evolution, including the evolution of photosynthesis [8,9], have been recognized. Various perspectives on evolution of photosynthesis have been reported in literature [8-25], whereas our understanding of transition from anaerobic to aerobic world is still fragmentary. The recent genomic report on *Cfl. aurantiacus* [5], along with previous physiological, ecological and biochemical studies, indicate that the anoxygenic phototroph bacterium *Cfl. aurantiacus* has many interesting and certain unique features in its metabolic pathways. The *Cfl. aurantiacus* genome contains numerous aerobic/anaerobic gene pairs and oxygenic/anoxygenic metabolic pathways in the *Cfl. aurantiacus* genome [5], suggesting numerous gene adaptations/replacements in *Cfl. aurantiacus* to facilitate life under both anaerobic and aerobic growth conditions. These

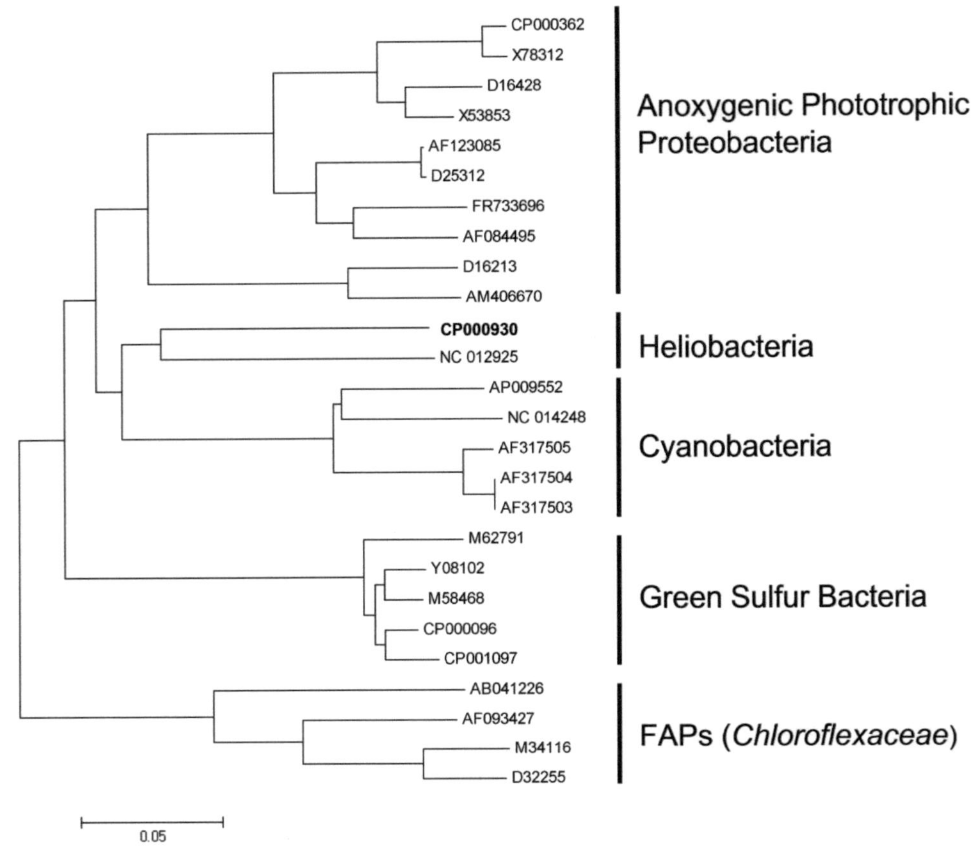

Fig. 1. Phylogenetic tree of photosynthetic bacteria.
The tree was constructed with un-rooted neighbor joining 16S rRNA dendrogram from five phyla of photosynthetic microbes, including cyanobacteria, heliobacteria, phototrophic anoxygenic Proteobacteria, green sulfur bacteria and filamentous anoxygenic phototrophs (FAPs). Bacterial names and accession numbers of 16S rRNA genes: (1) Phototrophic anoxygenic Proteobacteria: *Roseobacter denitrificans* OCh114 (CP000362), *Roseobacter litoralis* (X78312), *Rhodobacter capsulatus* (D16428), *Rhodobacter sphaeroides* 2.4.1 (X53853), *Rhodopseudomonas faecalis* strain gc (AF123085), *Rhodopseudomonas palustris* (D25312), *Rhodopseudomonas acidophila* (FR733696), *Rhodopseudomonas viridis* DSM 133 (AF084495), *Rubrivivax gelatinosus* (D16213); (2) heliobacteria: *Heliobacterium gestii* (AB100837), *Heliobacterium modesticaldum* (CP000930); (3) cyanobacteria: *Oscillatoria amphigranulata* str. 19-2 (AF317504), *Oscillatoria amphigranulata* str. 11-3 (AF317503), *Oscillatoria amphigranulata* str. 23-3 (AF317505), *Microcystis aeruginosa* NIES-843 (AP009552), *Nostoc azollae* 0708 (NC_014248); (4) green sulfur bacteria: *Chlorobaculum thiosulfatiphilum* DSM 249 (Y08102), *Pelodictyon luteolum* DSM 273 (CP000096), *Chlorobium limicola* DSM 245 (CP001097), *Chlorobaculum tepidum* TLS (M58468), *Chlorobium vibrioforme* DSM 260 (M62791); and (5) FAPs: *Chloroflexus aurantiacus* J-10-fl (M34116), *Chloroflexus aggregans* (D32255), *Oscillochloris trichoides* (AF093427), *Roseiflexus castenholzii* DSM 13941 (AB041226)

include duplicate genes and gene clusters for the alternative complex III (ACIII) [26,27], auracyanin (a type I blue copper protein) [28,29] and NADH:quinone oxidoreductase (complex I); and several aerobic/anaerobic enzyme pairs in central carbon metabolism (pyruvate metabolism and the tricarboxylic acid (TCA) cycle) and tetrapyrroles and nucleic acids biosynthesis [5]. Overall, genomic information is consistent with a high tolerance for oxygen that has been reported in the growth of *Cfl. aurantiacus*.

Phylogenetic analyses on the photosystems and comparisons to the genome and reports of other photosynthetic bacteria suggest lateral or horizontal gene transfers between *Cfl. aurantiacus* and other photosynthetic bacteria [3,30,31]. The *Cfl. aurantiacus* genome suggests possible evolutionary connections of photosynthesis. Here we probe some proposed lateral gene transfers using the phylogenetic analyses on important proteins/enzymes on chlorophyll biosynthesis, photosynthetic electron transport chain, and central carbon metabolism. Further, we also discuss the evolutionary perspectives on assembling photosynthetic machinery, autotrophic carbon assimilation and unique components on the electron transport chains of *Cfl. aurantiacus* and other phototrophic and non-phototrophic bacteria.

2. Results and discussion

a. Photosynthetic components

The photosystem of *Cfl. aurantiacus* is a chimeric system with contains a peripheral light harvesting complex chlorosomes and an integral membrane B808-866-type II RC (quinone-type) core complex. Chlorosomes are typically found in type I (Fe-S type) RC phototrophic organisms, such as green sulfur bacteria (GSBs) [32] and the recently discovered aerobic anoxygenic bacterium *Chloroacidobacterium thermophilium* [1], whereas the B808-866-RC core complex is arranged similarly to the LH-RC core complex in phototrophic Proteobacteria [33]. Thus, the *Cfl. aurantiacus* photosystem indicates little correlation between the RC type and light-harvesting antenna complexes in the assembly of the photosystem of anoxygenic phototrophic bacteria [8,34]. Two hypotheses, which are selective loss and fusion, for evolutionary of photosynthetic RCs have been proposed [8,35]. The phylogenic analyses and evolutionary perspectives of the integral membrane-RC core complex in *Cfl. aurantiacus* and other phyla of phototrophic bacteria are presented in several reports [8,36,37] for readers who are interested in further information. It is possible that during the evolution of photosynthesis chlorosomes were transferred between *Cfl. aurantiacus* and GSBs, which have larger chlorosomes and more genes encoding chlorosome proteins [38,39], and that the integral membrane core antenna complex and a type II RC in *Cfl. aurantiacus* were possibly transferred either to or from photosynthetic anoxygenic Proteobacteria.

b. Electron transfer complexes

Four copies of auracyanin genes have been identified in the *Cfl. aurantiacus* genome and two aurancyanin proteins have been characterized biochemically and structurally [28]. Auracyanin has also been biochemically characterized in *Roseiflexus castenholzii* [40], which only has one copy of aurancyanin gene in the genome [5]. The gene encoding a putative auracyanin has been identified in the genome of the non-photosynthetic aerobic thermophilic bacterium *Thermomicrobium roseum* DSM 5159, which is evolutionally related to *Cfl. aurantiacus* [41]. Genes encoding auracyanin may have been transferred either to or from

Thermomicrobium roseum. Further, higher plants, green algae and cyanobacteria operate the photosynthetic electron transport via a water-soluble mobile type I blue copper protein plastocyanin. Auracyanin may have evolved from or to plastocyanin in cyanobacteria.

Most of phototrophic bacteria use the cytochrome bc_1 or b_6/f complex for transferring electrons during phototrophic growth, whereas *Chloroflexi* species operate photosynthetic electron transport using a unique complex, namely alternative complex III (ACIII) [1,26,27]. Two sets of ACIII gene clusters, one containing seven genes and the other containing thirteen genes, have been identified in the *Cfl. aurantiacus* genome [5]. The seven subunit complex has been characterized biochemically [27]. In contrast, *Roseiflexus castenholzii*, which is a member of a familia *Chloroflexaceae* and phylogenetically closely related to *Cfl. aurantiacus* [42], contains only one copy of the ACIII operon with a six-gene cluster (Rcas_1462-1467) [5]. In addition to *Cfl. aurantiacus* and other members of *Chloroflexaceae*, genes encoding ACIII, which contains seven subunits [27], have also been identified in the *Chloroacidobacterium thermophilium* genome [1]. ACIII has also been identified in non-phototrophic bacterium *Rhodothermus marinus* [43] and suggested to wide-spread in prokayrotes [44]. Genes encoding ACIII may have been transferred either from or to evolved from or to *Chloroacidobacterium thermophilium* (and/or *Rhodothermus marinus*). Further, ACIII may have evolved from or to the cytochrome bc_1 or b_6/f complex.

NADH:quinone oxidoreductase (Complex I, EC 1.6.5.3) is known to be responsible for the electron transport in the respiratory chain. Two sets of the Complex I genes, one of which forms a gene cluster, have been identified in the *Cfl. aurantiacus* genome [5]. Two Complex I gene clusters have also been identified in some anaerobic anoxygenic phototrophic Proteobacteria (AnAPs), such as *Rhodobacter* [Rba.] *sphaeroides* and *Rhodopseudomonas* [Rps.] *palustris*, and gene expression profile in *Rba. sphaeroides* suggests that one of the gene clusters is responsible for photosynthetic electron transport during phototrophic and anaerobic growth and the other is required for the respiratory chain during aerobic and dark growth [45]. **Fig. 2** shows the phylogenetic trees constructed based on the amino acid sequences of the subunit F of Complex I (encoded by the *nuoF* gene) in phototrophic bacteria. The subunit F protein in **Fig. 2A** is encoded by the gene locus Caur_2901 in the gene cluster (Caur_2896 to Caur_2909), and the subunit F protein in **Fig. 2B** is encoded by the gene locus Caur_1185. No Complex I genes have been identified in the green sulfur bacteria, which cannot respire or grow in darkness. Note that one subunit F protein in *Cfl. aurantiacus* is more related to the protein in anoxygenic phototrophic Proteobacteria than to the protein in heliobacteria and cyanobacteria (**Fig. 2A**) and the other *Cfl. aurantiacus* subunit F protein is more related to the protein in heliobacteria and cyanobacteria than to the protein in anoxygenic Proteobacteria (**Fig. 2B**), suggesting different biological functions for two NADH:quinone oxidoreductase complexes found in the *Cfl. aurantiacus* genome.

c. (Bacterio)chlorophyll biosynthesis

AcsF (aerobic cyclase) and BchE (anaerobic cyclase) are suggested to be responsible for biosynthesis of the isocyclic ring of (bacterio)chlorophylls and conversion of Mg-protoporphyrin monomethyl ester (MgPMMe) to Mg-divinyl-protochlorophyllide *a* (PChlide) under aerobic and anaerobic growth conditions, respectively [46-51] (**Fig. 3A**). Both MgPMMe and PChlide are suggested to be photosensitizers of higher plants and green algae that produce reactive oxygen species in response to the excess light [52]. Both *acsF* (Caur_2590) and *bchE* (Caur_3676) are detected in the *Cfl. aurantiacus* genome [5]. AcsF has

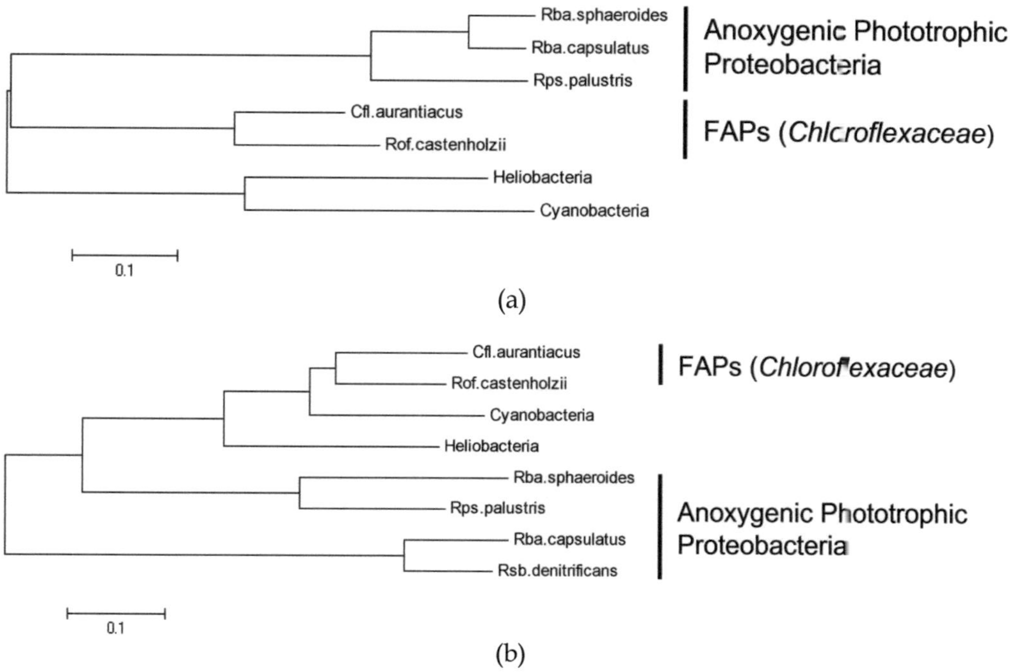

Fig. 2. Phylogenetic tree of the NADH:quinine oxidoreductase (Complex I) in phototrophic bacteria.
The subunit F proteins of *Cfl. aurantiacus*, *Roseiflexus* [Rof.] *castenholzii* (FAPs), *Rhodobacter* [Rba.] *sphaeroides* and *Rhodopseudomonas* [Rps.] *palustris* (anoxygenic Proteobacteria) in **Fig. 2A** and **2B** are encoded by different *nuoF* genes. Two Complex I are identified in *Cfl. aurantiacus*, *Rof. castenholzii*, *Rba. sphaeroides* and *Rps. palustris*, and one Complex I gene cluster is found in heliobacteria, cyanobacteria and some phototrophic anoxygenic Proteobacteria (e.g., *Rba. capsulatus* and *Roseobacter* [Rsb.] *denitrificans*). The trees are constructed based on amino acid sequences using the phylogenetic software MEGA5 [65] with un-rooted neighbor jointing method.

not been identified in any strictly anaerobic phototrophic bacteria (e.g., green sulfur bacteria and heliobacteria). In addition to Proteobacteria (including aerobic and anaerobic anoxygenic phototrophic Proteobacteria) and cyanobacteria, several non-phototrophic α-Proteobacteria also contain the *acsF* gene, including several facultative methotrophic bacteria (e.g., *Methylocella silvestris*, *Methylobacterium* [Mtb.] sp. 4-46, *Mtb. populi*, *Mtb. chloromethanicum*, *Mtb. radiotolerans* and *Mtb. extorquens*) and the environmental bacterium *Brevundimonas subvibrioides* (**Fig. 3B**). Roles of the gene encoding the putative AcsF in these non-phototrophic bacteria are unclear. AcsF has also been characterized for *Cfl. aurantiacus* grown under anaerobic conditions [50]. Together, the role of AcsF remains to be further understood. BchE is widely spread in all phyla of anoxygenic phototrophic bacteria (e.g., anoxygenic phototrophic Proteobacteria, green sulfur bacteria, heliobacteria and FAPs) and some facultative methyltrophic bacteria and cynaobacteria also contain the gene encoding the putative BchE (**Fig. 3C**). Experimental evidence indicates that the *bchE* genes in the cyanobacterium *Synechocystis* sp. PCC 6803 are important but do not contribute to the formation of the isocyclic ring of chlorophylls [47].

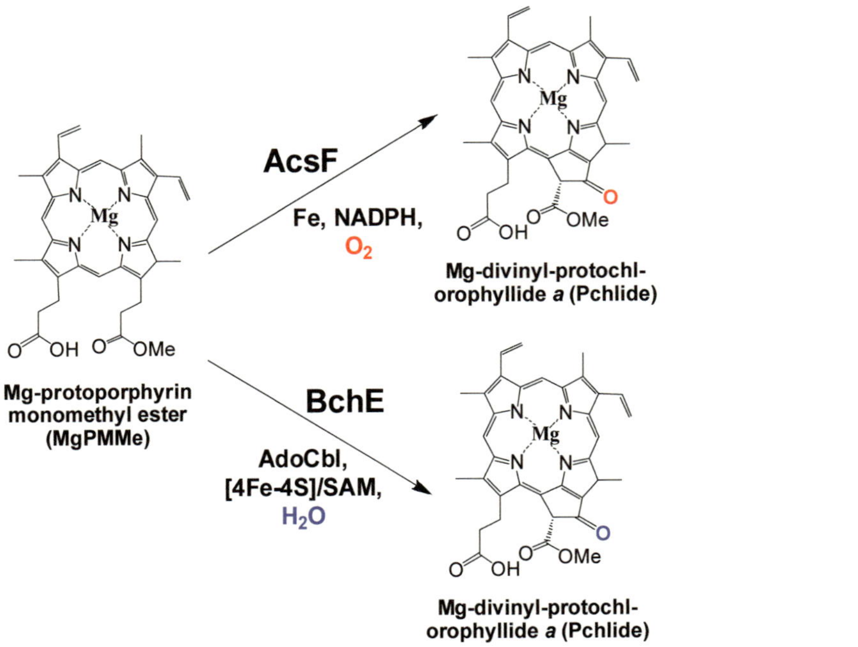
AcsF
Fe, NADPH,
O2
Mg-divinyl-protochl-
orophyllide a (Pchlide)
Mg-protoporphyrin
monomethyl ester
(MgPMMe)
BchE
AdoCbl,
[4Fe-4S]/SAM,
H2O
Mg-divinyl-protochl-
orophyllide a (Pchlide)

(a)

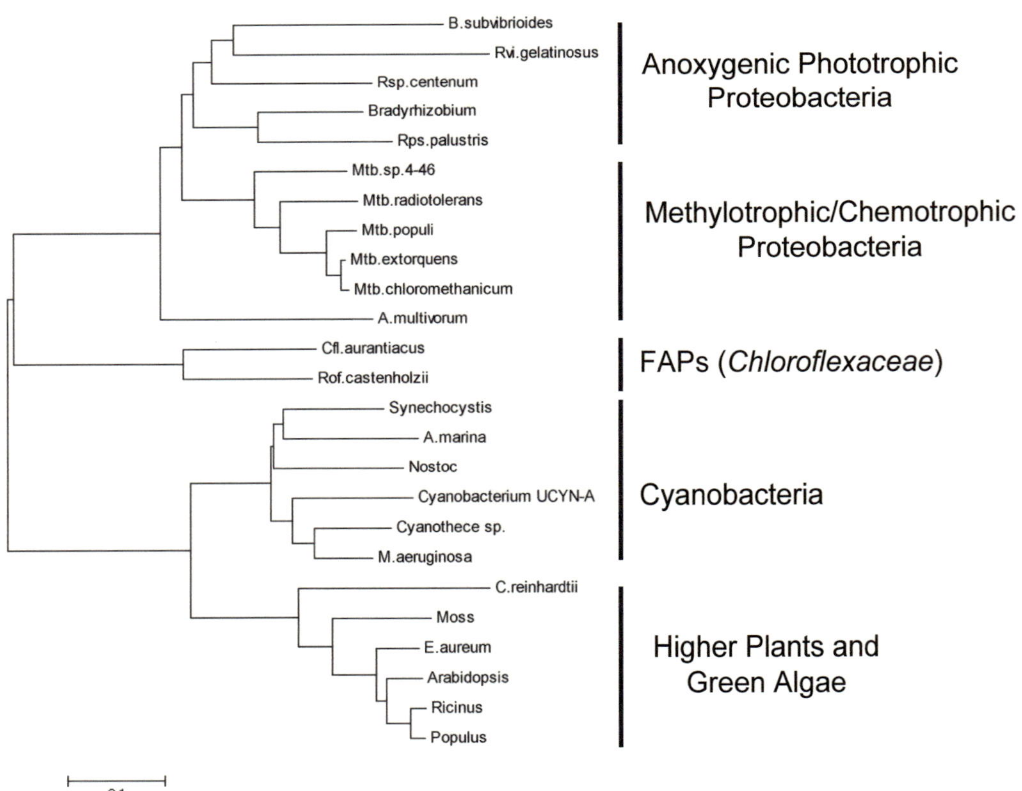
B.subvibrioides
Rvi.gelatinosus
Rsp.centenum
Bradyrhizobium
Rps.palustris
Mtb.sp.4-46
Mtb.radiotolerans
Mtb.populi
Mtb.extorquens
Mtb.chloromethanicum
A.multivorum
Cfl.aurantiacus
Rof.castenholzii
Synechocystis
A.marina
Nostoc
Cyanobacterium UCYN-A
Cyanothece sp.
M.aeruginosa
C.reinhardtii
Moss
E.aureum
Arabidopsis
Ricinus
Populus
0.1
Anoxygenic Phototrophic Proteobacteria
Methylotrophic/Chemotrophic Proteobacteria
FAPs (Chloroflexaceae)
Cyanobacteria
Higher Plants and Green Algae

(b)

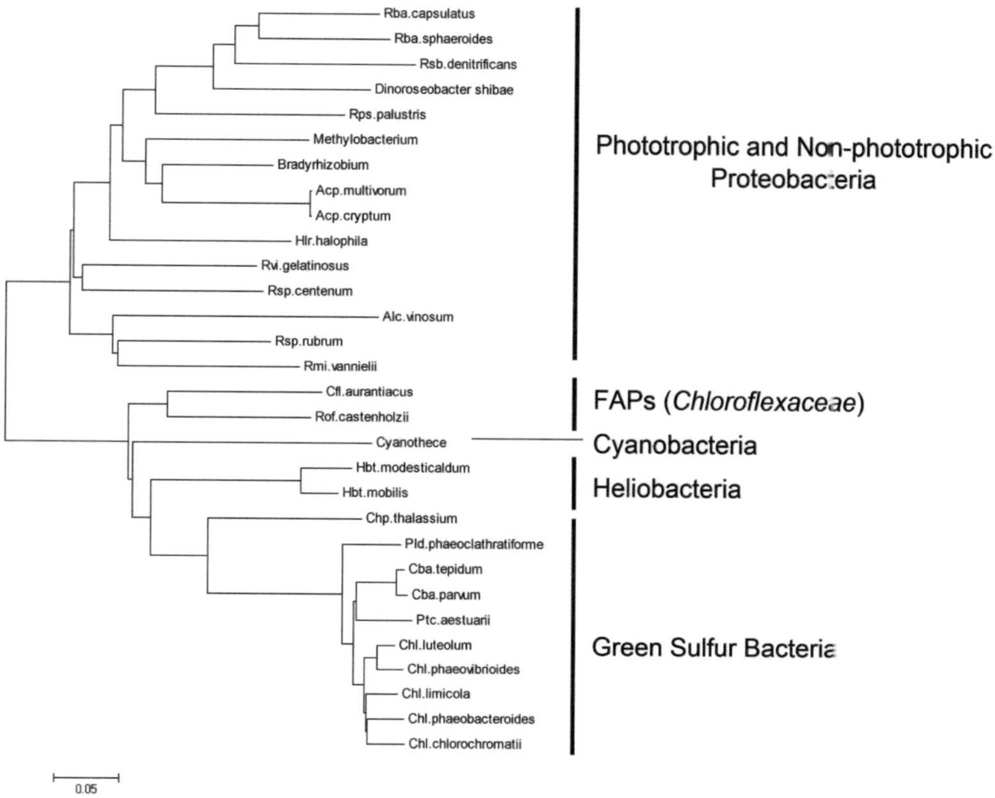

(c)

Fig. 3. Reactions of aerobic cyclase (AcsF) and anaerobic cyclase (BchE) and the phylogenetic trees.
Conversion of MgPMMe into PChlide is suggested to be catalyzed by AcsF and BchE under aerobic and anaerobic conditions, respectively (A). The phylogenetic relationships of AcsF (B) and BchE (C) are shown. The trees are constructed based on amino acid sequences using the phylogenetic software MEGA5 [65] with un-rooted neighbor jointing method.

Phylogenetic analyses suggest that the *acsF* gene in *Cfl. aurantiacus* and other *Chloroflexaceae* species are more evolutionarily related to the genes in anoxygenic phototrophic Proteobacteria than to the genes in oxygenic phototrophs (cyanobacteria, green algae and higher plants) (**Fig. 3B**), and that the *bchE* gene in *Cfl. aurantiacus* is more evolutionarily related to the genes in strictly anaerobic phototrophs (green sulfur bacteria and heliobacteria) than to the genes in phototrophic and non-phototrophic Proteobacteria (**Fig. 3C**). It is possible that the *Cfl. aurantiacus acsF* gene was transferred either to or from Proteobacteria, and the *Cfl. aurantiacus bchE* gene was transferred either to or from heliobacteria and green sulfur bacteria. The phylogenetic analyses of AcsF and BchE in **Fig. 3** likely suggest horizontal gene transfers among phototrophic bacteria and also between phototrophic and non-phototrophic bacteria.

d. Central carbon metabolism

Here we analyze enzymes/gene products for pyruvate metabolism, which takes place in every living organism, and the TCA cycle. In contrast to other phyla of phototrophic bacteria, *Cfl. aurantiacus* and other members of *Chloroflexaceae* are only bacteria containing both anaerobic and aerobic gene pairs for pyruvate and α-ketoglutarate metabolism: pyruvate/α-ketoglutarate dehydrogenase (aerobic enzymes) and pyruvate/α-ketoglutarate synthase (or pyruvate/α-keto-glutarate:ferredoxin oxidoreductase (PFOR/KFOR)) (anaerobic enzymes).

Fig. 4A shows the phylogenetic analyses of the E1 protein of α-ketoglutarate dehydrogenase (encoded by *sucA*) from FAPs and anoxygenic phototrophic Proteobacteria. Note that the *Cfl. aurantiacus* α-ketoglutarate dehydrogenase has higher sequence identities to many gram-(+) non-phototrophic *Bacillus* strains (~50%) than phototrophic anoxygenic Proteobacteria (~40%). Similar results also find in the sequence alignments of the E1 protein of pyruvate dehydrogenase, and the *Cfl. aurantiacus* enzyme has ~51-55% identities with *Thermobifida fusca, Streptomyces cattleya, Acidothermus cellulolyticus, Saccharopolyspora erythraea,* and *Sanguibacter keddieii* and ~38-44% or lower identities with the phosynthetic Proteobacteria and cyanobacteria (data not shown). These results support the horizontal gene transfer between microbial genomes. **Fig. 4B** shows the phylogenetic tree of the E1 protein of pyruvate dehydrogenase. The *Cfl. aurantiacus* enzyme is less related to cyanobacteria and anoxygenic phototrophic Proteobacteria.

Fig. 4C suggests that α-ketoglutarate synthase in *Cfl. aurantiacus* are more closely related to the enzyme in heliobacteria than in green sulfur bacteria. While the biochemical studies of the *Cfl. aurantiacus* α-ketoglutarate synthase have not been reported, the phylogenetic analyses of α-ketoglutarate synthase are consistent with the central carbon flow in these three phyla of photosynthetic bacteria: the green sulfur bacteria operate the reductive (reverse) TCA cycle, and *Cfl. aurantiacus* and heliobacteria have strong carbon flow via either a complete or a partial oxidative (forward) TCA cycle [34].

Fig. 4D suggests that pyruvate synthase in heliobcteria evolved prior to the enzymes in other phyla of photosynthetic bacteria, and that the enzyme in *Cfl. auranticus* is remotely related to the enzymes in GSBs and cyanobacteria, which are likely from the same origins, similar to the tree of the E1 protein of pyruvate dehydrogenase (**Fig. 4B**). Together, the phylogenetic analyses suggest pyruvate metabolism of anoxygenic phototrophic Proteobacteria is more related to cyanobacteria than to *Cfl. aurantiacus* (and perhaps FAPs). Compared to the experimental data, acetate can support the growth of *Cfl. aurantiacus* during anaerobic growth in the light and during aerobic growth in darkness [53], and acetate excretion has been reported during the pyruvate-grown heliobacteria [54,55] but not on other phyla of photosynthetic bacteria. *Cfl. aurantiacus* likely uses pyruvate synthase for assimilate acetyl-CoA. Since heliobacteria do not have pyruvate dehydrogenase, their pyruvate synthase is supposed to convert pyruvate to acetyl-CoA, which is then converted to acetate. Further, pyruvate synthase is essential for the growth of green sulfur bacteria because it is required to convert acetyl-CoA generated from the reductive TCA cycle to pyruvate, whereas the role of pyruvate synthase in oxygenic phototrophic bacteria (cyanobacteria) is not clear, as pyruvate synthase is sensitive to oxygen during biochemical characterization *in vitro.*

e. Autotrophic carbon assimilation

Cfl. aurantiacus can grow photoautotrophically and uses the 3-hydroxypropionate (3HOP) bi-cycle to assimilate inorganic carbon [5,56-58]. Both 3HOP bi-cycle and the widely distributed Calvin-Benson cycle can operate in both aerobic and anaerobic conditions.

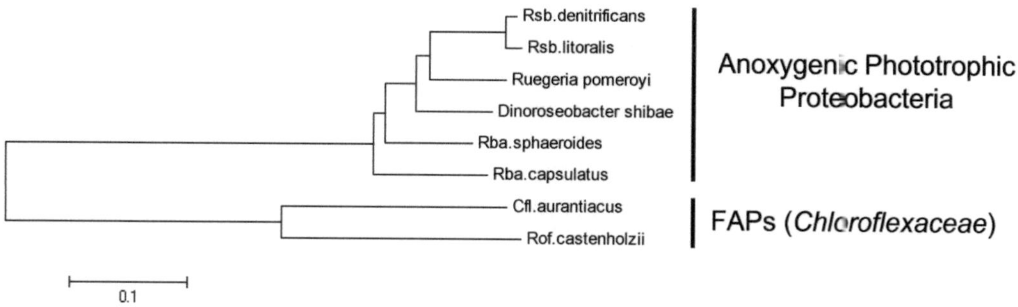
Rsb.denitrificans
Rsb.litoralis
Ruegeria pomeroyi
Dinoroseobacter shibae
Rba.sphaeroides
Rba.capsulatus
Cfl.aurantiacus
Rof.castenholzii
Anoxygenic Phototrophic Proteobacteria
FAPs (Chloroflexaceae)
0.1

(a)

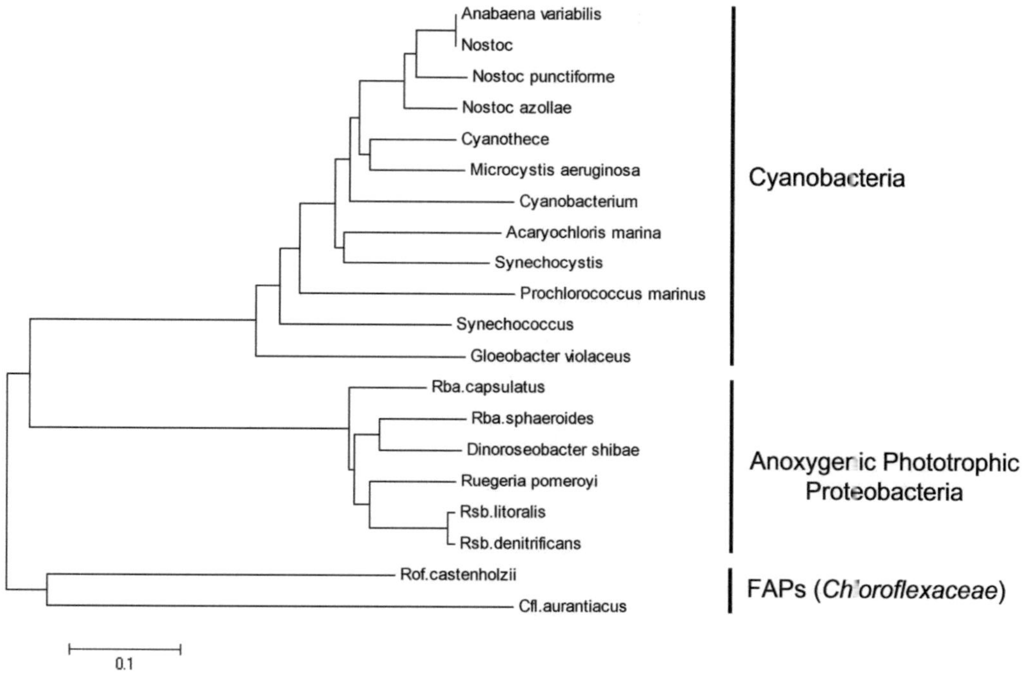
Anabaena variabilis
Nostoc
Nostoc punctiforme
Nostoc azollae
Cyanothece
Microcystis aeruginosa
Cyanobacterium
Acaryochloris marina
Synechocystis
Prochlorococcus marinus
Synechococcus
Gloeobacter violaceus
Rba.capsulatus
Rba.sphaeroides
Dinoroseobacter shibae
Ruegeria pomeroyi
Rsb.litoralis
Rsb.denitrificans
Rof.castenholzii
Cfl.aurantiacus
Cyanobacteria
Anoxygenic Phototrophic Proteobacteria
FAPs (Chloroflexaceae)
0.1

(b)

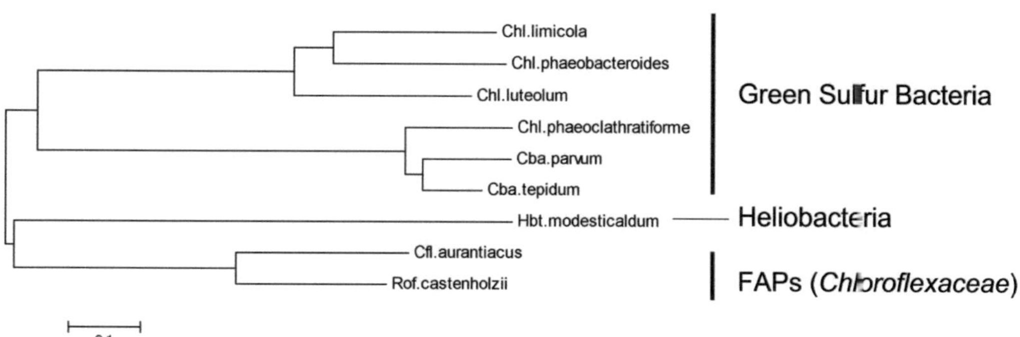
Chl.limicola
Chl.phaeobacteroides
Chl.luteolum
Chl.phaeoclathratiforme
Cba.parvum
Cba.tepidum
Hbt.modesticaldum
Cfl.aurantiacus
Rof.castenholzii
Green Sulfur Bacteria
Heliobacteria
FAPs (Chloroflexaceae)
0.1

(c)

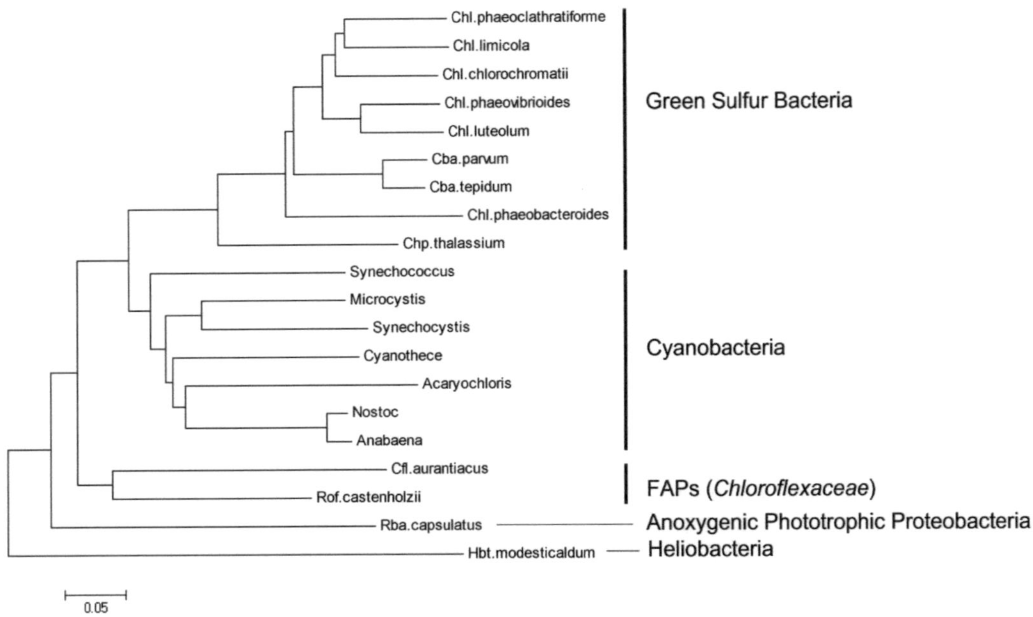

(d)

Fig. 4. The phylogenetic trees of α-ketoglutarate dehydrogenase (A), pyruvate dehydrogenase (B), α-ketoglutarate synthase (C) and pyruvate synthase (D).
The trees are constructed based on amino acid sequences using the phylogenetic software MEGA5 [65] with un-rooted neighbor jointing method.

However, one significant problem leading to low photosynthesis efficiency of higher plants and oxygenic phototrophs is photorespiration and energy waste resulting from the interactions of oxygen with RuBisCO (ribulose 1,5-bisphosphate carboxylase/oxygenase) [12], the carboxylase in the Calvin-Benson cycle. Different from the Calvin-Benson cycle, the 3HOP bi-cycle assimilates bicarbonate instead of CO_2 (**Fig. 5A**). The 3HOP bi-cycle, which operates in *Cfl. aurantiacus* and most likely in other members of *Chloroflexaceae* [57], is similar to 3-hydroxypropionate/4-hydroxybutyrate (3HOP/4HOB) cycle reported in several archaea [59,60] (**Fig. 5B**). Several enzymes operate in both 3HOP bi-cycle and 3HOP/4HOB cycle, including enzymes for assimilating inorganic carbon: acetyl-CoA carboxylase and propionyl-CoA carboxylase. 16S rRNA analyses suggest that Archaea developed earlier than the bacteria capable of using light as the energy sources [3], so the 3HOP bi-cycle may have evolved from the 3HOP/4HOB cycle.

Other horizontal gene transfers can be also found in the autotrophic carbon assimilation on other members of *Chloroflexales*. For example, several strains in the family of *Oscillochloridaceae* assimilate inorganic carbon via the Calvin-Benson cycle and have an incomplete TCA cycle [61]. In addition to oxygenic phototrophs, anaerobic anoxygenic phototrophic Proteobacteria (AnAPs) also operate the Calvin-Benson cycle. In contrast to oxygenic phototrophs, poor substrate specificity of RuBisCO should not be a serious concern for anoxygenic phototrophs like AnAPs and *Oscillochloridaceae*. It is possible that the genes in the Calvin-Benson cycle in may transfer between *Oscillochloridaceae*, AnAPs and cyanobacteria. Furthermore, *Dehalococcoides ethanogenes* strain 195, a Gram-positive non-phototrophic bacteria in the subphylum 2 of *Chloroflexi* [62], uses (*Re*)-citrate synthase [63] and has a branched TCA cycle [63,64]. Together, three members of the phylum *Chloroflexi*, *Cfl. aurantiacus*, *Oscillochloridaceae* and *Dehalococcoides ethanogenes* have distinct central carbon metabolic pathways.

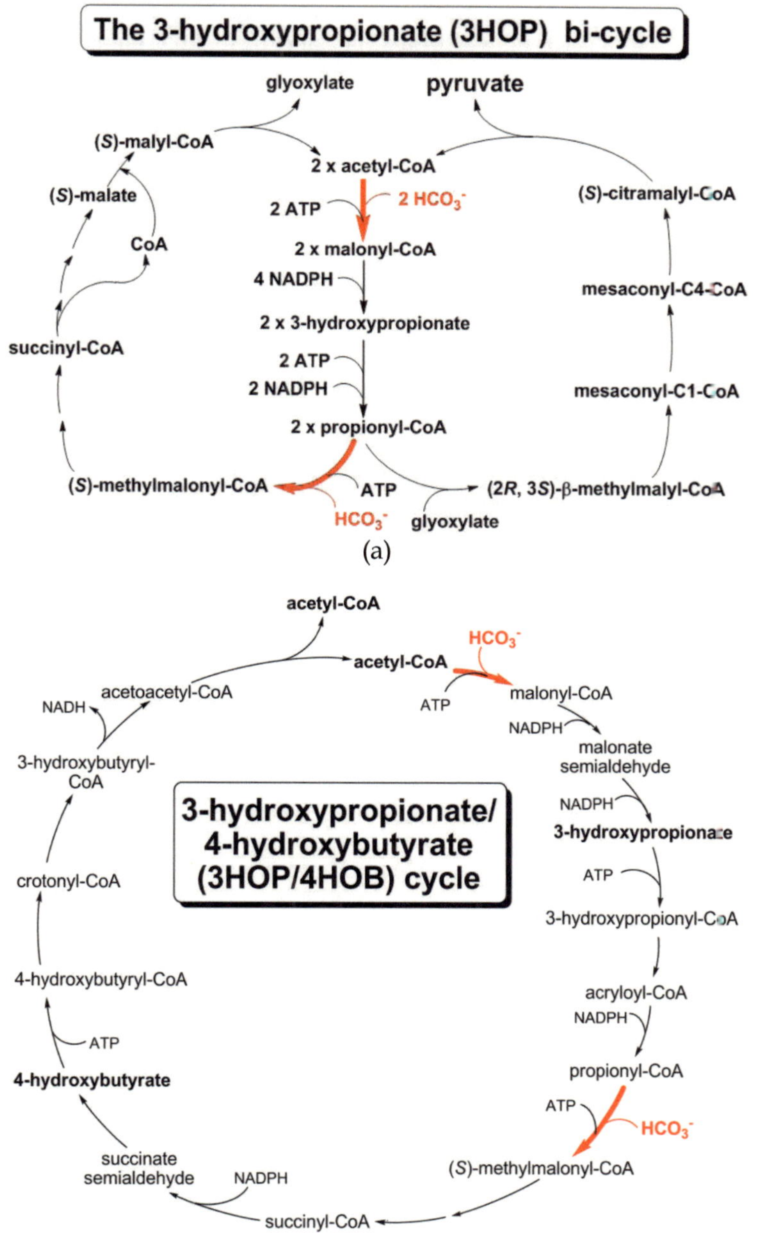

Fig. 5. Autotrophic carbon assimilations through 3-hydroxypropionate bi-cycle (A) and 3-hydroxypropionate/4-hydroxybutyrate cycle.
Several enzymes, including acetyl-CoA carboxylase and propionyl-CoA carboxylase, operate in both 3HOP bi-cycle and 3HOP/4HOB cycle.

3. Conclusions

Previous physiological, ecological and biochemical studies [4] as well as genomic analyses [5], indicate that the anoxygenic phototroph bacterium *Cfl. aurantiacus* has many interesting and certain unique features in its metabolic pathways. The evolutionary links of *Cfl. aurantiacus* and other phototrophic bacteria suggested from this report are summarized in **Table 1.** It has been recognized that the type II RCs were transferred between the *Chloroflexi* species (or FAPs) and the anoxygenic phototrophic Proteobacteria. Sequence alignments and phylogenetic analyses illustrated in this report suggest: (i) Some *Cfl. aurantiacus* enzymes in essential metabolic pathways are more related to the anoxygenic phototrophic Proteobacteria than other phototrophic bacteria, whereas other enzymes are more related to other phototrophic bacteria than anoxygenic phototrophic Proteobacteria; and (ii) some *Cfl. aurantiacus* enzymes in essential carbon metabolic pathways are more related to non-photosynthetic microbes than other phyla of phototrophic bacteria. Together, our studies support lateral/horizontal gene transfers among microbes, and suggest that photosynthesis is likely an adaption to the environments [9].

	Anoxygenic Phototrophic Proteobacteria	Green Sulfur Bacteria	Heliobacteria	Cyanobacteria
Photosynthetic Machinery				
Chlorosomes		√		
B808-866 core antenna complex	√			
Reaction center	√			
Electron Transport Chain				
Complex I -A[a]	more related		less related	less related
Complex I –B[b]	less related		related	more related
Auracyanin				√
Pigment Biosynthesis				
Aerobic cyclase (AcsF)	more related			less related
Anaerobic cyclase (BchE)	less related	more related	more related	?[c]
Central Carbon Metabolism				
Pyruvate dehydrogenase	√			√
Pyruvate synthase	related	more related	less related	more related
α-ketoglutarate dehydrogenase	√			
α-ketoglutarate synthase		less related	more related	

[a] The *Cfl. aurantiacus* complex I with clustered genes (**Fig. 2A**)
[b] The *Cfl. aurantiacus* complex I without clustered genes (**Fig. 2B**)
[c] The putative bchE genes in some cyanobacteria have been reported not to function as anaerobic cyclase.

Table 1. Evolutionary perspectives of selective proteins/enzymes/cellular complexes in *Cfl. aurantiacus* versus other phyla of phototrophic bacteria

4. Abbreviations

AnAPs:	anaerobic anoxygenic phototrophic Proteobacteria
AcsF:	aerobic cyclase
ACIII:	alternative complex III
BchE:	anaerobic cyclase
FAPs:	filamentous anoxygenic phototrophs (or green non-sulfur bacteria or green gliding bacteria)
GSBs:	green sulfur bacteria
3HOP bi-cycle:	3-hydroxypropionate bi-cycle
3HOP/4HOB cycle:	3-hydroxypropionate/4-hydroxybutyrate cycle
RC:	reaction center
RuBisCO:	ribulose 1,5-bisphosphate carboxylase/oxygenase
TCA cycle:	tricarboxylic acid cycle

5. Acknowledgements

The author thanks Dr. Robert E. Blankenship for introducing the author into the fields of photosynthesis and the financial support of start-up fund from Clark University.

6. References

[1] Bryant, D.A. et al. (2007). *Candidatus* Chloracidobacterium thermophilum: an aerobic phototrophic Acidobacterium. Science 317, 523-6.
[2] Giovannoni, S.J. and Stingl, U. (2005). Molecular diversity and ecology of microbial plankton. Nature 437, 343-8.
[3] Pace, N.R. (1997). A molecular view of microbial diversity and the biosphere. Science 276, 734-40.
[4] Hanada, S. and Pierson, B.K. (2006). The family chloroflexaceae. *The Prokaryotes*, 3rd Ed., Vol. 7, Springer, New York. pp. 815 – 842.
[5] Tang, K.H. et al. (2011). Complete Genome Sequence of the Filamentous Anoxygenic Phototrophic Bacterium *Chloroflexus aurantiacus*. BMC Genomics 12, 334.
[6] Jain, R., Rivera, M.C., Moore, J.E. and Lake, J.A. (2002). Horizontal gene transfer in microbial genome evolution. Theor Popul Biol 61, 489-95.
[7] Pierce, S.K., Massey, S.E., Hanten, J.J. and Curtis, N.E. (2003). Horizontal transfer of functional nuclear genes between multicellular organisms. Biol Bull 204, 237-40.
[8] Hohmann-Marriott, M.F. and Blankenship, R.E. (2011). Evolution of photosynthesis. Annu Rev Plant Biol 62, 515-48.
[9] Leslie, M. (2009). Origins. On the origin of photosynthesis. Science 323, 1286-7.
[10] Blankenship, R.E. (1992). Origin and early evolution of photosynthesis. Photosynth Res 33, 91-111.
[11] Blankenship, R.E. (2001). Molecular evidence for the evolution of photosynthesis. Trends Plant Sci 6, 4-6.
[12] Blankenship, R.E. (2002). *Molecular Mechanisms of Photosynthesis*. Blackwell Science Ltd, Oxford.
[13] Blankenship, R.E. and Hartman, H. (1998). The origin and evolution of oxygenic photosynthesis. Trends Biochem Sci 23, 94-7.

[14] Allen, J.F. (2005). A redox switch hypothesis for the origin of two light reactions in photosynthesis. FEBS Lett 579, 963-8.

[15] Allen, J.F. and Martin, W. (2007). Evolutionary biology: out of thin air. Nature 445, 610-2.

[16] Bjorn, L.O. and Govindjee. (2009). The evolution of photosynthesis and chloroplasts. Curr Sci 96, 1466-1474.

[17] Buick, R. (2006). When did oxygenic photosynthesis evolve? Phil Trans R Soc B 363, 2731-2743.

[18] De Marais, D.J. (2000). Evolution. When did photosynthesis emerge on Earth? Science 289, 1703-5.

[19] Gupta, R.S. (2010). Molecular signatures for the main phyla of photosynthetic bacteria and their subgroups. Photosynth Res 104, 357-72.

[20] Hengeveld, R. (2007). Two approaches to the study of the origin of life. Acta Biotheor 55, 97-131.

[21] Mulkidjanian, A.Y. (2009). On the origin of life in the zinc world: 1. Photosynthesizing, porous edifices built of hydrothermally precipitated zinc sulfide as cradles of life on Earth. Biol Direct 4, 26.

[22] Olson, J.M. and Blankenship, R.E. (2004). Thinking about the evolution of photosynthesis. Photosynth Res 80, 373-86.

[23] Olson, J.M. and Pierson, B.K. (1987). Evolution of reaction centers in photosynthetic prokaryotes. Int Rev Cytol 108, 209-48.

[24] Raven, J. (2009). Functional evolution of photochemical energy transformation in oxygen-producing organisms. Funct Plant Biol 36, 505-515.

[25] Shi, T. and Falkowski, P.G. (2008). Genome evolution in cyanobacteria: the stable core and the variable shell. Proc Natl Acad Sci U S A 105, 2510-5.

[26] Yanyushin, M.F., del Rosario, M.C., Brune, D.C. and Blankenship, R.E. (2005). New class of bacterial membrane oxidoreductases. Biochemistry 44, 10037-45.

[27] Gao, X., Xin, Y., Bell, P.D., Wen, J. and Blankenship, R.E. (2010). Structural analysis of alternative complex III in the photosynthetic electron transfer chain of Chloroflexus aurantiacus. Biochemistry 49, 6670-9.

[28] Lee, M., del Rosario, M.C., Harris, H.H., Blankenship, R.E., Guss, J.M. and Freeman, H.C. (2009). The crystal structure of auracyanin A at 1.85 A resolution: the structures and functions of auracyanins A and B, two almost identical "blue" copper proteins, in the photosynthetic bacterium Chloroflexus aurantiacus. J Biol Inorg Chem 14, 329-45.

[29] McManus, J.D., Brune, D.C., Han, J., Sanders-Loehr, J., Meyer, T.E., Cusanovich, M.A., Tollin, G. and Blankenship, R.E. (1992). Isolation, characterization, and amino acid sequences of auracyanins, blue copper proteins from the green photosynthetic bacterium *Chloroflexus aurantiacus*. J Biol Chem 267, 6531-40.

[30] Oyaizu, H., Debrunner-Vossbrinck, B., Mandelco, L., Studier, J.A. and Woese, C.R. (1987). The green non-sulfur bacteria: a deep branching in the eubacterial line of descent. Syst Appl Microbiol 9, 47-53.

[31] Woese, C.R. (1987). Bacterial evolution. Microbiol Rev 51, 221-71.

[32] Frigaard, N.-U. and Bryant, D.A. (2006). Chlorosomes: antenna organelles in photosynthetic green bacteria. *In* Shively JM (ed) Complex intracellular structures in prokaryotes. Springer, Berlin, pp 79–114.

[33] Tang, K.H., Urban, V.S., Wen, J., Xin, Y. and Blankenship, R.E. (2010). SANS investigation of the photosynthetic machinery of *Chloroflexus aurantiacus*. Biophys J 99, 2398-407.

[34] Tang, K.H., Tang, Y.J. and Blankenship, R.E. (2011). Carbon metabolic pathways in phototrophic bacteria and their broader evolutionary implications. Front Microbio 2, 165.

[35] Mathis, P. (1990). Compared structure of plant and bacterial photosynthetic reaction centers. Evolutionary implications. Biochim Biophys Acta 1018, 163-7.

[36] Blankenship, R.E. (2010). Early evolution of photosynthesis. Plant Physiol 154, 434-8.

[37] Mix, L.J., Haig, D. and Cavanaugh, C.M. (2005). Phylogenetic analyses of the core antenna domain: investigating the origin of photosystem I. J Mol Evol 60, 153-63.

[38] Bryant, D.A. et al. (2011). Comparative and functional genomics of anoxygenic green bacteria from the taxa *Chlorobi*, *Chloroflexi*, and *Acidobacteria*. *In* Burnap RL, and Vermaas W, (eds.), *Advances in Photosynthesis and Respiration, Functional Genomics and Evolution of Photosynthetic Systems*. Vol. 33, Springer, Dordrecht. in press.

[39] Frigaard, N.U., Chew, A.G.M., Maresca, J.A. and Bryant, D.A. (2006). Bacteriochlorophyll Biosynthesis in Green Bacteria. *In* Grimm B, Porra RJ, Rudiger W, and Scheer H (eds), *Chlorophylls and Bacteriochlorophylls*. Springer Academic Publishers, Dordrecht. pp 201-221

[40] Tsukatani, Y., Nakayama, N., Shimada, K., Mino, H., Itoh, S., Matsuura, K., Hanada, S. and Nagashima, K.V. (2009). Characterization of a blue-copper protein, auracyanin, of the filamentous anoxygenic phototrophic bacterium *Roseiflexus castenholzii*. Arch Biochem Biophys 490, 57-62.

[41] Wu, D. et al. (2009). Complete genome sequence of the aerobic CO-oxidizing thermophile *Thermomicrobium roseum*. PLoS One 4, e4207.

[42] Hanada, S., Takaichi, S., Matsuura, K. and Nakamura, K. (2002). *Roseiflexus castenholzii* gen. nov., sp. nov., a thermophilic, filamentous, photosynthetic bacterium that lacks chlorosomes. Int J Syst Evol Microbiol 52, 187-93.

[43] Refojo, P.N., Teixeira, M. and Pereira, M.M. (2010). The alternative complex III of *Rhodothermus marinus* and its structural and functional association with caa3 oxygen reductase. Biochim Biophys Acta 1797, 1477-82.

[44] Refojo, P.N., Sousa, F.L., Teixeira, M. and Pereira, M.M. (2010). The alternative complex III: a different architecture using known building modules. Biochim Biophys Acta 1797, 1869-76.

[45] Pappas, C.T. et al. (2004). Construction and validation of the *Rhodobacter sphaeroides* 2.4.1 DNA microarray: transcriptome flexibility at diverse growth modes J Bacteriol 186, 4748-58.

[46] Gough, S.P., Petersen, B.O. and Duus, J.O. (2000). Anaerobic chlorophyll isocyclic ring formation in Rhodobacter capsulatus requires a cobalamin cofactor. Proc Natl Acad Sci U S A 97, 6908-13.

[47] Minamizaki, K., Mizoguchi, T., Goto, T., Tamiaki, H. and Fujita, Y. (2008). Identification of two homologous genes, chlAI and chlAII, that are differentially involved in isocyclic ring formation of chlorophyll a in the cyanobacterium Synechocystis sp. PCC 6803. J Biol Chem 283, 2684-92.

[48] Ouchane, S., Steunou, A.S., Picaud, M. and Astier, C. (2004). Aerobic and anaerobic Mg-protoporphyrin monomethyl ester cyclases in purple bacteria: a strategy adopted to bypass the repressive oxygen control system. J Biol Chem 279, 6385-94.

[49] Pinta, V., Picaud, M., Reiss-Husson, F. and Astier, C. (2002). Rubrivivax gelatinosus acsF (previously orf358) codes for a conserved, putative binuclear-iron-cluster-containing protein involved in aerobic oxidative cyclization of Mg-protoporphyrin IX monomethylester. J Bacteriol 184, 746-53.

[50] Tang, K.H., Wen, J., Li, X. and Blankenship, R.E. (2009). Role of the AcsF protein in Chloroflexus aurantiacus. J Bacteriol 191, 3580-7.

[51] Tottey, S., Block, M.A., Allen, M., Westergren, T., Albrieux, C., Scheller, H.V., Merchant, S. and Jensen, P.E. (2003). Arabidopsis CHL27, located in both envelope and thylakoid membranes, is required for the synthesis of protochlorophyllide. Proc Natl Acad Sci U S A 100, 16119-24.

[52] Li, Z., Wakao, S., Fischer, B.B. and Niyogi, K.K. (2009). Sensing and responding to excess light. Annu Rev Plant Biol 60, 239-60.

[53] Madigan, M.T., Petersen, S.R. and Brock, T.D. (1974). Nutritional studies on *Chloroflexus*, a filamentous photosynthetic, gliding bacterium. Arch Microbiol 100, 97-103.

[54] Pickett, M.W., Williamson, M.P. and Kelly, D.J. (1994). An enzyme and 13C-NMR of carbon metabolism in heliobacteria. Photosynth. Res. 41, 75-88.

[55] Tang, K.H., Yue, H. and Blankenship, R.E. (2010). Energy metabolism of *Heliobacterium modesticaldum* during phototrophic and chemotrophic growth. BMC Microbiol 10, 150.

[56] Holo, H. (1989). *Chloroflexus aurantiacus* secretes 3-hydroxypropionate, a possible intermediate in the assimilation of CO_2 and acetate. Arch Microbiol 151, 252-256.

[57] Zarzycki, J., Brecht, V., Muller, M. and Fuchs, G. (2009). Identifying the missing steps of the autotrophic 3-hydroxypropionate CO_2 fixation cycle in *Chloroflexus aurantiacus*. Proc Natl Acad Sci U S A 106, 21317-22.

[58] Strauss, G. and Fuchs, G. (1993). Enzymes of a novel autotrophic CO2 fixation pathway in the phototrophic bacterium *Chloroflexus aurantiacus*, the 3-hydroxypropionate cycle. Eur J Biochem 215, 633-43.

[59] Hugler, M. and Sievert, S.M. (2011). Beyond the Calvin cycle: autotrophic carbon fixation in the ocean. Ann Rev Mar Sci 3, 261-89.

[60] Berg, I.A., Kockelkorn, D., Buckel, W. and Fuchs, G. (2007). A 3-hydroxypropionate/4-hydroxybutyrate autotrophic carbon dioxide assimilation pathway in Archaea. Science 318, 1782-6.

[61] Berg, I.A., Keppen, O.I., Krasil'nikova, E.N., Ugol'kova, N.V. and Ivanovskii, R.N. (2005). Carbon metabolism of filamentous anoxygenic phototrophic bacteria of the family *Oscillochloridaceae*. Microbiology 74, 258-264.

[62] Seshadri, R. et al. (2005). Genome sequence of the PCE-dechlorinating bacterium *Dehalococcoides ethenogenes*. Science 307, 105-8.

[63] Tang, Y.J., Yi, S., Zhuang, W.Q., Zinder, S.H., Keasling, J.D. and Alvarez-Cohen, L. (2009). Investigation of carbon metabolism in *Dehalococcoides ethenogenes* strain 195 by use of isotopomer and transcriptomic analyses. J Bacteriol 191, 5224-31.

[64] West, K.A. et al. (2008). Comparative genomics of *Dehalococcoides ethenogenes* 195 and an enrichment culture containing unsequenced *Dehalococcoides* strains. Appl Environ Microbiol 74, 3533-40.

[65] Tamura, K., Peterson, D., Peterson, N., Stecher, G., Nei, M. and Kumar, S. (2011). MEGA5: Molecular Evolutionary Genetics Analysis using Maximum Likelihood, Evolutionary Distance, and Maximum Parsimony Methods. Mol Biol Evol, in press.

16

Mechanism of Cargo Recognition During Selective Autophagy

Yasunori Watanabe[1,2] and Nobuo N. Noda[2]
[1]*Graduate School of Life Science, Hokkaido University*
[2]*Institute of Microbial Chemistry, Tokyo*
Japan

1. Introduction

Autophagy is an intracellular bulk degradation system conserved among eukaryotes from yeast to mammals. It is responsible for the degradation of cytosolic components and organelles in response to nutrient deprivation. There are three main types of autophagy: macroautophagy, microautophagy and chaperone-mediated autophagy (CMA). Microautophagy sequesters cytoplasmic components and delivers them for degradation by direct invagination or protrusion/septation of the lysosomal or vacuolar membrane (Mijaljica et al., 2011; Uttenweiler and Mayer, 2008). CMA targets specific cytosolic proteins that are trapped by the heat shock cognate protein of 70 kDa (hsc70) and, through interaction with lysosome-associated membrane protein type 2A (LAMP-2A), they are then translocated into the lysosomal lumen for rapid degradation (Orenstein and Cuervo, 2010). Macroautophagy, hereafter referred to as autophagy, is the most well characterized process of the three. During autophagy, double membrane structures called autophagosomes sequester a portion of the cytoplasm and fuse with the lysosome (or vacuole in the case of yeast and plants) to deliver their inner contents into the organelle lumen (Mizushima, 2007; Mizushima et al., 2010). Analyses of autophagy-related (Atg) proteins have unveiled dynamic and diverse aspects of mechanisms that underlie membrane formation during autophagy (Mizushima et al., 2010; Nakatogawa et al., 2009). As the contents of autophagosomes are indistinguishable from their surrounding cytoplasm (Baba et al., 1994), autophagy has long been considered a nonselective catabolic pathway. Recent studies, however, have provided evidence for the selective degradation of various targets by autophagy. In autophagy-deficient neuronal cells, intracellular protein aggregates accumulate and eventually lead to neurodegeneration, suggesting that autophagy selectively degrades harmful protein aggregates (Hara et al., 2006; Komatsu et al., 2006). Damaged or superfluous organelles, such as mitochondria and peroxisomes, and even intracellular infectious pathogens are also selectively degraded by autophagy (Goldman et al., 2010; Gutierrez et al., 2004; Manjithaya et al., 2010; Nakagawa et al., 2004; Noda and Yoshimori, 2009). In the budding yeast *Saccharomyces cerevisiae*, α-mannosidase and aminopeptidase I are selectively transported to the vacuole through autophagic pathways (Baba et al., 1997; Hutchins and Klionsky, 2001).

Although the precise molecular mechanisms of cargo selection by autophagy are yet to be established, an increasing number of autophagic receptors that are responsible for recognition of specific cargoes have been identified. These include Atg19 and Atg34 in the selective transport of vacuolar enzymes to the vacuole through autophagy (Leber et al., 2001; Scott et al., 2001; Suzuki et al., 2010), p62 and neighbor of BRCA1 gene 1 (NBR1) in the autophagic degradation of ubiquitinated protein aggregates (Bjorkoy et al., 2005; Kirkin et al., 2009), PpAtg30 in pexophagy (autophagic degradation of peroxisome) (Farre et al., 2008), and Atg32 and Nix1 in mitophagy (autophagic degradation of mitochondria) (Kanki et al., 2009; Novak et al., 2010; Okamoto et al., 2009). Most of these receptors interact directly with Atg8-family proteins, which are crucial factors in autophagosome biogenesis.

We have been studying the mechanisms of specific cargo recognition during autophagy, especially those of the selective delivery of vacuolar enzymes into the vacuole in yeast. We summarize here the current knowledge of such mechanisms as revealed by biochemical and structural studies.

2. Recognition of vacuolar enzymes by Atg19 and Atg34

2.1 Selective transport of vacuolar enzymes by autophagic pathways

In the budding yeast *S. cerevisiae*, α-mannosidase (Ams1) and a precursor form of aminopeptidase I (prApe1) are selectively delivered into the vacuole through the cytoplasm to vacuole targeting (Cvt) pathway under vegetative conditions, and via autophagy under starvation conditions. The Cvt pathway is topologically and mechanistically similar to autophagy (Lynch-Day and Klionsky, 2010); therefore, studies on the molecular mechanisms of cargo recognition in the Cvt pathway will provide insight into the basic mechanism of selective autophagy. prApe1, the primary Cvt cargo, is synthesized in the cytosol as a precursor form with a cleavable propeptide consisting of 45 amino acid residues at the N terminus (Klionsky et al., 1992) and assembles into a dodecamer. The prApe1 dodecamer further self-assembles into a higher order structure called the Ape1 complex. The existence of a specific receptor for prApe1 was proposed when it was observed that prApe1 transport to the vacuole by the Cvt pathway is both specific and saturable.

Two groups simultaneously discovered that Atg19 has all of the characteristics needed to be a receptor for prApe1 in Cvt transport (Leber et al., 2001; Scott et al., 2001). Characterization of the protein revealed that Atg19 is needed for the stabilization of prApe1 binding to the Cvt vesicle membrane, and that in *atg19Δ* cells, prApe1 maturation is inhibited while autophagy is not affected (Suzuki et al., 2002). In addition, Atg19 binds to prApe1 in a propeptide-dependent manner, suggesting that the propeptide region is responsible for the recognition of prApe1 by the Cvt pathway machinery (Shintani et al., 2002). A secondary-structure prediction suggested that the prApe1 propeptide forms a helix-turn-helix structure and that the first helix exhibits the characteristics of an amphipathic α helix (Martinez et al., 1997). Our previous study revealed that the region containing the first helix of the prApe1 propeptide (residues 1-20) is sufficient for interaction with Atg19 (Watanabe et al., 2010). This is consistent with a previous report showing that the first helix of the prApe1 propeptide is critical for prApe1processing (Oda et al., 1996). *In vitro* pull-down assays showed that the coiled coil domain of Atg19 (residues 124-253), which contains a predicted coiled coil between amino acids 160 and 187, directly interacts with the prApe1 propeptide. This is consistent with a previous report showing that the prApe1-binding site of Atg19 is located in the region between amino acid residues 153 and 191 (Shintani et al., 2002).

Ams1, another Cvt cargo, oligomerizes after synthesis and associates with the Ape1 complex through the action of Atg19. Atg19 has two stable domains, the N-terminal domain (residues 1-123) and the Ams1 binding domain (ABD; residues 254-367, see below for further details). Ams1 associates with Atg19 via the ABD that is distinct from the prApe1 binding site and therefore Atg19 can simultaneously interact with both prApe1 and Ams1. prApe1, Ams1, and Atg19 assemble into a large complex called the Cvt complex, which was identified as an electron-dense structure localized close to the vacuole by electron microscopy (Baba et al., 1997). Atg11 interacts with Atg19 to recruit the Cvt complex to the preautophagosomal structure (PAS), which plays a central role in autophagosome formation near the vacuole (Shintani et al., 2002; Suzuki and Ohsumi, 2010). Atg19 further interacts with Atg8, which is localized at the PAS and involved in the elongation of autophagosomes, using the Atg8 family-interacting motif (AIM; 412-WEEL-415) to induce formation of the Cvt vesicle (Noda et al., 2008). Atg8 is conjugated to phosphatidylethanolamine (PE) and associates with autophagosomes or the Cvt vesicle (Ichimura et al., 2000). This explains why the vesicle selectively surrounds only the cargo. After transport to the vacuole, the prApe1 propeptide is removed via a proteinase B-dependent reaction to generate mature Ape1 (mApe1), and the Ape1 complex disassembles back into dodecamers. Atg34, an Atg19 paralog, functions as an additional receptor protein for Ams1 but not prApe1 only under starvation conditions (Suzuki et al., 2010). Although Atg34, similar to Atg19, has the predicted coiled coil (residues 130-157), Atg34 is not capable of interacting with prApe1.

Recently, two cargoes that are selectively delivered to the vacuole have been identified: leucine aminopeptidase III (Lap3) (Kageyama et al., 2009) and aspartyl aminopeptidase (Ape4) (Yuga et al., 2011). Lap3 is transported to the vacuole for degradation only when it is overproduced under nitrogen starvation conditions. Lap3 forms a homohexameric complex of ~220 kDa, which further forms an aggregate independently of prApe1. Although this transport is partially mediated by Atg19, it remains to be determined whether Lap3 can interact with Atg19. Ape4 is the third Cvt cargo, which is similar in primary structure and subunit organization to Ape1. Ape4 lacks the N-terminal propeptide that is used by prApe1 for binding to Atg19. As the Ape4-binding site in Atg19 is located between the prApe1- and Ams1-binding sites (residues 204-247), these enzymes are unlikely to compete with each other for binding to Atg19. As Atg34 did not interact with Ape4, it might not be involved in Ape4 transport. More recently, Suzuki *et al.* elucidated that selective autophagy downregulates Ty1 transposition by eliminating Ty1 virus-like particles (VLPs) from the cytoplasm under nutrient-limited conditions (Suzuki et al., 2011). Although Ty1 VLPs are not vacuolar enzymes, they are targeted to autophagosomes by an interaction with Atg19. The N-terminal domain of Atg19 is specifically required for selective transport of Ty1 VLPs to the vacuole, though Atg19 is able to interact with Ty1 Gag without the N-terminal domain. Selective autophagy might safeguard genome integrity against excessive insertional mutagenesis caused during nutrient starvation by transposable elements in eukaryotic cells.

2.2 Structural basis for Ams1 recognition by Ag19 and Atg34

Scott *et al.* suggested that Ams1 is delivered to the vacuole in an Atg19-dependent manner (Scott et al., 2001). Ams1 was found to associate with Atg19, and a defect in the Ape1-Atg19 complex formation was shown to severely affect the import of Ams1 into the vacuole, whereas Ams1 was dispensable for transport of the Ape1-Atg19 complex. This suggests that Ams1 might exploit the prApe1 import system to achieve its own effective transport to the vacuole and that it is tethered to the Ape1-Atg19 complex through interaction with Atg19. In our recent study, we identified the Ams1 binding domain (ABD) in Atg19 and Atg34 by limited proteolysis of full-length Atg19, an *in vitro* pull-down assay as well as sequence alignment (Watanabe et al., 2010). In *atg19Δ* cells expressing Atg19$^{\Delta ABD}$, Ams1 transport to the vacuole was inhibited, suggesting that the Atg19 ABD is required for Ams1 transport to the vacuole through the Cvt pathway. In such cells, prApe1 transport to the vacuole is the normal process. These results indicate that the Atg19 ABD is specifically responsible for the transport of Ams1, but not prApe1, to the vacuole through the Cvt pathway.

The Atg19 and Atg34 ABD structures were determined in solution using NMR spectroscopy (Figure 1A and B) (Watanabe et al., 2010). Both ABDs comprise eight β-strands (A-H), of which A, B, E, and H form an antiparallel β-sheet; the surface of this sheet faces a second antiparallel β-sheet comprising C, D, F, and G, thus forming a typical immunoglobulin-like β-sandwich fold. The Atg19 and Atg34 ABD structures are similar to each other with a root mean square difference of 2.1 Å for 102 residues (Z-score calculated by the Dalilite program (Holm and Park, 2000) is 12.8). There are relatively large structural differences between the Atg19 and Atg34 ABDs in the loops located at the bottom of the immunoglobulin fold (the loop connecting strands A and B (AB loop), CD, EF, and GH loops). In contrast, the loops located at the top of the immunoglobulin fold (the BC, DE, and FG loops) have a similar conformation. Furthermore, the residues comprising the top loops, especially those of the DE loop, are more strongly conserved between the Atg19 and Atg34 ABDs than those comprising the bottom loops (Figure 1). In the DE loop, His-310/296, Glu-311/297, Ile-314/300, and Lys-315/301 of Atg19/Atg34 are exposed. Among these exposed residues, His-310/296 and/or Glu-311/297 of the Atg19/Atg34 ABD are essential for Ams1 recognition. Further analysis showed that in *atg19Δatg34Δ* cells expressing Atg19^{H310A} (substitution of His-310 with alanine) but not Atg19^{E311A}, transport of Ams1-GFP to the vacuole under autophagy-inducing conditions is inhibited. This indicates that the conserved His residue in the DE loop of the Atg19 ABD plays a critical role in Ams1 recognition and that Ams1 binding of the Atg19 ABD is essential for Ams1 transportation to the vacuole. Similar experiments using Atg34 mutants showed that His-296 of Atg34 ABD, which corresponds to His-310 of Atg19 ABD, also plays a critical role in Ams1 recognition.

The ABDs in Atg19 and Atg34 have a β-sandwich fold that is observed in a variety of immunoglobulins and immunoglobulin-like domains responsible for recognizing various proteins. Because antibodies generally recognize antigens using the hypervariable loops from both the VH and VL regions, their manner of antigen binding should differ from that of monomeric ABDs with Ams1. Interestingly, however, the ABD-Ams1 interaction resembles that observed between camelid antibody fragments and their antigens, as camelid antibodies lack a light chain and function as a monomer where hypervariable loops of the VH are responsible for antigen binding (Muyldermans, 2001). It also mimics the interaction of monobodies (artificially designed proteins that use a fibronectin type III domain as a scaffold) and their targets, as monobodies interact with their targets using similar loops in a monomeric immunoglobulin fold. Camelid antibody fragments and monobodies interact with their target proteins using loops clustered at one side of their immunoglobulin fold; these loops are topologically equivalent to the BC, DE, and FG loops of the Atg19 and Atg34 ABDs, one of which was shown to be crucial for Ams1 recognition as mentioned above. Therefore, they might recognize Ams1 using these loops in a similar manner with camelid antibodies and monobodies. In order to further elucidate the recognition mechanism of Ams1 by the ABD, structural determination of the Ams1-ABD complex by X-ray crystallography is needed. We have already succeeded in overexpressing *S. cerevisiae* Ams1 in *Pichia pastoris* and purifying it on a large scale (Watanabe et al., 2009). Crystallization and structural determination of the Ams1-ABD complex are now in progress.

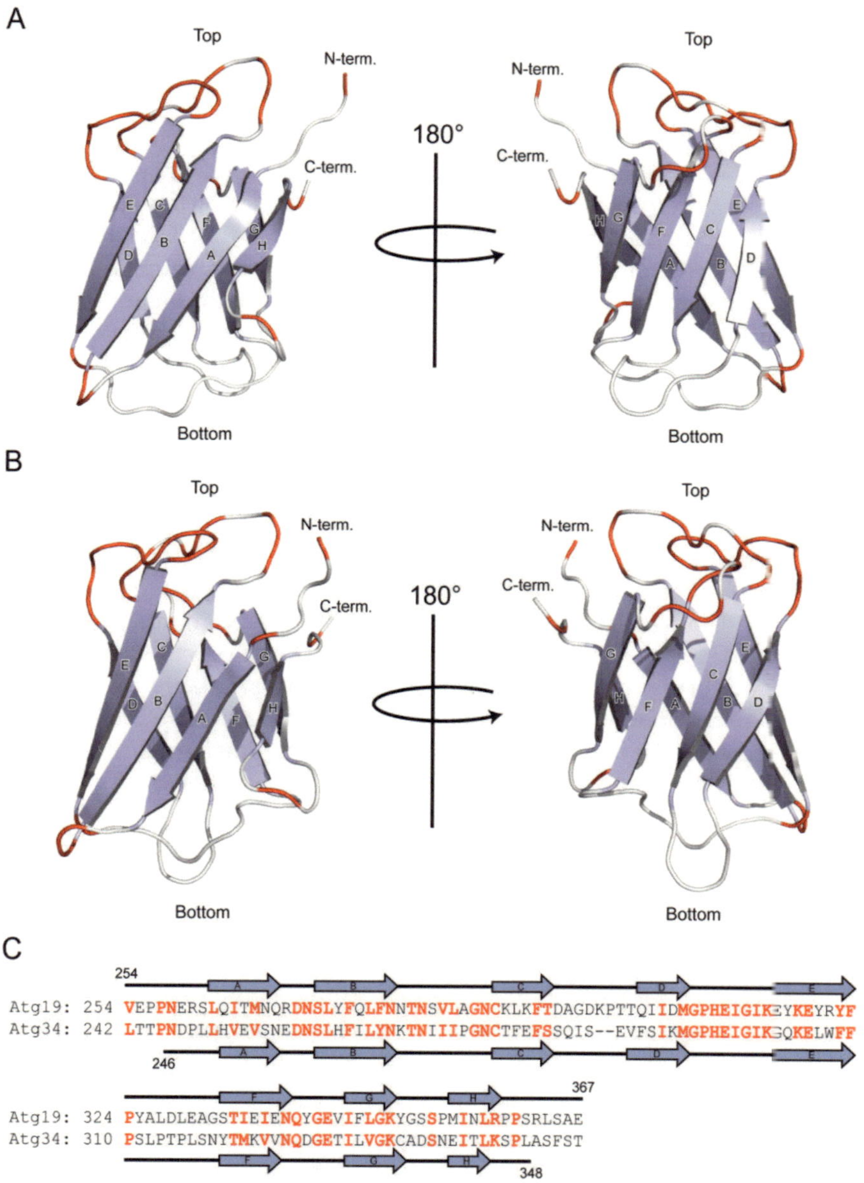

Fig. 1. Solution structures of Atg19 and Atg34 ABDs. (A), (B) Ribbon diagrams of the Atg19 ABD and Atg34 ABD structures, respectively. Strands are colored *light blue* and labeled. Loop residues conserved between Atg19 and Atg34 are colored *red*. *Left* and *right* are related by a 180° rotation along the *vertical axis*. (C) Sequence alignment between Atg19 and Atg34 ABDs. *Gaps* are introduced to maximize the similarity. Conserved or type-conserved residues are colored *red*. Secondary structure elements of the Atg19 and Atg34 ABDs are shown above and below the sequence, respectively.

3. Receptor proteins required for the selective degradation of organelles by autophagy

Damaged or superfluous organelles, such as mitochondria and peroxisomes, are also selectively degraded by autophagy. To date, several receptor proteins which function in these selective types of autophagy have been identified.

3.1 Receptor proteins in mitophagy

The mitochondrion is an organelle that produces energy through oxidative phosphorylation and simultaneously generates reactive oxygen species (ROS), causing oxidative damage to mitochondrial DNA, protein and lipids, and often inducing cell death. Therefore, an appropriate quality control of mitochondria is important to maintain proper cellular homeostasis. Selective degradation of mitochondria via autophagy, known as mitophagy, is the primary mechanism for mitochondrial quality control. Two groups independently identified Atg32 as a receptor protein for mitophagy in yeast (Kanki et al., 2009; Okamoto et al., 2009). Atg32 is a single-pass mitochondrial outer membrane protein, and its N- and C-terminal domains are oriented towards the cytoplasm and the intermembrane space, respectively. Mitochondria-anchored Atg32 binds Atg11 during mitophagy to recruit mitochondria to the PAS. When mitophagy is induced, Atg32 is phosphorylated, for which Ser-114 and Ser-119 of Atg32 are required. , The phosphorylation of Atg32 is required for Atg32-Atg11 interaction and mitophagy (Aoki et al., 2011). By controlling the activity and/or localization of the kinase that phosphorylates Atg32, cells may regulate the amount of mitochondria or remove damaged or aged mitochondria through mitophagy. Similarly to other receptor proteins, Atg32 binds to Atg8 using the AIM sequence, 86-WQAI-89, and this binding is required for the efficient sequestration of mitochondria by the autophagosome. The Atg32-Atg8 interaction may restrict autophagosome formation to the neighborhood of the targeted mitochondrion by gathering surrounding Atg proteins. Although no homolog of Atg32 has been identified in mammals, Nix was recently shown to be a mammalian mitophagy receptor protein (Novak et al., 2010). Similarly to Atg32, Nix has a transmembrane domain in its C terminus that can target the protein to the mitochondrial outer membrane and has a functional AIM, 35-WVEL-38, that directly interacts with mammalian Atg8 homologs. Thus Nix may fulfill the function of Atg32 in mammals.

3.2 Receptor protein in pexophagy

Peroxisomes have diverse functions, including the decomposition of hydrogen peroxide and the oxidation of fatty acids. The specific degradation of peroxisomes by autophagy, pexophagy, is conserved from yeast to humans and is triggered physiologically to allow cells to clear an excess of peroxisomes. The study of methylotrophic yeasts, particularly *P. pastoris* and *Hansenula polymorpha*, has led to the current understanding of the molecular mechanism governing pexophagy. Farre *et al.* identified Atg30 as a receptor protein for pexophagy in *P. pastoris* (Farre et al., 2008). PpAtg30 interacts with peroxisomes via two peroxisomal membrane proteins, Pex3 and Pex14, and with autophagy machinery via PpAtg11 and PpAtg17, which organize the PAS. Several residues on PpAtg30 are phosphorylated under pexophagy conditions and, similarly to Atg32, such phosphorylation, especially that of Ser-112, is required for PpAtg11 interaction. The isolation membrane then expands and surrounds the PpAtg30-localizing peroxisomes, in order to selectively degrade surplus peroxisomes. Unlike other receptor proteins, PpAtg30 has no AIMs, so that PpAtg30 is unable to interact directly with PpAtg8. It is important to understand how the isolation membrane expands around the peroxisome surface while excluding cytosolic contents, and this is speculated to involve the interaction of PpAtg30 with an unidentified protein in the isolation membrane, or the interaction of PpAtg8 with another protein on the peroxisome surface.

4. Receptor protein for selective autophagy in *C. elegans*

Germ granules are restricted to the germ cells of many higher eukaryotes and are believed to carry germ cell determinants (Strome and Lehmann, 2007). Germ granules in *Caenorhabditis elegans,* also known as P granules, are maternally contributed and dispersed throughout the cytoplasm of a newly fertilized embryo. During *C. elegans* embryogenesis, some P granules are left in the cytoplasm destined for the somatic daughter cell and these P granules are quickly disassembled and/or degraded (Hird et al., 1996). Recently, Zhang *et al.* provided evidence that the P granule components PGL-1 and PGL-3 that remain in the cytoplasm destined for somatic daughters are selectively removed by autophagy through the receptor protein SEPA-1 (Zhang et al., 2009). In autophagy-deficient somatic cells, PGL-1 and PGL-3 extensively accumulate in the P granules, and SEPA-1 mediates the accumulation of these P granules into aggregates, termed PGL granules, through its self-oligomerization and direct interaction with PGL-3. SEPA-1 can also directly interact with LGG-1, an Atg8 homolog. Thus, PGL granules associated with SEPA-1 could be incorporated into autophagosomes through the interaction of SEPA-1 with LGG-1. Because the expression of SEPA-1 is restricted to somatic cells, the selective exclusion of P granules is ensured only in these cells. An *in vitro* pull-down assay showed that the SEPA-1 fragment containing amino acids 39 to 160 is required for both self-oligomerization and interaction with PGL-3 and that the SEPA-1 fragment containing amino acids 289 to 575 is required for interaction with LGG-1. The SEPA-1 fragment that interacts with LGG-1 contains a canonical AIM sequence, 469-YQEL-472 (Noda et al., 2010). Thus, the YQEL sequence in SEPA-1 is a potential candidate for a functional AIM.

5. Recognition of ubiquitinated cargoes by p62 and NBR1

Autophagic degradation of ubiquitinated protein aggregates is important for cell survival. Defects in autophagy cause the accumulation of ubiquitin-positive protein inclusions, leading to severe liver injury (Komatsu et al., 2005) and neurodegeneration (Hara et al., 2006; Komatsu et al., 2006). The polyubiquitin-binding protein p62, also called sequestosome 1 (SQSTM1), is a common component of protein aggregates found in both the brain and the liver of patients suffering from protein aggregation diseases. These include Lewy bodies in Parkinson's disease, neurofibrillary tangles in Alzheimer's disease, and huntingtin aggregates (Kuusisto et al., 2001, 2002; Nagaoka et al., 2004; Zatloukal et al., 2002). In the liver, Mallory bodies, hyaline bodies in hepatocellular carcinoma, and α1 antitrypsin aggregate contain p62 (Zatloukal et al., 2002); all of these aggregates contain polyubiquitinated proteins. p62 interacts with ubiquitin via its C-terminal UBA domain (Vadlamudi et al., 1996) and self-assembles via its N-terminal PB1 domain (Ponting et al., 2002), thereby forming large aggregates containing ubiquitinated proteins. p62 further interacts with LC3, a mammalian homolog of Atg8, via the LC3 interacting region (LIR;

residues 321-342), so that ubiquitinated protein aggregates containing p62 are selectively degraded by autophagy (Bjorkoy et al., 2005; Komatsu et al., 2007; Pankiv et al., 2007). Therefore, it is implied that p62 functions as a receptor protein for ubiquitinated proteins to be degraded in lysosomes. It is also hypothesized that p62 functions as a receptor for organelles such as peroxisomes and mitochondria (Kim et al., 2008; Kirkin et al., 2009), and for intracellular bacteria (Dupont et al., 2009; Yoshikawa et al., 2009; Zheng et al., 2009). Recently, neighbor of BRCA1 gene 1 (NBR1) has been identified as another autophagy receptor (Kirkin et al., 2009). The structure of NBR1 is similar to that of p62, and NBR1 can bind both LC3 via the LIR and ubiquitinated proteins via the UBA domain. Like p62, NBR1 is sequestered into the autophagosome via LC3-interaction and/or p62-interaction and markedly accumulates in autophagy-deficient tissues.

To clarify the molecular mechanism of ubiquitinated cargo recognition by p62 and NBR1, it is necessary to elucidate which proteins conjugated with either K48-linked or K63-linked polyubiquitin chains are targeted to the autophagy/lysosomal degradation pathway. Classically, proteins conjugated with K48-linked polyubiquitin chains are recognized as the proteolytic substrate by the UBD-containing proteasomal receptors. Recently, K63-linked chains have been implicated in proteolytic degradation of misfolded and aggregated proteins (Olzmann et al., 2007; Tan et al., 2008; Wooten et al., 2008). Given the reported preference of the known ubiquitin-binding autophagy receptors for K63-linked ubiquitin chains, cargoes conjugated with K63-linked ubiquitin chains may be preferentially targeted to the autophagy/lysosomal degradation pathway. However, p62 has been shown to compete for ubiquitinated cargo with the classical proteasomal receptors. Accumulation of p62 resulting from inhibition of autophagy compromised degradation of proteasomal substrates, most likely due to the excessive interaction between p62 and substrates conjugated with K48-linked polyubiquitin chains (Korolchuk et al., 2009). It remains to be clarified how p62 distinguishes between K48-linked and K63-linked polyubiquitin chains.

To date, several structural studies have been performed on p62. Isogai *et al.* determined the crystal structure of the UBA domain of mouse p62 and the solution structure of its ubiquitin-bound form (Isogai et al., 2011). In crystals, the p62 UBA domain adopts a dimeric structure, which is distinct from that of other UBA domains. In solution, the domain exists in equilibrium between the dimer and monomer forms, and ubiquitin-binding shifts the equilibrium toward the monomer to form a 1:1 complex between the UBA domain and ubiquitin. The extreme C-terminal end of the p62 UBA domain is responsible for dimerization of the domain. Mutations that inhibit dimerization of the p62 UBA domain increase the affinity of p62 for ubiquitin. These results suggest an autoinhibitory mechanism in the p62 UBA domain to avoid self-degradation by the ubiquitin-proteasomal system. The interaction between the p62 LIR and LC3 is structurally and functionally well characterized (Ichimura et al., 2008; Noda et al., 2008). In the structure of the p62 LIR in complex with LC3, the tryptophan and leucine residues in the p62 LIR, DDDWTHL, interact with two hydrophobic pockets, the W-site and the L-site, where tryptophan and leucine residues respectively interact (Noda et al., 2010) on the surface of LC3. In agreement with the structure of the p62 LIR in complex with LC3, the tryptophan and leucine residues are involved in the turnover of p62 via autophagy. Recently, the structure of the NBR1 LIR in complex with GABARAP, a LC3 paralog, has also been determined (Rozenknop et al., 2011). Similar to the interaction between p62 LIR and LC3, the tyrosine and the third isoleucine residues in the NBR1 LIR, EDYIII, interact with GABARAP in a manner typical of the interaction of AIM with Atg8 homologs.

6. Receptor proteins required for the restriction of infectious bacterial growth by autophagy

Autophagy also serves as a cell-autonomous effector mechanism of innate immunity in the cytosol. It does this through restricting bacterial proliferation by separating bacteria from the nutrient-rich cytosol and delivering them into bactericidal autolysosomes. Several examples showing that these types of autophagy are mediated by cytosolic bacteria-recognizing receptor proteins have been recently reported. In *Drosophila melanogaster*, PGRP-LE, a receptor protein for bacterial peptidoglycans, induces autophagy of wild-type but not listeriolysin-deficient *Listeria monocytogenes*, suggesting that this pathway specifically selects cytosolic bacteria for autophagy (Yano et al., 2008). However, it is unknown how PGRP-LE induces the autophagic degradation of *L. monocytogenes* and whether PGRP-LE binds the *D. melanogaster* Atg8 orthologs.

Salmonella enterica Typhimurium (*S.* Typhimurium) typically occupies a membrane bound compartment, the Salmonella-containing vacuole (SCV), in host cells. In mammalian cells, *S.* Typhimurium and other bacteria enter the cytosol and are released from SCVs, then become coated with a dense layer of ubiquitin (Perrin et al., 2004) and are delivered to lysosomes via autophagy (Birmingham et al., 2006). It was reported that p62 functions as a receptor protein for delivering such ubiquitin-coated bacteria into autophagosomes (Dupont et al., 2009; Zheng et al., 2009). In addition to p62, two other receptor proteins were identified, NDP52 (Thurston et al., 2009) and Optineurin (Wild et al., 2011), both of which recognize ubiquitin-coated bacteria. NDP52 is recruited by ubiquitin-coated *S.* Typhimurium and binds both ubiquitin via a zinc-finger domain (residues 420-446) and LC3. Although NDP52 interacts with LC3, it has not been clarified whether NDP52 has an AIM. NDP52 also coordinates a signaling complex including Tank-binding kinase (TBK1), Sintbad and Nap1. *In vitro* binding studies revealed that a SKICH domain (residues 1-127) in NDP52 is required for direct Nap1 binding. Thereby, NDP52 recruits TBK1 to ubiquitin-coated *S.* Typhimurium. The ability of NDP52 to serve as an adaptor for TBK1 also seems to be critical in the cell-autonomous response, but it remains to be determined what role TBK1 plays in association with NDP52. OPTN is another autophagy receptor protein that binds and co-localizes with LC3 via an AIM (178-FVEI-181) and ubiquitin via its ubiquitin binding in ABIN and NEMO (UBAN) domains. The N-terminal region of OPTN (residues 1-127) also interacts with TBK1 (Morton et al., 2008). When TBK1 is recruited into *S.* Typhimurium via OPTN, it becomes activated and phosphorylates OPTN at Ser-177, one residue N-terminal to the OPTN AIM. The phosphorylation of Ser-177 increases the affinity between OPTN and LC3, which is consistent with the previous review showing that acidic residues are preferred at the N-terminal side of the AIM for higher affinity with Atg8 homologs (Noda et al., 2010). Although p62, NDP52 and OPTN target the same bacteria, NDP52 and OPTN localize to microdomains on the surface of ubiquitinated bacteria where p62 does not co-localize (Cemma et al., 2011; Wild et al., 2011). Because these autophagy receptors have their respective ubiquitin-binding domains, the distinct specificities for different ubiquitin chains may result in partitioning of the receptors to different subdomains on the bacterium. However, it is still unknown which type of ubiquitin chains are conjugated to the bacterial surface components.

7. Concluding remarks

Autophagy receptor proteins have two main functions: recognizing autophagic cargoes, and interacting with Atg8 homologs. Because of these functions, autophagy receptor proteins can tether autophagic cargoes to the isolation membrane (e.g., Atg19; Figure 2) so that the cargoes are selectively and efficiently engulfed by an autophagosome and transported into the vacuole/lysosome. Although the mechanism of direct interaction with Atg8 homologs

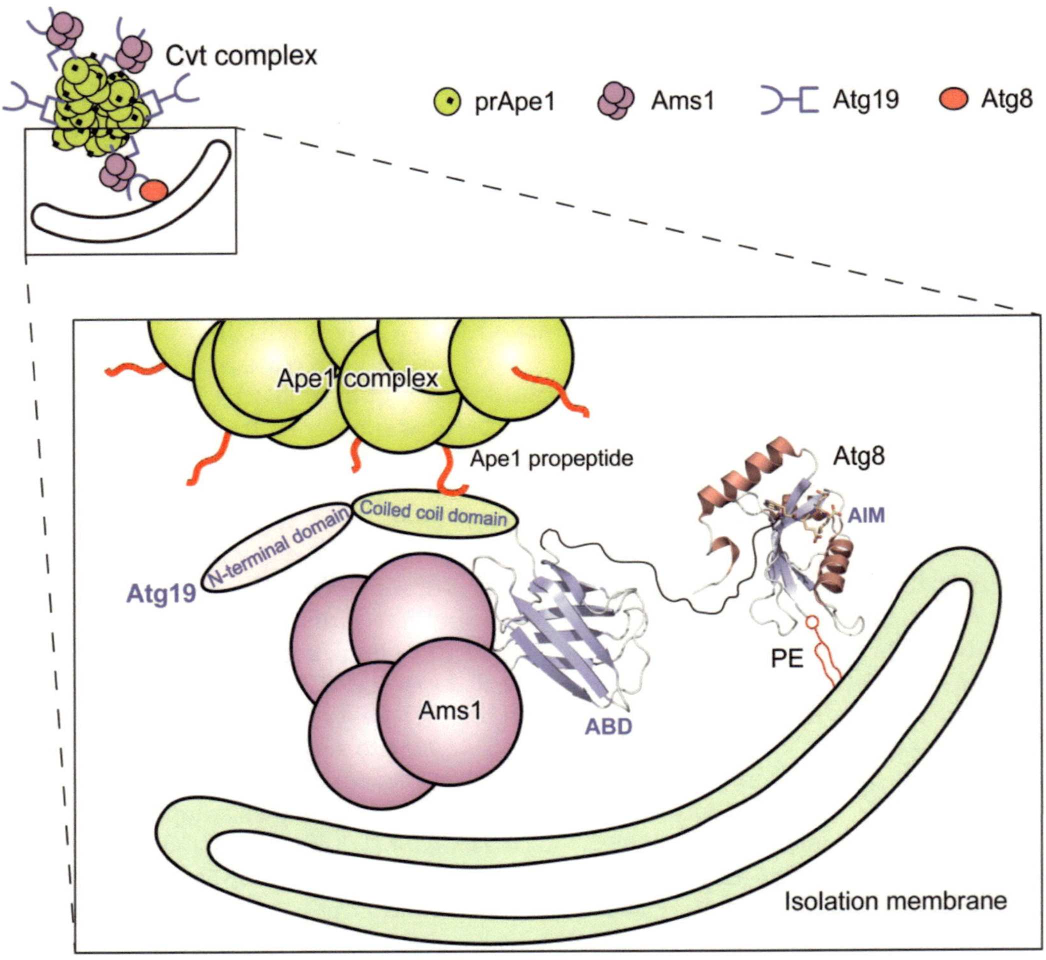

Fig. 2. Atg19 tethers autophagic cargoes to the isolation membrane. Atg19 interacts with prApe1 and Ams1 via the coiled coil domain and ABD, respectively. Atg19 also interacts with Atg8, which is conjugated to phosphatidylethanolamine (PE) and associates with the isolation membrane, via the AIM. Therefore, autophagic cargoes are tethered to the isolation membrane.

via the AIM is common among most autophagy receptor proteins, the recognition mechanisms of autophagic cargoes by autophagy receptor proteins are too divergent to be elucidated. We reported that Atg19 and Atg34 ABDs, similarly to the camelid antibody and monobody, recognize Ams1, an autophagic cargo, using the loops clustered at one side of their immunoglobulin fold. Recent proteomics analysis has identified proteins that are selectively degraded by autophagy (Onodera and Ohsumi, 2004). Although the recognition mechanism of these target proteins by autophagy has not been established, autophagy-specific receptor proteins possessing an ABD-like fold might be responsible. Identification and structural analysis of other autophagy receptor proteins are required for further clarification of the molecular mechanism of specific cargo recognition during autophagy.

8. Acknowledgment

This work was supported by Grants-in-Aids for Young Scientists (A) from the Ministry of Education, Culture, Sports, Science and Technology of Japan and for JSPS Fellows from Japan Society for the Promotion of Science.

9. References

Aoki, Y., Kanki, T., Hirota, Y., Kurihara, Y., Saigusa, T., Uchiumi, T. & Kang, D. (2011). Phosphorylation of Serine 114 on Atg32 mediates mitophagy. *Mol Biol Cell*, Vol.22, No.17, Sep, pp. 3206-3217, 1059-1524

Baba, M., Osumi, M., Scott, S.V., Klionsky, D.J. & Ohsumi, Y. (1997). Two distinct pathways for targeting proteins from the cytoplasm to the vacuole/lysosome. *J Cell Biol*, Vol.139, No.7, Dec 29, pp. 1687-1695, 0021-9525

Baba, M., Takeshige, K., Baba, N. & Ohsumi, Y. (1994). Ultrastructural analysis of the autophagic process in yeast: detection of autophagosomes and their characterization. *J Cell Biol*, Vol.124, No.6, Mar, pp. 903-913, 0021-9525

Birmingham, C.L., Smith, A.C., Bakowski, M.A., Yoshimori, T. & Brumell, J.H. (2006). Autophagy controls Salmonella infection in response to damage to the Salmonella-containing vacuole. *J Biol Chem*, Vol.281, No.16, Apr 21, pp. 11374-11383, 0021-9258

Bjorkoy, G., Lamark, T., Brech, A., Outzen, H., Perander, M., Overvatn, A., Stenmark, H. & Johansen, T. (2005). p62/SQSTM1 forms protein aggregates degraded by autophagy and has a protective effect on huntingtin-induced cell death. *J Cell Biol*, Vol.171, No.4, Nov 21, pp. 603-614, 0021-9525

Cemma, M., Kim, P.K. & Brumell, J.H. (2011). The ubiquitin-binding adaptor proteins p62/SQSTM1 and NDP52 are recruited independently to bacteria-associated microdomains to target Salmonella to the autophagy pathway. *Autophagy*, Vol.7, No.3, Mar, pp. 341-345, 1554-8627

Dupont, N., Lacas-Gervais, S., Bertout, J., Paz, I., Freche, B., Van Nhieu, G.T., van der Goot, F.G., Sansonetti, P.J. & Lafont, F. (2009). Shigella phagocytic vacuolar membrane remnants participate in the cellular response to pathogen invasion and are regulated by autophagy. *Cell Host Microbe*, Vol.6, No.2, Aug 20, pp. 137-149, 1931-3128

Farre, J.C., Manjithaya, R., Mathewson, R.D. & Subramani, S. (2008). PpAtg30 tags peroxisomes for turnover by selective autophagy. *Dev Cell*, Vol.14, No.3, Mar, pp. 365-376, 1534-5807

Goldman, S.J., Taylor, R., Zhang, Y. & Jin, S. (2010). Autophagy and the degradation of mitochondria. *Mitochondrion*, Vol.10, No.4, Jun, pp. 309-315, 1567-7249

Gutierrez, M.G., Master, S.S., Singh, S.B., Taylor, G.A., Colombo, M.I. & Deretic, V. (2004). Autophagy is a defense mechanism inhibiting BCG and Mycobacterium tuberculosis survival in infected macrophages. *Cell*, Vol.119, No.6, Dec 17, pp. 753-766, 0092-8674

Hara, T., Nakamura, K., Matsui, M., Yamamoto, A., Nakahara, Y., Suzuki-Migishima, R., Yokoyama, M., Mishima, K., Saito, I., Okano, H., *et al.* (2006). Suppression of basal autophagy in neural cells causes neurodegenerative disease in mice. *Nature*, Vol.441, No.7095, Jun 15, pp. 885-889, 0028-0836

Hird, S.N., Paulsen, J.E. & Strome, S. (1996). Segregation of germ granules in living Caenorhabditis elegans embryos: cell-type-specific mechanisms for cytoplasmic localisation. *Development*, Vol.122, No.4, Apr, pp. 1303-1312, 0950-1991

Holm, L. & Park, J. (2000). DaliLite workbench for protein structure comparison. *Bioinformatics*, Vol.16, No.6, Jun, pp. 566-567, 1367-4803

Hutchins, M.U. & Klionsky, D.J. (2001). Vacuolar localization of oligomeric alpha-mannosidase requires the cytoplasm to vacuole targeting and autophagy pathway components in Saccharomyces cerevisiae. *J Biol Chem*, Vol.276, No.23, Jun 8, pp. 20491-20498, 0021-9258

Ichimura, Y., Kirisako, T., Takao, T., Satomi, Y., Shimonishi, Y., Ishihara, N., Mizushima, N., Tanida, I., Kominami, E., Ohsumi, M., *et al.* (2000). A ubiquitin-like system mediates protein lipidation. *Nature*, Vol.408, No.6811, Nov 23, pp. 488-492, 0028-0836

Ichimura, Y., Kumanomidou, T., Sou, Y.S., Mizushima, T., Ezaki, J., Ueno, T., Kominami, E., Yamane, T., Tanaka, K. & Komatsu, M. (2008). Structural basis for sorting mechanism of p62 in selective autophagy. *J Biol Chem*, Vol.283, No.33, Jun 4, pp. 22847-22857, 0021-9258

Isogai, S., Morimoto, D., Arita, K., Unzai, S., Tenno, T., Hasegawa, J., Sou, Y.S., Komatsu, M., Tanaka, K., Shirakawa, M., *et al.* (2011). Crystal Structure of the Ubiquitin-associated (UBA) Domain of p62 and Its Interaction with Ubiquitin. *J Biol Chem*, Vol.286, No.36, Sep 9, pp. 31864-31874, 0021-9258

Kageyama, T., Suzuki, K. & Ohsumi, Y. (2009). Lap3 is a selective target of autophagy in yeast, Saccharomyces cerevisiae. *Biochem Biophys Res Commun*, Vol.378, No.3, Jan 16, pp. 551-557, 1090-2104

Kanki, T., Wang, K., Cao, Y., Baba, M. & Klionsky, D.J. (2009). Atg32 is a mitochondrial protein that confers selectivity during mitophagy. *Dev Cell*, Vol.17, No.1, Jul, pp. 98-109, 1534-5807

Kim, P.K., Hailey, D.W., Mullen, R.T. & Lippincott-Schwartz, J. (2008). Ubiquitin signals autophagic degradation of cytosolic proteins and peroxisomes. *Proc Natl Acad Sci U S A*, Vol.105, No.52, Dec 30, pp. 20567-20574, 0027-8424

Kirkin, V., Lamark, T., Sou, Y.S., Bjorkoy, G., Nunn, J.L., Bruun, J.A., Shvets, E., McEwan, D.G., Clausen, T.H., Wild, P., *et al.* (2009). A role for NBR1 in autophagosomal degradation of ubiquitinated substrates. *Mol Cell*, Vol.33, No.4, Feb 27, pp. 505-516, 1097-4164

Klionsky, D.J., Cueva, R. & Yaver, D.S. (1992). Aminopeptidase I of Saccharomyces cerevisiae is localized to the vacuole independent of the secretory pathway. *J Cell Biol*, Vol.119, No.2, Oct, pp. 287-299, 0021-9525

Komatsu, M., Waguri, S., Chiba, T., Murata, S., Iwata, J., Tanida, I., Ueno, T., Koike, M., Uchiyama, Y., Kominami, E., *et al.* (2006). Loss of autophagy in the central nervous system causes neurodegeneration in mice. *Nature*, Vol.441, No.7095, Jun 15, pp. 880-884, 0028-0836

Komatsu, M., Waguri, S., Koike, M., Sou, Y.S., Ueno, T., Hara, T., Mizushima, N., Iwata, J., Ezaki, J., Murata, S., *et al.* (2007). Homeostatic levels of p62 control cytoplasmic inclusion body formation in autophagy-deficient mice. *Cell*, Vol.131, No.6, Dec 14, pp. 1149-1163, 0092-8674

Komatsu, M., Waguri, S., Ueno, T., Iwata, J., Murata, S., Tanida, I., Ezaki, J., Mizushima, N., Ohsumi, Y., Uchiyama, Y., *et al.* (2005). Impairment of starvation-induced and constitutive autophagy in Atg7-deficient mice. *J Cell Biol*, Vol.169, No.3, May 9, pp. 425-434, 0021-9525

Korolchuk, V.I., Mansilla, A., Menzies, F.M. & Rubinsztein, D.C. (2009). Autophagy inhibition compromises degradation of ubiquitin-proteasome pathway substrates. *Mol Cell*, Vol.33, No.4, Feb 27, pp. 517-527, 1097-2765

Kuusisto, E., Salminen, A. & Alafuzoff, I. (2001). Ubiquitin-binding protein p62 is present in neuronal and glial inclusions in human tauopathies and synucleinopathies. *Neuroreport*, Vol.12, No.10, Jul 20, pp. 2085-2090, 0959-4965

Kuusisto, E., Salminen, A. & Alafuzoff, I. (2002). Early accumulation of p62 in neurofibrillary tangles in Alzheimer's disease: possible role in tangle formation. *Neuropathol Appl Neurobiol*, Vol.28, No.3, Jun, pp. 228-237, 0305-1846

Leber, R., Silles, E., Sandoval, I.V. & Mazon, M.J. (2001). Yol082p, a novel CVT protein involved in the selective targeting of aminopeptidase I to the yeast vacuole. *J Biol Chem*, Vol.276, No.31, Aug 3, pp. 29210-29217, 0021-9258

Lynch-Day, M.A. & Klionsky, D.J. (2010). The Cvt pathway as a model for selective autophagy. *FEBS Lett*, Vol.584, No.7, Apr 2, pp. 1359-1366, 0014-5793

Manjithaya, R., Nazarko, T.Y., Farre, J.C. & Subramani, S. (2010). Molecular mechanism and physiological role of pexophagy. *FEBS Lett*, Vol.584, No.7, Apr 2, pp. 1367-1373, 0014-5793

Martinez, E., Jimenez, M.A., Segui-Real, B., Vandekerckhove, J. & Sandoval, I.V. (1997). Folding of the presequence of yeast pAPI into an amphipathic helix determines transport of the protein from the cytosol to the vacuole. *J Mol Biol*, Vol.267, No.5, Apr 18, pp. 1124-1138, 0022-2836

Mijaljica, D., Prescott, M. & Devenish, R.J. (2011). Microautophagy in mammalian cells: revisiting a 40-year-old conundrum. *Autophagy*, Vol.7, No.7, Jul, pp. 673-682, 1554-8627

Mizushima, N. (2007). Autophagy: process and function. *Genes Dev*, Vol.21, No.22, Nov 15, pp. 2861-2873, 0890-9369

Mizushima, N., Yoshimori, T. & Ohsumi, Y. (2010). The Role of Atg Proteins in Autophagosome Formation. *Annu Rev Cell Dev Biol*, Oct 29, pp., 1081-0706

Morton, S., Hesson, L., Peggie, M. & Cohen, P. (2008). Enhanced binding of TBK1 by an optineurin mutant that causes a familial form of primary open angle glaucoma. *FEBS Lett*, Vol.582, No.6, Mar 19, pp. 997-1002, 0014-5793

Muyldermans, S. (2001). Single domain camel antibodies: current status. *J Biotechnol*, Vol.74, No.4, Jun, pp. 277-302, 0168-1656

Nagaoka, U., Kim, K., Jana, N.R., Doi, H., Maruyama, M., Mitsui, K., Oyama, F. & Nukina, N. (2004). Increased expression of p62 in expanded polyglutamine-expressing cells and its association with polyglutamine inclusions. *J Neurochem*, Vol.91, No.1, Oct, pp. 57-68, 0022-3042

Nakagawa, I., Amano, A., Mizushima, N., Yamamoto, A., Yamaguchi, H., Kamimoto, T., Nara, A., Funao, J., Nakata, M., Tsuda, K., *et al.* (2004). Autophagy defends cells against invading group A Streptococcus. *Science*, Vol.306, No.5698, Nov 5, pp. 1037-1040, 0036-8075

Nakatogawa, H., Suzuki, K., Kamada, Y. & Ohsumi, Y. (2009). Dynamics and diversity in autophagy mechanisms: lessons from yeast. *Nat Rev Mol Cell Biol*, Vol.10, No.7, Jul, pp. 458-467, 1471-0072

Noda, N.N., Kumeta, H., Nakatogawa, H., Satoo, K., Adachi, W., Ishii, J., Fujioka, Y., Ohsumi, Y. & Inagaki, F. (2008). Structural basis of target recognition by Atg8/LC3 during selective autophagy. *Genes Cells*, Vol.13, No.12, Dec, pp. 1211-1218, 1356-9597

Noda, N.N., Ohsumi, Y. & Inagaki, F. (2010). Atg8-family interacting motif crucial for selective autophagy. *FEBS Lett*, Vol.584, No.7, Apr 2, pp. 1379-1385, 0014-5793

Noda, T. & Yoshimori, T. (2009). Molecular basis of canonical and bactericidal autophagy. *Int Immunol*, Vol.21, No.11, Nov, pp. 1199-1204, 0953-8178

Novak, I., Kirkin, V., McEwan, D.G., Zhang, J., Wild, P., Rozenknop, A., Rogov, V., Lohr, F., Popovic, D., Occhipinti, A., *et al.* (2010). Nix is a selective autophagy receptor for mitochondrial clearance. *EMBO Rep*, Vol.11, No.1, Jan, pp. 45-51, 1469-3178 (Electronic)

Oda, M.N., Scott, S.V., Hefner-Gravink, A., Caffarelli, A.D. & Klionsky, D.J. (1996). Identification of a cytoplasm to vacuole targeting determinant in aminopeptidase I. *J Cell Biol*, Vol.132, No.6, Mar, pp. 999-1010, 0021-9525

Okamoto, K., Kondo-Okamoto, N. & Ohsumi, Y. (2009). Mitochondria-anchored receptor Atg32 mediates degradation of mitochondria via selective autophagy. *Dev Cell*, Vol.17, No.1, Jul, pp. 87-97, 1534-5807

Olzmann, J.A., Li, L., Chudaev, M.V., Chen, J., Perez, F.A., Palmiter, R.D. & Chin, L.S. (2007). Parkin-mediated K63-linked polyubiquitination targets misfolded DJ-1 to aggresomes via binding to HDAC6. *J Cell Biol*, Vol.178, No.6, Sep 10, pp. 1025-1038, 0021-9525

Onodera, J. & Ohsumi, Y. (2004). Ald6p is a preferred target for autophagy in yeast, Saccharomyces cerevisiae. *J Biol Chem*, Vol.279, No.16, Apr 16, pp. 16071-16076, 0021-9258

Orenstein, S.J. & Cuervo, A.M. (2010). Chaperone-mediated autophagy: molecular mechanisms and physiological relevance. *Semin Cell Dev Biol*, Vol.21, No.7, Sep, pp. 719-726, 1084-9521

Pankiv, S., Clausen, T.H., Lamark, T., Brech, A., Bruun, J.A., Outzen, H., Overvatn, A., Bjorkoy, G. & Johansen, T. (2007). p62/SQSTM1 binds directly to Atg8/LC3 to facilitate degradation of ubiquitinated protein aggregates by autophagy. *J Biol Chem*, Vol.282, No.33, Aug 17, pp. 24131-24145, 0021-9258

Perrin, A.J., Jiang, X., Birmingham, C.L., So, N.S. & Brumell, J.H. (2004). Recognition of bacteria in the cytosol of Mammalian cells by the ubiquitin system. *Curr Biol*, Vol.14, No.9, May 4, pp. 806-811, 0960-9822

Ponting, C.P., Ito, T., Moscat, J., Diaz-Meco, M.T., Inagaki, F. & Sumimoto, H. (2002). OPR, PC and AID: all in the PB1 family. *Trends Biochem Sci*, Vol.27, No.1, Jan, pp. 10, 0968-0004

Rozenknop, A., Rogov, V.V., Rogova, N.Y., Lohr, F., Guntert, P., Dikic, I. & Dotsch, V. (2011). Characterization of the interaction of GABARAPL-1 with the LIR motif of NBR1. *J Mol Biol*, Vol.410, No.3, Jul 15, pp. 477-487, 0022-2836

Scott, S.V., Guan, J., Hutchins, M.U., Kim, J. & Klionsky, D.J. (2001). Cvt19 is a receptor for the cytoplasm-to-vacuole targeting pathway. *Mol Cell*, Vol.7, No.6, Jun, pp. 1131-1141, 1097-2765

Shintani, T., Huang, W.P., Stromhaug, P.E. & Klionsky, D.J. (2002). Mechanism of cargo selection in the cytoplasm to vacuole targeting pathway. *Dev Cell*, Vol.3, No.6, Dec, pp. 825-837, 1534-5807

Strome, S. & Lehmann, R. (2007). Germ versus soma decisions: lessons from flies and worms. *Science*, Vol.316, No.5823, Apr 20, pp. 392-393, 1095-9203 (Electronic) 0036-8075 (Linking)

Suzuki, K., Kamada, Y. & Ohsumi, Y. (2002). Studies of cargo delivery to the vacuole mediated by autophagosomes in Saccharomyces cerevisiae. *Dev Cell*, Vol.3, No.6, Dec, pp. 815-824, 1534-5807

Suzuki, K., Kondo, C., Morimoto, M. & Ohsumi, Y. (2010). Selective transport of alpha-mannosidase by autophagic pathways: identification of a novel receptor, Atg34p. *J Biol Chem*, Vol.285, No.39, Sep 24, pp. 30019-30025, 0021-9258

Suzuki, K., Morimoto, M., Kondo, C. & Ohsumi, Y. (2011). Selective Autophagy Regulates Insertional Mutagenesis by the Ty1 Retrotransposon in Saccharomyces cerevisiae. *Dev Cell*, Vol.21, No.2, Aug 16, pp. 358-365, 1534-5807

Suzuki, K. & Ohsumi, Y. (2010). Current knowledge of the pre-autophagosomal structure (PAS). *FEBS Lett*, Vol.584, No.7, Apr 2, pp. 1280-1286, 0014-5793

Tan, J.M., Wong, E.S., Kirkpatrick, D.S., Pletnikova, O., Ko, H.S., Tay, S.P., Ho, M.W., Troncoso, J., Gygi, S.P., Lee, M.K., *et al.* (2008). Lysine 63-linked ubiquitination promotes the formation and autophagic clearance of protein inclusions associated with neurodegenerative diseases. *Hum Mol Genet*, Vol.17, No.3, Feb 1, pp. 431-439, 0964-6906

Thurston, T.L., Ryzhakov, G., Bloor, S., von Muhlinen, N. & Randow, F. (2009). The TBK1 adaptor and autophagy receptor NDP52 restricts the proliferation of ubiquitin-coated bacteria. *Nat Immunol*, Vol.10, No.11, Nov, pp. 1215-1221, 1529-2908

Uttenweiler, A. & Mayer, A. (2008). Microautophagy in the yeast Saccharomyces cerevisiae. *Methods Mol Biol*, Vol.445, pp. 245-259, 1064-3745

Vadlamudi, R.K., Joung, I., Strominger, J.L. & Shin, J. (1996). p62, a phosphotyrosine-independent ligand of the SH2 domain of p56lck, belongs to a new class of ubiquitin-binding proteins. *J Biol Chem*, Vol.271, No.34, Aug 23, pp. 20235-20237, 0021-9258

Watanabe, Y., Noda, N.N., Honbou, K., Suzuki, K., Sakai, Y., Ohsumi, Y. & Inagaki, F. (2009). Crystallization of Saccharomyces cerevisiae alpha-mannosidase, a cargo protein of the Cvt pathway. *Acta crystallographica Section F, Structural biology and crystallization communications*, Vol.65, No.Pt 6, Jun 1, pp. 571-573, 1744-3091

Watanabe, Y., Noda, N.N., Kumeta, H., Suzuki, K., Ohsumi, Y. & Inagaki, F. (2010). Selective transport of alpha-mannosidase by autophagic pathways: structural basis for cargo recognition by Atg19 and Atg34. *J Biol Chem*, Vol.285, No.39, Sep 24, pp. 30026-30033, 0021-9258

Wild, P., Farhan, H., McEwan, D.G., Wagner, S., Rogov, V.V., Brady, N.R., Richter, B., Korac, J., Waidmann, O., Choudhary, C., *et al.* (2011). Phosphorylation of the autophagy receptor optineurin restricts Salmonella growth. *Science*, Vol.333, No.6039, Jul 8, pp. 228-233, 0036-8075

Wooten, M.W., Geetha, T., Babu, J.R., Seibenhener, M.L., Peng, J., Cox, N., Diaz-Meco, M.T. & Moscat, J. (2008). Essential role of sequestosome 1/p62 in regulating accumulation of Lys63-ubiquitinated proteins. *J Biol Chem*, Vol.283, No.11, Mar 14, pp. 6783-6789, 0021-9258

Yano, T., Mita, S., Ohmori, H., Oshima, Y., Fujimoto, Y., Ueda, R., Takada, H., Goldman, W.E., Fukase, K., Silverman, N., *et al.* (2008). Autophagic control of listeria through intracellular innate immune recognition in drosophila. *Nat Immunol*, Vol.9, No.8, Aug, pp. 908-916, 1529-2908

Yoshikawa, Y., Ogawa, M., Hain, T., Yoshida, M., Fukumatsu, M., Kim, M., Mimuro, H., Nakagawa, I., Yanagawa, T., Ishii, T., *et al.* (2009). Listeria monocytogenes ActA-mediated escape from autophagic recognition. *Nat Cell Biol*, Vol.11, No.10, Oct, pp. 1233-1240, 1465-7392

Yuga, M., Gomi, K., Klionsky, D.J. & Shintani, T. (2011). Aspartyl aminopeptidase is imported from the cytoplasm to the vacuole by selective autophagy in Saccharomyces cerevisiae. *J Biol Chem*, Vol.286, No.15, Apr 15, pp. 13704-13713, 0021-9258

Zatloukal, K., Stumptner, C., Fuchsbichler, A., Heid, H., Schnoelzer, M., Kenner, L., Kleinert, R., Prinz, M., Aguzzi, A. & Denk, H. (2002). p62 Is a common component of cytoplasmic inclusions in protein aggregation diseases. *Am J Pathol*, Vol.160, No.1, Jan, pp. 255-263, 0002-9440

Zhang, Y., Yan, L., Zhou, Z., Yang, P., Tian, E., Zhang, K., Zhao, Y., Li, Z., Song, B., Han, J., *et al.* (2009). SEPA-1 mediates the specific recognition and degradation of P granule components by autophagy in C. elegans. *Cell*, Vol.136, No.2, Jan 23, pp. 308-321, 0092-8674

Zheng, Y.T., Shahnazari, S., Brech, A., Lamark, T., Johansen, T. & Brumell, J.H. (2009). The adaptor protein p62/SQSTM1 targets invading bacteria to the autophagy pathway. *J Immunol*, Vol.183, No.9, Nov 1, pp. 5909-5916, 0022-1767

Part 4

Regulatory Molecules

17

Role of Ceramide 1-Phosphate in the Regulation of Cell Survival and Inflammation

Alberto Ouro, Lide Arana, Patricia Gangoiti and Antonio Gomez-Muñoz
Department of Biochemistry and Molecular Biology
Faculty of Science and Technology, University of the Basque Country, Bilbao
Spain

1. Introduction

Cell and tissue homeostasis is essential for normal development of an organism. When this is altered, metabolic dysfunctions and disease are prone to occur. Therefore, the maintenance of an appropriate balance in the activation / inhibition of the different metabolic pathways and cell signaling systems is simply vital.

Many lipids, including simple sphingolipids, are known to regulate cell activation and metabolism (Gomez-Munoz et al., 1992; Gomez-Munoz, 1998; Gomez-Munoz, 2004; Gomez-Munoz, 2006; Hannun & Obeid, 2008; Chen et al., 2011; Hannun & Obeid, 2011) . Some of them, including sphingosine, ceramides and their phosphorylated forms sphingosine 1-phosphate (S1P) and ceramide 1-phosphate (C1P) have been described as crucial regulators of key processes that are essential for normal development, and have also been involved in the establishment and progression of different diseases (Gangoiti et al., 2008a; Arana et al., 2010). In particular, ceramides can induce cell growth arrest and cause apoptosis, when they are generated (Hannun et al., 1986; Kolesnick, 1987; Kolesnick & Hemer, 1990; Merrill & Jones, 1990; Merrill, 1991; Hannun, 1994; Kolesnick & Golde, 1994; Hannun & Obeid, 1995; Hannun, 1996; Spiegel & Merrill, 1996; Merrill et al., 1997; Kolesnick et al., 2000; Hannun & Obeid, 2002; Merrill, 2002). Nonetheless, although in general, ceramides are negative signals for cell survival, in neurons they can induce cell growth (Goodman & Mattson, 1996; Ping & Barrett, 1998; Brann et al., 1999; Song & Posse de Chaves, 2003; Plummer et al., 2005). Also, ceramides play important roles in the regulation of cell differentiation, inflammation, tumor development (Okazaki et al., 1990; Mathias et al., 1991; Dressler et al., 1992; Hannun, 1994; Kolesnick & Golde, 1994; Hannun & Obeid, 1995; Gomez-Munoz, 1998; Menaldino et al., 2003), bacterial and viral infections, and ischemia-reperfusion injury(Gulbins & Kolesnick, 2003). More recently, ceramides have been associated with insulin resistance through activation of protein phosphatase 2A and the subsequent dephosphorylation and inactivation of protein kinase B (PKB) (Schmitz-Peiffer, 2002; Adams et al., 2004; Stratford et al., 2004), and toll-like receptor 4 (TLR4)-dependent induction of inflammatory cytokines, a fact essential for TLR4-dependent insulin resistance (Holland et al., 2011).

Concerning ceramide generation, there are three different mechanisms by which these molecules can be synthesized in cells. Ceramides can be generated by i) *de novo* synthesis, which takes place in the endoplasmic reticulum (ER), ii) by the action of different sphingomyelinases (SMases) in the plasma membrane, lysosomes, or mitochondria, and iii) by reacylation of sphingosine, a pathway known as the salvage or recycled pathway (Hannun & Obeid, 2011). The biosynthetic and degradative pathways of ceramide are shown in figure 1, where further products of ceramide metabolism are also indicated.

Natural ceramides tipically have long *N*-acyl chains ranging from 16 to 26 carbons in length (Merrill, 2002; Pettus et al., 2003a; Merrill et al., 2005), and some times longer in tissues such as skin. Many studies have used a short-chain analog (N-acetylsphingosine, or C_2-ceramide) in experiments with cells in culture because it can be incorporated into cells more easily and rapidly than long-chain ceramides. Of note, although C_2-ceramide was suggested not to occur *in vivo*, recent studies demonstrated that C_2-ceramide does exist in mammalian tissues. In particular, C_2-ceramide was found in rat liver cells (Merrill et al., 2001; Van Overloop et al., 2007), and brain tissue (Van Overloop et al., 2007). Ceramide generation is also relevant because this sphingolipid is the precursor of important bioactive molecules that can also regulate cellular functions. For instance, stimulation of ceramidases results in generation of sphingosine (Fig. 1), which was first described as a physiological inhibitor of protein kinase C (PKC) (Hannun et al., 1986). There are numerous reports in the scientific literature showing that PKC is inhibited by exogenous addition of sphingosine to cells in culture. Moreover, Merrill and co-workers demonstrated that addition of the ceramide synthase inhibitor fumonisin B1 to J774.A1 macrophages to increase the levels of endogenous sphingoid bases, also inhibited protein kinase C (Smith et al., 1997). Further work showed that sphingosine can affect the activity of other important enzymes that are involved in the regulation of metabolic or cell signaling pathways such as the Mg^{2+} dependent form of phosphatidate phosphohydrolase (Jamal et al., 1991; Gomez-Munoz et al., 1992), phospholipase D (PLD) (Natarajan et al., 1994), or diacylglycerol kinase (DAGK) (Sakane et al., 1989; Yamada et al., 1993). Sphingosine, in turn, can be phosphorylated by the action of sphingosine kinases to generate S1P, which is a potent mitogenic agent and can also inhibit apoptosis in many cell types (Olivera & Spiegel, 1993; Wu et al., 1995; Spiegel et al., 1996; Spiegel & Merrill, 1996; Spiegel & Milstien, 2002; Spiegel & Milstien, 2003). More recently, we demonstrated that S1P stimulates cortisol (Rabano et al., 2003) and aldosterone secretion (Brizuela et al., 2006) in cells of the zona fasciculata or zona glomerulosa, respectively, of bovine adrenal glands, suggesting that S1P plays an important role in the regulation of steroidogenesis.

A major metabolite of ceramide in cells is ceramide-1-phosphate (C1P), which is formed directly through phosphorylation of ceramide by the action of ceramide kinase (CerK) (Fig. 1). There is increasing evidence suggesting that C1P can regulate cell proliferation and apoptosis (Reviewed in (Gomez-Munoz, 1998; Gomez-Munoz, 2004)), and Chalfant and co-workers have implicated C1P in inflammatory responses (Reviewed in (Chalfant & Spiegel, 2005; Lamour & Chalfant, 2005)). In addition, Shayman's group demonstrated that C1P plays a key role in phagocytosis (Hinkovska-Galcheva & Shayman; Hinkovska-Galcheva et al., 1998; Hinkovska-Galcheva et al., 2005).

The aim of the present chapter is to review and update recent progress on the regulation of cell survival and inflammation by C1P.

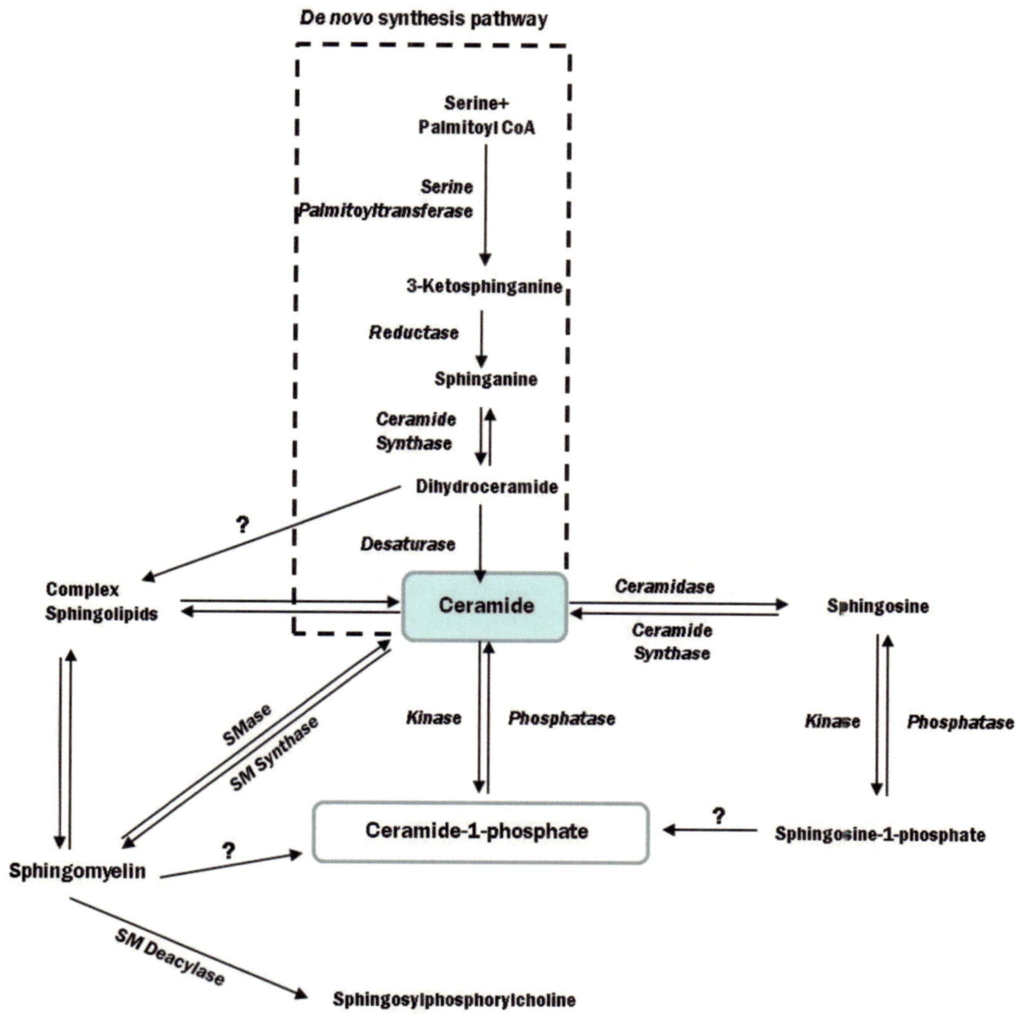

Fig. 1. Biosynthesis of simple sphingolipids in mammalian cells. Ceramide is the central core of sphingolipid metabolism. It can be produced by *de novo* synthesis through the concerted action of serine palmitoyltransferase and dihydroceramide synthase or by degradation of sphingomyelin (SM) through sphingomyelinase (SMase) activation. Ceramides can also be generated through metabolism of more complex sphingolipids. Phosphorylation of ceramide by ceramide kinase gives rise to ceramide-1-phosphate. The reverse reaction is catalyzed by ceramide-1-phosphate phosphatase, or by lipid phosphate phosphatases. Alternatively, ceramide can be degraded by ceramidases to form sphingosine, which can, in turn, be phosphorylated to sphingosine-1-phosphate by sphingosine kinases. The reverse reaction is catalyzed by sphingosine-1-phosphate phosphatases, or by lipid phosphate phosphatases. Sphingosine-1-phosphate lyase breaks down Sphingosine-1-phosphate to hexadecenal and ethanolamine phosphate, both of which can be recycled back to generate phosphatidylethanolamine. Sphingomyelin *N*-deacylase generates sphingosylphosphorylcholine, also known as lysosphingomyelin.

2. Biosynthesis of ceramide 1-phosphate. The essential role of ceramide kinase

At present, the only enzyme known to produce C1P in mammalian cells is ceramide kinase (CerK). This enzyme was first observed in brain synaptic vesicles (Bajjalieh et al., 1989), and was later found in human leukemia HL-60 cells (Kolesnick & Hemer, 1990). Cerk was first reported to be confined to the microsomal membrane fraction, but more recent studies indicate that it is mainly located in the cytosol (Mitsutake et al., 2004). These contradictory observations may arise from the different degrees of enzyme expression in different cell types, and it may also be possible that subcellular localization of this enzyme varies depending on cell metabolism. In this connection, Van Veldhoven and co-workers found that tagged forms of human CerK (FLAG-HsCerK and EGFP-HsCerK fusions), upon expression in Chinese Hamster Ovary (CHO) cells, were mainly localized to the plasma membrane, whereas no evidence for association with the ER was observed (Van Overloop et al., 2006). These findings are in agreement with those of Boath et al. (Boath et al., 2008) who showed that ceramides are not phosphorylated at the ER but must be transported to the Golgi apparatus for phosphorylation by CerK. When C1P is synthesized, it traffics from the Golgi network along the secretory pathway to the plasma membrane, where it can be back-exchanged into the extracellular environment and then bind to acceptor proteins such as albumin or lipoproteins (Boath et al., 2008). These observations are consistent with published work by Chalfant's group (Lamour et al., 2007), and it was demonstrated that CerK utilizes ceramide transported to the trans-Golgi apparatus by ceramide transport protein (CERT). In fact, downregulation of CERT by RNA interference resulted in strong inhibition of newly synthesized C1P, suggesting that CERT plays a critical role in C1P formation. However, Boat et al (Boath et al., 2008) reported that the transport of ceramides to the vicinity of CerK is not dependent upon CERT intervention. The reason for such discrepancy is unknown at the present time, but it is possible that the different experimental approaches used in those studies rendered different results. Specifically, whilst Lamour and co-workers used siRNA technology to inhibit CERT (Lamour et al., 2007), Boath and co-workers utilized pharmacological inhibitors (Boath et al., 2008). Also, it might be possible that different cell types may have different subcellular distribution of CerK, and / or that expression of this enzyme activity is not the same in all cell types.

With regards to the regulation of CerK, its ability to move intracellularly from one compartment to another and the dependency on cations (mainly Ca^{2+} ions) for activity seem to be well established. More recently, CerK has been proposed to be regulated by phosphorylation/dephosphorylation processes (Baumruker et al., 2005), and that it can be myristoylated at its N-terminus, a feature that is related to targeting proteins to membranes. Nonetheless, cleavage of the myristoylated moiety did not affect the intracellular localization of the enzyme. In addition, both CerK location and activity seem to require the integrity of its PH domain, which actually includes the myristoylation site, as deletion of this domain abolishes both the specific subcellular localization of the enzyme, as well as its activity (Baumruker et al., 2005).

Although CerK is thought to be the only enzyme for production of C1P, it was reported that bone marrow-derived macrophages (BMDM) from CerK-null mice (CerK-/-) still had significant levels of C1P (Boath et al., 2008). This observation suggests that there are other metabolic pathways, at least in mammals, capable of generating C1P independently of CerK.

Specifically, formation of C_{16}-C1P, which is a major species of C1P in cells, was not abolished in CerK-/- BMDM. Two alternative pathways for generation of C1P in cells might be: i) acylation of S1P by a putative acyl transferase that would catalyze the formation of a N-linked fatty acid in the S1P moiety to form C1P, and ii) cleavage of sphingomyelin (SM) by the action of a D-type SMase (SMase D), which would generate choline and C1P in an analogous manner to that of phospholipase D acting on phosphatidylcholine to produce choline and phosphatidic acid (PA). However, work from our own lab (Gomez-Munoz et al., 1995a) and that of others (Boath et al., 2008) demonstrated that acylation of S1P to form C1P does not occur in mammalian cells. Also, formation of C1P by the action of a putative SMase D has not yet been reported for mammalian cells. SMase D is a major component of the venom of a variety of arthropods including spiders of the gender *Loxosceles* (the brown recluse spider), such as *L. reclusa*. SMase D is also present in the toxins of some bacteria including *Corynebacterium pseudotuberculosis*, or *Vibrio damsela* (Truett & King, 1993). The bites of this spider result in strong inflammatory responses and may lead to renal failure, and occasionally lead to death (Lee & Lynch, 2005). Although we found no evidence for an analogous activity of SMase D in rat fibroblasts (Gomez-Munoz et al., 1995a), this possibility should be explored in more detail using different types of cells; so it is possible that SMase D may still be the cause for C1P generation in selective tissues.

Concerning regulation, mammalian CerK was demonstrated to be highly dependent on Ca^{2+} ions for activity (Van Overloop et al., 2006). More recently, it has been shown that treatment of human lung adenocarcinoma A549 cells and Chinese hamster ovary cells (CHO) with orthovanadate, a potent inhibitor of tyrosine phosphatases, increased CerK expression potently (Tada et al., 2010), suggesting a possible regulation of CerK by phosphorylation/dephosphorylation processes on tyrosine residues. Also, it has been suggested that CerK expression can be regulated through activation of Toll-like receptor 4 (TLR-4) by agonists such as the bacterial toxin lipopolysaccharide (Rovina et al., 2010).

The cloning of CerK (Sugiura et al., 2002) opened a new avenue of research that led to determination of important structural properties of this enzyme. The protein sequence has 537 amino acids with two protein sequence motifs, an N-terminus pleckstrin homology (PH) domain, and a C-terminal region containing a Ca^{2+}/calmodulin binding domain. Using site-directed mutagenesis, it was found that leucine 10 in the PH domain is essential for the catalytic activity of CerK (Kim et al., 2005). In addition, it was reported that the interaction between the PH domain of CERK and phosphatidylinositol 4,5-bisphosphate regulates the plasma membrane targeting and the levels of C1P (Kim et al., 2006). CERK also contains the five conserved sequence stretches (C1-C5) that are specific for lipid kinases (Reviewed in (Baumruker et al., 2005)).

With regards to substrate specificity, it was reported that phosphorylation of ceramide by CERK is stereospecific (Wijesinghe et al., 2005). The latter report also showed that a minimum of a 12-carbon acyl chain was required for normal CERK activity, whereas the short-chain ceramide analogues C_8-ceramide, C_4-ceramide, or C_2-ceramide were poor substrates for CERK. It was concluded that CERK phosphorylates only the naturally occurring D-erythro-ceramides (Wijesinghe et al., 2005). However, C_2-ceramide has been shown to also be a good substrate for CerK, especially when albumin is used as a carrier, and that C_2-ceramide can be converted to C_2-C1P within cells (Van Overloop et al., 2007). This raises the possibility that C_2-C1P is also a natural sphingolipid, capable of eliciting important biologic effects, as previously demonstrated (i.e. stimulation of cell proliferation (Gomez-Munoz et al., 1995a)). These observations suggested that substrate presentation is an important factor when testing CerK activity and that the use of different vehicles may result in different outcomes. Also, it should be borne in mind that CerK expression may not be the same in all cell types. The importance of CERK in cell signaling was emphasized in experiments using specific small interfering RNA (siRNA) to silence the gene encoding for CerK. Downregulation of CerK blocked the response of the enzyme to treatment with ATP, the calcium ionophore A23187, or interleukin 1-betta (Pettus et al., 2003b; Chalfant & Spiegel, 2005), and led to a potent inhibition of arachidonic acid release and PGE_2 formation in A549 lung adenocarcinoma cells. The relevance of CerK in cell biology was also highlighted in studies using CerK null mice; specifically, a potent reduction in the amount of neutrophils in the blood and spleen of these animals compared to their wild type counterparts was observed, whereas de amount of leukocytes, other than neutrophils, was increased in those mice. These observations suggested an important role of CerK in neutrophil biology (Graf et al., 2008). In addition to CerK, a ceramide kinase-like (CERKL) protein was identified in human retina (Tuson et al., 2004), and this was subsequently cloned (Bornancin et al., 2005). However, CERKL failed to phosphorylate ceramide or other related lipids, under conditions commonly used to measure CERK activity. Therefore, the role of this protein in cell biology is unclear at the present time.

CerK has also been reported to exist in dicotyledonous plants, where it was associated to the regulation of cell survival (Bi et al., 2011). Also, it has been recently found that a conserved cystein motif is critical for rice CerK activity and function (Bi et al., 2011). However, no reports on the possible existence of Cerk in monocot plants are available at the present time.

3. Catabolism of ceramide 1-phosphate

From the above discussion, it should be apparent that C1P is a bioactive metabolite, capable of altering cell metabolism rapidly and potently. So, the existence of enzymes capable of degrading C1P seemed to be feasible for regulation of C1P levels. The identification of a specific C1P phosphatase in rat brain (Shinghal et al., 1993), and hepatocytes (Boudker & Futerman, 1993), together with the existence of CerK suggested that ceramide and C1P are interconvertible in cells. C1P phosphatase is enriched in brain synaptosomes and liver plasma membrane fractions, and appeared to be distinct from PA phosphohydrolase, the phosphatase that hydrolyzes PA. Nonetheless, C1P can also be converted to ceramide by the action of a PA phosphohydrolase that is specifically located in the plasma membrane of cells (Waggoner et al., 1996). The latter enzyme belongs to a family of at least three mammalian lipid phosphate phosphatases (LPPs) (Brindley & Waggoner, 1998). LPPs have recently been shown to regulate cell survival by controlling the levels of intracellular PA and S1P pools (Long et al., 2005), and also to regulate leukocyte infiltration and airway inflammation (Zhao et al., 2005). Dephosphorylation of C1P might be a way of terminating its regulatory effects, although the resulting formation of ceramide could potentially be detrimental for cells. Controlling the levels of ceramide and C1P by the coordinated action of CERK and C1P phosphatases, may be of crucial importance for the metabolic or signaling pathways that are regulated by these two sphingolipids. It could be speculated that another possibility for degradation of C1P might be its deacylation to S1P, which could then be cleaved by lyase activity to render a fatty aldehyde and ethanolamine phosphate (Merrill & Jones, 1990), or to sphingosine by the action of S1P phosphatases (Fig. 1). However, no C1P deacylases or lyases have so far been identified in mammalian tissues, suggesting that the only pathway for degradation of C1P in mammals is through phosphatase activity.

4. Ceramide 1-phosphate and the control of cell growth and death

The first report showing that C1P was biologically active was published in 1995 (Gomez-Munoz et al., 1995a). C1P was found to have mitogenic properties as it stimulated DNA synthesis and cell division in rat or mouse fibroblasts (Gomez-Munoz et al., 1995a; Gomez-Munoz et al., 1997). Subsequent studies using primary macrophages, demonstrated that like for most growth factors, the mechanisms whereby C1P exerted its mitogenic effects implicated stimulation of the mitogen-activated protein kinase kinase (MEK)/Extracellularly regulated kinases 1-2 (ERK1-2), phosphatidylinositol 3-kinase (PI3-K)/protein kinase B (PKB, also known as Akt), and c-Jun terminal kinase (JNK) pathways (Gangoiti et al., 2008b). In addition, C1P caused stimulation of the DNA binding activity of the transcription factor NF-κB, and the selective inhibitors of MEK, PI3-K, and JNK (PD98059, LY290042, and SP600125), respectively) completely blocked NF-κB activation. Another major target of PKB is glycogen synthase kinase-3β (GSK-3 β), which expression was increased in the presence of C1P. This led to up-regulation of cyclin D1 and c-Myc, two important markers of cell proliferation that are targets of GSK-3β.

In addition, we found that C1P-stimulated macrophage proliferation, involved activation of sphingomyelin synthase (SMS), an enzyme that catalyzes the transfer of phosphocholine from phosphatidylcholine (PC) to ceramide to synthesize sphingomyelin (SM). The other by-product of this reaction is diacylglycerol (DAG), which is a well-established activator of protein kinase C (PKC). Conventional and novel PKC isoforms respond to DAG by translocating to the plasma membrane so that these enzymes can then express their activity and act on signaling events. In this connection, C1P stimulated the translocation and activation of the alpha isoform of PKC (PKC-α) in macrophages, and this resulted to be essential for stimulation of cell growth by C1P (Gangoiti et al., 2010c).

In a more recent report, it has been demonstrated that another essential kinase involved in the regulation of cell proliferation by C1P is the mammalian target of rapamycin (mTOR) (Gangoiti et al., 2010a). Activation of this kinase was tested my measuring the phosphorylation state of its downstream target p70S6K after treatment with C1P. Activation of mTOR/ p70S6K was dependent upon prior activation of PI3-K, as selective inhibition of this kinase blocked mTOR phosphorylation and activation. In addition, C1P caused phosphorylation of PRAS40, a component of the mTOR complex 1 (mTORC1) that is absent in mTORC2, and inhibition of the small G protein Ras homolog enriched in brain (Rheb), which is also a specific component of mTORC1, completely blocked C1P-stimulated mTOR phosphorylation, DNA synthesis and macrophage growth. C1P also caused phosphorylation of another Ras homolog gene family member, RhoA, and inhibition of its downstream effector RhoA-associated kinase (ROCK) also blocked C1P-stimulated mTOR and cell proliferation. It was concluded that mTORC1, and RhoA/ROCK are essential components of the mechanism whereby C1P stimulates macrophage proliferation. However, phospholipase D (PLD), and cAMP are not involved in the mitogenic effect of C1P (Gomez-Munoz et al., 1995a; Gomez-Munoz et al., 1997). Concerning intracellular calcium levels, which have also been implicated in the regulation of cell proliferation, the situation is controversial.

Although short-chain C1Ps failed to induce Ca^{2+} mobilization in fibroblasts (Gomez-Munoz et al., 1995a; Gomez-Munoz et al., 1997) or neutrophils (Rile et al., 2003), and natural C_{16}-C1P did not alter intracellular Ca^{2+} concentrations in A549 cells (Pettus et al., 2004), C_2-C1P- or C_8-C1P, caused intracellular Ca^{2+} mobilization in calf pulmonary artery endothelial (CAPE) cells (Gijsbers et al., 1999), thyroid FRTL-5 (Hogback et al., 2003), or Jurkat T-cells (Colina et al., 2005), suggesting that regulation of Ca^{2+} homeostasis may be cell type specific.

Finally, it should be pointed out that C1P has been recently shown to be a key mediator in the development and survival of retina photoreceptors, and to also play a critical role in photoreceptor differentiation (Miranda et al., 2011)

Apart from its mitogenic effect, another mechanism by which C1P controls cell homeostasis is by prevention of apoptosis (reviewed in (Gangoiti et al., 2010b)). We previously demonstrated that natural C1P blocked apoptosis in bone marrow-derived macrophages (Gomez-Munoz et al., 2004; Gomez-Munoz et al., 2005), and this was confirmed by Mitra and co-workers (Mitra et al., 2007) who found that down-regulation of CerK in mammalian cells reduced growth, and promoted apoptosis. Also, downregulation of CerK blocked epithelial growth factor-induced cell proliferation. However, in contrast to these observations, it was reported that addition of the cell-permeable C_2-ceramide to cells overexpressing CerK led to C_2-C1P formation and stimulation of apoptosis (Graf et al., 2007). This controversy can be explained by the fact that overexpression of CerK would substantially increase the intracellular levels of C1P, especially when cells are supplied with high concentrations of exogenous cell permeable C_2-ceramide; this action would cause overproduction of C_2-C1P inside the cells, which is toxic at high concentrations (Gomez-Munoz et al., 1995a; Gomez-Munoz et al., 2004).

When cells become apoptotic, their metabolism undergoes important changes from early stages. For example, apoptotic bone marrow-derived macrophages express high acid sphingomyelinase (A-SMase) activity and show high levels of ceramides compared to non-apoptotic cells (Gomez-Munoz et al., 2003; Hundal et al., 2003). Of interest, inhibition of A-SMase activation resulted to be one of the mechanisms by which C1P blocks apoptosis (Gomez-Munoz et al., 2004). C1P also blocked the activity of A-SMase in cell-free systems (in vitro), suggesting that inhibition of this enzyme takes place by direct physical interaction of C1P with the enzyme.

Recent work by our group (Granado et al., 2009a) showed that ceramide levels are also increased in alveolar NR8383 macrophages when they become apoptotic. However, A-SMase activity was only slightly enhanced in these cells under apoptotic conditions. This suggested the intervention of a different pathway for ceramide generation in these cells. In subsequent work we demonstrated that the mechanism whereby ceramide levels increased in apoptotic alveolar macrophages involved activation of serine palmitoyltransferase (SPT), the key regulatory enzyme of the de novo pathway of ceramide synthesis. Like for A-SMase, inhibition of SPT activation by treatment with C1P prevented the alveolar macrophages from entering apoptosis. These findings led to conclude that C1P promotes macrophage survival by blocking ceramide accumulation, and action that can be brought about through inhibition of either A-SMase activity, or SPT, depending on cell type.

The prosurvival effect of C1P was highlighted by the demonstration that intracellular levels of C1P were substantially decreased when the cells became apoptotic. It was hypothesized that depletion of intracellular C1P could result in the release of A-SMase from inhibition, thereby triggering ceramide generation an apoptotic cell death (Gomez-Munoz et al., 2004). Once generated, ceramides act on different intracellular targets to induce apoptosis. One of these targets is protein kinase B (or Akt), a kinase that lies downstream of PI3-K, a major signaling pathway through which growth factors promote cell survival. Using two different experimental approaches, it was demonstrated that PI3-K was also a target of C1P (Gomez-Munoz et al., 2005). On one hand, PI3-K activation was demonstrated by immunoprecipitation of the enzyme from whole cell lysates and assayed in vitro using ^{32}P-phosphatidylinositol. On the other hand, an in vivo approach provided evidence of phosphatidylinositol (3,4,5)-trisphosphate (PIP3) formation in intact cells that were prelabeled with ^{32}P-orthophosphate (Gomez-Munoz et al., 2005). PIP3 is a major product of PI3-K, and was shown to directly inhibit A-SMase (Testai et al., 2004). Therefore, it could be speculated that PI3-K activation might potentiate the inhibitory effect of C1P on A-SMase through generation of PIP3. C1P stimulated the phosphorylation of PKB, which was sensitive to inhibition by wortmannin or LY294002, thereby confirming that PI3-K was the enzyme responsible for its phosphorylation. These two PI3-K inhibitors also blocked the prosurvival effect of C1P, as expected (Gomez-Munoz et al., 2005). Another relevant finding was that C1P caused IkB phosphorylation and stimulation of the DNA binding activity of NF-kB in primary cultures of mouse macrophages (Gomez-Munoz et al., 2005). Of note, C1P up-regulated the expression of anti-apoptotic Bcl-XL, which is a downstream target of NF-kB. The latter results provided the first evidence for a novel biological role of natural C1P in the regulation of cell survival by the PI3-K/PKB/NF-kB pathway in mammalian cells (Gomez-Munoz et al., 2005).

As mentioned above, C1P can be metabolized to ceramide by different phosphatases, and then further converted to sphingosine and S1P by the coordinated actions of ceramidases and sphingosine kinases. Therefore, it could be speculated that the effects of C1P might be mediated through C1P-derived metabolites. However, usually ceramides and C1P exert opposing effects, (i.e. on PLD activation, adenylyl cyclase inhibition, or Ca^{2+} mobilization), and C1P is not able to reproduce the effects of S1P (Gomez-Munoz et al., 1995a; Gomez-Munoz et al., 1995b; Gomez-Munoz et al., 1997; Gomez-Munoz, 1998). Also, ceramides can decrease the expression of Bcl-XL (Chalfant & Spiegel, 2005), whereas C1P causes its up-regulation (Gomez-Munoz et al., 2005). Finally, no ceramidases capable of converting C1P into S1P have so far been reported to exist in mammalian cells, and S1P and C1P inhibit A-SMase through different mechanisms (Gomez-Munoz et al., 2003; Gomez-Munoz et al., 2004). Therefore, it can be concluded that C1P acts on its own right to regulate cell homeostasis. The above observations suggest that regulation of the enzyme activities involved in ceramide and C1P metabolism is essential for cell fate. Elucidation of the mechanisms controlling ceramide and C1P levels may help develop new molecular strategies for preventing metabolic disorders, or designing novel therapeutic agents for treatment of disease.

5. Ceramide 1-phosphate and the control of inflammation

Inflammation is, in principle, a beneficial process for protecting the organism against infection or injury. However, it can be detrimental when it becomes out of control. Apart from the classical signaling pathways and metabolites that are involved in the regulation of inflammation, it is now well accepted that ceramides are key elements in the inflammatory response (Lamour & Chalfant, 2005; Wijesinghe et al., 2008; Gomez-Munoz et al., 2010). For instance, it was reported that activation of A-SMase and the subsequent formation of ceramides play an important role in pulmonary infections as it facilitates internalization of bacteria into lung epithelial cells (Gulbins & Kolesnick, 2003). In this context, inhibition of A-SMase by C1P could be important to reduce or prevent infection in the lung.

Inflammatory mediators include chemokines, cytokines, vasoactive amines, products of proteolytic cascades, phospholipases, or lipids such as eicosanoids and sphingolipids. A major mediator of inflammation is PLA_2 activity. In particular, group IV cytosolic $cPLA_2$ (or $cPLA_2$-alpha) has been involved in receptor-dependent and independent production of eicosanoids, which are major components of inflammatory responses. Sphingolipids, including ceramides, have also been described as key mediators of inflammation (Hayakawa et al., 1996; Serhan et al., 1996; Manna & Aggarwal, 1998; Newton et al., 2000). More recently a role for ceramide in the development of allergic asthmatic responses and airway inflammation was established (Masini et al., 2008), and exogenous addition of C_2-ceramide to cultured astrocytes induced 12-lipoxigenase leading to generation of reactive oxygen species (ROS) and inflammation (Prasad et al., 2008). Also, A-SMase-derived ceramide was involved in platelet activating factor (PAF)-mediated pulmonary edema (Goggel et al., 2004). Subsequently, it was proposed that at least some of the pro-inflammatory effects of ceramides might in fact be mediated by its further metabolite C1P. The first report on the regulation of arachidonic acid (AA) release and the production of prostaglandins by C1P was from the laboratory of Charles Chalfant (Pettus et al., 2003b). This group demonstrated that C1P was able to stimulate AA release and prostanoid synthesis in A549 lung adenocarcinoma cells. In a follow up report, the same group showed that the mechanism whereby C1P stimulates AA release occurs through direct activation of $cPLA_2$ (Pettus et al., 2004). Subsequently, it was found that C1P is a positive allosteric activator of $cPLA_2$-alpha, and that it enhances the interaction of the enzyme with PC (Subramanian et al., 2005). In further work, the same group demonstrated that activation of $cPLA_2$-alpha by C1P is chain length-specific; in particular, C1P bearing acyl chains equal or higher than six carbons were able to efficiently activate $cPLA_2$-alpha in vitro, whereas shorter acyl chains (in particular C_2-C1P) were unable to activate the enzyme. It was concluded that the biological activity of C_2-C1P does not occur via eicosanoid synthesis (Wijesinghe et al., 2008). Also, C1P was shown to act in coordination with S1P to ensure maximal production of prostaglandins. Specifically, S1P was shown to induce cyclooxigenase-2 (COX-2) activity, which then uses $cPLA_2$-derived AA as substrate to synthesize prostaglandins (Pettus et al., 2005). Further details on the role of C1P in inflammatory response can be found in different reviews (Chalfant & Spiegel, 2005; Lamour et al., 2007; Wijesinghe et al., 2007), Wijesinghe et al., and recent work by Murayama and co-workers (Nakamura et al., 2011).

6. Ceramide 1-phosphate and the control of cell migration

Macrophage populations in tissues are determined by the rates of recruitment of monocytes from the bloodstream into the tissue, the rates of macrophage proliferation and apoptosis, and the rate of macrophage migration or efflux. Recently, our group demonstrated that exogenous addition of C1P to cultured Raw 264.7 macrophages stimulated cell migration

(Granado et al., 2009b). Interestingly, this action could only be observed when C1P was applied to the cells exogenously, and not by increasing the intracellular levels of C1P (i.e. through agonist stimulation of CerK, or by using the "caging" strategy to deliver C1P intracellularly (Lankalapalli et al., 2009)). This observation led us to identify a specific receptor through which C1P stimulates chemotaxis. This putative receptor seems to be located in the plasma membrane, has low affinity for C1P and has an apparent Kd of approximately 7.8 μM. The receptor is specific for C1P and is coupled to Gi proteins. Ligation of this receptor with C1P caused phosphorylation of ERK1-2, and PKB, and inhibition of either of these pathways completely abolished C1P-stimulated macrophage migration. Moreover, C1P stimulated the DNA binding activity of NF-kB, and blockade of this transcription factor resulted in full inhibition of macrophage migration. These observations suggest that MEK/ERK1-2, PI3-K/PKB (or Akt) and NF-kB are crucial signaling pathways for regulation of cell migration by C1P. It was concluded that this newly identified receptor could be an important drug target for treatment of illnesses in which cell migration is a major cause of pathology, as it occurs in atherosclerosis or in the metastasis of tumors.

7. Other relevant biological actions of C1P

In a previous report, Hinkovska-Galcheva et al (Hinkovska-Galcheva et al., 1998) showed that endogenous C1P can be generated during the phagocytosis of antibody-coated erythrocytes in human neutrophils that were primed with formylmethionylleucylphenylalanine. More recently, the same group demonstrated that C1P is a key mediator of neuthophil phagocytosis (Hinkovska-Galcheva et al., 2005). In addition, it was reported that C1P can be formed in neutrophils upon incubation with cell-permeable [^{3}H]*N*-hexanoylsphingosine (C_6-ceramide) (Rile et al., 2003), and Riboni and co-workers (Riboni et al., 2002) found that C1P can be generated in cerebellar granule cells both from SM-derived ceramide and through the recycling of sphingosine produced by ganglioside catabolism. C1P can be also generated by the action of interleukin 1-betta on A549 lung adenocarcinoma cells (Pettus et al., 2003b), or by stimulation of bone marrow-derived macrophages with macrophage-colony stimulating factor (M-CSF) (Gangoiti et al., 2008b). We found that C1P is present in normal bone marrow-derived macrophages isolated from healthy mice (Gomez-Munoz et al., 2004), and that C1P levels are substantially decreased in apoptotic macrophages. These observations are consistent with recent findings showing that CerK plays a key role in the stimulation of cell proliferation in A549 human lung adenocarcinoma cells (Mitra et al., 2007), and the induction of neointimal formation via cell proliferation and cell cycle progression in vascular smooth muscle cells by C1P (Kim et al., 2011).

8. Conclusion

The implication of simple sphingolipids in the regulation of cell activation and metabolism has acquired special relevance in the last two decades. Most attention was first paid to the effects elicited by ceramide because this sphingolipid turned out to be essential in the regulation of cell death, differentiation, senescence, and various metabolic disorders and diseases. However, its phosphorylated form, C1P, was thought not to be so important. However, C1P has emerged as a crucial bioactive sphingolipid, and this chapter highlights the relevance of C1P in cell biology. Specifically, C1P has now been established as key regulator of cell growth and survival, and its relevance in the regulation of cell migration is beginning to emerge. Also importantly, the discovery that C1P can act both intracellularly or as receptor ligand opens a broad avenue to investigate its implication in controlling cell metabolism. In addition to this, C1P has been postulated to be a potent proinflammatory agent, acting directly on $cPLA_2$ to trigger eicosanoid production. Therefore, C1P and CerK, the major enzyme responsible for its biosynthesis, may be key targets for developing new pharmacological strategies for treatment of illnesses associated to cell growth and death, and cell migration , such as chronic inflammation, cardiovascular diseases, neurodegeneration, or cancer.

9. Acknowledgement

Work in AGM lab is supported by Ministerio de Ciencia e Innovación (Madrid, Spain), Departamento de Educación, Universidades e Investigación del Gobierno Vasco (Gazteiz-Vitoria, Basque Country), and Departamento de Industria, Comercio y Turismo del Gobierno Vasco (Gazteiz-Vitoria, Basque Country).

10. References

Adams, J. M., 2nd, Pratipanawatr, T., Berria, R., Wang, E., DeFronzo, R. A., Sullards, M. C. and Mandarino, L. J. (2004). Ceramide content is increased in skeletal muscle from obese insulin-resistant humans. *Diabetes* 53, 1,(Jan, 2004) 25-31, 0012-1797

Arana, L., Gangoiti, P., Ouro, A., Trueba, M. and Gomez-Munoz, A. (2010). Ceramide and ceramide 1-phosphate in health and disease. *Lipids Health Dis* 9, 2010) 15, 1476-511X

Bajjalieh, S. M., Martin, T. F. and Floor, E. (1989). Synaptic vesicle ceramide kinase. A calcium-stimulated lipid kinase that co-purifies with brain synaptic vesicles. *J Biol Chem* 264, 24,(Aug 25, 1989) 14354-60, 0021-9258

Baumruker, T., Bornancin, F. and Billich, A. (2005). The role of sphingosine and ceramide kinases in inflammatory responses. *Immunol Lett* 96, 2,(Jan 31, 2005) 175-85, 0165-2478

Bi, F. C., Zhang, Q. F., Liu, Z., Fang, C., Li, J., Su, J. B., Greenberg, J. T., Wang, H. B. and Yao, N. (2011). A conserved cysteine motif is critical for rice ceramide kinase activity and function. *PLoS One* 6, 3,(2011) e18079, 1932-6203

Boath, A., Graf, C., Lidome, E., Ullrich, T., Nussbaumer, P. and Bornancin, F. (2008). Regulation and traffic of ceramide 1-phosphate produced by ceramide kinase: comparative analysis to glucosylceramide and sphingomyelin. *J Biol Chem* 283, 13,(Mar 28, 2008) 8517-26, 0021-9258

Bornancin, F., Mechtcheriakova, D., Stora, S., Graf, C., Wlachos, A., Devay, P., Urtz, N., Baumruker, T. and Billich, A. (2005). Characterization of a ceramide kinase-like protein. *Biochim Biophys Acta* 1687, 1-3,(Feb 21, 2005) 31-43, 0005-2736

Boudker, O. and Futerman, A. H. (1993). Detection and characterization of ceramide-1-phosphate phosphatase activity in rat liver plasma membrane. *J Biol Chem* 268, 29,(Oct 15, 1993) 22150-5, 0021-9258

Brann, A. B., Scott, R., Neuberger, Y., Abulafia, D., Boldin, S., Fainzilber, M. and Futerman, A. H. (1999). Ceramide signaling downstream of the p75 neurotrophin receptor mediates the effects of nerve growth factor on outgrowth of cultured hippocampal neurons. *J Neurosci* 19, 19,(Oct 1, 1999) 8199-206, 0270-6474

Brindley, D. N. and Waggoner, D. W. (1998). Mammalian lipid phosphate phosphohydrolases. *J Biol Chem* 273, 38,(Sep 18, 1998) 24281-4, 0021-9258

Brizuela, L., Rabano, M., Pena, A., Gangoiti, P., Macarulla, J. M., Trueba, M. and Gomez-Munoz, A. (2006). Sphingosine 1-phosphate: a novel stimulator of aldosterone secretion. *J Lipid Res* 47, 6,(Jun, 2006) 1238-49, 0022-2275

Colina, C., Flores, A., Castillo, C., Garrido Mdel, R., Israel, A., DiPolo, R. and Benaim, G. (2005). Ceramide-1-P induces Ca2+ mobilization in Jurkat T-cells by elevation of Ins(1,4,5)-P3 and activation of a store-operated calcium channel. *Biochem Biophys Res Commun* 336, 1,(Oct 14, 2005) 54-60, 0006-291X

Chalfant, C. E. and Spiegel, S. (2005). Sphingosine 1-phosphate and ceramide 1-phosphate: expanding roles in cell signaling. *J Cell Sci* 118, Pt 20,(Oct 15, 2005) 4605-12, 0021-9533

Chen, Y., Liu, Y., Sullards, M. C. and Merrill, A. H., Jr. (2011). An introduction to sphingolipid metabolism and analysis by new technologies. *Neuromolecular Med* 12, 4,(Dec, 2011) 306-19, 1535-1084

Dressler, K. A., Mathias, S. and Kolesnick, R. N. (1992). Tumor necrosis factor-alpha activates the sphingomyelin signal transduction pathway in a cell-free system. *Science* 255, 5052,(Mar 27, 1992) 1715-8, 0036-8075

Gangoiti, P., Arana, L., Ouro, A., Granado, M. H., Trueba, M. and Gomez-Munoz, A. (2010a). Activation of mTOR and RhoA is a major mechanism by which Ceramide 1-phosphate stimulates macrophage proliferation. *Cell Signal* 23, 1,(Aug 18, 2010a) 27-34,

Gangoiti, P., Camacho, L., Arana, L., Ouro, A., Granado, M. H., Brizuela, L., Casas, J., Fabrias, G., Abad, J. L., Delgado, A. and Gomez-Munoz, A. (2010b). Control of metabolism and signaling of simple bioactive sphingolipids: Implications in disease. *Prog Lipid Res* 49, 4,(Oct, 2010b) 316-34, 0163-7827

Gangoiti, P., Granado, M. H., Alonso, A., Goñi, F. M. and Gómez-Muñoz, A. (2008a). Implication of Ceramide, Ceramide 1-Phosphate and Sphingosine 1-Phosphate in Tumorigenesis. *Translational Oncogenomics* 3 2008a) 67-79, 1177-2727

Gangoiti, P., Granado, M. H., Arana, L., Ouro, A. and Gomez-Munoz, A. (2010c). Activation of protein kinase C-alpha is essential for stimulation of cell proliferation by ceramide 1-phosphate. *FEBS Lett* 584, 3,(Feb 5, 2010c) 517-24, 0014-5793

Gangoiti, P., Granado, M. H., Wang, S. W., Kong, J. Y., Steinbrecher, U. P. and Gomez-Munoz, A. (2008b). Ceramide 1-phosphate stimulates macrophage proliferation through activation of the PI3-kinase/PKB, JNK and ERK1/2 pathways. *Cell Signal* 20, 4,(Apr, 2008b) 726-36, 0898-6568

Gijsbers, S., Mannaerts, G. P., Himpens, B. and Van Veldhoven, P. P. (1999). N-acetyl-sphingenine-1-phosphate is a potent calcium mobilizing agent. *FEBS Lett* 453, 3,(Jun 25, 1999) 269-72, 0014-5793

Goggel, R., Winoto-Morbach, S., Vielhaber, G., Imai, Y., Lindner, K., Brade, L., Brade, H., Ehlers, S., Slutsky, A. S., Schutze, S., Gulbins, E. and Uhlig, S. (2004). PAF-mediated pulmonary edema: a new role for acid sphingomyelinase and ceramide. *Nat Med* 10, 2,(Feb, 2004) 155-60, 1078-8956

Gomez-Munoz, A. (1998). Modulation of cell signalling by ceramides. *Biochim Biophys Acta* 1391, 1,(Mar 6, 1998) 92-109, 0005-2736

Gomez-Munoz, A. (2004). Ceramide-1-phosphate: a novel regulator of cell activation. *FEBS Lett* 562, 1-3,(Mar 26, 2004) 5-10, 0014-5793

Gomez-Munoz, A. (2006). Ceramide 1-phosphate/ceramide, a switch between life and death. *Biochim Biophys Acta* 1758, 12,(Dec, 2006) 2049-56, 0005-2736

Gomez-Munoz, A., Duffy, P. A., Martin, A., O'Brien, L., Byun, H. S., Bittman, R. and Brindley, D. N. (1995a). Short-chain ceramide-1-phosphates are novel stimulators of DNA synthesis and cell division: antagonism by cell-permeable ceramides. *Mol Pharmacol* 47, 5,(May, 1995a) 833-9, 0026-895X

Gomez-Munoz, A., Frago, L. M., Alvarez, L. and Varela-Nieto, I. (1997). Stimulation of DNA synthesis by natural ceramide 1-phosphate. *Biochem J* 325 (Pt 2), Jul 15, 1997) 435-40, 0264-6021

Gomez-Munoz, A., Gangoiti, P., Granado, M. H., Arana, L. and Ouro, A. (2010). Ceramide 1-Phosphate in Cell Survival and Inflammatory Signaling. Sphingolipids as Signaling and Regulatory Molecules. C. Chalfant and M. D. Poeta. Austin (Tx), Landes Bioscience and Springer Science+Business Media: 118-130.

Gomez-Munoz, A., Hamza, E. H. and Brindley, D. N. (1992). Effects of sphingosine, albumin and unsaturated fatty acids on the activation and translocation of phosphatidate phosphohydrolases in rat hepatocytes. *Biochim Biophys Acta* 1127, 1,(Jul 9, 1992) 49-56, 0005-2736

Gomez-Munoz, A., Kong, J., Salh, B. and Steinbrecher, U. P. (2003). Sphingosine-1-phosphate inhibits acid sphingomyelinase and blocks apoptosis in macrophages. *FEBS Lett* 539, 1-3,(Mar 27, 2003) 56-60, 0014-5793

Gomez-Munoz, A., Kong, J. Y., Parhar, K., Wang, S. W., Gangoiti, P., Gonzalez, M., Eivemark, S., Salh, B., Duronio, V. and Steinbrecher, U. P. (2005). Ceramide-1-phosphate promotes cell survival through activation of the phosphatidylinositol 3-kinase/protein kinase B pathway. *FEBS Lett* 579, 17,(Jul 4, 2005) 3744-50, 1177-5793

Gomez-Munoz, A., Kong, J. Y., Salh, B. and Steinbrecher, U. P. (2004). Ceramide-1-phosphate blocks apoptosis through inhibition of acid sphingomyelinase in macrophages. *J Lipid Res* 45, 1,(Jan, 2004) 99-105, 0022-2275

Gomez-Munoz, A., Waggoner, D. W., O'Brien, L. and Brindley, D. N. (1995b). Interaction of ceramides, sphingosine, and sphingosine 1-phosphate in regulating DNA synthesis and phospholipase D activity. *J Biol Chem* 270, 44,(Nov 3, 1995b) 26318-25, 0022-2275

Goodman, Y. and Mattson, M. P. (1996). Ceramide protects hippocampal neurons against excitotoxic and oxidative insults, and amyloid beta-peptide toxicity. *J Neurochem* 66, 2,(Feb, 1996) 869-72, 0022-3042

Graf, C., Rovina, P., Tauzin, L., Schanzer, A. and Bornancin, F. (2007). Enhanced ceramide-induced apoptosis in ceramide kinase overexpressing cells. *Biochem Biophys Res Commun* 354, 1,(Mar 2, 2007) 309-14, 0006-291X

Graf, C., Zemann, B., Rovina, P., Urtz, N., Schanzer, A., Reuschel, R., Mechtcheriakova, D., Muller, M., Fischer, E., Reichel, C., Huber, S., Dawson, J., Meingassner, J. G., Billich, A., Niwa, S., Badegruber, R., Van Veldhoven, P. P., Kinzel, B., Baumruker, T. and Bornancin, F. (2008). Neutropenia with Impaired Immune Response to Streptococcus pneumoniae in Ceramide Kinase-Deficient Mice. *J Immunol* 180, 5,(Mar 1, 2008) 3457-66, 0022-1767

Granado, M. H., Gangoiti, P., Ouro, A., Arana, L. and Gomez-Munoz, A. (2009a). Ceramide 1-phosphate inhibits serine palmitoyltransferase and blocks apoptosis in alveolar macrophages. *Biochim Biophys Acta* 1791, 4,(Apr, 2009a) 263-72, 0005-2736

Granado, M. H., Gangoiti, P., Ouro, A., Arana, L., Gonzalez, M., Trueba, M. and Gomez-Munoz, A. (2009b). Ceramide 1-phosphate (C1P) promotes cell migration Involvement of a specific C1P receptor. *Cell Signal* 21, 3,(Mar, 2009b) 405-12, 0898-6568

Gulbins, E. and Kolesnick, R. (2003). Raft ceramide in molecular medicine. *Oncogene* 22, 45,(Oct 13, 2003) 7070-7, 0950-9232

Hannun, Y. A. (1994). The sphingomyelin cycle and the second messenger function of ceramide. *J Biol Chem* 269, 5,(Feb 4, 1994) 3125-8, 0021-9258

Hannun, Y. A. (1996). Functions of ceramide in coordinating cellular responses to stress. *Science* 274, 5294,(Dec 13, 1996) 1855-9,

Hannun, Y. A., Loomis, C. R., Merrill, A. H., Jr. and Bell, R. M. (1986). Sphingosine inhibition of protein kinase C activity and of phorbol dibutyrate binding in vitro and in human platelets. *J Biol Chem* 261, 27,(Sep 25, 1986) 12604-9, 0021-9258

Hannun, Y. A. and Obeid, L. M. (1995). Ceramide: an intracellular signal for apoptosis. *Trends Biochem Sci* 20, 2,(Feb, 1995) 73-7, 0968-0004

Hannun, Y. A. and Obeid, L. M. (2002). The Ceramide-centric universe of lipid-mediated cell regulation: stress encounters of the lipid kind. *J Biol Chem* 277, 29,(Jul 19, 2002) 25847-50,

Hannun, Y. A. and Obeid, L. M. (2008). Principles of bioactive lipid signalling: lessons from sphingolipids. *Nat Rev Mol Cell Biol* 9, 2,(Feb, 2008) 139-50, 0028-0836

Hannun, Y. A. and Obeid, L. M. (2011). Many ceramides. *J Biol Chem* 286, 32,(Aug 12, 2011) 27855-62, 0021-9258

Hayakawa, M., Jayadev, S., Tsujimoto, M., Hannun, Y. A. and Ito, F. (1996). Role of ceramide in stimulation of the transcription of cytosolic phospholipase A2 and cyclooxygenase 2. *Biochem Biophys Res Commun* 220, 3,(Mar 27, 1996) 681-6, 0006-291X

Hinkovska-Galcheva, V., Boxer, L. A., Kindzelskii, A., Hiraoka, M., Abe, A., Goparju, S., Spiegel, S., Petty, H. R. and Shayman, J. A. (2005). Ceramide 1-phosphate, a mediator of phagocytosis. *J Biol Chem* 280, 28,(Jul 15, 2005) 26612-21, 0021-9258

Hinkovska-Galcheva, V. and Shayman, J. A. Ceramide-1-phosphate in phagocytosis and calcium homeostasis. *Adv Exp Med Biol* 688, 131-40, 0065-2598

Hinkovska-Galcheva, V. T., Boxer, L. A., Mansfield, P. J., Harsh, D., Blackwood, A. and Shayman, J. A. (1998). The formation of ceramide-1-phosphate during neutrophil phagocytosis and its role in liposome fusion. *J Biol Chem* 273, 50,(Dec 11, 1998) 33203-9, 0021-9258

Hogback, S., Leppimaki, P., Rudnas, B., Bjorklund, S., Slotte, J. P. and Tornquist, K. (2003). Ceramide 1-phosphate increases intracellular free calcium concentrations in thyroid FRTL-5 cells: evidence for an effect mediated by inositol 1,4,5-trisphosphate and intracellular sphingosine 1-phosphate. *Biochem J* 370, Pt 1,(Feb 15, 2003) 111-9, 0264-6021

Holland, W. L., Bikman, B. T., Wang, L. P., Yuguang, G., Sargent, K. M., Bulchand, S., Knotts, T. A., Shui, G., Clegg, D. J., Wenk, M. R., Pagliassotti, M. J., Scherer, P. E. and Summers, S. A. (2011). Lipid-induced insulin resistance mediated by the proinflammatory receptor TLR4 requires saturated fatty acid-induced ceramide biosynthesis in mice. *J Clin Invest* 121, 5,(May 2, 2011) 1858-70, 0021-9738

Hundal, R. S., Gomez-Munoz, A., Kong, J. Y., Salh, B. S., Marotta, A., Duronio, V. and Steinbrecher, U. P. (2003). Oxidized low density lipoprotein inhibits macrophage apoptosis by blocking ceramide generation, thereby maintaining protein kinase B activation and Bcl-XL levels. *J Biol Chem* 278, 27,(Jul 4, 2003) 24399-408, 0022-9258

Jamal, Z., Martin, A., Gomez-Munoz, A. and Brindley, D. N. (1991). Plasma membrane fractions from rat liver contain a phosphatidate phosphohydrolase distinct from that in the endoplasmic reticulum and cytosol. *J Biol Chem* 266, 5,(Feb 15, 1991) 2988-96, 0021-9258

Kim, T. J., Kang, Y. J., Lim, Y., Lee, H. W., Bae, K., Lee, Y. S., Yoo, J. M., Yoo, H. S. and Yun, Y. P. (2011). Ceramide 1-phosphate induces neointimal formation via cell proliferation and cell cycle progression upstream of ERK1/2 in vascular smooth muscle cells. *Exp Cell Res* 317, 14,(Aug 15, 2011) 2041-51, 0014-4827

Kim, T. J., Mitsutake, S. and Igarashi, Y. (2006). The interaction between the pleckstrin homology domain of ceramide kinase and phosphatidylinositol 4,5-bisphosphate regulates the plasma membrane targeting and ceramide 1-phosphate levels. *Biochem Biophys Res Commun* 342, 2,(Apr 7, 2006) 611-7, 0006-291X

Kim, T. J., Mitsutake, S., Kato, M. and Igarashi, Y. (2005). The leucine 10 residue in the pleckstrin homology domain of ceramide kinase is crucial for its catalytic activity. *FEBS Lett* 579, 20,(Aug 15, 2005) 4383-8, 0014-5793

Kolesnick, R. and Golde, D. W. (1994). The sphingomyelin pathway in tumor necrosis factor and interleukin-1 signaling. *Cell* 77, 3,(May 6, 1994) 325-8, 0272-4340

Kolesnick, R. N. (1987). 1,2-Diacylglycerols but not phorbol esters stimulate sphingomyelin hydrolysis in GH3 pituitary cells. *J Biol Chem* 262, 35,(Dec 15, 1987) 16759-62, 0021-9258

Kolesnick, R. N., Goni, F. M. and Alonso, A. (2000). Compartmentalization of ceramide signaling: physical foundations and biological effects. *J Cell Physiol* 184, 3,(Sep, 2000) 285-300, 0021-9541

Kolesnick, R. N. and Hemer, M. R. (1990). Characterization of a ceramide kinase activity from human leukemia (HL-60) cells. Separation from diacylglycerol kinase activity. *J Biol Chem* 265, 31,(Nov 5, 1990) 18803-8, 0021-9258

Lamour, N. F. and Chalfant, C. E. (2005). Ceramide-1-phosphate: the "missing" link in eicosanoid biosynthesis and inflammation. *Mol Interv* 5, 6,(Dec, 2005) 358-67, 1543-2548

Lamour, N. F., Stahelin, R. V., Wijesinghe, D. S., Maceyka, M., Wang, E., Allegood, J. C., Merrill, A. H., Jr., Cho, W. and Chalfant, C. E. (2007). Ceramide kinase uses ceramide provided by ceramide transport protein: localization to organelles of eicosanoid synthesis. *J Lipid Res* 48, 6,(Jun, 2007) 1293-304, 0022-2275

Lankalapalli, R. S., Ouro, A., Arana, L., Gomez-Munoz, A. and Bittman, R. (2009). Caged ceramide 1-phosphate analogues: synthesis and properties. *J Org Chem* 74, 22,(Nov 20, 2009) 8844-7, 5163–5166

Lee, S. and Lynch, K. R. (2005). Brown recluse spider (Loxosceles reclusa) venom phospholipase D (PLD) generates lysophosphatidic acid (LPA). *Biochem J* 391, Pt 2,(Oct 15, 2005) 317-23, 0264-6021

Long, J., Darroch, P., Wan, K. F., Kong, K. C., Ktistakis, N., Pyne, N. J. and Pyne, S. (2005). Regulation of cell survival by lipid phosphate phosphatases involves the modulation of intracellular phosphatidic acid and sphingosine 1-phosphate pools. *Biochem J* 391, Pt 1,(Oct 1, 2005) 25-32, 0264-6021

Manna, S. K. and Aggarwal, B. B. (1998). IL-13 suppresses TNF-induced activation of nuclear factor-kappa B, activation protein-1, and apoptosis. *J Immunol* 161, 6,(Sep 15, 1998) 2863-72, 0022-1767

Masini, E., Giannini, L., Nistri, S., Cinci, L., Mastroianni, R., Xu, W., Comhair, S. A., Li, D., Cuzzocrea, S., Matuschak, G. M. and Salvemini, D. (2008). Ceramide: a key signaling molecule in a Guinea pig model of allergic asthmatic response and airway inflammation. *J Pharmacol Exp Ther* 324, 2,(Feb, 2008) 548-57., 0022-3565

Mathias, S., Dressler, K. A. and Kolesnick, R. N. (1991). Characterization of a ceramide-activated protein kinase: stimulation by tumor necrosis factor alpha. *Proc Natl Acad Sci U S A* 88, 22,(Nov 15, 1991) 10009-13, 1091-6490

Menaldino, D. S., Bushnev, A., Sun, A., Liotta, D. C., Symolon, H., Desai, K., Dillehay, D. L., Peng, Q., Wang, E., Allegood, J., Trotman-Pruett, S., Sullards, M. C. and Merrill, A. H., Jr. (2003). Sphingoid bases and de novo ceramide synthesis: enzymes involved, pharmacology and mechanisms of action. *Pharmacol Res* 47, 5,(May, 2003) 373-81, 1043-6618

Merrill, A. H., Jr. (1991). Cell regulation by sphingosine and more complex sphingolipids. *J Bioenerg Biomembr* 23, 1,(Feb, 1991) 83-104, 0145-479X

Merrill, A. H., Jr. (2002). De novo sphingolipid biosynthesis: a necessary, but dangerous, pathway. *J Biol Chem* 277, 29,(Jul 19, 2002) 25843-6, 0021-9258

Merrill, A. H., Jr. and Jones, D. D. (1990). An update of the enzymology and regulation of sphingomyelin metabolism. *Biochim Biophys Acta* 1044, 1,(May 1, 1990) 1-12, 0005-2736

Merrill, A. H., Jr., Schmelz, E. M., Dillehay, D. L., Spiegel, S., Shayman, J. A , Schroeder, J. J., Riley, R. T., Voss, K. A. and Wang, E. (1997). Sphingolipids--the enigmatic lipid class: biochemistry, physiology, and pathophysiology. *Toxicol Appl Pharmacol* 142, 1,(Jan, 1997) 208-25, 0041-008X

Merrill, A. H., Jr., Sullards, M. C., Allegood, J. C., Kelly, S. and Wang, E. (2005). Sphingolipidomics: high-throughput, structure-specific, and quantitative analysis of sphingolipids by liquid chromatography tandem mass spectrometry. *Methods* 36, 2,(Jun, 2005) 207-24, 1548-7091

Merrill, A. H., Jr., Sullards, M. C., Wang, E., Voss, K. A. and Riley, R. T. (2001). Sphingolipid metabolism: roles in signal transduction and disruption by fumonisins. *Environ Health Perspect* 109 Suppl 2, May, 2001) 283-9, 0091-6765

Miranda, G. E., Abrahan, C. E., Agnolazza, D. L., Politi, L. E. and Rotstein, N. P. (2011). Ceramide-1-phosphate, a new mediator of development and survival in retina photoreceptors. *Invest Ophthalmol Vis Sci* 52, 9,(2011) 6580-8, 1552-5783

Mitra, P., Maceyka, M., Payne, S. G., Lamour, N., Milstien, S., Chalfant, C. E. and Spiegel, S. (2007). Ceramide kinase regulates growth and survival of A549 human lung adenocarcinoma cells. *FEBS Lett* 581, 4,(Feb 20, 2007) 735-40, 0014-5793

Mitsutake, S., Kim, T. J., Inagaki, Y., Kato, M., Yamashita, T. and Igarashi, Y. (2004). Ceramide kinase is a mediator of calcium-dependent degranulation in mast cells. *J Biol Chem* 279, 17,(Apr 23, 2004) 17570-7, 0021-9258

Nakamura, H., Tada, E., Makiyama, T., Yasufuku, K. and Murayama, T. (2011). Role of cytosolic phospholipase A(2)alpha in cell rounding and cytotoxicity induced by ceramide-1-phosphate via ceramide kinase. *Arch Biochem Biophys* 512, 1,(Aug 1, 2011) 45-51, 0003-9861

Natarajan, V., Jayaram, H. N., Scribner, W. M. and Garcia, J. G. (1994). Activation of endothelial cell phospholipase D by sphingosine and sphingosine-1-phosphate. *Am J Respir Cell Mol Biol* 11, 2,(Aug, 1994) 221-9, 1044-1549

Newton, R., Hart, L., Chung, K. F. and Barnes, P. J. (2000). Ceramide induction of COX-2 and PGE(2) in pulmonary A549 cells does not involve activation of NF-kappaB. *Biochem Biophys Res Commun* 277, 3,(Nov 2, 2000) 675-9, 0006-291X

Okazaki, T., Bielawska, A., Bell, R. M. and Hannun, Y. A. (1990). Role of ceramide as a lipid mediator of 1 alpha,25-dihydroxyvitamin D3-induced HL-60 cell differentiation. *J Biol Chem* 265, 26,(Sep 15, 1990) 15823-31, 0021-9258

Olivera, A. and Spiegel, S. (1993). Sphingosine-1-phosphate as second messenger in cell proliferation induced by PDGF and FCS mitogens. *Nature* 365, 6446,(Oct 7, 1993) 557-60, 0028-0836

Pettus, B. J., Bielawska, A., Kroesen, B. J., Moeller, P. D., Szulc, Z. M., Hannun, Y. A. and Busman, M. (2003a). Observation of different ceramide species from crude cellular extracts by normal-phase high-performance liquid chromatography coupled to atmospheric pressure chemical ionization mass spectrometry. *Rapid Commun Mass Spectrom* 17, 11,(2003a) 1203-11, 0951-4198

Pettus, B. J., Bielawska, A., Spiegel, S., Roddy, P., Hannun, Y. A. and Chalfant, C. E. (2003b). Ceramide kinase mediates cytokine- and calcium ionophore-induced arachidonic acid release. *J Biol Chem* 278, 40,(Oct 3, 2003b) 38206-13, 0022-9258

Pettus, B. J., Bielawska, A., Subramanian, P., Wijesinghe, D. S., Maceyka, M., Leslie, C. C., Evans, J. H., Freiberg, J., Roddy, P., Hannun, Y. A. and Chalfant, C. E. (2004). Ceramide 1-phosphate is a direct activator of cytosolic phospholipase A2. *J Biol Chem* 279, 12,(Mar 19, 2004) 11320-6, 0021-9258

Pettus, B. J., Kitatani, K., Chalfant, C. E., Taha, T. A., Kawamori, T., Bielawski, J., Obeid, L. M. and Hannun, Y. A. (2005). The coordination of prostaglandin E2 production by sphingosine-1-phosphate and ceramide-1-phosphate. *Mol Pharmacol* 68, 2,(Aug, 2005) 330-5, 0026-895X

Ping, S. E. and Barrett, G. L. (1998). Ceramide can induce cell death in sensory neurons, whereas ceramide analogues and sphingosine promote survival. *J Neurosci Res* 54, 2,(Oct 15, 1998) 206-13, 1097-4547

Plummer, G., Perreault, K. R., Holmes, C. F. and Posse De Chaves, E. I. (2005). Activation of serine/threonine protein phosphatase-1 is required for ceramide-induced survival of sympathetic neurons. *Biochem J* 385, Pt 3,(Feb 1, 2005) 685-93, 0264-6021

Prasad, V. V., Nithipatikom, K. and Harder, D. R. (2008). Ceramide elevates 12-hydroxyeicosatetraenoic acid levels and upregulates 12-lipoxygenase in rat primary hippocampal cell cultures containing predominantly astrocytes. *Neurochem Int* 53, 6-8,(Nov-Dec, 2008) 220-9, 0197-0186

Rabano, M., Pena, A., Brizuela, L., Marino, A., Macarulla, J. M., Trueba, M. and Gomez-Munoz, A. (2003). Sphingosine-1-phosphate stimulates cortisol secretion. *FEBS Lett* 535, 1-3,(Jan 30, 2003) 101-5, 0014-5793

Riboni, L., Bassi, R., Anelli, V. and Viani, P. (2002). Metabolic formation of ceramide-1-phosphate in cerebellar granule cells: evidence for the phosphorylation of ceramide by different metabolic pathways. *Neurochem Res* 27, 7-8,(Aug, 2002) 711-6, 0364-3190

Rile, G., Yatomi, Y., Takafuta, T. and Ozaki, Y. (2003). Ceramide 1-phosphate formation in neutrophils. *Acta Haematol* 109, 2,(2003) 76-83, 0001-5792

Rovina, P., Graf, C. and Bornancin, F. (2010). Modulation of ceramide metabolism in mouse primary macrophages. *Biochem Biophys Res Commun* 399, 2,(Aug 20, 2010) 150-4, 0006-291X

Sakane, F., Yamada, K. and Kanoh, H. (1989). Different effects of sphingosine, R59022 and anionic amphiphiles on two diacylglycerol kinase isozymes purified from porcine thymus cytosol. *FEBS Lett* 255, 2,(Sep 25, 1989) 409-13, 0014-5793

Schmitz-Peiffer, C. (2002). Protein kinase C and lipid-induced insulin resistance in skeletal muscle. *Ann N Y Acad Sci* 967, Jun, 2002) 146-57, 0077-8923

Serhan, C. N., Haeggstrom, J. Z. and Leslie, C. C. (1996). Lipid mediator networks in cell signaling: update and impact of cytokines. *Faseb J* 10, 10,(Aug, 1996) 1147-58, 0892-6638

Shinghal, R., Scheller, R. H. and Bajjalieh, S. M. (1993). Ceramide 1-phosphate phosphatase activity in brain. *J Neurochem* 61, 6,(Dec, 1993) 2279-85, 0022-3042

Smith, E. R., Jones, P. L., Boss, J. M. and Merrill, A. H., Jr. (1997). Changing J774A.1 cells to new medium perturbs multiple signaling pathways, including the modulation of protein kinase C by endogenous sphingoid bases. *J Biol Chem* 272, 9,(Feb 28, 1997) 5640-6, 0021-9258

Song, M. S. and Posse de Chaves, E. I. (2003). Inhibition of rat sympathetic neuron apoptosis by ceramide. Role of p75NTR in ceramide generation. *Neuropharmacology* 45, 8,(Dec, 2003) 1130-50, 0028-3908

Spiegel, S., Foster, D. and Kolesnick, R. (1996). Signal transduction through lipid second messengers. *Curr Opin Cell Biol* 8, 2,(Apr, 1996) 159-67, 0955-0674

Spiegel, S. and Merrill, A. H., Jr. (1996). Sphingolipid metabolism and cell growth regulation. *Faseb J* 10, 12,(Oct, 1996) 1388-97,

Spiegel, S. and Milstien, S. (2002). Sphingosine 1-phosphate, a key cell signaling molecule. *J Biol Chem* 277, 29,(Jul 19, 2002) 25851-4, 0021-9258

Spiegel, S. and Milstien, S. (2003). Sphingosine-1-phosphate: an enigmatic signalling lipid. *Nat Rev Mol Cell Biol* 4, 5,(May, 2003) 397-407, 1471-0072

Stratford, S., Hoehn, K. L., Liu, F. and Summers, S. A. (2004). Regulation of insulin action by ceramide: dual mechanisms linking ceramide accumulation to the inhibition of Akt/protein kinase B. *J Biol Chem* 279, 35,(Aug 27, 2004) 36608-15, 0021-9258

Subramanian, P., Stahelin, R. V., Szulc, Z., Bielawska, A., Cho, W. and Chalfant, C. E. (2005). Ceramide 1-phosphate acts as a positive allosteric activator of group IVA cytosolic phospholipase A2 alpha and enhances the interaction of the enzyme with phosphatidylcholine. *J Biol Chem* 280, 18,(May 6, 2005) 17601-7, 0021-9258

Sugiura, M., Kono, K., Liu, H., Shimizugawa, T., Minekura, H., Spiegel, S. and Kohama, T. (2002). Ceramide kinase, a novel lipid kinase. Molecular cloning and functional characterization. *J Biol Chem* 277, 26,(Jun 28, 2002) 23294-300, 0021-9258

Tada, E., Toyomura, K., Nakamura, H., Sasaki, H., Saito, T., Kaneko, M., Okuma, Y. and Murayama, T. (2010). Activation of ceramidase and ceramide kinase by vanadate via a tyrosine kinase-mediated pathway. *J Pharmacol Sci* 114, 4,(2010) 420-32, 1347-8613

Testai, F. D., Landek, M. A., Goswami, R., Ahmed, M. and Dawson, G. (2004). Acid sphingomyelinase and inhibition by phosphate ion: role of inhibition by phosphatidyl-myo-inositol 3,4,5-triphosphate in oligodendrocyte cell signaling. *J Neurochem* 89, 3,(May, 2004) 636-44, 0022-3042

Truett, A. P., 3rd and King, L. E., Jr. (1993). Sphingomyelinase D: a pathogenic agent produced by bacteria and arthropods. *Adv Lipid Res* 26, 1993) 275-91, 0065-2849

Tuson, M., Marfany, G. and Gonzalez-Duarte, R. (2004). Mutation of CERKL, a novel human ceramide kinase gene, causes autosomal recessive retinitis pigmentosa (RP26). *Am J Hum Genet* 74, 1,(Jan, 2004) 128-38, 0002-9297

Van Overloop, H., Denizot, Y., Baes, M. and Van Veldhoven, P. P. (2007). On the presence of C2-ceramide in mammalian tissues: possible relationship to etherphospholipids and phosphorylation by ceramide kinase. *Biol Chem* 388, 3,(Mar, 2007) 315-24, 0264-6021

Van Overloop, H., Gijsbers, S. and Van Veldhoven, P. P. (2006). Further characterization of mammalian ceramide kinase: substrate delivery and (stereo)specificity, tissue distribution, and subcellular localization studies. *J Lipid Res* 47, 2,(Feb, 2006) 268-83, 0022-2275

Waggoner, D. W., Gomez-Munoz, A., Dewald, J. and Brindley, D. N. (1996). Phosphatidate phosphohydrolase catalyzes the hydrolysis of ceramide 1-phosphate, lysophosphatidate, and sphingosine 1-phosphate. *J Biol Chem* 271, 28,(Jul 12, 1996) 16506-9, 0021-9258

Wijesinghe, D. S., Lamour, N. F., Gomez-Munoz, A. and Chalfant, C. E. (2007). Ceramide kinase and ceramide-1-phosphate. *Methods Enzymol* 434, 2007) 265-92, 0076-6879

Wijesinghe, D. S., Massiello, A., Subramanian, P., Szulc, Z., Bielawska, A. and Chalfant, C. E. (2005). Substrate specificity of human ceramide kinase. *J Lipid Res* 46, 12,(Dec, 2005) 2706-16, 0022-2275

Wijesinghe, D. S., Subramanian, P., Lamour, N. F., Gentile, L. B., Granado, M. H., Szulc, Z., Bielawska, A., Gomez-Munoz, A. and Chalfant, C. E. (2008). The chain length specificity for the activation of group IV cytosolic phospholipase A2 by ceramide-1-phosphate. Use of the dodecane delivery system for determining lipid-specific effects. *J Lipid Res* 50, 10,(Dec 15, 2008) 1986-95, 0022-2275

Wu, J., Spiegel, S. and Sturgill, T. W. (1995). Sphingosine 1-phosphate rapidly activates the mitogen-activated protein kinase pathway by a G protein-dependent mechanism. *J Biol Chem* 270, 19,(May 12, 1995) 11484-8, 0021-9258

Yamada, K., Sakane, F., Imai, S. and Takemura, H. (1993). Sphingosine activates cellular diacylglycerol kinase in intact Jurkat cells, a human T-cell line. *Biochim Biophys Acta* 1169, 3,(Sep 8, 1993) 217-24, 0005-2736

Zhao, Y., Usatyuk, P. V., Cummings, R., Saatian, B., He, D., Watkins, T., Morris, A., Spannhake, E. W., Brindley, D. N. and Natarajan, V. (2005). Lipid phosphate phosphatase-1 regulates lysophosphatidic acid-induced calcium release, NF-kappaB activation and interleukin-8 secretion in human bronchial epithelial cells. *Biochem J* 385, Pt 2,(Jan 15, 2005) 493-502, 0264-6021

18

Cholesterol: Biosynthesis, Functional Diversity, Homeostasis and Regulation by Natural Products

J. Thomas[1], T.P. Shentu[1] and Dev K. Singh[2*]

[1]Department of Medicine, University of Illinois, Chicago

[2]Division of Developmental Biology, Department of Pediatrics

Children's Hospital of University of Illinois, University of Illinois at Chicago

USA

1. Introduction

Most of the discussions on cardiovascular diseases include the relative level of total plasma cholesterol whereas complete lipid profile is very important as clinical diagnotics for cardiovascular risk. For example, the status of triglycerides, low density lipoprotein cholesterol (LDL cholesterol), high density lipoprotein cholesterol (HDL cholesterol), thyroid functions, insulin and lipid peroxides etc as these lipids are related to cardiovascular disease either as markers of another underlying disturbance or along the same pathways of cholesterol metabolism. According to latest World Health Organization (WHO) report about 50% of the heart attacks occur in individual with high level of cholesterol. The most abundant sterol in animal system is in the form of cholesterol whereas plants lack cholesterol but they contain structurally similar other sterol and similar biosynthetic pathway exist both in plants and animals as well as some prokaryotes also synthesize some specific sterols. In 1948, the Framingham Heart Study - under the direction of the National Heart Institute (now known as the National Heart, Lung, and Blood Institute or NHLBI) - embarked on an ambitious project in health research. At the time, little was known about the general causes of heart disease and stroke, but the death rates for CVD had been increasing steadily since the beginning of the century and had become an American epidemic. The Framingham Heart Study became a joint project of the National Heart, Lung and Blood Institute and Boston University. The concern about cholesterol was largely fueled by this study and others that provided strong evidence that when large populations are observed, persons with higher than average serum total cholesterol have a higher incidence of coronary artery disease (CAD). Laboratory reports often mention two main types of cholesterol, the HDL cholesterol (often termed the "good" cholesterol) and LDL cholesterol (often mis-named the "bad" one. Even if the total LDL is lowered, the important fraction is actually the small LDL, which is more easily oxidized into a potentially atherogenic particle than its larger, more buoyant counterpart. Bacteria or other infectious agents are being looked at as part of the culprits as causative factors in initiating injury to the arterial wall. Cholesterol is then attracted to this 'rough' site on the blood vessel

* Corresponding Author

wall in an attempt to heal the wall so that blood will flow smoothly over the injured area. Cholesterol itself is not the cause of CAD. The blood cholesterol is rather only a reflection of other metabolic imbalances in the vast majority of cases. If we assume blood lipid (fats) status are associated with some risk of cardiovascular disease. The question is which lipids are the important markers, and even more important what should we do about them.

Lowering cholesterol too aggressively or in artificial circumstances or having too low total cholesterol is also undesirable. Plasma cholesterol level below 200mg% is desirable whereas as 200-239mg% and above 240mg% is considered as borderline and high level of cholesterol, respectively. Ingested cholesterol comes from animal sources (plants and prokaryotes do not contain cholesterol (Gylling and Miettinen, 1995) such as eggs, meat, dairy products, fish, and shellfish or biosynthesized from the breakdown of carbohydrates, lipids, or proteins available in the food. One study has estimated that the complete abolition of dietary cholesterol absorption would reduce plasma cholesterol by up to 62% (Gylling and Miettinen, 1995). About 50% of dietary cholesterol is absorbed through intestinal enterocytes, while the rest is excreted through feces (Ostlund et al., 1999). It is estimated that half of ingested cholesterol enters the body while the other half excreated in the feces. Normally, the more cholesterol we absorb, the less our bodies make. There is slight increase in plasma cholesterol with increase in the amount of cholesterol ingested each day, usually is not changed more than ± 15 percent by altering the amount of cholesterol in the diet. Although the response of individuals differs markedly. Cholesterol is integeral part of membranes and perform a number of vital functions in the cell and due to this property of cholesterol, each cell has the capability to biosynthesize cholesterol if it is required. Cholesterol is a component of steroid hormones, including pregnenolone, estrogens, progesterone, testosterone, vitamin D and bile acids. Bile acids are involved in lipid digestion, absorption, and excretion.

In this chapter, mode of intracellular and extracellular cholesterol transport through acceptors-donors and thereafter cholesterol trafficking pathways will be described in detail. Furthermore, we will discuss the regulation of cholesterol at enzymatic/transcriptional level and diverse functions of cholesterol in our body. Taken together, this book chapter will address recent advances in cholesterol metabolism both *in vitro* and *in vivo* models related to absorptions, biosynthesis, transport, excretion and therapeutic targets for new drugs and natural compounds.

1.1 Cholesterol biosynthetic pathway

As early as 1926, studies by Heilbron, Kamm and Owens suggested that squalene is precursor of cholesterol biosynthesis (Garrett & Grisham, 2007). In the same year, H.J. Channon, demonstrated first time that animals fed on shark oil produced more cholesterol in the tissues. In 1940, Bloch and Rittenberg, first time demonstrated that mice fed on radiolabled acetate showed significant radiolabeled cholesterol (Bloch et al., 1945; Kresge et al., 2005). In 1952, Konard Bloch and Robert Langdon showed conclusively that squalene as well as cholesterol are synthesized from acetate for which Fyodor Lynen and Bloch were awarded the Noble Prize in Medicine/Physiology in 1964. Cholesterol is biosynthesized from 2-carbon metabolic intermediate, acetyl-CoA hooked end to end involving a number of enzymatic reactions and finally get converted into the 27-carbon molecule of cholesterol. Metabolism (catabolism) of lipids, carbohydrates and proteins lead to the formation Acetyl-CoA. Proteins are generally are not catabolized for the purpose of energy and usually broken down into amino acids for denovo protein biosynthesis, under excessive protein consumption or during certain disease states, certain proteins can be catabolized to acetyl-CoA. Non-essential fatty acids, trans-fatty acids, and saturated fats, and refined carbohydrates are general source of excessive acetyl-CoAwhich pressurize our body to biosynthesize cholesterol. In other words, cholesterol is formed from excess calories which usually are generated most often from carbohydrates and fats.

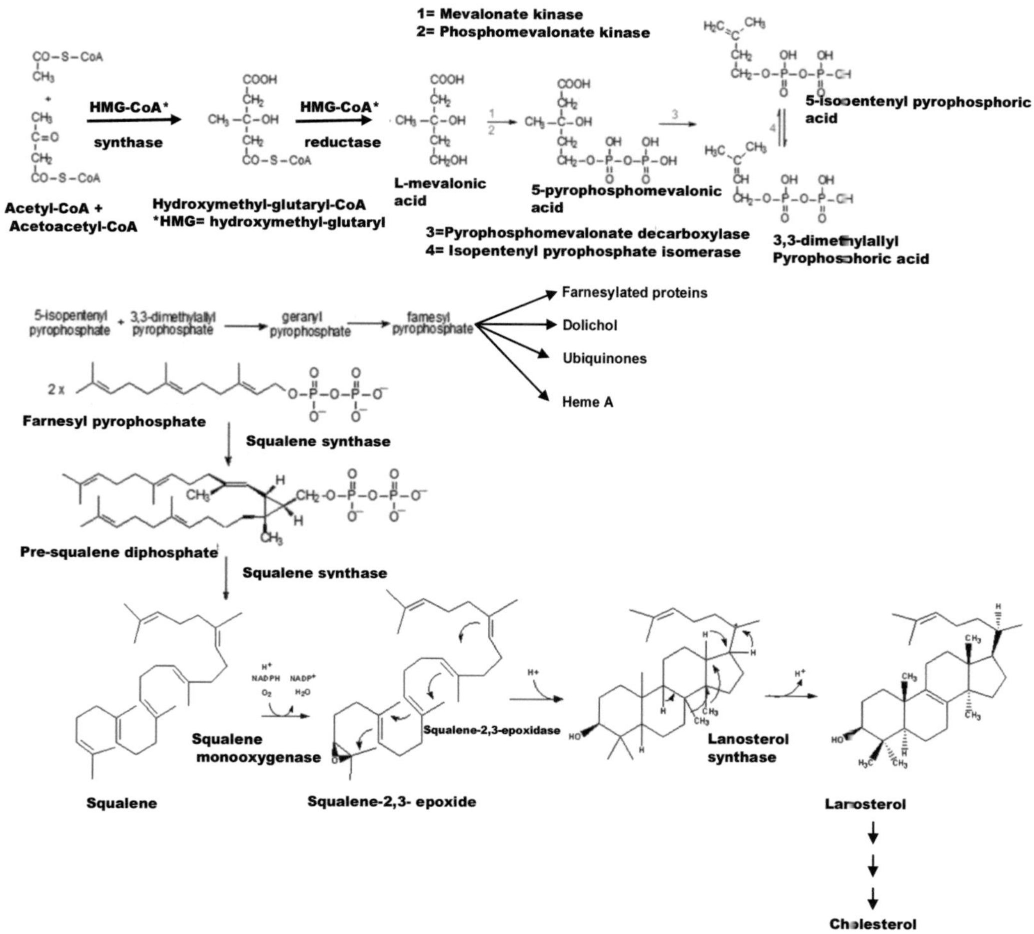

Fig. 1. Cholesterol Biosynthetic Pathway

The process of cholesterol synthesis has five major steps:

1. Acetyl-CoAs are converted to 3-hydroxy-3-methylglutaryl-CoA (HMG-CoA)
2. HMG-CoA is converted to mevalonate
3. Mevalonate is converted to the isoprene based molecule, isopentenyl pyrophosphate (IPP), with the concomitant loss of CO_2
4. IPP is converted to squalene
5. Squalene is converted to cholesterol.

Acetyl-CoA units are converted to mevalonate by a series of reactions that begins with the formation of **HMG-CoA (Figure 1).** Unlike the HMG-CoA formed during ketone body synthesis in the mitochondria, this form is synthesized in the cytoplasm. However, the pathway and the necessary enzymes are the same as those in the mitochondria. Two moles of acetyl-CoA are condensed in a reversal of the thiolase reaction, forming acetoacetyl-CoA. Acetoacetyl-CoA and a third mole of acetyl-CoA are converted to HMG-CoA by the action of HMG-CoA synthase. HMG-CoA is converted to mevalonate by HMG-CoA reductase, HMGR (this enzyme is bound in the endoplasmic reticulum, ER). HMGR absolutely requires NADPH as a cofactor and two moles of NADPH are consumed during the conversion of HMG-CoA to mevalonate. The reaction catalyzed by HMGR is the rate limiting step of cholesterol biosynthesis, and this enzyme is subject to complex regulatory controls which will be discussed in separate section of this book chapter.

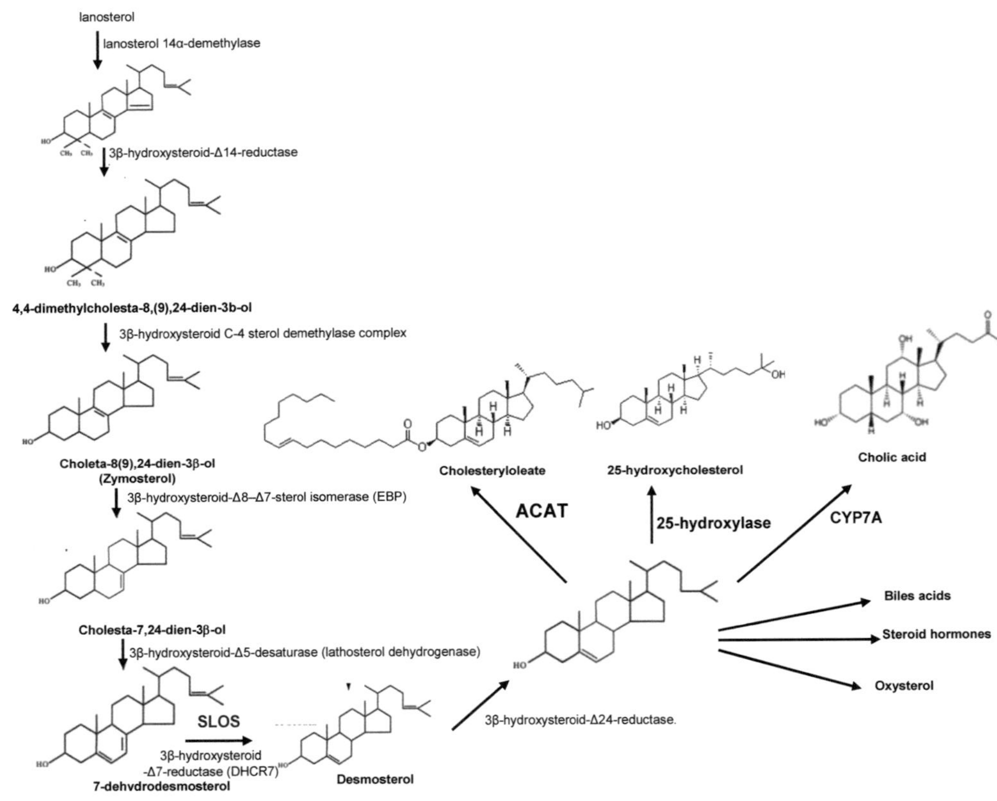

Fig. 2. Post squalene pathway of cholesterol and other sterol Biosynthesis

Mevalonate is then activated by three successive phosphorylations, yielding 5-pyrophosphomevalonate. Phosphorylation mevalonate and successive reactions maintain its solubility, since otherwise these are insoluble in water. After phosphorylation, an ATP-dependent decarboxylation yields isopentenyl pyrophosphate, IPP, an activated isoprenoid molecule. Isopentenyl pyrophosphate is in equilibrium with its isomer, dimethylallyl pyrophosphate, DMPP. One molecule of IPP condenses with one molecule of DMPP to generate geranyl pyrophosphate, GPP. GPP further condenses with another IPP molecule to yield farnesyl pyrophosphate, FPP. Finally, the NADPH-requiring enzyme, squalene synthase catalyzes the head-to-tail condensation of two molecules of FPP, yielding squalene (squalene synthase also is tightly associated with the endoplasmic reticulum). Squalene undergoes a two step cyclization to yield lanosterol catalyzed by sequalene monooxygenase and sequalene 2, 3 epoxidase enzymes. Sequalene mono oxygenase is the second committed step in cholesterol biosynthesis and lead to the formation squalene 2, 3 epoxide. This enzymatic reaction require supernatant protein factor (SPF) and NADPH as a cofactor to introduce molecular oxygen as an epoxide at the 2, 3 position of squalene. The activity of supernatant protein factor itself is regulated by phosphorylation/dephosphorylation (Singh et al., 2003). Through a series of 19 additional reactions, lanosterol is converted to cholesterol. The first sterol intermediate, lanosterol, is formed by the condensation of the 30 carbon isoprenoid squalene as explained above and figure 1., and subsequent enzymatic reactions define the 'post-squalene' half of the pathway figure 2.

The conversion of lanosterol to cholesterol involves the reduction of the C-24 double bond, removal of three methyl groups at the C-14 and C-4 positions, and 'migration' of the C-8(9) double bond (Figure 2) (for a review, see (Herman, 2003)). Some of the enzymatic reactions must occur in sequence; for example, Δ^8–Δ^7 isomerization cannot precede C-14α demethylation. The saturation of the C-24 double bond of lanosterol can occur at multiple points in the pathway, creating two immediate precursors for cholesterol, desmosterol [cholesta-5(6), 24-dien-3β-ol] and 7-dehydrocholesterol (7DHC), whose relative abundance may vary among different tissues. Desmosterol, in particular, appears to be abundant in the developing mammalian brain (Herman, 2003). Several post-squalene sterol intermediates serve additional cellular functions as well. The C-14 demethylated derivatives of lanosterol, 4,4-dimethyl-5α-cholesta-8,14,24-trien-3β-ol and 4,4-dimethyl-5α-cholesta-8,24-dien-3β-ol, have meiosis-stimulating activity and accumulate in the ovary and testis, respectively (Rozman et al., 2002). 7-Dehydrocholesterol is the immediate precursor for vitamin D synthesis. Selected human enzymes of post-squalene cholesterol biosynthesis have also been identified based on homology to sterol biosynthetic enzymes from *Arabidopsis thaliana* (Herman, 2003; Waterham et al., 2001). In addition, cholesterol and other sterol intermediates can be converted to oxysterols that can act as regulatory signaling molecules and bind orphan nuclear receptors such as LXRα (Fitzgerald et al., 2002).

Normal healthy adults synthesize cholesterol at a rate of approximately 1g/day and consume approximately 0.3g/day. A relatively constant level of cholesterol in the body (150 - 200 mg/dL) is maintained primarily by controlling the level of *de novo* synthesis which is partly regulated in part by the dietary intake of cholesterol. Cholesterol from both diet and synthesis is utilized in the formation of membranes and in the synthesis of the steroid hormones and bile acids. The greatest proportion of cholesterol is used in bile acid synthesis.

1.1.1 Regulation of cholesterol biosynthetic pathway

Regulation of the pathway includes sterol-mediated feedback of transcription of several of the genes, including the rate-limiting enzyme HMGR and HMG-CoA synthase, as well as a variety of post-transcriptional mechanisms. Recently, increased attention has been focused on the regulated degradation of HMGR in the ER. The cellular supply of cholesterol is maintained at a steady level by four distinct mechanisms:

1. Regulation of HMGR activity and levels (upstream regulation of cholesterol biosynthesis)
2. Regulation of sequalene mono-oxygenase activity by sequalene supernant protein factor (down stream biosynthesis of cholesterol). Once this step is taken place then it ends up biosynthesis of cholesterol.

3. Regulation of excess intracellular free cholesterol through the activity of acyl-CoA: cholesterol acyltransferase, ACAT (internalization of excessive cholesterol).
4. Regulation of plasma cholesterol levels via LDL receptor-mediated uptake and HDL-mediated reverse transport.

The first seven enzymes of cholesterol biosynthesis are soluble proteins with the exception of 3-hydroxy-3-methylglutaryl CoA reductase (HMGR), which is an integral endoplasmic reticulum (ER) membrane protein (Gaylor, 2002; Goldstein and Brown, 1990; Kovacs et al., 2002). HMG-CoA for cholesterol biosynthesis is generated within the cytosol. It is also synthesized within the mitochondria where its hydrolysis by HMG-CoA lyase generates ketones for energy during fasting (Herman, 2003). Reactions generating mevalonate can also occur within the peroxisome, and the subsequent reactions that result in the production of farnesyl-pyrophosphate are exclusively peroxisomal. The remainder of the biosynthetic reactions occurs in the ER with enzymes and substrates that are membrane-bound. Thus, cholesterol biosynthetic enzymes are compartmentalized in the cytosol, ER and/or peroxisome, adding another level of complexity to the regulation of this metabolic pathway. Particularly, direct regulation of HMGCoA reductase activity means for controlling the level of cholesterol biosynthesis. Recently, increased attention has been focused on the regulated degradation of HMGR in the ER.

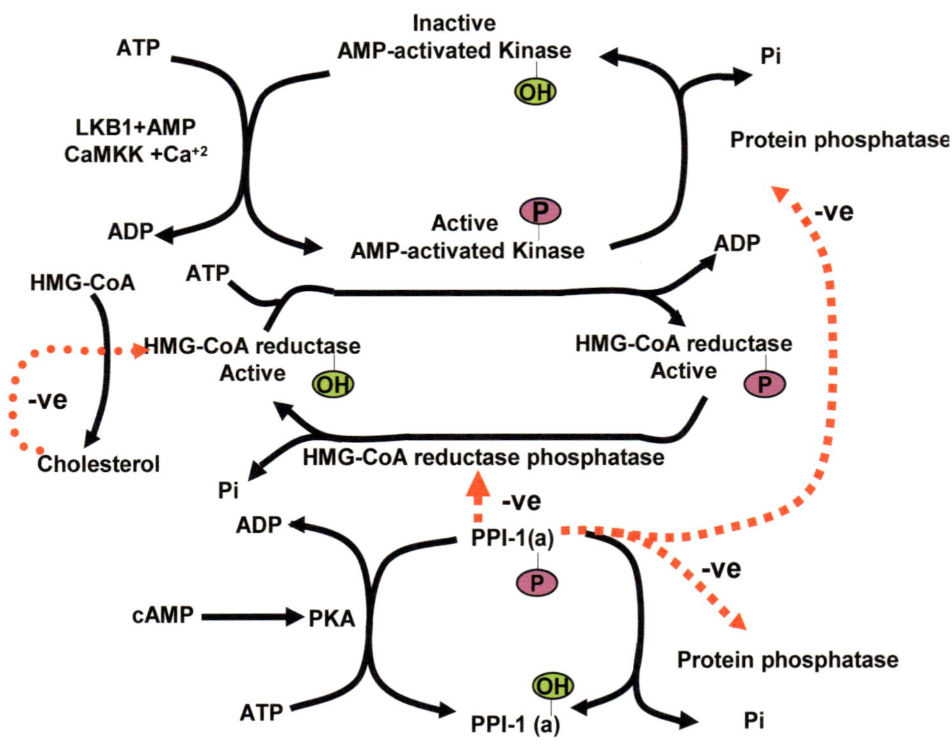

Fig. 3. Regulation of HMGCoA reductase

In vivo system, the activity of HMGCoA reductase is controlled by four distinct mechanisms: feed-back inhibition, control of gene expression, rate of enzyme degradation and phosphorylation-dephosphorylation. Cholesterol controls the first three mechanisms itself as cholesterol acts as a feed-back inhibitor of pre-existing HMGR as well as inducing rapid degradation of the enzyme. The latter is the result of cholesterol-induced polyubiquitination of HMGR and its degradation in the proteosome (explained in separate section in details). This ability of cholesterol is a consequence of the **sterol sensing domain, SSD** of HMGR. In addition, when cholesterol is in excess the amount of mRNA for HMGR is reduced as a result of decreased expression of the gene. By covalent modification of HMGreductase through the process phosphorylation/ dephosphorylation plays a very important role in regulation of HMGR. The enzyme is most active in its unmodified form. Phosphorylation of the enzyme decreases its activity (Figure 3). HMGR is phosphorylated by adenosine mono phosphate-activated protein kinase, **(AMPK)which is** itself activated via phosphorylation catalyzed by 2 enzymes. LKB1 is the primary kinase sensitive to rising AMP levels which was first identified as a gene in humans carrying an autosomal dominant mutation in Peutz-Jeghers syndrome, PJS and mutated in lung adenocarcinomas. The second AMPK phosphorylating enzyme is calmodulin-dependent protein kinase kinase-beta (CaMKKβ) which induces phosphorylation of AMPK due to increase in the level of intracellular Ca^{2+} as a result of muscle contraction. The activity of HMGR is additionally controlled by the cAMP signaling pathway (Figure 3). Enhanced level of cAMP lead to activation of cAMP-dependent protein kinase, PKA. In the context of HMGR regulation, PKA phosphorylates phosphoprotein phosphatase inhibitor-1 (PPI-1) leading to an increase in its activity. PPI-1 can inhibit the activity of numerous phosphatases including protein phosphatase 2C (PP2C) and HMG-CoA reductase phosphatase which remove phosphates from AMPK and HMGR, respectively. This maintains AMPK in the phosphorylated and active state, and HMGR in the phosphorylated and inactive state. As the stimulus leading to increased cAMP production is removed, the level of phosphorylations decreases and that of dephosphorylations increases. The net result is a return to a higher level of HMGR activity. The intracellular level of cAMP itself is regulated by hormonal stimuli, thus regulation of cholesterol biosynthesis is hormonally controlled. One of the most common hormone, insulin leads to a decrease in cAMP, which in turn activates cholesterol biosynthesis. Alternatively, glucagon and epinephrine increase the level of cAMP in turn lead to inhibition of cholesterol biosynthesis. The basic function of epinephrine and glucagon hormones is to control the availability and delivery of energy in all the cells in our body. Degradation of HMGR and inhibition of its biosynthesis, are the two long term regulatory processes for cholesterol biosynthesis as when the levels of cholesterol are high, resulted in reduction in the expression of the HMGR gene. On other hand, low levels of cholesterol activate its expression.

1.1.2 Proteolytic regulation of HMG-CoA

The stability of HMGR is regulated as the rate of flux through the mevalonate synthesis pathway changes. When the flux is high the rate of HMGR degradation is also high. When the flux is low, degradation of HMGR decreases. This phenomenon can easily be observed in the presence of the statin drugs. HMGR is localized to the ER and like SREBP contains a sterol-sensing domain, SSD. When sterol levels increase in cells there is a concomitant increase in the rate of HMGR degradation. The degradation of HMGR occurs within the proteosome, multiprotein complex dedicated for protein degradation.The primary signal directing proteins to the proteosome is ubiquitination. Ubiquitin is a 7.6kDa protein that is covalently attached to proteins targeted for degradation by ubiquitin ligases (Kimura and Tanaka, 2010). These enzymes attach multiple copies of ubiquitin allowing for recognition by the proteosome. HMGR has been shown to be ubiquitinated prior to its degradation. The primary sterol regulating HMGR degradation is cholesterol itself. As the levels of free cholesterol increase in cells, the rate of HMGR degradation increases.

1.2 Disorders of post-sequalene cholesterol biosynthesis

Seven disorders (Smith-Lemli-Opitz, desmosterolosis, X-linked dominant chondrodysplasis punctata, child syndrome, lathosterolosis and hydrops-ectopic calcification-moth-eaten skeletal dysplasia) of post-squalene cholesterol biosynthesis have been reported, only the most commonly disorder, Smith-Lemli-Opitz syndrome (SLOS) will be discussed in this section of the book as described and reviewed else where (Herman, 2003). SLOS was the first described disorder of post-squalene cholesterol biosynthesis and is by far the most common, with an incidence of approximately 1/40 000 to 1/50 000 in the USA (reviewed in (Kelley and Hennekam, 2000). It was initially described in 1964 as an autosomal recessive major malformation syndrome. In 1993, Irons *et al.(Irons et al., 1993)* detected decreased plasma cholesterol levels and elevated 7DHC in several patients with SLOS, suggesting an enzymatic efficiency of 7-dehydrocholesterol reductase (7DHCR).

1.3 Regulation of cholesterol biosynthesis at transcriptional level

As cells need more sterol they will induce their synthesis and uptake, conversely when the need declines synthesis and uptake are decreased. Regulation of these events is brought about primarily by sterol-regulated transcription of key rate limiting enzymes and by the regulated degradation of HMGR. Activation of transcriptional control occurs through the regulated cleavage of the membrane-bound transcription factor sterol regulated element binding protein (**SREBP**). As discussed earlier in this book chapter, degradation of HMGR is controlled by the ubiquitin-mediated pathway for proteolysis. Sterol control of transcription affects more than 30 genes involved in the biosynthesis of cholesterol, triacylglycerols, phospholipids and fatty acids. Transcriptional control requires the presence of an octamer sequence in the gene termed the sterol regulatory element-1 (SRE-1). It has been shown that SREBP is the transcription factor that binds to SRE-1 elements. It turns out that there are 2 distinct SREBP genes, SREBP-1 and SREBP-2. In addition, the SREBP-1 gene encodes 2 proteins, SREBP-1a and SREBP-1c/ADD1 (ADD1 is adipocyte differentiation-1) as a consequence of alternative exon usage. SREBP-1a regulates all SREBP-responsive genes in both the cholesterol and fatty acid biosynthetic pathways. SREBP-1c controls the expression of genes involved in fatty acid synthesis and is involved in the differentiation of adipocytes. SREBP-1c is also an essential transcription factor downstream of the actions of insulin at the level of carbohydrate and lipid metabolism. SREBP-2 is the predominant form of this transcription factor in the liver and it exhibits preference at controlling the expression of genes involved in cholesterol homeostasis, including all of the genes encoding the sterol biosynthetic enzymes. In addition SREBP-2 controls expression of the LDL receptor gene.

Regulated expression of the SREBPs is complex in that the effects of sterols are different on the SREBP-1 gene versus the SREBP-2 gene. High sterols activate expression of the SREBP-1 gene but do not exert this effect on the SREBP-2 gene. The sterol-mediated activation of the SREBP-1 gene occurs via the action of the liver X receptors (LXRs). The LXRs are members of the steroid/ thyroid hormone super family of cytosolic ligand binding receptors that migrate to the nucleus upon ligand binding and regulate gene expression by binding to specific target sequences. There are two forms of the LXRs: LXRα and LXRβ. The LXRs form heterodimers with the retinoid X receptors (RXRs) and as such can regulate gene expression either upon binding oxysterols (e.g. 22*R*-hydroxycholesterol) or 9-*cis*-retinoic acid. All 3 SREBPs are proteolytically activated and the proteolysis is controlled by the level of sterols in the cell. Full-length SREBPs have several domains and are embedded in the membrane of the endoplasmic reticulum (ER). The N-terminal domain contains a transcription factor motif of the basic helix-loop-helix (bHLH) type that is exposed to the cytoplasmic side of the ER. There are 2 transmembrane spanning domains followed by a large C-terminal domain also exposed to the cytosolic side. The C-terminal domain (CTD) interacts with a protein called SREBP cleavage-activating protein (**SCAP**). SCAP is a large protein also found in the ER membrane and contains at least 8 transmembrane spans. The C-terminal portion, which extends into the cytosol, has been shown to interact with the C-terminal domain of SREBP. This C-terminal region of SCAP contains 4 motifs called WD40 repeats. The WD40 repeats are required for interaction of SCAP with SREBP. The regulation of SREBP activity is further controlled within the ER by the interaction of SCAP with insulin regulated protein (**Insig**). When cells have sufficient sterol content SREBP and SCAP are retained in the ER via the SCAP-Insig interaction. The N-terminus of SCAP, including membrane spans 2–6, resembles HMGR which itself is subject to sterol-stimulated degradation (see above). This shared motif is called the sterol sensing domain (SSD) and as a consequence of this domain SCAP functions as the cholesterol sensor in the protein complex. When cells have sufficient levels of sterols, SCAP will bind cholesterol which promotes the interaction with Insig and the entire complex will be maintained in the ER.

The Insig proteins bind to oxysterols which in turn affects their interactions with SCAP. Insig proteins can cause ER retention of the SREBP/SCAP complex. In addition to their role in regulating sterol-dependent gene regulation, both Insig proteins activate sterol-dependent degradation of HMGR.

When sterols are scarce, SCAP does not interact with Insig. Under these conditions the SREBP-SCAP complex migrates to the Golgi where SREBP is subjected to proteolysis. The cleavage of SREBP is carried out by 2 distinct enzymes. The regulated cleavage occurs in the lumenal loop between the 2 transmembrane domains. This cleavage is catalyzed by site-1 protease, **S1P**. The function of SCAP is to positively stimulate S1P-mediated cleavage of SREBP. The second cleavage, catalyzed by site-2 protease, **S2P**, occurs in the first transmembrane span, leading to release of active SREBP. In order for S2P to act on SREBP, site-1 must already have been cleaved. The result of the S2P cleavage is the release of the N-terminal bHLH motif into the cytosol. The bHLH domain then migrates to the nucleus where it will dimerize and form complexes with transcriptional coactivators leading to the activation of genes containing the SRE motif. To control the level of SREBP-mediated transcription, the soluble bHLH domain is itself subject to rapid proteolysis.

1.4 Interaction of cholesterol with other lipids

Cholesterol plays a vital role in determining the physiochemical properties of cell membranes. However, the detailed nature of cholesterol–lipid interactions is a subject of ongoing debate. Cholesterol primarily serves as a structural component of cellular membranes. When incorporated into phospholipid bilayers, cholesterol aligns so that its polar hydroxyl group is near the interface with the aqueous environment while its hydrophobic body is buried in the bilayer (Ohvo-Rekila et al., 2002; Olsen et al., 2011a; Olsen et al., 2011b). The interaction of cholesterol with neighboring phospholipids alters membrane structure. The alignment and ordering of nearby phospholipid tails causes membrane condensation, decreasing the area of the membrane and increasing the thickness (Ohvo-Rekila et al., 2002). Cholesterol also broadens the liquid-to-solid phase transition, inducing an intermediate liquid-ordered phase that retains lateral mobility while increasing lipid order (Feigenson, 2007; Simons and Vaz, 2004; van Meer et al., 2008). These changes result in a mechanically stronger membrane with decreased permeability due to tighter packing among lipids (Ikonen, 2008; Simons and Vaz, 2004). The low activity pool consists of cholesterol that is sequestered within the phospholipids and relatively inaccessible to other molecules, while the high activity pool of cholesterol that is more accessible and mobile in the non-condensed phospholipids of the membrane (Olsen et al., 2011b). Distribution between the high and low activity pools is determined by the ability of the phospholipids to condense with cholesterol, which in turn is dependent on the phospholipid composition of the membrane. Thus, raising the plasma membrane cholesterol concentration can saturate the ability of the membrane to accommodate cholesterol in the condensed phospholipids, whereupon excess cholesterol transitions into the high activity pool where it is more available for trafficking to the ER. Sphingomyelin is other important lipid present in the biomembrane that interact with cholesterol (Garmy et al., 2005). Interaction between cholesterol and sphingomyelin has a very high biological significance as lipid-lipid interaction leads to the formation of ordered lipid domains in the plasma membrane of eukaryotic cells (Simons and Ikonen, 1997). Molecular association between cholesterol and sphingomyelin is very important as this constitute cholesterol enriched microdomains of plasma membrane and plays a very important role in cellular functions such as the control of signal transduction pathways. On the other hand, the formation of cholesterol-sphingomyelin molecular complexes in the intestinal lumen explains the mutual inhibitory effects of cholesterol and sphingomyelin on their intestinal absorption (Nyberg et al., 2000).

1.5 Cholesterol homeostasis

Every eukaryotic cell require cholesterol as its not only integral part of membrane as well as plays very important role in cell signalling pathways. That the reason, all eukaryotic cells, which have specialized methods of recruiting and synthesizing the cholesterol only when it is needed. While effectively maintaining intracellular cholesterol homeostasis, these processes leave excess circulating though the body, leading to atherosclerotic plaque development and subsequent coronary artery disease. Thus, levels of cholesterol and related lipids circulating in plasma are important predictive tools utilized clinically to diagnose the risk of a cardiovascular diseases (Daniels et al., 2009; Ikonen, 2008; Simons and Ikonen, 2000; Singh et al., 2007). Cholesterol is transported in the plasma predominantly as cholesteryl esters associated with lipoproteins and dietary cholesterol is transported from the small intestine to the liver in the form of chylomicrons. Cholesterol synthesized by the liver, as well as any dietary cholesterol in the liver that exceeds hepatic needs, is transported in the serum in the form of low density lipoproteins (LDLs). In the liver, VLDLs are biosyntheiszed and are converted to LDLs by endothelial cell-associated lipoprotein lipase. High density lipoproteins (HDLs)can extract Cholesterol found in plasma membranes and esterified by the HDL-associated enzyme LCAT. The cholesterol acquired from peripheral tissues by HDLs can then be transferred to VLDLs and LDLs via the action of cholesteryl ester transfer protein (apo-D) which is associated with HDLs. In humans, HDL levels are a very well known measurement of cardiac health due to their strong inverse relationship with coronary artery disease. Peripheral cholesterol is returned to the liver by the process called **reverse cholesterol transport by HDLs** and ultimately, cholesterol is excreted in the bile as free cholesterol or as bile salts following conversion to bile acids in the liver (Figure 4).

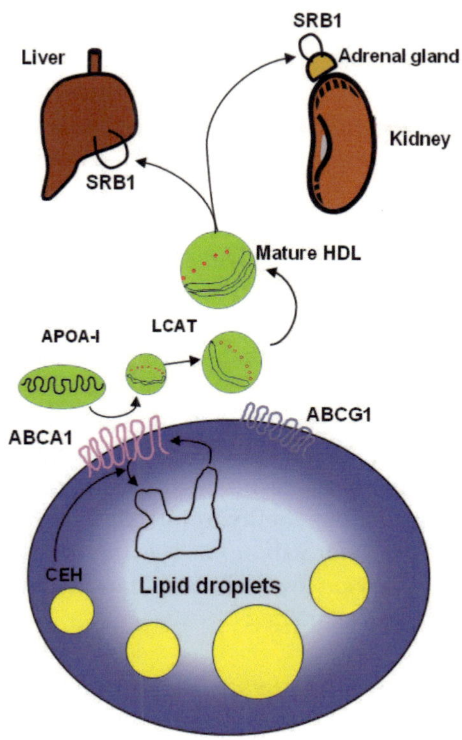

Fig. 4. Cholesterol homeostasis

1.6 Cholesterol and ion channels

A number of studies has demonstrated that the level of membrane cholesterol regulate the ion channel functions *(Levitan et al., 2010; Maguy et al., 2006; Martens et al., 2004)*. The impact of cholesterol on different types of ion channels is highly heterogeneous. In most of the cases cholesterol supresses channel activity that may include decrease in the open probability, unitary conductance and/or the number of active channels on the membrane. This effect was observed in several types of K^+ channels, voltage-gated Na^+ and Ca^{+2} channels, as well as in volume-regulated anion channels. However, there are also several types of ion channels, such as epithelial Na^+ channels (eNaC) and transient receptor potential (Trp) channels that are inhibited by the removal of membrane cholesterol. Finally, in some cases changes in membrane cholesterol affect biophysical properties of the channel such as the voltage dependence of channel activation or inactivation. Clearly, therefore more than one mechanism has to be involved in cholesterol-induced regulation of different ion channels.

Studies from our laboratory have shown that many major and important ion channels are regulated by changes in the level of membrane cholesterol (Epshtein et al., 2009; Levitan, 2009; Levitan et al., 2010; Rosenhouse-Dantsker et al., 2011; Singh et al., 2009b; Singh et al., 2011). In general three different mechanisms may be involved in regulation of ion channels by cholesterol: (i) specific interactions (ii) changes in the physical properties of the membrane bilayer and (iii) maintaining the scaffolds for protein-protein interactions. Furthermore, our recent data for the first time demonstrate a specific cholesterol-channel binding (Singh et al., 2011). One possibility is that cholesterol may interact directly and specifically with the transmembrane domains of the channels protein. Direct interaction between channels and cholesterol as a boundary lipid was first proposed in a "lipid belt" model by (Marsh and Barrantes, 1978) suggesting that cholesterol may be a part of a lipid belt or a "shell" constituting the immediate perimeter of the channel protein. Moreover, studies from our laboratory demonstrated that inwardly-rectifying K^+ channel are sensitive to the chiral nature of the sterol analogue providing further support for the hypothesis that sensitivity of these channels to cholesterol can be due to specific sterol-protein interactions (Romanenko et al., 2002). An alternative mechanism proposed by Lundbaek and colleagues (Lundbaek and Andersen, 1999) suggested that cholesterol may regulate ion channels by hydrophobic mismatch between the transmembrane domains and the lipid bilayer. More specifically, it was proposed that when a channel goes through a change in conformation state within the viscous medium of the lipid membrane it may induce deformation of the lipid bilayer surrounding the channel. If this is the case, then a stiffer less deformable membrane will increase the energy that is required for the transition (Levitan et al., 2010). It is important to note that these mechanisms are not mutually exclusive. A lipid shell surrounding a channel may also affect the hydrophobic interactions between the channels and the lipids and increase the deformation energy required for the transitions between closed and open states. Finally, obviously, cholesterol may also affect the channels indirectly through interactions with different signalling cascades.

1.7 Cholesterol and oxysterol as major oxidized component in oxidized LDL

Oxidation of LDL is considered as the major risk factors for the development of coronary artery disease (CAD) and plaque formation (reviewed in (Berliner et al., 2001; Levitan and Shentu, 2011). Indeed, elevated levels of oxLDL are associated with an increased risk of CAD (Toshima et al., 2000) and correlate with plasma hypercholesterolemia both in humans (van Tits et al., 2006) and in the animal models of atherosclerosis (Hodis et al., 1994; Holvoet et al., 1998). It is also well-known that exposure to oxLDL induces an array of proinflammatory and proatherogenic effects but the mechanisms that underlie oxLDL-induced effects remain controversial. Most of studies involving oxLDL are based on ex-vivo oxidation of LDL. The term oxidized LDL is used to describe LDL preparations which have been oxidatively modified *ex vivo* under defined conditions, or isolated from biological sources.

The most typical procedure of LDL oxidation *ex vivo* is incubation of LDL with metal ions, Cu^{2+} in particular, that leads to the generation of multiple oxidized products in the LDL particle, including oxysterols, oxidized phospholipids, and modified apolipoprotein B (reviewed in (Burkitt, 2001; Levitan and Shentu, 2011). The oxidized LDL preparations described in the literature are broadly divided into two main categories: "minimally modified LDL" (MM-LDL) and (fully or extensively) oxidized LDL (oxLDL) based on the degree of LDL oxidation. Cu^{2+} oxidation of LDL can generate both minimally modified and fully oxidized LDLs depending on the duration of the exposure and ion concentration. Two other procedures that are also used to generate oxLDL *ex vivo* are enzymatic oxidation by 15-lypoxygenase or myeloperoxidase or by incubating LDL with 15-lypoxygenase expressing cells (Boullier et al., 2006). It is important to note that while it is controversial whether Cu^{2+} oxidation occurs *in vivo,* it was shown that there are significant similarities between Cu^{2+} oxidized LDL and oxLDL found in atherosclerotic lesions (Yla-Herttuala et al., 1989). **However, oxidation of LDL is a complex process that yields an array of bioactive compounds with different biological properties and its composition depends on the degree of LDL oxidation** It is known that oxLDL includes various oxidized products that are divided into neutral lipids, a group that includes cholesterol and cholesterol ester, and phospholipids that include SM, PE, LPC and PC (Subbanagounder et al., 2000). Our recent studies have identififed five major oxysterols in strongly oxidized LDL (16 hours of oxidation data submitted for publication) by the GC analysis: 7α-hydroxycholesterol , 7β-hydroxycholesterol, cholesterol 5α,6α- and 5β,6β-epoxides, and 7-ketocholesterol similar to earlier studies (Brown et al., 1996). Our recent unpublished data showed that the most abundant oxysterol is 7β-hydroxycholesterol, followed by 7-ketocholesterol with 7α-hydroxycholesterol and cholesterol 5β,6β-epoxide representing minor oxysterols in oxLDL. Two oxysterols, 25-hydroxycholesterol and 27-hydroxycholesterol do not constitute significant components of oxLDL complex, but were found in human atherosclerotic lesions (Bjorkhem et al., 1994; Zidovetzki and Levitan, 2007). 27-hydroxycholesterols, specially is abundant in human atherosclerotic lesions and macrophage-derived foam cells (Brown and Jessup, 1999).

1.8 Impact of oxLDL on cholesterol-rich membrane rafts

Membrane rafts were originally described as cholesterol- and sphingolipid-rich microdomains that provide platforms for protein-protein interactions in multiple signaling cascades (Brown and London, 1998; Simons and Gerl, 2010; Simons and Ikonen, 1997). Membrane rafts are small (10–200 nm), heterogeneous, highly dynamic, sterol- and sphingolipid-enriched domains that compartmentalize cellular processes" (Pike, 2006). Most recently, Simons and Gerl (Simons and Gerl, 2010) further defined membrane rafts as "dynamic, nonoscale, sterol-sphingolipid-enriched, ordered assembles of proteins and lipids" that are regulated by specific lipid-lipid, protein-lipid, and protein-protein interactions (Simons and Gerl, 2010). The goal of this section of book chapter is to discuss the recent advances in our understanding of the impact of oxLDL on membrane rafts. Oxysterols are found in abundance in Cu^{2+}-oxidized LDL, in which cholesterol is oxidized preferably at 7 positions resulting in the generation of 7-ketocholesterol, 7*β*-hydroxycholesterol, and 7*α*-hydroxycholesterol (Brown and Jessup, 1999). In addition, 27-hydroxycholesterol, that is, generated *in vivo,* has been shown to accumulate in foam cells in atherosclerotic lesions (Brown and Jessup, 1999). Several studies have shown that oxysterols result in inhibition of cholesterol efflux in the mouse and it was suggested that impairment of cholesterol homeostasis by the inhibition of cholesterol efflux may be mechanism by which oxysterols affect cellular function (Kilsdonk et al., 1995; Terasaka et al., 2008). Interestingly, 7-ketocholesterol was shown to deplete cholesterol specifically from the raft domains in human macrophages (Gaus et al., 2004) and disrupt lipid packing of the immunological synapses in sterol-enriched T lymphocytes (Rentero et al., 2008) and in cholesterol-rich membrane domains in endothelial cells (Shentu et al., 2010). These observations suggest that incorporation of oxysterols may also play an important role in oxLDL-induced disruption of cholesterol-rich membrane domains.

1.9 Manipulation of cholesterol in cellular system (*in vitro* and *in vivo* models)

The physiological importance of cholesterol in the cell plasma membrane has attracted increased attention in recent years. Consequently, the use of methods of controlled manipulation of membrane cholesterol content has also increased sharply, especially as a method of studying putative cholesterol-enriched cell membrane domains (rafts). The most common means of modifying the cholesterol content of cell membranes is the incubation of cells or model membranes with cyclodextrins, a family of compounds, which, due to the presence of relatively hydrophobic cavity, can be used to extract cholesterol from cell membranes (Zidovetzki and Levitan, 2007). Under conditions commonly used for cholesterol extraction, cyclodextrins may remove cholesterol from both raft and non-raft domains of the membrane as well as alter the distribution of cholesterol between plasma and intracellular membranes. In addition, other hydrophobic molecules such as phospholipids may also be extracted from the membranes by cyclodextrins. Here, we discuss useful control strategies that may help to verify that the observed effects are due specifically to cyclodextrin-induced changes in cellular cholesterol.

The high affinity of methyl-β-cyclodextrin (MβCDs) for cholesterol can be used not only to remove cholesterol from the biological membranes but also to generate cholesterol inclusion complexes that donate cholesterol to the membrane and increase membrane cholesterol level. MβCD-cholesterol inclusion complexes are typically generated by mixing cholesterol suspension with a cyclodextrin solution as described earlier (Christian et al., 1997; Klein et al., 1995; Levitan et al., 2000; Zidovetzki and Levitan, 2007). The ratio between the amounts of cholesterol and cyclodextrin in the complex determines whether it will act as cholesterol acceptor or as cholesterol donor (Zidovetzki and Levitan, 2007). The efficiency of cholesterol transfer from MβCD inclusion complex to biological membranes depends on MβCD:cholesterol molar ratio, MβCD-cholesterol concentration, and duration of the exposure (Zidovetzki and Levitan, 2007). Thus, it is important to note that exposing cells to MβCD-cholesterol complexes that contain saturating amounts of cholesterol typically results not just in replenishing cholesterol to control levels but in significant cholesterol enrichment. Cholesterol enrichment is observed even if the cells were first depleted of cholesterol and then exposed to MβCD-cholesterol complexes. ApoE -/- mice cholesterol ester transfer protein (CETP) knock out mice are ideal *in vivo* animal model to study cholesterol biosynthetic pathway and its regulation. Our recent study revealed that WT and ApoE -/- mice fed on normal chow diet for 4 weeks, showing a significant increase in the level of cholesterol in ApoE -/- mice as compared to WT (Shentu et al., 2011).

1.10 Treatment of hypercholesterolemia

Reductions in circulating cholesterol levels can have profound positive impacts on cardiovascular disease, particularly on atherosclerosis, as well as other metabolic disruptions of the vasculature. Control of dietary intake is one of the easiest and least cost intensive means to achieve reductions in cholesterol. But in most of the alleviated levels of cholesterol cannot be controlled merely with exercise. Drug therapy is very important to avoid the cardiovascular effects of high cholesterol levels in such patients. In this section of the book chapter, we have discussed the regulation cholesterol by drugs as well as will review recent work from our laboratory for the regulation of cholesterol by natural products such as tea/green tea, policosanol and garlic compounds.

1.10.1 Regulation of cholesterol by natural products

In this section of book the regulation of cholesterol by natural products, tea, policosanol and garlic is discussed.

1.10.1.1 Tea and tea compounds

Epidemiological studies have indicated that tea consumption is associated with a lower risk of cardiovascular disease. This decreased risk is attributed to the ability of tea to lower serum cholesterol levels, and several clinical studies have demonstrated that black tea can lower serum total- and LDL-cholesterol (Davies et al., 2003; Maron et al., 2003). Green tea has been shown to be hypocholesterolemic in animal studies, with the bulk of evidence indicating that tea polyphenols reduce the absorption of dietary and biliary cholesterol and promote its fecal excretion (Koo and Noh, 2007)..

Feeding studies have been equivocal on the ability of green tea extract to inhibit cholesterol synthesis. Although a recent study by Bursill and colleagues (Bursill et al., 2007) showed a decrease in serum lathosterol (an indicator of whole body cholesterol synthesis) in rabbits fed a green tea extract, a similar study with rats by these investigators (Bursill and Roach, 2007) was unable to demonstrate a decrease in this serum sterol, despite significant reductions in hepatic cholesterol levels and an increase in LDL receptor expression. A feeding study by Chan *et al.* (Chan et al., 1999) was similarly unable to demonstrate an effect of green tea extract on hepatic HMG-CoA reductase activity. Measuring cholesterol synthesis *in vivo* is difficult, whereas *in vitro* studies are more tractable. In this regard, Gebhardt and colleagues reported that several common polyphenols (luteolin, quercetin) were able to decrease cholesterol synthesis when added to cultured hepatocytes or hepatoma cell cultures (Gebhardt, 2003). This inhibition appeared to occur at the level of HMG-CoA reductase. Tea polyphenols (Abe et al., 2000), as well as the simple polyphenol resveratrol have been shown to directly inhibit squalene monooxygenase, a rate-limiting downstream enzyme in cholesterol synthesis. Two studies by Bursill and colleagues (Bursill et al., 2001; Bursill and Roach, 2006) demonstrated an increase in HMG-CoA reductase and LDL-receptor mRNA in HepG2 cells incubated with green tea extract or its principal component, epigallocatechin gallate (EGCG), and a decrease in cellular lathosterol, indicating that cholesterol synthesis was inhibited in treated cells (Singh et al., 2009a). Together, these studies suggest that green tea polyphenols are inhibitory to cholesterol synthesis by inhibiting HMGCoA reductase (Singh et al., 2009a). Moreover, the effect of black tea extract, which consists predominantly of a diverse mixture of polymerized polyphenols termed theaflavins and thearubigins, has not been examined, despite the recent clinical evidence that black tea can modestly reduce serum cholesterol levels

1.10.1.2 Decrease in cholesterol level by policosanol

Policosanol, a mixture of very long-chain alcohols isolated from sugarcane, at doses of 10 to 20 mg/day has been shown to lower total and LDL cholesterol by up to 30%, equivalent to low-dose statin therapy (Gouni-Berthold and Berthold, 2002). In both short-term (≤12-week)

and long-term (up to 2-year) randomized, placebo-controlled, double-blind studies, policosanol lowered LDL-cholesterol in normocholesterolemic patients by an average of 33%, and in hypercholesterolemic patients by 24% (for review, see (Gouni-Berthold and Berthold, 2002; Varady et al., 2003). In normocholesterolemic patients, policosanol caused a small and generally insignificant increase in high-density lipoprotein-cholesterol, whereas in seven clinical studies of dyslipidemic patients high-density lipoprotein-cholesterol was increased by an average of 17%. Policosanol is also effective in rabbits and monkeys, where it lowers blood cholesterol and reduces the development of atherosclerotic plaques (Wang et al., 2003), but it was found not to be effective in hamsters (Wang et al., 2003).

The major components of policosanol are the primary alcohols octacosanol (C28; ~60%), triacontanol (C30; 12–14%), and hexacosanol (C26; 6–12%), with lesser amounts of other alcohols with chain lengths of 24 to 34 carbons (Singh et al., 2006) . The product has no evident toxicity and is available over-the-counter in many outlets. The active component(s) has not been established, but it has been shown that very long-chain alcohols can undergo oxidation to fatty acids with subsequent peroxisomal β-oxidation, which also yields chain-shortened metabolites (Singh et al., 1987). D-003, a mixture of very long-chain saturated fatty acids, also purified from sugarcane, similarly lowers LDL and total cholesterol in normocholesterolemic patients (Castano et al., 2005) and in normocholesterolemic and casein-induced hypercholesterolemic rabbits, and a more rapid onset of effects suggests that oxidation of policosanols to very long-chain fatty acids may be necessary for their hypocholesterolemic actions (Menendez et al., 2001; Menendez et al., 2004). Several studies have demonstrated that policosanol inhibits cholesterol synthesis in laboratory animals and cultured cells, and it is thought that this is the principal mechanism by which it lowers blood cholesterol levels. In the latter study, policosanol did not affect the incorporation of [^{14}C]mevalonate into cholesterol, indicating that policosanol was acting at or above mevalonate synthesis. However, policosanol did not inhibit HMG-CoA reductase (mevalonate synthase) when added to cell lysates, arguing against a direct interaction with this enzyme. The ability of policosanol to prevent the up-regulation of HMG-CoA reductase activity in these cells in response to lipid-depleted media suggested that policosanol suppresses HMG-CoA reductase synthesis or enhances enzyme degradation. Similar results were obtained with D-003 (Menendez et al., 2001), although neither study measured HMG-CoA reductase protein levels. Our studies explored that policosanol and identify the active component(s) of this natural product inhibits cholesterol synthesis by inhibiting HMGCoA reductase enzyme (Singh et al., 2006).

1.10.1.3 Inhibition of cholesterol biosynthesis by garlic

Garlic is rich in sulfur-containing compounds, principally S-allylcysteine and alliin, the latter of which is rapidly metabolized when garlic is crushed and alliinase is released. The highly reactive sulfenic acid that is formed from alliin condenses to allicin, which then rapidly recombines to various di- and tri-sulfides, depending on conditions. Ultimately these compounds are believed to yield allyl mercaptan and allyl methyl sulfide, which can react with cellular components or be eliminated on the breath. The organosulfur compounds formed in garlic are highly reactive with other sulfhydryl compounds, including cysteines found in proteins, and it is likely that the chemical modification of enzyme-sulfhydryls is responsible for the purported therapeutic effects of garlic. The question of which compounds are most important to the therapeutic effects of garlic remains unresolved, although several studies have shown that the diallyl disulfides, allyl mercaptan, and Salk(en)yl cysteines are effective inhibitors of cholesterol synthesis in cells (Gebhardt and Beck, 1996; Liu and Yeh, 2000; Singh and Porter, 2006). Similarly, the enzyme targets that mediate the effects of garlic have not been identified.

Our studies with hepatoma cells in which cholesterol and intermediates are radiolabeled and identified by coupled gas chromatography–mass spectrometry reveal that garlic causes the accumulation of sterol 4α-methyl oxidase substrates and that an allyl disulfide or allyl sulfhydryl group is necessary for inhibition by garlic-derived compounds (Singh and Porter, 2006).

1.10.2 Treatment of Hypercholesterolemia by available drug therapy

Drug treatment to lower plasma lipoprotein /or cholesterol is primarily aimed at reducing the risk of athersclerosis and subsequent coronary artery disease that exists in patients with elevated circulating lipids. Drug therapy usually is considered as an option only if non-pharmacologic interventions (altered diet and exercise) have failed to lower plasma lipids.

1.10.2.1 Members of statins family

These are fungal HMG-CoA reductase (HMGR) inhibitors **from members of statins family. Atorvastin (Lipitor), simvastatin (Zocor) and lovastatin (Mevacor) belongs to this family, are widely used for lowering the plasma cholesterol. During the course of treatment, cellular uptake of LDL from plama is significantly increased**, since the intracellular synthesis of cholesterol is inhibited as cells are dependent on extracellular sources of cholesterol. Important isoprenoid compounds require mevalonate as the precursor as a result long term treatment carry some risk of toxicity. A component of the natural cholesterol lowering supplement, red yeast rice, is in fact a statin-like compound. Other beneficial effects of statins other than lowering blood cholesterol levels via their actions on HMGR are ability to reduce the prenylation of numerous pro-inflammatory modulators. Thus, inhibition of this post-translational modification by the statins interferes with the important functions of many signaling proteins which is manifest by inhibition of inflammatory responses.

1.10.2.2 Fibrates compounds

Second group drugs belongs to a series of compounds are derivatives of fibric acid and although known since 1930 but identified as cholesterol lowering drugs very recently. Fibrates are activators of the peroxisome proliferator activated receptor-α (PPARα) class of proteins and are classified as nuclear receptor co-activators. Fibrates result in activation of β-oxidation and thereby decreasing the level of triacyl glycerol and cholesterol rich VLDL in liver as well as enhances the clearance of chylomicrons remnants, and increase in the level of HDLs. These drugs are also known to increase the lipase activity which in turn promotes rapid VLDL turnover. Gemifibrozil (Lopid) and Fenofibrate (Tricor) are two therapeutic drugs available in the market.

1.10.2.3 Cholestyramine or colestipol

Cholestyramine or colestipol (resins) compounds are nonabsorbable resins that bind bile acids which are then not reabsorbed by the liver but excreted. The drop in hepatic reabsorption of bile acids releases a feedback inhibitory mechanism that had been inhibiting bile acid synthesis. As a result, a greater amount of cholesterol is converted to bile acids to maintain a steady level in circulation. Additionally, the synthesis of LDL receptors increases to allow increased cholesterol uptake for bile acid synthesis, and the overall effect is a reduction in plasma cholesterol. This treatment is ineffective in homozygous FH patients since they are completely deficient in LDL receptors.

1.10.2.4 Ezetimibe

This drug is sold under the trade names Zetia® or Ezetrol® and is also combined with the statin drug simvastatin and sold as Vytorin® or Inegy®. Ezetimibe functions to reduce intestinal absorption of cholesterol, thus effecting a reduction in circulating cholesterol. The drug functions by inhibiting the intestinal brush border transporter involved in absorption of cholesterol. This transporter is known as Niemann-Pick type C1-like 1 (NPC1L1). NPC1L1 is also highly expressed in human liver. The hepatic function of NPC1L1 is presumed to limit excessive biliary cholesterol loss. NPC1L1-dependent sterol uptake is regulated by cellular cholesterol content. In addition to the cholesterol lowering effects that result from inhibition of NPC1L1, its' inhibition has been shown to have beneficial effects on components of the metabolic syndrome, such as obesity, insulin resistance, and fatty liver, in addition to atherosclerosis. Ezetimibe is usually prescribed for patients who cannot tolerate a statin drug or a high dose statin regimen. There is some controversy as to the efficacy of ezetimibe at lowering serum cholesterol and reducing the production of fatty plaques on arterial walls. The combination drug of ezetimibe and simvastatin has shown efficacy equal to or slightly greater than atorvastatin (Lipitor®) alone at reducing circulating cholesterol levels.

1.10.2.5 New approaches for the treatment of hypercholesterolemia

In the last decade, a number of epidemiological and clinical studies have demonstrated a direct correlation between the circulating levels of HDL cholesterol and a reduction in the potential for atherosclerosis and coronary heart disease (CHD). Individuals with low levels of HDL (below 40mg/dL) are at higher risk of coronary heart disease CHD) then individual with level above 50mg/dL. Clinical studies have demonstrated that infusion of HDL component, apolipoprotein A-1 (apoA-1) in patients, significantly increases the level of HDL. The newest strategies are targeted to up regulate the level of HDL cholesterol instead of decreasing the level of total cholesterol. Cholesterol ester transfer protein (CETP) is secreted primarily from the liver and plays a critical role in HDL metabolism by facilitating the exchange of cholesteryl esters (CE) from HDL for triglycerides (TG) in apoB containing lipoproteins, such as LDL and VLDL. The activity of CETP directly lowers the cholesterol levels of HDLs and enhances HDL catabolism by providing HDLs with the TG substrate of hepatic lipase. Thus, CETP plays a critical role in the regulation of circulating levels of HDL, LDL, and apoA-I. It has also been shown that in mice naturally lacking CETP most of their cholesterol is found in HDL and these mice are relatively resistant to atherosclerosis. CETP inhibitors have failed in the clinical trials as their use has increased negative cardiovascular events and death rates in test subjects.

2. Conclusion

Cholesterol is an essential component in cell membrane, as a precursor for the synthesis of steroid hormones vitamin D, and bile acids that aid in digestion and cellular signal transduction. Half of the cholesterol is *de novo* synthesized in liver and is transported through various lipoprotein. Dysfunction in cholesterol metabolism can lead to hypercholesterolemia which is a major factor in the development of atherosclerosis. Mode of intracellular and extracellular cholesterol transport through acceptors-donors and thereafter cholesterol trafficking pathways are highly co-ordinated with each other, regulated at enzymatic/transcriptional level and diverse functions of cholesterol in our body. Taken together, this book chapter addressed recent advances in cholesterol metabolism related to absorptions, biosynthesis, transport, excretion and therapeutic targets for new drugs and natural compounds.

3. Acknowledgment

We thank Professor (Dr) Papasani V. Subbaiah, Associate Professor (Dr) Irena Levitan, Professor (Dr) Todd Porter and Assistant Professor (Dr) Ramachandran Ramaswamy for their multiple critical discussions. Special thanks are to Software Professional, Mr Ravi Kesavarapu, for help in making the figures for the manuscript.

4. References

Abe, I., T. Seki, K. Umehara, T. Miyase, H. Noguchi, J. Sakakibara, and T. Ono. 2000. Green tea polyphenols: novel and potent inhibitors of squalene epoxidase. *Biochemical and Biophysical Research Communications*. 268:767-771.

Berliner, J.A., G. Subbanagounder, N. Leitinger, A.D. Watson, and D. Vora. 2001. Evidence for a role of phospholipid oxidation products in atherogenesis. *Trends in Cardiovascular Medicine*. 11:142-147.

Bjorkhem, I., O. Andersson, U. Diczfalusy, B. Sevastik, R.J. Xiu, C. Duan, and E. Lund. 1994. Atherosclerosis and sterol 27-hydroxylase: evidence for a role of this enzyme in elimination of cholesterol from human macrophages. *The Proceedings of National Academy Sciences, U S A*. 91:8592-8596.

Boullier, A., Y. Li, O. Quehenberger, W. Palinski, I. Tabas, J.L. Witztum, and Y.I. Miller. 2006. Minimally oxidized LDL offsets the apoptotic effects of extensively oxidized LDL and free cholesterol in macrophages. *Arteriosclerosis, Thrombosis and Vascular Biology*. 26:1169-1176.

Brown, A.J., R.T. Dean, and W. Jessup. 1996. Free and esterified oxysterol: formation during copper-oxidation of low density lipoprotein and uptake by macrophages. *Journal of Lipid Research*. 37:320-335.

Brown, A.J., and W. Jessup. 1999. Oxysterols and atherosclerosis. *Atherosclerosis*. 142:1-28.

Brown, D.A., and E. London. 1998. Functions of lipid rafts in biological membranes. *Annual Review of Cell and Developmental Biology*. 14:111-136.

Burkitt, M.J. 2001. A critical overview of the chemistry of copper-dependent low density lipoprotein oxidation: roles of lipid hydroperoxides, alpha-tocopherol, thiols, and ceruloplasmin. *Archieves of Biochemistry and Biophysics*. 394:117-135.

Bursill, C., P.D. Roach, C.D. Bottema, and S. Pal. 2001. Green tea upregulates the low-density lipoprotein receptor through the sterol-regulated element binding Protein in HepG2 liver cells. *Journal of Agricultural Food Chemistry*. 49:5639-5645.

Bursill, C.A., M. Abbey, and P.D. Roach. 2007. A green tea extract lowers plasma cholesterol by inhibiting cholesterol synthesis and upregulating the LDL receptor in the cholesterol-fed rabbit. *Atherosclerosis*. 193:86-93.

Bursill, C.A., and P.D. Roach. 2006. Modulation of cholesterol metabolism by the green tea polyphenol (-)-epigallocatechin gallate in cultured human liver (HepG2) cells. *Journal of Agricultural Food Chemistry*. 54:1621-1626.

Bursill, C.A., and P.D. Roach. 2007. A green tea catechin extract upregulates the hepatic low-density lipoprotein receptor in rats. *Lipids*. 42:621-627.

Castano, G., R. Mas, L. Fernandez, J. Illnait, S. Mendoza, R. Gamez, J. Fernandez, and M. Mesa. 2005. A comparison of the effects of D-003 and policosanol (5 and 10 mg/day) in patients with type II hypercholesterolemia: a randomized, double-blinded study. *Drugs under Experimental and Clinical Research*. 31 Suppl:31-44.

Chan, P.T., W.P. Fong, Y.L. Cheung, Y. Huang, W.K. Ho, and Z.Y. Chen. 1999. Jasmine green tea epicatechins are hypolipidemic in hamsters (Mesocricetus auratus) fed a high fat diet. *Journal of Nutrition*. 129:1094-1101.

Christian, A.E., M.P. Haynes, M.C. Phillips, and G.H. Rothblat. 1997. Use of cyclodextrins for manipulating cellular cholesterol content. *Journal of Lipid Research*. 38:2264-2272.

Daniels, T.F., K.M. Killinger, J.J. Michal, R.W. Wright, Jr., and Z. Jiang. 2009. Lipoproteins, cholesterol homeostasis and cardiac health. *International Journal of Biological Sciences*. 5:474-488.

Davies, M.J., J.T. Judd, D.J. Baer, B.A. Clevidence, D.R. Paul, A.J. Edwards, S.A. Wiseman, R.A. Muesing, and S.C. Chen. 2003. Black tea consumption reduces total and LDL cholesterol in mildly hypercholesterolemic adults. *Journal of Nutrition*. 133:3298S-3302S.

Epshtein, Y., A.P. Chopra, A. Rosenhouse-Dantsker, G.B. Kowalsky, D.E. Logothetis, and I. Levitan. 2009. Identification of a C-terminus domain critical for the sensitivity of Kir2.1 to cholesterol. *The Proceedings of National Academy Sciences, U S A*. 106:8055-8060.

Feigenson, G.W. 2007. Phase boundaries and biological membranes. *Annual Review of Biophysics and Biomolecular Structure*. 36:63-77.

Fitzgerald, M.L., K.J. Moore, and M.W. Freeman. 2002. Nuclear hormone receptors and cholesterol trafficking: the orphans find a new home. *Journal of Molecular Medicine (Berl)*. 80:271-281.

Garmy, N., N. Taieb, N. Yahi, and J. Fantini. 2005. Interaction of cholesterol with sphingosine: physicochemical characterization and impact on intestinal absorption. *Journal of Lipid Research*. 46:36-45.

Gaus, K., L. Kritharides, G. Schmitz, A. Boettcher, W. Drobnik, T. Langmann, C.M. Quinn, A. Death, R.T. Dean, and W. Jessup. 2004. Apolipoprotein A-1 interaction with plasma membrane lipid rafts controls cholesterol export from macrophages. *The FASEB Journal*. 18:574-576.

Gaylor, J.L. 2002. Membrane-bound enzymes of cholesterol synthesis from lanosterol. *Biochemical Biophysical Research Communications*. 292:1139-1146.

Gebhardt, R. 2003. Variable influence of kaempferol and myricetin on *in vitro* hepatocellular cholesterol biosynthesis. *Planta Medicine*. 69:1071-1074.

Gebhardt, R., and H. Beck. 1996. Differential inhibitory effects of garlic-derived organosulfur compounds on cholesterol biosynthesis in primary rat hepatocyte cultures. *Lipids*. 31:1269-1276.

Goldstein, J.L., and M.S. Brown. 1990. Regulation of the mevalonate pathway. *Nature*. 343:425-430.

Gouni-Berthold, I., and H.K. Berthold. 2002. Policosanol: clinical pharmacology and therapeutic significance of a new lipid-lowering agent. *American Heart Journal*. 143:356-365.

Gylling, H., and T.A. Miettinen. 1995. The effect of cholesterol absorption inhibition on low density lipoprotein cholesterol level. *Atherosclerosis*. 117:305-308.

Herman, G.E. 2003. Disorders of cholesterol biosynthesis: prototypic metabolic malformation syndromes. *Human Molecular Genetics*. 12 Spec No 1:R75-88.

Hodis, H.N., D.M. Kramsch, P. Avogaro, G. Bittolo-Bon, G. Cazzolato, J. Hwang, H. Peterson, and A. Sevanian. 1994. Biochemical and cytotoxic characteristics of an *in vivo* circulating oxidized low density lipoprotein (LDL-). *Journal of Lipid Research*. 35:669-677.

Holvoet, P., G. Theilmeier, B. Shivalkar, W. Flameng, and D. Collen. 1998. LDL hypercholesterolemia is associated with accumulation of oxidized LDL, atherosclerotic plaque growth, and compensatory vessel enlargement in coronary arteries of miniature pigs. *Arteriosclerosis, Thrombosis and Vascular Biology*. 18:415-422.

Ikonen, E. 2008. Cellular cholesterol trafficking and compartmentalization. *Nature Reviews Molecular Cell Biology*. 9:125-138.

Irons, M., E.R. Elias, G. Salen, G.S. Tint, and A.K. Batta. 1993. Defective cholesterol biosynthesis in Smith-Lemli-Opitz syndrome. *Lancet*. 341:1414.

Kelley, R.I., and R.C. Hennekam. 2000. The Smith-Lemli-Opitz syndrome. *Journal of Medical Genetics*. 37:321-335.

Kilsdonk, E.P., D.W. Morel, W.J. Johnson, and G.H. Rothblat. 1995. Inhibition of cellular cholesterol efflux by 25-hydroxycholesterol. *Journal of Lipid Research*. 36:505-516.

Kimura, Y., and K. Tanaka. 2010. Regulatory mechanisms involved in the control of ubiquitin homeostasis. *The Journal of Biochemistry*. 147:793-798.

Klein, U., G. Gimpl, and F. Fahrenholz. 1995. Alteration of the myometrial plasma membrane cholesterol content with beta-cyclodextrin modulates the binding affinity of the oxytocin receptor. *Biochemistry*. 34:13784-13793.

Koo, S.I., and S.K. Noh. 2007. Green tea as inhibitor of the intestinal absorption of lipids: potential mechanism for its lipid-lowering effect. *Journal of Nutritional Biochemistry*. 18:179-183.

Kovacs, W.J., L.M. Olivier, and S.K. Krisans. 2002. Central role of peroxisomes in isoprenoid biosynthesis. *Progress in Lipid Research*. 41:369-391.

Levitan, I. 2009. Cholesterol and Kir channels. *IUBMB Life*. 61:781-790.

Levitan, I., A.E. Christian, T.N. Tulenko, and G.H. Rothblat. 2000. Membrane cholesterol content modulates activation of volume-regulated anion current in bovine endothelial cells. *Journal of General Physiology*. 115:405-416.

Levitan, I., Y. Fang, A. Rosenhouse-Dantsker, and V. Romanenko. 2010. Cholesterol and ion channels. *Subcell Biochemistry*. 51:509-549.

Levitan, I., and T.P. Shentu. 2011. Impact of oxLDL on Cholesterol-Rich Membrane Rafts. *Journal of Lipids*. 2011:730209.

Liu, L., and Y.Y. Yeh. 2000. Inhibition of cholesterol biosynthesis by organosulfur compounds derived from garlic. *Lipids*. 35:197-203.

Lundbaek, J.A., and O.S. Andersen. 1999. Spring constants for channel-induced lipid bilayer deformations. Estimates using gramicidin channels. *Biophysical Journal*. 76:889-895.

Maguy, A., T.E. Hebert, and S. Nattel. 2006. Involvement of lipid rafts and caveolae in cardiac ion channel function. *Cardiovascular Research*. 69:798-807.

Maron, D.J., G.P. Lu, N.S. Cai, Z.G. Wu, Y.H. Li, H. Chen, J.Q. Zhu, X.J. Jin, B.C. Wouters, and J. Zhao. 2003. Cholesterol-lowering effect of a theaflavin-enriched green tea extract: a randomized controlled trial. *Archieves in Internal Medicine* 163:1448-1453.

Marsh, D., and F.J. Barrantes. 1978. Immobilized lipid in acetylcholine receptor-rich membranes from Torpedo marmorata. *The Proceedings of National Academy of Scences, U S A*. 75:4329-4333.

Martens, J.R., K. O'Connell, and M. Tamkun. 2004. Targeting of ion channels to membrane microdomains: localization of KV channels to lipid rafts. *Trends in Pharmacological Sciences*. 25:16-21.

Menendez, R., A.M. Amor, I. Rodeiro, R.M. Gonzalez, P.C. Gonzalez, J.L. Alfonso, and R. Mas. 2001. Policosanol modulates HMG-CoA reductase activity in cultured fibroblasts. *Archieves of Medical Research*. 32:8-12.

Menendez, R., R. Mas, J. Perez, R.M. Gonzalez, and S. Jimenez. 2004. Oral administration of D-003, a mixture of very long chain fatty acids prevents casein-induced endogenous hypercholesterolemia in rabbits. *Canadian Journal of Physiology and Pharmacology*. 82:22-29.

Nyberg, L., R.D. Duan, and A. Nilsson. 2000. A mutual inhibitory effect on absorption of sphingomyelin and cholesterol. *Journal of Nutritional Biochemistry*. 11:244-249.

Ohvo-Rekila, H., B. Ramstedt, P. Leppimaki, and J.P. Slotte. 2002. Cholesterol interactions with phospholipids in membranes. *Progress in Lipid Research*. 41:66-97.

Olsen, B.N., P.H. Schlesinger, D.S. Ory, and N.A. Baker. 2011a. 25-Hydroxycholesterol increases the availability of cholesterol in phospholipid membranes. *Biophysical Journal*. 100:948-956.

Olsen, B.N., P.H. Schlesinger, D.S. Ory, and N.A. Baker. 2011b. Side-chain oxysterols: From cells to membranes to molecules. *Biochimica et Biophysica Acta. (in press)*

Ostlund, R.E., Jr., M.S. Bosner, and W.F. Stenson. 1999. Cholesterol absorption efficiency declines at moderate dietary doses in normal human subjects. *Journal of Lipid Research*. 40:1453-1458.

Pike, L.J. 2006. Rafts defined: a report on the Keystone Symposium on Lipid Rafts and Cell Function. *Journal of Lipid Research*. 47:1597-1598.

Rentero, C., T. Zech, C.M. Quinn, K. Engelhardt, D. Williamson, T. Grewal, W. Jessup, T. Harder, and K. Gaus. 2008. Functional implications of plasma membrane condensation for T cell activation. *PLoS One*. 3:e2262.

Romanenko, V.G., G.H. Rothblat, and I. Levitan. 2002. Modulation of endothelial inward-rectifier K+ current by optical isomers of cholesterol. *Biophysical Journal*. 83:3211-3222.

Rosenhouse-Dantsker, A., D.E. Logothetis, and I. Levitan. 2011. Cholesterol sensitivity of KIR2.1 is controlled by a belt of residues around the cytosolic pore. *Biophysical Journal*. 100:381-389.

Rozman, D., M. Cotman, and R. Frangez. 2002. Lanosterol 14alpha-demethylase and MAS sterols in mammalian gametogenesis. *Molecular and Cellular Endocrinology*. 187:179-187.

Shentu, T.P., I. Titushkin, D.K. Singh, K.J. Gooch, P.V. Subbaiah, M. Cho, and I. Levitan. 2010. oxLDL-induced decrease in lipid order of membrane domains is inversely correlated with endothelial stiffness and network formation. *American Journal Physiology: Cell Physiology*. 299:C218-229.

Simons, K., and M.J. Gerl. 2010. Revitalizing membrane rafts: new tools and insights. *Nature Reviews Molecular Cell Biology*. 11:688-699.

Simons, K., and E. Ikonen. 1997. Functional rafts in cell membranes. *Nature*. 387:569-572.

Simons, K., and E. Ikonen. 2000. How cells handle cholesterol. *Science*. 290:1721-1726.

Simons, K., and W.L. Vaz. 2004. Model systems, lipid rafts, and cell membranes. *Annual Review of Biophysics and Biomolecular Structure*. 33:269-295.

Singh, D.K., S. Banerjee, and T.D. Porter. 2009a. Green and black tea extracts inhibit HMG-CoA reductase and activate AMP kinase to decrease cholesterol synthesis in hepatoma cells. *Journal of Nutritional Biochemistry*. 20:816-822.

Singh, D.K., L.R. Gesquiere, and P.V. Subbaiah. 2007. Role of sphingomyelin and ceramide in the regulation of the activity and fatty acid specificity of group V secretory phospholipase A2. *Archieves of Biochemistry Biophysics*. 459:280-287

Singh, D.K., L. Li, and T.D. Porter. 2006. Policosanol inhibits cholesterol synthesis in hepatoma cells by activation of AMP-kinase. *Journal of Pharmacology and Experimental Therapeutics*. 318:1020-1026.

Singh, D.K., V. Mokashi, C.L. Elmore, and T.D. Porter. 2003. Phosphorylation of supernatant protein factor enhances its ability to stimulate microsomal squalene monooxygenase. *Journal of Biological Chemistry*. 278:5646-5651.

Singh, D.K., and T.D. Porter. 2006. Inhibition of sterol 4alpha-methyl oxidase is the principal mechanism by which garlic decreases cholesterol synthesis. *Journal of Nutrition*. 136:759S-764S.

Singh, D.K., A. Rosenhouse-Dantsker, C.G. Nichols, D. Enkvetchakul, and I. Levitan. 2009b. Direct regulation of prokaryotic Kir channel by cholesterol. *Journal of Biological Chemistry*. 284:30727-30736.

Singh, D.K., T.P. Shentu, D. Enkvetchakul, and I. Levitan. 2011. Cholesterol regulates prokaryotic Kir channel by direct binding to channel protein. *Biochimica et Biophysica Acta*. 1808:2527-2533.

Singh, H., N. Derwas, and A. Poulos. 1987. Very long chain fatty acid beta-oxidation by rat liver mitochondria and peroxisomes. *Archieves of Biochemistry Biophysics*.259:382-390.

Subbanagounder, G., A.D. Watson, and J.A. Berliner. 2000. Bioactive products of phospholipid oxidation: isolation, identification, measurement and activities. *Free Radical Biology and Medicine*. 28:1751-1761.

Terasaka, N., S. Yu, L. Yvan-Charvet, N. Wang, N. Mzhavia, R. Langlois, T. Pagler, R. Li, C.L. Welch, I.J. Goldberg, and A.R. Tall. 2008. ABCG1 and HDL protect against endothelial dysfunction in mice fed a high-cholesterol diet. *The Journal of Clinical Investigation*. 118:3701-3713.

Toshima, S., A. Hasegawa, M. Kurabayashi, H. Itabe, T. Takano, J. Sugano, K. Shimamura, J. Kimura, I. Michishita, T. Suzuki, and R. Nagai. 2000. Circulating oxidized low density lipoprotein levels. A biochemical risk marker for coronary heart disease. *Arteriosclerosis, Thrombosis and Vascular Biology*. 20:2243-2247.

van Meer, G., D.R. Voelker, and G.W. Feigenson. 2008. Membrane lipids: where they are and how they behave. *Nature Reviews Molecular Cell Biology*. 9:112-124.

van Tits, L.J., T.M. van Himbergen, H.L. Lemmers, J. de Graaf, and A.F. Stalenhoef. 2006. Proportion of oxidized LDL relative to plasma apolipoprotein B does not change during statin therapy in patients with heterozygous familial hypercholesterolemia. *Atherosclerosis*. 185:307-312.

Varady, K.A., Y. Wang, and P.J. Jones. 2003. Role of policosanols in the prevention and treatment of cardiovascular disease. *Nutrition Review*. 61:376-383.

Wang, Y.W., P.J. Jones, I. Pischel, and C. Fairow. 2003. Effects of policosanols and phytosterols on lipid levels and cholesterol biosynthesis in hamsters. *Lipids*. 38:165-170.

Waterham, H.R., J. Koster, G.J. Romeijn, R.C. Hennekam, P. Vreken, H.C. Andersson, D.R. FitzPatrick, R.I. Kelley, and R.J. Wanders. 2001. Mutations in the 3beta-hydroxysterol Delta24-reductase gene cause desmosterolosis, an autosomal recessive disorder of cholesterol biosynthesis. *The American Journal of Human Genetics*. 69:685-694.

Yla-Herttuala, S., W. Palinski, M.E. Rosenfeld, S. Parthasarathy, T.E. Carew, S. Butler, J.L. Witztum, and D. Steinberg. 1989. Evidence for the presence of oxidatively modified low density lipoprotein in atherosclerotic lesions of rabbit and man. *The Journal of Clinical Investigation*. 84:1086-1095.

Zidovetzki, R., and I. Levitan. 2007. Use of cyclodextrins to manipulate plasma membrane cholesterol content: evidence, misconceptions and control strategies. *Biochimica et Biophysica Acta*. 1768:1311-1324.

19

Stobadine – An Indole Type Alternative to the Phenolic Antioxidant Reference Trolox

Ivo Juranek, Lucia Rackova and Milan Stefek
Institute of Experimental Pharmacology and Toxicology, Slovak Academy of Sciences
Slovakia

1. Introduction

Treatment of free radical pathologies by antioxidants has been substantiated by studies in animal models of diseases. However, so far the therapy of oxidative stress-related diseases has not found satisfactory application in clinical practice. This may be due to an insufficient efficacy of the antioxidants available, their unsuitable pharmacokinetics, lack of selectivity, presence of adverse side effects, their toxicity, etc. Thus, new antioxidants have to be identified. In numerous studies searching for novel antioxidant compounds, trolox, a water soluble analogue of alpha-tocopherol, is commonly utilized as a reference antioxidant. Chemically, trolox represents a carboxylic acid chromane (Fig. 1A). Due to its good water solubility, this antioxidant has been broadly used as a standard when screening antioxidant efficacy of other prospectively active compounds in studies involving chemical, subcellular, cellular and tissue models of oxidative stress mediated injury (Aruoma, 2003; Huang et al., 2005; Prior et al., 2005). On the other hand, a fairly large and specific group of substances with beneficial antioxidant effects is derived from indole structure (Suzen, 2007). This puts a demand on the reassessment of the suitability of a phenol-type reference trolox, particularly in studies on screening of nitrogen heterocyclic antioxidants.

The pyridoindole stobadine (Fig. 1A) has been found out to be an effective chain-breaking antioxidant scavenging a variety of reactive oxygen species (Kagan et al., 1993; Steenken et al., 1992; Stefek & Benes, 1991). The input of stobadine into the literary data on indole-type antioxidants comprises more than two hundred PubMed references. Several comprehensive reviews cover stobadine action as determined in a variety of models including simple chemical systems, biological models at subcellular, cellular or organ level, followed by extensive studies in vivo in a number of free-radical disease models, and that also in comparison with other drugs (Horakova et al., 1994; Horakova & Stolc, 1998; Juranek et al., 2010; Stolc et al., 1997; Stefek et al., 2010). The main goal of the present paper is to provide an overview on the current data of the indole-type antioxidant stobadine and to compare them with those of the phenol-type antioxidant trolox.

2. Comparison of stobadine with trolox

Structural features, physicochemical properties, mechanism of action and efficiency of stobadine in various models of free radical damage are summarized and compared with those of trolox. Consequently, stobadine may be highlighted as a promising reference antioxidant, which may readily be utilized as a standard in studies testing antioxidative efficacy of other indole-type substances.

A)

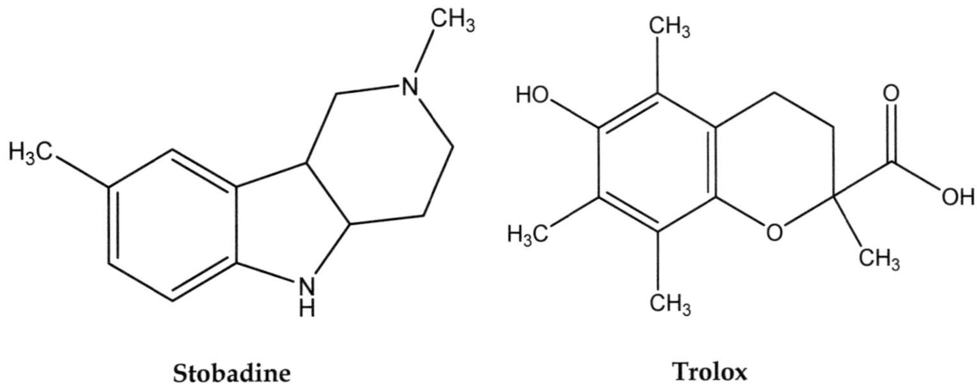

B)

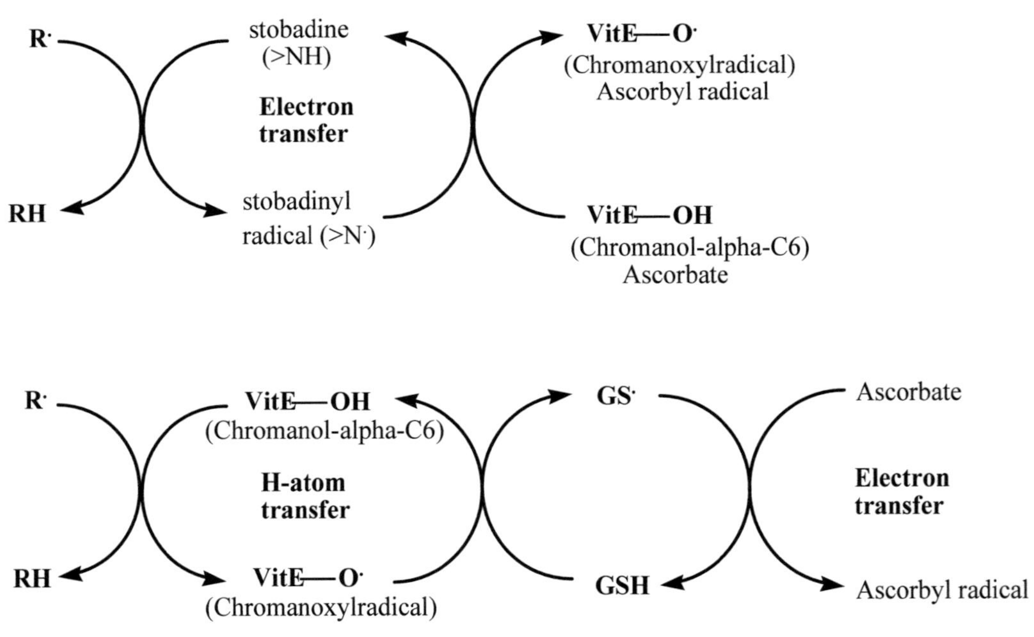

Fig. 1. Structures of stobadine and trolox (A) and possible mechanisms of free radical scavenging by stobadine and vitamin E/trolox (B).
Biologically relevant coupled reactions that might recycle stobadine (Kagan et al., 1993) and vitamin E/trolox (Davies et al., 1988) are depicted.

2.1 Physico-chemical properties

Trolox, 6-hydroxy-2,5,7,8-tetramethylchroman-2-carboxylic acid, has the character of organic acid (Nonell et al., 1995), while stobadine, (-)-cis-2,8-dimethyl-2,3,4,4a,5,9b-hexahydro-1H-pyrido[4,3b]indole, is an organic base. Despite the fact that trolox is more lipophilic than stobadine (log P values 2.83 and 1.95, respectively), at the physiological pH 7 stobadine preferentially distributes into lipid compartment, while trolox preferentially resides in the water phase. The acidobasic behaviour accounts for this apparent discrepancy. With the pKa value of the carboxyl group 3.89 (Barclay & Vinqvist, 1994), trolox undergoes virtually complete dissociation at physiological pH (99.92% of COO^- form of trolox). On the other hand, stobadine with the pKa value of the tertiary nitrogen 8.5 (Stefek et al., 1989) has around 92% of the basic nitrogen in protonated form at pH 7. As a result of the acidobasic equilibrium, the corresponding distribution ratios at pH 7 of trolox and stobadine, D = 0.33 (Barclay et al., 1995) and 3.72 (Kagan et al., 1993), respectively, clearly favour partitioning of stobadine into the lipid phase, yet not that of trolox. This may explain the profound drop of the apparent antioxidant efficiency of trolox in experimental models involving membranous systems (Horakova et al., 2003; Juskova et al., 2010; Rackova et al., 2002; Rackova et al., 2004; Stefek et al., 2008).

2.2 Redox properties

Early pulse radiolysis studies indicated differences with regard both to the centre of antioxidant activity, residing in the indolic nitrogen or phenolic moiety of stobadine or trolox, respectively, and to deprotonation mechanism following the oxidation of the parent molecules (Fig. 1B). It was demonstrated that one-electron oxidation of stobadine leads to the formation of its radical cation (Steenken et al., 1992). That deprotonates from the indolic nitrogen and gives a resonance stabilised nitrogen-centred radical. With regard to the pKa value of approx. -5 of trolox-derived phenoxyl radical cation (Davies et al., 1988) and its expected extremely rapid deprotonation, no spectral evidence for generation of the trolox radical cation was obtained. However, depending on the reaction conditions, electron transfer followed by proton shift or even sequential proton loss and electron transfer (SPLET) has been suggested as a radical scavenging mechanism of phenolic antioxidants involving trolox and alpha-tocopherol (Musialik, 2005; Svanholm et al., 1974).

As shown in Table 1, stobadine and trolox are characterised by comparable rate constants of their interactions with the majority of individual reactive oxygen species tested. The major differences concern the second order rate constants of their reactions with superoxide and hydroxyl radicals. A considerably higher $k_{superoxide}$ value was found for trolox (Nishikimi & Machlim, 1975) than that of stobadine (Kagan et al., 1993) (Table 1). On the other hand, the study by Bielski (1982) showed a notably low second order rate constant of trolox for its reaction with superoxide ($k_{superoxide}$ <0.1 $M^{-1}.s^{-1}$), while Davies et al. (1988) reported an apparent absence of trolox reaction with superoxide.

Regarding the hydroxyl radical scavenging, Davies et al. (1988) reported the value $k_{.OH^{\cdot}}$ for trolox to be comparable with that of stobadine (Table 1). Nonetheless, according to the study of Aruoma et al. (1990), the second order rate constant of trolox for scavenging $HO^{\cdot}$ radicals is about one order higher than that of stobadine. These findings are in a good agreement with our data obtained in a study where the efficacy of stobadine and trolox in inhibition of hydroxyl-radical-induced cross-linking of bovine serum albumin were assessed (Kyselova et al., 2003).

Reactive oxygen species	Rate constant ($M^{-1}.s^{-1}$)	
	Stobadine	Trolox
HO·	7 x 10^9 15.9 x 10^9 (Steenken et al., 1992; Stefek & Benes, 1991)	8.5 x 10^{10} (Aruoma et al., 1990)
CH_3COO· Cl_3COO·	< 5x10^6 6.6 x 10^8 (Steenken et al., 1992)	2.5 x 10^6 3.7 x 10^8 (Simic, 1980; Davies et al., 1988)
DPPH·	4.9 x 10^2 (Rackova et al., 2004)	1.6 x 10^3 (Rackova et al., 2004)
C_6H_6O·	5.1 x 10^8 (Steenken et al., 1992)	4.1 x 10^8 (Davies et al., 1988)
$O_2^{·-}$	7.5 x 10^2 (Kagan et al., 1993)	1.7 x 10^4 <0.1 (Nishikimi & Machlin, 1975; Bielski, 1982)
$^1O_2$	1.3 x 10^8 (Steenken et al., 1992)	pH6 3.5x10^8 (Nonell et al., 1995)

Table 1. Second-order rate constants of stobadine and trolox interaction with reactive oxygen species and 1,1`-diphenyl-2-picrylhydrazyl (DPPH) stable free radical.

Redox potential of stobadine (E = 0.58 V) (Steenken et al., 1992) is more positive than that of vitamin E (E = 0.48 V) (Neta & Steenken, 1982). Hence, at pH 7, the stobadine radical, formed as a consequence of stobadine free-radical-scavenging activity, may subtract proton from the trolox molecule resulting in regeneration of the parent stobadine molecule. Indeed, Steenken et al. (1992) in their pulse radiolytic study demonstrated the ability of trolox to recycle stobadine from its one-electron oxidation product, to give a corresponding trolox phenoxyl radical. When stobadine and trolox were present simultaneously in oxidatively stressed liposomes, trolox spared stobadine in the system in a dose-dependent manner (Rackova et al., 2002). Direct interaction of trolox with the stobadinyl radical resulting in the recovery of parent stobadine molecule appears to be a plausible mechanism. Thus, under physiological conditions, the antioxidant activity of stobadine may be potentiated by vitamin E. In a good agreement with this idea, Horakova et al. (1992) showed that the antioxidant action of stobadine was profoundly diminished in tocopherol-deficient rat liver microsomes.

Analogically, in biological systems, vitamin E (E = 0.48 V) can be regenerated from its phenoxyl radical via the interaction with ascorbate (Davies et al., 1988), which possesses a more negative redox potential (E = 0.30 V) (Neta & Steenken, 1982); depicted in Figure 1B. In a similar way, the stobadinyl radical was shown to be quenched by ascorbate, as demonstrated by the increased magnitude of the ascorbyl radical ESR signal generated in the presence of stobadine in the system of lipoxygenase + arachidonate (Kagan et al., 1993). Hence, one may expect that in biological systems, the antioxidant potency of both trolox and stobadine may be modulated by their mutual interactions with other lipid- or water-soluble antioxidants.

2.3 Antioxidant efficacies in various assay systems

In a homogeneous system, antioxidant activity stems from an intrinsic chemical reactivity towards radicals. In membranes, however, the reactivity may differ as there are additional factors involved, such as a relative location of the antioxidant and radicals, ruled predominantly by their distribution ratios between water and lipid compartments. As already mentioned, a notably lower distribution ratio of trolox compared to that of stobadine may account for their different efficacies in systems involving lipid interface (membranes) in comparison to homogenous units (true solutions).

In the ethanolic solution, trolox scavenged 1,1`-diphenyl-2-picrylhydrazyl (DPPH) radical more efficiently than did stobadine, based on the initial velocity measurements (Rackova et al., 2002). The finding was corroborated by the respective rate constants (Rackova et al., 2004) as shown in Table 2.

In the models of oxidative damage comprising soluble proteins in buffer solutions, the water-soluble antioxidants stobadine and trolox have free access both to free radical initiator and to protein-derived radicals. Stobadine inhibited the process of albumin cross-linking due to the oxidative modifications induced by the Fenton reaction system of $Fe^{2+}/EDTA/H_2O_2/$ascorbate less effectively than did trolox (Kyselova et al., 2003). The experimental IC_{50} values correlated well with the reciprocal values of the corresponding second order rate constants for scavenging OH radicals.

Trolox, in comparison with stobadine, was found to be a more potent inhibitor of 2,2'-azobis-2-amidinopropane (AAPH)-induced precipitation of the soluble eye lens proteins (Stefek et al., 2005). In contrary, production of free carbonyls due to protein oxidation was more efficiently inhibited by stobadine. Both stobadine and trolox showed comparable efficacies in an experimental glycation model in preventing glycation-related fluorescence changes of bovine serum albumin as well as in lowering the yield of 2,4-dinitrophenylhydrazine-reactive carbonyls as markers of glyco-oxidation (Table 2) (Stefek et al., 1999).

On the other hand, trolox was found to be much less effective in inhibiting AAPH-induced peroxidation of di-oleoyl-phosphatidylcholine (DOPC) liposomes with respect to stobadine (Rackova et al., 2004; Rackova et al., 2006; Stefek et al., 2008), as exemplified by the respective IC_{50} values 25.3 and 93.5 µM, shown in Table 2. Stobadine, in comparison with trolox, more effectively prolonged the lag phase of Cu^{2+}-induced low-density lipoprotein (LDL) oxidation measured by diene formation (Horakova et al., 1996). The same pattern of efficacy in prevention of the lipid oxidation boost was shown in the system of tissue homogenate. Stobadine showed a more potent inhibitory effect than trolox on lipid peroxidation in rat brain homogenates exposed to Fe^{2+}/ascorbate as documented by thiobarbituric acid reactive substances (TBARS) levels (Table 2; Horakova et al., 2000). Interestingly, in the case of alloxan-induced lipid peroxidation of heat denaturized rat liver microsomes, the inhibitory efficacy of stobadine and trolox was comparable (Stefek & Trnkova, 1996). This finding may indicate that the critical competition of the scavengers with the alloxan-derived initiating reactive oxygen species takes place outside the membrane in the bulk solution.

Assay system		Parameter measured	Stobadine	Trolox
AAPH induced LPO in DOPC liposomes (Rackova et al., 2002)		IC_{50} (μmol/L)	25.3 ± 14.6	93.5 ± 8.5
BSA cross-linking induced Fe^{2+}/EDTA/H_2O_2/ascorbate (Kyselova et al., 2003)		IC_{50} (μmol/L)	0.65 ± 0.08	0.13 ± 0.02
AAPH-induced oxidative modification of soluble eye lens proteins (Stefek et al., 2005)	Inhibition of protein precipitation	IC_{50} (μmol/L)	121 ± 15	79 ± 8
	Inhibition of protein oxidation		44 ± 8	131 ± 20
Oxidative modification of BSA in an experimental glycation model (Stefek et al., 1999)	Glucose attachment into the molecule of BSA	Amadori product (with respect to 8.2 ± 0.4 nmol/mg BSA for control without inhibitor)	8.1 ± 0.5 (0.25 mmol/L)	7.4 ± 0.7 (0.25 mmol/L)
	Glycation-induced fluorescence changes of BSA	Relative fluorescence (with respect to 11.2 ± 0.7 nmol/mg BSA for control without inhibitor)	7.9 ± 0.7 (0.25 mmol/L)	6.5 ± 0.4 (0.25 mmol/L)
	Formation of DNPH-reactive carbonyl groups in BSA	Carbonyl groups (with respect to 5.6 ± 0.4 nmol/mg BSA for control without inhibitor)	3.4 ± 0.5 (0.25 mmol/L)	3.3 ± 0.2 (0.25 mmol/L)
Cu^{2+}-mediated oxidation of LDL (Horakova et al., 1996)		Δt_{lag} (min) (The increase in lag time given by one stobadine molecule per single LDL particle)	1.5	0.38
Fe^{2+}/ascorbate induced oxidative damage of rat brain homogenate (Horakova et al., 2000)	Inhibition of TBARS production	IC_{50} (μmol/L)	35	98

Assay system	Parameter measured	Stobadine	Trolox
AAPH-induced haemolysis of rat erythrocytes (Juskova et al., 2010)	t_{lag} (min) (88.6 ± 2.2 for control erythrocytes)	> 300 (100 µmol/L)	144 (100 µmol/L)

LPO, lipid peroxidation; DOPC, dioleoyl phosphatidylcholine; BSA, bovine serum albumin; AAPH, 2,2`-azobis (2-amidinopropane)hydrochloride; DNPH, dinitrophenylhydrazine; TBARS, thiobarbituric acid reactive substances.

Table 2. Summary of antioxidant and protective efficacies of stobadine and trolox in experimental models of oxidative damage.

In the cellular system of intact erythrocytes exposed to peroxyl radicals generated by thermal degradation of the azoinitiator AAPH in vitro, stobadine, in comparison to trolox, protected more powerfully erythrocytes from haemolysis, as shown (Table 2) by the respective lag phase prolongations (Juskova et al., 2010). In another cellular model, stobadine increased the viability of hydrogen-peroxide treated PC12 cells more effectively than did trolox, while both compounds reduced the content of malondialdehyde with a comparable efficiency (Horakova et al., 2003).

3. Conclusion

On balance then, the present paper, by summarizing the current data on both trolox and stobadine, underscores the structural and physicochemical differences between the two compounds as respective representatives of phenolic- and indole-type antioxidants. The structural variance explains their different mechanisms of antioxidant action and variable efficacies in the range of assay systems studied. Considering a plethora of studies reported on stobadine antioxidant action, physicochemical properties, and on a variety of its other biological activities, stobadine may represent a pertinent indole-type reference antioxidant. Hence, in studies of indole compounds, stobadine antioxidant standard may be utilized as a more relevant alternative to structurally diverse trolox.

4. Acknowledgements

Financial support by grants of the EU COST Programme (Action CM1001) and the National Agencies VEGA (2/0083/09, 2/0067/11 and 2/0149/12) and APVV (51-017905) is gratefully acknowledged.

5. References

Aruoma, O.I. (2003). Methodological considerations for characterizing potential antioxidant actions of bioactive components in plant foods, *Mutation Research* Vol. 523-524: 9–20.

Aruoma, O.I., Evans, P.J., Kaur, H., Sutcliffe, L., & Halliwell, B. (1990). An evaluation of the antioxidant and potential pro-oxidant properties of food additives and of Trolox C, vitamin E and probucol, *Free Radical Research Communication* Vol. 10: 143–157.

Barclay, L.R., Artz, J.D., & Mowat, J.J. (1995). Partitioning and antioxidant action of the water-soluble antioxidant, Trolox, between the aqueous and lipid phases of phosphatidylcholine membranes: ^{14}C tracer and product studies, *Biochimica et Biophysica Acta* Vol. 1237: 77–85.

Barclay, L.R.C., & Vinqvist, M.R. (1994). Membrane peroxidation: inhibiting effects of watersoluble antioxidants on phospholipids of different charge types, *Free Radical Biology & Medicine* Vol. 16: 779–788.

Bielski, B.H.J. (1983). Evaluation of the reactivities of $HO_2\cdot/O_2^{-}\cdot$ with compounds of biological interest. In Oxygen radicals and their scavenger systems, Vol I (ed. G. Cohen and RE Greenwald), Amsterdam: Elsevier, pp. 1–7.

Davies, M.J., Fornit, L.G., & Willson, R.L. (1988). Vitamin E Analogue Trolox C. E.S.R. and Pulse-Radiolysis Studies of Free-Radical Reactions, *The Biochemical Journal* Vol. 255:513–522.

Horakova, L., Briviba, K., & Sies, H. (1992a). Antioxidant activity of the pyridoindole stobadine in liposomal and microsomal lipid peroxidation, *Chemico–Biological Interactions* Vol. 83, 85–93.

Horakova, L., Giessauf, A., Raber, G., & Esterbauer, H. (1996). Effect of stobadine on Cu $^{(++)}$ -mediated oxidation of low-density lipoprotein, *Biochemical Pharmacology* Vol. 51(10): 1277–1282.

Horakova, L., Licht, A., Sandig, G., Jakstadt, M., Durackova, Z., & Grune, T. (2003). Standardized extracts of flavonoids increase the viability of PC12 cells treated with hydrogen peroxide: effects on oxidative injury, *Archives of Toxicology* Vol. 77(1): 22–9.

Horakova, L, Ondrejickova, O., Bachrata, K., & Vajdova, M. (2000). Preventive effect of several antioxidants after oxidative stress on rat brain homogenates, *General Physiology and Biophysics* Vol. 19(2): 195–205.

Horakova, L., Sies, H., & Steenken S. (1994). Antioxidant action of stobadine, *Methods in Enzymology* (Edited by Packer L.), Part D, Academic Press, San Diego, Vol. 234, pp. 572–580.

Horakova, L., & Stolc, S. (1998) Antioxidant and pharmacodynamic effects of pyridoindole stobadine, *General Pharmacology* Vol. 30: 627–638.

Huang, D., Ou, B., & Prior, R.L. (2005). The chemistry behind antioxidant capacity assays, *Journal of Agricultural and Food Chemistry* Vol. 53: 1841–1856.

Juranek, I., Horakova, L., Rackova, L., & Stefek, M. (2010). Antioxidants in Treating Pathologies Involving Oxidative Damage: An Update on Medicinal Chemistry and Biological Activity of Stobadine and Related Pyridoindoles, *Current Medicinal Chemistry* Vol. 17: 552–570.

Juskova, M., Snirc, V., Krizanova, L., & Stefek, M. (2010). Effect of carboxymethylated pyridoindoles on free radical-induced haemolysis of rat erythrocytes *in vitro*, *Acta Biochimica Polonica* Vol. 57(2): 153–6.

Kagan, V.E., Tsuchiya, M., Serbinova, E., Packer, L., & Sies, H. (1993). Interaction of the pyridoindole stobadine with peroxyl, superoxide and chromanoxyl radicals, *Biochemical Pharmacology* Vol. 45(2): 393–400.

Kyselova, Z., Rackova, L., & Stefek, M. (2003). Pyridoindole antioxidant stobadine protected bovine serum albumin against the hydroxyl radical mediated cross-linking in vitro, *Archives of Gerontology and Geriatrics* Vol. 36: 221–229.

Musialik, M. & Litwinienko, G. (2005). Scavenging of dpph• Radicals by Vitamin E Is Accelerated by Its Partial Ionization: the Role of Sequential Proton Loss Electron Transfer, *Organic Letters* Vol. 7(22): 4951–4954.

Neta, P., & Steenken, S. (1982). One electron redox potentials of phenols, hydroxy- and aminophenols and related compounds of biological interest, *The Journal of Physical Chemistry* Vol. 86: 3661–3667.

Nishikimi, M., & Machlin, L.J. (1975). Oxidation of alpha-tocopherol model compound by superoxide anion, *Archives of Biochemistry and Biophysics* Vol. 170(2):684–689.

Nonell, S., Moncayo, L., Trull, F., Amat-Guerri, F., Lissi, E.A., Soltermann, A.T., Criado, S., & Garcia, N.A. (1995). Solvent influence on the kinetics of the photodynamic degradation of trolox, a water-soluble model compound for vitamin E, *Journal of Photochemistry and Photobiology B: Biology* Vol. 29: 157–162.

Prior, R.L., Wu, X., & Schaich, K. (2005). Standardized methods for the determination of antioxidant capacity and phenolics in foods and dietary supplements, *Journal of Agricultural and Food Chemistry* Vol. 53: 4290–4302.

Rackova, L., Majekova, M., Kostalova, D., & Stefek, M. (2004). Antiradical and antioxidant activities of alkaloids isolated from Mahonia aquifolium. Structural aspects, *Bioorganic & Medicinal Chemistry* Vol.12: 4709–4715.

Rackova, L., Snirc, V., Majekova, M., Majek, P., & Stefek, M. (2006). Free Radical Scavenging and Antioxidant Activities of Substituted Hexahydropyridoindoles.Quantitative Structure-Activity Relationships, *Journal of Medicinal Chemistry* Vol. 49; 2543–2548.

Rackova, L., Stefek, M., & Majekova, M. (2002). Structural aspects of antioxidant activity of substituted pyridoindoles, *Redox Report* Vol. 7: 207–214.

Simic MG. (1980). Peroxyl radical from oleic acid. In Simic, MG. (ed.), *Autoxidation in Food and Biological Systems*. Plenum, New York, pp. 17–26.

Steenken, S., Sunquist, A.R., Jovanovic, S.V., Crockett, R., & Sies, H. (1992). Antioxidant activity of the pyridoindole stobadine. Pulse radiolytic characterization of one-electron-oxidized stobadine and quenching of singlet molecular oxygen, *Chemical Research in Toxicology* Vol. 5: 355–360.

Stefek, M., Benes, L., & Zelnik, V. (1989). N-oxygenation of stobadine, a gamma-carboline antiarrhythmic and cardioprotective agent: the role of flavin-containing monooxygenase. *Xenobiotica* Vol. 19: 143–50.

Stefek, M., & Benes, L. (1991). Pyridoindole stobadine is a potent scavenger of hydroxyl radicals, *FEBS Letters* Vol. 294: 264–266.

Stefek, M., & Trnkova, Z. (1996). The pyridoindole antioxidant stobadine prevents alloxan-induced lipid peroxidation by inhibiting its propagation, *Pharmacology & Toxicology* Vol. 78(2):77–81.

Stefek, M., Krizanova, L., & Trnkova, Z. (1999). Oxidative modification of serum albumin in an experimental glycation model of diabetes mellitus in vitro: effect of the pyridoindole antioxidant stobadine, *Life Sciences* Vol. 65(18–19):1995–1997.

Stefek, M., Kyselova, Z., Rackova, L., & Krizanova, L. (2005). Oxidative modification of rat lens proteins by peroxyl radicals *in vitro*: Protection by the chain-breaking antioxidants Stobadine and Trolox. *Biochimica et Biophysica Acta–Molecular Basis of Disease* Vol. 1741(1-2): 183–190.

Stefek, M., Snirc, V., Djoubissie, P.-O., Majekova, M., Demopoulos, V., Rackova, L., Bezakova, Z., Karasu, C., Carbone, V., & El-Kabbani, O. (2008). Carboxymethylated pyridoindole antioxidants as aldose reductase inhibitors: Synthesis, activity, partitioning, and molecular modelling, *Bioorganic & Medicinal Chemistry* Vol. 16: 4908–4920.

Stefek, M., Gajdosik, A., Gajdosikova, A., Juskova, M., Kyselova, Z., Majekova, M. Rackova, L., Snirc, V., Demopoulos, V.J. & Karasu, Ç. (2010). Stobadine and related pyridoindoles as antioxidants and aldose reductase inhibitors in prevention of diabetic complications. In Stefek, M. (ed.), *Advances in molecular mechanisms and pharmacology of diabetic complications*. Transworld Research Network, Kerala, pp. 285-307.

Stolc, S., Vlkolinsky, R., & Pavlasek, J. (1997). Neuroprotection by the pyridoindole stobadine: A minireview, *Brain Research Bulletin* Vol. 42: 335–340.

Suzen, S. (2007). Antioxidant Activities of Synthetic Indole Derivatives and Possible Activity Mechanisms, *Topics in Heterocyclic Chemistry* Vol. 11: 145–178.

Svanholm, U., Beckgaard, K., & Parker, V.D. (1974). Electrochemistry in media of intermediate activity. VIII. Reversible oxidation products of α-tocopherol model compound. Cation radical, cation, and dication, *Journal of the American Chemical Society* Vol. 96: 2409–2413.